Y0-BRY-334

THE CARNIVOROUS PLANTS

Darlingtonia californica drawn from a field specimen. By kind permission of Shahid Naeem, Berkeley, California, USA

THE CARNIVOROUS PLANTS

Dr B.E. Juniper

Department of Plant Sciences
University of Oxford
South Parks Road
Oxford
OX1 3RA

Dr R.J. Robins

Agricultural & Food Research Council
Institute of Food Research
Colney Lane
Norwich
NR4 7UA

Dr D.M. Joel

Department of Weed Research
Agricultural Research Organization
Newe-Ya'ar Experiment Station
Israel
31999

ACADEMIC PRESS

Harcourt Brace Jovanovich, Publishers

London San Diego New York Berkeley Boston
Sydney Tokyo Toronto

LIBRARY
COLBY-SAWYER COLLEGE
NEW LONDON, NH 03257

OVER
QK
917
.J854
1989

ACADEMIC PRESS LIMITED
24–28 Oval Road
London NW1 7DX

United States Edition published by
ACADEMIC PRESS, INC.
San Diego
CA 92101

103081

Copyright © 1989 by
ACADEMIC PRESS LIMITED

All rights reserved. No part of this book may be reproduced or transmitted in any form or by any means, electronic or mechanical, including photocopy, recording, or any information storage and retrieval system without permission in writing from the publisher.

British Library Cataloguing in Publication Data

Juniper, B.E. (Barrie Edward)
The carnivorous plants.
1. Carnivorous plants
I. Title II. Robins, R.J. III. Joel, D.M.
581

ISBN 0–12–392170–8

Text design by Eric Drewery

Typeset in Great Britain by
Photo·graphics, Honiton, Devon.

Printed in Great Britain at
The Printing House, Oxford.

To our wives

Bridget, Carol and Alma

who have suffered much

Contents

Preface

Plants, as is now becoming widely recognized, exploit animals in almost as many ways as animals use plants. Yet only rarely do they eat animals in the sense of catching, holding and devouring prey. However, the manner in which they function as carnivores grants insights into plant form, function and evolution not otherwise readily available. The diversity of morphological, biochemical and commensal features generates both the lay and the scientific interest in this diverse group.

The carnivorous plants exhibit features that are common to many other non-carnivorous plants. However, the extent to which these features have developed and the combination of different features in small organs is unique and can therefore be exploited by using such plants as models for scientific research. This unique character is perhaps best emphasized where secretory and transport mechanisms are concerned; *Dionaea* for example can be triggered to synthesize and secrete both ions and digestive enzymes by simple chemical stimulation. This simplicity contrasts with almost all other secretory and transport activities in 'conventional' plants which are the result of interior control.

Some of the subjects of this book may have evolved into carnivorous plants from having initially developed sophisticated, external defences against insects. As the defence of crops against insects by chemical means becomes both less effective and more expensive it may not be unrealistic to think that some of the findings in this book may lead to the future breeding of less vulnerable crops, with their own defensive mechanisms. The carnivorous plants are not, therefore, just a botanical curiosity.

Throughout this book we have used the word 'carnivorous' as opposed to the more common but incomplete term 'insectivorous' plant. The diet of animal-eating plants lies well outside the great class INSECTA, even outside the phylum ARTHROPODA. Their diet may include MOLLUSCA, such as the slug *Agriolimax* in *Dionaea*, earthworms, spiders and woodlice (*Armadillidium*) in the pitchers of *Nepenthes* and *Sarracenia*, *Paramecium*, tadpoles and small fish in *Utricularia* (Carpenter, 1884) and even, anecdotally, small rabbits in *Byblis* and *Drosera gigantea*. However, it is our belief that, with the exception of the specialized fungi that prey solely upon nematodes and the few algae which graze upon bacteria (Bird and Kalff, 1986), both of which lie outside the scope of this book, carnivorous plants in general have adapted and refined their various mechanisms to an arthropod diet. The rest are a gastronomic bonus.

Since the precise academic study of carnivorous plants began somewhere about the beginning of the seventeenth century, there have been just two major scientific books on the subject in English: by Darwin (1875) and Lloyd (1942). Darwin's work is intense, detailed, still regarded as a classic text and shows characteristic strokes of genius and insight. Sixty per cent of his book is devoted to the sundews (*Drosera*). Lloyd's book is intense, detailed and virtually comprehensive; he combined wit, wisdom and a polished literary style with an insatiable curiosity and a voluminous knowledge. In these more harried times there probably never can be another Darwin or Lloyd and in no sense is our book intended to be a replacement for their texts. We shall be happy if the serious student of plants has all three books on the shelf: Darwin for his inimitable, limpid style and simple, practical approach; Lloyd for his comprehensive knowledge; and ourselves for the insight into mechanisms provided by modern science.

The stimulus to produce this book came principally from the association in one laboratory of the three authors, each of whom had, separately, developed an interest in particular aspects of the world of carnivorous plants. In contrast with the two previous authors we have adopted a layout that concentrates on mechanisms rather than taxonomic groupings. We have embraced this approach not only because our particular interest lies in that direction, but also because there has only been a limited advance in the taxonomy of the phylogenetic relationships of the various groups since Lloyd's time. The taxonomy is briefly summarized, but is not critically treated. Nor do we make any effort to deal with the rapidly expanding field of the horticulture of the carnivorous plants.

Acknowledgements

Darwin wrote *Insectivorous Plants* at Down House in Kent, England. Lloyd wrote most of his *Carnivorous Plants* after he had retired from the Professorship of Botany at McGill University, Canada. Much of the present text was written by the three authors, over the last eight years, in the Botany School (now the Department of Plant Sciences), Oxford. We would like to thank Professor F.R. Whatley, FRS, Head of the Department, for his tolerance and support throughout this whole period.

By an odd coincidence, part of the present text, on the ecology and the commensal inhabitants of carnivorous plants, was completed by B.E. Juniper in Tucson (the Department of Ecology and Evolutionary Biology, University of Arizona). Francis Lloyd, early in his career 1906–1909, also spent some time in Tucson, at the Carnegie Institute Desert Laboratory working on his book *The Physiology of Stomata*. We are most grateful to Professor Conrad Istock, the Chairman of the above department, for his hospitality and help with parts of the manuscript.

We would also like to thank Dr H.C. Bennet-Clark (Oxford), Dr B.S. Cox (Oxford), Dr A. Dafni (Haifa), Dr R.H. Disney (Cambridge), Professor J.L. Harper, FRS (Bangor), Dr P.A. Rea (Rothamsted) and Dr T. King (Abingdon), who have kindly read parts of the manuscript. We accept, however, full responsibility for any mistakes, omissions or misinterpretations therein.

Mrs Rosemary Wise, Botanical Artist, Oxford University, prepared, with her consummate skill almost all the line diagrams and drawings used in this text. Mr J.K. Burras, the Superintendent of the Botanic Gardens, Oxford grew all the plants for our experimental and depictive purposes and was a rich source of information and contacts. Mr P. Taylor (Kew) provided us with preserved material of *Genlisea*, and was very helpful in many different matters concerning *Utricularia*. Miss C.J. Cresswell and Miss J.M. Humphries were invaluable with their skills in word-processing and reference chasing, and, in addition, we would like to thank all those innumerable librarians and secretaries throughout the world who so kindly answered our time-wasting queries and sought out obscure references.

History

Some of the earliest observations on *Sarracenia* were made by Robert Morison (1699), the first Professor of Botany in Oxford. Burdon Sanderson's epoch-making discoveries on the electrical signals of *Dionaea* were begun at University College, London, in 1873, but continued by him after he was appointed to the Chair of Physiology in Oxford in 1882. The research on *Dionaea* continued in Oxford until 1899. Much further work was done here by S.H. Vines, the Sherardian Professor of Botany, in the period 1897–1901. We would like to think that we are carrying on, in this city, an established tradition of careful observation, experiment and interpretation.

The Future

Many carnivorous plants are on the verge of extinction. We hope that this book may assist in convincing authorities all over the world that the preservation of these plants is a matter of importance.

This book is dedicated to all those past, present and future, who investigate the improbable with an open mind, thereby discovering new wonders.

B.E.J.
R.J.R.
D.M.J.
Oxford, 1988

Glossary

The Carnivorous Genera, Authorities and Origin of the Names

Aldrovanda Linnaeus, 1753. In honour of Ulisse Aldrovandi, 1522–1605, founder of the Botanic Garden in Bologna (1567).

Brocchinia J.H. Schultes, 1830. For Giovanni Battista Brocchi, 1772–1825, director of the Botanic Garden of Brescia, Italy.

Byblis R.A. Salisbury, 1808. From the Greek meaning 'nymph'.

Catopsis Grisebach, 1864. Greek for 'view', possibly because they grow on the tops of trees.

Cephalotus La Billardière, 1806. From the Greek *kephalatos* meaning 'headed', which refers to the filaments of the stamens.

Darlingtonia J. Torrey, 1851. Named after Dr William Darlington, 1782–1863, botanist of West Chester, Pennsylvania, USA.

Drosophyllum Link, 1805. Greek for 'dew-leaf'. A description of the stalked glands on leaves giving the impression of being covered with dew.

Drosera Linnaeus, 1753. Greek for 'dewy'. An allusion to the leaves appearing to be covered in dew.

Dionaea J. Ellis, 1773. A surname of Venus, goddess of love and daughter of Jupiter and Dione.

Genlisea H.G.L. Reichenbach, 1828 and Auguste Sainte-Hilaire, 1833. Named after the Contesse Stéphanie Félicité de Genlis.

Heliamphora George Bentham, 1840. From the Greek 'sun-pitcher'. But Bentham got it slightly wrong; he meant to call it 'marsh-pitcher' (Gk. *helos*: *amphora*), but somehow *helios*, 'sun', got recorded, which has led to confusion.

Ibicella Van Eseltine, 1929 (*Martynia lutea*, Lindley). *Ibicella* a diminutive of *Ibex*, with reference to the horned fruits, hence the common name 'Devil's claws'. *Martynia* in honour of John Martyn FRS, 1699–1768, Professor of Botany at Cambridge, England.

Nepenthes Linnaeus, 1753. Greek for 'no mind' or 'without care', an allusion to the story in Homer's *Odyssey* where Helen mixed the wine with the drug Nepenthe so that by drinking it man would be free from care and grief.

Pinguicula Linnaeus, 1753. A diminutive from the Latin for 'fat', *pinguis*, a description of the greasy texture of the leaves.

Sarracenia Linnaeus, 1753. In honour of Dr Michel Sarrazin, 1659–1734, a physician in Quebec, Canada, who sent *Sarracenia purpurea* to Tournefort.

Triphyophyllum Airy-Shaw, 1952. Greek for 'three different types of leaf'.

Utricularia Linnaeus, 1753. Latin for 'little pouch'.

Abbreviations

2,4-D 2,4-dichlorophenoxyacetic acid
4PA *p*-fluorophenylalanine
ABA abscissic acid
ATP adenosine triphosphate
***p*-CMB** *p*-chloromercuribenzenesulphonic acid
DCMU 3-(3,4-dichlorophenyl)-1,1-dimethylurea
DIPFP diisopropylfluorophosphonate
DEAE diethylaminoethyl
DNP dinitrophenol
EDAX energy dispersive X-ray analyser
EDTA ethylenediaminetetraacetic acid
EM electron microscope
EMMA electron microscope microanalyser
ER endoplasmic reticulum
rER rough endoplasmic reticulum
sER smooth endoplasmic reticulum
FC fusicoccin
IAA indoleacetic acid
LAMMA laser microprobe mass analyser
NMR nuclear magnetic resonance
PA-TCH-SP periodic acid-thiosemicarbazide-silver proteinate
PAS periodic acid Schiff
PCIB 4-chlorophenoxyisobutyric acid
PCP pentachlorophenol
PEG polyethyleneglycol
PMSF phenylmethanesulphonyl fluoride
PTA phosphotungstic acid
RNA ribonucleic acid
Rubisco ribulose-1,6-bisphosphate carboxylase
SEM scanning electron microscope
TEM transmission electron microscope
TIBA triiodobenzoic acid
Tlc thin layer chromatography
TMV tobacco mosaic virus
UV ultraviolet
ZIO zinc iodide osmium

PART I

Introduction: The Syndrome and the Habitat

CHAPTER 1

The Carnivorous Syndrome

Introduction

Carnivorous plants (Latin: *Carnis* – flesh; *Vorare* – devour or swallow) are those plants which use entrapped animal tissues for their nutrition. There is no doubting the carnivorous habit of many plants but, surprisingly to the layperson, there is no general agreement on whether certain other plants are indeed carnivorous. *Roridula* (Darwin, 1875) and the teazle *Dipsacus* (Christy, 1923) were at one time considered to be carnivorous, but Lloyd (1942) did not recognize them as such. *Brocchinia reducta* and *Catopsis berteroniana*, respectively a terrestrial and an epiphytic bromeliad, are now fully recognized as true carnivores. Other bromeliads in the same family may soon follow and others in the same order must be considered very close to true carnivory; yet other plants, not having all the 'necessary' carnivorous mechanisms, are just on the borderline, evolving, it can be speculated, either into or out of the carnivorous habit (Fig. 1.1).

In this first chapter we shall consider what complex of characters comprises the carnivorous 'syndrome' and thus attempt a definition of what is a carnivorous plant, recognizing always that the boundaries of carnivory are not universally agreed and are shifting all the time. Later, in Chapters 17 to 19, we shall review the fossil record and, as far as it goes, attempt to trace the path of specific features of carnivory, such as the gland, throughout the plant kingdom and study the very few instances where part of a specific sequence of evolution may still exist in the higher plants.

Insect and Plant Interactions

Plants and animals have been interacting presumably ever since the distinction between the autotrophic and heterotrophic forms was established. There is good fossil evidence of insect and plant interactions from the Carboniferous period onwards (Scott *et al.*, 1985), and more limited data as far back as the Devonian (Scott and Taylor, 1983): lycopod plant debris, for example, has been found in the gut of the large Devonian fossil arthropod *Arthropleura* (Rolfe and Ingham, 1967). There can be no doubt that plants were developing defence mechanisms against a constant insect assault long before that time. Whether any such mechanisms took the further step of evolving into a carnivorous device we cannot say – carnivorous plants, in the generally accepted sense, appear to be restricted to the angiosperms and certain fungi but there are suggestions that primitive plants (see page 294) may have moved further down this path than commonly supposed. However, virtually all such records seem to have been lost; only the angiosperm evolutionary sequence, fragmentary though it is, can be studied with any completeness.

The Occurrence of Carnivorous Taxa in the Plant Kingdom

In the absence of an adequate fossil record and in view of the minimal available cytological analysis (Kress, 1970; Kondo, 1984), we can only turn in the first instance to the occurrence of carnivory within the well-established taxa of the angiosperms, glean what we can from their associations and, as we shall discuss later, see what can be learned from a comparison of structural features.

Figure 1.2 shows a proposed phylogenetic 'tree' of the angiosperms, modified from Takhtajan (1969). While all such trees are subject to dispute, particularly in their drawings of precise boundaries between the proposed sub-classes, no other such tree will

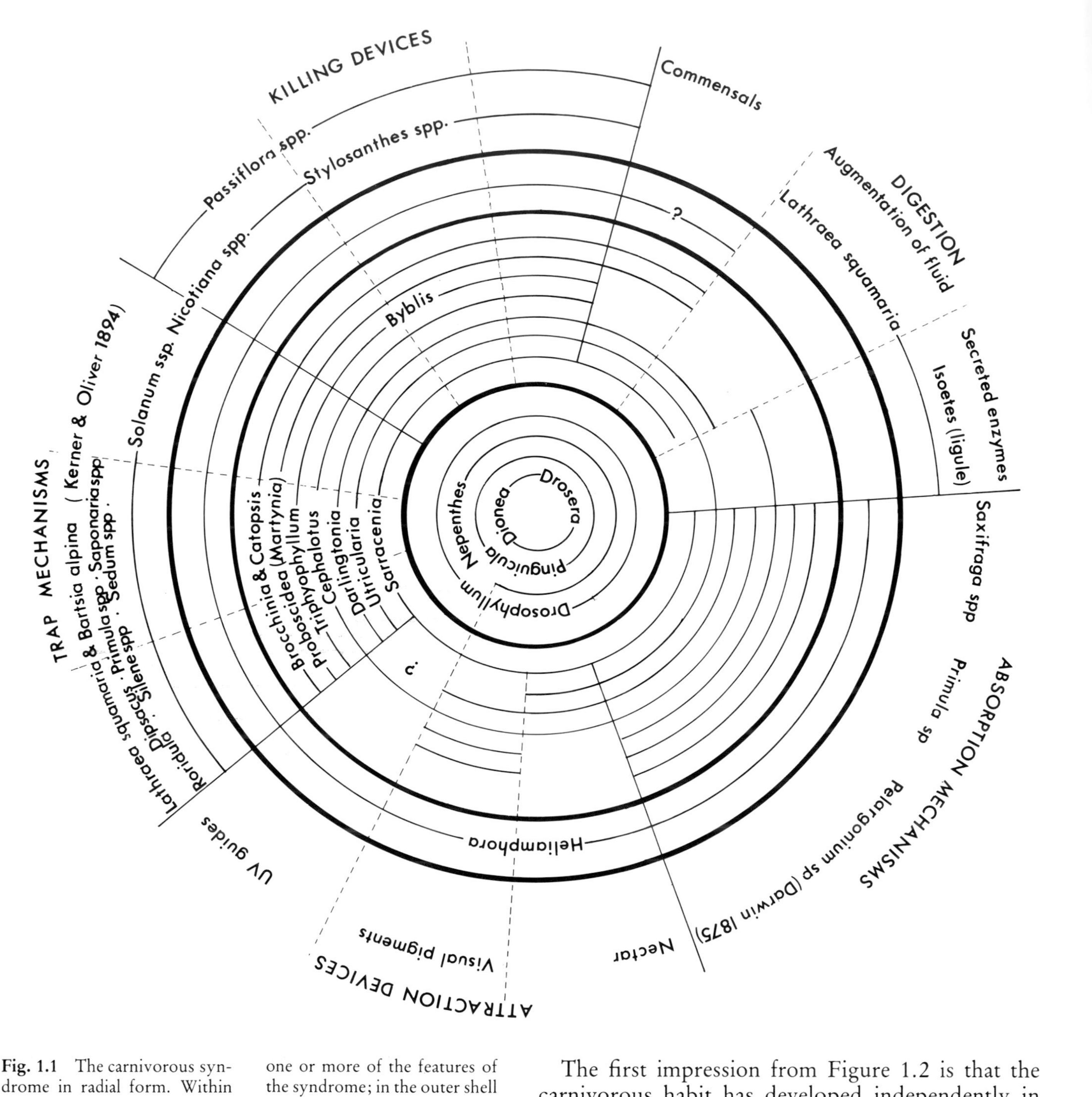

Fig. 1.1 The carnivorous syndrome in radial form. Within the inner ring are all those genera possessing every feature of the carnivorous syndrome. In the first 'shell' are those genera whose carnivory is widely accepted, but which lack one or more of the features of the syndrome; in the outer shell are those whose carnivory is in doubt; beyond the outer ring are examples of genera and species of plants with a few features of carnivory.

alter the first conclusion drawn below. This figure indicates where unique carnivorous orders are said, by Takhtajan, to occur or where carnivorous plant families or sub-tribes are thought to lie within more diverse orders.

The first impression from Figure 1.2 is that the carnivorous habit has developed independently in the plant kingdom at least six times. This may be more apparent than real, as we shall see later (cf. Fig. 17.3 and see Ch. 17–19). More recent ideas on the origin of flowering plants suggest that such rigid boundaries, as are suggested between groups in Figure 1.2, cannot represent the fluidity and reticulation that must have been, and probably still is, a feature of plant evolution.

It is interesting that this number of six is also the minimum number of different evolutionary events arrived at by Thompson (1981) leading to myrme-

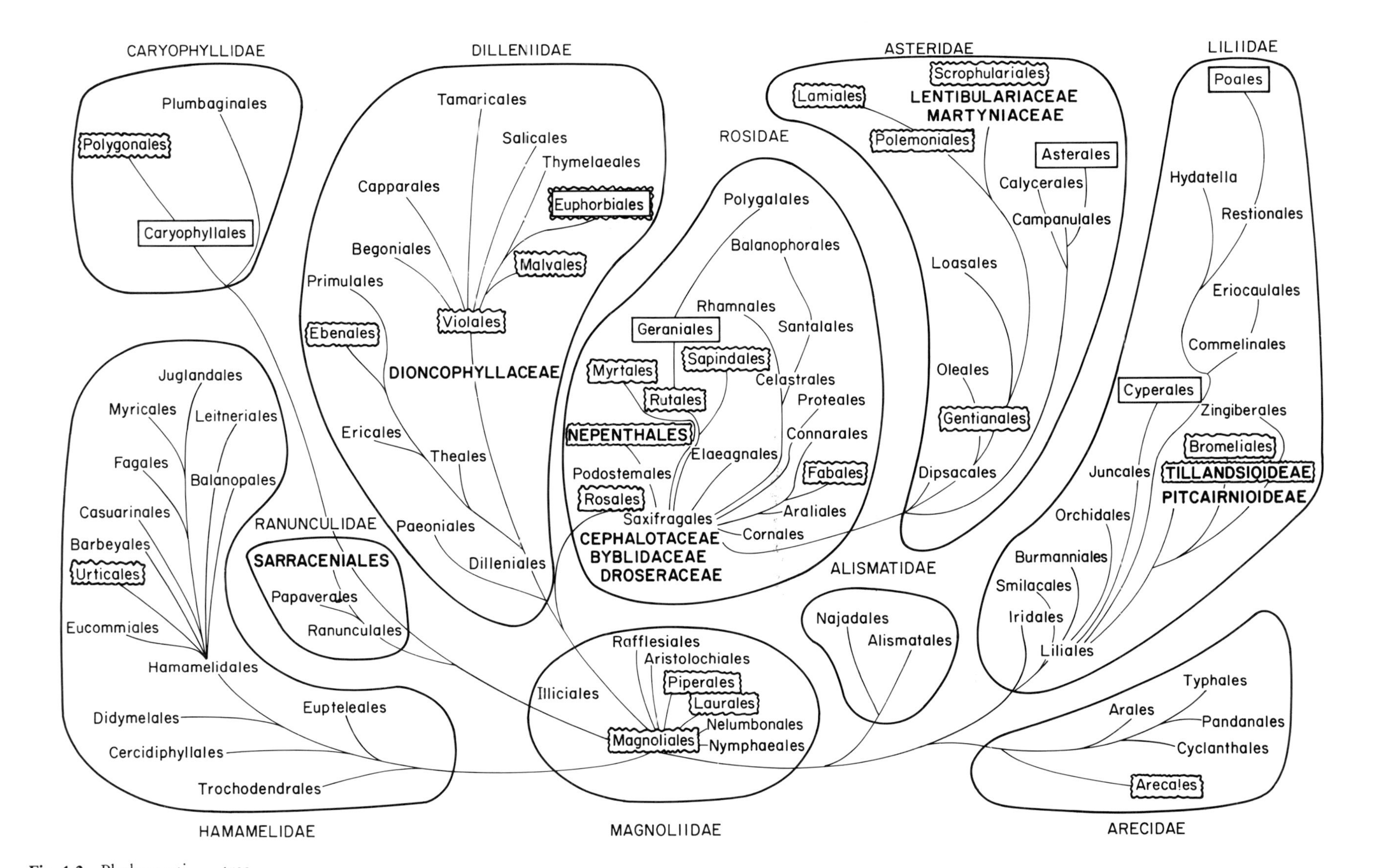

Fig. 1.2 Phylogenetic tree, redrawn with several modifications, from Takhtajan (1969). The carnivorous plant orders or families are shown in bold capitals, e.g. **BYBLIDACEAE**. C-4 plant families are shown [Poales]. Ant-plant orders and families are shown {Piperales}. Sometimes these syndromes overlap.

cophily, the 'ant-plants'. Like carnivory, myrmecophily is an adaption to exploit nutrient-poor habitats (Huxley, 1980; Thompson, 1981). It is interesting, too, that the number of species of both carnivorous and myrmecophilous plants is virtually the same, namely about 500 (Huxley, C.R. pers. comm., 1987). A base number as small as six may indicate that, given the catalogue of the earth's plant biota, the number of potential evolutionary solutions to the problems of low nutrient availability is limited.

Carnivorous Plants in the Angiosperms

It is not our intention in this text to deal critically with the taxonomy of the angiosperm carnivorous plants. Nevertheless, in Table 1.1 we provide a list of all the known carnivorous plants in their families,

Table 1.1 All the conventionally recognized carnivorous plants in families and orders, their approximate numbers of species and world-wide distribution

Order	Family and Genera	No. of Species[a]	Geographical Distribution
Sarraceniales	Sarraceniaceae		
	Heliamphora	5	Venezuela, Guyana and Brazil
	Sarracenia	8	eastern North America
	Darlingtonia	1	northern California and southern Oregon, USA
Nepenthales	Nepenthaceae	68	Eastern Tropics and westwards to Ceylon, the Seychelles and Madagascar
	Nepenthes		
	Droseraceae		
	Dionaea	1	North and South Carolina, USA
	Aldrovanda	1	Ubiquitous
	Drosophyllum	1	Portugal, Spain, Morocco
	Drosera	110	Ubiquitous, but about half the species confined to southwestern Australia
Violales	Dioncophyllaceae		
	Triphyophyllum	1	Sierra Leone, Liberia, southwestern Ivory Coast
Saxifragales	Byblidaceae		
	Byblis	2	Australia from northwest to southwest
	Cephalotaceae		
	Cephalotus	1	Australia, extreme southwest
Scrophulariales	Lentibulariaceae		
	Pinguicula	52	Northern and southern Hemisphere
	Utricularia	200	Ubiquitous
	Biovularia	2	One species in Brazil and one in Australasia
	Polypompholyx	2	Tropical Australia and South America
	Genlisea	15	Brazil, Guyana, Surinam, French Guiana, Cuba, West Africa
	Martyniaceae (related genera: *Proboscidea, Martynia*)		
	Ibicella	1+?	Southern USA and Mexico
Bromeliales	Bromeliaceae		
	Brocchinia reducta	1+?	Southeastern Venezuela and Guyana
	Catopsis berteroniana	1	Southern Florida to southeastern Brazil

[a]All species numbers, except *Ibicella* and the Bromeliaceae, from Schlauer (1986).

Table 1.2 Taxonomic and nomenclatural authorities for all the genera and sub-genera of carnivorous plants. All authorities in bibliography

Taxon	Authorities				
Darlingtonia	Torrey, 1850–1854				
Heliamphora	Maguire, 1978				
Sarracenia	Bell, (1949, 1952, 1954) Bell and Case, 1956 McDaniel, 1971 Case and Case, 1974, 1976 Schnell and Krider, 1976 De Buhr, 1975b, 1977a Schnell, 1978	Macfarlane, 1908a Uphof, 1936 Thanikaimoni and Vasanthy, 1972 Shetler, 1974a			
Nepenthes section 1 Aneurosperma	Schmid-Hollinger, 1979	Macfarlane, 1908b			
Nepenthes section 2 Eunepenthes		Danser, 1928			
Aldrovanda				Markgraf, 1954 Schmid, 1964	
Dionaea					
Drosera		Diels, 1906 Laundon, 1959	Shetler, 1974		
Drosera sub-genus *Ergaleium*	De Buhr, 1977b Kondo, 1976 Kondo and Lavarack, 1984				
Drosophyllum					De Candolle, 1873 Schlauer, 1986
Triphyophyllum	Airy-Shaw, 1951 Metcalfe, 1951	Carlquist, 1976a, and b, 1981 Cronquist, 1968, 1981			
Byblis					
Cephalotus					
Biovularia					
Pinguicula	Godfrey and Stripling, 1961 Ernst, 1961 Casper, 1966				
Utricularia	Lloyd, 1935 Reinert and Godfrey, 1962 Taylor, 1967 Robins and Subramanyam, 1980	Taylor, 1988			
Genlisea	Taylor, 1967 Fromm-Trinta, 1984 Fromm-Trinta and Taylor, 1986				
Polypompholyx	Taylor, 1964				
Ibicella (Proboscidea: Martynia) lutea	Van Eseltine, 1929 Hevly, 1969				
Brocchinia reducta	Smith and Downs, 1974–1979 Benzing *et al.*, 1985				
Catopsis berteroniana					

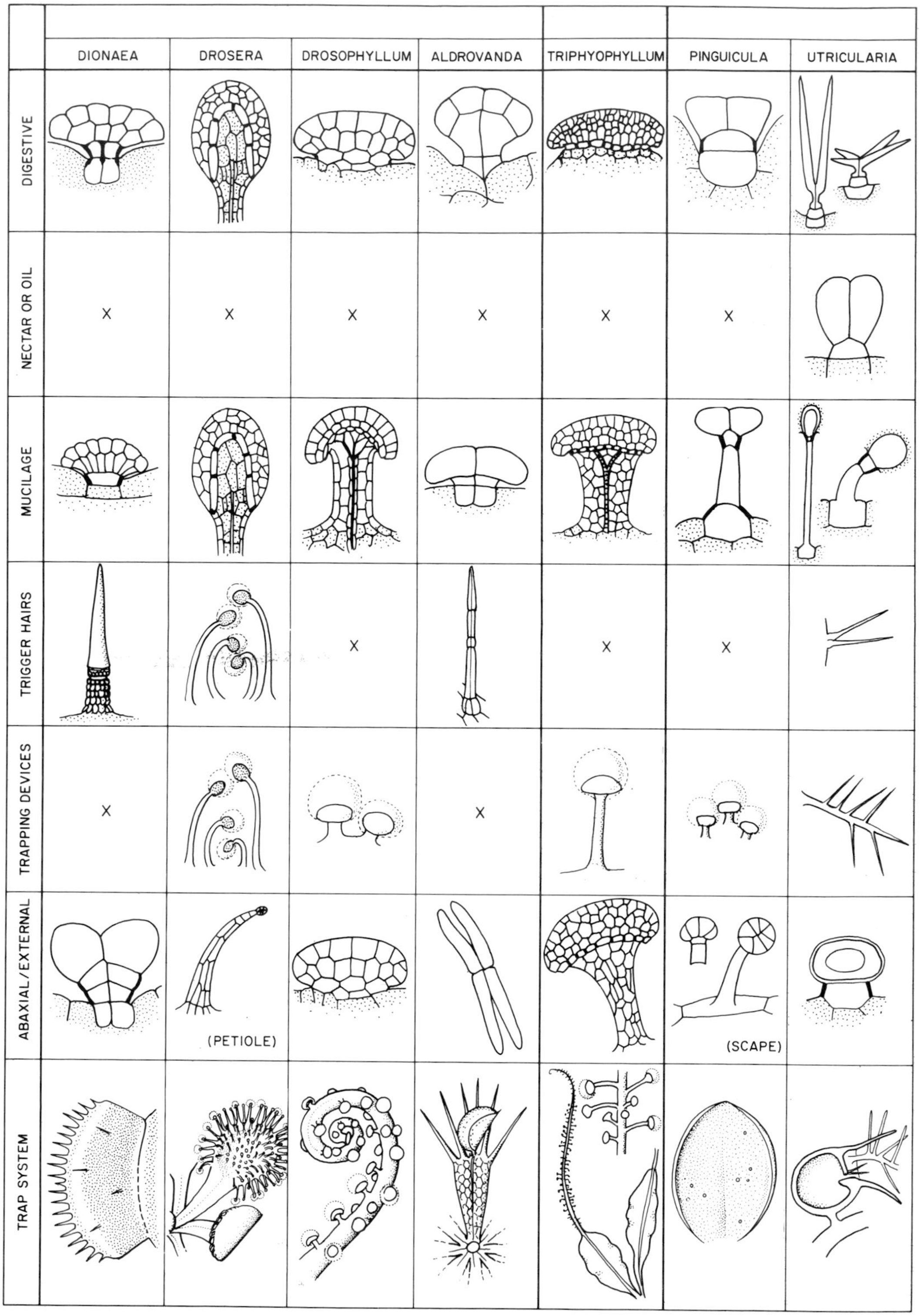

Fig. 1.3 Glandular structures and trap mechanisms of all known carnivorous plant genera.

	DIONAEA	DROSERA	DROSOPHYLLUM	ALDROVANDA	TRIPHYOPHYLLUM	PINGUICULA	UTRICULARIA
DIGESTIVE							
NECTAR OR OIL	X	X	X	X	X	X	
MUCILAGE							
TRIGGER HAIRS			X		X	X	
TRAPPING DEVICES	X			X			
ABAXIAL/EXTERNAL		(PETIOLE)				(SCAPE)	
TRAP SYSTEM							

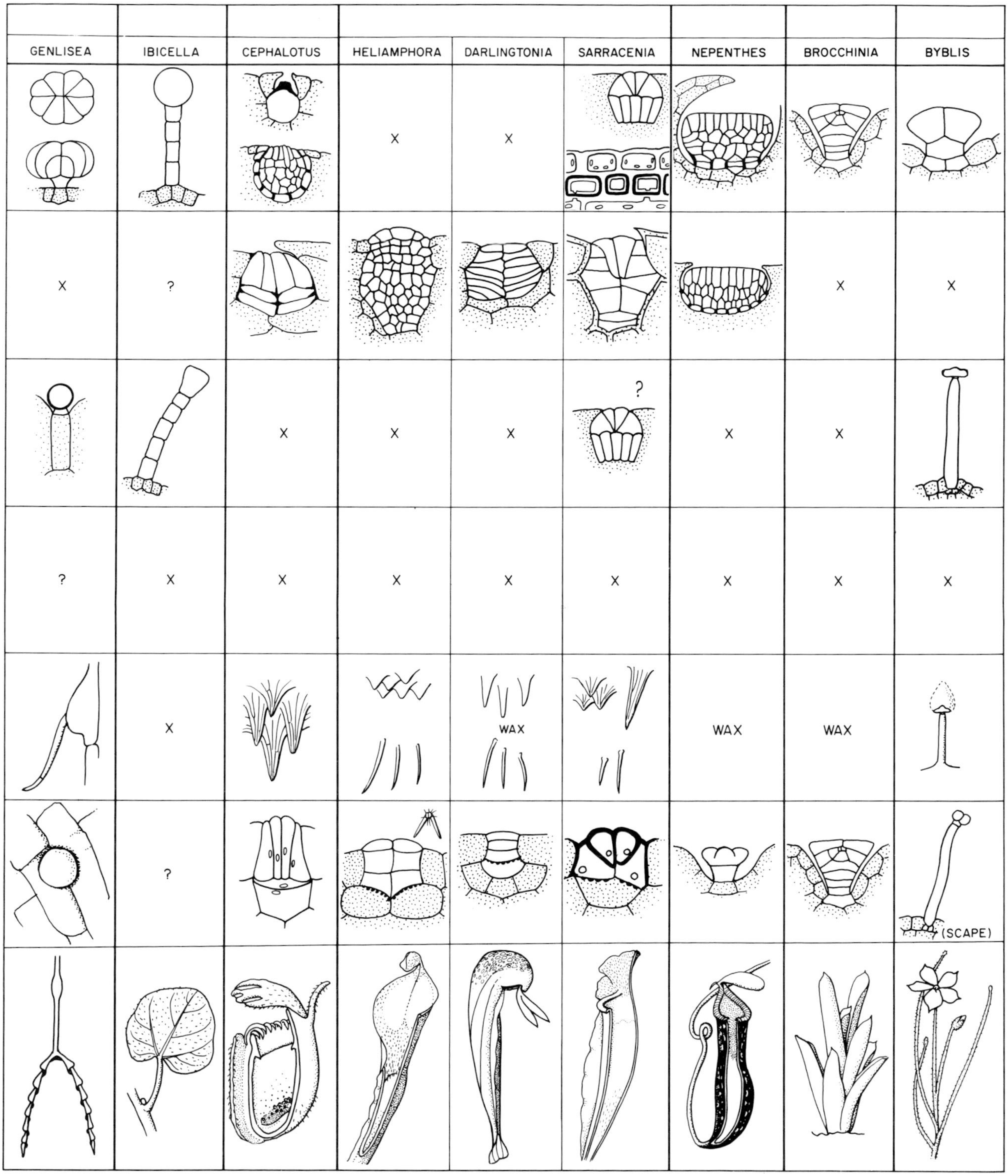

GENLISEA	IBICELLA	CEPHALOTUS	HELIAMPHORA	DARLINGTONIA	SARRACENIA	NEPENTHES	BROCCHINIA	BYBLIS
			X	X				
X	?						X	X
		X	X	X	?	X	X	
?	X	X	X	X	X	X	X	X
	X			WAX		WAX	WAX	
	?							(SCAPE)

with generally accepted species numbers and names according to Schlauer (1986). We also provide geographical distributions where relevant, but more detailed maps of certain species or genera are provided in Chapter 3. In the final columns of Table 1.2 we provide, uncritically, the most prominent taxonomic authorities relating to each genus or family.

The Concept of a Carnivorous Syndrome

All plants, above the level of the prokaryotes, can be looked upon as possessing clusters of characters (a syndrome) whereby the genotype is translated into the superficial features which confront the environment. Sometimes, as in C4 Fig. 1.2 or CAM plants, this syndrome is most familiar at the enzyme level. At other times, as in epiphytic plants or carnivores, the most obvious manifestations are anatomical and morphological. One of the most striking instances is that of the carnivorous plant, whose essential syndrome is summarized in Table 1.3.

Table 1.3 The carnivorous syndrome

1. (Attract)	4. Kill
2. Retain	5. (Digest)
3. Trap	6. Absorb useful substances

Most of the single features which together build up the sophisticated carnivorous nature of a plant, are pronounced to a greater or lesser extent in almost all seed plants. It is not necessary to assume that any one of these individual features has a common origin in evolution. Jeffree (1986) has shown that even a simple structure like a trichome can definitely be polyphyletic. A few of these features from Table 1.3 will now be exemplified.

Features of the Carnivorous Syndrome in Mesophytic Plants

Means of attraction are highly developed in the flowers of many, but not all, angiosperms. The secretion of mucilage is almost universal in root caps. This mucilage is produced in a virtually identical way to the 'fangschleim' of carnivorous glands, (see Ch. 6) but, in roots, serves principally to lubricate the passage of the root tip through the soil. Many flower systems retain, albeit temporarily, their pollinators. Movement is a common capability of many plants. Many of these movements are of course very slow, but others, not necessarily insect related, are very rapid (Hill and Findlay, 1981). The secretion of digestive enzymes is common to many germinating seeds and virtually every surface of a plant is capable of absorbing both large and small molecules. We shall expand several of these points later in Chapters 17 and 18.

The simple occurrence of these features in carnivorous plants is therefore of no surprise. The important point is that several such features develop in a certain limited area of the plant and that these features are expressed there to an extreme extent (Fig. 1.3). One analogy, which we have used in part before, is with a deck of cards: the cards are the same to all players; the expert player may assemble his or her 'hand' to the greatest advantage. This strong and integrated performance leads to the special capability, expressed as carnivory.

The fact that certain mechanisms are expressed in the extreme makes them, *inter alia*, a convenient and powerful model system for such basic plant studies as the study of transport systems, the study of sensory mechanism and the study of movement in plants.

The Evolution of the Carnivorous Syndrome

It is often a source of wonder to those not familiar with the diversity of the plant kingdom that the carnivorous habit could have evolved at all. The counter-argument runs as follows: since all the features of the carnivorous syndrome need great clusters of gene products and are required in precise synchrony, how can they have come about by normal evolutionary and selective processes? This facile argument ignores the fact that all the features of the syndrome that we consider essential to this style of life are, as we shall show, common and occur uniquely or occasionally grouped together in plants that do not rely on carnivory. However, the successful carnivorous plant must achieve a co-ordinated syndrome of characteristics. Not all carnivorous plants possess all the syndrome, the level of evolution is bound to vary from family to family, even from genus to genus, and some genera and species may be on the edge of carnivory. We have attempted to summarize this diversity in Figure 1.1, which presumes, in a centripetal fashion, to synthesize the various levels of development. The 'nucleus' and the inner 'shell' comprise those genera about whose carnivorous tendencies we believe there to be no dispute. The outer shell (e.g. *Heliamphora*) purports to show that, although a plant appears to possess much, if not most, of the syndrome, its ability as a carnivore may be weak (see page 16). Beyond the outer shell lie just a few examples of those genera and species with some extraordinary features, but which are not generally regarded as true carnivores. Every boundary line is certain to be contentious.

The Evolution of Detailed Features of the Carnivorous Syndrome

As already stated, for a plant to be carnivorous it should develop at least two distinctive and rare devices: traps and digestive organs. The addition of the strategem of attraction is, of course, of great advantage to a carnivore, although when too specific it might limit the range of prey and induce an unnecessary dependence on the limited availability of certain insects. It may come as some surprise to discover, as we shall see on page 291, how widespread is the phenomenon of indiscriminate insect-killing by conventional plants.

The attraction of insects for killing as opposed to pollination or dispersal

An initial step in the trapping procedure is, of course, the attraction of prey. This complex includes coloured markings, reflection and/or transmission of sunlight, odours, nectar, etc. (see Ch. 5). Nectar, when present, would seem to be the strongest and final temptation to prey.

By themselves most of these characteristics are, as we emphasize, of no particular note within the higher plant kingdom as a whole. These features are highly and widely developed in flowers mainly for the purposes of pollination, occasionally for seed dispersal (Van der Pijl, 1982), but sometimes also for defence, as via extra-floral nectaries in conjunction with ants. How carnivorous plants attempt to solve the paradox between attracting and killing prey, yet attracting and rewarding pollinators, we shall address later on (page 273).

An interesting, but not quite unique, feature of many a carnivorous plant's attractive armament is the glistening globule. Greenbottles, bluebottles and fleshflies all apparently head for glistening objects immediately they have hatched out. A similar device may be the basis of the pollination mechanism of *Parnassia*, where the sterile stamens form glistening knobs. To these conventional devices can be added a feature familiar to students of pollination, the ultraviolet (UV) guide. As has now clearly been shown, many, but not all, carnivorous plants have what appears to be a system of UV guides converging on the focus of the trap system (Joel *et al.*, 1985); these devices will be described in more detail in Chapter 5. From this initial step of attraction follow the equally important steps of trapping and retention, followed by digestion and absorption. Not only must all of these phenomena be assembled, they must be actuated in a precise sequence. What is the nature then of these important devices? The terms *trap* and *digestive gland* do not, in any sense, represent single characters. Each is built, as we shall see, of a series of structures, and biochemical and physiological mechanisms integrate the functions of trapping and digestion. Are they inflexible and immovable in their developmental sequences? We shall suggest possible answers to these questions in Chapters 17, 18 and 19.

CHAPTER 2

The History of the Study of Carnivorous Plants

Although occasionally not without unintentional humour, early studies of carnivorous plants do not, on the whole, reflect much credit on the observational powers of the early naturalists. There seems often to have been a marked reluctance on the part of otherwise highly respected scientists to admit to a carnivorous role for plants. This reluctance has a parallel in the far more significant inhibitions of many botanists, roughly over the same period, to admit to the sexual functions of flowers or even to the existence of sexual systems in plants at all (Delaporte, 1982; Mayr, 1983). There can also be no doubt that many competent artists failed, in their beautiful depictions, to record the dead insects on the sticky surfaces of, for example, *Pinguicula*, *Drosera* and *Drosophyllum*. Only Gerard (1633) left a few midges or gnats on his drawings of sundews, but he failed to realize their significance.

In the brief historical account that follows we have sought to indicate not only when the individual genus or species was first described but, more important to our argument, when its remarkable nature was first recognized. This understanding may come very much later than the actual discovery of the plant, and the year or spread of years is given after the generic name in the sub-headings.

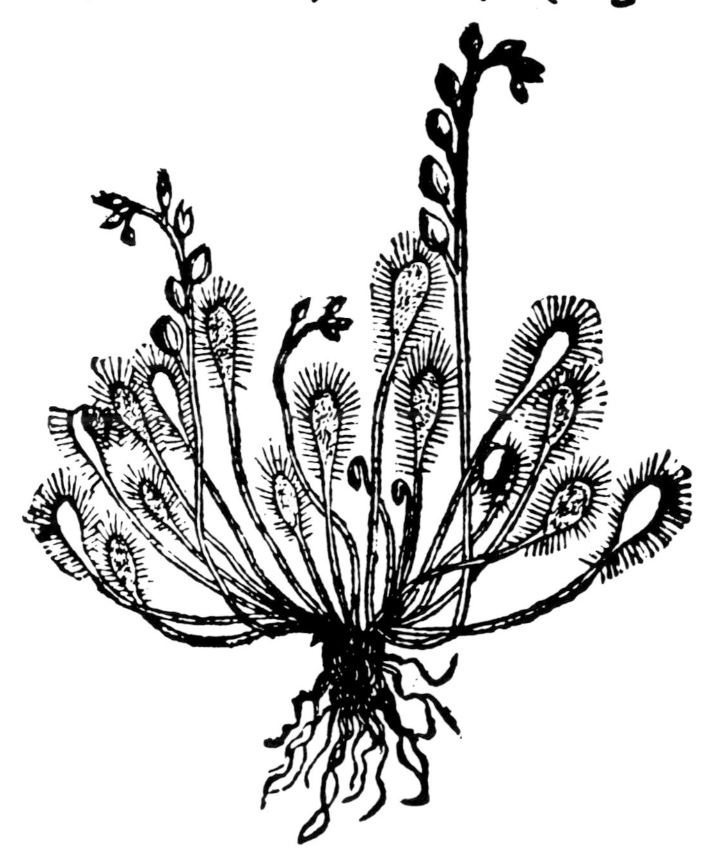

Fig. 2.1 A *Drosera* species, either *D. anglica* or *D. intermedia* from Dodoens' 'Cruydeboeck', the earliest illustration of a carnivorous plant. The first edition of this text was in 1548, but is not readily available. In later editions e.g. the third from 1618, the pagination changes and this illustration will be found on page 774. The title reads (in old Flemish) 'Sundew or Loopichweed' – the latter is roughly translated as 'Glandwort' – a medicinal herb with supposed efficacy against colds, i.e. moisture out of the ear or nose!
The same illustration will be found in Lyte (1578).

Drosera 1780–1875 (The sundews)

In our chronological study the large and widespread genus *Drosera* was not only the first to be described, but the first also to be understood. The first edition of Dodoens' 'Cruydeboeck' (1554) has, under the section on mosses, an excellent illustration of what is possibly *Drosera intermedia* or *D. anglica*. Further editions of the Dodoens, in various languages, such as the more readily available third Dutch edition (1618), changed the pagination and in this later version the *Drosera* appears on page 774. Henry Lyte translated the L'Écluse version of the Dodoens into English in 1578 and in this text the same figure, reproduced here as Figure 2.1, was copied yet again. In Lyte's 'A Nievve Herball' there is a verbatim translation of the original Dutch:

> The fourth kind (of moss) called *Rosa Solis*, hath reddishe leaves, somewhat rounde, hollowe, rough, with long stemmes almost fashioned lyke little spoones, amongst the whiche commeth up a short stalke, crooked at the toppe, and carrying little white flowers. This herbe is of a very strange nature and marvelous: for although that the Sonne do shine hoate, and a long time thereon, yet you shall finde it always moyst and bedewed, and the small leaves thereof upon this herbe, so much the moystier it is, and the more bedewed and for that cause it was called *Rosa Solis* in Latine, whiche is to say in Englishe, The dew of the Sonne, or Sonnedewe.

John Tradescant (senior) recognized it during his plant-collecting travels in northern Russia in 1618: 'Also Rosasollis I found theare . . .', he wrote; it was obviously familiar to him. (Tradescant, J., Ashmole M.S. 824; Leith-Ross, 1984).

In 1780 a Mr Gardom, an amateur botanist from Derbyshire, England, a London surgeon, Mr Whately, and, at almost exactly the same time, a Herr Roth in Germany noticed flies imprisoned within the contracted leaves of *Drosera rotundifolia* (see Withering, 1887). Some very simple experiments by these observers, subsequently confirmed by Withering (1887), showed that the *Drosera* leaves had the powers of curvature, the peripheral glands were able to bend (Fig. 2.2) and the gland heads to secrete upon the receipt of a stimulus. By 1790, Sowerby was able to write that, '. . . glandular bristles, each tipped with a clear viscid globule, which ornament the margin and surface the leaves, and which, by a degree of irritability lately observed by some naturalists, are thought to contract, and to imprison insects, like the Canadian (*sic*) *Dionaea*, a plant of the same natural order. This irritability, if it exists, is only to be detected in hot sunny days, nor indeed do the flowers expand in any other weather.' Nevertheless, in 1855 Trécul was prepared to deny emphatically that any such movement occurred, despite its corroboration by many botanists. In a standard text, as late as 1867 Duchartre, another French botanist, was prepared to admit the movement, but sought to deny that any plant structure could kill and absorb insect material. The elegant and all-embracing experiments of Charles Darwin on *Drosera* (1875) provided such overwhelming proof that sceptics received little further credence (see Fig. 2.2).

It is worth noting that, at first, Darwin was self-deprecating concerning his carnivorous plant experiments. He wrote to J.D. Hooker on 29 July 1860 '. . . latterly I have done nothing here; but at

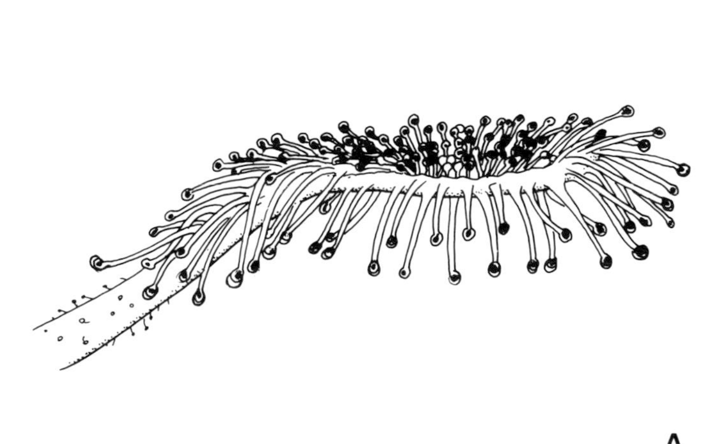

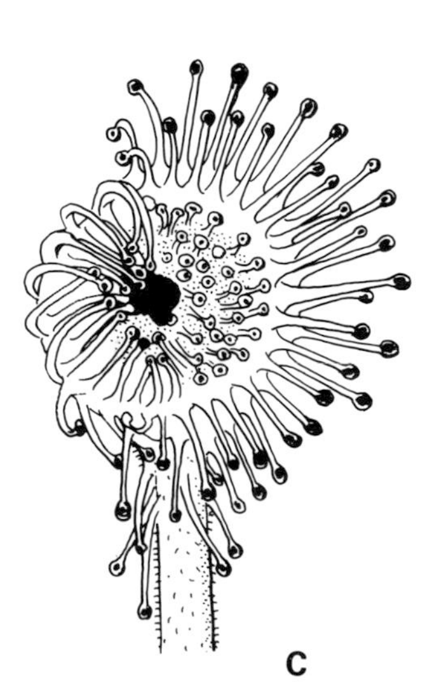

Fig. 2.2 *Drosera rotundifolia* from Darwin (1875). In the first edition these are Darwin's Figs. 2, 4 & 5. They are reproduced here with his original legends. (*A*) (left) 'Old leaf viewed laterally: enlarged about five times.' (*B*) 'Leaf (enlarged) with all the tentacles closely inflected, from immersion in a solution of phosphate of ammonia (one part in 87,500 of water).' (*C*) 'Leaf (enlarged) with the tentacles on one side inflected over a bit of meat placed on the disc.' (Redrawn by kind permission of John Murray, London.)

first I amused myself with a few observations on the insect-catching powers of *Drosera*, and I must consult you some time whether my "twaddle" is worth communicating to the Linnean Society . . .' (Darwin, 1887). Apart from Hooker, Darwin's other major stimulus in his monumental work on these plants was David Moore of the Glasnevin Botanic Gardens in Dublin who both supplied him with carefully grown carnivorous plants, which other sources were not often able to do, and communicated with him by letter with his own observations (Nelson and Seaward, 1981).

Sarracenia 1737–1791 (Pitcher-plant or Trumpet-leaf)

An early and quite accurate drawing of a *Sarracenia* followed close on the heels of Dodoens' picture of *Drosera*. Clusius (otherwise L'Ecluse or L'Escluse, 1601) published an excellent illustration of the vegetative parts of what was almost certainly a specimen of *S. purpurea*, from a sketch which in some way travelled from eastern North America, via Lisbon to Paris. According to Parkinson (1640) (who incidentally directly copied L'Éscluse's sketch and added spurious flowers) it was John Tradescant, junior this time (Leith-Ross, 1984), who was the first to send living plants home to Britain. His garden specimens do not seem to have survived his death nor did Morison's (see below); and the genus, in the form of cultivated plants, did not reappear in Britain until they arrived at Kew Gardens about 1773.

A puzzle is the incorrectly labelled sketch of what must be a *Sarracenia* on page 1435 of Gerard (1633). The plant can only be *S. minor* or *S. psittacina*, but how such a rare specimen found its way from Florida or thereabouts to England at this early date is unknown. An excellent sketch, with flowers, appeared in John Josselyn's 'New-England Rarities' of 1672 (Fig. 2.3). His ecological observations along with a depiction of its very limited root development were sound. He called it, however, *Hollow Leaved Lavender*. He was probably following L'Éscluse who in 1601, as we saw above, called his plant *Limonium peregrinum*; Gerard (1633), who copied him, called it *Limonio congener*. Limoniums are Plumbaginaceae, but are quite close, in evolutionary terms, to the Sarraceniaceae. Our present generic name comes from a Dr Sarrazin who, some time after Tradescant, sent material from Quebec to Paris.

Morison (1699) attempted a serious description of the genus, which he called *Coilophyllum* (Gk. hollow leaf) *virginianum*. Morison wrote 'D.J. Banister sent our plants from Virginia, with drawings of both species; like countless other marsh plants they seem not to take too kindly to cultivation' (see page 37). He mistook too the function of the terminal lobe which he described as '. . . serving as a lid, capable of moving like a hinge, frequently used to cover the mouth of the cavity . . .' and, of the second species, '. . . a lip or covering which in these species serves as a lid which seems to have been designed by divine providence to shield and protect the plant from damage by rain, and which bears a striking resemblance to the lid of a box, or the cover of a wooden jar' [our translation]. This 'hinge' idea was copied by many subsequent authors. In fact, the terminal lobe never moves, nor indeed does the pitcher elongate or grow further after the hood has opened. Like those of Josselyn's, Morison's ecological observations were, however,

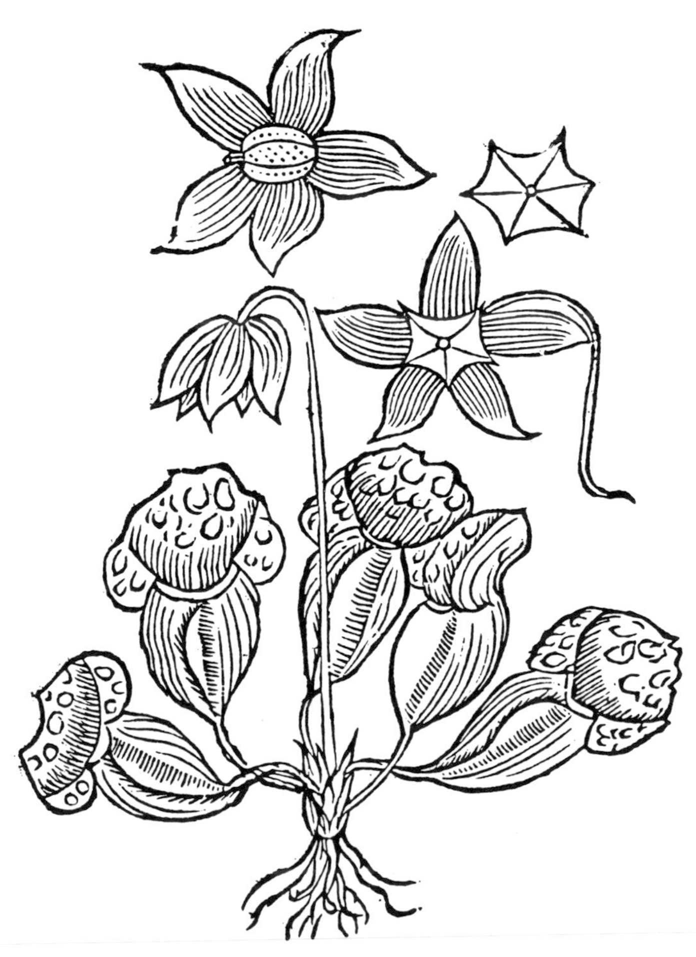

Fig. 2.3 The 'hollow-leaved lavender' (*Sarracenia purpurea*) from John Josselyn's *New England's Rarities Discovered*, 1672.

sound: 'It likes cold, wet places, not the spongy ground among cranberry' [our translation].

Catesby (1737–54) who depicted *S. purpurea* very accurately, along with *S. minor*, but called the former *Sarracena canadensis*, spotted that the pitchers were often full of insects and suggested that these permanently filled reservoirs might act as a refuge or habitat for certain insects from the frogs that preyed on them (also carefully depicted in the above work). Later research has demonstrated that the pitchers can contain a complex micro-environment (see Ch. 14), but that this is not its principal function. Collinson (1765; see Henrey, 1975) got very close to the truth when he remarked in a letter to Linnaeus that '. . . many poor insects lose their lives by being drowned in these cisterns of water . . .', but it was Bartram (1791) who finally realized that *Sarracenia* both catches and kills insects, and evidently digests them in some way. Any ideas of enhanced nutrition were far in the future.

Macbride (1818), studied *Sarracenia minor (adunca)* with great care and appreciated that there was a very sophisticated trapping mechanism 'The fly which has thus changed its situation, will be seen to stand unsteadily [on the edge of the pitcher], it totters for a few seconds, slips, and falls to the bottom of the tube where it is either drowned, or attempts in vain to ascend against the points of the hairs.' The essential truth eluded him, however, and he concludes: 'What purpose beneficial to the growth of these plants may be effected by the putrid masses of insects, I have never ascertained.' However, in the next sentence, Macbride made an observation of extraordinary prescience:

> '. . . but I learn from a hint given in the article *Dionaea*, in Rees's *Cyclopaedia*, that it has been discovered that the air evolved is wholesome to plants. I once entertained a suspicion that this air might be of such a deleterious nature as to cause the precipitation of insects exposed to it, but I have long since relinquished it as entirely groundless.' (See pages 252 and 269; Cameron *et al.*, 1977; Joel and Gepstein, 1985).

It was not until 1829 (Burnett) that the suggestion was made that *Sarracenia* could digest insects in a way analogous to that of a human stomach. Burnett again repeated that the lids of *Nepenthes* and *Sarracenia* could open and close as lids, but in an otherwise light-hearted, almost frivolous article, made some telling observations. He wrote that he had observed '. . . the sundew . . . fattening on the delicates that it had caught . . .', and, noting that there were some animals not digested but apparently using the pitchers of *Nepenthes* and *Sarracenia* as habitats, observed that '. . . even this simple digestive apparatus is not free from intestinal worms.' (see page 251)

Dionaea 1769–1834 (Venus's Fly-trap)

The extraordinary properties of *Dionaea* were, in strict chronological terms, the next to be appreciated and again the names of Bartram (1791) and Collinson appear. In 1769 Ellis, who achieved other fame from his better known text 'Directions for Bringing Over Seeds and Plants from the East Indies' (1770) sent Linnaeus a drawing and a good description of a plant to which he had given the name of *Dionaea*; the name being derived from one of the synonyms for Venus. He stated that he had received the specimen from Peter Collinson (see Rendle, 1925: Blanche Henrey, 1975) who had been given it by John Bartram of Philadelphia, botanist to the late king. Ellis wrote to Linnaeus:

> '. . . Nature may have some views towards its nourishment in forming the upper joint of its leaf like a machine to catch food: upon the middle of this lies the bait for the unhappy insect that becomes its prey . . . the two lobes rise up, grasp it fast, lock the rows of spines together, and squeeze it to death . . . three small erect spines are fixed near the middle of each lobe, over the glands, that effectually put an end to all its struggles.'

The overall description is so accurate as to indicate, at least second-hand, that it was based on field observation; although certain details, in particular the role of the spines, were erroneous. Enormous interest was generated by this extraordinary little plant and even the President of the United States, Thomas Jefferson, reported in his correspondence that he had collected and distributed *Dionaea* plants from Charleston in 1787 (see Boyd, 1955). Linnaeus, however, with all the facts in front of him, but only dried specimens to study, refused to believe the obvious. He wrote (see Smith, 1821 and Darwin, 1888, p. 243) that as soon as the insects ceased to struggle the leaf lobes opened up and let them go. It was a botanical draughtsman, a Mr Warner of Winchester, who in 1804 spotted (see Edwards, 1804) that the spines on the lobes were the only sensitive part of the plant. Not until 1834 did the Rev. Dr Curtis of Wilmington, North Carolina, publish the first completely accurate description of the operation of the *Dionaea* lobes. Research on this extraordinary plant more or less stood still until the 1870s. Darwin demonstrated that, as with *Drosera*, the lobes of *Dionaea* move slowly in response to chemical stimulation, though much of his work on *Dionaea* remains unpublished and can be found in Jones (1923).

The next major event was Burdon Sanderson's epoch-making discoveries of 1873–74. In the teeth of opposition, particularly from some members of the botanical establishment such as Sachs (1887), he was able to show, in a series of elegant experiments, that an action potential very similar to that observed in stimulated animal muscle, could be propagated across the leaf lobe from a stimulated trigger hair. Again the established critics were trying to deny the animal nature of plants, though Burdon Sanderson's 'live' demonstrations published by the Royal Society were quickly accepted.

Cephalotus follicularis 1800–1823 (Australian pitcher-plant)

According to Lloyd (1942), *Cephalotus follicularis*, the Western Australian pitcher-plant, was first collected by Archibald Menzies, naturalist to the Vancouver Expedition of 1791. In December 1792 (Willis, 1965), the naturalist with the expedition of Bruny d'Entrecasteaux, La Billardière, landed somewhere in the area, but the locations as Willis (1965) and Carr and Carr (1976) point out, are confused, and it is not at all clear that he collected the plant himself. In fact, Willis (1965) is of the opinion that La Billardière never got within 200 miles of any of the known locations of *Cephalotus*. According to Willis, he must have received dried samples from some other voyager and he could not have seen Brown's samples, since Brown did not reach England until October 1805. La Billardière described the plant in detail in a communication to the French Institute in July 1805. As the source of the material Willis favours Antoine Guichenot, one of the gardeners with Leschenault's expedition, who certainly collected the material in the King George's Sound area in February 1803, and who returned to France in 1804.

Who first realized this plant's carnivorous nature? It must have been Robert Brown in December of 1800 (see Mabberley, 1985, his p. 84) who, in the King George Sound area with Flinders' expedition again collected the plant. Flinders wrote:

> Amongst the plants collected by Mr Brown and his associates, was a small one of a novel kind, which we commonly call the pitcher plant. Around the root leaves are several little vases lined with spiny hairs, and these were generally found to contain a sweetish water, and also a number of dead ants. It cannot be asserted that the ants were attracted by the water, and prevented by the spiny hairs from making their escape; but it seemed not improbably, that this was a contrivance to obtain the means necessary either to the nourishment or preservation of the plant.

The plant was painted by the artist on board, the gifted Ferdinand Bauer, and is reproduced in Mabberley (1985). Whatever the honours of discovery, La Billardière did describe it for the first time in 1805 (see Dakin, 1919; Grieve, 1961; Willis, 1965), hence his authority *C. follicularis* Labill. By 1823, when it was introduced to Kew Gardens by Philip Parker King, W.J. Hooker was well aware of its diet and noted, as Brown and Flinders had originally, that the genus appeared mainly to trap ants (Hooker, 1831).

Heliamphora 1846 (Sun-pitcher? or Marsh-pitcher: see Glossary)

Heliamphora was discovered by Sir R. Schomburgk on Roraima, in what was British Guiana, in 1838 (see Brown, 1901). Bentham (1840) described it in detail but cast doubts on its carnivorous ability. Lindley (1846) noted its secretory nature and virtually accepted its carnivorous habit by reason of its similarity to *Sarracenia* and the recently discovered *Darlingtonia*. Oddly enough J.D. Hooker does not mention *Heliamphora* in his comprehensive review of the carnivores in 1874 (Hooker, 1875) and in his description of *Heliamphora* in the Curtis's 'Botanical Magazine' of 1890 does not comment on its carnivory either. It may be that, both as we have observed in the Botanic Garden in Oxford and by personal observations of recent collectors in the Roraima area (Burras, J.K., pers. comm., 1979), *Heliamphora* is not a particularly avid collector of flies (see page 42 for a comment on the proposed evolution of the habit).

Byblis 1846–48 (Rainbow plant)

In Australia, there are two known species of *Byblis*: *B. liniflora* from the north, in the Gulf of Carpentaria area; and *B. gigantea* from the extreme west around Perth, separated, in fact, by almost the complete span of the Australian continent (Bentham and Mueller, 1864). Lindley noted their existence in 1846 and Planchon (1848) described them in detail. There is no specific mention in either of these reports of their carnivorous habit, but the extensive comparisons with both *Drosera* and *Dionaea*, whose heterotrophic mode of nutrition was recognized by this date, suggest that their peculiar diet was well appreciated. Oddly enough, as with *Heliamphora* above, *Byblis* was not mentioned by Hooker (1875) in his general review.

Darlingtonia californica 1853 (Cobra plant: Cobra lily)

Darlingtonia californica (Frontispiece), closely related to *Sarracenia*, restricted to a few counties of North California and South Oregon, was discovered in October 1841 by Mr J.D. Brackenridge, Assistant Botanist of the US Exploring Expedition under Captain Wilkes (Lloyd, 1942). He found it somewhere in the Mount Shasta region; the exact location is unclear – not surprising since he was being pursued by Indians at the time. John Torrey described it in 1850 (his manuscript was not published until 1854) considering it sufficiently different to justify a new genus and naming it after 'my highly esteemed friend Dr. William Darlington, of West Chester, in Pennsylvania, whose valuable botanical works have contributed so largely to the scientific reputation of this country.' It was not brought into cultivation until 1869 (Canby, 1874), it first flowered in the eastern United States in 1870, and, as reported by J.D. Hooker, in 1871 in England (Cheatham, 1976). It was not studied in detail until 1875–77 by Austin (Ch. 3 and 14) whose pioneering field work has not previously been published (see Appendix 1).

Nepenthes 1858 (Pitcher-plant: Monkey's cup, Dutchman's pipe)

The genus was being described as early as the middle of the seventeenth century by Governor Flacourt of Madagascar (cited by Lloyd, 1942). The new plant was called *Amramitico* in 1658, but the name did not persist (Mellichamp, 1979). For a fuller account of these early discoveries and observations see Veitch (1898).

Linnaeus gave the plant its present name in 1753, in allusion to the story in Homer's *Odyssey* where Helen mixed the wine with the drug *Nepenthe* (Gk. literally 'no mind') so that by drinking it man might be freed from care and grief. The shape of the pitchers in some species resembles the Greek rhincton or drinking horn.

However, as we have seen, Linnaeus was not inclined to accept the carnivorous nature of plants and, despite many beautiful depictions by Fitch in Curtis's Botanical Magazine in the first half of the nineteenth century, not until 1858 was the first tentative suggestion made as to its carnivorous proclivities. Hooker W.J. (1858) writes '. . . the pitcher . . . no doubt is a great provision of nature for decoying and for the destruction of insects.' He was writing of *N. villosa*, recently collected from Kina-Baloo (*sic*), Borneo.

Aldrovanda 1861 (Waterwheel plant)

Aldrovanda, originally named *Aldrovandia*, has a very limited literature. According to Lloyd (1942), this odd little plant, which Darwin called a miniature aquatic *Dionaea* (Darwin, 1875, his Ch. IV; and see p. 53) was first observed by Plukenet in India in 1696. But it was not until 1861 that De Sassus (cited by Darwin, 1875) first called attention to the irritability of the leaves. Because of its small size and the relative difficulty of keeping it in cultivation, very much less work has been done on this monotypic genus than the closely related *Dionaea*.

Drosophyllum 1869 (Dewy-pine: Portuguese sundew)

Drosophyllum was known as long ago as 1661 (Grisley) and in 1689 Tournefort described it in his Portuguese flora as *Ros solis lusitanicus maximus*. The relationship to *Drosera* was obviously appreciated. It was beautifully depicted by W. Fitch in Hooker (1869) and Hooker, like Tournefort, observed that this odd carnivore was very similar to *Drosera*. He wrote '. . . the glandular hairs, . . . have rigid pedicels that are not endowed with the motive power of those of English sundews which curve towards their prey when once it is entangled.' There seems no doubt that he comprehended its carnivorous nature at this date. Darwin (1875) made a limited study of *Drosophyllum lusitanicum*, which he states was well known as a fly-catcher to the Portuguese around Oporto, who hung it up in their cottages for that purpose. Lloyd (1942) casts doubt on this statement and it is certainly true that although *Drosophyllum* is a very effective catcher of small flies in the living state, it soon desiccates and becomes inoperative when dead.

Pinguicula 1873? (Butterwort)

Pinguicula, in this case *P. vulgaris*, seems to have been brought to the attention of Charles Darwin as a possible carnivorous plant by a Mr W. Marshall of Cumberland. It is not completely clear from Darwin's account which year this observation took place, but it must have been shortly before 1875. Even then some very observant workers were still not totally convinced. In his entertaining little book *Flowers and their Unbidden Guests* Kerner (1878) wrote (of *Pinguicula alpina*): 'The primary function of the glandular trichomes on the leaves of *Pinguicula*, and numerous other plants, is certainly therefore, to keep off those creeping insects whose bodily dimensions are so small that their visits would not bring about allogamy; but this, of course, does not exclude the possibility of such insects as get caught and remain adherent being digested, and serving as welcome, if *not very luxurious food*' [our italics].

Genlisea 1874

According to Lloyd (1942), *Genlisea* was first discovered by Auguste de Saint-Hilaire in Brazil in 1833. Other species were discovered in Africa, such as *G. africana* in Angola. The Cuban species found by C. Wright about 1866 (cited by Lloyd, 1942) has never been found again. *G. guianensis* and *G. roraimensis* were found on Roraima by McConnell, Quelch in 1894 (Brown *et al.*, 1901). Incidentially this expedition also found *Brocchinia reducta* (see page 19). *Genlisea's* carnivorous nature does not seem to have been properly appreciated until it was studied by a Danish worker, E. Warming (1874), cited by Darwin (1875).

Utricularia 1876 (Bladderwort)

Sowerby (1797) came very close to comprehending the nature of the extraordinary *Utricularia* trap, but made the understandable error of thinking the bladders were flotation devices. He wrote: '. . . bristly leaves bearing little compressed curved bladders, open and bearded at the tip, containing a bubble of air, and a drop of watery fluid, in which, when highly magnified, Dr Withering observed a quantity of extremely minute solid particles. *Aquatic insects frequently take up lodgings in these bladders*' [our italics].

In 1867, Duchartre, who as we have already seen was generally unconvinced of the carnivorous habit of plants, was still talking about the bladders as if they were flotation devices 'they are, at first, filled with a somewhat gelatinous liquid which weighs them down and then they serve to keep the plant on the bottom. Soon the four-branched hairs, which line the interior, secrete a gas which accumulates while the liquid diminishes; in this way the plant becomes more buoyant and since it is not restrained by a root system detaches itself from the mud and rises slowly to the surface . . .' [our translation]. Darwin (1875) made one of his very rare mistakes in observation when studying the *Utricularia* trap by believing that the aquatic animals forced their way into the trap. According to Lloyd (1942), it was an American botanist, Mrs Mary Treat, who, having studied *Utricularia* for some years, reported (Treat, 1875, 1876) that in *U. purpurea* the prey was suddenly engulfed as if drawn into a partial vacuum. She observed that the engulfed larvae were digested within 48 hours and concluded that the little bladders were like so many stomachs, digesting and assimilating food.

Ibicella lutea [syn. *Martynia* & *Proboscidea*] 1875?: 1916 (Devil's Claws)

This genus, until recently rarely included in a list of carnivorous plants, was first tentatively suggested as a carnivore by Beal (1875). He carried out only a little experimental work and his contribution was missed by Lloyd (1942). Mameli (1916), who seems not to have known of the work by Beal and was, in her turn, not noticed by Lloyd, established its carnivorous nature. Mameli's study of the capture, degradation and absorption of insect prey on the surface of *Ibicella lutea* (Lindl.) (which she called *Martynia*) is comprehensive. She writes, in her summary, 'However I can state now that the principal result of this study demonstrates the capacity of *Ibicella* (*Martynia*) *lutea* effectively to catch insects and to absorb albuminous substances which naturally or experimentally come in contact with its leaves' [our translation]. We have grown *I. lutea* (Fig. 2.4) under greenhouse conditions, and Cornelius Muller (pers. comm., 1982) confirms (see page 309) that in its native habitat the leaf surfaces are viscid (Fig. 2.4) and often covered with trapped insects, particularly gnats and small dipterans. However, to our knowledge, no further work on this genus in respect of its carnivorous habit has been attempted since 1916.

Triphyophyllum peltatum 1965–1979

Triphyophyllum peltatum is a recently recognized example of the evolution of the carnivorous habit within a family many members of which have only some of the features of carnivory (Schmid, 1964). It was first discovered in the Ivory Coast by A. Chevalier in 1907 and named *Ouratea glomerata*. It is now known to exist, precariously, in Liberia and Sierra Leone. Airy-Shaw (1951) described it in detail, and its affinities with the Nepenthaceae and Droseraceae were noted (see page 284) by both he and Metcalfe (1951) in his paper on the anatomy of the Dioncophyllaceae. Yet neither author, despite pointing out the very remarkable morphological and anatomical similarities between the tentacles of *Triphyophyllum* and those of *Drosophyllum*, suggest a carnivorous habit for this plant.

Again no mention was made of its carnivorous habit by Schmid (1964), although the affinities with the other carnivorous genera were noted. Menninger (1965) speculated tentatively that one of the leaf forms might have a carnivorous role and Marburger (1979) strongly supported this view with detailed comparisons between the glandular structures of *T. peltatum* and those of *Drosera* and *Drosophyllum*. But it was not until the definitive field observations of Green *et al.* (1979), combined with their laboratory work at Kew, that the carnivorous nature was established.

Brocchinia reducta, ? *B. hectioides* and *Catopsis berteroniana* 1974

There is no reason to suppose that the total list of the world's carnivorous plants is closed. Many

Fig. 2.4 *Ibicella lutea*, a native of southern California and northern Mexico. (Original by Rosemary Wise.)

Fig. 2.5 *Brocchinia reducta* of the wet but infertile Venezuelan savannah. (Original by Rosemary Wise.)

plants, as we shall see in Chapters 17 and 18, may retain insects. These insects may drown and decay within the plants' cavities, but the plant does not fall within our precise definition (see Ch. 1) because it lacks some feature of the carnivorous syndrome.

It was inevitable that members of the Bromeliaceae should have been investigated since their occupancy of often arid or oligotrophic habitats, together with possession of an imbricate impounding leaf form is suggestive of a carnivorous habit. Three species so far, *Brocchinia reducta* (Fig. 2.5) (Smith and Downs, 1974; Givnish *et al.*, 1984), the closely related and possibly con-specific *B. hectioides* (Benzing, pers. comm., 1986) and *Catopsis berteroniana* (Fish, 1976; Benzing *et al.*, 1976; Frank and O'Meara, 1984; Benzing, 1986) have been found to fulfil all the requisites. All three are slightly anomalous bromeliads. The brocchinias are Pitcairnioideae, and are always terrestrial on acid, wet, sandy soil in full sun in the Guyana Highlands. *Catopsis* is a Tillandsioideae bromeliad, but is a little different from the conventional epiphyte in growing in full sun at the top of red mangroves (*Rhizophora mangle*) in Florida. It is yellow-green in colour with an upright habit and is covered with a thick, slippery, white, epicuticular wax. It has been suggested that the most stringent limiting factor for the conventional shade-living epiphytic bromeliad is water, not nutrients (Thompson, 1981; Givnish *et al.*, 1984; Benzing *et al.*, 1985; and see page 144); hence the apparent absence of true carnivory amongst most epiphytic groups of plants.

A few more species of carnivorous plants will undoubtedly be found. Likely areas are desert fringes, cf. *Ibicella* and *Drosophyllum*, and amongst growth forms further candidates are almost certain to emerge from the aroids, bromeliads and similar tank-forming genera such as *Dipsacus*. Nor is it beyond the realms of speculation that completely new forms of carnivorous plant might arise, based on mutant epiascidiate-leaf-forming species, so readily does this striking aberration arise, as we shall see in Chapter 4.

CHAPTER 3

The Habitats of Carnivorous Plants and their Regional Distribution

The General Habitat Preference and Strategy of Carnivorous Plants

The carnivorous plants, in general terms, substituted leaves for roots as they exploited low-nutrient environments. They have taken advantage of the insect fauna within these environments for their normal nutrition, but also seem, for the most part, to have lost the ability to compete in a normal, calcium-rich mesophytic habitat. The net primary production has been partitioned (see Ch. 7) and, having diverted part of their biomass into the trap mechanisms, they have left themselves as marginal performers in terms of photosynthesis and conventional soil-nutrient scavenging. It is now generally agreed that carnivorous plants are able to exist satisfactorily as mesophytes and, up to a point, use soil nutrients, but do best when supplied with insects. This still contentious issue will, however, be discussed fully in Chapter 7. Givnish *et al.* (1984) have discussed a cost/benefit model, with particular reference to the bromeliads, for the general restriction of carnivorous plants to moist habitats which are poor in nutrients, but which enjoy high light-intensities (page 145). As we shall see below, this restriction is not wholly true.

The genus *Utricularia* and its relatives, although well-attested carnivorous plants, are so different from the general run, in growth form, habitat and life cycles, that they will be considered separately (page 43).

A Tabulated Summary of Ecological Preferences

Summaries are inclined to conceal, behind crude generalizations, more than they reveal. However, if we attempt to tabulate the commonly identified characteristics of carnivorous plants, the exceptions to this overall pattern may reveal other truths.

Carnivorous plants are:

(i) generally weak-rooted perennials of varying longevity of variable habit and life form, but often forming huge clonal colonies by stolons or rhizomes;
(ii) generally calcifuges;
(iii) intolerant of plant competition, almost invariably that of other flowering plants, but sometimes even that of other carnivorous plants and mosses;
(iv) generally tolerant of low-nutrient environments;
(v) intolerant of shade or of low-light conditions;
(vi) tolerant of and frequently opportunistically dependent on low-temperature fires;
(vii) tolerant of temporary or permanent waterlogging.

We shall attempt to exemplify these broad headings by reference to certain specific carnivorous plants on page 25 onwards.

The Carnivorous Plant and its Root System

Most carnivorous plants have either weak root systems, e.g. most *Drosera* species (well depicted by Nitschke, 1860), *Dionaea*, *Pinguicula*, *Cephalotus*, *Sarracenia* and most *Nepenthes*, or no roots at all, e.g. *Utricularia*, *Genlisea* and *Aldrovanda*. There are, however, four totally unrelated species, *Drosophyllum lusitanicum*, *Byblis gigantea*, *Ibicella lutea* and *Triphyophyllum peltatum*, that have substantial if not necessarily phreatophyte-type root systems.

It may be more than a coincidence that all of them live in acid (low-nutrient) habitats which are seasonally very dry. They are all 'fly-paper' types of trap, but none of them possesses any power of movement. The four tend to occur as individual plants, dependent mostly, if not entirely, on seed distribution and not, so far as is known, forming clonal colonies by any form of vegetative reproduction. In contrast, all the weakly-rooted 'fly-paper' type carnivorous plants, except for the tuberous droseras which we shall describe in detail on page 26, have at least some power of movement. We shall discuss the significance of this apparent allocation of resources in Chapter 7.

The seedling stage

Regardless of whether or not the carnivorous plants in question subsequently develop a vestigial or vigorous root system, it appears that all *seedling* carnivorous plants studied have feeble root systems. *Drosophyllum*, although it subsequently develops a conventional root, is very slow to anchor to the soil and, long after germination, even when the seedling is 2–3 cm high, can still be displaced or pushed over with ease. Green (1967) writes of *Nepenthes gracilis*:

> 'The seedling root is a very meagre one, and often does not appear until after the cotyledons; wind can uproot the small plants as well as dry them. If the humidity is high, the seedlings survive even in a seemingly dry soil, though it is not clear how this is achieved. A bunch of long, tough root hairs appears outside the seed case even before the tip of the root shows, but these are not enough for anchorage and they are doubtfully enough for absorption.'

[Stern (1917) [cited by Green] suggested that the seed coat was important, supplying water by capillary action along its furrows.

The General Distribution and Growth Form of Carnivorous Plants

Carnivorous plants are world-wide, but each carnivorous species occupies a restricted micro-habitat. They can be found from low to high altitudes and latitudes; in extremes of cold and aridity, in the semi-shade of tropical forests (e.g. several *Drosera* species); and in 'toxic' streams from serpentine rocks (e.g. *Darlingtonia* and several *Nepenthes* species). Detailed distribution maps of certain species or genera will be given (see Figs. 3.5, 3.6, 3.7, 3.8, 19.5, 19.6 and 19.7). However, it would serve no purpose to give distribution maps for many genera such as *Drosera, Utricularia* and *Aldrovanda* which are widespread. So far as is known, there are no totally halophytic examples of carnivorous plants although *Dionaea*, *Darlingtonia*, perhaps *Sarracenia purpurea* (Butler, 1985) and certain species of *Nepenthes* are known to be salt-tolerant.

In growth form, carnivorous plants range from minute rosettes, (*Drosera occidentalis* is 6 mm across and *D. pygmaea* 10 mm across), to tropical forest climbers (*Triphyophyllum peltatum* and some *Nepenthes* species) most of which, as we have seen, are perennial to a greater or lesser degree. Some carnivorous plants, e.g. *Drosera erythrorhiza* (Dixon and Pate, 1980), are known from tuber sheaths to be over 50 years old, and other clonal colonies, e.g. *Darlingtonia* (see page 38) are known to have occupied the same area for over 100 years. Others are more peripatetic. Individual species of many of the genera may persist over a winter or a dry season as tubers or turions (see page 27), many Western Australian and South African droseras come into this category. On the other hand, some such as *Darlingtonia* and *Sarracenia purpurea* endure extremes of low temperature, and the trap systems, plus their commensal fauna, emerge from the snow to continue their existence in a new spring. Just a few species are able to complete their life cycle during one single season, such as those species of *Utricularia*, *Genlisea africana* and *Drosera* growing in the warm and wet West African summer (Green *et al.*, 1979). For the rest of the season these annual species persist as seed.

The Carnivorous Plant as an Epiphyte

Until comparatively recently, the existence of obligate epiphytic carnivorous plants could be disputed. It is true that several species of *Utricularia* (Slack, 1986) can be found growing in the wet moss at the bases of the dripping trunks of tropical forest trees (Thanikaimoni, 1966; Sugden and Robins, 1979). On the South American mainland and Cuba respectively, *Pinguicula moranensis* and *P. lignicola* and *P. casabitoana* (*cladophila*) in the mossy forests in the mountains of Haiti can be found actually growing in the trees of the tropical forest. Even more convincingly, *Utricularia nelumbifolia*, in an unspecified *Brocchinia* or *Tillandsia* (see page 261), and *U. humboldtii* in *Brocchinia cordylinoides* appear to be obligate phytotelmatic inhabitants (see page 260) of the pitchers of these non-carnivorous bromeliads. Certain *Nepenthes* species (see page 34) definitely lose contact with the ground entirely. However, most convincing of all is the newly discovered carnivorous plant, the obligate epiphyte, *Catopsis berteroniana*. This tillandsioid bromeliad ranges from southern Florida through many islands of the Caribbean to the Greater Antilles and, on the mainland, to southern Mexico and eastern Brazil. It is an epiphyte over all its range, commonly reported as living high in the branches of the mangrove *Rhizophora mangle*, in other mangroves, e.g. *Conocarpus* and *Avicennia*, and, away from the mangrove rim, in conventional forest trees to 1200 m in the mountain forests.

The Carnivorous Plant as a Calcifuge

Carnivorous plants are generally believed to be calcifuges; the detailed position is not so clear. Some, e.g. *Drosophyllum* and many *Drosera* and *Sarracenia* species, are strictly so; however, *Drosera falconeri*, the newly discovered species from Northern Australia (Tsang, 1980; Kondo, 1984) is not, nor is *D. erythrorhiza* (see page 26). *D. anglica* (see page 27) is normally found in strictly calcifuge sites, but here and there it can be found on soils of a basic pH, such as calcareous tufa (Bussy, 1974). Similar habitats have been noted for many *Pinguicula* species (Hadač, 1977, and see Ch. 19; see Studnička, 1981; Diaz Gonzales *et al.*,1982; Taylor and Cheek, 1983). Lime-rich soils generally imply a luxuriant vegetation. However, *P. vallisneriifolia* (Diaz Gonzales *et al.*, 1982) evades the competition which seems to restrict so many carnivorous plants (see page 24), as we shall see, by growing in the precarious habitat of wet limestone cliffs in southern Spain (Fig. 3.1). Another limestone-cliff dweller, totally unrelated but avoiding the competition in a similar fashion, is *Nepenthes clipeata*, a non-climbing species from Borneo (Slack, 1986). Most other *Nepenthes* seem restricted to arid soils (see page 35). In a sense, these two totally unrelated species seem to be behaving like the epiphytes discussed on page 34. Most *Nepenthes*, other than *N. clipeata* seem to be restricted to arid soils (see page 35). Some *Utricularia* species will tolerate a degree of calcium, but the epiphytic bladderworts seem to be strict calcifuges (Slack, 1986).

However, it should be noted that any consideration of the calcifuge/calcicole phenomenon must take into account not only the quantity and availability of the calcium ion in the soil but, as Rychnovska-Soudkova (1953, 1954) points out, also the pH of the soil. This phenomenon is discussed on page 134.

Fluctuations in the availability of calcium

Most carnivorous plants, then, grow in calcium-deficient environments. As we shall see later, where that factor has been studied, they do not seem to depend upon their prey as a calcium source (Ch. 7). The calcium content of most of their prey, e.g. dipterans, hemipterans, arachnida, must be negligible (see Gibson, 1983a) but snails are commonly trapped in *Nepenthes* pitchers as are tree frogs and lizards in *Sarracenia* (Jones, 1921) and, very occasionally, fish in *Utricularia* (Simms, 1884; Grudger, 1947). In the very acidic environment of the trap, the shells and skeletons must be partly degraded. Do certain carnivorous plants have selective ion carriers for Ca^{++} in their plasma-membranes as are known for other ions (see Chs. 9 and 19)?

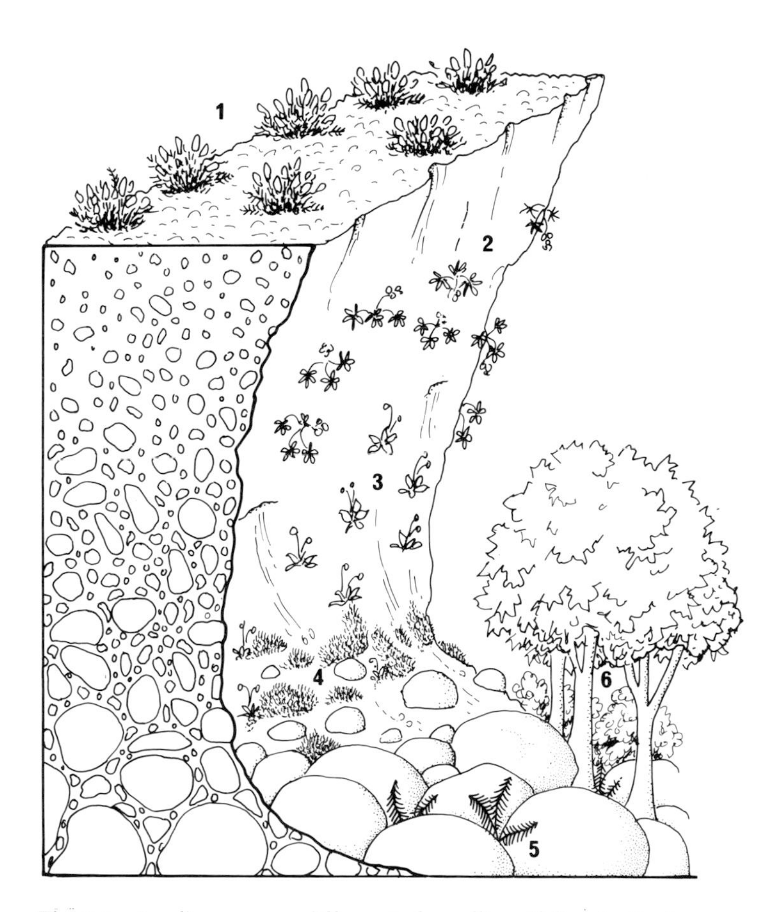

Fig. 3.1 A limestone cliff cross-section through the Salto del Caballo (Sierra Tejeda, Granada, Spain). Altitude 1700 m. 1. *Erinus alpinus*; 2. *Potentilla caulescens*; 3. *Pinguicula vallisnerifolia*; 4. *Eucladium verticillatum*; 5. *Dryopteris/Cystopteris*; 6. *Daphne/Acer* woodland. (Redrawn from Gonzales *et al.*, 1982, with permission.)

Calcium-intolerance and former vegetation patterns

This general inability to penetrate high pH soils dominated by mesophytes may sometimes offer a clue to past vegetation patterns. *Drosera rotundifolia*, a strict calcifuge over its immense range, is still found in the Lebanon Mountains on the Nubian sandstone and many hundred of kilometres south of its present range, suggesting a much wetter Lebanon in the Wurm 1 period (Feinbrun, 1942).

General Features of the Tolerance of Carnivorous Plants for Other Plant Species

Carnivorous plants of many different species, genera, growth forms and trap mechanisms may grow side by side. In fact, other carnivores, as well as *Sphagnum* moss and a few species of orchids, are virtually the only plants generally found growing in the immediate vicinity (Knight *et al.*, 1970; Nash, 1973; Beckner, 1979; Hindley, 1980; O'Connell, 1981; Fowlie, 1985). It may be that *Sphagnum*, as Bell (1959) has pointed out, is capable of markedly altering the relative concentrations of cations of different valencies in natural solutions flowing through it. In this way the moss alters the pH, reduces the list of potential competitors and thus paves the way for invasion of the carnivorous plant. The only exception to this apparent association with *Sphagnum* seems to be *Aldrovanda vesiculosa* which occurs only in places that are more or less continuously flooded with water, but not in those with a dense *Sphagnum* cover (Fijalkowski, 1958).

Few carnivorous plants appear able to compete successfully with other forms of vegetation. This inability is exemplified by Gibson's (1983a) study of the carnivorous plants of the bogs of the south-eastern United States. As his graph shows (Fig. 3.2), there is an almost perfect inverse relationship between the number of carnivorous plants per square metre and the weight in grams of the competing conventional plants. This lack of competitive ability can be expressed in another way. As Schwaegerle (1983) showed with *Sarracenia purpurea*, also in the eastern United States, pitcher-plants have a relatively low intrinsic rate of increase. In other words, compared with normal herbaceous plants, their rate of population increase, even under apparently ideal conditions, is slow. There are, however, certain very interesting exceptions to this competition rule, as we shall see on pages 36, 42 and 44.

By trapping and concentrating large quantities of protein and phosphate-rich prey, terrestrial carnivorous plants will tend to enrich the soil in which they live, so providing a basis for competitive invasion. This phenomenon may account, in part, for their failure to continue to occupy a given locality for long periods. The exceptions (see pages 27, 38 and 39) are interesting.

General Features of Carnivorous Plants' Preference for Soil Type

In general, carnivorous plants exploit infertile soils, often arenaceous or granitic, that are predominantly wet or seasonally wet. However, the widely held belief that carnivorous plants only occupy damp or wet acid environments is erroneous. *Drosophyllum*

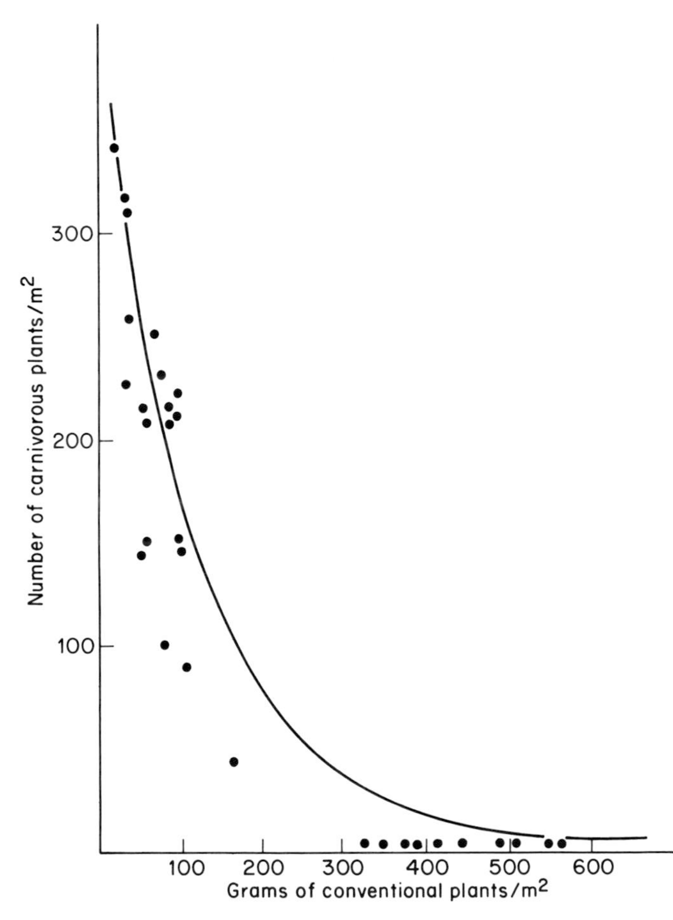

Fig. 3.2 Maximum density of all carnivorous plant series as a function of conventional plant density. The maximum number of carnivorous plants/square metre decreases as the density of conventional bogplant biomass increases, mostly grass. Carnivorous plants tend to occur only in the sparsest areas within a bog. This relationship includes over 11 species (*D. brevifolia*, *D. capillaris*, *D. filiformis*, *S. psittacina*, *S. rubra*, *S. flava*, *S. leucophylla*, *S. purpurea*, *P. planifolia*, *P. lutea*, *P. caerulea*). (Redrawn with permission from Gibson, 1983a; Fig. 34, p. 127.)

(see page 30), around the Iberian peninsula, and *Ibicella*, in the Sonoran Desert of the south-west United States, continue to function as effective carnivores on arid sandstone soils throughout many months of total drought. Many species of the tuberous droseras (see page 27) avoid the long, hot, dry months by aestivation. On the other hand, many carnivorous plants that are not normally aquatic, e.g. *Drosera burmanni*, *Dionaea* and some *Sarracenia* species, can endure total submersion for a while (see Ch. 19 for a comment on the evolution of *Aldrovanda*).

In a similar manner to the ant-plants (Huxley, 1980), an exploitation of insect protein has enabled this group to penetrate nutrient-poor habitats. The whole nutrition argument will be reviewed in Chapter 7 but, in brief, Chandler and Anderson (1976c) have shown that certain *Drosera* species, when supplied with the dipteran *Drosophila* as a prey, grow better with low soil nitrogen levels than do similar plants with a full inorganic nutrient supply but no flies. The same may be true of *Dionaea* (see Ch. 7 and Roberts and Oosting, 1958). Thus, we shall argue that the carnivorous habit generally confers most advantage in a nutrient-poor habitat where, with their particular insect-scavenging properties, carnivorous plants will, at least temporarily, be at an advantage. It seems to be generally true, although not tested specifically for carnivorous plants, that a species adapted to a low-nutrient habitat cannot compete where the nutrient supply is high (Grime, 1979). Carnivory, in a brutal analysis, is just one of the innumerable ruts into which evolution has forced certain groups of the plant kingdom. It puts severe restraints on where any particular species can live. On the other hand, it has been suggested by Aldenius *et al.* (1983) that an insect diet, perhaps by supplying a broad range of trace elements as well as the macro-nutrients, enables the carnivorous plant to increase its efficiency of nutrient uptake from the soil. It is worth noting too, that so far as we can ascertain from our own observations and from the literature, no member of this group forms a mycorrhizal association with its roots (Peyronel, 1932; Dudderidge, J., 1984, unpublished observations on *D. capensis*). MacDougal (1899) confirmed the general absence of mycorrhizal associations with carnivorous plants, but did note one instance of a vesicular–arbuscular-like invasion of the roots of a *Sarracenia purpurea* which otherwise showed no disease symptoms.

There may be no external or vesicular–arbuscular mycorrhizal association known in carnivorous plants, but we should note the odd, and as yet unexplained, phytotelmatous association between a *Mucor* species and a nematode in a *Sarracenia* pitcher (see page 258).

General Ecological Features of Carnivorous Plants as Exemplified by Certain Genera and Species

Although the carnivorous plants are generally tolerant of nutrient-deficient soils, they possess three features of intolerance which we will examine in some detail: they are unable to cope generally with *competition*, *desiccation* and *shading*. The properties developed in response to these pressures are diverse.

In this brief summary of the distribution and ecology of some of the carnivorous species it is not our intention to be comprehensive; very little indeed is known of the ecology of many species. However, we have selected some species of the genera *Drosera* and *Nepenthes*; some genera of the *Sarraceniaceae*, the species of *Byblis*; and three of the monotypic genera, *Dionaea*, *Drosophyllum* and *Triphyophyllum*, for deeper consideration. They illustrate, in their various growth forms and habitat preferences, certain attributes common to all in the group. A summary appears on page 45 and the special case of *Utricularia* on page 43.

The Distribution and Ecology of Certain *Drosera* Species

Drosera must be considered, at least in species and growth-form terms, one of the largest and most diverse of the carnivorous plant genera, with over 90 known species and outnumbered only by *Utricularia*. As we shall see in Chapter 19, the *Drosera* group (Fig. 19.1), or its ancestors, has almost certainly given rise to the very different genera of *Aldrovanda* and *Dionaea*.

The centre of diversity of the genus would appear to be Australia and New Zealand, where the greatest range of growth forms and habitats is observed (DeBuhr, 1977b). Over 40 species are found in a limited area of the southwest of Western Australia (Fig. 19.7). Southern Africa too is a rich source, whereas only a few species occur in the north temperate zone. Of these, two are circumboreal, *D. anglica* and *D. rotundifolia*; one nearly so, *D. intermedia*; and, in temperate North America, there are four additional 'good' species: *D. brevi-*

folia, *D. capillaris*, and *D. filiformis* of the southeastern United States (see Fig. 19.6), and *D. linearis* of the northern States and Canada (Shetler, 1974b; Schnell, 1976).

In growth form, *D. gigantea* of Australia may, in exceptional specimens, reach over a metre tall, and *D. filiformis* of the United States Gulf Coast almost half that height. On the other hand, *D. pygmaea* and *D. occidentalis* of Australia hug the ground and have leaves less than 1 cm across.

Generally, all species of *Drosera* observe the calcifuge rule of carnivorous plants, but *D. linearis*, which grows from Labrador, west around the Great Lakes and south to the State of Michigan in the USA, is notable both for colonizing alkaline marl bogs (Wood, 1955; Schnell, 1980, 1982b) and its extreme lack of competitive ability. Very frequently it is accompanied here by *Sarracenia purpurea* which, as we shall see later on, can tolerate a wide range of soils from acid to alkaline.

As we saw earlier, almost all carnivorous plants are intolerant, to a greater or lesser degree of competition, except that of *Sphagnum* species and a few other mosses and, commonly, other carnivorous plants. O'Connell (1981) noted that, in an embryonic raised bog in Ireland, *Sphagnum contortum* and *Drosera rotundifolia* were almost universal and the only other plant of a high leaf-area index was *Menyanthes trifoliata*. Apart from *Sphagnum* species, Lavarack (1981b) noticed that *Byblis liniflora* in Australia was frequently found with *Drosera indica* with which it is often confused and more loosely is associated with *D. petiolaris*, *D. spathulata*, *D. burmanni*, *Utricularia chrysantha*, and, in the northern part of its range, *Nepenthes mirabilis*.

Some other angiosperm associations have also been noted. Hindley (1980) observed that two species of lady-slipper orchids, *Cypripedium reginae* and *C. pubescens* grow in association with *Drosera rotundifolia* in bogs in Vermont, USA. The three species were comparatively rare in the 8 hectare (20 acre) bog, but the sundews were always found in association with the orchids, although apparently similar habitats were abundant throughout the whole area. This association with, or tolerance of, orchids was also noticed in a study by Nash (1973) of one of the tuberous droseras, *D. macrantha* (*planchonii*) of South Australia. The specialist orchid growers, too, are aware of the tolerance of orchids for carnivorous plants, particularly *Drosera* and *Pinguicula* (Beckner, 1979). Certain species of *Nepenthes* (see page 36) and *Darlingtonia* (see page 41) also appear to tolerate orchids. Most of these observations seem to be suggesting that the carnivorous plants tolerate orchids. The tolerance may be reciprocal. On the other hand Tsang (1980) noticed that the newly discovered *Drosera falconeri*, grew only in dense grassy areas; but since this species seemed tolerant of relatively alkaline conditions (pH 8) it may also be an exception to our general competition rule.

This general inability to tolerate competition, except that of *Sphagnum* (as we shall see again below with *Dionaea*), was noted as long ago as 1633 by Gerard (p. 1557). He wrote (of *Drosera rotundifolia*): 'They grow in desert, sandie and sunny places, but yet waterie, and seldome other where than among the white marsh which groweth on the ground and also upon bogs.'

Tolerance of shade

Drosera is also generally intolerant of shade and requires a specific combination of temperature and light intensity for successful flowering and colony establishment (Pissarek, 1965). This light-intensity/temperature requirement would seem to be a further limitation, but again there are exceptions.

There are three species peculiar to the tropical northeast of Australia: *D. adelae*, *D. schizandra* and *D. prolifera*. All three are unusual in that they grow in hot, humid, rain-forest conditions and show what appear to be conventional responses to low-light conditions, e.g. the normally rather small leaves of most members of the Droseraceae have been replaced with large broad leaves (Lavarack, 1979). Could it also be that, as in so many cases carnivorous plants seem to have exchanged roots for leaves, these larger leaves serving to intercept nutrient-rich foliar leachings under the canopy of the tropical forest? As we shall see in Chapter 7, on extraordinary features of nutrition, some *Drosera* species may benefit from trapping pollen grains and spores on their leaves; they are not obligate carnivores.

The Habitat Preferences of North Temperate *Droseras*

The sundew species in northern latitudes are not very abundant nor, for the most part, has their ecology been studied in any great detail. This is in contrast to the more numerous southern droseras whose ecology (see page 27) and nutrition (see Ch. 7) have been studied in depth. There is, however, a detailed study of ombrotrophic (rain-nourished) mires of northern Britain in which, *inter*

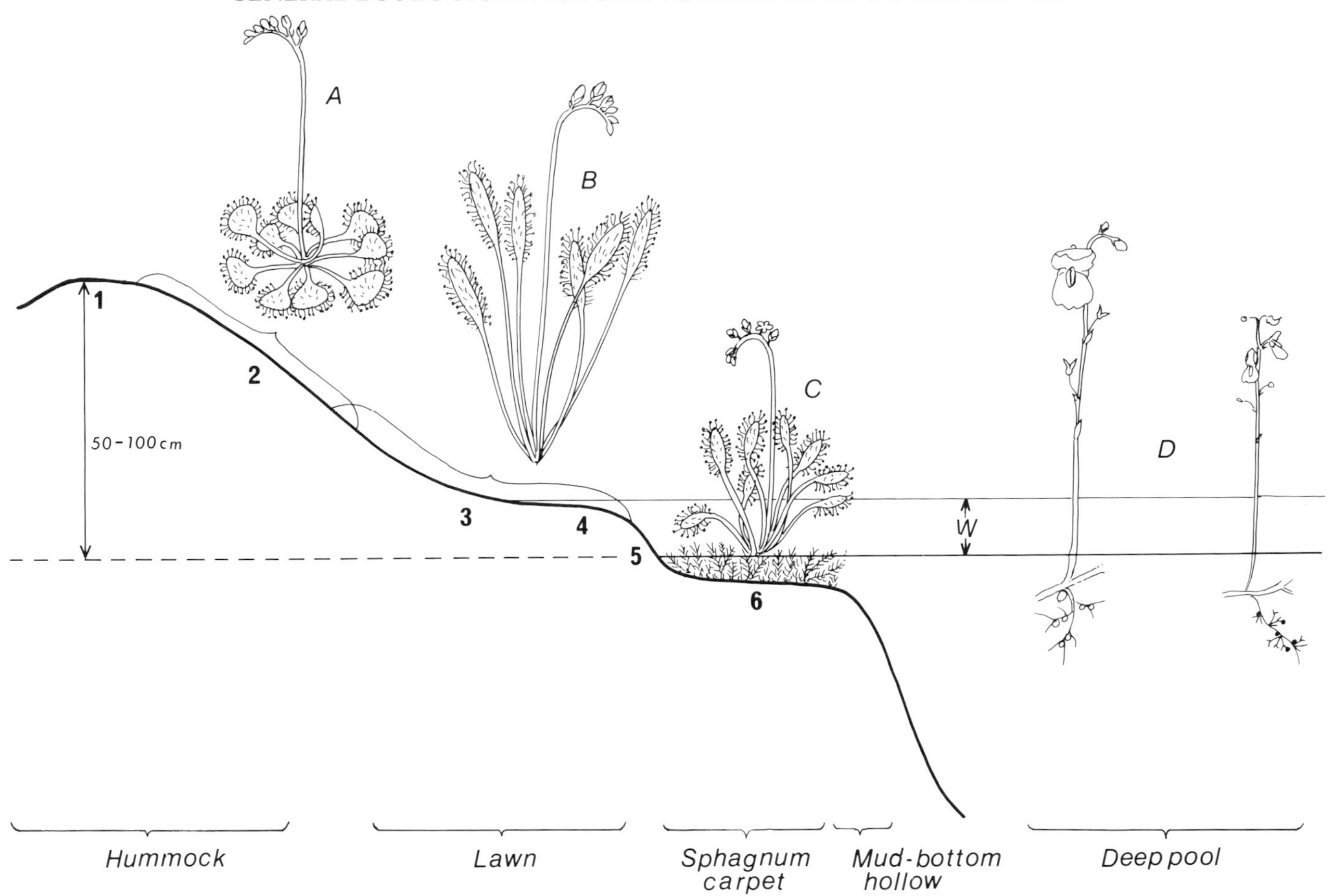

Fig. 3.3 The local distribution of northern European droseras in a blanket bog. (Data from Lindsay 1983, 1985; Caroline Aldridge, pers. comm., 1986). (*A*) *Drosera rotundifolia*; (*B*) *D. anglica*; (*C*) *D. intermedia*; (D) *Utricularia vulgaris* and *U. minor*. 1. *Sphagnum imbricatum;* 2. *S. rubellum;* 3. *S. magellanicum*; 4. *S. papillosum*; 5. *S. pulchrum*; 6. *S. cuspidatum*. N.B. The water level (W) may fluctuate considerably, but the long-leaved sundew (*D. intermedia*) appears not only to tolerate partial submergence, but to continue catching insects during this period.

alia, Lindsay and his colleagues (1983, 1985) have studied the microtopographic distribution of the three British species of *Drosera*. On belt transects, 10 cm^2 quadrats showed that on the 'hummock' (at least 10 cm above the mire water table) was *D. rotundifolia* (Fig. 3.3). At the 'lawn' level, *D. rotundifolia* overlapped with *D. anglica*. *Drosera intermedia* was restricted to the '*Sphagnum* carpet' or 'mud-bottom hollow'. Below the droseras, in the water-filled hollows, were also found *Utricularia vulgaris* and *U. minor*. To what extent these zonations reflect prey-partitioning by the different species will be discussed on page 141.

The Ecology of a Tuberous Sundew *Drosera erythrorhiza*

On the coastal sand-plains of southwestern Australia, there are 20 species of droseras that form tubers. In this diverse group some species are climbers, e.g. *D. macrantha* and *D. sulphurea* (Pate and Dixon, 1982). The tuberous habit is also found in some South African droseras and in a few species of the genus *Utricularia* (Lloyd, 1935), particularly those that grow on tree boles where desiccation may be a problem. *D. erythrorhiza* is one of the few tuberous carnivorous plant species to have been studied in detail (Dixon and Pate, 1978; Pate and Dixon, 1978). Like *D. linearis* of North America it is rare amongst carnivorous plants in being able to colonize an alkaline environment, in this case calcareous sand. Vegetative growth is from March to October, followed by aestivation by underground tubers from November to February during the hot months. The new tubers form during July and August. Daughter tubers are produced on the ends of horizontal stolons or rhizomes: up to 13 daughter tubers may form per plant, but 2–5 is the more usual complement (Pate and Dixon, 1982). Summer fires are common between October and February, although they do not occur every year. As with other *Drosera* species, *Sarracenia* (page 38), *Darlingtonia* (page 41), *Byblis* (DeBuhr, 1975a; page 42) *Cephalotus* (DeBuhr, 1976) and *Dionaea*, this species of

Drosera is tolerant of low-temperature burning. Pate and Dixon (1982) in fact consider that, after a fire, flowering is enhanced in many of these tuberous species.

The colonies occur as loosely structured clones (cf. *Darlingtonia*, page 40), some apparently of great size and complexity. Although flowers are sometimes formed and seeds set, particularly after fire, no plants were ever seen by the above authors to develop from these seeds. The seed establishment of colonies must therefore be a very rare event. After the new tubers are formed in July and August, a positively geotropic axillary shoot (dropper) develops from the stem base and penetrates the epidermal sheath of the parent tuber to form a replacement tuber within the emptying storage tissue of its parent (see Fig. 3.4), a form of direct biomass recycling. Rhizomes are produced higher up the stem and then extend radially. Each of these swells terminally to produce a daughter tuber. The tuberous and almost self-parasitic habit of the species embodies the potential for the carry-over to the next season of a substantial fraction of the nutrient resources of one plant population. The leaf rosette serves as the major storage organ for the minerals obtained from the parent tuber and rooting medium. Efficiency of carry-over was P (77%), N (79%), dry matter (71%), Mg (63%), K (56%), but Ca, Na and Zn (25–39%). Insects would seem to be an adequate source of nitrogen and phosphate, and the ash of the frequent fires the principal source of potassium (Dixon and Pate, 1978; Pate and Dixon 1978) (see 'nutrition', Ch. 7).

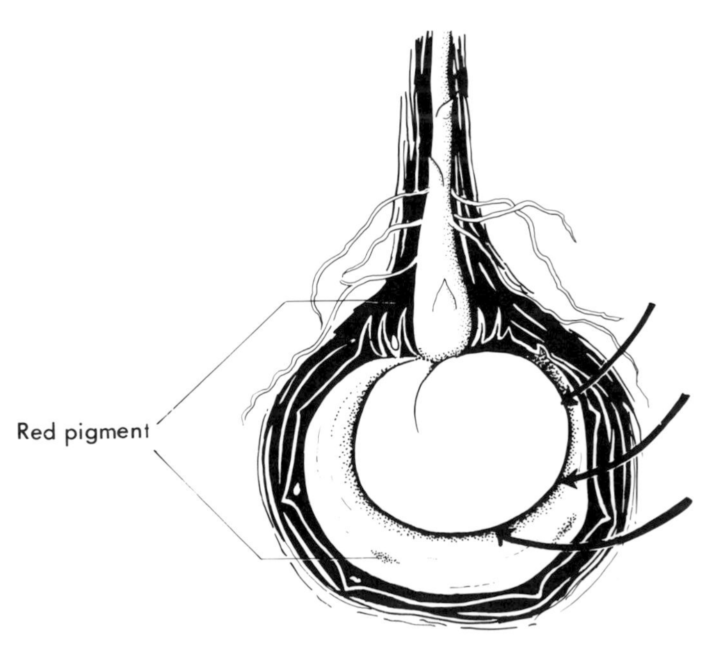

Fig. 3.4 The perennating organ, tuber, of *Drosera erythrorhiza* showing a new tuber (broad arrows) developing in the living ground tissues of the current year's tuber. A red pigment develops both at the base of the stem and, in a granular form, in the ground tissue of the current tuber. (Redrawn from Russell, 1959.)

The Successive Degradation of the Tuber and the Release of Free Pigment from Western Australian Sundews

A feature common to many of the tuber-forming sundews, and particularly to *D. erythrorhiza* as its name implies, is the accumulation of a free red pigment around the tuber (see below). The resting bud of the current year's tuber (Fig. 3.4) lies within a ring of scale leaves near the scar of attachment of the old stem. The bud shoots in early autumn to form a new stock; then, within the scale leaves the new stem swells basally, pressing the surrounding scale leaves outwards to form a funnel-like structure around the swollen base. From a site on the under surface of this swollen base a bud drops within the funnel of scale leaves until it presses on the current tuber. The epidermis of the current tuber ruptures around the stem base allowing the dropping bud to penetrate the living tissues of the current tuber where it inverts and swells to form a new tuber (Russell, 1959). It might be expected that the new tuber would become deformed and the current tuber wither rapidly as it is penetrated, but the exposed ground tissue appears to degrade only at the rate necessary to allow the expansion of the new tuber. Degradation seems to occur both by the extraction of reserve material through the intact vascular system still linking current tuber and stem and by autolysis of parenchyma cells at the exposed surface (Fig. 3.4, broad arrows). The new tuber finally comes to occupy the whole of the space inside the persistent old epidermis. A concentration of the red pigment (Fig. 3.4, upper narrow arrow) is seen at the point of entry of the dropping bud; sometimes it forms a damp red plug. Russell concludes that the release of this free pigment is a direct result of the breakdown of the epidermis and exposure of the underlying cells. Red pigment-granules also appear scattered generally in the cavity beneath the epidermis and throughout the body of the current tuber. These red granules may concentrate and solidify into a red amorphous form or, rarely, as yellow needle-like crystals. Large *D. erythrorhiza* tubers may comprise up to 12 more or less complete layers of annual tuber peripheries. Russell speculates that

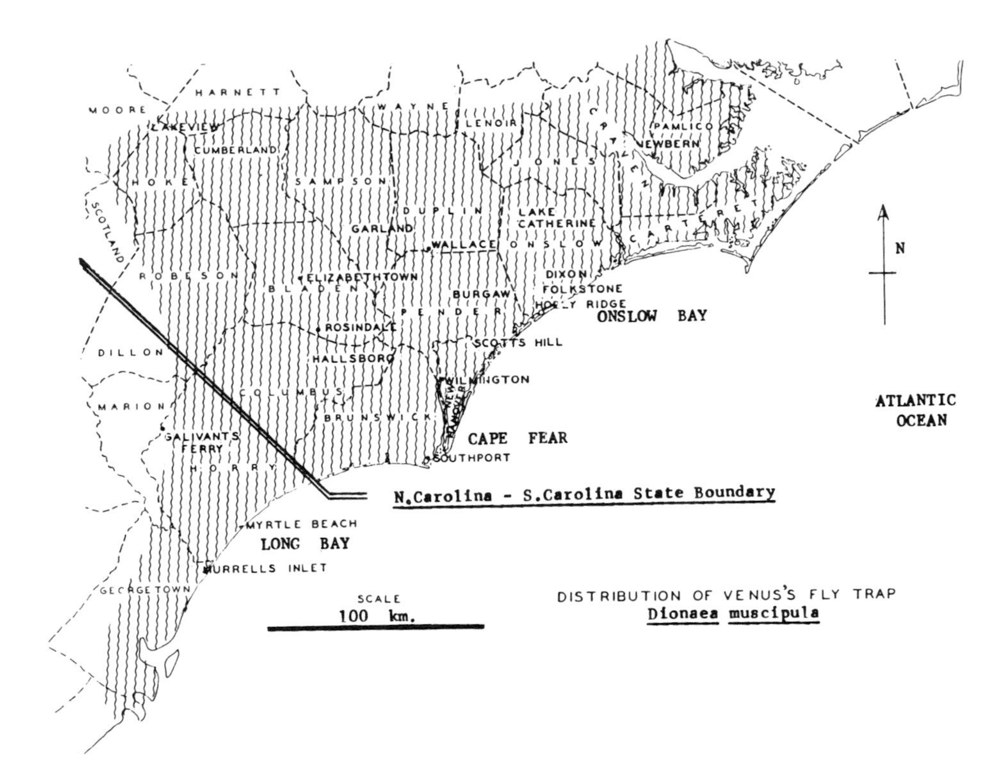

Fig. 3.5 The distribution of *Dionaea muscipula* along the eastern seaboard of the United States. This map indicates the widest known distribution, now, in part, reduced by urban development, agriculture and collecting. (Data from Coker, 1928; Roberts and Oosting, 1958; cf. Fig. 19.6*A*).

these layers may form some protection against desiccation and also, since the degrading tissue always appears healthy and not diseased, have some antibiotic action. The *D. erythrorhiza* pigment is a naphthaquinone, whose properties are discussed on page 242.

The Distribution Pattern and Ecology of *Dionaea*

Dionaea is considered by many to be the doyen of carnivorous plants, although in evolutionary terms it is relatively recent (see Ch. 19), monotypic and grows in a very limited area (Fig. 3.5). Nevertheless, it illustrates most of the ecological characteristics described above for other species. In cultivation, *Dionaea* is moderately tolerant both of nutritional neglect and irregular irrigation, although it does not flourish if dry and ill nourished, but it is completely intolerant of the competition offered by flowering plants, moss (*Polytrichum*) or liverwort (*Lunularia*), in whose presence, the *Dionaea* plants regress and rapidly disappear. Only *Sphagnum* species in the same pan cause no regression (cf. *Drosera* above). Interestingly enough, *Dionaea*'s very close relative *Aldrovanda* (see Ch. 19), about which very little indeed is known but which is observed to grow in totally flooded conditions, cannot tolerate the presence of *Sphagnum* moss (Fijalkowski, 1958).

These recent observations on *Dionaea* in cultivation are supported by an earlier detailed ecological study in its native habitat which is the central southeastern coastal plain of North America (Roberts and Oosting, 1958).

The Distribution Pattern of *Dionaea*

The species ranges in the United States from Beaufort County of North Carolina south to Charleston County, South Carolina, a distance of about 320 km (200 miles), although its actual distribution is now severely constricted and dissected by agriculture, human habitation and plant collectors. This coastal plain is principally pre-Pleistocene sand. There is no fossil evidence either for the pre-Pleistocene distribution or possible migration of *Dionaea*, within this rich carnivorous-plant area (Figs 3.5 and 19.6), but it seems likely that favourable conditions for *Dionaea* existed in the area at least as early as the Pliocene period (Roberts and Oosting, 1958).

Typically, *Dionaea* grows in a semi-pocosin* or semi-savannah area, i.e. between the wet evergreen-shrub bogs and the dry, sandy *Pinus palustris* and wire grass (*Aristida stricta*) savannahs. The climatic conditions vary, exposed ground temperatures ranging from a minimum of −12°C (10°F) in winter to

* Pocosin = acid, sometimes brackish swamps of the southeastern United States dominated by leathery-leaved shrubs.

a brief maximum of 46°C (155°F) in the long, hot summers. These potentially desiccating conditions are modified by all-season rain. The rainfall in the *Dionaea* area is about 128 cm (50 in.) per year with peaks in June, July and August. In one extended period of observation by Roberts and Oosting (1958) the longest dry period was only 12 days. *Dionaea* generally occupies the level areas where water does not collect, although it is occasionally inundated (cf. *Aldrovanda*; see pages 53 and 95) and flooded traps have even been seen to catch aquatic animals (Schnell, 1976). It commonly grows on undisturbed ground, but it may colonize railway embankments where weed suppression by controlled burning is rigorous.

The soils dry to a depth of 5–10 cm (2–4 in.) only during the late spring and early summer and water is generally always available to the *Dionaea* roots. The soil of the whole area is very infertile, which has at least helped to preserve the remaining *Dionaea* stands from destruction through agriculture. It is typically a sand, mixed with the darker fragments of charred vegetation, from the numerous fires, and traces of organic matter, with pHs in the range 3.5 to 4.9. The levels of calcium, manganese and nitrate in the Roberts and Oosting study were below the limits of estimation, with ammonia at 2 ppm, and iron, magnesium and phosphate about 1 ppm. The concentration of potassium was 2 ppm, which is equivalent to an agricultural rating of 'medium'.

The Fire-Climax and *Dionaea*

Apart from the almost continuous and ever-flowing water supply contributed by the seasonal rains, frequent light showers and high humidity, the other striking feature of the *Dionaea* habitat is fire. Most areas where the *Dionaea* grows are burned fairly frequently; there is a sparse cover of ground vegetation and virtually no superficial humus. The soil surface is generally completely bare between individual plants, although at the pocosin edge, *Sphagnum* spp. of which, as we have seen, *Dionaea* is tolerant, form a dense mat. The plants in the artificially burned or cleared areas, in the Roberts and Oosting (1959) experiments, all flowered earlier with a higher percentage of flowers. Those in the burned areas looked healthier, according to these authors and, two years after burning, the plants of the burnt areas were more vigorous than those from areas where burning was prevented. Roberts and Oosting also carried out some artificial shading experiments and these showed that a reduction of the light intensity by about 50% both reduced the incidence of flowering and caused the plants to become etiolated and more prone to insect attack.

Dionaea thus shows all the characteristics of a carnivorous plant in ecological terms: tolerance of infertile soils and an intolerance of desiccation, shading and competition, but with the interesting, although not unique feature of an ability to withstand both flooding and, like many *Drosera* species (see page 28 and Whitehead, 1973), *Darlingtonia* (Austin, see Appendix 1) and *Byblis* (DeBuhr, 1975a), low temperature ground fires.

The Distribution and Ecology of *Drosophyllum*

Drosophyllum is a marked exception to the general run of the carnivorous plants and deserves, in its ecology, this slightly more extended treatment. *Drosophyllum* is found around the southwestern European Mediterranean coast (see Fig. 3.6) and just at the northern tip of Morocco, mostly in areas that may be consistent with dependence on sea mist (see below). Along the Atlantic coast of Portugal it stretches as far north as Oporto and, although the sea mists extend farther north than this, Oporto may represent its natural temperature-dependent northern limit. Coming down the Portuguese coast, *Drosophyllum* is found both north and south of the Tagus and extends east to the Portuguese–Spanish border. There is a gap in its distribution over much of the Gulf of Cadiz region. However, this stretch may be a recently drowned coastline. It appears again just south of Cape Trafalgar in the Zohara de los Atunes region, extends south and eastwards past the Straits of Gibraltar and reaches into the Mediterranean about as far as Estepona. It seems likely that this area is the eastern limit of effective coastal mists. The warmer waters of the Mediterranean proper are probably at this point not sufficiently different in temperature to the mainland to generate the correct meteorological conditions.

Habitat Preference

D. lusitanicum is a strict calcifuge growing over virtually all its range on the rather poor sandstone-derived soils of the Iberian Miocene. Most commonly, the plant will be found on steep boulder-strewn slopes, usually south or southwest facing

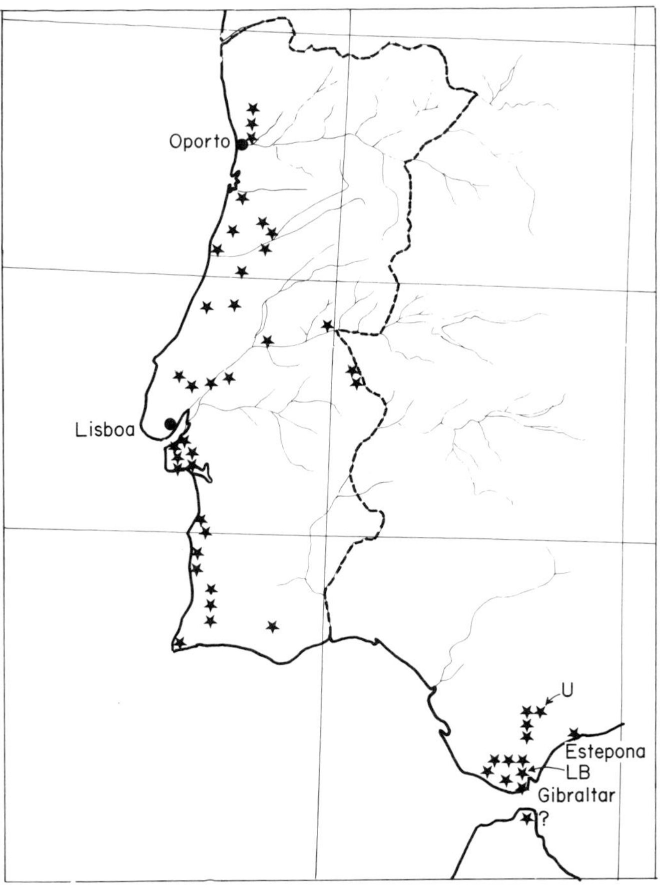

Fig. 3.6 The distribution of *Drosophyllum lusitanicum* in southwestern Europe and northern Africa. LB, Los Barrios, Province of Cadiz; *U* = Province of Ubrique, see text p. 30. (Data from: Jahandiez and Maire, 1931; Fernandes, 1941a, b; Tutin *et al.*, in *Flora Europaea*, 1964; Franco, 1971; B.E., Smythies, pers. comm., 1978; E. Allen, pers. comm., 1979; B.E., Juniper, pers. obsv., 1978.)

and generally not far from the sea. Some colonies at least must be salt-tolerant. These slopes are free-draining and, in a normal year will be baked dry from the end of April to early October.

D. lusitanicum is a colonizer. It can be looked for wherever the soil surface and vegetation cover have been broken, either directly by human hand or indirectly by fire or landslip, or, more rarely, in the stable but very arid and open Portuguese maritime pine macchia (Harshberger, 1925). It may not be a coincidence that it grows, over the whole of its distribution, in an earthquake zone, on the boundary between the European and African tectonic plates. This area is one in which minor shakes are relatively common and major earthquakes, e.g. Lisbon 1755, not unknown. Apart from earthquakes, the building boom in hotels and holiday villas along the whole of the Iberian coast, combined with extensive activities by the military of both Spain and Portugal, all of which match the distribution of *D. lusitanicum*, have provided yet more sites for it to invade. However, *Drosophyllum* can occasionally be found, as we have seen, in the undisturbed but relatively bare ground under established but open stands of *Pinus pinea* (the Stone Pine or Umbrella Pine), where these grow on acidic sand.

On a fresh site with no competition *D. lusitanicum* will be large and free-flowering, measuring up to 40 cm across and 60 cm tall. As the competing vegetation moves in, *Drosophyllum*, which is normally cauline in habit, responds by growing up through the competing herbs and shrubs and forming fresh vegetational tufts and flowering shoots above the invaders. We have seen *Drosophyllum* emerging between the stems of *Calluna vulgaris*, *Erica* spp. and *Stauracanthus* spp. It would seem, in this growth habit, to have a preference for the spiny open-textured shrubs, possibly because it is easier for the *Drosophyllum* to retain its upright habit amongst the thorns or spikes and possibly too because there is less internal shading than in a heavier-leaved type of shrub such as the widespread *Cistus* spp.

At first sight *Drosophyllum* growing in this fashion resembles some very vigorous epiphyte or even a semi-parasite, but a close examination reveals a long, thin, woody, perennating stem stretching down through the interstices of the host shrub, eventually disappearing under a neighbouring boulder. The dead but persistent leaves of previous years' growth give the whole system the appearance of a contorted flue brush. This perennating structure, entangled in the host plant (cf.*Drosera*, see page 49), no doubt contributes to its resistance to wind on an exposed site. How many years this escalating growth habit can be maintained is not known. It would take many years of continuous observation to determine these data on a site which was fortunate enough to escape burning for a while. The superficial appearance of the persistent leaf bases of the older plants would suggest that this process can continue for several years. On the other hand, Schwartz (1974) suggests that, without plants of similar size nearby, *Drosophyllum* normally behaves as a biennial. The senior author (Juniper, 1978) took stem samples from an established *Drosophyllum* site where the mature plants had been killed by fire. Stem sections 1.1 cm in diameter, from growth ring counting, suggested an age of 6–7 years. *Drosophyllum*, unlike *Dionaea*, *Darlingtonia* and some *Sarracenia* species, apparently cannot endure burning, but may persist as seed. Finally, when the competition becomes too vigorous *Drosophyllum* is ousted, probably establishing seedlings on a freshly exposed site. Although there is not direct evidence for this, seeds may remain ungerminated until the next burn clears away the competition. *Drosophyllum* in the Zohara area of southwestern Spain was found growing in company, apart from the host plants already mentioned, with several *Cistus* species, e.g. *C. ladanifer*, *C. albidus*, and *C. salvifolius*, as well as *Tuberaria lignosa*, *T. guttata*, *Anagallis monelli*, *Lavandula stoechas* and *Ulex* species; cf. *Heliamphora* (Bogner, 1976). *Drosophyllum* was never found in the most vigorous stands of *Cistus ladanifer*, which can grow up to 150 cm tall with the individual plants touching each other.

Drosophyllum and Water Conservation

Drosophyllum has, for a carnivorous plant, a fairly well-developed root system as can be seen from França's elegant drawing (1925). As compared with *Dionaea* or many *Drosera* species (see pages 21 and 50), it is not atrophied. However, particularly in the seedling stage, even after several leaves have developed and stalked glands are well-formed, the root system is vestigial. Yet it seems able to survive without any obvious means of water conservation, on south-facing exposed hillsides through at least 5 months of completely dry and often very hot Mediterranean weather (Allen, E., pers. comm., 1978). During most of this period it is in an apparently functioning, insectivorous state. Its glistening arrays of stalked tentacles have, moreover, only a vestigial cuticle to prevent overall water loss. As Mazrimas (1972) and other workers have suggested, it may benefit from the coastal mists that are generated by the cool Atlantic ocean meeting the warmer Mediterranean coast. Many other plants are known to receive at least marginal benefit from mist (Juniper and Jeffree, 1983). The mechanism of taking advantage of the mist in this plant is possibly similar to that of an osmometer. The sugar-rich mucilage over the gland heads, which serves to attract, entrap and suffocate an insect, may absorb water vapour from the saturated air of the early mornings. Once bound into the gel of sugars the water will not readily be lost into the desiccating air of a hot, dry day. Nevertheless, one should be cautious about accepting the mist hypothesis uncritically since some flourishing colonies of *Drosophyllum* e.g. south of Ubrique Province of Cadiz (Smythies, B.E., pers. comm., 1978) and in the Los Barrios area, Province of Cadiz (Allen, E., pers. comm., 1978) (see 'U' and 'L.B.' on Fig. 3.6) are too far inland to benefit from the occasional misty morning. However, the Ubrique area enjoys, for Spain, a generous winter, spring and autumn rainfall, and there are numerous underground streams in the area providing, at least locally, an adequate ground water supply.

The Distribution and Ecology of *Triphyophyllum:* the Part-Time Carnivorous Plant

If *Drosophyllum* is the exception to our general rule that carnivorous plants only colonize damp or seasonally damp environments, then *Triphyophyllum peltatum* of West African forests seems to be the most obvious exception to the shade rule. *T. peltatum* is a well-rooted liane and in this way alone it differs from most other carnivorous plants (see page 21). Sometimes the liane may be up to 40 m in length (Menninger, 1965), and it produces leaves of three different types. Two of these appear cyclically on the juvenile short shoots and the third only at maturity after the internodes have begun extending and the plant has become a vigorous climber. The mature leaves have apical hooks which assist the scrambling habit even to the level of the tree canopy. In this sense it behaves rather like a giant *Drosophyllum* (cf. page 31). One of the two leaf types on the short shoots confers the carnivorous habit. The lamina is reduced so that the leaf is virtually cylindrical and the whole surface is studded with larged stalked glands (Green *et al.*, 1979; Marburger, 1979). These leaves with their glands are totally different from any other structure on the plant and more closely resemble a *Drosophyllum* leaf.

The species is restricted to a limited area of West African rain forest with a roughly uniform climate (Airy-Shaw, 1951). Probably most of the area where it is now found is secondary forest, but nevertheless of a relatively dense shade, with a dry season of 6–7 months from mid-November to late May or early June. Peak rainfall (July–August up to 150 cm per month) intercalates with heavy showers in June and from September to early November. The rock is mostly archaean gneiss. The soil is principally a decomposed laterite which varies from being waterlogged in the wet season to arid in the dry.

LIBRARY COLBY-SAWYER COLLEGE NEW LONDON, NH 03257

The pH is about 4.2 and the soil generally nutrient deficient (Green *et al.*, 1979). Under natural conditions, the juvenile plant starts by producing about 10–15 aglandular leaves followed by 1–6 glandular leaves, the latter tending not to develop until the short shoots are 400–500 mm high. These glandular leaves of the juvenile plants are produced just before the height of the rainy season. Could this be an adaptation selected to synchronize with the maxima of prey insects? The life cycle of the species is then apparently completed with the production of a rapidly extending perennial liane which will finally bear flowers and seeds at the level of the forest canopy. *Triphyophyllum* seems to be the only evergreen climbing carnivorous plant on the African continent and, as such, has obvious parallels with *Nepenthes* which fills a mostly similar niche in southeastern Asia. There seems to be no similar genus in the Americas. Nevertheless, *Triphyophyllum* differs from *Nepenthes* in penetrating the substantial shade of at least the secondary forest, whereas *Nepenthes* in its various forms tends to occur on the forest edge, in clearings, or even on the seashore, and does not commonly attempt to gain the light above the first canopy by climbing. That there is nothing odd in the ecology of the West African tropical rain-forests in this respect is evinced by the fact that several species of *Utricularia*, *Drosera indica* and *Genlisea africana*, growing nearby, occupy temporarily inundated lateritic pans. These species complete a cycle of growth and flowering to survive the dry season as seeds or in a vegetatively dormant form (Green *et al.*, 1979).

The Distribution and Ecology of *Nepenthes*

Nepenthes, the eastern tropical pitcher-plant, grows from sea level, even sometimes within the spray zone (see page 34), to about 3000 m for *N. rajah*, up to 3500 m for *N. villosa* (both found on Mt. Kinabalu in Borneo, Burbidge, 1897) and up to 3520 m for *N. vieillardii* in Irian Jaya. The centre of origin appears to be somewhere in Borneo (see Fig. 3.7) where there may be as many as 28 species out of the world total of about 80 (Smythies, 1963). The genus extends westward to the isolated population of *N. khasiana* on the Khasia Hills in Assam, with a single species each in the Seychelles and Sri Lanka (Chapman, 1947 but see Wijewantha, 1952) and two in Madagascar, but it is unknown on the African continent. Danser (1928) has speculated on this odd distribution in relation to continental drift. Northwards it reaches southern China and to the south it has colonized the tropical forests of the Cape York Peninsula of Queensland, Australia, reaching as far south as Coen. To the southeast it has reached New Caledonia.

Some species have a very wide distribution. *N. mirabilis* (Fig. 3.7) ranges from southeastern China to New Guinea, Western Java and south to Queensland, Australia. *N. ampullaria*, the commonest of the ground-living species, has a very odd, disjunct distribution (Fig. 3.7). To the west, it occupies what is essentially the Sunda-shelf. To the east, and separated by 1600 km, it is found widely in what is now the western half of Papua New

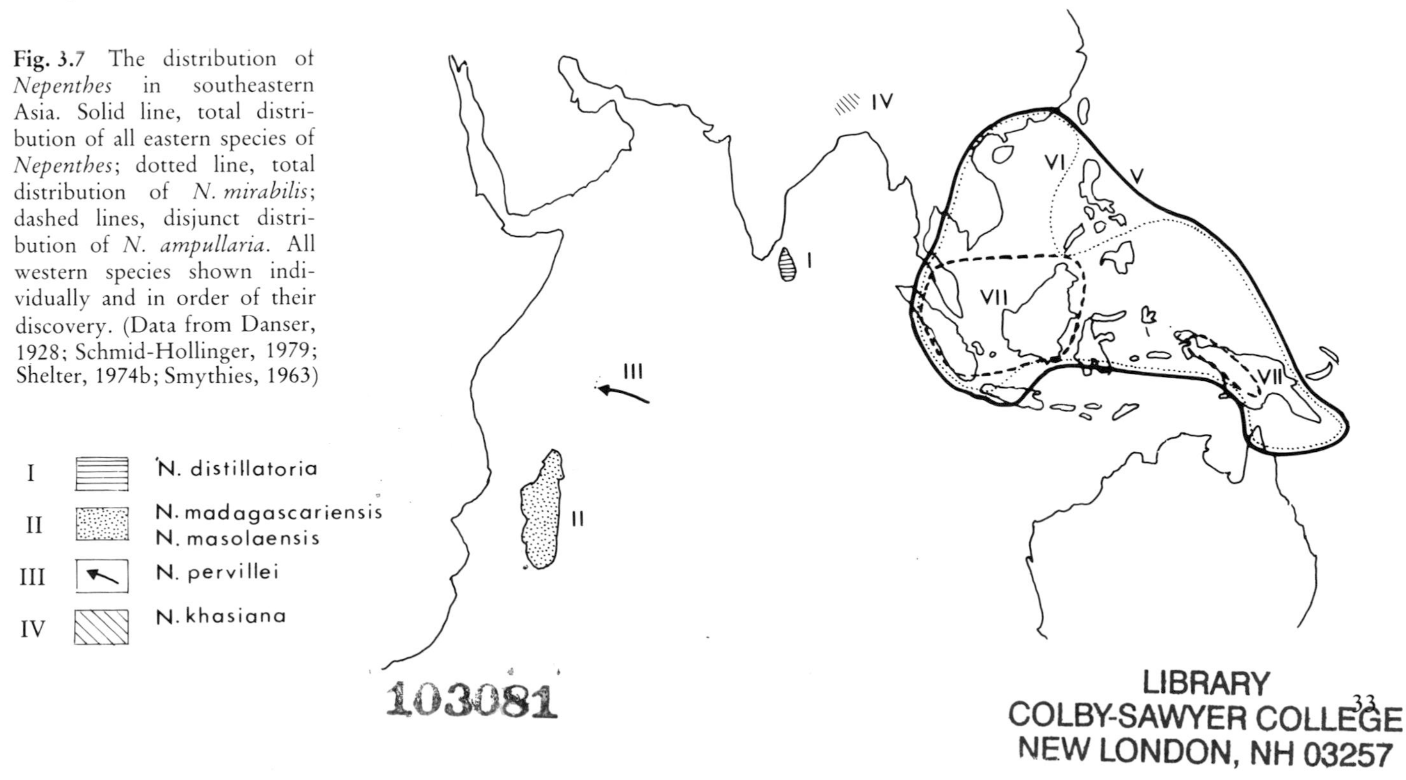

Fig. 3.7 The distribution of *Nepenthes* in southeastern Asia. Solid line, total distribution of all eastern species of *Nepenthes*; dotted line, total distribution of *N. mirabilis*; dashed lines, disjunct distribution of *N. ampullaria*. All western species shown individually and in order of their discovery. (Data from Danser, 1928; Schmid-Hollinger, 1979; Shelter, 1974b; Smythies, 1963)

103081

LIBRARY
COLBY-SAWYER COLLEGE
NEW LONDON, NH 03257

Guinea (approximately what was Dutch New Guinea).

Many species seem, on the other hand, to be very closely restricted endemics (Shetler, 1974b). As Holttum (1954) has pointed out, the high-ridge mountain species of Malaya, surrounded as they are by dense forests that they cannot penetrate, are as cut off and potentially endemic as if they were on an island. The true number of species must probably always remain in doubt since the genus forms hybrids very readily (Kurata and Toyoshima, 1972). According to Smythies (1963), hybrids seem to be a rare occurrence in the wild, but they are common in cultivation (Veitch, 1906). On the other hand, no fewer than 25 species (Danser, 1928; Smythies, 1963) have been found on only one spot on one mountain, a fact which may be attributed to the comparative youth of the genus.

Nepenthes as a Halophyte

The conventional habitat of a *Nepenthes*, on the edge of, but rarely in, the forest (see page 35), does not suggest a group with halophytic tendencies. Nevertheless, *N. albomarginata*, *N. reinwardtiana* and *N. treubiana* have all been recorded by Danser (1928) as growing along the shore line, sometimes well within the spray zone and more recently R. and L. Lewin (pers. comm., 1986) have seen *N. mirabilis* growing in the sand by the sea shore on Palau. All of these observations lend support to Speir's (1981) speculation that *Nepenthes* had a coastal distribution during the Cretaceous period.

The Growth Form of *Nepenthes* Species

As one would expect from such a widespread and apparently rapidly evolving genus, the growth form of *Nepenthes* is varied.

A young plant has a rosette of leaves spreading on all sides of the short stem and each leaf normally bears a pitcher at the tip. In these very young plants, the bases of the pitchers are round and rest on the ground with their mouths facing inwards (see below). *N. ampullaria* persists in this form and never develops aerial pitchers (Green, 1967; see below). Only when the stem produces leaves of full size does it begin to grow rapidly in length, when it may produce pitchers of a different shape with slender curved bases (Holttum, 1954). The dimorphism of the pitchers can be very marked in some species, e.g. the western species *N. distillatoria* in Sri Lanka, *N. pervillei* in the Seychelles and *N. madagascariensis* and *N. masoalaensis* in Madagascar (Schmid-Hollinger, 1979). This extreme dimorphy of the pitchers and also of the leaves often leads to confusion and new species have, in fact, been described on account of this dichotomy. For a review of this phenomenon see Jebb (1987; Appendix 2).

The dimorphic change is of a common pattern in almost all species. Interestingly enough, *Darlingtonia* (Hooker, 1875; see Ch. 4) also passes through a similar and discontinuous dimorphy. The transition in *Nepenthes* is part of the syndrome of changes from a short rosette plant to a climber. The internodes lengthen, the tendrils become coiled, apparently adapted to grasp other vegetation; the leaf blade becomes shorter and blunter; and the pitchers are transformed. The upper pitchers are more tubular or infundibulate, as opposed to the avoid or globose lower pitchers. The mouth of the pitcher now faces away from the plant axis (cf. *Darlingtonia* and *Sarracenia*, Ch. 5 and Figs 5.7 and 5.8), the wings along the front of the pitcher become much reduced, even absent in some species, the peristome is narrower, and the spur at the apex of the pitcher is now finer and more slender. This dimorphism is most marked in such species as *N. ampullaria* which does not form true upper pitchers at all. It may be that the extreme form of the lower pitchers in this species precludes the formation of upper pitchers due to a restricted morphological amplitude, these being reduced to a mere swelling on the tendril tip (Jebb, 1987).

The flowering phase of *Nepenthes*

The switch from lower to upper pitcher type appears usually to coincide with the onset of flowering (cf. *Triphyophyllum, page* 32; Menninger, 1965). Experience with cuttings in botanic gardens suggests that this phase shift is not irreversible. Cuttings from flowering material seem commonly to revert to non-flowering, lower-pitcher-type-producing plants.

Nepenthes as an epiphyte

Nepenthes, which always, so far as we can discover, germinates in the ground and commonly but not universally in its mature phase, has the habit of a climber, may adopt an epiphytic habit. On the other hand, *Triphyophyllum* (see page 32), unlike most carnivorous plants, has a robust root system and never seems to lose its connection with the ground (Green *et al.*, 1979). Richards (1952) has pointed out: 'No sharp line can be drawn between climbers and epiphytes. Many epiphytes have climbing stems and some root climbers . . . start life rooted in the ground, but eventually lose their connection with it and become epiphytes.' Burbidge (1882) wrote (of *N. veitchii*): 'This is a true epiphyte. I never met with it on the ground anywhere, but in great quantity 20–100 feet high on tree trunks. Its distichous habit is unique, I fancy, and then some of the leaves actually clasp around the tree just as a man would fold his arms around it in similar circumstances.' Smythies likewise (1963) wrote: 'Near the summit of Mt. Santubong (Borneo) there is a small colony of *N. veitchii* which I recently examined . . . The plants are growing on ten or a dozen trees of *Tristania anomala* and I failed to find a single plant that had any present connection or showed any evidence of past connection with the ground.' Burbidge (1896) also saw *N. villosa* (*edwardsiana*) as a pure epiphyte on *Casuarina*, *Dacrydium* and *Rhododendron*. *N. reinwardtiana* sometimes, as we saw earlier, grows by the sea, but both Smythies (1963) and Hanna (1975) have observed this species well inland growing apparently as a pure epiphyte high up on the branches of *Dipterocarpus oblongifolius*. Smythies concludes that, in agreement with Holttum, the normal habit of a *Nepenthes* is a climber (cf. *Triphyophyllum* and *Drosophyllum*). Those excellent Victorian growers like Veitch and Burbidge knew how weak-rooted many carnivorous plants were. In 1897, Burbidge wrote: '. . . all insect-catching plants were characterised by their sparse or weakly root system . . .'. This statement is generally (see page 21) but not completely true. The switch from terrestrial living to being a facultative epiphyte, under wet tropical-forest conditions, with good stem-drainage and a thick bark moss layer, may not be too significant. Is it possible that, as a *Nepenthes* makes the switch from being a terrestrial to an epiphytic plant, the individual in such circumstances could split up physically and spread clonally in the branches? Could the plants that Smythies observed in 1963 have been the members of such a clone?

When *Nepenthes* does appear as a scrambling shrub or epiphyte, Smythies speculates that there may be some inhibiting factor at work such as lack of light, restricting its proper development. He notes that there are a number of common species, e.g. *N. ampullaria*, *N. rafflesiana* and *N. mirabilis*, which are never found as pure epiphytes, but he has not observed any Borneo species, with the possible exception of *N. rajah*, that is incapable of climbing. Often, he observes, climbers will be quite vigorous, using their tendrils, which are extensions of the mid-vein of the leaf, to enfold a twig, but not producing any pitchers. He surmises that a nitrogen-rich soil may inhibit pitcher formation. However, the *Nepenthes* of the Seychelles (see page 36) is probably growing in a nitrogen-rich habitat and is abundant in pitcher formation.

The Ecology of *Nepenthes* Species

In Borneo (Smythies, 1963) *Nepenthes* is found from sea level to 3500 m (11000 ft.). The plants are most conspicuous on the higher mountains in the Montane Rain Forest (Richards, 1952) sometimes known as the Mossy Forest or Mountain Ericaceous Forest. They are rare or absent from the Mountain Oak forests which are two-storied and shady. M. Jebb, (pers. comm., 1986) saw *N. maxima* and *N. neoguineensis* in New Guinea commonly on the narrow ridge tops where the canopy is generally thin and low and the soil sparse. Such isolated ridges could, as we have already suggested, be a rich source of new species. Below about 1000 m (3000 ft elevation) they are almost completely absent from the Dipterocarp Forest, but common in the Tropical Heath Forest and Peatswamp Forest. The latter forest types are edaphic climaxes of the Tropical Rain-forest. Physiologically, 'Tropical Heath' and 'Peatswamp' are similar. The soils are deficient in nutrients, the drainage is impeded and they are predominantly acidic. The *Nepenthes* floras of the two types are similar. Smythies suggests that the difference between the *Nepenthes* penetration into these two forest types, and its relative exclusion from the *Dipterocarpus* type, is based on two factors: light and soil conditions. He notes that, although some *Nepenthes* can grow vegetatively in fairly heavy shade, no flowers are produced. In addition, due to the relative fertility of the dipterocarp soil, large areas are temporarily cleared for agriculture. Here you would expect to find *Nepenthes* seedlings since there is no shortage of light and the seeds are minute, filiform and wind-borne. Seedlings are never seen. He speculates that a deficiency of nitrogen in the soil may be a prerequisite for the growth of *Nepenthes* under natural conditions but this needs experimental confirmation. We shall see a suggestion of this inhibitory factor in Chandler and Anderson's work on *Drosera* (Ch. 7).

Holttum (1954) states that *Nepenthes* species, at least in Malaya, are climbers of *open* country, and persist only in the early stages of regeneration of the forest. They are not, in his estimation, normally found in the primitive high forest which formerly covered most of Malaya. They are found in Malaya and, as we have seen elsewhere, on exposed mountain ridges where the washed-out soil is poor and will not support large trees. Further support for this general pattern of ecological tolerance of *Nepenthes* species comes from Gibson's (1983b) observations on *N. rajah*. *N. rajah* grows on the serpentine soils (cf. *Darlingtonia*, page 38) of Mount Kina-Balu, between 1650 and 2650 m elevation and receives a rainfall of about 150 cm per year. At these elevations, long periods of full sunshine are rare and the humidity high the whole year round. *N. rajah* seedlings are to be seen only on recent landslides or other areas of disturbed ground. As they penetrate into the stunted forest their large leaves are exposed to the full skylight. Elsewhere on Kina-Balu, between 800 and 1000 m, *N. villosa (edwardsiana)* is also found on serpentine soils, (Kaul, 1982) (cf. *Darlingtonia*, page 39; Fowlie, 1985). We shall discuss the serpentine phenomenon more fully in connection with the distribution of *Darlingtonia* but, briefly, it would seem that serpentine endemics are characterized by being able to obtain sufficient calcium even at the low concentrations characteristic of these sites. The heavy metal ion concentrations may serve only to keep away the opposition, not to provide a specific need for the endemics (Kruckerberg, 1954). Further evidence of the tolerance, if not preference, of *Nepenthes* for nutrient-poor soils, this time of a man-made type, comes from Sri Lanka, where Chapman (1947) observed *Nepenthes* growing in the 'patana', areas of poor grassland which are regularly burnt by the local farmers to provide short-term grazing. Elsewhere in Sri Lanka (Bridget Juniper, pers. comm., 1982) *N. distillatoria* could be observed, like *N. pervillei* in the Seychelles, growing only on exposed granite rocks often beside new road cuttings rooting into crevices (see page 308) and flopping down the slopes away from the forest edge.

The Species Associations of *Nepenthes*

As with many other species of carnivorous plants, different species of *Nepenthes* may often be seen growing side-by-side (Kurata and Toyoshima, 1972). Whether, as is proposed with *Sarracenia* (see page 142) different species are prey-partitioning where they are co-habiting, thus avoiding competition, is not known. They often form hybrids in the wild, as Kurata and Toyoshima observed, but whether they failed to persist, as with *Sarracenia*, where the field hybrids are mostly formed as a result of disturbance of the habitat (see page 38) has not been ascertained.

There are, as we have seen and shall see (see pages 26 and 41) scattered observations suggesting that carnivorous plant species in general associate more often than would be expected by chance, with certain non-carnivorous species. As with many other carnivorous plants, *Nepenthes* is sometimes found in association with orchids – in this case an unknown *Nepenthes* with *Paphiopedilum volonteanum* (syn. *P. hookerae*) (Fowlie and Lamb, 1981) and *N. villosa (edwardsiana)* with *Paphiopedilum dayanum*, *Malaxis* and *Phalaenopsis* species (Fowlie, 1985). But this may only reflect the low competitive ability of the two diverse groups. It could be that the two plants are also carrying out a form of resource-partitioning, since the orchids are almost certainly mycorrhizal, whereas the *Nepenthes* probably relies on its insect prey as a source of phosphate (see page 131).

A theory has been proposed (Baker, 1877), and elaborated by Vesey-Fitzgerald (1940) that *Randia sericea*, a rare endemic of Mahé in the Seychelles, is an obligate associate of *Nepenthes pervillei*. Procter (1974) finds no support for this proposal and adds that the most that can be said is that they appear to have similar ecological requirements. In confirmation, there is one very healthy site for *N. pervillei* at Salazie, below Morne Seychellois, on Mahé. As with Procter's lists, *Dillenia ferruginea*, *Dianella ensifolia*, *Erythroxylon seychellarum* and *Gleichenia dichotoma* were present, but there was no sign of *Randia sericea* (Juniper, pers. obs., 1984).

As we shall discuss on page 308, *N. pervillei* may be anomalous amongst the *Nepenthes* group in that, although it grows on superficially arid soils, it may enjoy the benefits of nitrogen fixation and a substantial faecal rain from seabirds. Rao *et al.* (1969) made a study of the plant associations of *Nepenthes khasiana* growing in Assam. After an extensive species analysis on different sites, they came to the conclusion that *Nepenthes* had no definite association with any other plant. Otherwise, as we have observed elsewhere, in almost all its characteristics *Nepenthes* seems to be observing the general rules of intolerance of shade and a tolerance of, if not an obligate requirement for, areas of low competition on acidic, minimal-nutrient soils. In addition there are reports (Chapman, 1947; Lavarack, 1981a) that certain species of *Nepenthes* in, respectively, Sri Lanka and Australia can tolerate low-temperature burning and successfully recolonize the habitat.

The Distribution and Ecology of Certain Sarraceniaceae

Sarracenia, *Darlingtonia* and *Heliamphora* are hydrophyte genera that commonly inhabit bogs; seeps, particularly the Laracetum phase of Dansereau's bog succession (Dansereau and Segados-Vianna, 1952); swamps; wet, sandy meadows; pocosins; or savannahs where the soils are water-saturated (Schnell, 1982a). The genus *Sarracenia* occupies a broad band of the eastern seaboard of North America and *S. purpurea* extends beyond the Arctic Circle and west into the great plains (Figs 19.5 and 19.6). Excellent distribution maps of all the species are provided by McDaniel (1971). *S. purpurea* is not only the widest-ranging species, from the Gulf States to Labrador, and the most adaptable (Butler, 1985) but has also successfully become naturalized in Ireland (Kertland, 1968; Foss and O'Connell, 1984) and Sweden (Almborn, 1983). The soils this genus invades are moderately to intensely acid, although local colonies of *S. purpurea* can tolerate alkaline flushes (Wherry, 1929; Mandossian, 1965; Kertland, 1968; Schnell, 1976). The soils they inhabit are deficient in all nutrients, particularly nitrate and phosphate, but may be high in aluminium (Plummer, 1963; Pullen and Plummer, 1964). Some species (cf. *Dionaea*, page 30 and *Heliamphora*, Bogner, 1976) can tolerate standing water, as can *Pinguicula ionantha* (Folkerts, 1977), in contrast to *Darlingtonia* (see page 40). A few, e.g. *Sarracenia psittacina*, may even thrive under water for a time (Shetler, 1974a). In fact, in the Davenport Lake near Toms River, New Jersey, MacFarlane and Steckbeck (1933) reported that *S. purpurea* var. *stolonifera* was often to be found in floating or semi-floating masses of vegetation, its roots tangled into a mass of *Sphagnum*. Again we can note the toleration of *Sphagnum* species. As MacFarlane wrote, '. . . piled up masses of pitchers, seven to twelve feet in width and mostly floating on the water . . . In the sheltered bays of this lake floating beds of the purple bladderwort (*Utricularia purpurea*) in full bloom, alternated with sunken logs that were carpeted with *Drosera rotundifolia* and *D. intermedia*.' Schnell (1980) noticed *S. oreophila* in association with *Drosera capillaris*, *D. brevifolia*, *Utricularia subulata* and *U. cornuta*. Another but rarer habitat for *S. oreophila* was noted by Schnell in the same paper. Usually *S. oreophila* was to be found in grassy-sedge seeps or on the edges of bogs, but very occasionally in riparian sites, sandy or gravelly spits on the margins of erratic water courses. The plants here were not conspicuously vigorous and had apparently suffered secondary dispersal by flooding (see page 38). As with most other carnivorous plants, several species of *Sarracenia* may occur side-by-side (Dennis, 1980; Gibson 1983a). *Sarracenia purpurea* is often found with *S. flava*, and *S. flava* with *S. minor* (Schnell, 1976), comingling with droseras and *Dionaea*.

Folkerts (1982) reports that no fewer than 13 species of four different genera of carnivorous plants were found in a single *Sarracenia* bog. McMillan (1978) noted five species of *Sarracenia* in a single bog (see page 38 for the study of hybrid swarms but see Ch. 7 for Gibson's, 1983a, comments on interspecific competition). Also McDaniel (1971) noted that the different sizes and shapes of these species when associated together may suggest different roles in the environment. Givnish *et al.* (1984) noted that the close *Sarracenia* relative *Heliamphora* along with the 'new' carnivore *Brocchinia reducta* (see page 19) as well as *Drosera*, *Utricularia* and *Genlisea* species all grow near to one another in the sodden, wet, nutrient-poor, tepuis region of southeastern Venezuela (see page 42). All of these are so different in their mechanisms that, again, McDaniel's comments may apply.

Like many other carnivorous plants, *Sarracenia* is generally not tolerant of competition or particularly of shade (Schnell, 1980, 1982a) and *Sphagnum* species again seem to be the only broadly accepted associate. In alkaline bogs the mosses *Thuidium* and *Campylium* grow nearby (Mandossian, 1965). However, this lack of aggression only seems to apply in native habitats. On one bog in Ireland (Foss and O'Connell, 1984) where *S. purpurea* has been naturalized for 15 years, the invader is spreading successfully by seeding. In the centre of the colony the *Sarracenia* forms 100% of the vegetation and at the margins, where it is penetrating established *Calluna*, *Erica tetralix*, *Vaccinium oxycoccus* and *Andromeda* bog, the pitcher-plant already forms 50% of the vegetational cover. In none of the habitats studied does there seem to be any evidence of any consistent angiosperm association with *Sarracenia* in either acid or alkaline conditions (Mandossian, 1965). The only exceptions to the general intolerance of shade seem to be *S. rubra*, which can endure the shade of mixed woodland in the Carolina Mountains of the USA (Shetler, 1974a) and some populations of *S. oreophila*. The latter species seems able to survive in the understory of *Quercus* woodland in places (McDaniel, 1971), but not in every location (Dennis, 1980). Schnell (1980) notes specifically that, as *Quercus* spp. come to dominate a habitat, *Sarracenia oreophila* disappears.

Sarracenia and Local Ground Fires

Again like *Dionaea*, *Byblis*, *Darlingtonia* and some *Drosera* from all over the world, *Sarracenia* species can survive ground fires. Even if burned to the ground, they regenerate rapidly, possibly benefiting from the reduced competition (Plummer, 1963; Pullen and Plummer, 1964). Plummer and Kethley (1964), working with *S. flava*, noticed the vigorous competition in some stands from the wiregrass, *Sporobolus teretifolius*. The pitcher-plants appeared to be at some disadvantage until fire eliminated quantities of undecomposed leaves. They then grew and flowered luxuriantly. But, the authors noticed that during periods of severe competition insect entrapment was undiminished, suggesting that nutrition was not the limiting factor. Eleuterius and Jones (1969) also consider that *S. flava*, along with some native bog orchids, is probably fire-dependent and certainly is more vigorous after burning cf. *Dionaea*, *Drosera* and *Byblis* (DeBuhr, 1975a).

Hybrid Swarms in *Sarracenia*

As we have seen, several species of *Sarracenia* may occur side-by-side in the same bog, possibly by different syndromes of attraction and trapping (see Gibson, 1983a and Fig. 3.2), partitioning the available resources of insect prey (see Ch.7) and yet rarely, if ever, hybridizing effectively (see page 144). Hybrids are occasionally found and virtually every possible hybrid between the 14 or so species of *Sarracenia* has been achieved in the greenhouse, where they seem to be as vigorous and fertile as the parental forms. Hybrids in the wild seem to be particularly prevalent after disturbance (McDaniel, 1971). Evidence consistent with this comes from Schnell (1980) who noticed small, and admittedly not very vigorous, populations of *Sarracenia oreophila* on sandy and gravelly spits by river banks. These small groups had obviously been washed into these riparian sites by flooding. Torn by stream surges from their usual seeps or bog margins and co-mingled with other species, they may at certain times and places have achieved hybridization. Folkerts (1982) noticed that where he and his graduate students were working in bogs in Southern Alabama and Mississippi, hybrids are now present at points where their activities disturbed the soil. There is also anecdotal evidence that storm surges on the east coast of the United States, that have battered natural habitats of *Sarracenia* species, have also resulted in temporary hybrid swarms (see also Fig. 19.6). These observations might be consistent with Schnell's evidence above. Levin has suggested (Levin, 1976; Levin and Paine, 1974) that disturbance which reduces the sizes of a species patch makes hybridization more likely. It is suggested that a foraging pollinator habituated to one type of flower, but finding few available in one patch may visit several species. Such an hypothesis would not, however, explain Folkert's observations on disturbance.

The reason for the lack of long-term success of viable hybrids may be that resource partitioning (see Ch. 7) is probably a much more sophisticated and critical factor in *Sarracenia* ecology than at present realized, and intermediates between species may not attract a sufficiently broad spectrum of prey for effective nutrition.

The Distribution and Ecology of *Darlingtonia*

D. californica, were it not separated from *Sarracenia* by most of the width of the American continent, is so similiar in the view of some taxonomists to certain species of *Sarracenia* that it would probably not deserve generic distinction. It enjoys somewhat similar habitats, mostly montane meadows with small fresh streams amongst conifer and hardwood forests and a similar commensal fauna (see Ch. 14). As McDaniel (1966) and Debuhr (1977a) speculate, this similarity may be because the two genera have split from a common, more *Heliamphora*-like primitive stock, and separated in the northern hemisphere (Figs 19.5 and 19.6). *Darlingtonia* may have moved across the interior of what is now the United States through the mesic forests present during the tertiary period. *Sarracenia* subsequently may have speciated in what is now the southeastern United States (Fig. 19.5), and *S. purpurea*, much later, extended its range northwards and westwards.

In California (DeBuhr, 1973), *Darlingtonia* is found in the Sierra Nevada from El Dorado through Plumas and Nevada counties. Northwest in the Klamath Mountains it is distributed through southwest Siskiyou county, northwest Shasta and northeast Trinity counties. In northwest California it is found in Del Norte and Siskiyou counties. Over the border in Oregon, again in the Klamath Mountains it is found in Josephine and Curry county. Along

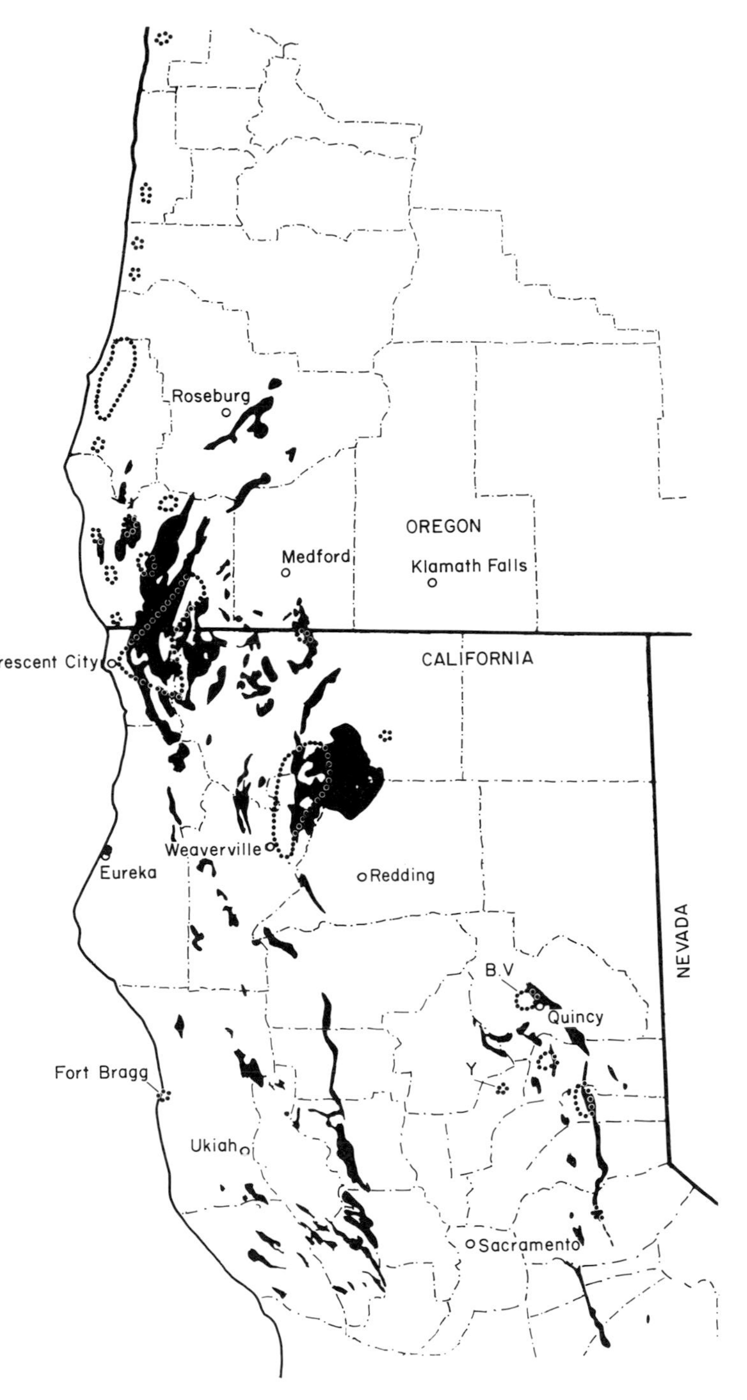

Fig. 3.8 The distribution of *Darlingtonia californica* in northern California and southern Oregon (. . .) in relation to the outcrops of serpentine (ultramafic) rocks of the region. Most of the colonies can be located on or near such outcrops, i.e. within the drainage areas, except perhaps the coastal colonies of Koos, Lane, Lincoln and Tillamook Counties of Oregon, the coastal site just north of Fort Bragg and a central site in Yuba County (Y). Townships with open circles, e.g. Redding, for general location only. B.V. = Butterfly Valley. (Data from De Buhr, 1973; Cheatham, 1976; Stansell, 1980; geological data, Irwin, 1966, 1977.)

the coast, and sometimes very close to the sea, it occurs in Coos, Lane, Lincoln and Tillamook counties and inland again in Yamhill, Linn and Lincoln counties.

The Distribution of *Darlingtonia* and Serpentine Rocks

The Klamath Mountain areas are rich in ultramafic rocks (Irwin, 1966) which contain many serpentines (Fig. 3.8; see below and cf. some *Nepenthes* species, page 36). In the Sierra Nevada, the locations in the Plumas (see below) and Nevada counties lie on the Shoo Fly formation, which has greenstone in the upper layers of rock (McMath, 1966). Greenstone is an altered basic igneous rock also containing serpentine minerals.

Butterfly Valley, Plumas County (BV in Fig. 3.8), a rich area for *Darlingtonia* and other rare species, lies at the northern end of the Sierra Nevada, at an altitude of about 1000 m and near serpentine rocks. Other colonies of *Darlingtonia* may be found, as we have seen, in similar habitats down to sea level and up to 2500 m (Fig. 3.8). The extremes of climate which these differences suggest, are obviously not a limiting factor. A study of the literature and an examination of the descriptions of all the known sites (Stansell, 1980) suggest that *Darlingtonia* may be dependent upon a continuous stream flow through serpentine rocks, or rocks having a similar high metal content (Ziemer, 1973). We should note, however, that with reference to Figure 3.8 not quite every site is located precisely over serpentine.

This discrepancy may be explained for the most part by underground streams, heavily charged with metal ions, emerging a long way from source. Just a few sites, not apparently associated in any way directly or indirectly with serpentines, have been identified so far. The coastal sites of central Oregon seem to be a long way from ultramafics, so too is one site in Yuba County, California, and that of Inglenook Fen, 10 km north of Fort Bragg in Mendocino County (see arrows in Fig. 3.8).

The availability of ions on serpentine soils

Kruckerberg (1954), in his study of serpentine soils, points out that calcium and all the other major plant nutrient levels are very low, but the plants in these habitats must tolerate high levels of magnesium, nickel and chromium. In general terms, plants on serpentine soils appear to be characterized by their ability to obtain sufficient calcium, even at the low concentrations peculiar to serpentine soils. Usually, serpentine endemics grow better on non-serpentine soils (Kruckerberg, 1954; Proctor and Woodell, 1975), thus they do not 'need' the high levels of heavy metals. It would seem that most serpentine endemics have adjusted to the high metal levels and have developed a high calcium-scavenging ability. In turn, they have generally lost the ability to compete in a calcium-rich, low-metal environment: *Darlingtonia* appears to fall into this category. The underlying strata of Butterfly Valley are mixed Paleozoic rocks, but contain no granites, larvas or other volcanics. Serpentine rocks, although not exposed at the site, do occur just outside the region and from the analyses it seems likely that the stream flows are charged by passage through these magnesium-rich deposits. A water analysis in Butterfly Valley showed the following ionic load: magnesium varying from 1.0–2.5 mg per litre, equivalent to 0.083–0.206 meq per litre, strongly suggesting a serpentine origin. However, the calcium levels in the area were not negligible, usually about half the relatively high magnesium levels, and the pH ranged from 7.5 to 9.5.

Darlingtonia and its Water Supply

All the *Darlingtonia* sites so far found (Fig. 3.8) are in areas of high rainfall. A major factor pointed out by DeBuhr (1973) which must regulate the distribution of *Darlingtonia* is not only available but ground water. All the seeps are continuous, drawing their supplies from these areas of high rainfall, from 100 cm in the Sierra Nevada (e.g. Butterfly Valley) up to 200 cm on the Oregon coast. This rain falls mainly in the winter: of no direct use to *Darlingtonia*, but serving to recharge the water table.

The temperature of the groundwater

At the time of collection at the Butterfly Valley site on 12 October 1985, a marked disparity was noticed between the air temperature at mid-day (25°C) and the water temperature (9°C). All the plants were in perfect condition and growing strongly and an even greater divergence of air and water temperature might be expected in the spring or early summer at that site. The altitude distribution, over 2000 m, suggests an even greater tolerance range than this. It has been noticed that, in cultivation, *Darlingtonia* grows better when the ground-water temperature is artificially held down. Could it be that the root system of *Darlingtonia* is indifferent to its ambient temperature (Ziemer, 1973), serving only to provide anchorage, water and perhaps calcium, but that its selective ion-pumps cease to operate effectively as the temperature rises?

Running water and the maintenance of the *Darlingtonia* habitat

All the colonies seen by DeBuhr (1973) grow in streamlets or in a seepage area. They are not immersed aquatics and none grows in standing water. In this respect they are similar to the unrelated Australian pitcher *Cephalotus* (Lowrie, 1978), but differ from *Heliamphora* (Bogner, 1976) which appears to be tolerant of standing water. The finest colonies in Butterfly Valley were growing, in almost pure stands (see below) in about 50 mm of steadily moving water. There is good evidence that some of the clonal colonies of the *Darlingtonia* in the Butterfly Valley area have maintained their position for at least 100 years (Austin, 1875–77; see Appendix 1). One might expect that the improvement of the soil, resulting from the sustained capture of insects, would result in the competitive invasion of more vigorous species. That this is not the case with *Darlingtonia* may be due in part to the dynamic nature of its habitat, forever removing the nutrient-rich humus that it produces. In this respect *Darlingtonia* differs from its near relative *Sarracenia* (see page 38) and many other carnivorous plants, which continually 'destroy' their own habitats. It resembles *Utricularia* which moves erratically but slowly in uncongested water (see page 43), or in its epiphytic forms on tree trunks where it is washed with continuous stem drainage. It resembles, too, *Aldrovanda* which occupies the slow-moving water in otherwise vegetation-free reed beds (Fijalkowski, 1958). We shall review the whole problem on page 45.

The orientation of *Darlingtonia* habitats

One further aspect of the local distribution of *Darlingtonia* has been noted, has no obvious explanation and requires further examination. Austin (1875–77; see Appendix 1) noticed that all the colonies of *Darlingtonia* that she discovered, in fact in three small valleys within a mile and a half (2.4 km) of each other and containing ten separate patches of plants, all faced south or southwest. The opposite faces in the same valleys, although apparently with similar ecological conditions, were unoccupied. This orientation, as we shall see in Chapter 4 on the patterning of the traps, brings about almost an exact and consistent north–south axis for the first pair of traps. No satisfactory explanation has yet been offered for this strict localization, but as we have already seen with *Drosera* (page 135) some carnivorous plants have strict requirements of temperature and light intensity to achieve flowering (Pissarek, 1965) and, as a result, adequate seed-set and colonization.

In this micro-habitat requirement, *Darlingtonia* bears a close resemblance to the south to southwest pattern of distribution for *Pinguicula lusitanica* over part of its range. Slack (1966) noticed that *P. lusitanica*, on the extreme edge of its range in parts of northern Scotland, had the conventional requirements of a suitable flush: on a seepage, but unlikely to be flooded, commonly acid, but not extremely so; and a limited choice of weakly competitive associates such as other carnivorous plants, saxifrages and small grasses and sedges. Where there was taller vegetation, e.g. *Myrica gale* and *Pteridium aquilinum*, *P. lusitanica* was found in the gaps. In addition, unlike *P. vulgaris* in comparable habitats (Gimingham and Cormack, 1964), which has some preference for north-facing slopes, *P. lusitanica* almost exclusively faces south or southwest. These 'sun-trap' localities have no correlation with rock type, and Slack (1986) suggests they may be correlated with the summer-flowering and autumn seed-set of *P. lusitanica*. In this respect *P. lusitanica* is similar to *Darlingtonia*, whereas *P. vulgaris* completes its seed-set in the late summer. It would be useful to test this hypothesis of carnivorous plants on the extreme edge of their range. Many carnivorous plants are marginal performers in terms of photosynthesis (see Ch. 7). Regardless of species, genera and even mechanism, as they reach the photosynthetic margin of their distribution, these light/temperature requirements may further limit their choice of habitat.

The Plant Associations of *Darlingtonia*

Darlingtonia may be found associated with *Sphagnum* moss, as are many carnivorous plants, but in many locations, e.g. Butterfly Valley, it grows in pure stands with virtually no other vegetation in great clusters ('islands') of what are at least partially, if not wholly, clonal colonies 5–30 m across. Thus an island of numerous plants may be a single genetic individual. Could it also be that, as with some other clonal colonies, e.g. certain forest trees, there is root or rhizome grafting and a sharing out of irregular food supply?

There is, at this site, virtually no humus accumulation and no *Sphagnum*, and the tangled root masses grow directly into the water of the many springs, seepages or streamlets. These islands grow in the light gaps of the mixed forest of the 'Arid Transition Zone' or 'Yellow Pine Forest' (Knight *et al.*, 1970). The principal trees are: *Pinus ponderosa* and *P. lambertiana*; *Abies concolor*; *Pseudotsuga menziesii*; *Calocedrus decurrens*; *Quercus kelloggi*; *Acer macrophyllum* and *Cornus nuttallii*. Few plants of the *Darlingtonia* islands grow into the shade of the canopy trees and where they do they are reduced in stature. The general impression of the *Darlingtonia* plant islands is of pure stands, but a small number of other plants are found in association, notably *Drosera rotundifolia*, which must be one of the most frequently observed co-associating carnivorous plants. Also commonly present are *Myrica californica*, *Vaccinium oxycoccus*, *Ledum glandulosum*, a few liliaceae such as *Camassia leichtlinii* and *Narthecium californicum*, the orchid *Habenaria dilatata* and *Rhododendron occidentale*. In addition, a few flowering plants of even greater rarity also are regularly found such as *Cypripedium californicum*, *Gentiana bisetaea*, *Rudbeckia californica* var. *glauca*, *Viola lanceolata* ssp. *occidentalis*. However, as with almost all other stands of carnivorous plants, there is generally little association with any other flowering plants except at the boundaries of the islands.

Fire Tolerance of *Darlingtonia*

The forests and chapparal of virtually the whole of California burn regularly. Authorities differ as to the frequency of such fires in the 'Yellow Pine Forest', but it is generally accepted that they must have burned up to recent historical time once every 3–20 years (Barbour and Major, 1977). Crown fires are rare or non-existent and the low-intensity ground fires appear to have little or no effect on the *Darlingtonia* colonies (Dakan, W., pers. comm., 1986) except to remove some of the competing vegetation.

The Distribution and Ecology of *Heliamphora*

The genus *Heliamphora* is restricted to the middle and upper levels of the Roraima Sandstone which form the high, isolated plateaux of the area of South America at the junction of Venezuela, Brazil and former British Guiana (Guyana). Geologically, the Guiana Shield is an ancient Pre-Cambrian land mass of long continental evolution. The tabular Proterozoic quartzitic sandstones and shales of the Roraima Formation, which comprise most of the basal rocks of the *Heliamphora* area, are probably about 1700 million years old. However, very much more recent successions of uplift and rapid erosion have given rise to the landscapes that we see today. The plains are punctuated by plateaux 600–1000 m high, covered with dense tropical rain forest. Large rivers cascade over spectacular falls such as the Kaietur and Amaila, which in fact lie just outside this area, the tallest over 1000 m high. Several high mesa-like 'tepui' rise above the general level, culminating in Mount Roraima (2730 m) at the point where Brazil, Venezuela and Guyana meet (McConnell, 1975). These tepuis are the homes of the various species of *Heliamphora*. But the precise periods when these tablelands became isolated is not clear. Three cycles of uplift and erosion are recognized (Krook, 1975), and it seems probable that the time of isolation of these species took place in the late Tertiary (Pliocene), by which time it is safe to assume that the basic *Heliamphora* form was well established. The subdivision from the *Darlingtonia/Sarracenia* group (see page 305 and Fig. 19.5) had almost certainly already taken place, and only the small species distinctions that we at present see were effected after connections between these remote high mountain tops were severed.

Very little is known of the ecology of *Heliamphora*, indeed its taxonomy is still in doubt (Maguire, 1978; Mazrimas, 1979; Steyermark and Maguire, 1984). As Maguire tersely puts it: 'six species are, with some difficulty, recognised for the genus . . .'. Ecological observations are rare; ecological experiments non-existent. Bogner (1976) saw *H. heterodoxa* on the Auyān-tepuí of Estado Bolivar, Venezuela. The Ayuān-tepuí is one of the largest in the Pan-tepui area of southern Venezuela, typical in its flat plateau summit, with an area of about 700 km^2 and a height of 2400 m above sea level. On the summit, *H. heterodoxa* grows in full sunlight in humus depressions, often with standing water, which have formed in the eroded sandstone. The hollows, locally known as 'oricangas', were probably formed in the semi-arid Pleistocene period (Bakker, 1957). The soil is saturated acid peat, poor in nutrients. The middle year temperature has a maximum of 14–15°C and a minimum of 4°C with frequent foggy conditions in the dry season. The *Heliamphora* grows in great masses with *Stegolepis*, *Paepalanthus*, *Tepuia*, *Xyris*, *Cyrilla*, *Tofieldia*, *Nietneria*, *Cottendorifia* and *Brocchinia*. At a different location on the Pan-tepui of southern Venezuela, Brewer-Carias (1973) noticed that the *Heliamphora* there grew in two different forms: one form, in marshy ground, was deeply embedded in acid peat, so submerged that its aperture was level with the surface; the other grew in slightly higher ground and the pitcher reached about 1.5 m.

Mazrimas (1979) records that one species, *H. tatei*, is dendroid, forming dense populations over extensive moor-like savannas. These plants may have been the ones seen by Brewer-Carias (1973). However, in marginal areas or in open sparse woodlands, Mazrimas notes that certain individuals may become partly scandent and will reach a height of 4 m. These ecological variants might suggest a separation in prey. However, *Heliamphora* may be a relatively ineffective carnivore (Burras, J.K., pers. comm., 1979); nor does it normally catch insects in the greenhouse. Nevertheless, unlike most other carnivores, it does seem able to stand some shade and competition. Although continuing to grow in low-light conditions, Burras, K. (pers. comm., 1979) noticed that *Heliamphora* ceased to form a functional pitcher. The epiascidiate pattern (see page 56) of growth was reduced to an almost vestigial pouch at the base of a virtually conventional flat lamina.

At the same site in Venezuela, Brewer-Carias (1973) also noticed *Utricularia humboldtii* growing in the water accumulating in the phytotelm of a giant terrestrial bromeliad, *Brocchinia*. Were this particular *Brocchinia*, like *B. reducta* demonstrated to be carnivorous (see page 19) the mutualistic/commensal relationships would become even more complex.

The ecology of *Byblis gigantea*

Byblis, like many Australian carnivorous plants that we have already seen, enjoys a sustained wet period in winter, but a long, hot, dry summer, during which it dies back to a rootstock. As reported by DeBuhr (1975a), *Byblis gigantea* in the Cannington Swamp near Perth, Western Australia grows in close association with other carnivorous plants such as *Utricularia (Polypompholyx) multifida*, *Drosera menziesii* and *Utricularia inaequalis (hookeri)*. The soil is primarily a white sand, with some accumulation of humus. Most of the plants in southwestern Australia are adapted to fire and *Byblis* (Carlquist, 1976), like its near neighbour the monotypic *Cephalotus* (DeBuhr, 1976) is no exception. Not only is the rootstock apparently totally resistant, there is also evidence that fire may be a requirement for seed germination (DeBuhr, 1976). Thus far *Byblis* is in general agreement with the standard ecological requirements of a carnivore. But in another location *Byblis gigantea* diverges from the general rules. 225 km north of Perth, and 56 km inland DeBuhr found *Byblis* growing in dense scrub which was up to a metre tall, and yet it was found there in greater concentration than the *Byblis* at the more conventional site of Cannington Swamp. DeBuhr points out that *Byblis*, like certain Australian *Drosera* species, often grows in tall, vigorous vegetation. A possible explanation of this apparent greater competitive ability may lie in the fact that *Byblis* species, unlike almost every other carnivorous plant studied save *Triphyophyllum*, has a thick and deeply penetrating rootstock (Schwartz, 1974). The stem has a significant quantity of secondary thickening (Carlquist, 1976) and can grow up to 0.5 m tall. At least in southwestern Australia, excessive soil moisture is more of a limiting factor than soil fertility, more species of plants being able to tolerate low fertility than can endure aquatic or semi-aquatic habits.

The Ecology of *Utricularia*

The genus *Utricularia*, including for convenience the now probably archaic genus *Polypompholyx*, is the most ubiquitous of any of the carnivorous plants and comprises over 275 species, with almost certainly more to be discovered. The flower structure of *Utricularia* is very similar to that of *Pinguicula*, yet it is difficult to imagine two more dissimilar genera within a family and the divergence of this sophisticated aquatic from what appears to be a primitive 'fly-paper' type of carnivorous plant has yet to be explained. Although it is not too difficult to speculate on the origin of a *Pinguicula*-type carnivore (Kerner, 1878), no adequate evolutionary sequence can yet be constructed even to present a speculative path for the origin of what appears to be a relatively homogeneous group.

Seed Production in *Utricularia*

So little work has yet been done on *Utricularia* that even its seed production has yet to be elucidated. Many species produce flowers of two types. One is an apparently normal, open chasmogamous type which is presumably insect pollinated, although evidence is sketchy. However, many species also produce cleistogamous flowers, often in the northern hemisphere, in the early spring. Although pollinators cannot, apparently, enter these flowers, they do produce viable seed.

The General Habitat of the Genus: the Aquatics

Many species of *Utricularia* are aquatic. They are found as strands or mats of plants floating in quiet, pollution-free acid ponds and bog-associated waters. Their flowers, as we shall see in Chapter 15, are borne on often tall, narrow scapes well above the water surface and the inflorescence may vary from a single to 15 flowers, depending upon the species.

A small community of *U. exoleta* in the Hula Valley in Israel occupies two distinct niches and deserves special mention. The first population grows in quiet, slow streams of cold water; the second in small ditches or puddles. Whilst many carnivorous plants are the first to establish themselves after fire (see *Dionaea* and *Sarracenia* p. 30 & p. 38) *U. exoleta* exploits the small puddles which are produced in the mud by cows and other large mammals. Therefore the presence of this species is dependent on its coexistence with a limited number of animals. By grazing, these animals also limit the competition of neighbouring plants.

The Terrestrial and Possible Epiphytes

The second habitat of the genus, found more commonly in sub-tropical regions, is terrestrial. These species most commonly grow in damp, sandy or peaty acid soils, with the main part of the plant at or below soil level. Some may also be found in *Sphagnum* mats and hummocks, marl bogs, sand and peat muck, the damp boles of tropical trees and even in the moss-covering on the tree itself. They have been seen amongst the moss and foliose Jungermanniae on the trunks of *Abies webbiana*, growing as high as 3550 m (11000 ft.) (Compton, 1909).

Many of these terrestrial species have narrow, flattened, pointed, green leaf-like structures, almost like seedling grass-blades, arising from the stem or base of the scape and projecting 1–5 mm above the ground. They apparently have a photosynthetic function (see Ch. 7).

The Turion or Resting Bud

Many aquatic species of *Utricularia* form a type of

hibernaculum (sometimes incorrectly called a winter bud) commonly known as a *turion*. In many northern species of the genus, as winter approaches and animal life in the upper layers of water disappears, the leaves at the extremities of the floating stems are enlarged and form buds. The older parts of the stems, together with their leaves, die. Their cavities, which were previously filled with air and buoyant, fill with water and sink to the bottom of the pond or lake, drawing down with them the turions. In some species, however, the turion may float free during the whole of the resting period and become frozen in by ice. After a northern winter, these buds elongate, detach themselves from the old stems, rise to the surface and there develop two rows of lateral branches in rapid succession.

This description suggests that turions are principally adapted to a summer/winter cycle. Turions are, however, able to resist drying. As Maier (1973a, b) has shown, the relative water content of the turion is lowered by its stored assimilates. When in a potentially desiccating situation, turions may be able to resist drying by their abundance of mucilage. Turions are, apparently, able to resist several months of dry air, losing only about 6% of their dry weight. At the end of winter, or the dry period, just before they sprout again the turions become sensitive to desiccation. Such a degree of protection would not only protect against frost damage, but also might protect sub-tropical or tropical species against a protracted dry season.

The Stolon and Tuber

Stolons, another leaf modification, arise from the base of the scape in the axils of the bracts. They may consist of short, fusiform, simple or compound appendages densely covered with mucilage glands, e.g. *U. uliginosa (affinis)* and *U. lloydii*, or the dissected leaf may develop as a short claw-like structure. These projections have been suggested to be anchoring structures or further capture mechanisms, but the latter never seems to be the case. In two species *U. neottioides* (Brazil) and *U. rigida* (West Africa) the stolons do anchor the plants by glueing them to a hard rock surface. As Lloyd (1935) points out, like their analogues in the genus *Podostemon*, the holdfasts are algal in appearance and texture, recalling some marine forms of life. In a few species, e.g. *U. nelumbifolia*, the stolons become stout aerial projections reaching from one water-filled bromeliad rosette, in which the plant lives, to another. Darwin (1875) gives a description of this remarkable plant:

> 'it . . . is only to be found growing in the water which collectes in the bottom of the leaves of a large *Tillandsia*, that inhabits abundantly an arid rocky part of the mountain, at an elevation of about 5000′ [1600 m] above the level of the sea. Besides the ordinary method by seed, it propagates itself by runners, which it throws out from the base of the flower-stem; this runner is always found directing itself towards the nearest *Tillandsia* when it inserts its point into the water and gives origin to a new plant, which in its turn sends out a new shoot. In this manner I have seen not less than six plants united.'

Tuberization of these stolons may occur in some species. Minute spherical or oval tubers resembling strings of beads were described by Goebel (see Lloyd, 1935) for *U. striatula (orbiculata)*, they were also seen by Compton (1909) in *U. brachiata* and by Darwin (1875) in *U. alpina (montana)*. All these species live in wet moss on rock surfaces or on tree trunks where they may be in danger of drying out. When well developed these tubers may be up to 26 mm in length (Darwin, 1875), lying commonly near the surface, but sometimes buried up to 5 cm deep. On the surface they may develop chlorophyll. Unlike most of the rest of the plant of *Utricularia* they do not contain any air, and they sink in water. They cannot be storage organs, as has been suggested and is somewhat implied by their name, as Darwin showed that they contain no starch. Darwin concluded that they contribute to water storage and showed that a plant with tubers in a dry soil could resist desiccation for many days: water was slowly withdrawn from the shrinking tubers. He writes:

> . . . I know of no case, besides the present one, of such organs [on *U. montana*] having been solely developed for this purpose. Prof. Oliver informs me that two or three other species of *Utricularia* are provided with these appendages; and the group containing them has in consequence received the name of *orchidoides*. All the other species of *Utricularia*, as well as closely related genera, are either aquatic or marsh plants; therefore, on the principle of nearly allied plants having a similar constitution, a never failing supply of water would probably be of great importance to our present species. We can thus understand the meaning of the development of its tubers, and of their number on the same plant, amounting in one instance to at least twenty.

The Ecology of the Carnivorous Plant: a Summary

Carnivorous plants are generally mesophytic plants exploiting low-nutrient habitats and possessed of relatively low competitive ability. Most of these habitats are permanently or temporarily wet-acid. Where they are dry-acid (e.g. *Drosophyllum*, *Ibicella*, *Triphyophyllum* and *Byblis*), there are hints that the problem of the rootstock probably not reaching the water table may be overcome by osmometer-like trichome water-scavenging, as is the case in *Drosophyllum*. So little is known about *Ibicella* that even speculation is impossible. It may only resemble a conventional, short-cycling, desert-fringe mesophyte, with a limited carnivorous capacity.

Triphyophyllum is a part-time carnivorous plant, i.e. is exploiting a nutrient-poor habitat in *time* as opposed to in *space*. *Byblis* seems to be the exception to every rule: deep-rooted, competitive and with an insect-catching ability spread throughout a substantial part of the year. Is it truly carnivorous? There is very little evidence either way.

The Characteristics of Plants Associated with Carnivorous Species

In previous sections, we have drawn attention to real or supposed plant associations. Most commonly, when examined closely, these associations do not seem to be anything more than fortuitous. We might expect that the non-carnivores in carnivorous plant habitats, where they exist, might have special features. With the general exception of the orchids, which are mostly if not universally mycorrhizal, this does not seem to be so. There seems, for example, to be no consistent association whatsoever with any nitrogen-fixing plant, of whatever type. The carnivorous plants, limited as they are by their commitment to trap-mechanisms, whose 'penalties' are discussed in Chapter 5, seem to be autonomous scavengers of the earth's varied sources of nutrients.

Carnivorous Plants and the Destruction of the Habitat

Carnivorous plants, by concentrating the protein-rich insect diet in a nutrient-poor habitat, would seem likely to augment the habitat for a competitive opposition. Epiphytism is one escape from this apparent dilemma. Moving slowly, by wind or current, in large bodies of water, like some *Utricularia* species (page 43) and *Aldrovanda*, or on stream sides or savannas where the water table is constantly moving like *Genlisea* (Magnussen, 1982), is another.

Living on the bole or trunk of a tree with a constant stream flow, like other species of *Utricularia* (e.g. *U. alpina*; Sugden and Robins, 1979) and some species of *Pinguicula*, is a third method. A habitat which is a constant stream of water and which is not only part of one's dispersal system, but is also toxic to most other plants is another, used by *Darlingtonia*. Lastly, and most commonly, the carnivorous plant will exploit the habitat as long as it can endure the competition, either rising above the encroaching opposition (as e.g. with *Drosophyllum*) or relying on the frequency of fires or grazing (e.g. *Dionaea* or *U. exoleta*) to reduce encroachment (as e.g. *Dionaea*), at the same time spreading abroad seeds (see Ch. 15) in random dispersion to exploit new habitats.

A Failure of Exploitation

As we have seen, the carnivorous plants have penetrated many habitats but, with the exception of a few species of *Utricularia* and perhaps some *Genlisea* species (of whose habitat requirements we know very little but which can live in very wet mud), they have not exploited the rich hunting grounds of the soil. The common toothwort (*Lathraea squamaria*) has, from time to time, been suggested as a carnivorous plant (Lloyd, 1942; Studnicka, 1982) but, although apparently having some of the features necessary to enable it to trap the soil fauna, convincing evidence is lacking.

In fact, despite the apparent advantages and the abundance of insect life, there are very few carnivorous plants. There is none, with the exception of a few fungi, below the angiosperms; there are no herbs, no shrubs or trees and there are large areas of the phylogenetic table (see Fig. 1.2) where carnivory does not seem to have evolved at all. Could it be that, as we shall review in Chapter 7, carnivory exacts so high a price in terms of the diversion of biomass into the trap component, that only the richest of insect-containing sites, and then only at certain times of the year can, effectively be exploited?

PART II

Attraction and Trapping

CHAPTER 4

General Morphological and Anatomical Features of the Traps of Carnivorous Plants

In this chapter we shall review all those general features of the *trapping leaves* of carnivorous plants, excluding those particular details of the trapping mechanism itself, which are the province of Chapter 6. Without regard for possible taxonomic affinities (see pages 6–7) we shall amalgamate for review 'Adhesive' or 'Mucilage' Traps; 'Snap' Traps; 'Pitchers'; and the 'Suction' Traps.

Adhesive Traps: General Considerations

Numerous stalked glands, each holding a large drop of sticky mucilage on its top, form the main trapping device of the *adhesive traps* (flypaper traps).

The secretion of viscid substances is very widespread in the plant kingdom, as we saw in Chapter 1 and will discuss again in Chapters 17–18. Such substances can be divided into three main groups: the aromatic, the resinous and the mucilaginous. Only the latter can, but does not invariably, carry an enzyme. The mucilaginous secretions, with or without enzymic addition, are by no means confined to the carnivorous plant, but are almost ubiquitous and involved in diverse roles.

The Adhesive Traps of the Droseraceae

As in other adhesive traps, those of the Droseraceae secrete their trapping mucilage from special stalked glands (see page 50). The Droseraceae are unique amongst the families of carnivorous plants in that some speculative sequence may be observed (Fig. 19.1) – from a non-moving mucilage trap, *Drosophyllum*, through the great, diverse and widespread genus of *Drosera* with its range from slow to perceptive movement of its tentacles. The Droseraceae progress too, almost as it were in a normal curve – from the dry habitat, monotypic *Drosophyllum* through the very numerous and habitat-diverse *Drosera* to the totally submerged and again monotypic *Aldrovanda* (see Ch. 3).

Drosophyllum

The monospecific genus *Drosophyllum* is a large plant, almost a shrub, sometimes reaching 1.6 m in height (Fig. 5.5).

The leaf is linear with a deep furrow along the upper side. There are three vascular bundles, a median and two laterals, arising from a single bundle entering at the base. Branchlets from each of these three bundles terminate in the stalked mucilage glands and under the sessile digestive glands (see Chs 6 and 8). The developing leaves have a reverse circination (Fig. 6.2), the rolled leaf-tip facing outwards, whereas in the related *Drosera*, the opposite is the case. Another characteristic feature is the persistence of leaves (see pages 31 and 50) giving, as we describe it, the appearance of a flue-brush. This persistence is normal and is not, as some authors (cited by Lloyd, 1942) have suggested, a pathological condition.

Drosera

The commonest type of *Drosera*, i.e. the north temperate form typified by *D. rotundifolia*, consists of a erect plant with a slender stem crowded by a rosette of leaves with flowering scapes growing in the leaf axils. The seedling arises from an ephemeral tap root which persists only in the formation of the earliest rosette of leaves. As the plant grows the old foliage dies off behind. Growing in other vegetation such as *Sphagnum*, in a strongly seasonal climate, one sees successive dead rosettes clinging to the stem and ending above in a living rosette. This growth form is clearly depicted by Nitschke (1860) and bears some resemblance to the perennial state described above for *Drosophyllum*.

Unlike *Drosophyllum* and *Byblis* (page 49), in most droseras the upper face of the young blade is applied to the petiole (Lloyd, 1942; his Fig. 16.17), in others, e.g. *D. filiformis* and *D. regia*, there is a true circination. Such differences may be due to the need to protect the tentacles on the upper surface in *Drosera* and on the lower side of *Drosophyllum*, but there seems no totally satisfactory explanation for these distinctions at the moment.

An interesting feature of, as far as is known, all *Drosera* leaves is an absence of a definite palisade layer, the mesophyll being made up of just a number of courses of more or less rounded cells (Fig. 6.1B). Moreover all the leaf cells, including the epidermis, contain chloroplasts. Despite this apparently anomalous anatomy there would appear, on the limited evidence quoted by Lloyd (1942), to be no evidence of ineffective photosynthesis as compared to normal non-carnivorous plants. Stomata occur on both surfaces.

The upper leaf surface of *Drosera* (Fig. 2.2) is covered with numerous stalked glands, the *tentacles*. These tentacles vary considerably in size, the peripheral are very long, becoming gradually shorter as they approach the leaf centre. This gradual difference in length is associated with a gradual difference in their ability to move: the longer the tentacle the faster its movement (see Ch. 6). The peripheral long tentacles increase the effective trapping area. The ontogeny of the tentacle is complex. The outermost layer (Oc) of the tentacle head (Fig. 4.1) is epidermal in origin. The layer 'lc' is parenchymatous and the bell (Ec) is partly epidermal and partly parenchymatous. Those cells which come to the surface at the base of the bell (B) are epidermic. The enclosed mass of tracheids, and reticulate, annular or spiral vessels connect to the leaf's vascular system.

In addition to these movable or inflexible stalked glands there are very numerous, small sessile glands (Fig. 6.1B). These are found on both leaf surfaces, on the stalks of the glands (Fig. 19.1) and on petioles and scapes. These latter are purely epidermal in origin.

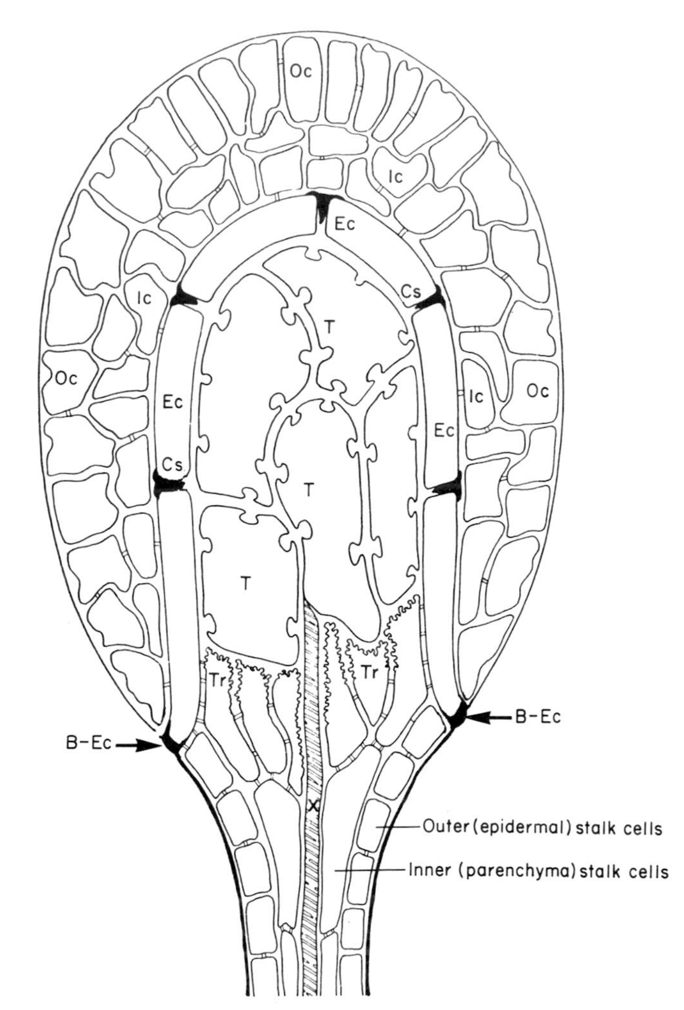

Fig. 4.1 Diagram of a stalked gland of *Drosera capensis*. Outer and inner stalk cells are indicated; X = xylem, T = tracheid, Ec = endodermoid cell, Cs = Casparian strip, Ic = inner gland cell, Oc = outer gland cell. A thin, perforate cuticle covers all of the outer gland cells and a more complete cuticular layer the base of the endodermoid bell (B–Ec, arrows) and the stalk (× 500).

Triphyophyllum

Triphyophyllum is a great woody climber from West Africa. Most carnivorous plants have no conventional leaves. Some, such as a few species of *Sarracenia*, have ephemeral phyllodia; only *Cephalotus* and *Nepenthes*, often inhabiting the partial shade of the edge of the tropical forest, has a clear distinction between the trap and the photosynthetic system.

Triphyophyllum, unique in this respect, has three different sorts of leaf for three occasions. The first (Menninger, 1965) is conventional, and obviously primarily photosynthetic. The second, like many other vines, is a tendril. The third, produced on juvenile shoots and timed just before the height of the rainy season (Green *et al.*, 1979) is carnivorous. The midrib is stout, the blade much reduced or absent and the whole structure from 15 to 25 cm long. The leaf unfolds with a reverse circination as in *Drosophyllum* and the stalked glands themselves, of two sorts, are anatomically similar, although very much larger than those of any other carnivorous plant being up to 3 mm long (Fig. 4.19). Apart from their heights, these two sets of glands, which do not overlap in their size ranges, are apparently identical. These glandular leaves are held erect, unlike their photosynthetic colleagues, and survive in an active state only a matter of weeks before being shed. As in *Drosophyllum*, there are numerous sessile glands on these leaves too. The stalked gland head is almost identical albeit twice the width, to that of *Drosera* (cf. Figs 4.19 and 6.2). Two to four layers of outer secretory cells overlie an endodermoid bell similar to that of Figure 4.1. Where *Drosera* (Fig. 4.1) and *Drosophyllum* have simple tracheids, *Triphyophyllum* is internally more elaborate. The vascular tracheids are extensively branched and ensheathed in parenchyma whose cells contain numerous amyloplasts. Unlike *Drosera*, there is no clear 'neck' zone in *T. peltatum*, and Green *et al.* (1979) speculate that living parenchymatous cells take over the function of the specialized cells in *Drosera*. As in *Drosophyllum* the stalked glands develop in sequence, secreting as they develop and secreting a new droplet if the first is removed. The sessile glands, again as in *Drosophyllum*, remain dry until stimulated. Both sets are reddish in colour contrasting vividly with the green background.

Stimulated glands do not move and in this respect they are similar to the stalked glands of *Drosophyllum*. Stimulation, as in *Drosophyllum*, leads to further release of secretion from stalked glands and this is reinforced by secretion from the sessile glands clustered at their bases.

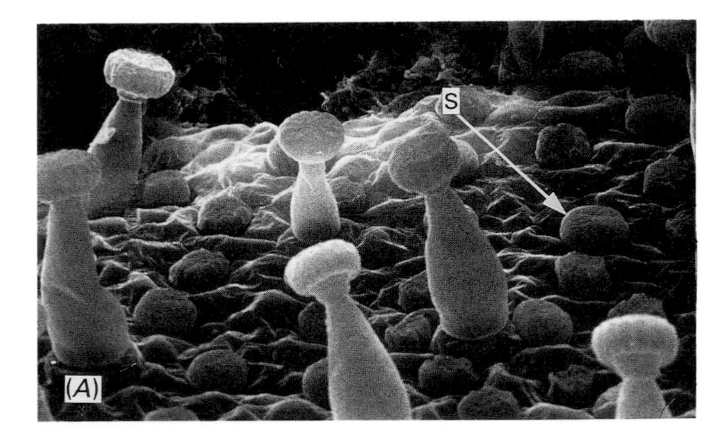

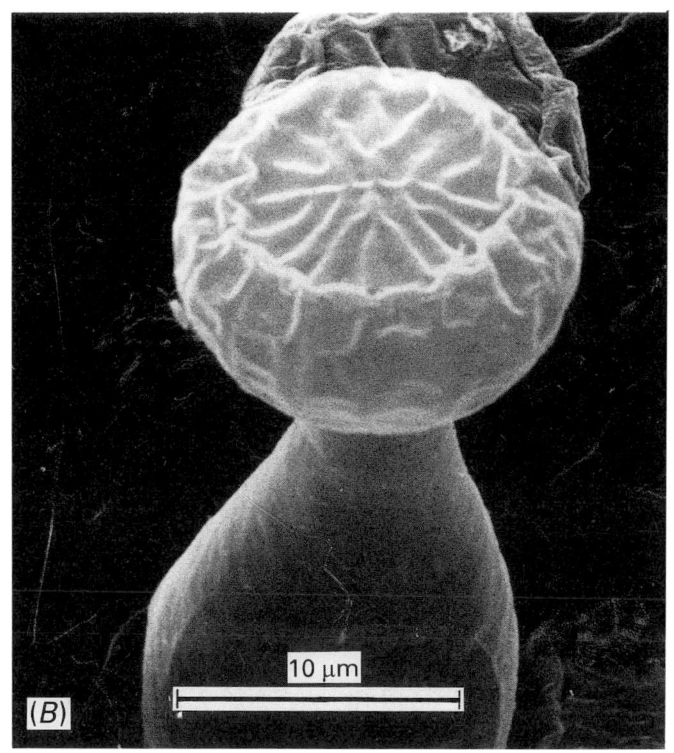

Fig. 4.2 The stalked and sessile glands of *Pinguicula moranensis*. (*A*) The sessile glands with their eight-celled heads cluster around the stalked glands with single-celled domes. As can clearly be seen in the lower micrograph (*B*), these single domed cells support a capital of sixteen radiating cells, cf. Figs 6.1*A* and 10.5 (SEM by G. Wakley, Oxford).

Pinguicula

A short vertical stem gives rise to a compact rosette of leaves which, in almost every species, lies flat on the ground. Most *Pinguicula* plants in full light (cf. *Catopsis*) are a pale, often lemon-green, but deepening in the shade. The leaves are soft, easily bruised, and greasy or shiny to the touch, hence their Latin generic. The entire leaves and peduncles are densely covered with stalked and sessile glands (see Figs 4.2 and 8.7C) with a much smaller number of sessile glands, with four-celled capitals, on the lower surface. The latter may be hydathodic in character and seem to have no role in the carnivorous process. The stalked glands end in a single domed cell supporting a capital of 16 radiating cells (Fig. 4.2). The sessile glands are similar, but with only eight cells in the head and the base of the gland lies flush with the epidermis (S in Fig. 4.2). In addition, some four rows of cells along the thin leaf margin, like the stalked glands, secrete mucilage (Lloyd, 1942).

Byblis

Like *Triphyophyllum*, but unlike almost every other carnivorous plant, *Byblis* species have substantial perennating rootstocks. From these rootstocks arise secondary branches, leaves and long-peduncled flowers held amongst the traps. All these aerial parts, even the flowering stems, are clothed with sessile and stalked glands (Figs 4.20 and 4.21). Stalked glands are so ubiquitous they can even be found on the ovary wall. *B. gigantea* has no apparent circination, but *B. liniflora* has a reverse circination similar to that of *Drosophyllum* (cf. Figs 4.20 and 6.2). The rigid, triangular in cross-section, thick-walled, yellow-green leaves (cf. *Pinguicula* and *Catopsis*) have a well-developed vascular system and mesophyll composed of only one type of cell (cf. *Drosera*, Fig. 6.1B). The stalked glands are permanently mucilaginous, but the sessile glands only throw out a weakly viscous mucilage when stimulated. After four to six hours these glands become dry again.

The 'Snap' Traps: *Dionaea* and *Aldrovanda*

The Trap of *Dionaea*

The perennial leaf rosette of *Dionaea* arises from a short unbranching rhizome which is clothed with the bases of the leaves. Each leaf has a succulent base which lies under soil surface and a green petiole which ends with a pair of trap lobes, together forming the leaf blade (Figs 4.3 and 6.12). The central vascular bundle, which runs along the petiole, extends to the midrib of the leaf blade and serves as the geometrical axis of the trap.

While the lower surface of the lobes (the trap outside) bears many stomata (Yaguchi and Kondo, 1981; Kondo and Yaguchi, 1983), their upper surface, which is slightly concave, supports various

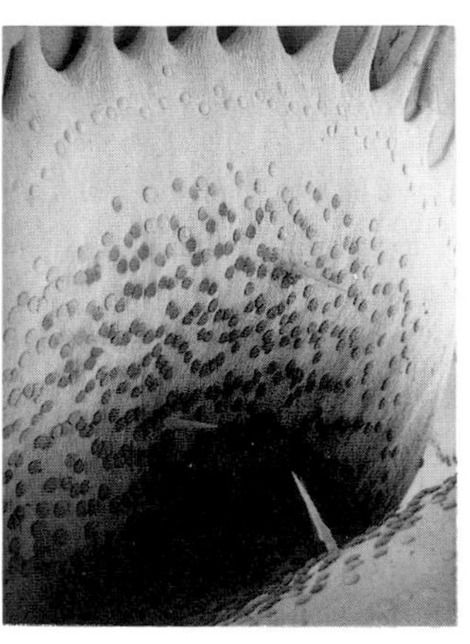

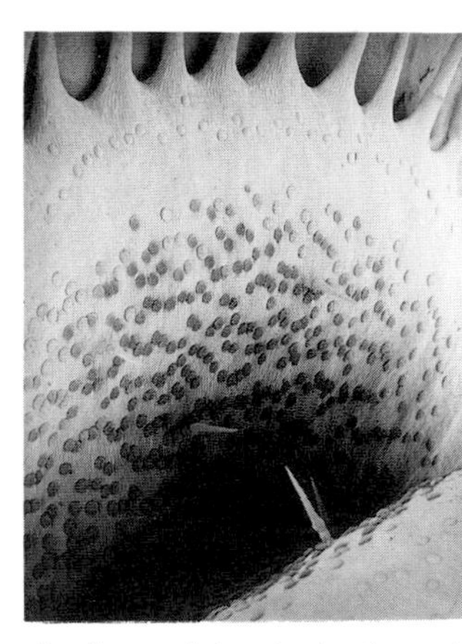

Fig. 4.3 SEM stereo-pair of the inside (adaxial) faces of the trap lobes of *Dionaea*. Five of the six trigger hairs, the sessile glands, peripheral glands and the base of the teeth are visible (SEM by D. Kerr, Oxford).

Fig. 4.4 The postures of *Dionaea*, after mechanical and chemical stimulation. (A) Unstimulated; (B) immediately after mechanical stimulation; (C) after chemical stimulation the peripheral rims of the trap lobes begin to come together; (C') lobe margins close completely. (D) peripheral rims now in tight contact and teeth cease to interlock and become more or less parallel (D').

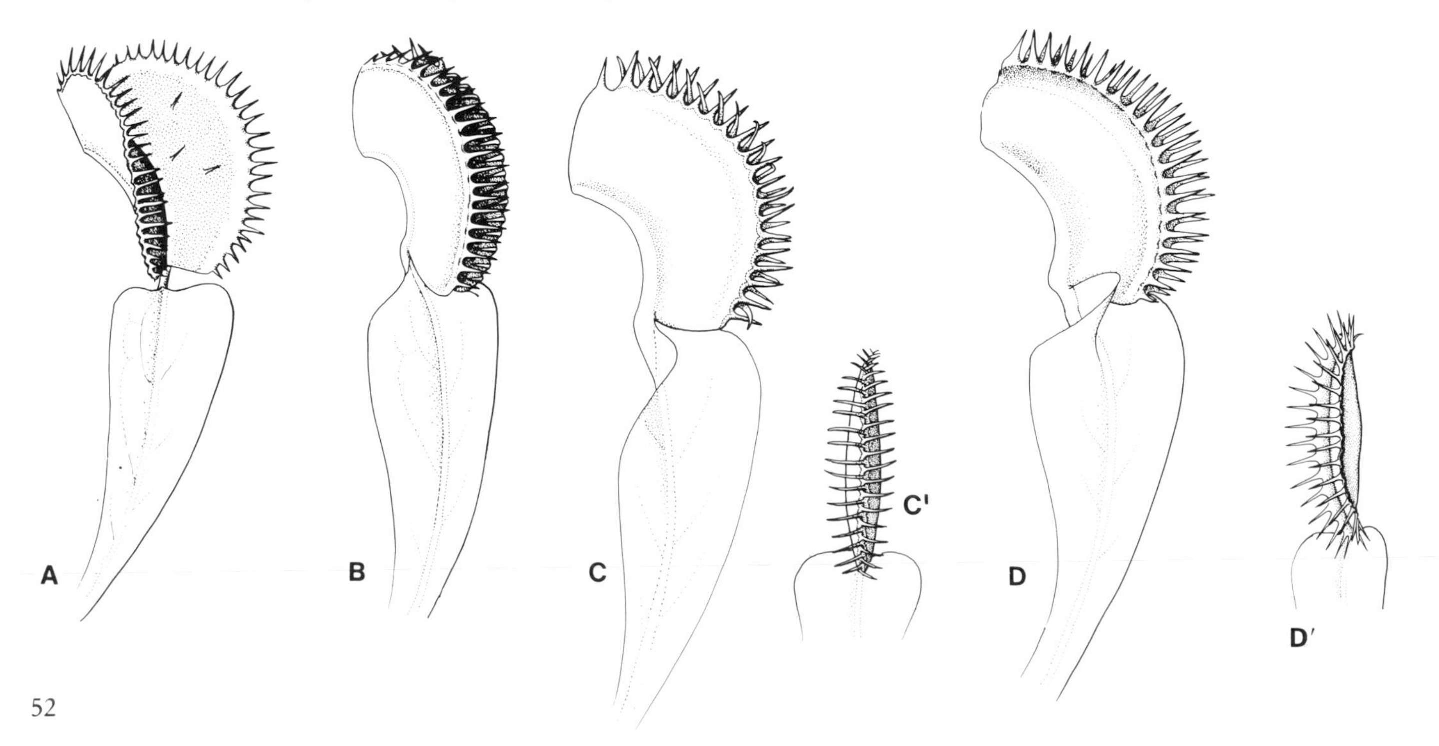

different structures serving in attraction, retention, sensing and digestion of the prey.

The trap of *Dionaea* can be divided into three principal zones (Figs 4.3 and 6.12):

(i) *The marginal teeth*: 14–20 marginal teeth are borne in the periphery of the trap, slightly bending upward (Fig. 4.4*A*; cf. *Drosera burmanni*, page 303). The teeth of the opposite lobes interlock when the trap shuts (Fig. 4.4*B*).

(ii) *The peripheral band*: this narrow region surrounds the upper side of the trap. It bears only a few small sessile glands which are normally colourless and are known to secrete carbohydrates. Each peripheral gland lies in a small depression, which can just be seen in Figure 4.3 (see also Fig. 8.2*B*), so that it is not damaged when the two lobes touch each other forcibly when the trap shuts and narrows (Fig. 4.4*D*,*E*). The peripheral band is UV-absorbing (see Fig. 5.2*C*).

(iii) *The digestive surface* (Fig. 6.12): The whole large central zone of the trap, which is limited by the peripheral band, is covered with numerous pigmented digestive glands lying on a flat epidermis. The digestive glands are larger than the peripheral glands and can easily be seen with the naked eye. When the trap closes this glandular region serves as the wall of a temporary digestive pool (Fig. 4.4*C*,*D*,*E*).

Amongst the digestive glands, and in symmetrical positions, are long pointed bristles (Figs. 4.4*A* and 6.12), usually numbering three on each lobe but sometimes as many as seven or nine. These bristles are the trigger hairs which serve as touch receptors (see page 99 onwards and 172).

External Glands in *Dionaea*

The whole outer surface of the trap, as well as the surfaces of the marginal teeth, are covered with stellate external glands (Fig. 4.5). Each external gland has a pair of basal cells lying between the epidermal cells. In addition, there are two endodermoid neck cells (see Fig. 4.1 for comparison and page 151), two glandular layers, the inner of which is made of two small cells and the outer layer of 2–6 elongated cells radiating from the centre. These external glands have a dense cytoplasm when the trap is developing. At maturity the endodermoid neck cells seem to remain highly active, accumulating chloride ions, while the outer glandular cells collapse (Fig. 4.5; Joel and Juniper, 1988). Thus they closely resemble the external glands of the *Utricularia* trap (see page 69).

The Trap of *Aldrovanda vesiculosa*

Aldrovanda is an aquatic plant. The leaves of *Aldrovanda*, only a few millimetres long, are arranged in whorls. A whorl is composed of eight leaves, evenly arranged around the stem (Fig. 4.6 and see p. 19 in Overbeck, 1982). The older internodes and leaf whorls die away successively as new ones are formed at the apex.

Fig. 4.5 External glands of a very young, developing and unopened trap of *Dionaea*. At this stage of the development of the trap, some gland collapse can be seen, whilst other glands are still showing living head cells (SEM, liquid nitrogen frozen, by S. Clarke, Oxford).

The petiole of each leaf is thin and broad and terminates in a bi-lobed leaf blade and 4–6 bristles (Fig.4.6). A vascular bundle runs down the middle of the petiole and extends to the midrib of the leaf blade. The leaf blade is twisted to the left by more than 100°, its midrib making an obtuse angle with the petiole in both the apical and the lateral view of the leaf whorl; thus its two lobes are situated in a tilted position and, hence, were sometimes termed the 'upper' and the 'lower' lobes (Ashida, 1934; Fenner, 1904). The description of *Aldrovanda* traps in this book is based mainly on the observations made by Ashida (1934, 1935), who published a most precise and comprehensive study of this unique plant.

Each leaf blade consists of a pair of concave lobes that are connected by the midrib. Each lobe consists of two portions: the marginal and the central zone. The marginal zone (M of Fig. 6.10) is very thin, formed of only two layers. The central zone (C of Fig. 6.10) is thicker, comprising three cell layers; it forms the digestive cavity when the trap shuts. Ashida (1934) further divided each zone into three regions (Fig. 6.10).

(i) **The marginal zone.**

a. *The rim*, which is the infolded margin of the lobe, comprises one peripheral line of cells, each showing a short tooth-like projection in its centre directed towards its parallel in the other lobe.

b. *The quadrifid region*, which is the main region of the marginal zone, bears quadrifid hairs on its inner surface (Fig. 4.7*A*,*B*) and smaller glands, normally with two arms, on the outside. This area closely resembles the trap wall of *Utricularia* in both comprising two thin cell layers and bearing quadrifids on the inner side together with external glands (see page 68). This gland distribution is further discussed when we consider the trapping mechanisms in (see pages 95 and 104).

c. *The hairless region* lies between the quadrifid region and the central zone, and bears only external two-armed glands.

(ii) **The central zone.**

d. *The digestive region* is densely covered with small digestive glands inside and has two-armed glands outside. This zone also bears about 15 thin trigger hairs (on each lobe) which are scattered between the glands.

e. *The detention region* lies in the middle of the central zone and bears only few glands and almost no trigger hairs. This is the region where entrapped animals are retained when trapped and digested.

f. *The midrib region* comprises the narrow zone around the midrib which carries densely packed digestive glands beween which 6–10 trigger hairs (Fig. 6.10) are also located.

Fig. 4.6 The waterwheel plant, *Aldrovanda vesiculosa* showing the whorled arrangement of traps on the stem. In its natural habitat the plant will lie, almost submerged in water, with the stem of the first five whorls, approximately, parallel to the water surface. One or two bristles from each whorl will project above the surface.

Pitcher Traps

Despite their diverse taxonomy (see Ch. 19), the pitcher traps have many features in common. Not only are they all forms of epiascidiate development (page 56) and this includes *Utricularia* and *Genlisea*, most of them employ ultraviolet light absorption patterns as a possible insect guide (page 75) and most of them orientate their traps in specific directions for what would appear to be a maximization of attraction (Ch. 5). *Sarracenia* and *Heliamphora* form more or less symmetrical circles facing inwards, whereas *Darlingtonia*, as a result of the tubular twist (see Fig. 4.9E), and *Cephalotus*, on its flexible petioles, face *outwards*. What is more, as Austin (1875D, see Appendix 1) observed, not only are the first pair of *Darlingtonia* pitchers usually orientated north and south (Fig. 5.7), but they are also significantly taller than any subsequent

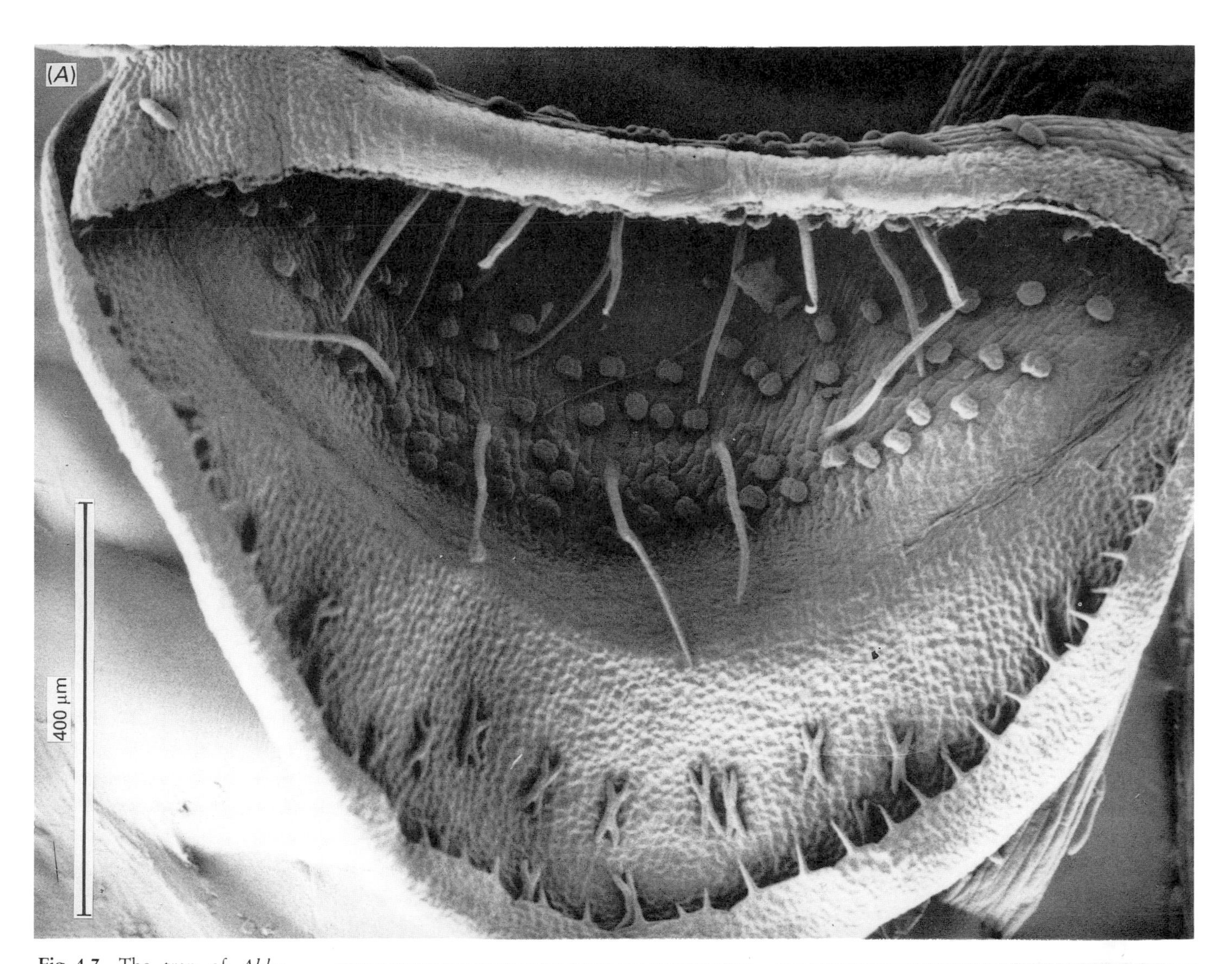

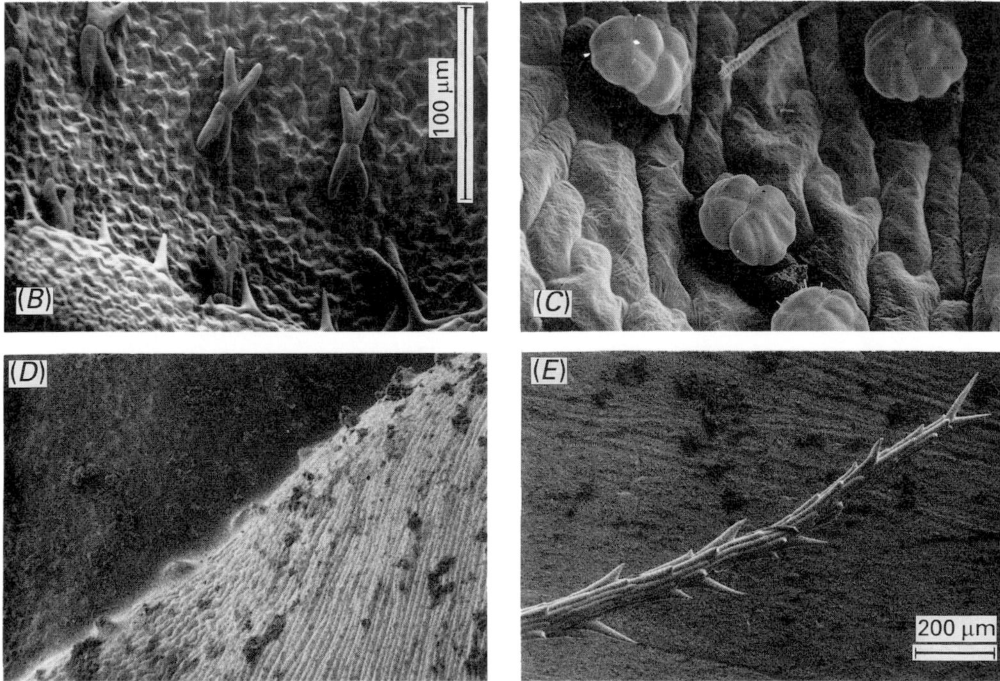

Fig. 4.7 The trap of *Aldrovanda vesiculosa*. (*A*) The whole trap in the open position. Note the numerous trigger hairs (cf. *Dionaea*) some of which may have been bent by the specimen preparation. Note the recurved teeth on the margin of the trap. The quadrifid glands (cf. *Utricularia* in Fig. 4.13*B*) are seen in higher magnification in (*B*). (*C*) The digestive glands at higher magnification (cf. Figs 6.13, 6.17 of *Dionaea*). (*D*) The sessile external glands. (*E*) The bristles of the whorl (see Fig. 6.10) of which the tips of the first few whorls will project above the surface. For a low magnification photograph of the whole plant see Fig. 4.6 (SEMs by C. Merriman, Oxford).

leaves. The same is true of *Sarracenia* species (page 85) and, to a lesser extent, of *Heliamphora*, and, at least in the juvenile stage, of *Nepenthes* (page 84 and Fig. 5.8). All these devices maximize presentation and reduce internal interference.

In addition to these features, all pitcher (deadfall) traps (except *Brocchinia*) use trichomes as part of the trap mechanism, and the totally unrelated *Nepenthes*, *Darlingtonia* and *Brocchinia*, at least in part, use detachable wax to prevent arthropod escape.

The morphology, venation, ontogeny and details of the pitcher-trap mechanisms, have been dealt with in great detail by many authors, particularly MacFarlane (1889, 1893), Lloyd (1942), Franck (1975, 1976) and Adams and Smith (1977). Such detail will not be repeated here, but only summarized and added to in the light of the latest observations.

Epiascidiate Leaves

All pitcher-type carnivorous plants are based upon the epiascidiate leaf. The word 'epiascidiate' derives from De Candolle (1898) where epi- and hypo- are used to determine the position of the original adaxial surface. Thus the adaxial surface of an epiascidiate leaf forms the inner face of a pitcher tube (Fig. 4.8). Such modified leaves are found amongst the species of several families, e.g.

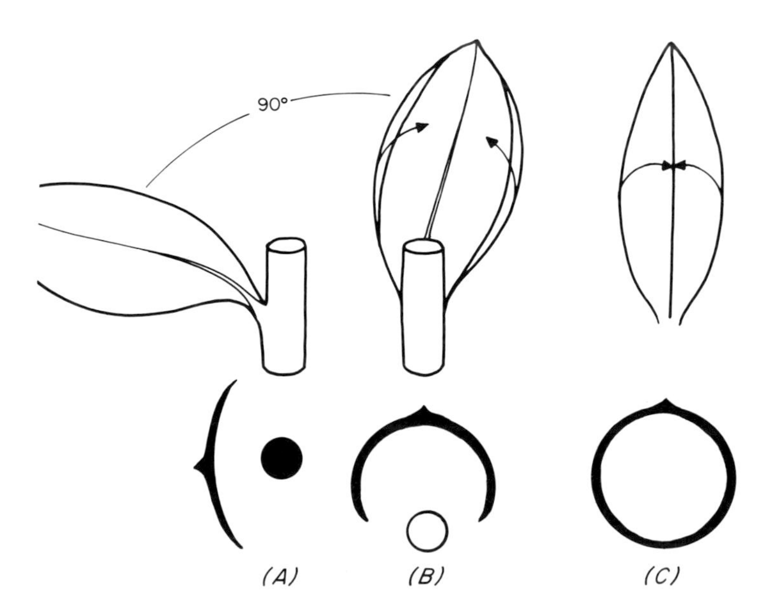

Fig. 4.8 The sequence of development of an epiascidiate leaf. (*A*) a normal leaf insertion; normal adaxial surface uppermost; (*B*) margins begin to roll over the adaxial surface; (*C*) fusion of margins complete; suppression of axis.

Sarraceniaceae
- *Sarracenia*
- *Darlingtonia*
- *Heliamphora*

Cephalotaceae
- *Cephalotus*

Bromeliaceae
- *Brocchinia*
- *Catopsis*

Nepenthaceae
- *Nepenthes*

Lentibulariaceae
- *Utricularia*
- *Polypompholyx*
- *Genlisea*

Note that in the Lentibulariaceae, the trap mechanisms of all but perhaps *Genlisea* are active bladders, not pitchers, but they are epiascidiate in ontogeny all the same.

Epiascidiate forms are constantly arising, but hypoascidiate leaf forms are just as common in the plant kingdom, although leading to other growth forms, not pitchers. The evolution of these structures is considered in Chapter 18.

The epiascidiate leaf as typified by *Darlingtonia*

The early primordial development of all the leaves of all pitcher-plants is remarkably similar (see Fig. 4.8). This text focuses on their development in *Darlingtonia californica* according to Franck (1975, 1976), with reference to other species where necessary.

The apex of *D. californica* comprises 1–4 tunica layers and a corpus with a central meristem 4–5 cells high and 6–7 cells wide. The apex swells from approximately 120 μm to 270 μm in diameter with periclinal divisions in the second and third layers of the tunica on the highest point of the apical slope to form a buttress. From its initiation the primordium is bifacial and accentuated by marginal growth of the basal zone, which forms a horseshoe sheath (equivalent to Fig. 4.8B).

Growth is upright and a dorsal median vascular bundle is established early (before the primordium is 100 μm tall). At this point an adaxial bulge forms periclinal division in the sub-epidermal layers and develops as the 'Querzone'. Up to about 200 μm tall, the margins of the primordia are free but further divisions of the 'Querzone' unite them. At this point a small surface pit is visible and the junction of the margins is still clear.

After the primordium, still growing parallel to the short axis, reaches a height of about 300 μm, adaxial growth from the Querzone and margins ceases and is redirected parallel to the main primordial axis excepting the pit region. Further divisions and cell elongation cause the formation of a tube. It should be noted, at this stage, that the heel may sometimes be in evidence.

After about this point different species tend to diverge in their further development. *D. californica* is particularly notable for the torsional twist that brings the morphologically ventral face to a dorsal position. It has been postulated that this makes the opening more accessible and prevents other pitchers growing into the mouth and blocking it. Such blocking, nevertheless, still occurs in the field (Naeem, S, pers. comm., 1985). A second distinctive feature of *Darlingtonia* is the fish-tail appendage commencing when the pitcher reaches a height of about 2.5 mm. Initially forming as two bulges on the distal point of the primordium the two flanges develop through apical growth. In cross-section the flanges are rolled and when mature carry nectaries within this edge (see Ch. 5).

The rest of the pitcher can be sub-divided into a broadly winged bifacial sheathing base (up to 45% max. of the length), the tubular region and keel (50–70% length), and the hood plus fish-tail appendage. The keel, although a single entity, is bifid from inception and carries two rows of alternate vascular bundles. It develops from periclinal divisions of the sub-epidermal layers continuous with the projected position of the 'old' leaf margins. The hood encloses the top of the pitcher leaving a gap accessible only from underneath. With the onset of maturity the central cells become hyaline and form a transparent 'window'.

Juvenile leaves of this species may lack the keel and have a simple beak instead of the fish-tail apparatus. This might suggest, according to Goebel (1891), that their development has been repressed due to the rapid, early maturation of the pitchers. It also implies that the development of the keel is not a direct consequence or continuation of the Querzone development.

The epiascidiate leaf form in other genera

In *Sarracenia purpurea*, whilst the lid does not occlude the mouth of the pitcher (but see *S. psittacina*) and lacks the fish-tail apparatus, the keel is outwardly single. Internally it has the alternate–opposite bundle arrangement similar to *D. californica*, but only one ventral suture bundle. Arber (1941) infers that this actually represents the agglomeration of two bundles as, just below the rolled rim of the aperture it divides into two lateral bundles. These bundles pass round the rim and, along with other bundles, converge towards the median bundle in the lid.

Heliamphora, the third of the Sarraceniaceae, appears relatively simpler, possessing an obvious double keel supplied with individual vascular bundles and having a poorly developed lid and no rolled collar on the rim.

In *Nepenthes*, the pitchers superficially closely resemble those of the Sarraceniaceae, but have several features that set them apart. Both the sheathing basal region and the petiolar region in *Nepenthes* pitchers are greatly elongated: there are two separated keels, the aperture has both an inwardly and outwardly turned collar and the junction of the lid and pitcher bears a small spike.

The separate keels imply that the adaxial surface has not completely disappeared and the pitcher is not completely unifacial (or epiascidiate). The relevance of the other characters is discussed in Chapter 18, dealing with the evolutionary development of pitchers in general.

Cephalotus represents a different means of attaining the same end. The mature plant possesses two forms of leaves: normal leaves and pitchers. (Occasionally intermediate forms can also be found.) The pitchers bear a double keel along the median suture and two separate ventral keels on either side. Problems in homology arise, however, as the well-developed petiole arises at 90° to the junction of the lid and tube, not at the base of the pitcher. At this junction there is another pair of wings which presumably marks the actual delimitations of the adaxial surface which include regions of the lid.

Early theories of the origin of epiascidiate leaf forms

The earlier theories, which have now been largely superseded by the modified peltation theory of Troll (1932), tended to be incomplete with no elucidation of the spur and were also unsupported by evidence gained from later studies on venation and development. Before the concept of *Oberblatt* and *Unterblatt* were introduced, again by Troll, theories tended to be restricted by the current view lacking this distinction. The 'Foliar' theory fails in that the evidence for the purported pinnae is lacking in terms of developmental structure and a mature venation pattern. Even with Bower's (1889a, b) reappraisal of this type of fused structure, using the concept of a phyllopodium rather than the misleading term 'pinna', the foliar theory lacks any real evidence and is of little predictive value. Hooker's appendicular theory (1859) is now universally disregarded as it has no developmental basis and appears to mix the original idea of a specialized gland (the pitcher) as an appendage of the leaf with the spur being the true apex and the lid as fused lateral pinnae. Roth's views (1953, 1954, 1957) have been questioned, and indeed refuted by many authors, and serve more as a challenge to the concept of unifaciality and peltation adhered to by Troll, if not correct in themselves. The new terminology introduced by Croizat (1961), following perhaps in Bower's phyllopodium footsteps, does not appear to have been developed sufficiently to warrant classification as a theory as such, at least not until some of its tenets have been tested.

Whilst peltation and unifaciality at present appear to be the most feasible solution, as Troll himself noted, they are not completely correct: the petiole actually lacks the required radial symmetry and is not therefore truly unifacial. If reference is made at this point to the pitchers of the Sarraceniaceae, it can be seen that similar trends also exist. If the wing structure(s) does represent the congenital fusion of the leaf margins, then it also seems reasonable to suppose that the sequence, *Heliamphora, Darlingtonia* to *Sarracenia* represents progressive stages tending towards complete fusion. *Nepenthes* can therefore be interpreted as an earlier stage where the two wings are still separated by a region of the original adaxial surface. This incomplete peltation is reflected in the incompleted unifaciality of the petiole region. The only problem with this analysis is why, in *Nepenthes*, the margin should fuse to form a tube; normally, incomplete unifaciality results in a normal dorsiventral leaf rather than in a peltate one. If this description is correct it implies that peltation is not an absolute criterion for pitcher development and the fundamental axioms of that theory are not universally applicable to pitcher formation.

Cephalotus follicularis has to be viewed differently from the above pitchers owing to its different orientation. Eichler (1881) observed a similar early development to *Nepenthes*. The lid is interpreted as a ligule and arises from what should develop into the ventral face of a *Nepenthes* pitcher. Troll called the lid a transverse pinna (Querfider) established by an outgrowth (Querwidst) on the upper surface of the presumptive leaf.

The petiole and non-ascidiate leaves all indicate a unifacial development of the pitcher and its peltate homology. The wings observed on the pitcher, unlike those of other families, are not related to the ad–ab suture and presumably originate by different means. That they are present suggests that wings or keels have a particular function and are not simply remnants of the evolutionary processes leading to the epiascidate leaf. It would be interesting to test the mechanical support these structures offer the pitcher considering their homology to 'T' girders in cross-section. Perhaps significantly both *Nepenthes* and *Cephalotus* possess more than one such structure and neither is normally ground-supported. In a similar light both have far more elaborate collar regions, with that of *Nepenthes* being a dual roll-over.

In summary, current views typologically relate the epiascidiate leaf to more conventional leaves via a modification of the peltation theory proposed by Troll. While this view appears at present to be the most satisfactory, it is obvious that more work needs to be done in this area.

Ontogenetic theories on the origin of the epiascidiate leaf

As discussed above, and in Table 4.1 ontogenetic theories on this leaf form abound and there is a wide divergence of opinion on the homologies of the parts. Most of the theories proposed can be examined with reference to *Nepenthes* and their application to the other pitcher-plants then discussed. *Nepenthes* poses particular problems in interpretation owing to the unique spur and the presence of a large photosynthetic blade and petiole intercalated with the pitcher. Table 4.1 lists the various segments imposed on the 'leaf' by each theory.

The Morphology and Zonation of all Epiascidiate 'Pitcher' Forms, Excluding *Utricularia*

The Morphology and Zonation of the *Nepenthes* Pitcher

The wall of the *Nepenthes* pitcher (Fig. 4.9*A*), although thin and frequently translucent, is remarkably strong; in striking comparison to the very fragile *Heliamphora* (see Ch. 6) and the floral pitchers of *Aristolochia* species. This strength is achieved through a thick-walled epidermis on both the outer and inner surfaces and heavily sclerenchymatized veins. In addition to these conventional strengthening features, there are also non-vascular idioblasts – large, spindle-shaped cells with spiral thickening. Similar structures are found in the walls of the *Dischidia* 'pitcher' (see Ch. 18). The selective

Table 4.1 Principal ontogenetic theories of the origin of epiascidiate leaves

Theory and main proponent	Flat photosynthetic portion	Radial tendril	Ascidium	Keel	Lid	Spur
Phyllodial theory (Link, 1824)	Winged petiole	Petiole	Hollowed	—	True lamina	—
Laminar theory (Wunshmann, 1872 see Lloyd, 1942)	Laminar	Midrib	Peltate laminar	—	—	—
Foliar theory (Macfarlane, 1889)	One pair of basal leaflets	Rachis	3–4 pairs lateral	Fused pinnae	—	—
Appendicular theory (Hooker, 1859)	Normal	Midrib	Developed gland	—	Fused pinnae	Apex
Modified peltation theory (Troll, 1932)	Unterblatt	Oberblatt (peltate)		—	Second peltation	Querfider (lateral pinna)
Bifasciate theory (Roth, 1953, 1954, 1957)	Development is sympodial not monopodial so theories based on peltation cannot be correct.					
Foliar runner (Croizat, 1961: see Heads *et al.*, 1984)	'Leaf' of *Nepenthes* neither comparable to normal dorsiventral leaves or stems but is new structure = foliar runner					

value of this strengthening is obviously to hinder the unorthodox escape of powerful-jawed leaf-cutting insects through the sides of the pitcher. Carpenter wasps (*Xylocopa* and *Ceratina* genera) are known occasionally to escape from various pitchers in this way.

Lloyd (1942) claims that the coarse and extensive vascularization of the walls of the pitcher suggest that the structure is produced by the expansion chiefly of the abaxial layer of the leaf. The importance of the vascular system is shown by the fact that young pitchers, when detached, lose their internal fluid, dehydrate and crumple very quickly.

The edge of the mouth of the pitcher is peculiar. In transverse section it looks like a T-shape, but with the horizontal bar of the 'T' generally sloping inwards. The two arms may be of different widths, not only according to the species but also according to the position of the pitcher on the plant. Generally, the outer margin is turned down and may sometimes, as in *N.lowii*, be very much longer than the inner one. On the other hand, in *N. ampullaria*, which develops a rhizome and from which the normal leaves grow up in the air but the pitcher leaves in clusters on the ground, the outer edge is very narrow and strongly reflexed with the inner very broad. *N. ampullaria* pitchers grow as rosettes sunk in the humus on the forest floor (Troll, 1932), set in tight clusters with the peristome edges often fitting closely together (cf. *Cephalotus*). The whole assembly would constitute a group of pitfall traps, each with a broad overhanging edge which would hinder the escape of large walking insects very effectively.

The complete T-section is very robust, strengthened by a thick cuticle and given stiffness by the marked corrugations of the surface (Zone 1 in Fig. 4.9A). On the inner edge of the peristome these corrugations end in minute teeth. Between each pair of teeth there is an opening leading down to a sunken nectary gland. The nectar oozes out in a drop that is held between the teeth, accessible to insects that stand on the upper surface of the peristome and reach down. This ring of nectar globules, combined with the nectar glands underneath the lid, constitute Hooker's 'attractive zone' (1859). The hard, almost polished, outer surface of the peristome is not smooth or unclimbable, at least to a normal insect: ants can often be found freely walking over the surface. Most of the teeth around

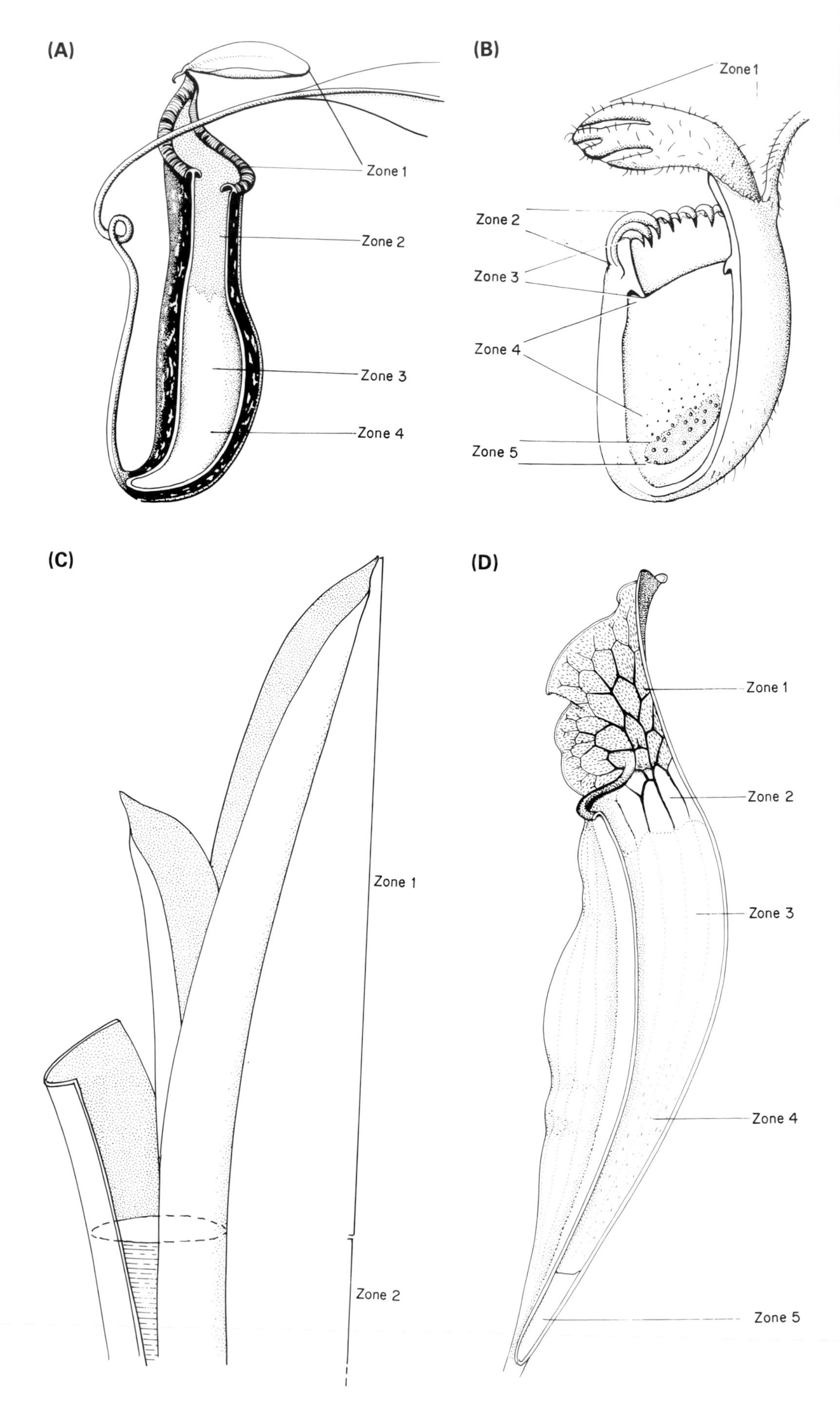

Fig. 4.9 Longitudinal sections through the 'traps' of the various 'pitcher' plants. Specific differences where they exist, e.g. *Nepenthes* and *Sarracenia*, are not recorded here, but are noted in the text.

(*A*) An aerial pitcher of *Nepenthes*: Zone 1, the lid and peristome with its associated nectaries; the 'attractive zone' of Hooker (1859). Nectaries between the teeth (cf. *Cephalotus*) attract insects to the peristome rim where they may slip onto the conductive zone. Zone 2, the wax-coated, slippery, conductive zone. Zone 3, an apparently functionless zone (see Fig. 8.12) and Zone 4, the principle absorptive zone.

(*B*) *Cephalotus follicularis*: Zone 1, the fenestrated lid which has nectaries on its underside; Zone 2, the peristome, with nectaries both between and occasionally on the teeth; Zone 3, the funnel-shaped collar with its imbricate hairs, a conductive zone; Zone 4, the smooth, gland-free region of the pitcher lumen; Zone 5, the gland-patch, with increasing size of gland towards the centre.

(*C*) *Brocchinia reducta*: Zones 1 and 2 are both glanduliferous, and differ only in the presence of detachable wax on the adaxial surface above the water-level and its absence below this point (see Fig. 6.43).

(*D*) *Sarracenia* (non-hooded species): Zone 1, the flap or hood of the pitcher, with numerous nectaries, downward-pointing hairs pigment and UV patterns; Zone 2, more or less colourless region of downward-pointing hairs and nectaries; Zone 3, smooth, gland-free zone, with occasional undeveloped stomata-like pits; Zone 4, the region of long, curved, virtually unridged, downward-pointing hairs, the 'eel-trap' zone of some authors; Zone 5, a further short, sometimes absent, gland-free, hair-free zone.

(*E*) *Darlingtonia californica*: two views to show the foliar twist, Zone 1, the 'fish-tail' or tongue, with nectaries; Zone 2, the peristome rim with further nectar glands; Zone 3, the areolar or 'window', fenestrated zone; Zone 4, the first conductive zone with both downward-pointing hairs and wax patches; Zone 5, the main, downward-pointing hair conductive zone; Zone 6, the smooth, aglandular zone.

(*F*) *Heliamphora*: Zone 1, the 'spoon' with two types of nectar glands on its underside; Zone 2, the first conducting zone of downward-pointing hairs which surrounds the whole of the smooth, attractive area of Zone 3; Zone 4, a smooth aglandular zone; Zone 5, the second conductive zone, with longer, more or less smooth, downward-pointing hairs, steadily increasing in density towards the base.

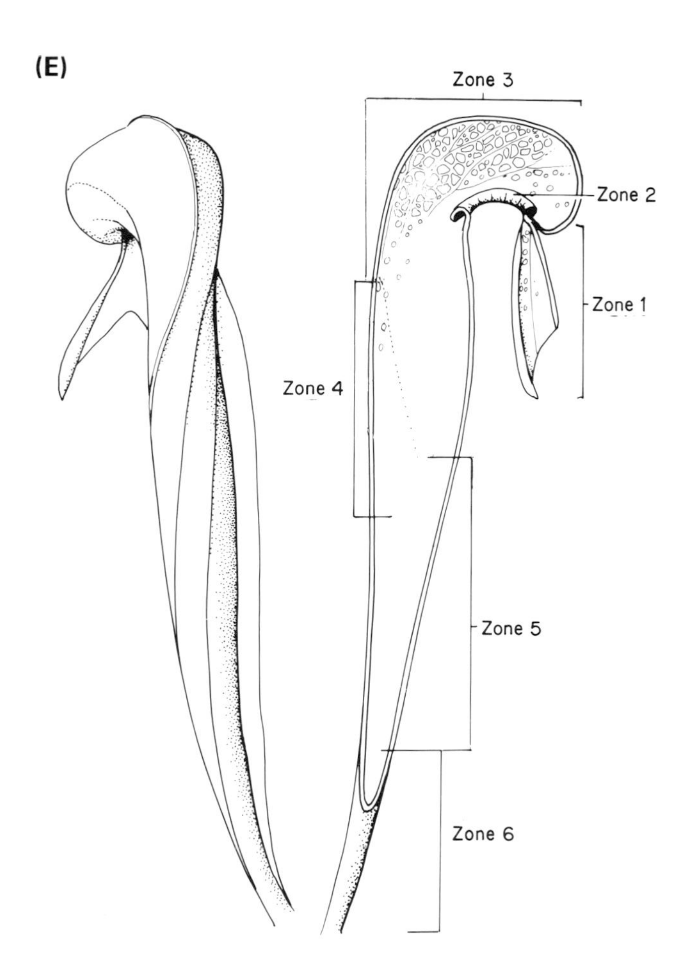

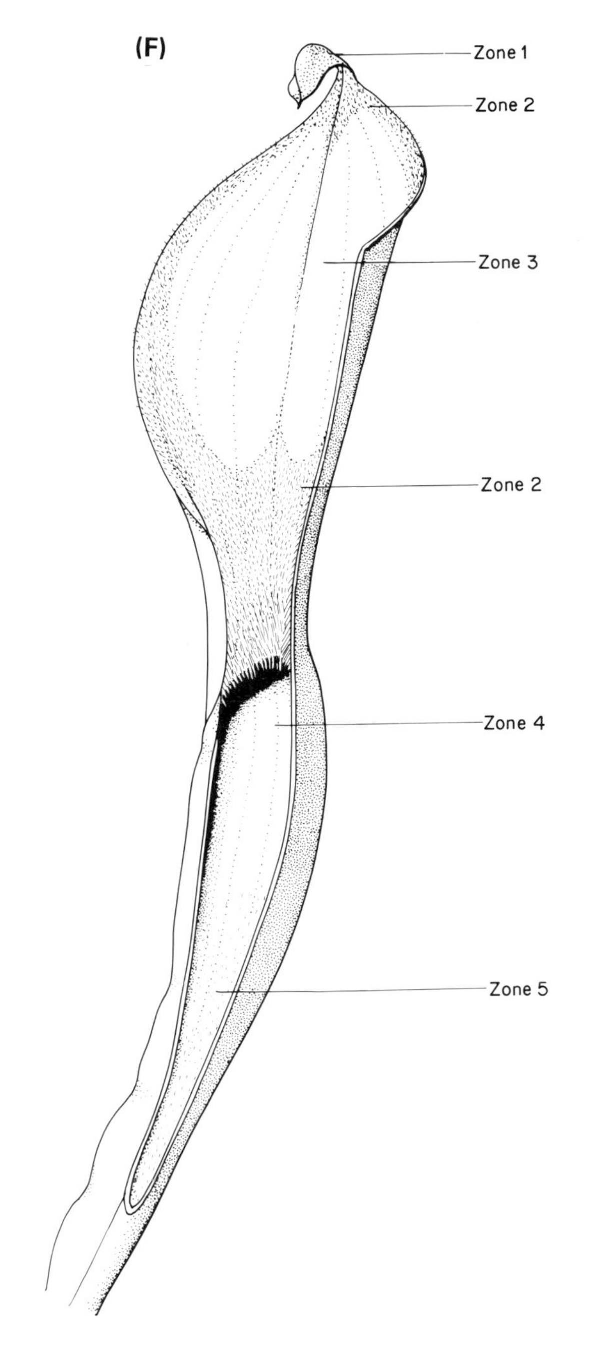

the base of the peristome are small, but in *N. bicalcarata* and *N. intermedia*, two teeth at the base of the lid are strongly developed (Lloyd, 1942, his Plate 7–23). The function of these long, curved, sharp thorns, which resemble the canine teeth of a cat, is obscure (but see page 261). The lid appears to play no part whatsoever in the trap mechanism itself. Its posture at maturity varies from overhanging the aperture to curled right back and is never known to move under stimulus as previously believed. A possible function in some species, but not others, maybe to prevent rain-water from entering the pitcher. One should remember that *Nepenthes* usually grows in regions of heavy rains. The entrance of rain-water would dilute the secreted digestive fluid and might raise the fluid to an ineffective level for insect trapping. But some species have no such protection and yet are effective.

Within the pitcher, and just below the nectar-secreting glands under the peristome lip, is the wax-secreting or conductive zone (Zone 2 in Fig. 4.9A). This broad zone is white, faintly iridescent and just slimy to the touch. The epidermal cells here are simply polygonal with the exception of the large number of slightly projecting lunate ones, which are oriented so that their concave edges are turned downwards (Fig. 4.10). Stomatal lips are not available for a claw to grab, but they cannot by themselves play a significant role in the loss of adhesion. They look like half-stomata but, according to Lloyd (1942), have no pore, and their function is unknown. They are certainly not wax-secreting glands. Over the entire epidermal cell surface is developed a coherent epicuticular wax structure which is described in detail on page 114. The wax chemistry of this layer is given in Table 6.2. Although a very wide range of surfaces from different species of pitchers has not yet been studied, apart from the variation in the slippery zone's band width, no other significant interspecific differences have been seen.

Just below the slippery zone is the digestive zone. This zone is usually glossy, in contrast with the glaucous green of the conductive zone. It may occasionally be red in colour (as e.g. in *N. ventricosa*). Although not an obvious trapping zone, the digestive zone's slimy surface would not be easy to climb, because of proteolytic secretion and the covering of wax scales which most insects' pad or claws would by now have from sliding over the slippery zone. Escape is also hindered by the fact that the liquid in the base of the pitcher is surface active. This soapiness may result in part from proteins discharged into the water, possibly from pitcher-derived polysaccharides also discharged into the water or, in older pitchers, from the various decay products (see also page 242 and 313). The effect, from whatever cause, is to coat, suffocate and drown very quickly even small insects which fall onto the liquid surface.

Like the asymmetric stomata in the slippery conductive zone, the construction of the glands in the digestive zone also hinders escape. The hood covering each gland may prevent insects from using them as footholds and simultaneously protect the secretory cells from damage by an insect's feet (Lloyd, 1942; Adams and Smith, 1977, their Fig. 5).

Fig. 4.10 The inner surface of the pitcher of *Nepenthes* × *benecti* Hort, showing the detachable wax and the asymmetrical half-stomata; (cf. Figs. 6.39–6.42; SEM by G. Wakley, Oxford).

The Pitcher of the Genus *Cephalotus*

The pitchers of *Cephalotus* vary from about 3–5 cm in length and about 2 cm in diameter. The mouth of the pitcher, unlike that of the other pitcher genera, is markedly ovoid and has been described (see Lloyd, 1942) as a slipper with the heel turned over to form a lid. The mouth of the pitcher is surrounded by a corrugated rim (Figs 4.9*B*, 6.35 and 6.36) each corrugation forming a claw-like tooth extending inward and downwards, much like an exaggerated form of the *Nepenthes* rim, except that the teeth are coarser. Lloyd stated that glands were not present, but gland-like structures are present as seen first in the scanning micrographs of Adams and Smith (1977). There are usually about 24 such teeth, not quite symmetrically arranged on either side. Both the surface of the rim and the teeth are covered, around the glands, with sharp-pointed imbricate cells (Fig. 6.35).

Within the pitcher, inside the conductive 'funnel', to use the terminology of Adams and Smith (1977), the surface is covered with very fine, downward-pointing hairs (Fig. 6.36). Below this region is a general glandular region (Zone 4) and the glandular patch (Zone 5 of Fig. 4.9*B* and Fig. 6.37).

The Tank of the Genus *Brocchinia*

The leaf rosette of ***Brocchinia*** forms a tall cylinder (Fig. 2.5). Its tightly overlapping leaf bases can impound water, thus serving as a 'tank'.

The tank of *Brocchinia* can be considered the simplest of the pitcher group. As Figure 4.9C shows, only two clearly demonstrated zones are apparent: the richly waxy and glandular Zone 1 (Fig. 6.43) and the eceriferous Zone 2. There is no evidence of different glandular function in either area and, according to Benzing *et al.* (1985) Benzing

(1986) both surfaces are capable of absorbing nutrients.

The Morphology and Zonation of the Pitchers of the Genus *Sarracenia*

The terminal lobe or flap of the *Sarracenia* pitcher not only looks like a lid but was interpreted as such by Morison (1699), in one of the first descriptions of *Sarracenia*, and by many after him (see page 14). In fact, after the semi-mature pitcher opens, the 'lid' does not move again and apparently has no trap or protective function whatsoever. The whole of the outer surface of the mature pitcher is covered with scattered nectar glands (Fig. 5.1) and rendered rough and hairy with trichomes. It is presumed that the external glands attract potential prey and lead them to the mouth. The internal surface has a distinct zonation, as shown by Hooker (1875) and subsequently used in a slightly modified form by Lloyd (1942), although the proportions, as we said in Chapter 3, vary. We shall use a similar classification, again slightly modified.

The zonation of the *Sarracenia* pitcher

For convenience, the inner surface of a *Sarracenia* pitcher can be divided into five zones (Fig. 4.9*D*). Just below the mouth or peristome of the pitcher (Zone 1) are numerous stomata and a large number of strongly developed, rigid, downward-pointing hairs. The second zone is characterized by the fact that each cell of the surface is prolonged downwards in a short, longitudinally striated process. Zone 3 below is devoid of hairs and processes and is almost entirely smooth. Zone 4, by far the longest, is crowded with long sinuous hairs, not as thick or robust as those found near the entrance. This is the so-called 'eel-trap' zone of some authors (see Fig. 26 of Heide-Jørgensen, 1986 and Fig. 4.11).

Zone 1 includes the cordate emarginate flap (Fig. 4.9*D*) and is called by Lloyd (1942) the 'attractive' zone. The margin of the flap is ciliated with more or less blunt hairs. The hairs are long, slightly curved, with the curve usually bringing the tip in towards the cuticle, and taper to a point. They are several millimetres long, but their most striking feature is the fine ribs which run their length from base to tip (see Figs 6.28, 6.29, 6.30 and pages 111–112).

Zone 2 is Lloyd's 'conducting' zone, but it is likewise attracting because of the quantity of nectar secreted. The attracting role of this zone has been demonstrated by the sequences taken by Oxford

Fig. 4.11 The 'eel-trap' or 'Reuse' (Ger.) zone of *Sarracenia*; Zone 4 of Fig. 4.9*D*.

Scientific Films, which show insects avidly feeding while demonstrating that their foothold is extremely precarious. In Zone 2 all the epidermal cells become trichomatous, and the abrupt transition is shown in Figure 4.9*D*. Instead of being long and finely ribbed, the trichomes are now short, still finely ribbed but so close together as to create the spurious impression that they overlap, the ribs on each trichome narrowing down to a point (see Fig. 6.30). There are apparently no stomata in this zone and glands are numerous, probably more numerous than in Zone 1. At normal temperatures and with an adequate water supply, the whole surface of this zone will be bathed in mucilage and the problems of adhesion for any type of insect can be appreciated (see Ch. 6).

Zone 3 is the glandular zone. The transition from Zone 2 is abrupt (Fig. 6.32) and the whole of Zone 3 is astomatous and glassy smooth; the epidermal cells are wavy-margined and thick-walled and there are numerous glands (Fig. 6.30), up to 15 or 20 per square millimetre. This is the secretory zone.

Zone 4 is an abrupt change from Zone 3. The epidermal walls change from being sinuous to plain and the cells are strictly isodiametric in dimensions. This is Lloyd's (1942) 'retentive and absorptive' zone (see Ch. 10). The hairs which are typical to this Zone are long, slightly curved and taper to a sharp point (Fig. 4.11) but are completely smooth and bear no sign of the ribs which are a striking feature of the hairs in Zones 1 and 2. The water level in *Sarracenia purpurea* varies a great deal depending on how much rain may have penetrated (Kingsolver, 1981; cf. *Darlingtonia* below). The hairs in Zone 4 will lie at, just above or sometimes reaching down just below, the water level. They will therefore provide an obstacle to insects attempting to climb up the epidermal walls, or to those in the water which attempt to effect a foothold on the pitcher sides.

Zone 5 is a relatively narrow band below Zone 4 in which the surface is smooth and glossy. There are no glands on this surface, nor are there any hairs. (See Ch. 8 for further discussion of the Zones.)

The Morphology of the *Darlingtonia* Pitcher

Unlike the pitchers of *Sarracenia*, the pitchers of *Darlingtonia* are typically twisted so that each dome is turned about 180° from the main axis. By this orientation all the pitchers are then turned to face outwards. *Darlingtonia californica* would thus seem to maximize its insect-catching capacity by presenting each new trap aperture in a different orientation (Fig. 4.9*E*). Field observations suggest that this phenomenon is universal for older pitchers in the wild (Austin, 1875–77; see Appendix 1). This phyllotaxy seems first to have been observed in *Darlingtonia*. Apart from possibly increasing the chance of prey attraction, this reorientation may also serve to reduce the chance of an elongating, but as yet unopened, pitcher growing into the aperture of a maturer neighbouring leaf. This phenomenon is not unknown in the wild (Naeem,S., pers. comm., 1985; and see page 84)

The reorientation seems to be true in the large pitchers reaching the height of 50–100 cm and growing thickly in large clumps. In the smaller leaves, which lie more or less prostrate, the typical twist allows the 'fish-tail' appendages (see below and Frontispiece) to lie on the surface of the ground, thus forming a ramp which may lead small creeping arthropods directly to the pitcher mouth (Lloyd, 1942). At its top, the tube of the pitcher spreads abruptly and bends sharply forward to form a dome, bringing the 'mouth' into a horizontal position underneath.

Morphology and Zonation of the *Heliamphora* Pitcher

The leaves of the six species of *Heliamphora* range in size from a few centimetres to over one metre. Although as suggested on pages 16 and 41 this genus may not be a very effective carnivorous plant, it does, nevertheless, have a very complex trap mechanism. It is more complex than suggested by Lloyd (1942) who probably did not have the advantage of abundant live material. We have divided the trap into five zones (Fig. 4.9*F*). Zone 1 is the 'spoon', possibly with an attracting function; Zone 3 is attractive and the nectar glands extend down into and overlap with the long, flexible, downward-pointing hairs of the conducting Zone 2 (Fig. 6.32) which completely surround Zone 3. Like *Sarracenia* (Figs 6.28–30) these hairs are hollow, but are more crowded. The ridges of the conducting surface are most pronounced (Fig. 6.32) and these hairs are relatively longer, more slender and distinctly flexible, the more so in the lower part of the conductive surface. The transition zone is shown in Figure 6.33. The hairs of Lloyd's detentive surface (Zone 4, Fig. 6.34) are short, very thick-walled and rigid, and are obviously designed to make escape difficult.

However, whereas the basal parts of other pitchers, particularly those of *Sarracenia* and *Darlingtonia*, are noticeably robust and obviously designed to prevent the escape of muscular-jawed insects through the walls, the walls of a *Heliamphora* pitcher are not so reinforced and collapse and crumple noticeably when flaccid.

The Trap of *Utricularia*

A General Description of the *Utricularia* Trap

A lucid and comprehensive description of the morphology and anatomy of the genus *Utricularia*, including the genera *Biovularia* and *Polypompholyx*, is given in Lloyd's review of 1935. What is more,

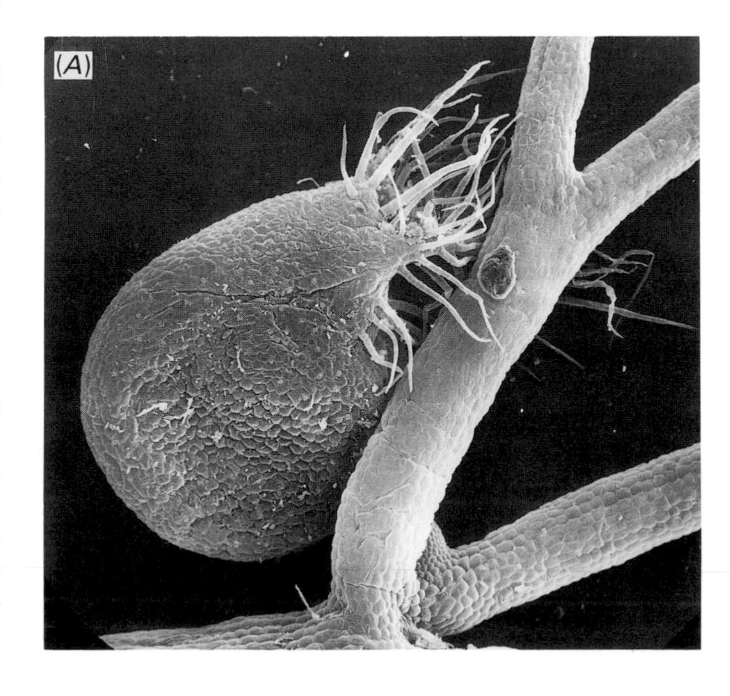

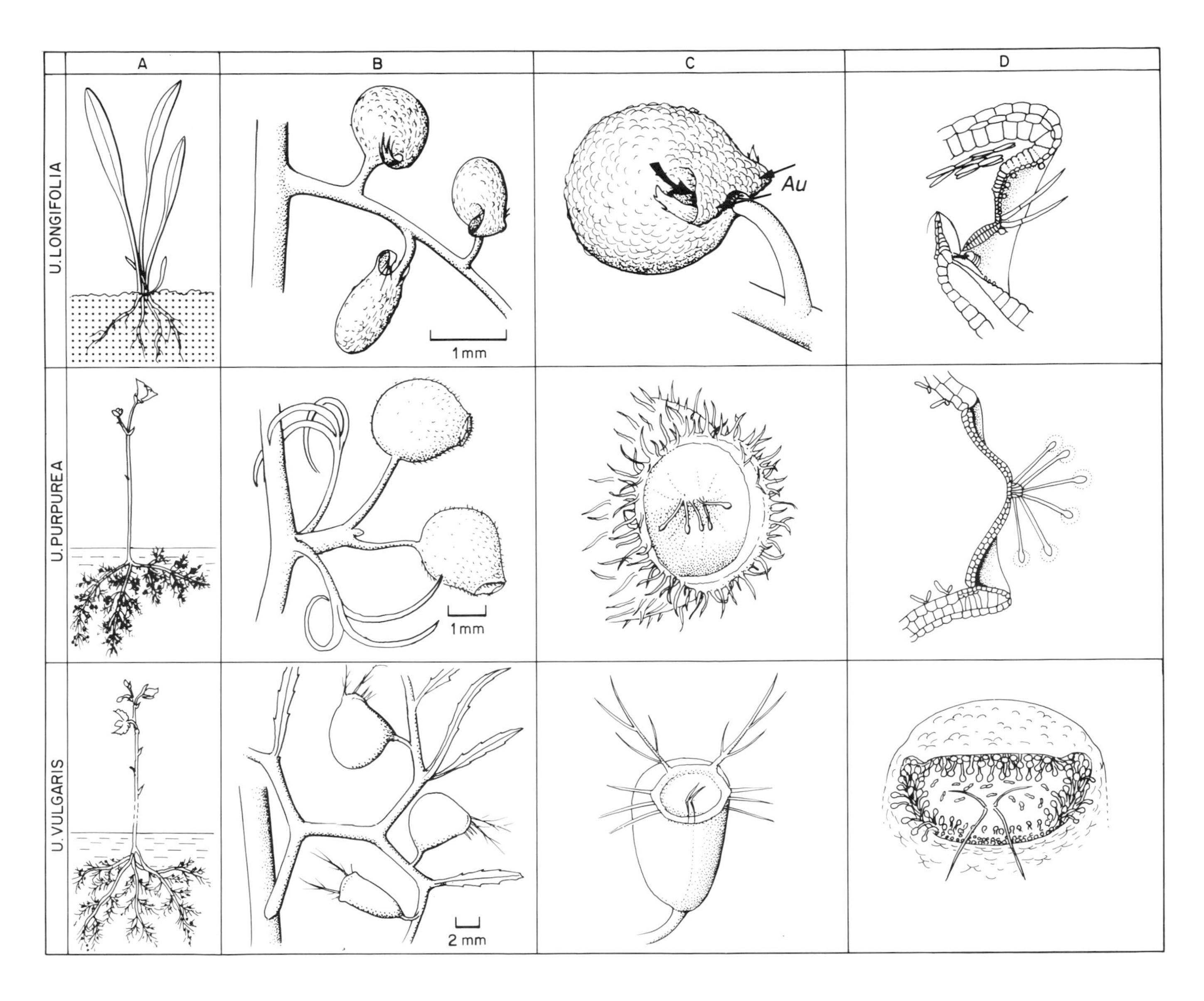

Fig. 4.12(*A*) Left. A trap of *Utricularia neglecta*, which is taxonomically close to *U. vulgaris*, showing the general form of the trap and its tentacles, cf. Figs. 4.12*B*–4.15 (SEM from Joel, 1982).

Fig. 4.12(*B*) Above. Showing something of the range of general habits, trap sizes and details of three different species of *Utricularia*.

U. longifolia. A mud or 'muck' living, sometimes semi-epiphytic species from Brazil. The striking blue/purple flowers on very tall scapes often, as Lloyd points out, rival the orchids in beauty. The traps, however, are minute and the submerged leaves, unlike *U. purpurea* and *U. vulgaris* are non-existent, whereas the aerial leaves are well developed. The globose traps are markedly tubercular, and the 'antennae' short and blunt-tipped. Two well-developed 'auricles' (Au) almost occlude the trap entrance and only very small, free-swimming or crawling prey could enter such a trap, perhaps under the auricles (curved arrow). The door is conventional except for having only two trigger hairs and a steeply-sloping vestibule.

U. purpurea ranges from North to South America including the larger Caribbean islands. The flowers are similar to those of *U. longifolia*, unlike *U. vulgaris* which are yellow, but there are no aerial leaves. The submerged leaves of this New World group are markedly curved (B) and grouped in whorls. The traps are larger than *U. longifolia* with short antennae, particularly round the trap (C). The trigger hairs are very striking, with globose, mucilaginous tips.

U. vulgaris is broadly distributed across the northern hemisphere, reaching northern Africa and temperate Asia. However, closely related species, e.g. *U. australis*, range from Japan to Australasia. The traps are large, up to 4 mm across, with very long antennae (B–C) around the trap. The submerged leaves (A–B) are well-developed and often almost obscure the traps. Air shoots with 'mussel-shaped leaves with stomata' (Lloyd, 1942) may arise in some forms of this large group. The trap mouth is large (C–D) with well-developed 'antennae' 'guiding' the prey to a mouth which is surrounded by mucilaginous stalked glands. (Fig. 4.15*C*). There are four trigger hairs tapering to a point.

Lloyd provides a general survey of the variation within the genus, the embryology, the anomalous behaviour at germination and the formation of the rootless seedling and the formation of turions, the resting or winter buds (page 43). No purpose would be served by repeating all his observations or his conclusions, few if any of which need qualification even after the passage of over 50 years. The following account briefly describes the trap and is mainly concerned with such new information as the latest techniques and more recent research have revealed.

The *Utricularia* trap: size and form

Similar to the other vegetative parts of this plant, the disposition of the traps around the plant shows a lack of organization (Fig. 4.12B). In almost all species, the traps arise from both the laminae and petioles of the leaves, as well as from the stolons and their branches (Fig. 4.12). Exceptions are *U. manni* and *U. striatula*, both epiphytic species in which traps apparently never occur on the leaves (Taylor, 1964).

The shape and size of the traps, as well as the position of the mouth, appendages and door, differ considerably between species and are often used as an aid for species identification. In many species the traps differ considerably in size (Fig. 4.12*B*). In others the traps are di- or trimorphic, with relatively large differences in size coupled with a disproportionate variation in the appendages. Some species also show variation within populations. Similarly, the size and number of trap glands exhibit great variations (Taylor, 1964). The traps range in size from 0.2 to 6 mm (Taylor, 1976). A new undescribed species from Australia develops even larger traps. In addition to tiny traps and ordinary traps, this new species develops giant traps as large as 10 mm (Taylor, P., pers. comm., 1986). These traps are effective in capturing prey and therefore seem most suitable for experimental work concerning trapping and digesting mechanisms.

General structure of the trap

The trap of *Utricularia*, a water-filled bladder also known as the utricle, is among the more sophisticated structures in the plant kingdom (Fig. 4.12*B*).

The trap comprises a thin-walled sac (Fig. 4.13*A*,*B*), mostly two cells thick, with a doorway of intricate structure obstructed by a door which, in due course, opens inwardly. The trap is supported by a footstalk joining its ventral side to the shoot or leaf (Figs. 4.12*B* and 4.13*A*).

Fig. 4.13(*A*) Schematic transverse section through a trap of *Utricularia*. (1) door; (2) pavement epithelium; (3) velum; (4) trigger hairs; (5) stalked mucilage glands; (6) antennae; (7) inner chamber with quadrifid glands; (8) spherical sessile glands; (9) bifid glands; (10) rostrum. For details of the trap entrance see Fig. 4.14. (Redrawn after Joel, 1982).

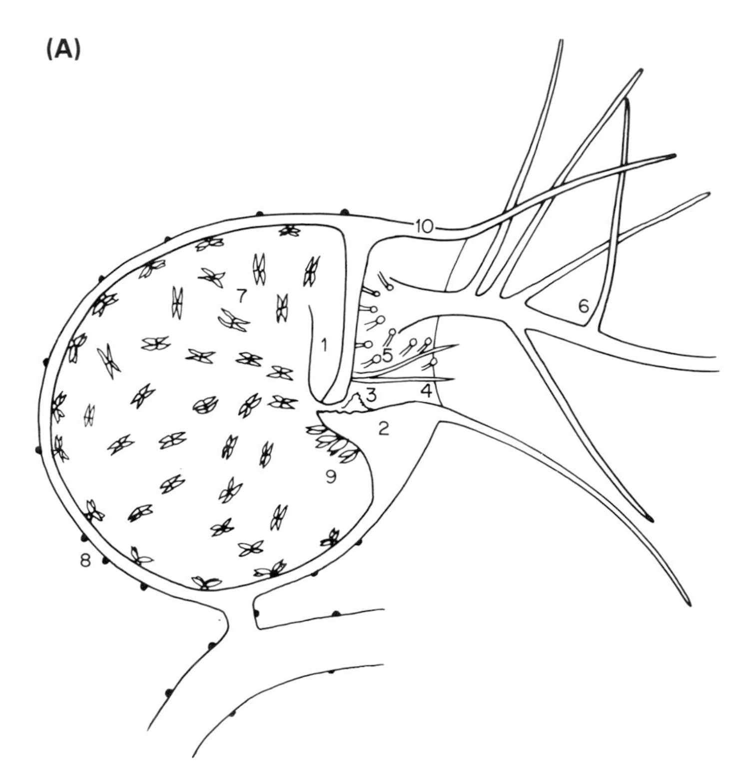

In most species, the edge of the doorway carries a pair of branched hairs; in addition, some elongated bristles may also be present around the doorway. With its bristles and hairs the trap resembles a crustacean animal in shape, hence the long branched hairs are frequently termed 'antennae', though they do not act as sensory organs (Fig. 4.12*B*).

The trap of *Utricularia* captures small aquatic animals. Darwin (1875) conjectured that the appendages around the doorway serve to direct potential prey towards the trapping mechanism by forming a sort of funnel leading to the door. This idea was clearly demonstrated to be correct in an experimental study by Meyers and Strickler (1979), who showed that the removal of antennae and bristles around the trap door decreases the rate of capture.

The ontogeny and structure of the *Utricularia* trap

The bladderwort trap is, like all other carnivorous pitchers, an epiascidiate leaf (see page 56). The opening of the trap arises in the very young plants as a slit caused by the invagination of the rounded

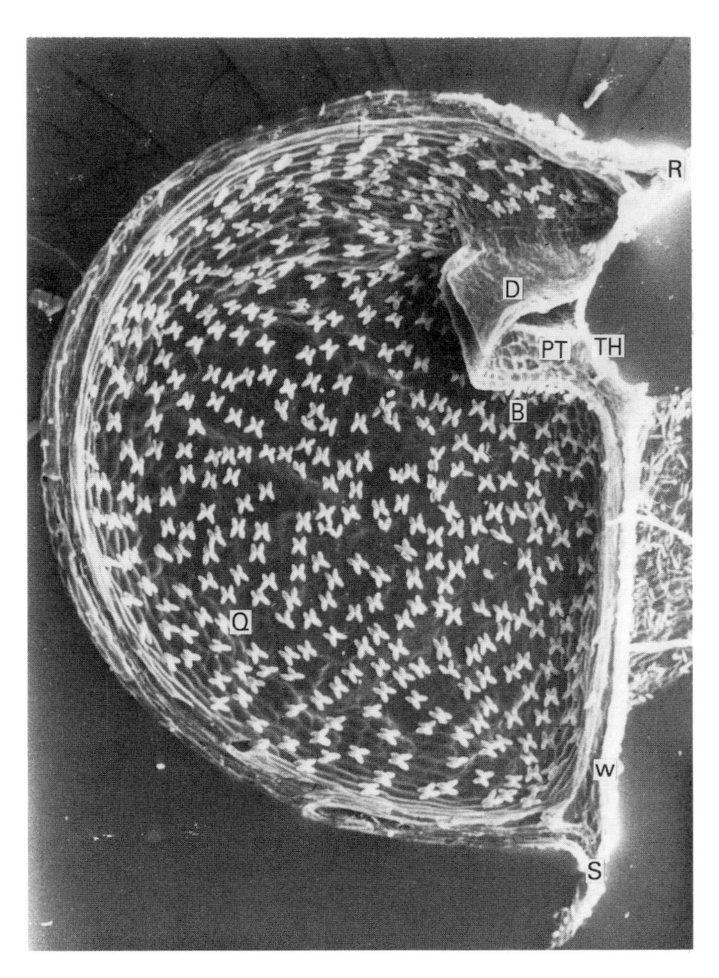

Fig. 4.13(*B*) A median section through the trap of *Utricularia monanthos*, cf. Fig. 4.13*A*. The quadrifid hairs (Q) are visible in face view and in profile (cf. Figs. 4.16*A*,*B*) as are the bifid hairs (B) under the threshold. The two-celled thick wall of the trap (W) is continuous with the rostrum (R) and the stalk (S). The door (D) partly obscures the threshold (TH) and in front of this can be seen the pavement epithelium (PE). (Reproduced by permission of Dr. B.A. Fineran, New Zealand, unpublished micrograph.)

Fig. 4.14 Detail of the trap region of a bladder of *Utricularia*. Note the mixture of long and short-stalked trichomes on the outer face of the door. Only two triggers of bristles (T) are shown in this section. The pavement epithelium (PE) secretes mucilage to achieve, along with the velum (V) a sealing of the lower edge of the trapdoor. Note that the outer layer cells of the two-celled walls of the bladder are smaller than those of the inner layer. Both exposed surfaces are cutinized. Both contain chloroplasts. Bifid hairs occur only on the inner face of the threshold. Lloyd suggests these may serve to discourage entrapped and creeping prey reaching the threshold. The two distinct layers of the door are shown, the inner capable of rapid expansion and contraction, the outer more or less rigid. The precise detail of this complex construction is shown in Lloyd, 1942, Plate 29, 1–3.

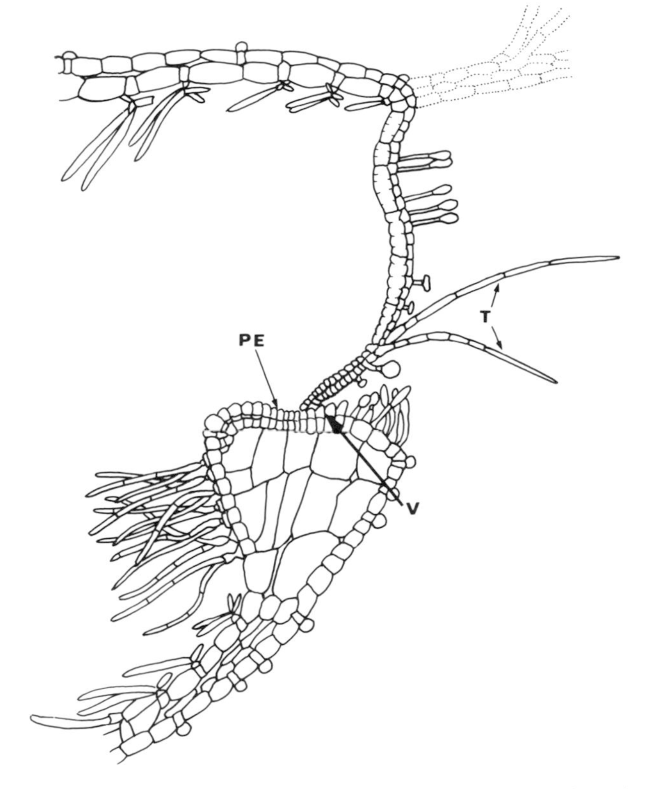

primordium. The lips of the slit turn inwards, the upper becoming the trap *door* while the lower lip becomes the *threshold* (Meierhofer, 1902). The sides of the entrance then extend, moving the lips apart to produce a funnel-shaped vestibule around the opening.

Although contiguous at their lateral extremities, the door and the threshold differ in their anatomical characters (Figs. 4.12*B* and 4.14). The threshold becomes semi-circular and forms a massive thickening, which preserves the shape of the opening and resists any crumpling when the trap is set. The side walls of the trap are thinner where they articulate with the threshold so that no distortion is exerted on the surrounding tissue when the side walls bend as the trap sets (Fig. 4.10*B*). In transverse section, the threshold is roughly triangular and continuous with the lower trap wall. The upper, free surface of the threshold shows at least three distinct regions: the inner zone, made of epidermal cells, forming a shelf projecting into the interior of the trap; the outer zone, doorstep, continuous with the lateral walls of the entrance; and between them is the middle zone, which is the '*pavement epithelium*' on which the door lies when the trap is closed. The pavement epithelium fits precisely the shape of the lower door edge and is provided with special devices which prevent leakage of water into the trap when the trap is set.

The three-dimensional form of the threshold varies in different species and is related to the characteristics of the door (Lloyd, 1942), although all show essentially the same kind of surface characters. This surface does not consist of a proper epidermis, but of a closely packed array of short glandular trichomes with their terminal cells touching each other. They superficially resemble cobblestones and are thus named '*pavement epithelium*' (Fineran, 1985).

The velum

The cuticle of the terminal cells of the glandular trichomes forming the threshold detaches as one piece when the trap matures, forming the *velum* – a cuticular structure preventing water from leaking into the trap when it is set (Lloyd, 1942; Broussaud and Vintéjoux, 1982; Fineran, 1985; Heide-Jørgensen, 1981; Figs 4.13A and 4.15B). The mechanism by which the cuticle separates from the surface of the head cells is not clear, but it is assumed that some secretion pushes it away from the cell walls. It has further been suggested that the labyrinthine wall typical of the head cells, on the side facing the door, helps to pump ions into the space formed beneath the velum and thus creates a higher hydrostatic pressure underneath it (Heide-Jørgensen, 1981). This pressure might keep the velum in a proper position, build an intermediary compartment between the low inner pressure and the surrounding medium, and thus prevent the velum from being pressed flat when water is sucked in as the trap is activated (see page 124).

The structure and properties of the door

The door which blocks the entry to the trap is flat, two cells thick, and fixed to the upper lip of the trap opening along a U-shaped line almost at right angles to the threshold plane (Figs 4.14 and 4.15). The door is roughly semi-circular, the shorter side

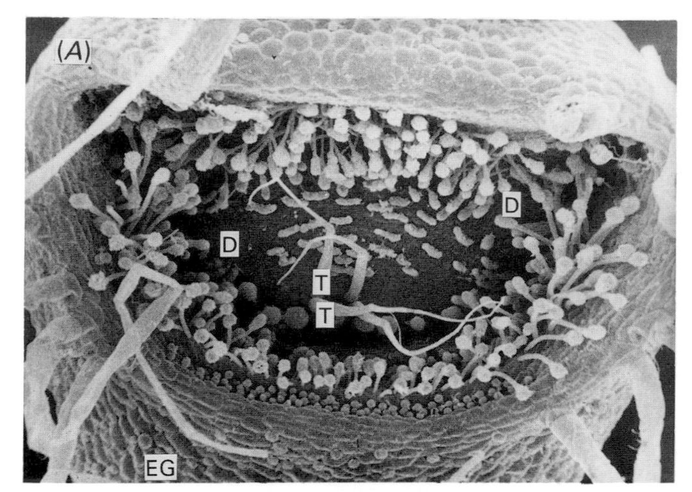

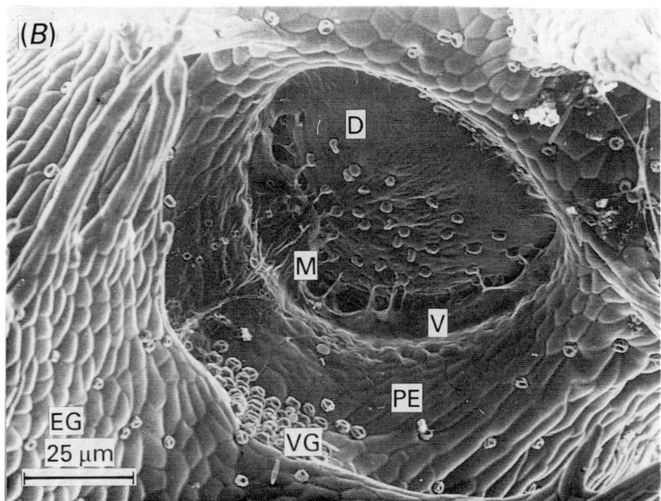

Fig. 4.15 The door and doorway of *Utricularia* traps; *A*,*B*-exterior views. (*A*) *U. australis* (taxonomically close to *U. vulgaris*); note the four trigger hairs (T) in this species and the door (D) which is surrounded by vestibule glands with stalks of varying length. (*B*) *U. monanthos*, a New Zealand species, has no trigger hairs on the door (D) and only a few sessile glands. The edge of the door is covered by the velum (V) and mucilage (M) coats both surfaces. The pavement epithelium (PE) carries vestibule glands which are all sessile in this species, as are the external glands (EG) on both of these species. (C) The stalked, mucilage-coated glands around the entrance of the trap of *Utricularia vulgaris* and a few other species. Their true function is unknown, but they may have a role in attracting certain types of prey. (SEMs *A*, Unpublished micrographs by kind permission of Dr H.S. Heide-Jørgensen, Copenhagen, Denmark; *B*, Dr B.A. Fineran, Christchurch, New Zealand; *C*, Liquid-nitrogen frozen by S. Clarke, Oxford.)

being its free edge facing the threshold. The shape of its free edge is maintained by turgor (Lloyd, 1942).

According to Lloyd (1942), there are four main regions in the door, forming a continuous structure. When the door opens, a strong reverse flexure occurs in a wide zone around the edge of attachment to the trap wall. This is the *hinge region* surrounding the lenticular *middle area* (Figs. 4.14, 4.15 and 6.45). While the above mentioned hinge region serves the movement of the door as a whole, the movement of the door edge is facilitated by another hinge, the *central flexible region*, which is a small, circular, thin patch of the door. The region lying below the central hinge (Figs 4.14 and 4.15) is thick and strong and often bears several trigger hairs which are believed to be responsible for activating its movement. This area is convex, in contrast to the upper hinge which is concave.

The central hinge, being very thin, allows the thickened edge of the door to bend up and inwardly. The very rapid movement of the door is based on unique wall structures which render the door highly flexible. In general, the outer wall layer of the door is thin and the inner is thick. In the region of the upper hinge and in the middle area, the walls of the outer cell layer lie athwart those of the inner cell layer. The outer, smaller cells are flat and approximately one-third the size of the cells of the inner layer. This ratio between the cells of the two layers changes along the door. In the lower side of the door, the inner cells are equal in size with the outer cells, and both have thick walls. While the cells of the outer layer are arranged in parallel rows, the elongated cells of the inner layer radiate from the centre of the door to both the top and the sides of the door. The closer to the centre the shorter these cells are, so that in the centre itself they are isodiametric (Fig. 4.16*A*).

The anticlinal walls of the outer cells possess buttresses; similar buttresses can also be seen in the inner cells, their maximal size occurring in the stiffest regions, mainly in the lower door edge. On the other hand, numerous smaller buttresses are typical of the region of maximum bending. In these regions, the hinges and neighbouring areas, the walls show a zigzag profile, presumably allowing the bending movement. At the same time, the inner wall layer is composed of highly flexible cells, each being constricted at regular intervals. For this reason each cell looks like a *group* of cells. The constrictions render the inner cells compressible, like bellows, and the two cell layers together form a structure of

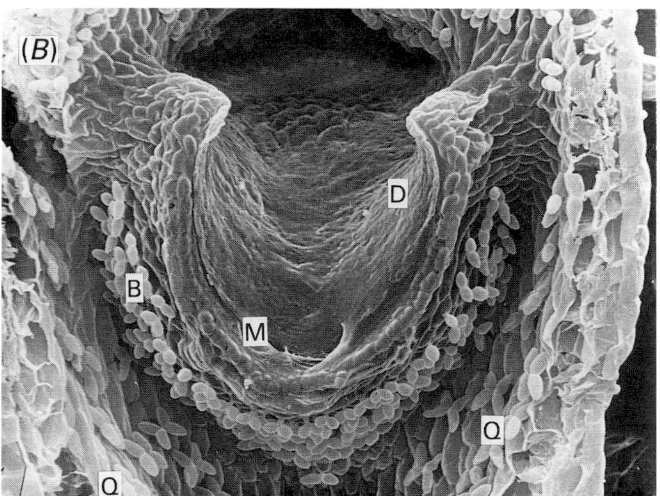

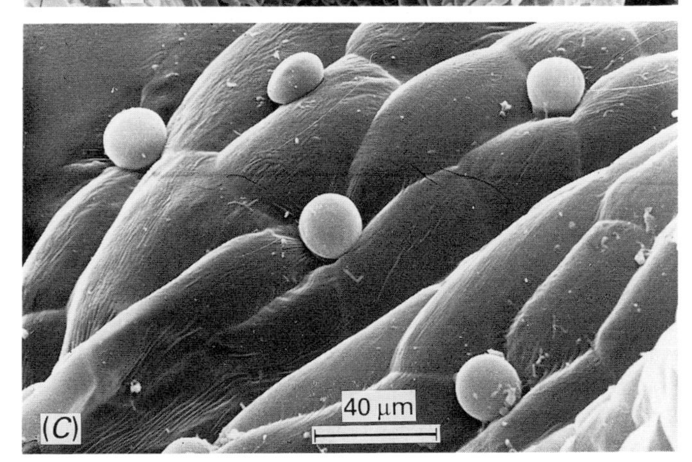

Fig. 4.16 Interior views of the trap: (*A*) *U. vulgaris* showing the concentric, articulated nature of the door and the long, thin bifids (B) which lie just inside the threshold. (*B*) *U. monanthos* showing how the 'U'-shaped threshold into which the door (*D*) fits and is sealed by mucilage (M). Bifid glands (B) cover the underside of the threshold and quadrifids (Q) the rest of the inside of the bladder.

(*C*) Outer surface of a trap of *U. monanthos* showing the external glands, cf. Fig. 6.49. (SEMs *A*. Liquid-nitrogen-frozen by S. Clarke; *B*. Unpublished micrograph by Dr. B.A. Fineran, Christchurch, New Zealand; *C*. Unpublished micrograph by Dr. B.A. Fineran, Christchurch, New Zealand).

great flexibility combined with robustness and rapid response (Lloyd, 1942; Heide-Jørgensen, 1981).

The tension of the door

When the door is freed from the trap surround, it springs outwards, and can be brought back only by plasmolysis. This induced loss of water indicates that it is held under pressure because of the turgor in its cells. Whilst the cells of both layers are equally turgid, only the inner cells are capable of appreciable expansion due to the above mentioned constrictions. This distinction causes a different index of expansion between the two layers and eventually leads to the internal pressure holding the door firmly in position on the threshold.

The trigger hairs

The trigger hairs, where they are present (but see Fig. 4.15*B*) are placed immediately below the central hinge. They are stiff, tapering, sharply pointed bristles, obliquely oriented and extending upwards (Figs. 4.12*B*, 4.13*A*, 4.14 and 4.15*A*) (Lloyd, 1942). Normally the *Utricularia* traps possess four hairs, but in some species, e.g. *U. flexuosa* there are (cf. *Dionaea*, page 52) six hairs and in others there are no trigger hairs at all. In these latter species, e.g. *U. cornuta*, *U. monanthos* (Fig. 4.16*B*) and *U. bifida*, the tripping apparatus consists of a group of glandular hairs, either stalked or sessile (Lloyd, 1942). In contrast to the trigger hairs of the Droseraceae (see page 304 and Fig. 19.1), in *Utricularia* there seem to be no endodermoid components in any part of the hair, and there is also no bending site along the hair.

It is possible that the trigger hairs have evolved from glandular structures situated on the door (e.g. in *U. cornuta*) (cf. Fig. 4.16*C*), in a manner similar to that believed to have happened in the Droseraceae (see Ch. 19).

The inner side of the trap

The inner surfaces of the traps of all *Utricularia* species (Figs 4.13*B* and 4.16*B*) are covered with typical divided glands, bearing two or four arms – the bifids (cf. Fig. 4.7*B*) and the quadrifids respectively – and in some rare cases having only a single arm (Taylor, P., pers. comm., 1986). In traps of the *U. vulgaris* type, and in most other species, the quadrifids are scattered all over the lateral walls, whereas the bifids are restricted to the underside of the threshold (Fig. 4.16*A*). While each quadrifid is surrounded by ordinary epidermal cells and is therefore separated from its neighbouring quadrifid, the bifids are compact because each epidermal cell supports a gland (Fineran, 1985).

There are some morphological differences between the bifids and the quadrifid, mainly in the orientation of the arms (Fig. 4.15) but no ultrastructural differences have been found (Fineran and Lee, 1975). For further consideration of their structure see Chapter 8.

The glands of the internal surface of *Utricularia* traps seem to have three main functions:

(i) the removal of water during the resetting phase (see page 123);
(ii) the secretion of digestive enzymes (see Ch. 9);
(iii) the absorption of digested products (see Ch. 10).

Some authors have assumed that water is absorbed mainly by the bifids (Sasago and Sibaoka, 1985a; Sydenham and Findlay, 1975), indicating that glands of different morphology may perform different functions *viz* the bifids are transferring water and the quadrifids are purely digestive. This distinction, however, has still to be confirmed.

The outer side of the trap

As in other digestive organs in carnivorous plants (Joel, 1985), the *Utricularia* trap bears typical external glands (Fig. 4.15*A*,*B*). First Kruck (1931) and then Nold (1934) suggested that these external glands are the sites of water extrusion from the trap. This is the opinion also held by some present-day authors and is supported by theoretical models suggested by Sydenham and Findlay (1975). Developmental and ultrastructural studies of these glands (Fineran and Lee, 1980; Fineran, 1980; Fineran, 1985) have revealed that the external glands might carry out successive functions during their development. According to these studies, the external glands of the young, developing traps take the place of roots as sites for solute absorption from the external medium, together with the similar external glands covering the other shoot organs (see Fig. 6.49 and Ch. 7).

At an early stage of development, the terminal cell differentiates as a transfer cell (Fineran, 1980). Tracer experiments have shown that the cuticle covering the terminal cells is permeable, and that the terminal cells of the external glands are the only permeable sites on the trap surface (Fineran and Gilbertson, 1980; Fineran, 1985; cf. *Drosophyllum*, pages 49–50).

At about the time when the traps begin to capture

prey, the wall ingrowths of the terminal cell are progressively buried under layers of secondary wall material. At the same time, the protoplast of the terminal cell becomes reduced and often senescent (Fineran, 1980, 1985), in a manner resembling that of the terminal cells of the hydropoten in other water plants (Lüttge and Krapf, 1969). Similar to the structural framework of hydropoten, the sub-terminal cells, frequently termed 'pedestal cells' look active, at least if judged by the structure of their cytoplasm. These cells show large labyrinthine wall ingrowths on the wall facing the terminal cell (Lüttge and Krapf, 1969; Fineran, 1980, 1985) and their lateral walls are completely impregnated, thus being endodermoid (see Ch. 8). They are probably involved in some glandular activity that needs to be further investigated.

Genlisea

All the species of *Genlisea* comprise small rosette plants with two kinds of leaves, both of which arise, in two dense clusters, from a slender rhizome. The epigeic green foliage leaves are linear or spatulate, while the thread-like submerged leaves, the traps, are forked and apparently have no chlorophyll (Fig. 4.17*A*). Lloyd (1942) believed that only the inflorescence appeared above water, since on herbarium material he found algae on the surfaces of the true leaves and noted that the leaves were astomatous. But this does not appear to be so from plants grown under cultivation (Slack, 1986) and the presence of algae may simply be a feature of a plant normally growing under very warm and wet conditions.

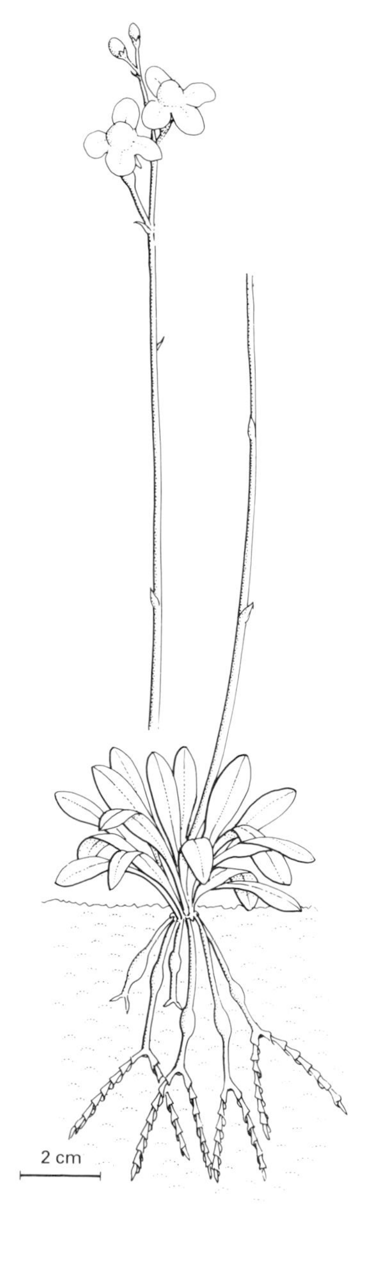

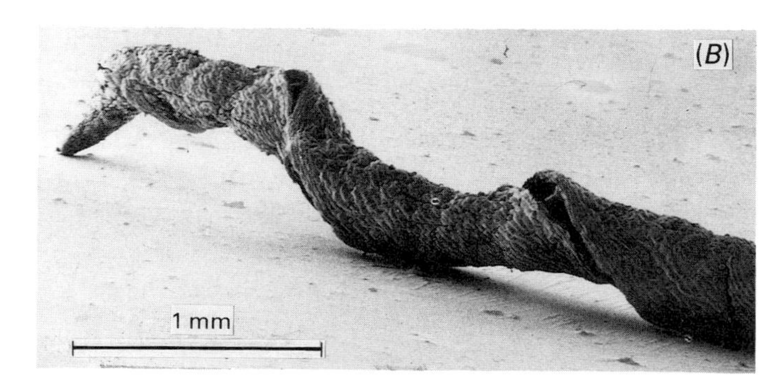

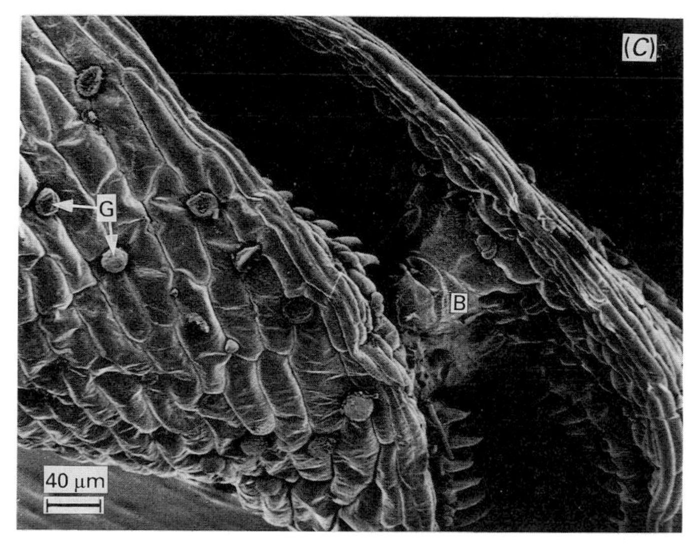

Fig. 4.17 The *Genlisea* trap. (*A*) The general habit of a *Genlisea* plant. The specimen from which this drawing was made had a scape 27 cm long (see p. 273).
(*B*) One 'arm' of the *Genlisea* trap (SEM). (For a general habit drawing see Lloyd, 1942, his plate 11/5 and Figs. 1.3 and 4.17A). (SEM by T. Scaysbrook, Oxford.)
(*C*) Exterior of the arm of a *Genlisea* trap, showing globular external glands (G), and the bridge (B) holding apart the two lips of the arm. Several rows of downward-pointing hairs are seen inside the trap. SEM by DMJ.

Genlisea does not develop any roots. However, when first observed, the whitish thin traps that occupy the underside of the foliage crown can easily be mistaken for roots. These structures were, therefore, often regarded as *Wurzelblätter* (Schmucker and Linnemann, 1959).

A General Description of the Trap

The trap of *Genlisea*, which is positively geotropic, is carried on a long stalk (Fig. 4.17*A*). The basal part of the trap, on top of the downwardly directed stalk, is a swollen utricle (bag called the 'bulb' by Slack, 1986) 1–4 mm long. This utricle is thought to serve as a digestive cavity.

A tubular channel, approximately 5–20 mm long, leads to this utricle. The free end of the tube terminates in a 'mouth' with two helical arms, each 10–50 mm long which serves as the main trapping device (Figs 4.17*B* and 4.18). The helical arms, the tubular channel and the utricle together form a continuous 'alimentary canal'.

The development of the trap

Despite its unique configuration, the development of the *Genlisea* trap closely resembles that of *Utricularia*: both are epiascidiate leaves, which have already been described on page 56.

Marginal growth leads to the formation of an invagination in the tip of the cylindrical primordium of the trap. In contrast to the spherical invagination in the primordial trap of *Utricularia* (see page 66), the invagination in *Genlisea* is tubular, arising from an extended marginal growth.

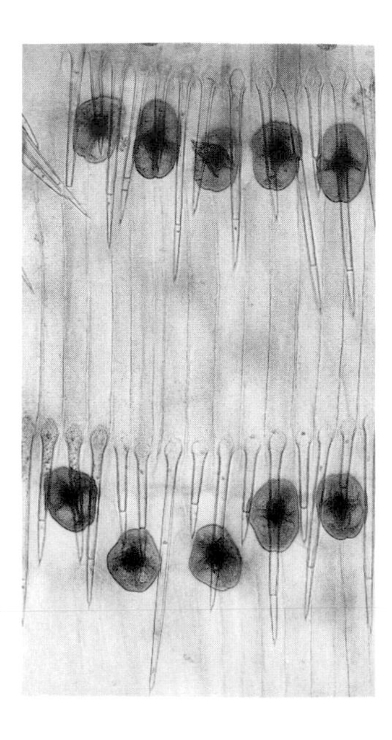

Fig. 4.18 Two transverse rows of glands situated immediately below dentative hairs on the inside of the tube of a trap of *Genlisea*. (Light micrograph made by D.M.J. from preserved material kindly supplied by P. Taylor, Kew.)

While the basal portion of the invagination develops into the sub-spherical hollow utricle, its neck forms a long tube which gradually widens towards the mouth where it forms a transverse slit with two lips (see Fig. 4.17C), a shorter ventral inner lip (homologous to the trap door in *Utricularia*) and a dorsal lip which is arched over the opening (homologous to the threshold in *Utricularia* traps).

In the last phase of the trap development, cell divisions are restricted to the two sides of the mouth only, where the lips are connected. These cell divisions lead to the elongation of the two parallel lips. Two arms thus develop on top of the tube, circinate when young and fully expanded when mature. The two parallel lips form spirals along the arms. The helical arms, with their spirals turning in opposite directions (Fig. 4.17) are interconnected on top of the tube. In this manner the slit, which extends in both arms, forms an extremely long and narrow mouth through which small animals can enter the trap (Fig. 4.17*A*, *B*).

The inner structure of the trap

Special trichomes bridge the two lips at short intervals, holding them apart at a constant distance (Fig. 4.17*B*). Each trichome consists of a row of three cells, the middle much enlarged. The inner surface of the arm is broken up into a series of many transverse ridges, each supporting a row of stiff trichomes directed inwards and towards the base of the arm (Figs. 4.17, *B,C* and 4.18). A ridge is frequently seen beginning and ending near one of the bridging trichomes (Plate 11–6 in Lloyd, 1942). The ridge passes obliquely inward and for a certain distance toward the apex of the arm and then bends back sharply at the middle of the arm. The ridge approaches both lips in harmonic curves, almost parallel to their edges (Plate 11, Lloyd, 1942). In this manner, a series of trichomes can easily be seen along the helical slit between neighbouring bridging trichomes (Fig. 4.17*B*).

One or two transverse rows of glands are always situated immediately below each ridge, hidden underneath the detentive hairs (Fig. 4.18). This repeating structure of hairy ridges and glandular rows extends along both arms and leads into the tube (Plate 11–11 in Lloyd, 1942). The inner surface of the tube is similarly broken up into a series of some 30–50 transverse trichomatous circular ridges, with rows of glands hidden underneath.

The utricle, which is wider than the tube, does not bear any hairs; instead it is crowded with many glands. Two longitudinal ridges, extending from its

Fig. 4.19, Right. *Triphyophyllum peltatum*, light micrographs. (*A*) Portion of a mature 'carnivorous' leaf showing both stalked and sessile (SS) glands. (*B*) Immature leaf tip showing the 'crozier' circinate venation; (cf. Fig. 6.2). (Unpublished light micrographs by kind permission of Dr Y. Heslop-Harrison.)

Fig. 4.20, Below. Young stalked and sessile glands at the circinate tip of *Byblis liniflora* (SEM by C. Merriman, Oxford).

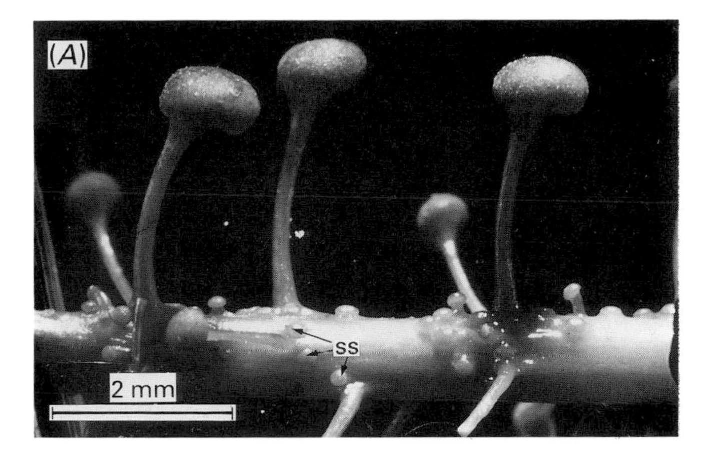

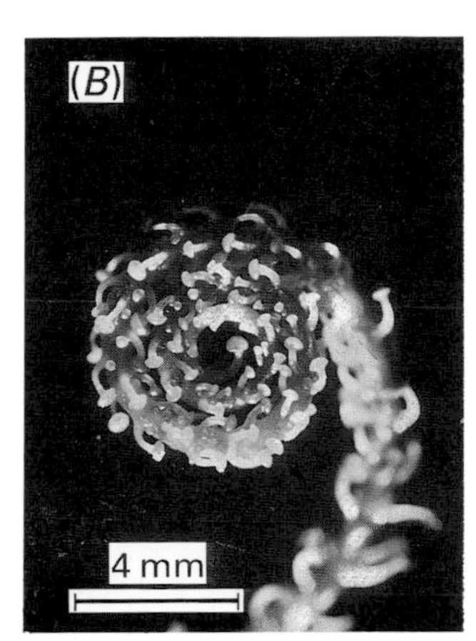

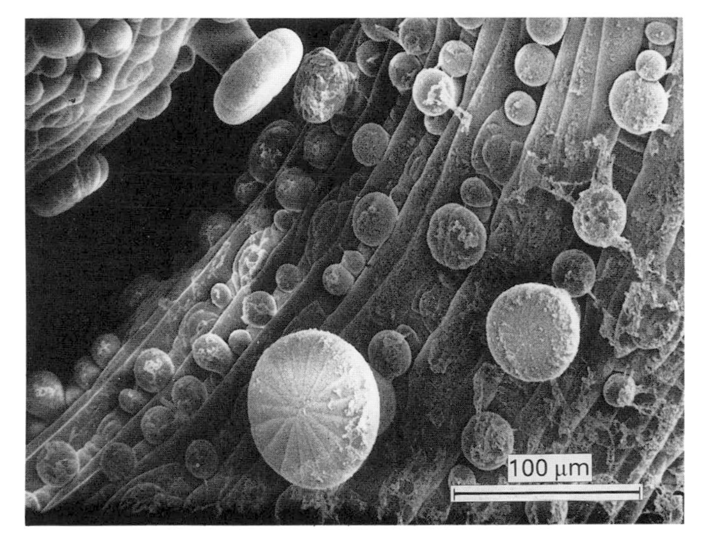

Fig. 4.21, Above. Mature stalked and sessile glands of *Byblis liniflora*. Inset: one sessile gland at higher magnification. (SEM by C. Merriman.)

base on the ventral wall and on the dorsal, serve as the main seat of glands in this part of the trap (Fig. 12–10 in Lloyd, 1942). In addition, similar glands are also scattered on the rest of the inner surface of the utricle. Each glandular ridge is provided with a single strand of conductive elements (see below). The glands that develop inside the various parts of the trap are built up of an upper layer of 2–8 cells which together form a cap-like glandular layer. A short, endodermoid neck cell connects the glandular cap to a basal cell, which is anchored in the epidermis.

The trap wall is thinnest in the twisted arms which comprise only two cell layers, as in the traps of *Utricularia* (see page 67). The wall becomes thicker in the tube and even more so towards the base of the utricle.

The vascular system of the trap consists of two main vascular bundles, a dorsal and ventral, each composed of a single spiral vessel accompanied by a thin strand of phloem (Lloyd, 1942). These two strands branch twice, first giving rise to one short vascular bundle to each of the glandular ridges in the wall of the utricle and then, near the mouth of the tube, each strand branches again, giving one branch to every trap arm.

The vascular bundles of each arm run along the two sides of the spiral in close proximity with the bridging hairs. All trap surfaces are studded with simple external glands (Fig. 4.17C) which resemble those of *Utricularia* (see page 69).

CHAPTER 5

Attracting and Retaining Systems of Carnivorous Plants

Introduction

The efficiency of carnivory is apparently aided by special mechanisms of attraction which entice insects to approach a baited area in which they are trapped. Yet no behavioural study has examined this seemingly obvious fact. There are, as will be shown, a number of parallels that may be drawn between the proposed mechanisms by which traps allure prey and those used by flowers. Thus, traps can produce visual, tactile and olfactory stimuli and may also offer a reward in the form of nectar. The possession of attractive mechanisms is of adaptive advantage to the plant because it potentially increases the frequency of capture. Not one carnivorous species has been shown to be specific in choosing its prey, although there are powerful indications of prey-partitioning (discussed in Chapter 7). May the traps of certain carnivorous species deliberately deceive insects by mimicking flowers? This question is discussed in Chapter 16, where we consider mimicry in terms of exploitation and mutualism.

Visual Stimuli

In many carnivorous species, and especially in certain pitcher-plants, typical colour patterns can be seen in the traps. Flavonoids, naphthoquinones and anthocyanins are responsible for this coloration (see Ch. 11), which appears red or purple to the human eye. Almost all these trap surfaces also show patterns of ultraviolet (UV) absorption, probably resulting from their naphthoquinone or flavonoid content. The pitchers of certain *Sarracenia* species are displayed with coloured veins that serve as nectar guides leading to the nectar pools at the peristome. This form of nectar guide stands out in most species by contrasting with a white, pale-green, greenish-yellow or other brightly coloured interspace. Contrary to the nectar guides of flowers, the nectar guides of *Sarracenia* are not only colour marks but nectariferous lines (Joel *et al.*, 1986; Fig. 5.1).

There is considerable information on the vision of bees and other insects which are known to serve as pollinators of flowers, but only a little is known of the spectral sensitivity of insects known to be trapped and its possible implications on the effectiveness of colour patterns on their attraction by traps. The best we can offer is the linkage between general findings on insect vision and some behavioural information together with morphological details of the traps.

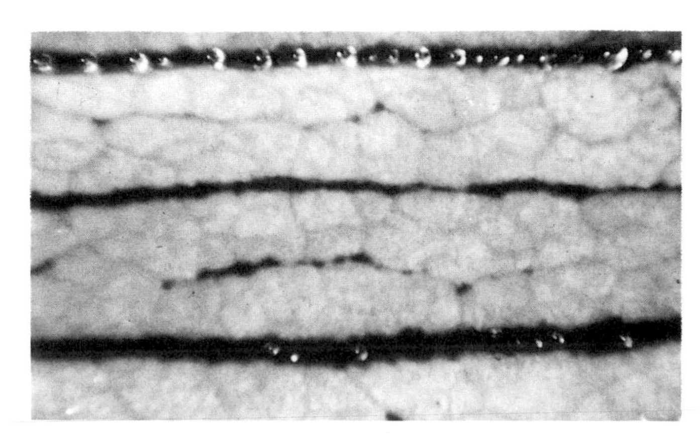

Fig. 5.1 Nectar droplets along nectar guides on the outer side of a *Sarracenia* pitcher (from Joel, 1986, with permission).

Ultraviolet Patterns

Ultraviolet patterns and insect vision

Arthropods are, in the main, sensitive to and strongly attracted by UV light (Menzel, 1979). Honey bees, flies and many other insects possess sensitivity to UV 100 times greater than to green light. The visible light (visible, that is, to the human eye) becomes relatively more attractive to these insects only in high intensities.

The more advanced insects, e.g. Hymenoptera, have the ability to distinguish the three primary insect colours: yellow (500–650 nm), blue (410–480 nm) and UV (300–390 nm). Others are unable to distinguish between these wavebands, their vision being based only on differences in light intensity. These latter insect groups show a preference for UV. UV is therefore a primary colour highly stimulating and most attractive to almost all arthropods under natural conditions, including those that are effectively colour-blind. Hence the considerable interest in UV patterns of insect-pollinated flowers (Daumer, 1958; Kevan and Baker, 1983; Vogel, 1983) and our own interest in UV patterns on the traps of carnivorous plants (Joel *et al.*, 1985, and see below).

Ultraviolet patterns of flowers and traps

Similar to many flowers, certain traps show conspicuous UV patterns. We believe that the various UV patterns found in the carnivorous traps represent a selective adaptation which facilitates the attraction of different groups of insects (Joel *et al.*, 1985). UV patterns appear to be significant in several ways:

(i) They seem to serve as orientation cues as in *Heliamphora* and *Brocchinia reducta*, where the entrance becomes conspicuous to visiting insects against the background by reflecting UV light while the outer surface is UV absorbing (Fig. 5.2*C*).

(ii) They give a definite colour pattern, for example, to *Drosophyllum*, delimiting the trap from neighbouring leaves. (The old dead leaves of this plant reflect UV and serve as a bright background to the UV-absorbing, green, living traps (Fig. 5.2*A*; see also Fig. 5.4).

(iii) They break up the colour pattern, as in *Dionaea*, where the glandular band at the periphery of the trap is UV absorbing and delimits the trapping site (Fig. 5.2*B*).

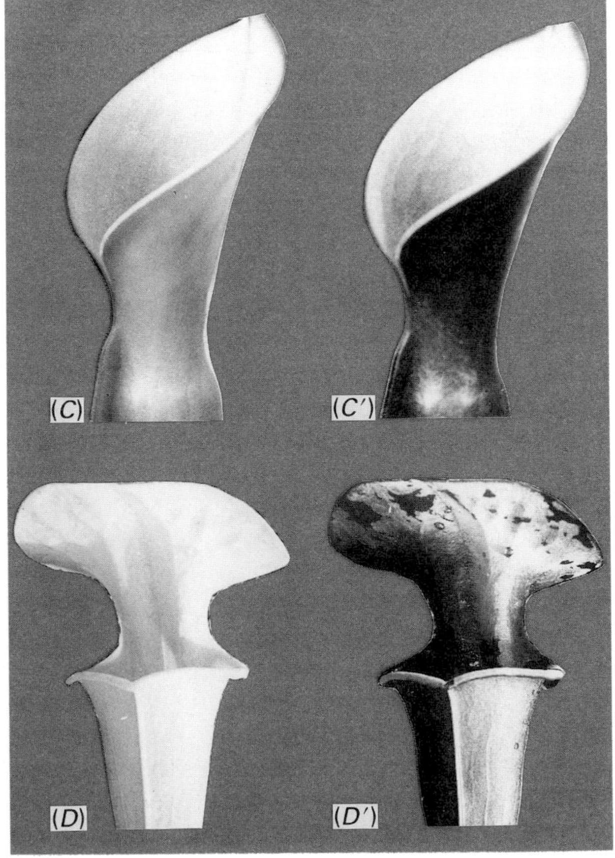

Fig. 5.2 Photographs taken through a visible light absorbing filter, showing UV patterns. (*A*) *Drosophyllum lusitanicum* with UV reflecting dead leaves serving as a bright background to the UV absorbing young leaves. (*B*) *Dionaea muscipula* with peripheral UV absorbing band, delimiting the central trapping site. (*C*) *Heliamphora nutans* showing UV absorption on outer trap surface, and UV reflection on inner side of the pitcher. (*D*) *Sarracenia flava* with conspicuous UV absorbing nectar.

(iv) They make the nectar of *Sarracenia* conspicuous to visiting insects. The nectar shows significant UV absorption that produces a visible contrast with the nectar-free trap surfaces (Fig. 5.2*D*).

(v) They seem to determine whether *Pinguicula* traps deter or attract insects, by changes in UV absorption of the leaf in accordance with the secretion of digestive pools on the trap surface (Fig. 8.16). The secretion is UV-absorbing probably as the result of the presence of amino acids and proteins (digestive enzymes).

The distribution of these UV patterns in traps of carnivorous plants resembles that of floral UV patterns, which are discussed by Jones and Buchmann (1974). The existence of UV patterns in and on the various traps further supports the view that the landing of insects on traps is often not a casual event but the result of a specific attraction (Joel *et al.*, 1985).

Reflection of UV may be the main signal or, more likely, one of a group of signals.

Colour Changes and Development in Traps

Pigmentation of many traps is not apparent until the trap is ready to capture insects. This is emphasized in most pitcher-plants, where the prominent coloured venation often develops only when the trap is open. *Sarracenia* leaves, for example, capture most of their insects during a relatively short period, apparently only when they are, in pigment terms, attractive to insects. In *S. purpurea*, the newest leaves are pale in colour; they acquire their typical pattern of coloured venation during their first days as mature leaves. During the developmental period, the capture ability of the leaf is low. When the development of pigmentation is complete, the capture ability suddenly peaks (Fish and Hall, 1978). Similarly, certain obligate inhabitants of the pitcher (see Ch. 14) enter the trap at approximately the same time. When the traps become older, the whole pitcher reddens and looks darker, resembling certain senescing flowers. The nectar guides are no longer prominent and apparently no longer attractive. The largest quantities of insect remains can be found in pitchers that are richest and deepest in colour (Edwards, 1876).

Pigmentation is often affected by such environmental factors as illumination and temperature but, according to Schnell (1976), the differences in colour do not affect the trapping capability of the pitchers of *Sarracenia*. *Dionaea* produces darker digestive glands when exposed to stronger light. The digestive glands of *Dionaea* are colourless when young, but they gradually acquire a reddish shade when exposed to sunlight and then often change to a deep-red coloration. Mature glands growing in weak sunlight do not ever go dark-red, indicating that the cyanidin production is light-dependent. While originally it may have been as a protective agent, observation indicates that it also improves capture.

Colour Change and 'Aggregation'

Similarly, the stalked glands of both *Drosera* and *Drosophyllum* usually develop a red pigment, especially when exposed to sunlight. This coloration changes when the digestive activity of the gland begins. Darwin (1875) showed that, when the glands of *Drosophyllum* were red, they instantly darkened and soon became very black. This is because of changes in the vacuolar composition in the glandular head cells, a process termed 'aggregation' by Darwin (see page 180). In this process the large, pigmented vacuoles split, while the total volume of vacuole decreases, consequently increasing the concentration of pigment in the smaller vacuoles arising from subdivision and resulting in a darker appearance of the pigment. Whilst this process seems to be associated with transport activities of the digestive cells (see Ch. 10), we do not know if it also affects the potential attractiveness of the trap as a whole. It is conceivable, however, that these colour changes serve to reorient potential prey towards unactivated regions of the trap, avoiding metabolically active tentacles which are devoid of trapping mucilage. In this respect the active glands of certain adhesive traps and the older pitchers of *Sarracenia* resemble senescing flowers which change their coloration after having been pollinated. This alteration of pattern might thus serve to optimize prey capture.

Glistening Droplets

In some flowers the watery exudate of stigmatic papillae provides a glistening attractant for dipterous visitors and it is well known that many insects, and Diptera in particular, are attracted by glistening drops (Kugler, 1956). In the spadix of *Anthurium hookerianum*, for example, the stigmatic exudate produces a series of glistening sugary droplets which are sipped by drosophilid flies (Vogel, 1983). Similarly, a series of droplets visited by insects is also produced by dense inflorescences composed of numerous flowers, each bearing an open nectary. Accordingly, therefore, the glistening droplets of mucilage of the adhesive traps, which are known to trap many flying insects (particularly Diptera), should act as an attractant.

The mucilage found on the adhesive traps of *Drosophyllum*, *Drosera*, *Byblis*, *Triphyophyllum* and *Pinguicula* is secreted by stalked glands and is involved both with trapping (see Ch. 6) and with digestion (see Chs 8–10). It is easily detected from a distance as it reflects both direct and indirect light of all colour bands, including UV. Any background which absorbs any one band or more of the visible light will therefore provide an adequate contrast to the glistening droplets, rendering them visible from all directions (see Figs 5.1, 5.4 and 5.5).

The ranks and patterns of mucilage droplets in the flypaper traps resemble the open nectar drops of those plants in which they are offered as a true reward to visiting insects which either act as pollinators (floral nectaries) or act in the defence of the plant (extra-floral nectaries). In the adhesive traps, however, no reward is obtained, in contrast to the pitcher traps, in which nectaries are found (see page 59). Whether in this the plants show deliberate mimicry is unknown (see Ch. 16).

Olfactory Stimuli

In addition to visual stimulation, insects in the vicinity of carnivorous plants may receive olfactory allurement. Insects are extremely sensitive to certain chemicals, in particular the pheromones used in mate location. They are also capable of discriminating between fragrances. Some insects are attracted only by very specific mixtures of a precise combination of certain volatiles, while others are less specific and therefore attracted by a wide range of flowers. This latter group is more likely to be exploited by carnivorous plants.

While optical cues are based on learned signals, chemical triggers act directly on the insect's instinct. Volatiles may act to attract insects to the general area of the traps, after which visual signals become more important. Once in close proximity to the traps, however, olfaction may once again become a major stimulant, releasing behavioural reactions originally necessary for pollination. These two triggers may act together to keep the potential prey on the trap, extending the time of each visit and increasing the probability of successful capture.

Volatile Organic Attractants

Despite much speculation, there is no direct experimental evidence for chemical attraction of potential prey by pitcher-plants. Nevertheless, volatiles are produced by these plants (see Chs 11 and 12). In *Sarracenia flava*, for example, 1.7% of the total oil extracted from its leaves was phenylacetaldehyde (Miles *et al.*, 1975), a compound with the scent of lilac or hyacinth that attracts moths to the flowers of *Araujia sericifera* (Cantelo and Jacobsen, 1979). n-Pentadecane, which comprises 0.6% of the total oil of *S. flava*, is one of the fragrance compounds released by flowers of *Magnolia* (Thien *et al.*, 1975). Other volatiles, (see Chs 11 and 12) might also serve as attractants to insects.

Surprisingly, the only information concerning insects attracted by chemical means to pitcher traps, is related to insect-inhabitants rather than to prey.

The adult female of *Wyeomyia smithii* (see Ch. 14 and Fig. 14.1) selects the youngest open leaves of *S. purpurea*. The behavioural pattern of the mosquito at these traps is a typical response to olfactory rather than to visual stimuli: the female lands in the trap while in flight, exhibiting a typical up and down bobbing flight pattern, which may constitute a period of final leaf assessment by the female before she lays her eggs (Bradshaw, 1983). Istock *et al.* (1983) provide evidence that a water-soluble chemical is responsible for this attraction of *W. smithii*. At the same time they ruled out the possibility that this compound is attractive to other insects as part of a trapping mechanism, because the trap does not seem to attract visits from mosquitoes other than *W. smithii*.

Utricularia may produce water-soluble attractant chemicals (Cohn, 1875), though the nature of these is unknown. Thus, sessile rotifers seem to settle on particular regions of the trap (see page 267) while minnows are able to discriminate *Utricularia* from other water plants (see Ch. 12).

Honey Scent

Several *Sarracenia* species that are known to produce nectar (see page 80) also release sweet attracting odour from their fresh tall pitchers (Gibson, 1983a; Slack, 1979). These species (*S. leucophylla*, *S. alata*, *S. rubra* (*alabamensis*), *S. oreophila* and *S. minor*) all show positive correlation between insect availability and pitcher density (Gibson, 1983a). The chemical composition of this odour is not known but merits investigation.

Nectar scent is also typical of certain *Heliamphora* traps in which the cap ('spoon') secretes nectar with a very strong scent and sweetness (Brewer-Carias, 1973). So too, but seasonally, does *Darlingtonia* (Austin, 1975G, Appendix 1).

Brocchinia reducta which grows in the vicinity of *Heliamphora* does not secrete any nectar at all, but nevertheless it does emit a sweet nectar-like odour, not unlike that produced by the nectaries of its neighbour, *H. heterodoxa* (Givnish *et al.*, 1984). Such a scent not only does not occur in the non-carnivorous species of *Brocchinia* but is apparently unique in the Bromeliaceae. The possibility that *Brocchinia* is mimicking *Heliamphora* is considered in Chapter 16.

All *Nepenthes* species, as well as *Cephalotus* and many species of *Sarracenia*, which produce nectar do not seem to produce any attractive scent at all. This is not surprising. A similar lack of correlation between nectar supply and the release of attractive odours is also found in insect pollinated flowers.

Drosophyllum is perhaps the only member of the large group of adhesive traps which is known to attract insects by olfactory means. It releases a very strong honey scent which is readily detected from a distance. This unique feature of *Drosophyllum* led the villagers in Oporto (Portugal) to use it as a 'fly-catcher' by hanging it up in their cottages (Darwin, 1875). By doing so they both achieved a very delicate sweet smell in their rooms and captured flies which entered their houses.

Attraction by the Smell of Putrefaction

Many pitcher traps were formerly believed to attract carrion and dung flies by releasing the odour of putrefaction, either through the decomposition of insect remains or through synthesizing the odours in their tissues. This belief can be rejected since we know that, if putrefaction odour is released by a pitcher it is the result, not the cause of trapping. It occurs, if at all, only when the trap has finished its trapping period. In each plant of *Sarracenia purpurea* one leaf normally contains more fresh insect remains than any of the others. Only this leaf, the oldest in the plant and the one which no longer catches any prey, could sometimes be characterized by considerable turbidity and a notable putrid odour (Fish and Hall, 1978).

Beaver (1983), expressing a similar view dealing with *Nepenthes* traps, has shown that observations on pitchers of *N. ampullaria* did not reveal larvae of detritus feeders (Ceratopogonidae) and predators (Chaoboridae) until 5–7 weeks after the pitcher had opened. Only on those rare occasions when a pitcher catches so many insects that they cannot all be digested will the corpses decay. Then the usual inhabitants of the pitcher are killed and opportunistic carrion feeders are attracted. In these cases the pitcher soon dies and does not profit from the visit of those insects attracted by putrefaction.

Rewards and Retention

Once potential prey has been lured to the entrance of a trap it needs either to be captured or retained within the vicinity in order to increase the likelihood of capture. In the active and adhesive traps (see Ch. 6) the former strategy is adopted. The pitchers adopt the latter and two mechanisms may be identified: (i) nectar is provided as a sugary reward and, (ii) where transparent windows called *areoles* (see page 82) have been developed to confuse the insect over the exit route (Fig 5.4).

Neither of these mechanisms occurs in non-pitchered carnivorous plants, with the exception of *Dionaea*, which produces nectar. In this snap-trap the potential prey is free to move around the trap surface until the snap-trap is activated by two or more contacts with the trigger hair (see page 98). Hence, it is advantageous to retain the insect on the lobes and alluring glands are located around the margins (Fig. 6.12).

Nectar

Nectar is basically a watery solution of sugar and well known as the dominant floral reward – the 'junk-food' of the plant kingdom. Insects suck nectar or lick it for their own nourishment and water supply (Vogel, 1983). All pitcher-plants offer this reward to insect visitors, thus tempting them to prolong their stay around the trapping site until some of them accidentally stumble and fall into the digestive cavity.

Structure and location of the nectaries

All pitcher-plants have nectaries in the peristome vicinity. Most species also exhibit nectaries on the outer surface of their traps (see Fig. 5.1).

In *Heliamphora*, two kinds of nectaries develop on the pitcher: large glands which can be seen only

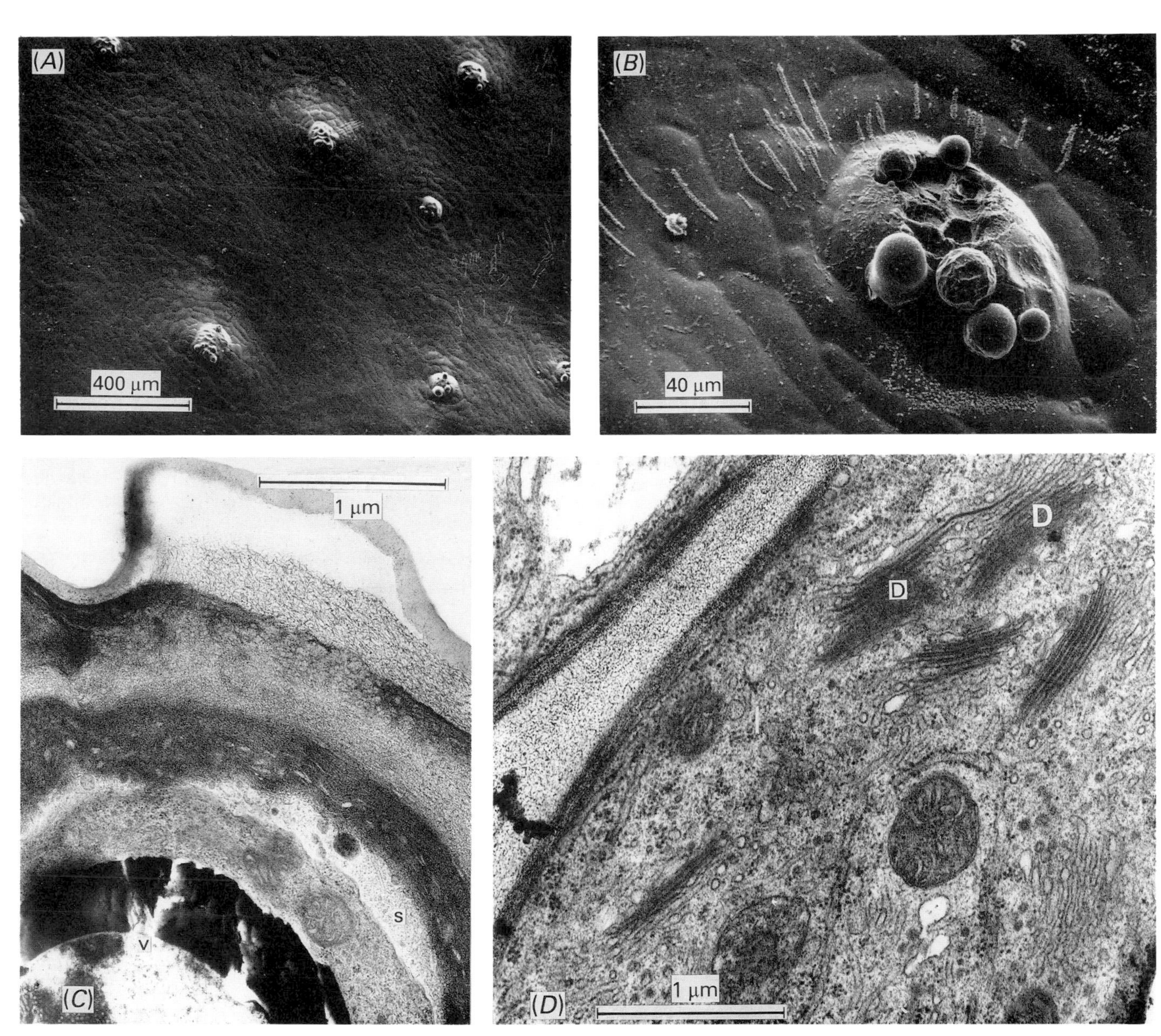

Fig. 5.3 Low (*A*) and intermediate (*B*) magnifications of the nectar glands from the 'spoon' of *Heliamphora nutans* (SEM by C. Merriman).
(*C*) and (*D*) TEM's showing a peristome nectary of *Nepenthes khasiana*. In (*C*) note the complex multi-layered wall just below the distended cuticle which is about to rupture. (s) is probably a slime layer recently exported through the plasma-membrane and (v) a vacuole with an electron-dense margin. In (*D*) the apparently very active dictyosomes of the peristome nectary can be seen and at 'D' a dictyosome may be dividing. (Original micrographs by permission of Drs A.E. Vassilyev and Lyudmila E. Muravnik, Leningrad, USSR.)
(*E*) Part of a secreting cell from a peristome nectary of an open pitcher of *Nepenthes khasiana*. The ER is closely aligned to the plasmamembrane with, it appears, fewer ribosomes on the outward face. The apparently active dictyosome is surrounded by numerous vesicles. Very prominent in these cells are what seem to be autophagic vacuoles (AV) with adjacent and well-developed mitochondria (M). (Unpublished micrograph by permission of Drs A.E. Vassilyev and Lyudmila E. Muravnik, Leningrad, USSR.)

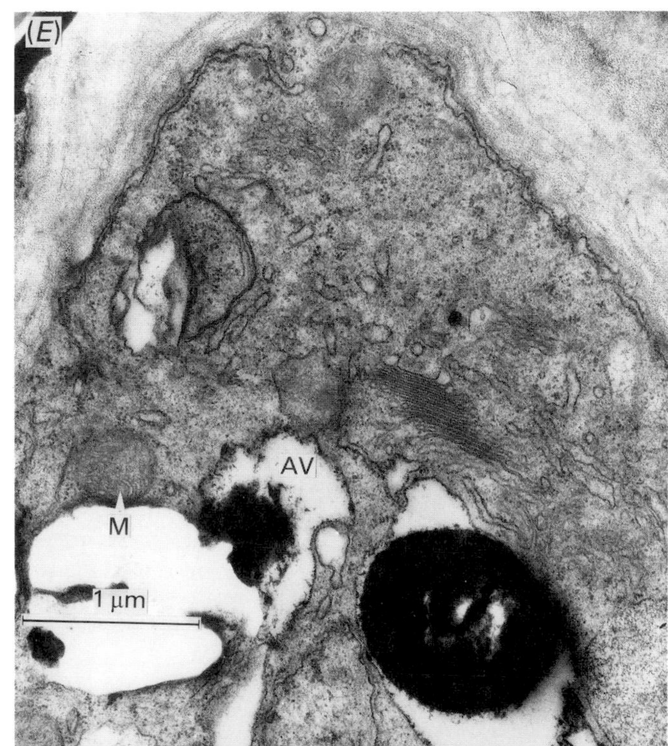

on the inner surface of the spoon (Fig. 5.3*A*, *B*) and smaller nectaries which are scattered on the outer surfaces of the pitcher, including the wings, and also extend inside the pitcher down below the rim into Zone 2, which is a band at the upper third of the pitcher covered with hairs (Adams and Smith, 1977). The structure of these glands has not been recently described or studied and is therefore only briefly mentioned in this section, following the description given by Lloyd (1942).

The smaller sunken glands appear in surface view to consist of six cells: two central 'cover' cells which sit on top of four glandular cells, the periphery of which can be seen around the free edge of the cover cells. A second inner course of four cells sits underneath them, with two further cells lying at the bottom of the gland. Cuticularized walls surround the gland and presumbly act like all endodermoid walls so typical and important to secretory activities (see Ch. 8). In their general structure these glands resemble the nectary glands of other Sarraceniaceae.

About 20 large nectaries are located in a more lateral position on the inside of the pitcher, the largest forming a ring around the hood apex (Adams and Smith, 1977). In principle, these glands show the same structural composition as the smaller ones, but each course of cells is built of a larger number of cells, so that, for example, there are more than two cover cells and about four courses of thick-walled cells underneath them. The periphery of the glandular structure is normally irregular. So large are these glands that they sometimes contain about 1000 cells. These large glands are in close contact with the vascular tissue.

In *Sarracenia*, nectaries are found in the peristome where they are most abundant. Nectaries do, however, develop on both the outer and the inner surfaces of the pitcher. On the outer surface they are often arranged along the pigmented nectar guides (Fig. 5.1). We therefore find that, contrary to all other nectar guides in non-carnivorous plants, the nectar guides on the outside of *Sarracenia* pitchers are not just colour marks but also nectariferous lines (Joel, 1986).

All glands in this genus are similar (the specific *Sarracenia*-type of Goebel, 1891) and closely resemble the small nectaries in *Heliamphora*. A characteristic of the nectaries in *Sarracenia*, however, is that the cover cells are conical and are wedged in between the four cells underneath, partly covering them. The walls of those cells that lie around the gland show dense protoplasts and cuticular wall impregnations, indicating their endodermoid nature. Methylene blue easily permeates the gland (Lloyd, 1942), indicating that the cuticle covering the gland is incomplete.

In contrast, in *Darlingtonia* the nectar roll carries large nectaries, compound glands built of two tiers of flat cells. Similar compound glands are also seen, to a lesser extent, on the forward interior face of the dome (Lloyd, 1942). The glands on the outer surface are much smaller.

The glands of *Darlingtonia* superficially resemble those of *Sarracenia*, but on the surface they look like ordinary epidermal cells. Being 'sunken', the main body of the gland is much wider than its outer cell. This inner part of the gland is composed of a row of flat cells which were evidently derived by periclinal divisions of an original protodermal cell (Lloyd, 1942).

The nectar glands of *Cephalotus* are scattered on the outside pitcher surface, on the inside surface of the lid and, most abundantly,on the surface of the recurved teeth of the peristome (Parkes, 1980). Each gland consists of six secretory cells seated on several endodermal cells. The secretory cells show a dense cytoplasm and large osmiophilic vacuoles. The endodermal cells show translucent cytoplasm and their transverse walls are typically lignified (Parkes, 1980).

In *Nepenthes*, nectaries are found on the shoot axis, petiole, lamina and tendril, as well as on their pitchers and flowers. Those of the pitcher develop on the underside of the lid and between the teeth of the inner edge of the peristome. The glands of the lid are homologous in structure to the digestive glands, both being multicellular, disc-shaped glands, featuring a row of columnar head cells at the surface, two to three layers of internal secretory cells and an endodermoid layer (Parkes, 1980). The nectaries of the inside edge of the peristome are very large, and may sometimes have an extracellular cavity. In *N. maxima* the nectar gland extends far back into the rim, and possibly is connected with neighbouring glands, thus forming a ring of secretory tissue right around the rim. In these very long glands an internal cavity is prominent. Apart from being surrounded by an endodermoid layer, these glands are also supplied with a ring of vascular bundles (Parkes, 1980).

Ultrastructure of nectaries

Nectar secretion and the ultrastructure of nectaries is rarely studied in the carnivorous plants. There is only one detailed description of the fine structure of a nectar secreting gland, that of the peristomal nectaries of *Nepenthes* by Vassilyev (1977 and Fig. 5.3*E*), which shows these to conform to our general knowledge of some floral nectaries (Schnepf, 1974; Vassilyev, 1977; Fahn, 1979).

All glandular cells of the peristomal nectaries of *Nepenthes* (Fig. 5.3*C, D*) show a large nucleus with numerous small nucleoli. The vacuome of the secretory cells is well developed. Several large vacuoles or a large central vacuole are clearly seen in the cells, all containing an electron-dense substance which is thought to be proteinaceous. The glandular cells are rich in dictyosomes. Though many dictyosomes are also present in the sheath cells, the most active dictyosomes are seen in the glandular cells. The numerous mitochondria are elliptical, sometimes long and branched. Their long cristae are densely packed (M in Fig. 5.3*E*). The endoplasmic reticulum is only slightly developed, showing solitary, long, rough cisternae. The plastids are polymorphic, sometimes cup-like and commonly devoid of any starch. Their stroma are dense, showing tubular and vesicular textures, very often containing groups of plastoglobuli. Microbodies and multivesicular bodies are often seen in these cells.

The cell walls are relatively thick, with many plasmodesmata, but do not show any ingrowths. The outer wall of each glandular cell is thicker, about 1.5 μm, and its peripheral zone accumulates some homogenous substance. Similarly, the irregular space between the plasmalemma and the cell wall (Fig. 5.3*C*) is filled with a granular substance, presumably mucilage, in which membranous elements are often seen (Vassilyev, 1977).

The composition of the nectar

Hepburn *et al.* (1927) analysed the nectar secreted by *Sarracenia flava* and *Darlingtonia californica* and came to the conclusion that both contain fructose. Unfortunately, no further analysis has been carried out on any of the nectars secreted by the different species of the pitcher genera. Nevertheless, there are suggestions that in some cases amino acids might also be included in the secreted nectar, as is known for floral nectaries (Baker and Baker, 1973). According to Joel *et al.* (1985), the pools of the viscid thick nectar on the surfaces of *S. flava* pitchers show a significant UV absorption (see page 75), probably due to phenolics and aromatic amino acids. Amino acids may, on the one hand, affect the 'taste' of the pitcher nectar and on the other may enable the pitcher to build relationships with a particular insect species (see page 78), as is known for a number of non-carnivorous associations in which nectar is supplied. This can be effective since many insects are known to be able to discriminate through taste, not only between the three primary sugars of nectar, i.e. glucose, fructose and sucrose, but also between different combinations of sugars with amino acids. The nectar provided by the pitchers of *Nepenthes* has not been studied, but the discovery of some protease in the nectar produced by its flowers (Daumann, 1930) is remarkable. The nectaries in *Nepenthes* show structural and ontogenetic patterns which resemble those of the digestive glands, and both are thought to have a common origin (see page 295 and Ch. 18). We should therefore expect to find traces of digestive enzymes in nectar which is secreted by at least some species of *Nepenthes*. Similarly, Parkes (1980) proposed that all glands in *Sarracenia* have a common origin (see Ch. 18). This argument was based particularly on evidence for hydrolytic enzyme activity in nectary glands on the outside of the pitchers similar to that of the digestive glands (see Ch. 9).

How do insects find the nectar?

The visual and olfactory cues discussed above, together with tongue guides, lead experienced insects to the nectariferous sites. However, the presence of nectar must be learned by probing. Nectar is recognized by taste when the sensory organs of the insect contact it directly. Only then, when a real reward is obtained by a naive insect visiting a pitcher, and visual and olfactory signals have been learned, will these signals help guide the experienced insect to the pitcher in their search for nectar.

The strategy is manifestly successful. Beaver (1983) reports how ants regularly visit the nectaries of the pitchers of *Nepenthes*. Inevitably, a proportion fall in. Presumably the nectaries found elsewhere on the stem, tendrils and leaf surfaces are comparable to those on the outer surface of *Sarracenia* pitchers in providing a trail of liquid reward leading to the pitcher rim. While it has been suggested that nectar may contain a narcotic to disorient or disable the prey, there is no evidence of this. The presence of coniine in the tissues of fresh pitchers of *Sarracenia flava* may, however, be significant in this context (see Ch. 12).

Light-transmitting Fenestrations (Areoles)

Insects, being positively UV phototropic, find their way out of closed spaces by following the direction of UV light. These wavelengths apparently signify skylight or 'open space' to insects (Mazokhin-Porshnyakov, 1969).

Nepenthes, and most species of *Sarracenia*, 'lose' many visiting insects which find their way out of the trap. Other pitcher-plants, notably *Darlingtonia* and a few species of *Sarracenia* have developed a special device both to exhaust the prey and to direct the would-be escaping insects deep into the trap by deceptive 'windows', termed *areoles* (Frontispiece).

The formation and distribution of the areoles

These 'windows' appear as a transparent fleck, a few millimetres wide, often surrounded with red 'mullions'. The areole is composed of several layers of rectangular cells, densely packed without intercellular spaces, the cells lacking chlorophyll and other pigments. According to Troll (1939), the white flecks of *Sarracenia leucophylla (drummondii)* are not fully transparent and show well-pronounced intercellular spaces, while in *Darlingtonia* they lack any sort of intercellular space, rendering the areoles fully transparent (Fig. 5.4). The whole structure is also devoid of vascular tissue and shows a smooth surface with no stomata, trichomes or glands. These features minimize both absorption and internal reflection of light, and render the areoles transparent. Similar 'window panes' surrounded by darkly pigmented mullions are also found in certain deceptive flowers (Dafni, 1984) as well as in the spathes of certain Araceae like *Arisaema laminatum* (Van der Pijl, 1953).

The 'windows' are often hardly conspicuous from the outside (except for a total absence of colour) while almost luminous and clearly evident to insects within the trap. They are transparent to UV (Fig. 5.4), making the windows highly attractive to insects attempting to escape. As the expected exit through the areoles does not materialize, the insects continue to pace or to fly up and down from one window to another, until they fall exhausted into the trap.

The translucent trap-door of *Utricularia*

A possible parallel case can be seen in the trap of *Utricularia* where the trap wall is chlorophyllous or darkly pigmented, while the door is remarkably transparent. Darwin (1875) was of the opinion that this was not accidental and that the spot of light thus formed may serve as an internal guide. However, although water-inhabiting invertebrates like flagellates, ciliates and free-living nematodes, demonstrate a phobic reaction to UV light (Menzel,

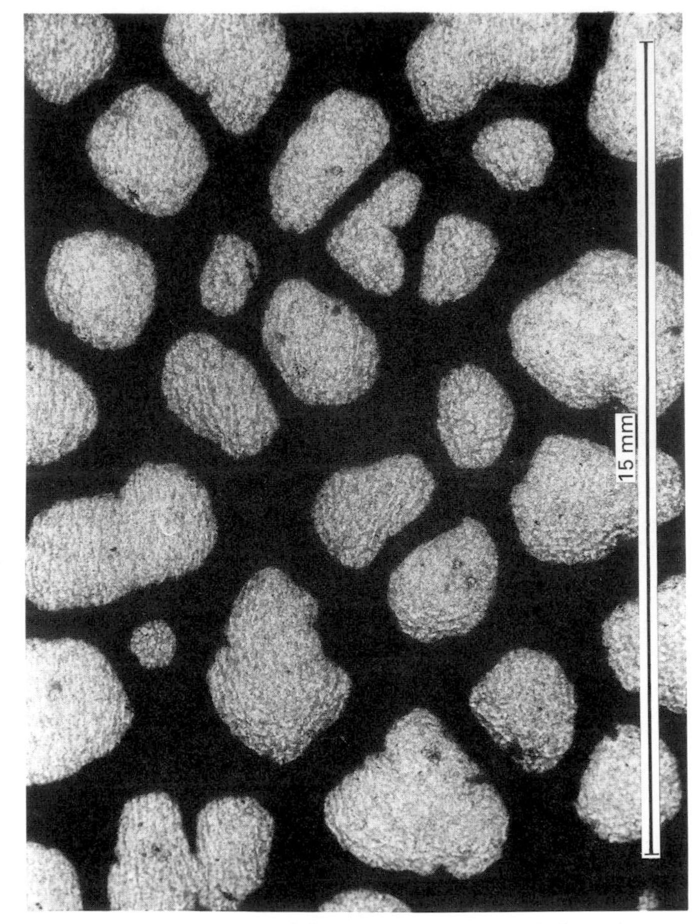

Fig. 5.4 Fly's-eye view of the areolae (windows) in the hood of *Darlingtonia californica*. Photographed from inside the hood with UV light and a UV transmitting filter. (Photomicrograph by Tony Allen, Oxford, UK.)

1979), these wavelengths do not penetrate turbid water. Thus, the spot of light described by Darwin might comprise wavelengths other than UV.

Tactile Stimuli

In *Utricularia* a strategy has apparently evolved in which a seemingly everlasting and abundant supply of food is provided for potential prey to consume.

The filamentous extensions (antennae and bristles) that arise from the traps of many *Utricularia* species (see Ch. 4 and Fig. 4.15C), whose structure closely resembles filamentous algae, may attract prey such as crustaceans. Crustaceans, like *Chydorous sphaericus* and chironomids, use two feeding modes when in contact with filamentous algae (Meyers and Strickler, 1979), *viz*. grazing on epiphytes while traversing an algal strand, and filter feeding while grasping a strand and remaining stationary. The antennae and bristles of the *Utricularia* traps, by mimicking filamentous algae, serve as feeding pathways drawing potential prey towards the entrance. Experimentally, it was shown that the antennae of *Utricularia* traps are twice as important as the bristles in enhancing the capture rate of crustaceans. Furthermore, the percentage of prey landings that resulted in capture was significantly higher for crustaceans landing on antennae than on either bristles or the trap body itself. These crustacean prey were never seen to swim directly into the trapdoor (Meyers and Strickler, 1979).

No carnivorous plant has yet been experimentally shown to produce touch stimuli. Nevertheless, stimuli of this type might play some role in directing potential prey into the traps of certain pitcher-plants. In this sense the imbricated cells of the epidermis at the entrance of a *Sarracenia* trap may represent a special adaptation, providing tactile stimuli which lead insects to the trapping site.

Selectivity and Allurement

Exact descriptions of insect capture in nature are lacking for many carnivorous species and therefore any proposal of what induces insects to enter traps must be speculative and circumstantial. Non-specific insect trapping, which is common to many carnivorous traps, is an indication of casual landing of prey on the trap. Williams (1976) reported that small insects tend to land on non-carnivorous plants surrounding *Drosera* at about the same rate as they land on *Drosera*. Accordingly, the catch made by *Drosera* species, as suggested by Achterberg (1973) depends primarily on two factors: the species of arthropod in its habitat and the abundance and activity of these species (see page 141 and Table 7.5). In *Dionaea*, the diversity of arthropods found in closed traps points to non-specific prey-preference (Lichtner and Williams, 1977) (but see page 140 for the possible changes of prey over time).

However, when the behaviour of single flies visiting the *Dionaea* traps is followed, they seem to be attracted to the outer trap margin (Jones, 1923), then known to secrete carbohydrates and now found also to be UV-absorbing (see page 75). This specific attraction pattern has been clearly demonstrated in the film produced by Oxford Scientific Films 'The Tender Trap'. *Dionaea* seems therefore to exploit the insects (and other arthropods) landing randomly on its traps together with insects attracted by UV light. Similarly, *Drosera* and *Drosophyllum*, which

Fig. 5.5 A *Drosophyllum lusitanicum* plant photographed (*A*) by normal photography; (*B*) by UV photography, using a visible-light absorbing filter. Note the mucilage droplets; visible light and UV reflecting and made more obvious in (*B*) by the UV absorbing background of the leaf (see Fig. 5.2).

show the typical glistening known to be attractive to flies (see page 83), catch various other arthropods which land randomly on the traps, together with those flies attracted by the glistening. The European droseras catch 50–95% flies when they are available (Achterberg, 1973). The percentage of Diptera amongst the entrapped fauna of *Drosera erythrorhiza* in Australia varies from one month to another, according to changes in the availability of the various species in the vicinity of the traps (see Ch. 7). It is, however, interesting to see that while this percentage changes from 5% in June to 70% in August (Dixon *et al.*, 1980), the total number of Diptera caught by the traps does not change and remains steady throughout the season. This phenomenon seems to result from the attractive nature of the glistening droplets. The total number of Collembola, on the other hand, does change through the season, being at its highest in June when Collembola are massively abundant. In contrast to Diptera, Collembola therefore seem to be caught randomly without being affected by any special attractive device.

The Orientation and Age of the Trap in Pitcher Genera

As the traps of *Sarracenia*, *Darlingtonia*, *Heliamphora*, *Cephalotus* and juvenile *Nepenthes* develop they orientate themselves radially, each successive trap generally at a large angle to the previous individual (Uphof, 1936, his Fig. 438; Fish and Hall, 1978, their Fig. 1; and Figs. 5.6, 5.7, 5.8 and Frontispiece). All the pitcher orifices, except *Darlingtonia*, face inwards. *Darlingtonia* achieves this precise orientation, and as Austin (1876E, Appendix 1) observed, precisely boxes the compass, in spite of the fact that it rotates as it develops, variably in direction, through 90°. These orientations would seem to be an adaptation to ensure that insects approaching the traps' cluster, from whatever angle, are lured by whatever set of stimuli (see above) are presented with the maximum possible signal. They should also ensure a trap of the right age (see Fig. 7.2). They also avoid the possibility in the case of *Darlingtonia* (Naeem, S., pers. comm., 1985 and see Frontispiece) that the young developing pitcher might grow, as sometimes still happens, into the orifice of a mature pitcher. The same problem, presumably, might also arise with the 'hooded' species of *Sarracenia*, *S. psittacina* and *S. minor*.

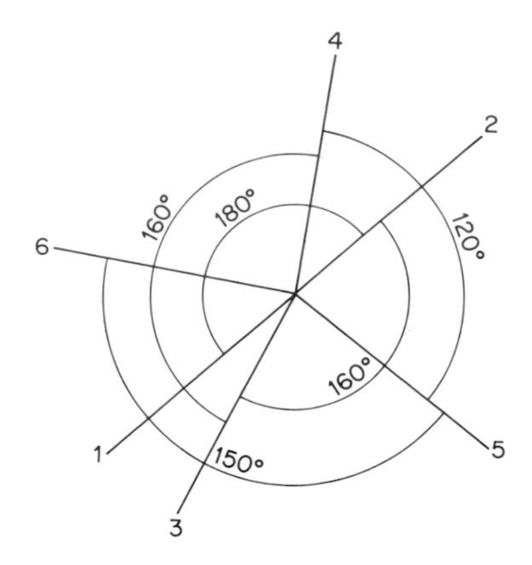

Fig. 5.6 The relative angles of insertion of successive pitchers of *Heliamphora nutans*, e.g. first pitcher 1 followed by pitcher 2 at 180° and 3 at 160°. Greenhouse-grown material.

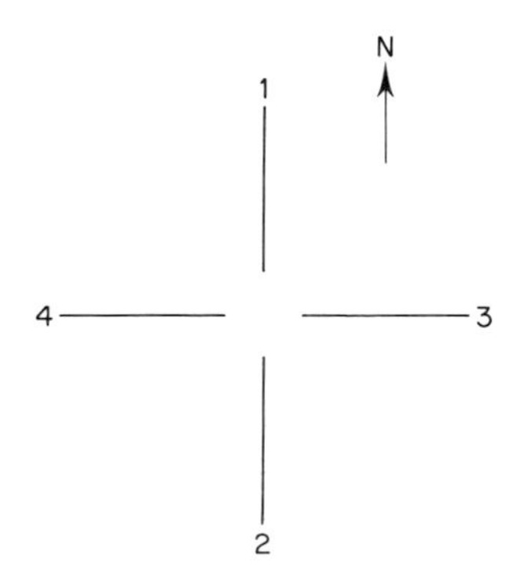

Fig. 5.7 The emergence of successive pitchers of *Darlingtonia californica*. The first pitcher (1) according to Austin (1875d, 1876e) is always north-facing and always, as the frontispiece shows, larger than the succeeding pitchers.

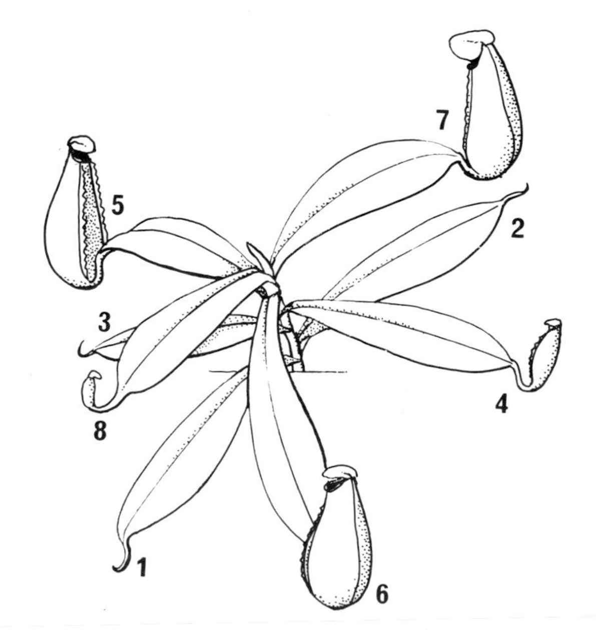

Fig. 5.8 The orientation of successive leaf tendrils or true pitchers in a young seedling of *Nepenthes pervillei*. Greenhouse-grown material.

Mature, climbing stem-traps of *Nepenthes* do not appear to have any obvious orientation, but it is interesting that the juvenile terrestrial forms of *Nepenthes* (see Fig. 5.8; Kerner and Oliver, 1895, their Fig. 23) also orient themselves in the same manner as the *Sarracenia* above.

The Height of the Trap

In addition to the orientations commonly seen in 'pitcher'-type traps (above), most pitcher forms achieve a gradation of height as well. A *Sarracenia* pitcher does not alter its height once the terminal lobe has opened. Thus, after the short period (see page 137) in which the pitcher matures to its maximum trapping ability, its height and orientation will not change. The following heights were taken by B.E.J. from a plant of *Sarracenia leucophylla* in the carnivorous plant collection at Oxford (U.K.):

Oldest leaf	⟶				*Youngest leaf*
607	577	545	514	466	420

(heights in mm)

Such a gradation can also be seen in the traps of *Darlingtonia* (Austin, 1875D, Appendix 1) where, both in the wild and in cultivation, the first pitcher is always markedly larger than the rest. It is possible that capture of insects might be enhanced by this linear exploitation of flying space.

The Radial Distribution of Traps in Other Carnivorous Species

The radial patterning of traps, possibly an adaptation to the maximization of trapping efficiency, is not restricted to pitcher types. Many Droseraceae, particularly the rosette-formers of the Western Australian flora (e.g. *Drosera zonaria* and *D. erythrorhiza*) form radial patterns. This is particularly well illustrated in the almost perfect radial symmetry shown in Russell's figure of *D. zonaria* (Russell, 1958).

The Inherent Complexity and Long-term Variability of the Attraction System

Studying attraction is fraught with difficulties and significant effects may occur only at specific seasons or with specific groups of insects. Again, it may be that only traps at a particular stage issue specific signals. As indicated, a combination of factors is probably involved, all making a proportionately different contribution to the process, the proportionality being dependent on the characteristics of the prey. Studies indicate that the prey captured reflects what is available, in contrast to that composing the commensals (see Ch. 14), so indicating that, in the main, the plants accept whatever is on offer. The way the prey is retained is the subject of the next chapter.

CHAPTER 6

Trapping Mechanisms

Introduction: Summary of Mechanisms Involved in Trapping

Four principle trapping strategies are employed by the carnivorous plants. Prey can: be glued to the trap; be caught in a snap trap; stumble into pitchers; be sucked into bladders. We have accordingly four principal groups of traps:

(i) adhesive traps (*Byblis*, *Drosophyllum*, *Drosera*, *Ibicella*, *Pinguicula* and *Triphyophyllum*);
(ii) snap traps (*Aldrovanda* and *Dionaea*);
(iii) pitchers (*Brocchinia*, *Catopsis*, *Cephalotus*, *Darlingtonia*, *Heliamphora*, *Nepenthes* and *Sarracenia*);
(iv) suction traps (*Utricularia* and, possibly, *Genlisea*).

All these mechanisms not only help to catch animals but also play a part in leading them to the site of digestion. The digestive strategies (see Ch. 8) are not always correlated with the trapping strategies, and different combinations and/or mercenary armies of commensal dipteran larvae are employed.

Though it is tempting, and of some advantage, to discuss each species separately, we shall, however, consider the trapping mechanisms in their broad sense after presenting, below, the general features of the trapping organs of each of the above mentioned groups.

Plants belonging to one family are often grouped in different classes of traps. In this respect certain members of both the Lentibulariaceae and the Droseraceae have developed similar traps. *Pinguicula* resembles *Drosera* in both its adhesive nature and in its movable leaves; whereas *Utricularia* resembles *Aldrovanda* not only in habitat, but also in the structure of some of its glands. These features, together with others, are dealt with in this chapter. The evolutionary problems they present are dealt with in Chapters 17–19.

Adhesive Traps

Adhesive traps bear stalked glands, each carrying one large drop of a trapping mucilage. Small animals which alight on such traps stick to the drops at once and cannot move any further. Larger animals creep on to the trap surface. At first they touch only one or two drops of mucilage which adhere to their legs but does not prevent further movement. As they creep along the trap they touch an increasing number of mucilage drops which accumulate on their bodies until they are heavily loaded with a mucilage that finally covers their legs and prevents them moving further. At first the lower side of the insect body touches the mucilage drops that are secreted on the top of stalked glands; then gradually, the insect's body is covered all over with mucilage, either because the insect falls from the top of hairs to the space on the leaf surface itself between the hairs, as in *Drosophyllum*, or because neighbouring hairs bend over and touch it from all directions, as in *Drosera*.

The advantage of raised glands lies not only in their ability to cover the upper side of the prey with mucilage, but also in the way they prevent the mucilage from being spread over the whole leaf surface. Adhesive traps normally employ restricted trap areas as temporary digestive pools (see page 169). If mucilage were spread over the whole leaf surface, a limited digestive pool around a single small prey could never be employed. In these traps energy is consumed in the secretion of mucilage and, in certain species, also in trapping movement.

Snap Traps

Snap traps are those active traps which rapidly shut when triggered by a visiting animal. The terrestrial *Dionaea* and the aquatic *Aldrovanda* adopted this trapping strategy and developed two important systems: sensory organs which translate touch stimuli into electric signals; and motile zones which are activated by the electric signals and respond in a rapid closure movement. In these traps, energy is consumed in the trapping movement.

Pitchers

All pitchers are passive traps, relying entirely on gravity to serve as the driving force leading prey into a digestive cavity. The digestive cavity is therefore always located in a position directly below an attracting zone, where small animals are tempted to stay by various means (see Ch. 5). Energy is therefore consumed in these traps mainly in producing attractions such as nectar. Several mechanisms have developed which lead prey to lose its foothold and prevent its escape. These include:

(i) smooth surfaces (*Sarracenia*) or unassailable waxy surfaces (*Brocchinia*, *Darlingtonia* and *Nepenthes*);
(ii) various geometrical structures which lead walking insects to horizontal positions where they easily stumble (the peristome of *Cephalotus*, *Sarracenia* and *Nepenthes*);
(iii) longish vertical tubes with narrow necks to force flying insects downwards;
(iv) covers of different shapes and structure which prevent the escape of those flying insects which have a strong vertical component in their flight, sometimes combined with transparent fenestrations (see Ch. 5).

Generally, pitchers produce nectar which not only attracts and serves as a reward, but also retains insects at the sites of trapping (see Chs 5 and 16).

There are two main types of pitchers: those adapted to trap *creeping creatures*, and those adapted to trap *flying insects*; some species, e.g. some *Nepenthes*, overlap in their function. The former normally lie on the ground or have developed walking routes for creeping insects. These traps are commonly quite short and often hairy. The traps that are adapted to catch flying insects are taller, much longer and normally associated with sophisticated attraction means. In *Sarracenia* certain species are adapted to creeping insects (e.g. *S. psittacina*), others to flying insects (e.g. *S. flava*). In *Nepenthes*, two types of traps are commonly found on the same plant: at first the plant produces its lower pitchers which trap mainly creeping insects; later on it develops upper traps which capture mainly flying insects (see Ch. 3).

Suction Traps

A closed cavity with negative hydrostatic pressure can trap small animals when its door is triggered to open. A quantity of water is sucked in, engulfing any small animals or soil particles which lie in its path. In this system, energy is consumed at the preparatory phase, when the traps are set (see page 118), not at the trapping movement itself. This strategy is employed by *Utricularia*. It is not clear whether *Genlisea* also employs a similar strategy. Clearly there is no door in *Genlisea*; however, the presence of soil particles in the cavity of its trap might indicate that the prey does not necessarily always force itself into the trap, but is sucked in, perhaps by water currents.

Mucilage-Secreting Glands

All glands that secrete trapping mucilage are raised on stalks. The stalk can be unicellular as in *Pinguicula* (Fig. 6.1*A*) and *Byblis* (Fig. 4.20), or multicellular as in *Drosophyllum* (Fig. 6.2), *Drosera* (Fig. 6.1*B*), *Ibicella* (Fig. 6.3) and *Triphyophyllum* (Fig. 4.18). The multicellular stalk is commonly provided with xylem (Fig. 6.4) (Lloyd, 1942; Green *et al.*, 1979; Joel, 1986) and according to Green *et al.* (1979) it also includes phloem in *Triphyophyllum* and in *Drosophyllum*.

The structure of the gland on top of the stalk closely resembles that of the neighbouring sessile digestive glands of each respective species: both are composed of external glandular layers and intermediate endodermoid cells. The general features of these glands will therefore be described in Chapter 8 with the description of other digestive glands.

The Composition of the Mucilage

The chemistry of the mucilage of *Drosera capensis* and related species

A typical trapping mucilage (*Fangschleim*) has been described for *Drosera capensis* (Rost and Schauer, 1977). The 4% aqueous solution of acidic polysacchar-

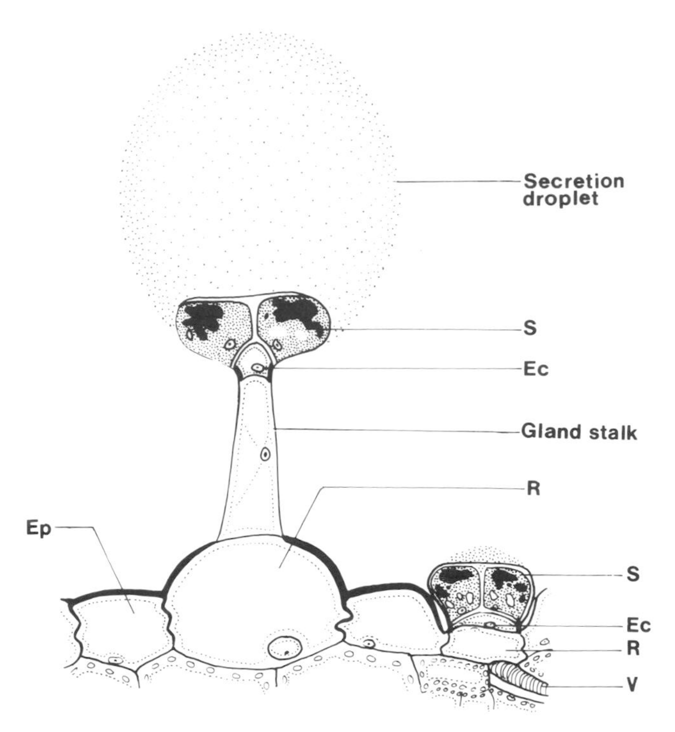

Fig. 6.1(*A*) The stalked and sessile glands of *Pinguicula vulgaris*; cf. Fig. 4.2. The adjoining epidermal cells (Ep) have thicker cuticles than those of the secretory cells of the gland heads. S, secretory cells; Ec, endodermoid cell; R, reservoir cell; V, vessel communicating with general vascular system of the leaf (Redrawn with permission from Heslop-Harrison and Knox, 1971).

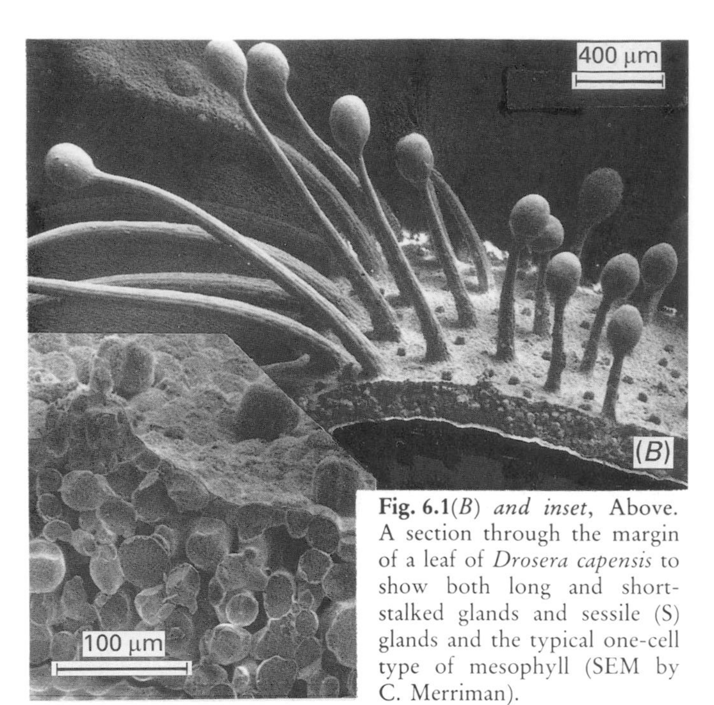

Fig. 6.1(*B*) *and inset*, Above. A section through the margin of a leaf of *Drosera capensis* to show both long and short-stalked glands and sessile (S) glands and the typical one-cell type of mesophyll (SEM by C. Merriman).

Fig. 6.2, Below. The unfolding 'crozier' tip of a young leaf of *Drosophyllum lusitanicum*, showing the mucilage, (M) beginning to emerge on the surface of the glands, and the marked furrow (F) on the adaxial surface (SEM by G. Wakley, Oxford).

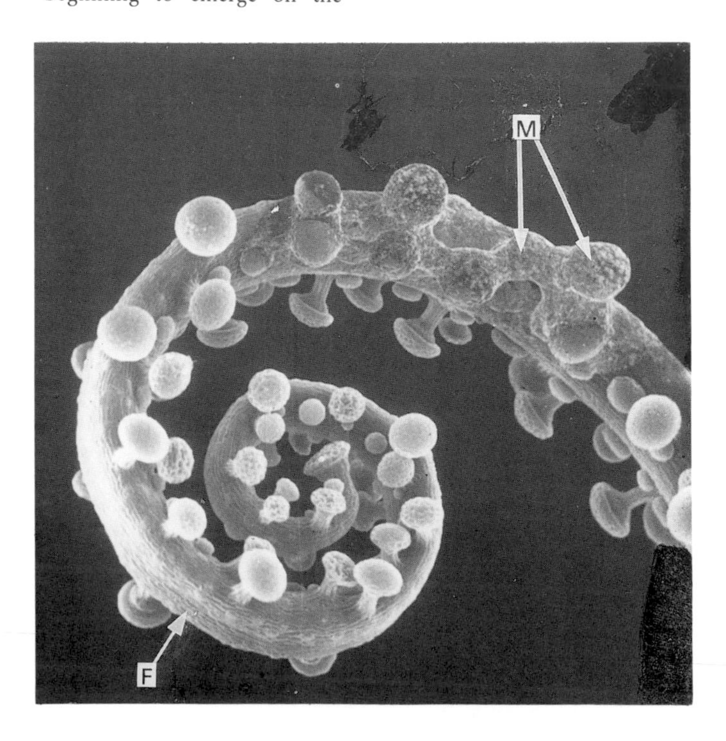

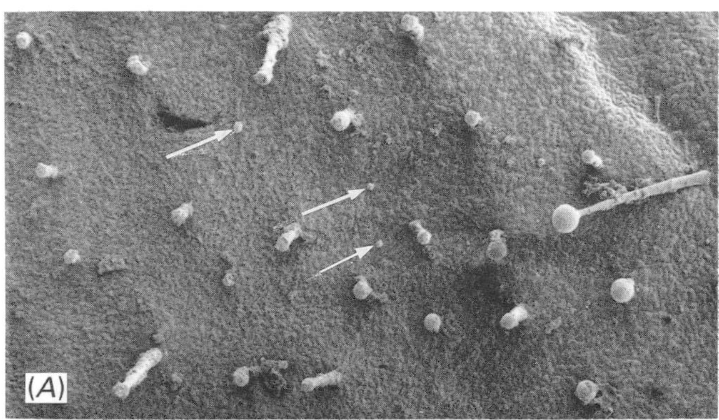

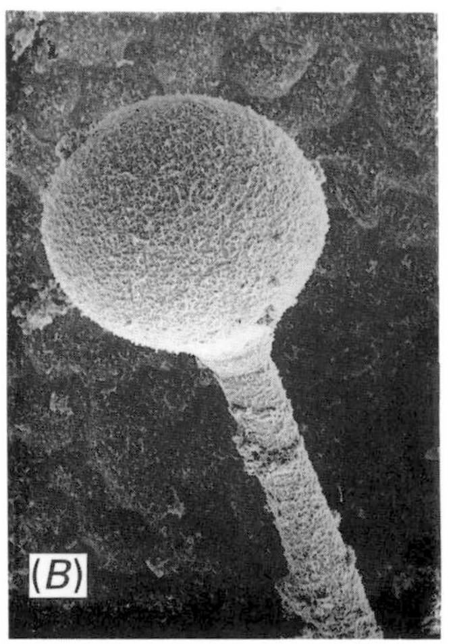

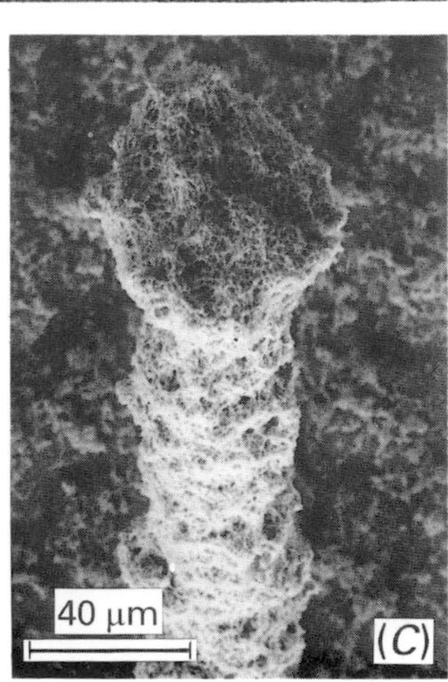

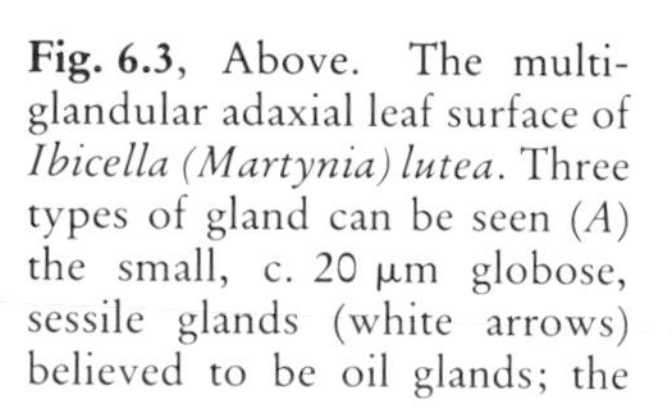

Fig. 6.3, Above. The multiglandular adaxial leaf surface of *Ibicella (Martynia) lutea*. Three types of gland can be seen (*A*) the small, c. 20 µm globose, sessile glands (white arrows) believed to be oil glands; the tall, (*B*) globose, stalked glands; (*C*) the richly mucilage coated, shorter-stalked tabular glands; (cf. Fig. 2.4 and see Mameli, 1916, her Fig. 14). (SEM by C. Merriman.)

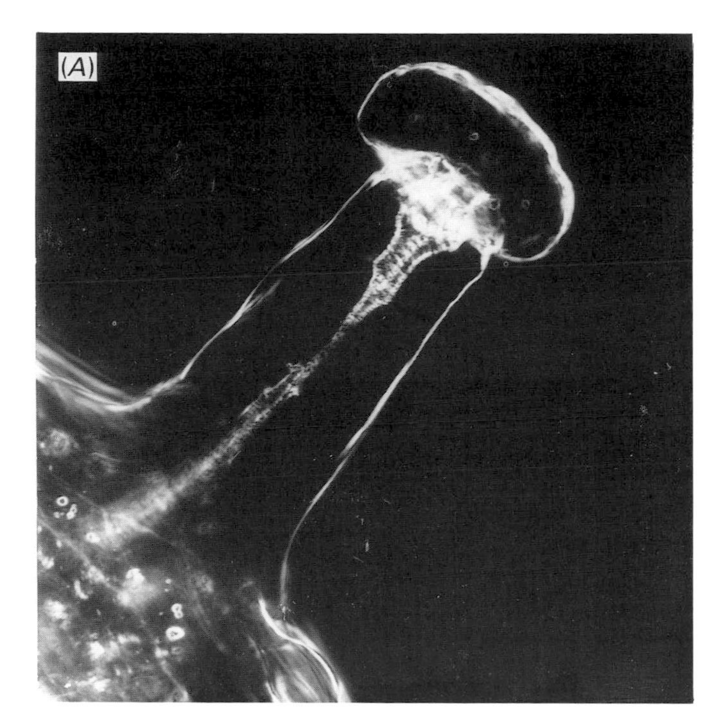

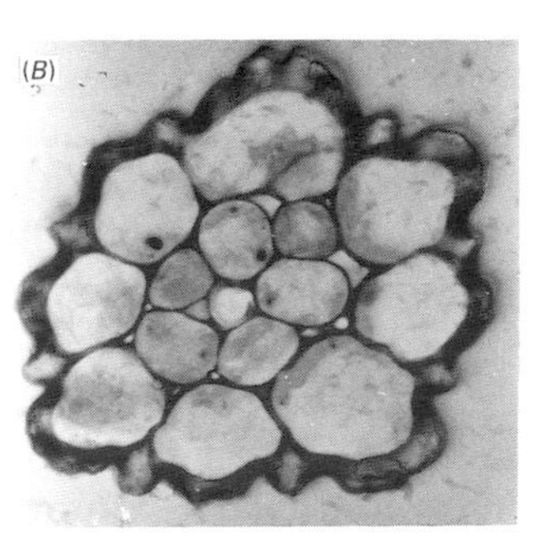

Fig. 6.4 The stalked gland of *Drosophyllum*. (*A*) Dark field micrograph showing the xylem in the centre of the stalk. (*B*) Cross-section of a stalk showing the thick cuticle, two layers of stalk cells and a central vessel element.

ides has a pH of 5. The freshly secreted mucilage is so viscous that it can be drawn out into threads several centimetres long. Maximum viscosity is observed at about pH 5 and it decreases irreversibly when the pH is either raised or lowered. A temperature rise from 0–8°C results also in an irreversible decrease in viscosity by about tenfold and at a nearly linear rate. Freezing, thawing and lyophilization also reduce the viscosity. The mucilage contains 22 mM Ca^{++}, 19 mM Mg^{++}, 0.9 mM K^{+} and 0.2 mM Na^{+}. Structured protein was absent from this secretion and there appears to be little significant difference between the product in *D. capensis* and the mucilage from related species (Rost and Schauer, 1977; Heinrich, 1984). The polysaccharide of the trapping mucilage of *Drosera* comprises L-arabinose, D-xylose, D-galactose, D-mannose and D-glucuronic acid. The molar ratio of these sugars in *D. capensis* is 3.6:1.0:4.9:8.4:8.2, respectively (Gowda *et al.*, 1983) and in *D. binata* 8.4:1.0:9.6:18.3:17.1 (Gowda *et al.*, 1982). Partial hydrolysis of the native and carboxyl-reduced polysaccharides of *D. capensis* gave various oligosaccharides that were characterized, and suggested a structure containing a D-glucurono D-mannan backbone (Fig. 6.5) having the disaccharide repeating unit of -β-D-Glc*p*A-(1→2)-α-D-Man*p*-(1→4)-. L-Arabinose and D-xylose are present as non-reducing furanosyl and pyranosyl end-groups, respectively, both attached α(1→3) to the D-glucuronic acid residues of the backbone. D-galactose is present as a non-reducing pyranosyl end-group linked α(1→3) to the D-mannose residues (Gowda *et al.*, 1982, 1983). The polysaccharide of *D. binata* is very similar to that of *D. capensis* the only difference being that the D-xylose content in this case is one-half that in *D. capensis* (Gowda *et al.*, 1982).

The mucilage of *Drosophyllum lusitanicum*

The mucilage from the related *D. lusitanicum* (Fig. 6.6A) does not seem to be significantly different, except that it is slightly more acidic with a pH in the range of 2.5–3.0 (Heslop-Harrison, 1976). The dry weight is 0.23% of which 0.19% is polysaccharide, and the remainder mainly ascorbate. As in *Drosera*, no protein is present and the sugars are xylose, gluconic acid, galactose, arabinose and traces

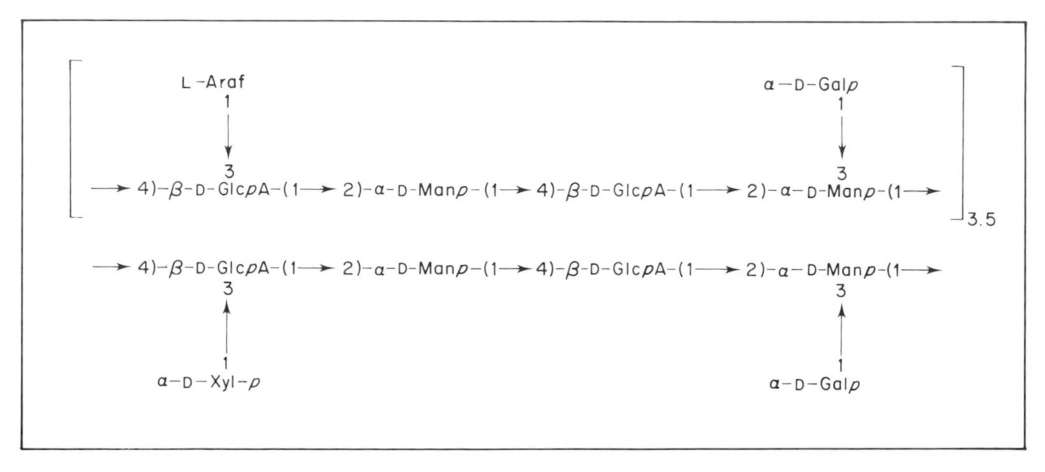

Fig. 6.5 Mucilage secreted by *Drosera capensis* (Reproduced by permission from Gowda *et al.* 1983.)

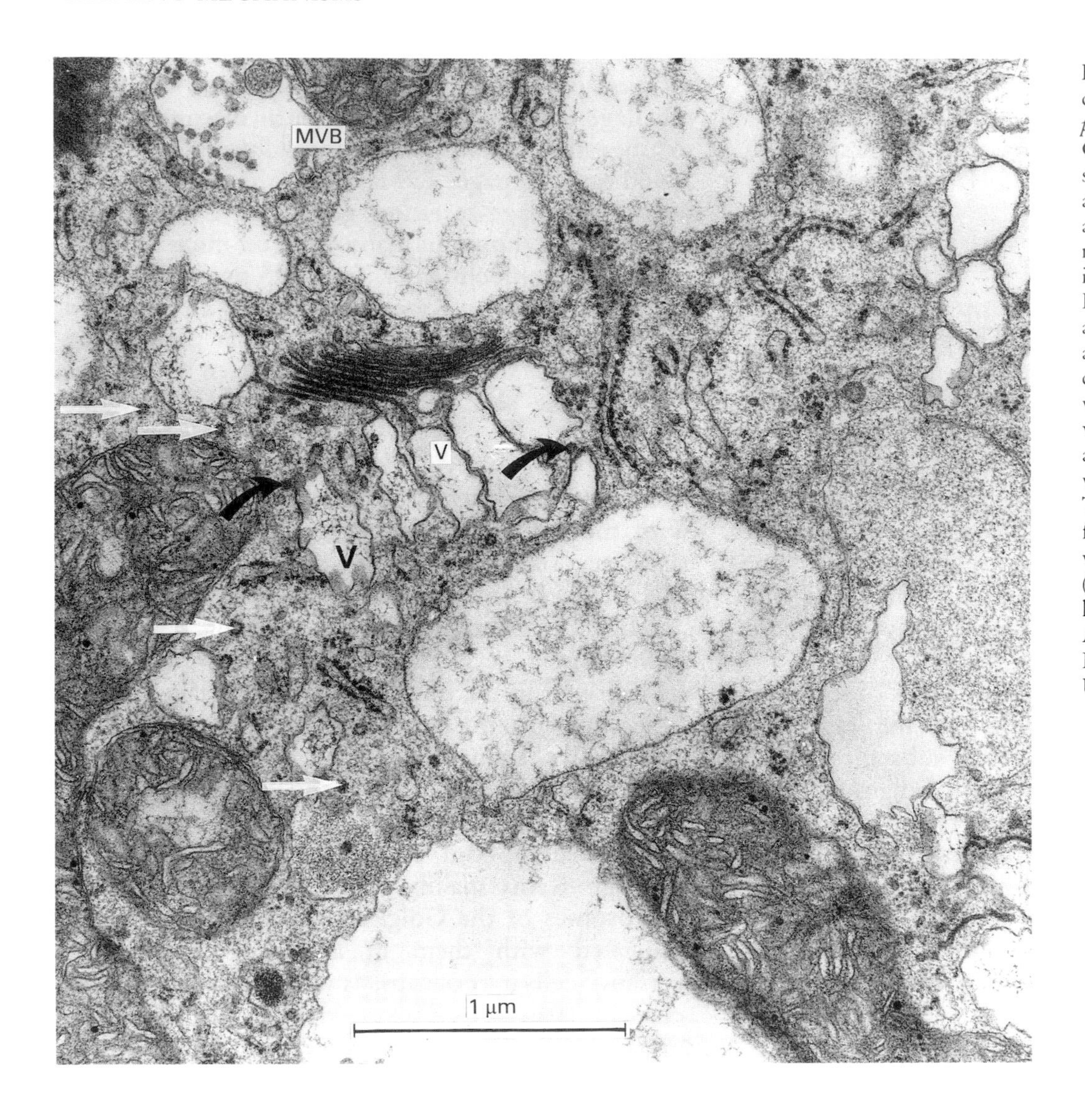

Fig. 6.8 An outer secretory cell of a stalked gland of *Drosophyllum* 2 h after stimulation. Compared to the unstimulated state (cf. Fig. 8.12) there is a significant reduction in the amount of tubular, non-ribosomal ER. There is an increase in the cisternal, ribosome-rich ER. The dictyosomes now appear to be producing fibrillar-rich vesicles (V) and the cytoplasm is rich in coated vesicles (white arrows) some of which, (black arrows) may be arising from the fibrillar-rich vesicles of the dictyosome. There are translucent as well as flocculent vacuoles and multivesicular bodies (MVB). (Unpublished micrograph by kind permission of Drs A.E. Vassilyev and Lyudmila E. Muravnik, Leningrad, USSR.)

glands, yet another indication of the muciformic nature of the dictyosomes in the stalked glands.

A final phase of the production of mucilage is the migration of the secreted substance across the cell wall and its exudation at the surface of the gland (Dexheimer, 1976).

After maturation, each gland has a full complement of external mucilage. The gland itself appears quiescent and is not further active in mucilage-secretion and only a few vesicles can be seen in its glandular cells (Outenreath and Dauwalder, 1982).

Schnepf, who was one of the first to study the ultrastructural aspects of mucilage secretion in traps of carnivorous plants, showed that dictyosomes are involved in the production of mucilage not only in *Drosera*, but also in *Pinguicula* and *Drosophyllum* (Figs 6.8 and 8.12).

Schnepf (1961b) has calculated the volumes of Golgi-complex-derived vesicles in this system and found that a dictyosome can produce about 3 vesicles per minute, each containing c. 0.02 μm^3 Each vesicle seems to exist for only about 2.5 min.

The production of mucilage is energy-dependent and various metabolic inhibitors affect the different stages of its secretion (Schnepf, 1963b).

A consistent development in the morphology of the cell walls occurs during mucilage secretion. Wall ingrowths develop in both the inner and the outer glandular cells. In the inner cells, the wall material which constitutes the in-growths appears darker, while in the outer cells it appears loosely fibrillar. Dexheimer (1976) suggests that the wall protuberances represent the remains of mucilage vesicles (see Ch. 8).

With autoradiography, using D-[1,3-^{3}H] galactose and labelling periods of 5–85 min Outenreath and Dauwalder (1982) found that the heaviest incorporation of radioactive tracers was associated with the plasma-membrane and with the outer periclinal cell walls of the outer cells. No specific labelling of any particular organelles was seen. However, the examination of secreted mucilage, in their experiments, generally showed only negligible labelling. This absence might indicate that the synthesis of mucilage does not permit a rapid incorporation of external D-galactose, which is assumed to represent a non-reducing pyranosyl end-group liked to $\alpha(1\rightarrow3)$ to D-mannose residues in the repeating unit of the mucilage polymers (see page 89). It is very likely that the synthesis of the repeating units of trapping mucilages, in the secretory cell, does not take place during the phase of secretion so far studied and, as Schnepf (1961a) puts it, we do not know whether the Golgi apparatus condenses or synthesizes mucilage. Biochemical studies aimed at elucidating the biosynthetic pathway of trapping mucilages are needed prior to further radio-labelling studies of the sites of mucilage manufacture, packing and release.

Schnepf (1972) has shown that the secretion of mucilage in *Drosophyllum* can be substantially inhibited by cycloheximide, indicating a role of protein synthesis, although puromycin was ineffective. While this may reflect different levels of uptake, further work is required to clarify this contradictory result.

Secretion of mucilages in *Utricularia*

The mucilages that are secreted by various glands in the *Utricularia* trap do not function as trapping mucilages (*Fangschleims*). The composition of these mucilages is not known. Cohn (1875) suggested that some of these glands secrete an attractive mucilage. Fineran and Lee (1980) found no evidence in *U. monanthos* that the vestibule glands secrete mucilage. This absence might be true for other species as well, although most species of *Utricularia* do seem to secrete some mucilage. Two main sites of mucilage release are known in the *Utricularia* traps namely the threshold epithelium; and the stalked glands which surround the trap door from the outside and are often also located on the outer side of the door itself (Fig. 4.15C).

The secretion of mucilage by these glandular structures was extensively studied by Vintéjoux (Broussaud and Vintéjoux, 1982; Vintéjoux, 1973a, b; Vintéjoux, 1974; Vintéjoux, 1976), who could clearly demonstrate that the dictyosomes and their vesicles are the main sites of mucilage production and transport in the secretory cells. In this respect, the secretion of mucilage in the stalk glands of *Utricularia* resembles the secretion of trapping mucilage in other carnivorous plants (see page 87 onwards).

Mechanisms of Movement

Introduction

A most attractive feature of carnivorous plants and one that has served as the origin of many legends, is the ability of many traps to move in response to external touch stimuli. *Dionaea*, which Darwin called 'the most wonderful plant in the world' (see Jones, 1923) is well known for its rapid closure movement. The most rapid movement is that of the trapdoor of *Utricularia*: high-speed cinematography has not yet determined the precise speed. Closure movements are also found in *Aldrovanda*, *Dionaea*, all the species of *Drosera* (with varying speeds) and, to a lesser extent, in the leaf margins of *Pinguicula* (Fig. 6.7).

Do these movements have underlying common principles? The available information is still too fragmentary to permit generalizations to be made and we are constrained to discussing separately what is known of the different mechanisms.

Movement in the Traps of the Droseraceae

The traps of the various Droseraceae, though so different from each other when viewed superficially, have many anatomical and physiological features in common (see Fig. 1.3 and Fig. 19.1). Variations in the composition of the leaf glands can be correlated with the development of the various trapping mechanisms (Williams, 1976), and will be discussed in Chapter 18.

In the perception of mechanical stimuli by traps, it has been shown that the moving Droseraceae, but not the primitive *Drosophyllum*, are capable of producing electrical signals by which they transmit the stimulus of the presence of prey to special motor cells that are responsible for the trap movement. One might expect the trigger hairs of the rapidly responding *Utricularia* to produce similar signals. However, there is to date no evidence of any such electrical mechanisms in this plant (see page 124).

When an insect lands on the leaf of *Dionaea* or *Aldrovanda*, it stimulates, mechanically, a rapid closure movement of the trap. Similarly, if it lands on a *Drosera* leaf, it stimulates a rapid movement of leaf tentacles. This rapid movement, in *Drosera* as in *Dionaea* and *Aldrovanda*, is followed by slower movements if, in addition to the mechanical stimuli, chemical stimuli are also applied to the leaf (Darwin, 1875).

The movements in *Drosera* spp.

When an insect alights on a *Drosera* leaf, it cannot avoid touching the trapping mucilage which covers the top of each tentacle (Fig. 6.9). In trying to remove the sticky substance from its legs and in an effort to escape the mucilaginous trap, the insect struggles toward the leaf edge, pulling strands of the secretion not only from the tentacles upon which it has landed, but from more and more tentacles (Fig. 6.9). Eventually, it smears much of the sticky polysaccharide over its body. As the tentacles are pulled back and forth by the insect, they are individually stimulated and begin to bend towards the centre of the leaf (see Fig. 2.2). The tentacles which carry the insect are the first to inflect, rapidly bearing it from the leaf margin to the gap between those unbent tentacles that obstruct its way to escape. Then all the neighbouring tentacles slowly bend, commonly to the centre of the leaf where the insect is held. However, should the insect land away from the centre, the tentacles are, nevertheless, accurately directed to the correct focus, as is shown in Figure 26.3 of Kerner and Oliver (1895). This slow tentacle movement is associated with a slow inflection of the whole surface of the leaf itself.

A few hours after the insect has landed on a *Drosera* leaf it lies in the centre of an incurved leaf between many tentacles which totally surround it. Most insects die from suffocation, their tracheae occluded with the mucilage.

The first movement of the tentacles, which is known to take place only in those tentacles that are directly touched by prey (cf. *Dionaea* and *Aldrovanda*) is regarded as the *rapid movement*. The other ensuing movements are designated the *slow movements* (Williams, 1976).

Darwin (1875) showed that the rapid response of each single tentacle can be triggered artificially by a purely mechanical stimulus delivered to each tentacle head. This rapid response starts shortly after the onset of a mechanical stimulation (10 s to several minutes) and requires several minutes, depending on the species (commonly 3–20 min) to be completed. The time required by the plant to complete the rapid movement depends upon the intensity of stimulation, on the plant's condition and on such external factors as temperature (Darwin, 1875; Williams and Pickard, 1972a, b, 1979). Since there is a predetermined direction (toward the leaf axis) this rapid movement of *Drosera* tentacles can be regarded as nastic (Behre, 1928).

The studies of *Drosera* by Darwin (1875), elucidating the principal modes of the behaviour of tentacles, and the studies of Burdon-Sanderson (1873, 1911) on *Dionaea* (see page 98 onwards) provide us with the basic knowledge of the rapid movements in the Droseraceae, where similar principles apply not

Fig. 6.9 A nematocerous fly trapped on the surface of *Drosera capensis*. Note the marked dimorphism of the peripheral, movable long-stalked glands and the shorter, almost immobile, central glands. Some of the peripheral glands (arrows) are beginning to respond to the presence of the insect. (SEM by D. Kerr.)

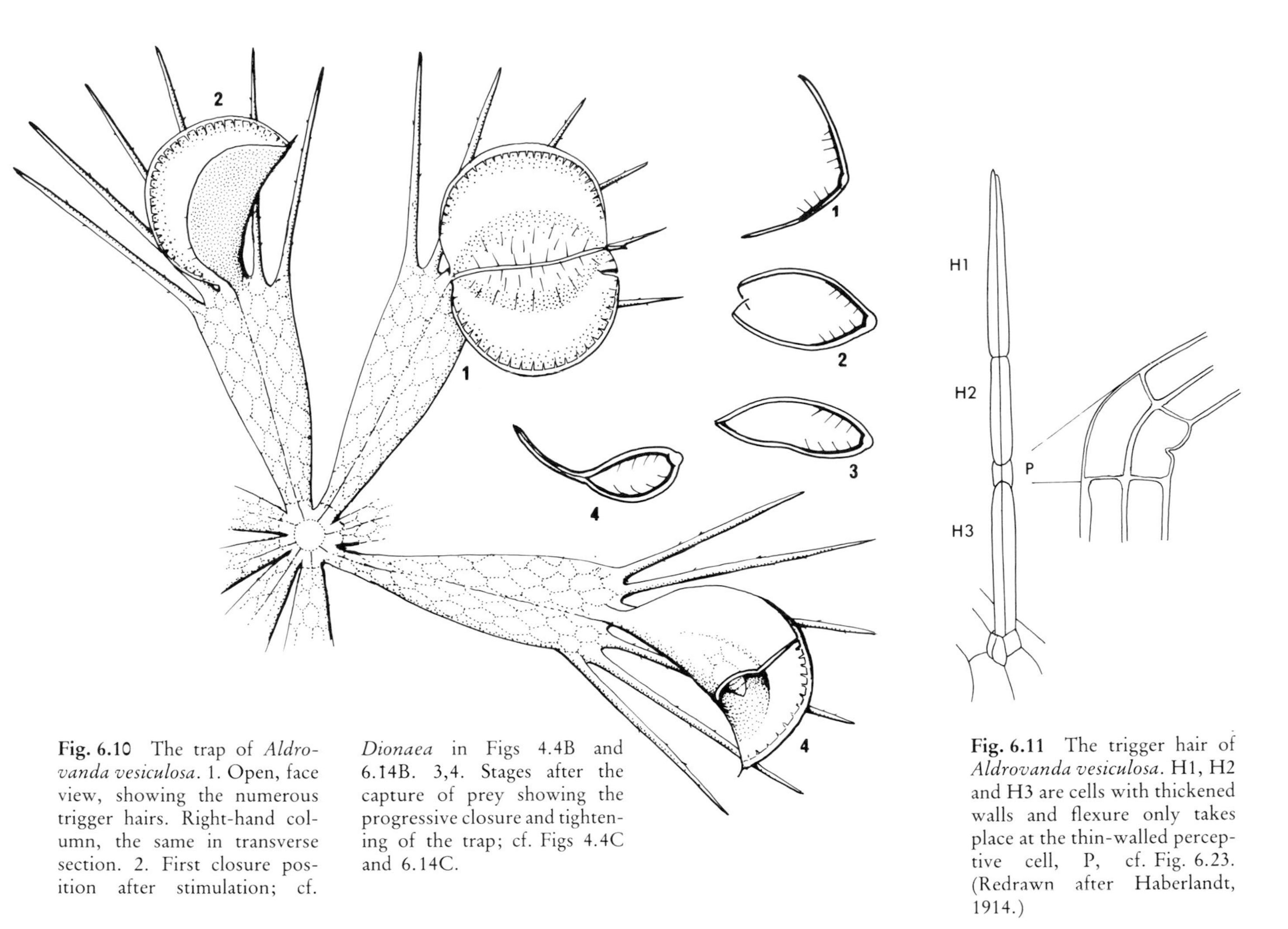

Fig. 6.10 The trap of *Aldrovanda vesiculosa*. 1. Open, face view, showing the numerous trigger hairs. Right-hand column, the same in transverse section. 2. First closure position after stimulation; cf. *Dionaea* in Figs 4.4B and 6.14B. 3,4. Stages after the capture of prey showing the progressive closure and tightening of the trap; cf. Figs 4.4C and 6.14C.

Fig. 6.11 The trigger hair of *Aldrovanda vesiculosa*. H1, H2 and H3 are cells with thickened walls and flexure only takes place at the thin-walled perceptive cell, P, cf. Fig. 6.23. (Redrawn after Haberlandt, 1914.)

only to *Drosera* or *Dionaea*, but also to *Aldrovanda*.

The action of the *Aldrovanda* trap

The fully mature trap is considered to be in:

(i) *The open phase*. The two leaf lobes stand apart (Fig. 6.10). When a trigger hair (Fig. 6.11) is touched the trap shuts. A single mechanical stimulus delivered to any of its mechanosensory trigger hairs will cause trap closure in a young healthy trap (cf. *Dionaea*). In somewhat older traps, two stimuli are frequently required to effect closure, but this response varies in different traps: some old traps might respond immediately to the first touch, while others will only respond to a sequence of many stimuli or will not respond at all.

The motile zone lies at the detention region, the bending movement of the trap being especially pronounced in a narrow portion of this region, nearer the midrib region.

(ii) *The closed phase*. As the result of the application of the first stimulus the trap shuts and the lobes come into contact only at their rims with the minute teeth inter-crossed (Fig. 6.10.2). In some cases the shutting movement goes on further until the lobes press more strongly upon each other, but even then they contact each other only at the rim. The midrib, which is nearly straight in the open phase, bends upwards when the trap shuts (Fig. 6.10.4).

At the closed phase, captured animals are contained but free to move within the cavity which is formed by the two trap lobes; then gradually the space narrows and digestive activity takes place.

(iii) *The narrowed phase*. It takes several minutes for the trap to narrow. This process is based on further bending of the lobes at the motile zone, which brings about the formation of a very small cavity, a narrow sickle-shaped space in the central zone.

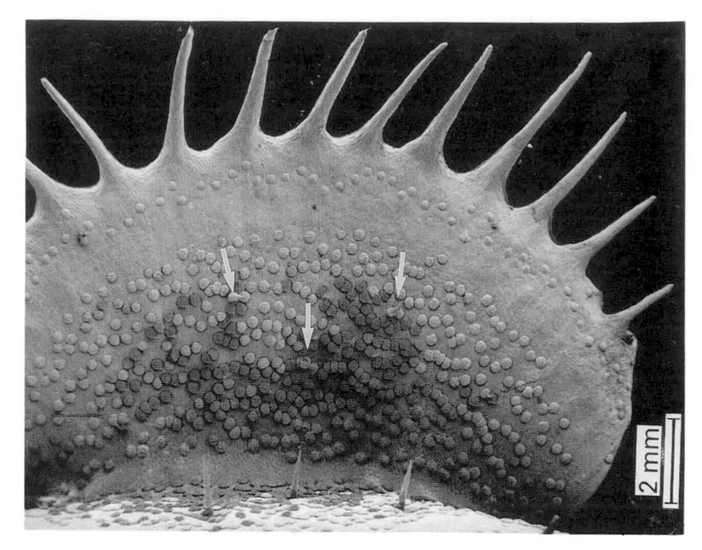

Fig. 6.12 The inner faces of the lobes of *Dionaea muscipula* showing the six trigger hairs, the three on the facing lobe are arrowed, the digestive glands surrounding them, and the nectaries around the rim (SEM).

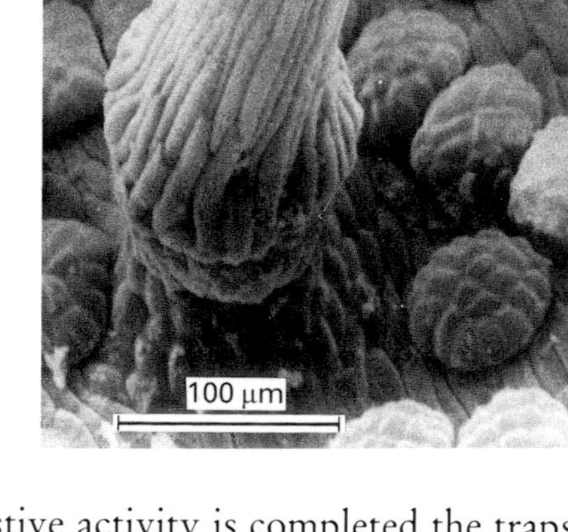

Fig. 6.13 Trigger hair of *Dionaea*. Sessile glands at the base. (SEM by G. Wakley.)

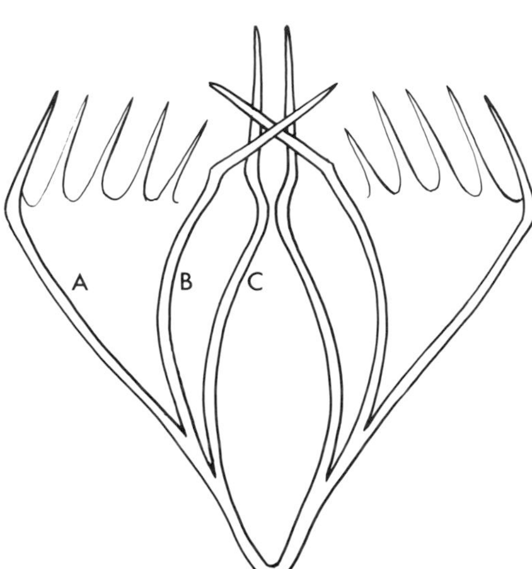

Fig. 6.14 The leaf-lobe positions of *Dionaea muscipula*: A, open; B, mechanically stimulated; C, chemically stimulated.

The marginal zone of each lobe is everted and applied closely to the corresponding region of the opposite lobe, thus preventing leakage of digestive fluid (cf. *Dionaea*) from the trap and to prevent penetration of external aqueous medium into the active trap. The narrowed state of the trap is maintained by the sharp bending of the lobes at the motile zone.

The quadrifids on the peripheral zone may remove water from the space between the very thin lobe margins in a manner which resembles the removal of water from the *Utricularia* trap (see page 119). This pumping would form a negative hydrostatic pressure at this zone which would enable the trap to hold the very thin periphery of its two lobes tightly together. The pumping would thus temporarily seal the trap when it is active in digestion (Joel, 1986).

When the digestive activity is completed the traps reopen. First it swells, again rebulges, and only then returns to the open phase.

The action of the trap of *Dionaea*

When the trigger hairs (Figs 6.12 and 6.13) are touched sufficiently to generate a stimulus (see page 98) the *open* trap rapidly shuts (Williams and Bennett, 1983). In this *closed phase* the marginal teeth interlock but still leave a certain space between (Figs 4.4*B* and 6.14*B*). The movement leading to the closed phase is regarded as the *rapid* movement. There are 14–21 teeth (Figs 6.12 and 6.14) along each margin. As the trap closes into the first position most of these interdigitate, but the accuracy is not total, nor on the second or third stimulation does the pattern precisely repeat itself. Only one trap in five repeats, completely accurately, the tooth pattern of its previous closure (Dossett, S., pers. comm., 1987).

Further stimulation of the trap (mainly chemical stimuli exerted by the prey [Darwin, 1875]) induces further closure (Figs 4.4*C* and 6.14*C*) which leads to a *narrowing* of the gap between the margins of the two trap lobes until the peripheral bands meet all the way along their margins forcibly sealing the digestive cavity thus formed (Fig. 4.4*D*).

This third state is the beginning of the digestive phase. Traps in the 'narrowed' phase not only have sealed cavities but also show parallel marginal teeth (versus interlocking teeth in the closed phase; see Fig. 4.4*D* and 6.14). In addition, the two concave trap lobes flatten, exerting pressure on the inside. This pressure was found to show a diurnal rhythm (Williams, R., pers. comm. 1982) which begins on the second day after stimulation, peaking at or just after midday (Fig. 6.15). This rhythm is dramatically shown in time-lapse cinematography.

At the same time mucilage is secreted by the small peripheral glands, and appears to form a 'sealing gasket' which prevents the leakage of digestive fluid from the trap when it is active in digestion (Joel, 1986). Narrowing is further discussed on page 105.

When the digestion of prey is completed, the trap reopens. This movement is very slow because it is based on real growth (Ashida, 1934). The reopening movement can be divided into a faster opening stage and a very slow 'rebulging' growth. A reopened trap can easily be identified from the upright position of its marginal teeth. It also differs from a virgin trap by the open cuticle covering its digestive glands (see page 164).

When large prey is captured, the trap may remain in its closed position for several weeks and then die without reopening. With small prey the trap can perform several digestive cycles during its lifetime.

The Perception of External Stimuli

What is the mechanism which enables a *Drosera* tentacle to respond to touch stimuli? And what is the mechanism which enables a *Dionaea* trap to snap shut in response to touching its trigger hairs? Darwin (1875) showed that the tentacle responds only to the stimulation of its glandular head, but the movement itself is carried out by bending of the basal part of the tentacle. There is, therefore, some distance between the touch receptor, or site of perception, and the site of response. Darwin (1875) suggested that an *impulse* mediates between these two sites and was convinced that this impulse must travel quite rapidly to enable the rapid response

Fig. 6.15 The diurnal rhythm of pressure changes between the trap lobes of *Dionaea muscipula* after mechanical stimulation and feeding. Note that the diurnal rhythm did not commence until about 36 h had passed, but continued for at least 8–9 days (Williams, R.A. pers. obs., 1983). The observed rhythm may, Williams suggests, be due to nyctinastic changes at the junction of the lobes with the petiole or changes in the turgor pressure of the cells of the abaxial faces of the lobes. The effect, in either case, is to redistribute the trapped prey and bathing enzyme secretion.

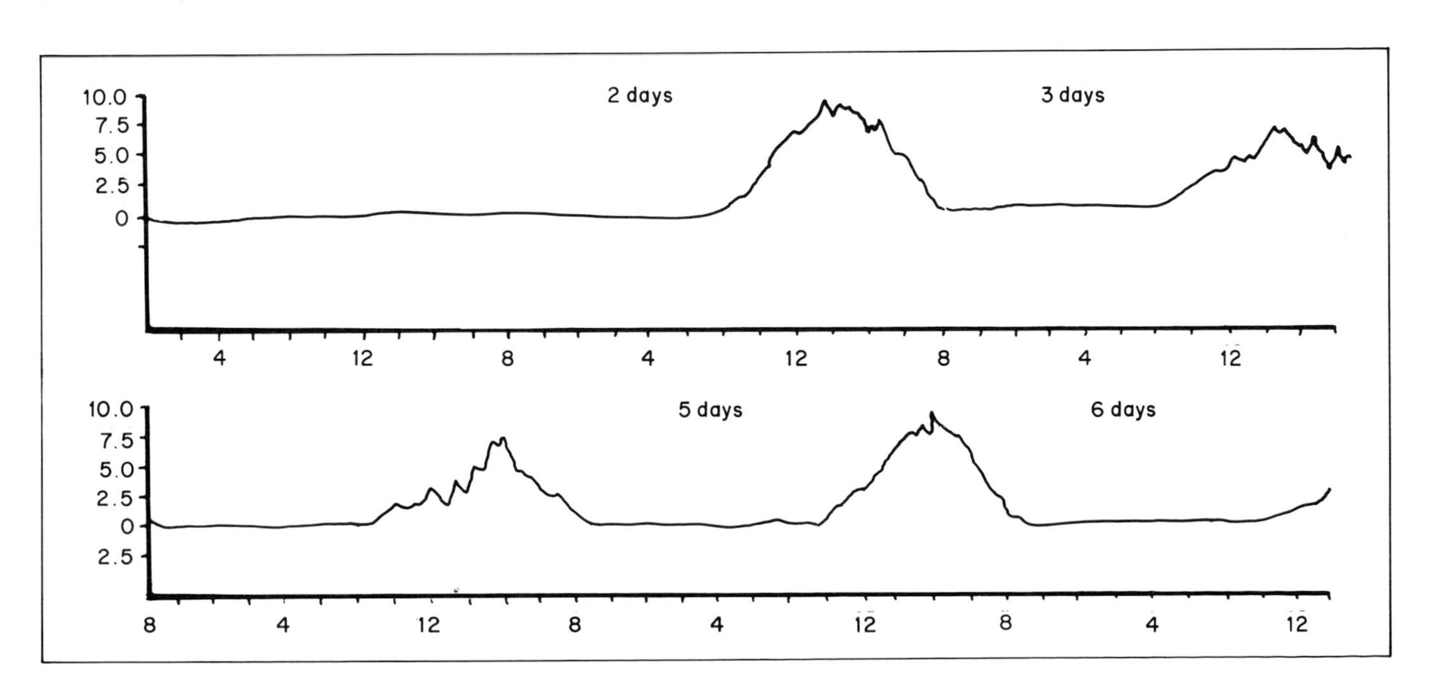

observed. In response to his request, Sir John Burdon Sanderson, a leading British physiologist of his time, examined the electrical activity of *Dionaea* traps and discovered that this impulse was electrical and resembled the principal signal of animal nerve and muscle (Burdon Sanderson, 1873, A full bibliography of these important papers can be found in Williams, 1973b).

Receptor potential and action potentials

The electrical system mediating between the touch receptor and the site of movement response has been studied in *Drosera* by Williams and Spanswick (1972) and Williams and Pickard (1972a). Once the head of a *Drosera* tentacle is touched, the surface voltage measured at the tentacle head typically drops (Fig. 6.16). This drop in potential is termed the *receptor potential.* The amplitude of the receptor potential is roughly proportional to the intensity of the stimulus, i.e. the receptor potential is a graded potential change. Accordingly, a light touch produces a small, brief receptor potential, a steady pressure delivered to the tentacle head produces a higher and longer negative receptor potential and a vigorous stroking yields a still more negative receptor potential of higher amplitude and of considerable duration.

As soon as the receptor potential attains a certain *threshold*, a series of short electrical pulses can be detected (Fig. 6.16), each rapidly travelling along the tentacle towards its base. These electrical pulses, which appear to belong to the all-or-nothing type, are the *action potentials.* Action potentials recur until the receptor potential drops below the original threshold.

After a sufficient number of action potentials have passed to the motile region at the base of the tentacle, the tentacle bends (Figs 6.1*B* and 6.9). This bending activity lasts as long as action potentials are provided.

The outermost large peripheral tentacles of each leaf produce action potentials of the greatest amplitude and duration (Williams and Spanswick, 1972; Williams and Pickard, 1979). These powerful action potentials result in a stronger bending movement of the tentacle. The tentacles of the inner portion of the leaf need an extremely strong stimulus to raise their receptor potential high enough to elicit a few action potentials. This difference in the electrical properties of the tentacles correlates with the difference in their ability to bend: the innermost tentacles bend but to a significantly lesser or negligible degree. Most studies have therefore been performed on the large, outermost tentacles.

The induction of the receptor potential

Two external factors might induce the formation of a receptor potential: a mechanical stimulus, as already mentioned above, and a chemical one. Darwin noticed that when the tentacle head is loaded with an extremely slight weight, or is struck

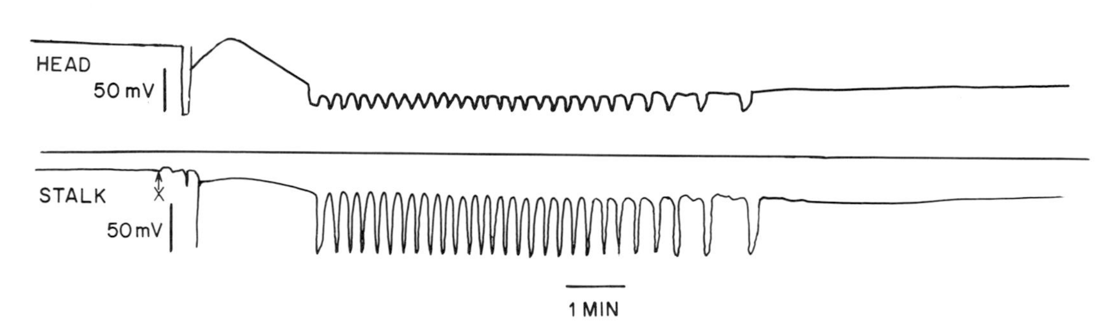

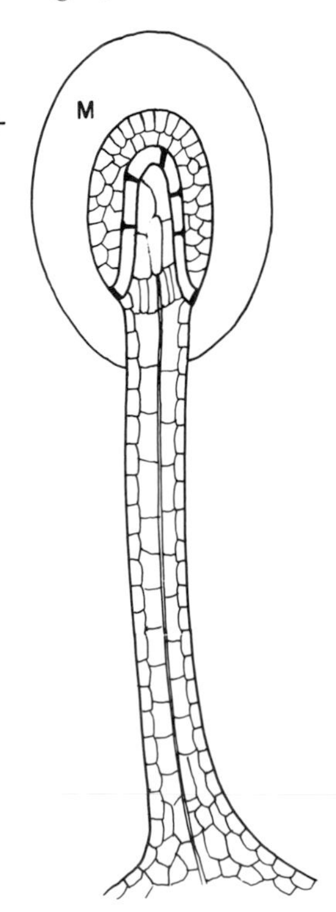

Fig. 6.16 Simultaneous recordings of a receptor potential and action potential from the head and upper stalk, and from the lower stalk of an outer tentacle of *Drosera rotundifolia.* The head region was initially stimulated by flexing it with the recording electrode. The stimulation continued until the tentacle bent and pulled away from the stimulating object. Electrical contact was maintained throughout the observation through the mucilage surrounding the receptor (M). The electrode was placed on the head of the tentacle at time 'X', and the preamplifier connected to the electrode recording from the head was turned on at the point indicated by the arrow (↑) on the trace. Redrawn with permission from Williams and Pickard (1972 a, b, 1974) and Williams and Spanswick (1976).

several times with a fine camel hair brush, the tentacle bends (Darwin, 1875, and letter to J.D. Hooker, 1862: in 'Letters', 1887). In addition, he found that various chemicals can cause the same effect. Balotin and Di Palma (1962) suggested that salt solutions induce osmotic changes which lead to the formation of action potentials in *Dionaea*. However, this possibility can be dismissed since osmotic agents such as sugar were later proved to be ineffective (Williams and Pickard, 1972a). Various ions are known to be able to induce changes in the potential which lead to the creation of action potentials. Insects do not seem to stimulate the movement of *Drosera* traps in this way since the ionic concentrations released by them when trapped are too low to activate the membranes. The influence of ions on the production of a receptor potential may nevertheless be indicative of the possible ionic nature of this process.

The Site of Mechanoreception

The site where receptor potentials are initiated in the tentacle or trigger hair (see page 100) is believed also to be the site of *mechanoreception*. It seems likely that this site is located in the region most easily deflected (Haberlandt, 1906: Figs 6.13 and 6.17).

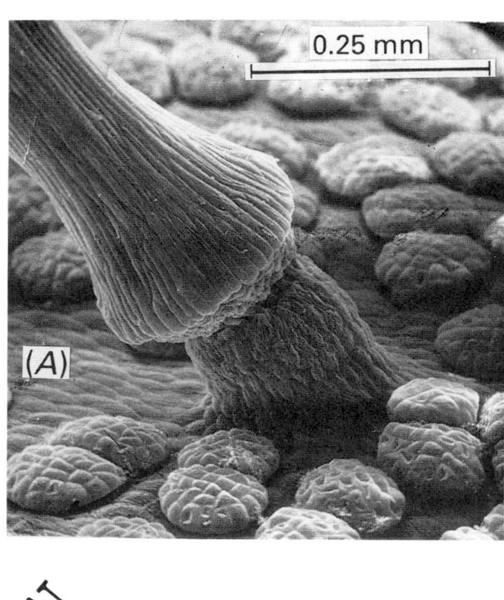

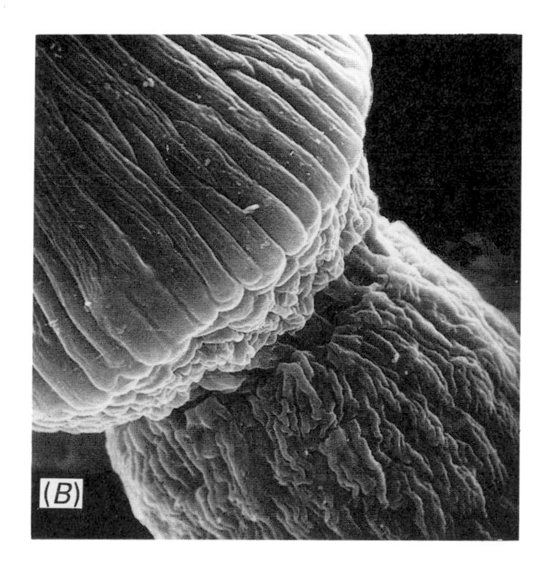

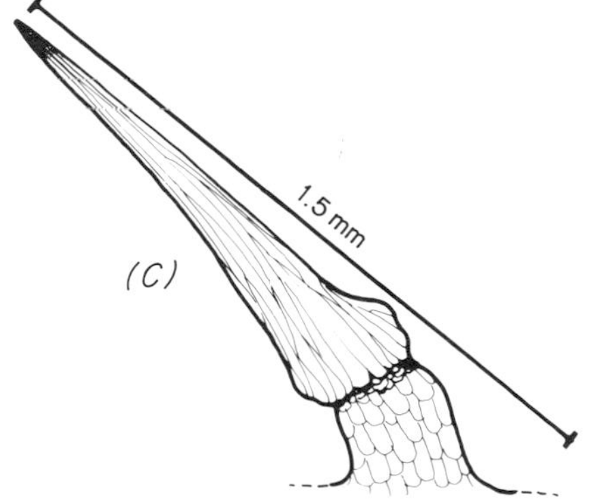

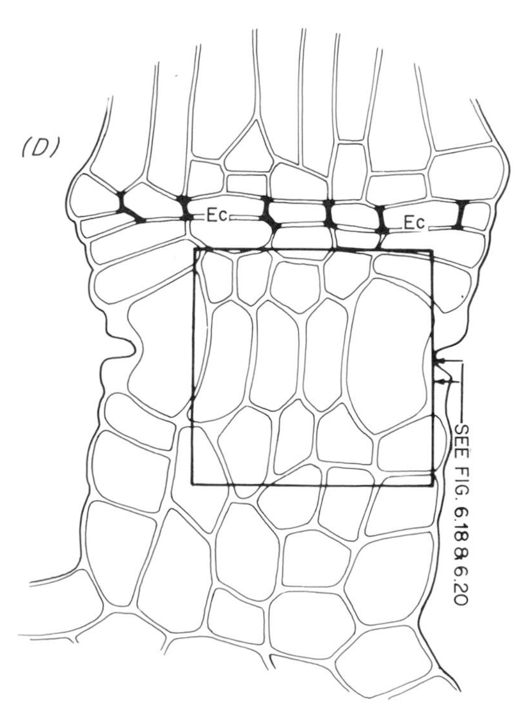

Fig. 6.17 The trigger hair of *Dionaea*. (*A*) The base of a trigger hair and several digestive glands on the surface of a lobe of *Dionaea muscipula*. (*B*) At higher magnification to show detail of the flexure zone. (*C, D*) The trigger hair of *Dionaea* on artist's longitudinal section to show the cellular distribution and in particular that of the endodermoid layer (Ec) and the microscopical structure at the point of flexure (SEMs by G. Wakley .)

In *Dionaea*

Munk (1876) removed surgically small portions of the trigger hair of *Dionaea*, from the tip towards the base, without affecting its sensitivity until he cut near the indented cells where, on the receipt of the cutting stimulus, the leaf immediately shut. A similar experiment, using isolated trigger hairs, was conducted by Benolken and Jacobsen (1970), who showed that when the indented cell layer of the *Dionaea* trigger hair was destroyed, the electric response of the hair to mechanical stimulation vanished. Further, they also demonstrated that, when a finely localized stimulus was applied from point to point along the hair, the response was negligible until the stimulus was administered at the level of the indented cells. All these results agree with the nineteenth-century notion that the sensory cells are located at the indentation of the hair, where bending strains are relatively pronounced upon stimulation (Fig. 6.17).

It should, however, be remembered that while touch stimuli lead to trap closure only when administered to the trigger hairs, almost the whole trap surface is sensitive to electric and to wounding stimuli. Both factors cause rapid closure if applied to any point on the trap other than its margins.

Interestingly, the upper portion of the trigger hair, as demonstrated above, does not respond to either electric or wounding stimuli. In this respect it closely resembles the marginal teeth.

The inexcitability of the trap margin is of an obvious advantage: herbivores cause trap closure only when they are located in a proper position to be captured.

In *Drosera*

In *Drosera*, the cells structurally homologous to the sensory cells of *Dionaea* are the epidermal cells of the upper part of the tentacle stalk (Williams, 1976). These stalk cells are located at a position where the stalk is most narrow and consequently is readily bent when force is applied to the head. This view replaces the traditional hypothesis that the receptors of *Drosera* tentacles are located in the head itself (Darwin, 1875; Haberlandt, 1906; Williams and Pickard, 1974).

The fine structure of the receptor cells

The sensory indented cells of *Dionaea* (Fig. 6.18) are characterized structurally by the presence of concentric endoplasmic reticulum (ER) complexes in a whorled configuration which surround large vacuoles containing polyphenolic deposits (Figs 6.18, 6.19 and 6.20 and cf. page 158). These ER complexes are confined to both poles of the sensory cells (Williams and Mozingo, 1971; Buchen

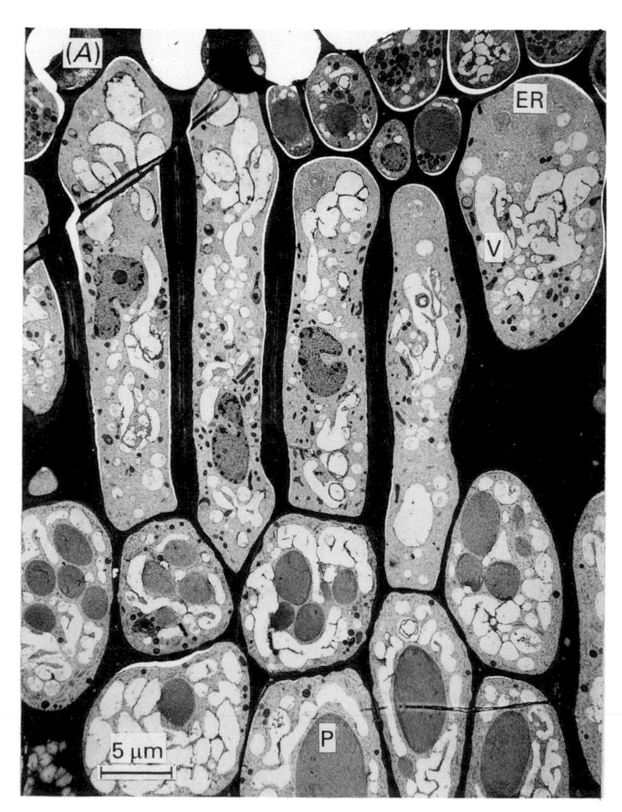

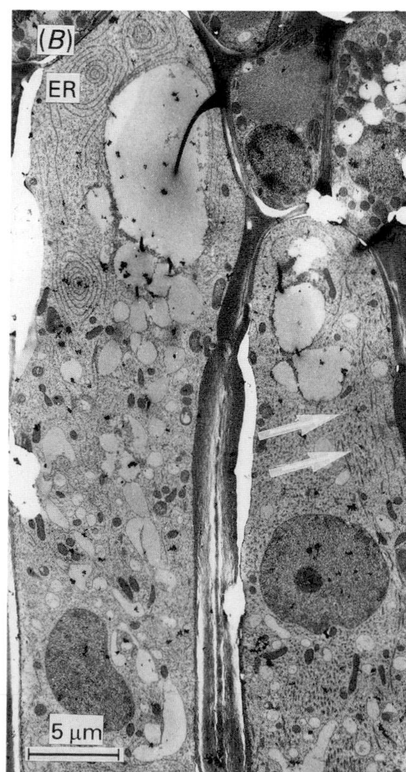

Fig. 6.18 (cf. Fig. 6.17) Freeze-substituted *Dionaea* sensory cells. (*A*) The polar arrangement of ER, vacuoles (V) and polyphenolic bodies (P). (*B*) The concentric-ring arrangement of the ER (ER) and thicker, fibrillar elements (arrows) passing along the cell. (By kind permission of Buchen and Schröder, 1986.)

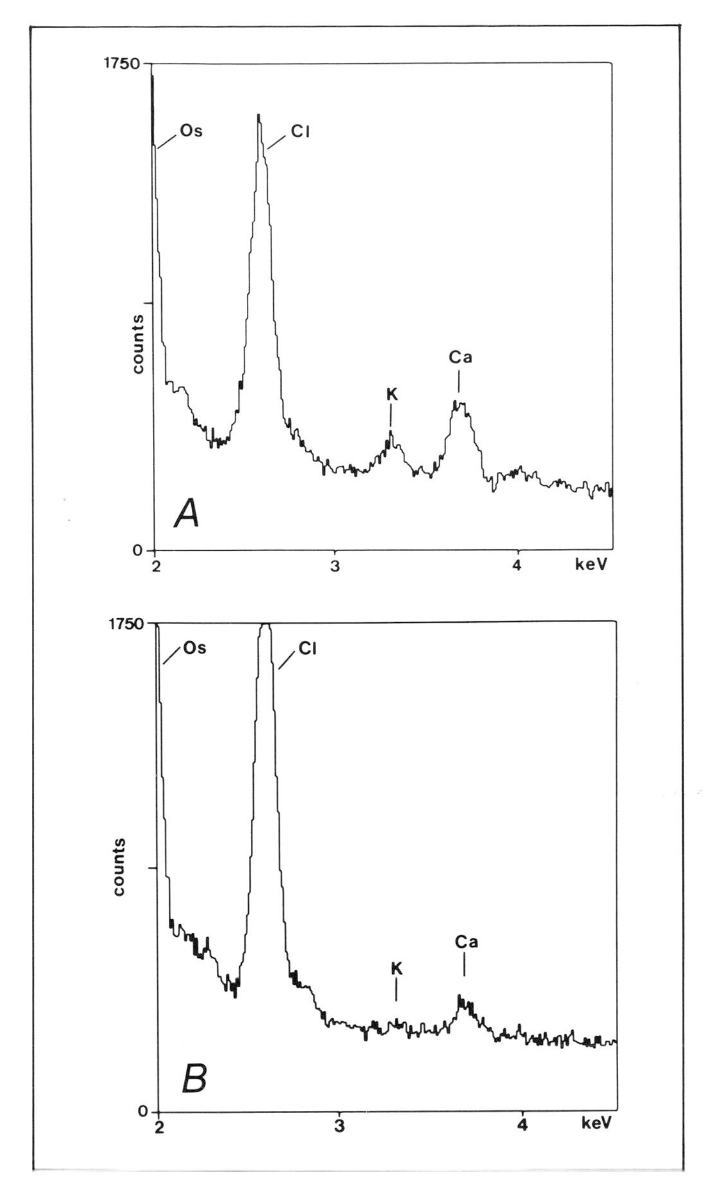

Fig. 6.19 EDAX spectra of the cell wall (*A*) and cytoplasm (*B*) of the freeze-substituted sensory cell of *Dionaea*. (Reproduced by kind permission of Buchen and Schröder, 1986.)

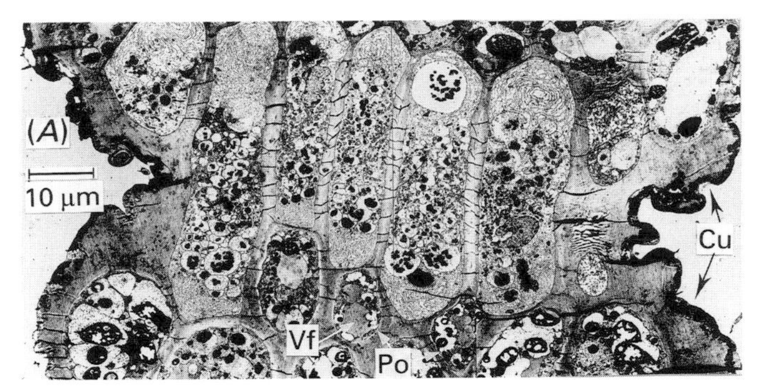

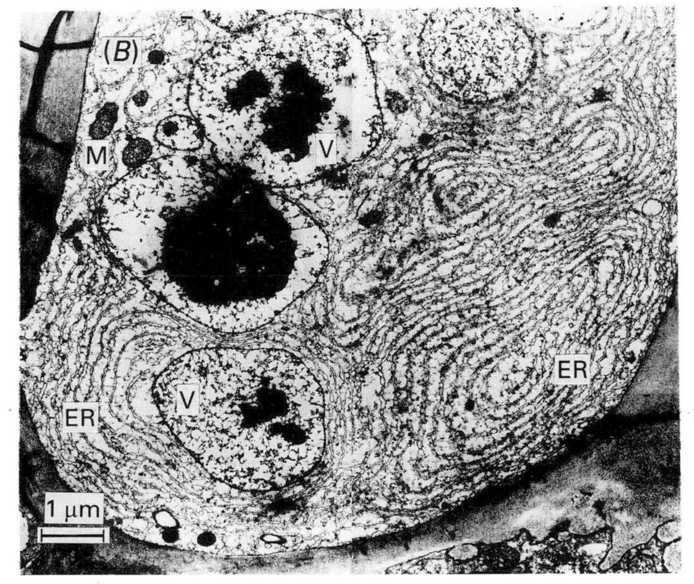

Fig. 6.20 Tangential length sections through sensory cells of the indentation zone of *Dionaea* at different magnifications. (*A*) some cells are cut in median section. Both cell poles contain numerous ER cisternae (ER), large vacuoles, sometimes with polyphenolic contents (V). In a podium cell (Po) there is a distinctive vacuole (Vf) with a fine-fibrillar content. Cu = cuticle. The dark lines traversing most of the cell walls are probably artefacts of preparation. (*B*) Details of the concentric ER and vacuoles with their polyphenolic bodies and mitochondria (M) can also be seen. (By kind permission Buchen *et al.*, 1983.)

et al., 1983) and resemble the distal ER complex in some root statocytes which act as gravity receptors (Juniper and French, 1973). These ER-vacuole complexes might serve as pressure transducers (Williams and Mozingo, 1971). When freeze-substituted material was examined under the electronmicroscope, long elements with a diameter of 120 nm or more were found in the cytoplasm of the sensory cells of *Dionaea* (Buchen and Schröder, 1986). They are longitudinally arranged and curved near the nucleus and towards the poles. These fine fibrillar long elements of medium electron density, might in some way be involved in the perception of stimuli or in the bending and rebending of the trigger hair (Buchen and Schröder, 1986). The role of the polyphenolic compounds in the vacuoles surrounded by the ER may, it can be speculated, be considered as storage, binding and release of ions necessary for electrical processing of stimulus transduction (Buchen *et al.*, 1983).

A change in the membrane potential can only be measured when structural polarity is established in these sensory cells. Similarly, short action potentials which are necessary for trap closure can be fired in the young traps only if the typical ER complex is already developed in the cell poles (Casser *et al.*,

1985; Buchen and Schröder, 1986). According to Hodick and Sievers (1988) the plasma-membrane of the sensory cells does not differ from that of other excitable cells (e.g. mesophyll cells) of the same trap. This was demonstrated by freeze fracturing of glutaraldehyde-fixed trap tissues.

Ionic content of resting sensory cells

Jacobson (1974) suggested that the membrane resting potential in *Dionaea* traps is K^+ mediated, whereas the action potential seemed to him to be Ca^{++} mediated. This assumption has recently been supported by a microscopical study using both histochemistry and energy dispersive X-ray microanalysis as well as LAMMA measurements, showing high peaks for Ca^{++} and for K^+ in the resting sensory cells (Buchen and Schröder, 1985). The Ca^{++} concentrations in the sensory cells were estimated in this study to be in the range of milliosmoles, and significantly higher than in neighbouring cells of the trigger hair and of the trap lobes (Buchen and Schröder, 1985). It is assumed that the Ca^{++} is compartmentalized in the sensory cells (Fig. 6.20). However, it is not yet clear what role each compartment plays in this process. The content of each individual compartment should first be established before coming to any conclusion. The ionic nature of the perception and transduction of mechanical stimuli therefore remains obscure.

Hodick and Sievers (1986) recently revealed experimentally that action potentials in *Dionaea* predominantly depended on the extracellular concentration of calcium. Peak depolarization was shifted by 27 mV per tenfold increase in Ca^{++} concentration, 1 mM $LaCl_3$ or EGTA can irreversibly block excitability, while 0.5 mM DNP and 2 mM NaN_3 block it reversibly. It was therefore proposed that the influx of Ca^{++} during excitation triggers changes in turgor pressure that result in movement (see page 103).

These findings in *Dionaea* closely resemble those obtained by Iijima and Sibaoka (1985) for *Aldrovanda*. The resting potential in the latter is composed of both diffusion and electromagnetic potentials; the first dependent mainly on K^+ and Na^+ and the latter on proton pumping.

The parallel between the ER whorls in these cells, and the sarcoplasmic reticulum in mammalian muscle is striking. This recent evidence for a key role of Ca^{++} in the mechanism is most exciting. Could it be that the Ca^{++} moves in and out of the ER cisternae in response to flexure, thereby generating an electrical impulse? Thus a mechanism of movement $\rightarrow$ Ca^{++} redistribution $\rightarrow$ current impulse, — the reverse of a muscle — might be evoked.

The 'memory' of the *Dionaea* trap

The trap lobes of *Dionaea* will normally not respond if the trigger hair is touched only once. This lack of response to the first touch benefits the plant by preventing accidental closure through wind-blown sand, raindrops etc. The trap will, however, snap shut if a second mechanical stimulus is delivered either to the same hair or to any other hair of the trap within about 20 or 40 s of the first stimulus (Burdon Sanderson 1876 and page 99).

At longer intervals after the first, the trap will respond to each of a series of touches in a pattern shown in Figure 6.21. When the stimuli are delivered at 1 min intervals the trigger hairs have to be touched six times to close the trap. The response to each successive touch after the first is greater

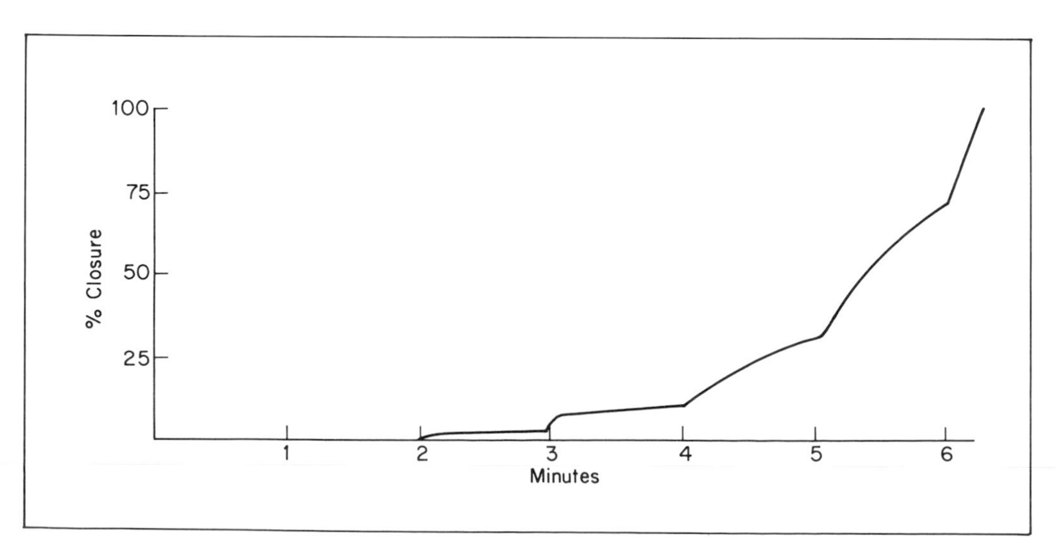

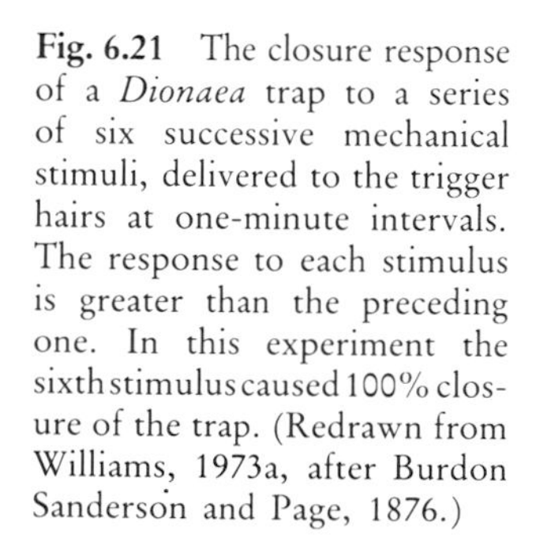
Fig. 6.21 The closure response of a *Dionaea* trap to a series of six successive mechanical stimuli, delivered to the trigger hairs at one-minute intervals. The response to each stimulus is greater than the preceding one. In this experiment the sixth stimulus caused 100% closure of the trap. (Redrawn from Williams, 1973a, after Burdon Sanderson and Page, 1876.)

than to the one preceding it, and the movement in response to each touch sums up to close the trap.

Single touches, spaced at 2 min intervals, do not cause the trap to move until the trigger hair had been deflected 10 times. In this case the trap does not fully close until the hair is touched 27 times (Burdon Sanderson, 1876; Williams, 1973a). The average number of touches needed to close a trap increases when the time intervals between the touches grow larger (Brown, 1916; Williams, 1973a).

Hodick and Sievers (1988) propose that a threshold concentration for Ca^{++} or some Ca^{++}-activated regulatory complexes must be reached before movement sets in. Accordingly, a single action potential does not allow enough Ca^{++} to enter the cytoplasm. Thus, at least two action potentials are necessary to reach this threshold. They further suggest that this increase in cytoplasmic Ca^{++} serves as the signal triggering turgor changes that serve as the driving force for trap closure (but see page 105).

Based on tracer experiments it was concluded by Iijima and Sibaoka (1985) that Ca^{++} has an important function in the membrane excitation of the trap cells in *Aldrovanda* and that the action potential is accompanied by calcium ion entry into the excitable cells. They suggest that the plasma-membrane of *Aldrovanda* trap cells acts as a calcium electrode at the peak of the action potential, in a manner similar to that of *Nitella* cells.

The Mode of Transport of the Potential

Action potentials are believed, admittedly on very little evidence, to be transmitted along plasma-membranes and via plasmodesmata. In plants (which lack neurones) the plasma-membrane constitutes the only low-resistance partially insulated continuum in which circuit patterns can readily be laid down by plasmodesmatal patterns (Juniper and Barlow, 1969; Gunning, 1978). It is therefore reasonable to deduce that action potentials are also generated on or in the plasma-membrane rather than on any other cell membrane (e.g. vacuole and/or ER; cf. Williams and Mozingo, 1971). The abundance of mitochondria in the centre of each sensory cell in *Dionaea*, near the cell indentation (Buchen *et al.*, 1983), supports the hypothesis that the mechanical bending of the cell wall is transduced to a physiological signal in this site on the plasma-membrane itself. Under these circumstances, the plasma-membranes of each receptor cell, but not the cell organelles or vacuoles, would suffer major shear and compression stresses.

Electrical coupling between excitable cells

The action potentials propagated from the mechanoreceptor is the signal operating the motor cells of the trapping system in the Droseraceae which show movement. How does the action potential pass from the sensing cells to the motile cells? To conduct action potentials a layer of cells must meet three conditions:

(i) the cells or their membranes must be excitable;
(ii) there must be a low resistance pathway between the cells;
(iii) there must be a coherent pathway for the transmission of the stimulus (Williams and Spanswick, 1976).

In *Drosera* where action potentials are transmitted along the stalk of the tentacle, from top to bottom, the two outer cell layers were shown to be excitable (Williams and Spanswick, 1976). Similarly in *Dionaea* and in *Aldrovanda* where action potentials are transmitted from the trigger hairs to the motile zone of the trap, both the lower and the upper epidermal cell layers seem to be excitable. In *Aldrovanda* traps that show only three cell layers, all three layers have been shown to be excitable (Iijima and Sibaoka, 1981). Similarly, the marginal cells of the trap are also excitable (Iijima and Sibaoka, 1981) and therefore might form a link between the three cell layers (see below).

The pathway and speed of the stimuli in *Dionaea* and *Aldrovanda*

The pattern of the spreading of an action potential in *Aldrovanda* is similar to that of *Dionaea* (Sibaoka, 1966) and both show a similar velocity of about 10 cm per second. The time required for the transmission of an action potential between two points on the trap seems to depend on the number of cells between them (Fig. 6.22). Since the long axis of each of the excitable cells in the trap lobes is perpendicular to the midrib, propagation of the action potential in this direction is faster than in the direction parallel to the midrib (Sibaoka, 1980).

There is strong circumstantial evidence that the low resistance required for the intercellular transmission of action potentials is provided by the plasmodesmata connecting excitable cells. While the periclinal walls separating the three cell layers of the *Aldrovanda* trap have only a few plasmodesmata, numerous plasmodesmata are found in the lateral and end walls of these cells. This pattern is consistent with general observations of root plasmodesmata

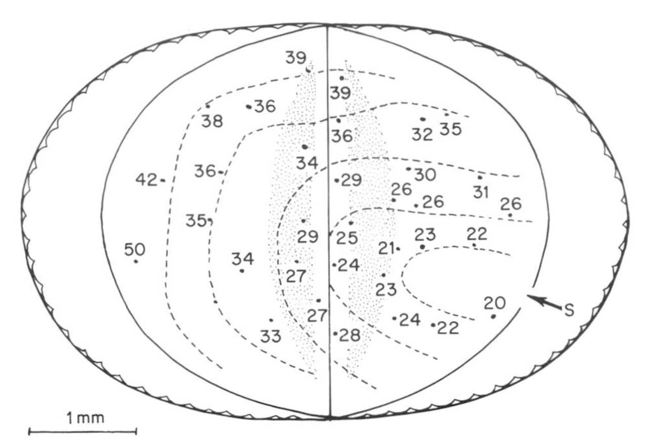

Fig. 6.22 Propagation of the action potential in *Aldrovanda*. Propagation of action potentials over the trap lobes. Electrical stimuli were applied to point S (close to the base of a sensory hair). Small open circles indicate the location of the recording micro-electrodes. The numerals beside each circle are the times (msec) between stimulation at S and recording of an action potential at these points. The broken lines are equitime curves, and the shaded areas represent the motor zones. (Redrawn with permission from Iijima and Sibaoka, 1982.)

(Juniper and Barlow, 1969; Gunning, 1978). This distribution of plasmodesmata in the excitable cells is also consistent with the measurements of the electrical coupling ratio of an epidermal cell to a middle-layer cell, which was 0.047 compared with the coupling ratio between adjacent middle-layer cells which was 20 times greater (0.80). An analysis of the results of an experiment in which a cell of the middle layer was exposed to an electric current revealed that the electrical resistance of the plasmamembrane, tonoplast and junction between two neighbouring cells were 11.1, 4.7 and 0.56 MΩ, respectively (Iijima and Sibaoka, 1982).

The pathway of the stimuli in *Drosera*

In a similar manner, the plasmodesmata in the outer tentacle cells of *Drosera* are concentrated in the end walls between outer stalk cells (Fig. 10.9) and are only sparsely distributed over the rest of these cells. Thus the distribution of the plasmodesmata is far denser along the path that the action potential would take, were it to pass down a column of cells (Williams and Spanswick, 1976). It must, however, be stressed that the presence of plasmodesmata is not, by itself, sufficient for the conduction of action potentials. The abundance of plasmodesmata between head and stalk of the *Drosera* tentacles is more likely to be associated mainly with solute transport, since there is no evidence of excitability of the head cells.

Comparison of possible different plasmodesmatal functions in the Droseraceae

In general, the presence of plasmodesmata is not given equal significance by the different authors studying the trapping mechanisms of carnivorous plants. Those studying glandular mechanisms regard plasmodesmata as a symplastic route through which transport is facilitated (Fig. 8.4*A–D*). Those studying the electrical response of traps regard plasmodesmata as possible symplastic routes through which electrical signals are propagated. Surprisingly, no comparative study has yet been done on the correspondence of plasmodesmatal structure to its electrical versus transportive function and an answer has yet to be given whether all, or even any, plasmodesmata are capable of conducting action potentials. A possible tissue for this study might be the tentacle of *Drosera*, where action potentials are restricted only to the tentacle itself, whereas absorbed digested matter is transported not only along the tentacle but also from the tentacle to the leaf tissues.

The Mechanism of the Rapid Movement in Snap Traps

Rapid leaf movements are usually assumed to be the result of changes in turgor pressure following a loss of solute (Hill and Findlay, 1981). Studies on *Dionaea* and *Aldrovanda* have led to two contradictory views being put forward as to how rapid movements occur in these species. While that proposed by Iijima and Sibaoka (1983) for *Aldrovanda* involves changes of turgor pressure due to some bulk solute movement in a similar manner to the probable mechanism in *Mimosa pudica* (Itill and Findlay, 1981), that put forward by Williams and Bonnett (1982) for *Dionaea* involves a radically new mechanism.

In *Aldrovanda*

The motor tissues in the trap lobes of this plant are located on both sides near the midrib, in what is called the motor zone (Ashida, 1934). The active motor cells are the inner epidermal cells of this zone which seem to lose their turgor upon receiving the trigger of an action potential. This turgor loss allows the other cell layers to bend the lobes (Ashida, 1934).

Two opposing forces seem to participate in trap closure.:

(i) turgor of the inner epidermal cells in the motor zone (motor cells);

(ii) turgor of the outer epidermal cells and of the middle layer.

An action potential triggers a rapid turgor loss in the motor cells, leading to an imbalance of the two forces and to trap closure. The rapid shutting in *Aldrovanda* is accompanied by a marked K^+ migration into the trap lumen (Fig. 6.23), possibly associated with ATP-ase activity (Iijima and Sibaoka, 1985) and accompanied by Cl^- as a counter ion. It was suggested that the influx of Ca^{++}, which is associated with membrane excitation by action potential, activates the membrane ATP-ase which is responsible for K^+ transport (Iijima and Sibaoka, 1985). The K^+ efflux is believed to result in a rapid decrease of turgor in the active motor cells. When immersed in 200 mM manitol, which is nearly isotonic to the epidermal cells, the trap did not shut. These results were interpreted by Iijima and Sibaoka (1983) as suggesting that solute leakage from the active motor cells requires the pressure inside the cells and is induced by bulk flow between the vacuole and the outside of the cell.

In *Dionaea*

The loss of turgor observed in the outer epidermis of *Dionaea* traps is due to an increase in wall plasticity (Williams and Bennett, 1982). Based on an experimental study of the effect of pH on trap movement (Table 6.1), it was suggested that the changes in wall plasticity are due to 'acid growth'. This leads to an irreversible expansion of the outer epidermis, which leads to trap closure. Support for this concept comes from the observations of Stuhlman (1948a, b) who showed such expansion to occur, while Yaguchi and Kondo (1981) showed water to move into the cells of the outer epidermis during trap closure. During the 1–3 s which are required for closure the trap cells lose about 30% of their ATP which is likely to be consumed in rapid transport of protons from the motor cells, thus leading to increased acidity of the cell walls and to 'acid growth' (Williams and Bennett, 1982).

Slow Movements

Narrowing in *Dionaea* traps

Once the *Dionaea* trap has closed and an insect is captured the trap must be further stimulated into the narrowed phase (Figs 6.14*C* and 4.4*C–E*) so that a closed digestive cavity is formed and the digestive cycle can start. Experimentally, narrowing and digestive juice secretion can be accomplished

Fig. 6.23 One trap of *Aldrovanda vesiculosa* in the closed position. Note the external glands and the barbed 'bristles' of the whorl to the right; cf. Figs 4.7A–E. (Liquid-nitrogen frozen SEM by S. Clarke, Oxford, UK.)

Table 6.1 Leaf closure in the *Dionaea* flytrap. The response of *Dionaea* trap lobes, whose extracellular spaces were perfused with 50 mM acetate of 2-(N-morpholino)ethanesulfonic acid buffers in the absence of mechanical stimulation, and in the presence of stimulation of trigger hairs at 6 s intervals with a camel-hair brush. Sap was expressed from the cut petiole in a pressure chamber and the leaves were rehydrated in the appropriate buffer solution. All buffer solutions were adjusted with sorbitol to approximately 70 mM per kilogram. Values are means ± standard errors, N.D., not determined.

pH[a]	Number of stimuli necessary to cause trap to close 50%	Closure rate of unstimulated traps (% per min)
Control[b]	7 ± 3	0.0 ± 0
3.0	N.D.	0.27 ± 0.01
4.0	N.D.	1.38 ± 0.20
4.5	11 ± 5	0.32 ± 0.16
4.75	60 ± 24	N.D.
5.0	90 ± 7[c]	0.0 ± 0
6.0	N.D.	0.0 ± 0

[a]The buffering capacity was 30 mM above the pH listed.
[b]Perfusion with 70 mM sorbitol.
[c]Traps requiring more than 100 stimuli for 50% closure were recorded as >100 and averaged with the data as 100.
(Redrawn with permission from Williams and Bennett, 1982.)

by two types of stimuli, mechanical and chemical (see page 195).

Insects are normally not crushed, at least not at first, when the trap shuts so no chemical stimuli can arise directly from the release of haemolymph, but they sometimes excrete or defecate copiously and struggle to escape. At this early point after the trap has shut both the chemical excretions and the mechanical movements stimulate the trap to start narrowing. These stimulations also lead to the secretion of a digestive juice (Robins, 1976 and see page 175) which penetrates the prey and releases digestion products. The chemical products released from a digested prey further stimulate the trap to narrow. When they are all absorbed no further stimulation is perceived by the trap and it reopens.

Mechanical stimulation

Affolter and Olivo (1975) found that the struggle by the entrapped prey against the trigger hairs in the closed trap results in the generation of action potentials which cease to occur when the prey stops moving. These stimuli result in greater force being exerted by the opposing lobes upon each other, with further tightening of the trap.

Lichtner and Williams (1977) used nylon bristles to stimulate the trigger hairs. They elicited five to ten action potentials every 15 min for 5 h or more and measured the extent or absence of narrowing after each interval in accord with the occurrence of action potentials. Fifty per cent of the traps were narrowed in this experiment after 5 h of periodic stimulation, indicating that the action potentials which were reported by Affolter and Olivo (1975) could be the result of the mechanical stimulation by prey of the trigger hairs. Although this is possible and very likely to be true, we should be aware of the possibility that the narrowing phase developed as the result of some wounding, as with the secretion of digestive fluid (see page 172), rather than of mechanical stimulation.

When containing digestible material the trap of *Dionaea* usually remains closed for 7–10 days, long after the prey has died and ceased to stimulate action potentials. This sustained closure indicates that, when trap-narrowing is chemically induced, it must be mediated by a mechanism other than action potentials (Lichtner and Williams, 1977). A hormonal mechanism seems the most likely, perhaps in a manner similar to that described for *Drosera* leaves (see below).

Chemical stimulation

Both sodium and ammonium ions were shown to induce narrowing in *Dionaea* traps at very low concentrations (Darwin, 1875). Concentration of 50 mM caused narrowing in more than 80% of the examined traps. Urea, glycine and L-lysine also induced narrowing, although to a lesser extent. Glucose- and salt-free egg albumen had no effect on the traps (Lichtner and Williams, 1977). Some effect may, though, be seen in response to high concentrations of different substances as glucose and potassium salts. (Balotin and Dipalma, 1962), presumably as a result of their being an osmoticum, rather than of direct chemical stimulation (Lichtner and Williams, 1977). The stimulation with sodium, which was experimentally induced by Lichtner and Williams (1977), was within the range of sodium concentrations found in insect haemolymph. Ammonium might be expected to be released in more than sufficient quantity if proteins were hydrolysed and deaminated during the digestion process. Glycine from haemolymph and from hydrolysed protein would perhaps have some stimulating effect on trap narrowing. Stimulation of secretion is discussed in Chapter 9.

Hormonal Regulation of the Leaf Movements in *Drosera*

Two different slow movements occur in *Drosera* traps as the result of chemical stimulation of their tentacles; the bending of neighbouring tentacles, and the inflection of the leaf itself. In both cases the site of stimulation is far from the site of reaction and therefore some sort of stimuli must move from the stimulated tentacle to the moving neighbouring tentacle and to the bending leaf region.

There is abundant evidence that these slow movements are not normally mediated by action potentials (see page 105 for *Dionaea*). In *Drosera* no action potential has ever been detected beyond the individual tentacle that receives mechanical stimulation. The cells on the surface of the *Drosera* leaves are not excitable and it is therefore impossible to induce action potentials by providing electric current pulses to the surface of the trap between the tentacles (Williams and Pickard, 1972b).

If electrical signals are excluded as possible mediators of the slow movements, the next most likely candidate would appear to be a plant growth regulator which is capable of inhibiting growth of *Drosera* tentacles on the side nearest the insect, or inhibiting the growth of the inner epidermis of *Dionaea* (Williams, 1971, 1976).

Some insight into the regulatory process of leaf blade bending in *Drosera capensis* and the possible participation of hormones in this process is provided by Bopp and Weber (1981; Fig. 6.24). They used pieces of 'Holländischer Edamer' cheese (a low-fat cheese), placing them on the leaf in different locations. As expected, they found that the leaf bent at the point of application. The bending angle depends on the site of stimulation and decreases from the tip of the leaf to the base. Thus, a 40° bending was observed 48 h following the application of cheese to a site 1 cm from the leaf base. When applied to the middle of the trap a 200° bending was observed, and when applied 1 cm from the tip of the trap a 260° bending was observed. Longer treatments often caused the leaf tip to roll right up.

The speed of bending is markedly enhanced when IAA is applied together with the cheese, but the total bending after 24 h is not significantly different. IAA alone also causes a bending reaction in the site of application. This reaction is, however, weaker, and reaches a smaller angle of bending. No stimulation could cause bending in the opposite direction, but IAA was active both when applied to the upper or to the lower sides of the leaf.

Experiments with different substances which are antagonistic to IAA (TIBA, PCIB and ABA) show a reversible inhibition of IAA reaction. These may demonstrate that IAA or other internal auxin-like hormones are involved in a growth stimulation of the leaf which leads to its bending.

The reaction chain appears to include a longitudinal transport of auxin. A ring of TIBA, which interrupts the transport of auxin to the bending point, leads to the accumulation of auxin just above the TIBA ring. The supposed stream of auxin, which is blocked in the bending point, seems to be induced primarily by the presence of prey, as TIBA alone does not induce leaf curvature.

Bopp and Weber (1981) thus came to the conclusion that a signal exerted by prey has to travel to the tip of the leaf to induce a reverse flow of an auxin-like substance. This flow is blocked at the position of the prey, by which it induces the growth of the adaxial (under)side of the leaf, resulting in leaf bending.

It should be added that, in addition to the development of concavity on the trap upper surface, which was discussed by Bopp and Weber, the part of the trap of *Drosera capensis* between prey and the leaf base often becomes convex as the result of

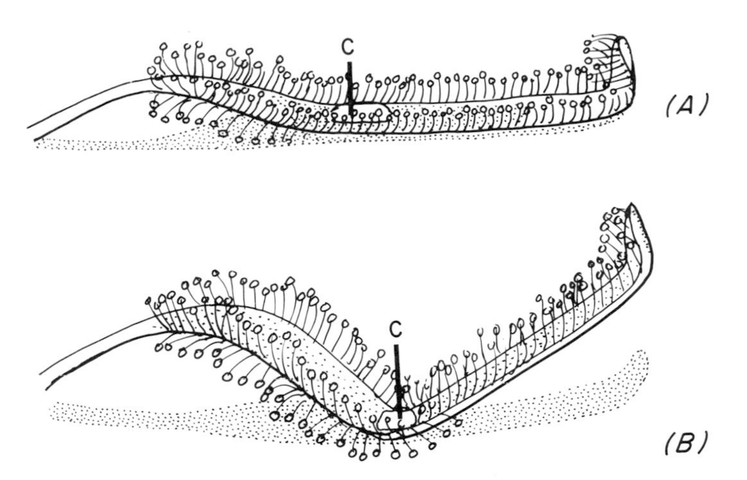

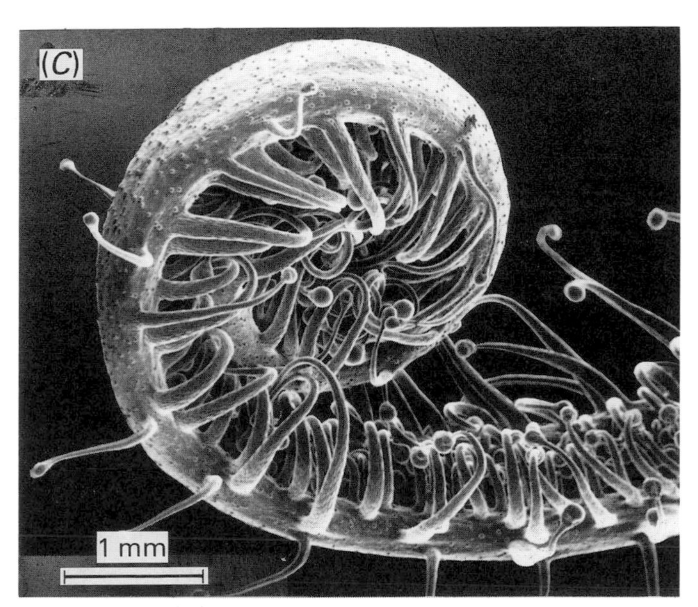

Fig. 6.24 (*A,B*) The visible response of a leaf of *Drosera capensis* to prey capture. In *D. capensis* the bending is enhanced by a simultaneous application of IAA either to the prey or to the leaf tip, and IAA alone can induce a curvature of the leaf. The curvature is always introrse (towards the axis) regardless of the point of application of the growth promoter. PCIB and TIBA inhibit the bending reversibly. ABA inhibits irreversibly. TIBA between the prey and the leaf tip reduces the bending reaction whereas TIBA on the basal part of the leaf has no effect.

(*A*) With cheese (C) as prey; bending with the reaction inhibited by a TIBA barrier between the prey and the leaf tip. The TIBA was applied between the tip and the middle of the leaf. The cheese was applied 24 h later and the picture recorded 24 h after feeding.

(*B*) The same as *A* but 20 μl of 1 mM IAA was applied to the leaf tip at the same time as the cheese. (Redrawn, with permission, from Bopp & Weber, 1981.)

(*C*) The tip of a tentacle of *D. capensis* responding to the capture of a fly. Note the marked circination as well as the incurving of most of the marginal trichomes. A few of the marginal trichomes curve in the 'wrong' direction. The sessile glands on the abaxial surface and on the margin of the leaf are prominent. For a similar circination, cf. Fig. 6.2B. (SEM by D. Kerr.)

prey capture (Fig. 6.24*B*). TIBA not only prevents the bending at the site of prey, but also inhibits this basal curvature.

We assume that, if auxin is indeed involved in the bending of the leaf, its differential distribution in the leaf, which is caused by the presence of prey, leads to growth movements of different directions in the distal and in the proximal leaf portions.

The Slow Movement in the Trap of *Pinguicula*

Little attention has been paid to the movement in the leaves of *Pinguicula* since it was first described and experimentally studied by Darwin (1875). Following the capture of prey by the sticky mucilage secreted by the stalked glands, the leaf margins adjacent to the trapping site curve upwards, covering the captured prey and forming a closed space around it (Figs 6.7 and 8.16*B*), where secreted digestive fluid accumulates and acts on the trapped animal. This curving movement is a growth movement slowly proceeding as the result of slight continued pressure or the absorption of nitrogenous matter. When the digestive activity is over, the leaf margins unfold. The whole process lasts from a few hours to several days.

Most experiments were performed on *P. vulgaris* and it is not known whether all species react in the same manner.

Darwin thought that the rolling of the leaf margin brings more glands into contact with the prey, and in some cases pushes it into new positions further away from the margin. It is now quite clear that the upward-curved leaf margins help to hold the secretion in place (see Figs 6.7 and 8.16B), as suggested by Lloyd (1942).

The Pitcher Traps: General Considerations

Trapping by pitcher-plants is based on two main elements:

- (i) stumbling devices that cause prey to fall into the digestive cavity, and
- (ii) retaining devices that prevent animals from escape.

The stumbling devices are complex structures at the opening of the trap, employing both geometrical and surface characteristics which, on the one hand, bring the potential prey to an inescapable position and, on the other hand, reduce its foothold by smooth or waxy surfaces. Waxy surfaces also prevent animals from ascending the trap wall. Another common retaining structure is the downwardly directed hair (see page 111). In general, pitchers feeding on ants are shorter, with a relatively wide opening, whereas pitchers trapping flying insects develop a long and narrow neck (Appendix 2). These differences seem to be correlated with the specific needs the trap faces when treating flying as opposed to creeping arthropods. A winged arthropod will start flying shortly after it stumbles. If the space around its wings is narrow, the air currents which are produced by its wings will force it downward. The distance between the peristome and the tube-narrowing seems to determine whether the insect will escape or be trapped. Some *Sarracenia* species lie horizontally; they also specialize in creeping insects. When a *Darlingtonia* trap lies on the ground, its nectariferous 'fish-tail' serves as a bridge for creeping insects. These insects will accumulate at the base of the trap due to the slight difference in height between the entrance of the pitcher and its base. Tall vertical *Darlingtonia* pitchers do not offer a bridge. Instead, they attract flying insects which are trapped by the long and narrow construction of the pitcher.

The Trapping Mechanism of *Darlingtonia*

The roles of the 'fish-tail' and the areolae

From the front of the mouth, i.e. from the free margin of the dome-shaped pitcher lid, a prominent forked appendage hangs down with a forward curve (Frontispiece). This appendage, frequently termed the 'fish-tail' (Zone 1 in Fig. 4.9*E*), is covered with nectaries, mainly on the ventral surface. In addition, many stiff, thick blunt hairs, directed towards the pitcher entrance, are also present on the fish-tail (Adams and Smith, 1977; Fig. 6.25). The roof of the dome and the back of the upper part of the pitcher and extending some way down Zone 3 of the tube are mottled with numerous white flecks or areolae, which are light-transmitting (Fig. 5.4). The transparency of the areolae is based on the entire absence of chlorophyll and other pigments in its tissue, and a lack of intercellular spaces (Lloyd, 1942). Each areole lies in a mesh of the vascular tissue and shows no stomata or external hairs. So complete is the UV light transmission through the areoles that they appear quite glassy and highly translucent. These apparent exits act to confuse insects, preventing their escape (see page 81). Similar trapping areoles are also found in three species of *Sarracenia*, (*S. minor*, *S. psittacina* and *S. leucophylla*) and similar structures of a less well-developed form are found in the lid of the *Cephalotus* trap (Lloyd, 1942).

The trap entrance of *Darlingtonia*

At the entrance of each pitcher the leaf edge rolls inwards forming the 'nectar roll' (Zone 2 of Fig. 4.9*E*). The nectar roll encircles the entrance and is connected both to the fish-tail appendage and to the keel, which is a ventral wing running externally along the adaxial side of the pitcher. This marked keel is one of the few morphological features clearly distinguishing *Darlingtonia* from *Sarracenia*.

Insects that are attracted by the nectaries, either along the fish-tail or elsewhere on the surface of the pitcher, are assisted by hairs to approach the pitcher entrance. There, along the gradient of gradually increasing density of nectaries, as well as increasing quantity of nectar, they are led directly to the nectar roll, where the heavy exudation of nectar (Austin, 1875H: see Appendix 1) serves as the strongest temptation for insects. Once inside the hood, the insect faces both the slippery surfaces and the false windows (areoles), leading them eventually to the bottom of the pitcher.

The slippery zone of the *Darlingtonia* pitcher

The slippery nature of the inside of the *Darlingtonia* pitchers is due partly to a detachable waxy layer, similar in function to that of *Nepenthes* species and partly to the imbrications of the epidermal cells, each being elongated into a sharp downwardly pointed hair which offers no foothold (Fig. 6.25). The wax layer, called the 'velvety portion' by Austin (1875C: see Appendix 1) occupies the first few cm of the pitcher tube (Fig. 6.26), the top of Zone 4

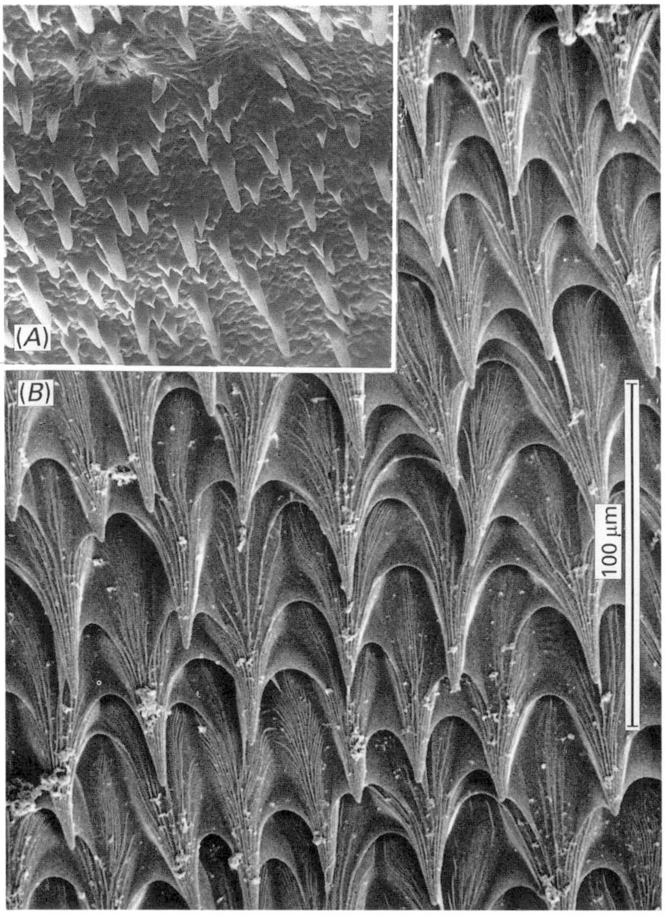

Fig. 6.25 (*A*) The trapping hairs inside the pitcher of *Darlingtonia*. (*B*) shows the blunt, *upward*-pointing hairs on the 'fish-tail'. The main area shows the structures of *A*, Zone 1 in Fig. 4.9, grade into the *downward*-pointing and more markedly ridged hairs of Zones 3–4 of Fig. 4.9 (cf. *Sarracenia*: Figs. 6.30–31 and *Cephalotus*, Fig. 6.36).

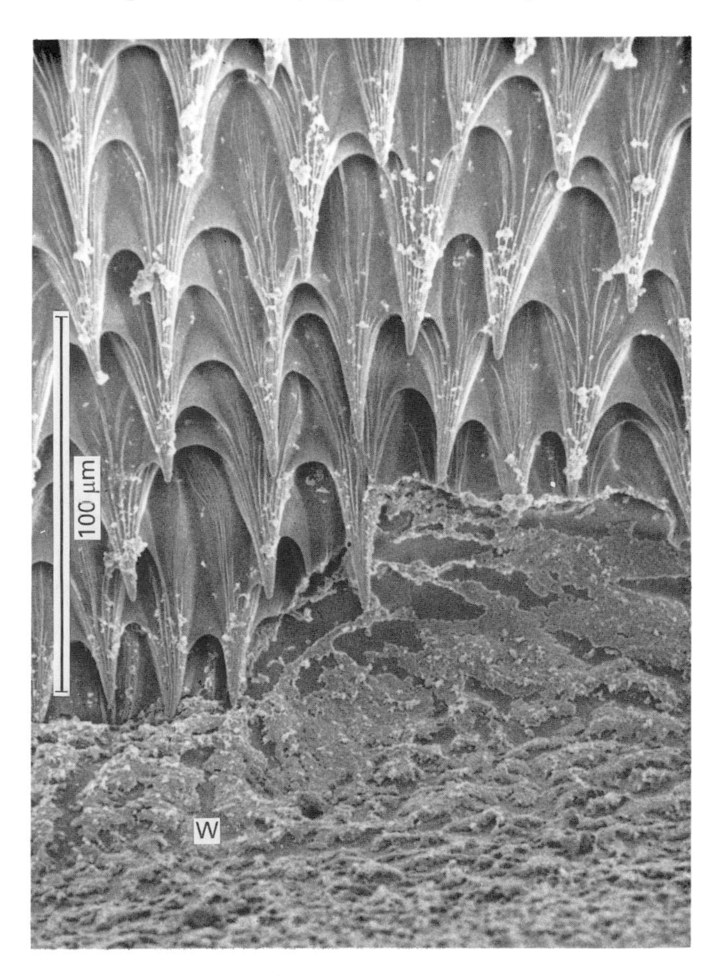

Fig. 6.26 Figure 6.25 continued into the next area of a wax patch (W); Zone 4 of Fig. 4.9. (SEM by S. Clarke, Oxford.)

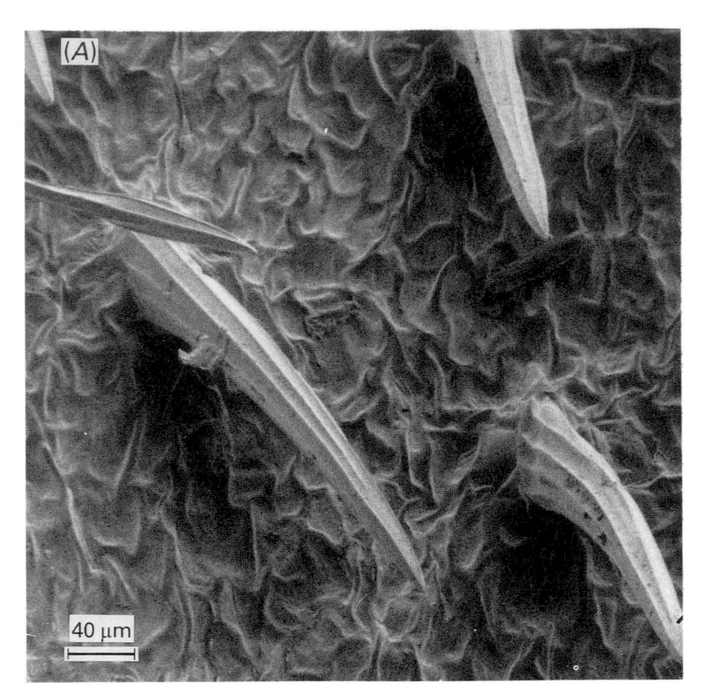

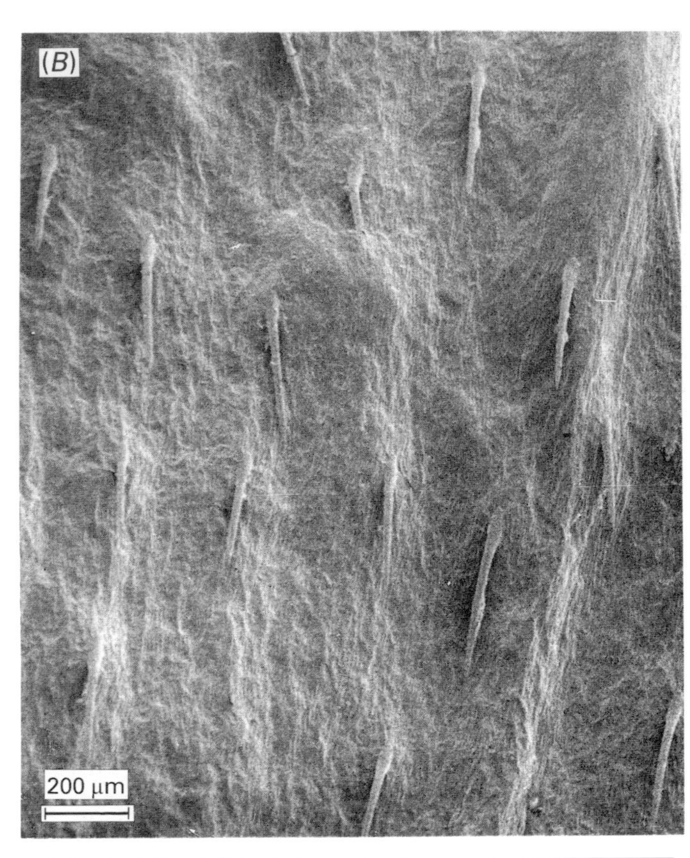

Fig. 6.27 (*A*) The hairs of Figs 6.25–26 elongate and become more detached from the epidermis, and grade (see Zones 5–6 in Fig. 4.9) into (*B*) the smooth, thin hairs. (SEM 6.25 by G. Wakley; 6.25*B*–6.27*B* by S. Clarke, Oxford.)

in Figure 4.9*F*. Unlike that in *Nepenthes* (Fig. 4.9*A*) the lower margin of this waxy zone is not clearly defined and precisely annular, but descends down in irregular embayments over the downward-pointing imbrication cellular teeth of Zone 4 of Figure 4.9*F*. Again, unlike *Nepenthes* (see page 113 and Figs 6.39–42), this waxy zone is only detected in pitchers grown under the very highest light intensities, i.e. the best field conditions. The imbricating teeth and the special 'windows' occur along the entire upper half of the pitcher, intermingling with nectaries at the hood, but not further beneath it. The imbricating teeth (Fig. 6.25–26) do not change in character as they pass over the areoles. These teeth are continued often half-way down the pitcher and overlap with the longer curved hairs of Zone 5 (Fig. 6.27). These hairs increase in density down the tube and are absent only in the very bottom of some pitchers (Zone 6). The bottom zone of the pitcher epithelium possesses no special glands but is capable of absorption (Lloyd, 1942). This permeability is a result of the discontinuity of the cuticle (see page 162). In contrast to most species of *Sarracenia, Darlingtonia* does not show any digestive glands in the pitcher (Goebel, 1893). Digestion of prey only takes place, so far as is known, by the associated fauna (Chs 8 and 14).

The Trapping Mechanism of *Sarracenia*

The epiascidiate leaves (see page 56) of *Sarracenia* form hollow pitchers which may be either erect or semi-recumbent. The posture may relate to features of prey selection (Chs 7 and 15). As we saw in Chapter 4, the pitcher may be divided into five zones (Fig. 4.9*D*) the first three of which concern us in the trapping function. The individual development of each zone will vary with the species. Zone 1, as we saw, is covered with numerous nectar glands (see page 60) and stiff, downward-pointing hairs. Each hair, as those of the shorter-haired imbricate Zone 2 and the long slender and curved hairs of Zone 4, is acutely ridged (Figs 6.28 and 6.29). These sharp ridges are present not only on the hairs themselves, but also on the connections between the hair and the epidermal wall and the epidermal walls themselves. These features can clearly be seen in TEM sections (Fig. 6.29).

Under normal conditions of growth these hairs will be coated with mucilage and although an insect's claw can gain a foothold by encircling the fine hair, the shape, the surface detail and the slipperiness prevent an effective grip. Between these hairs are numerous glands (G in Fig. 6.28). Zone 1 appears velvety and is usually highly coloured – red along

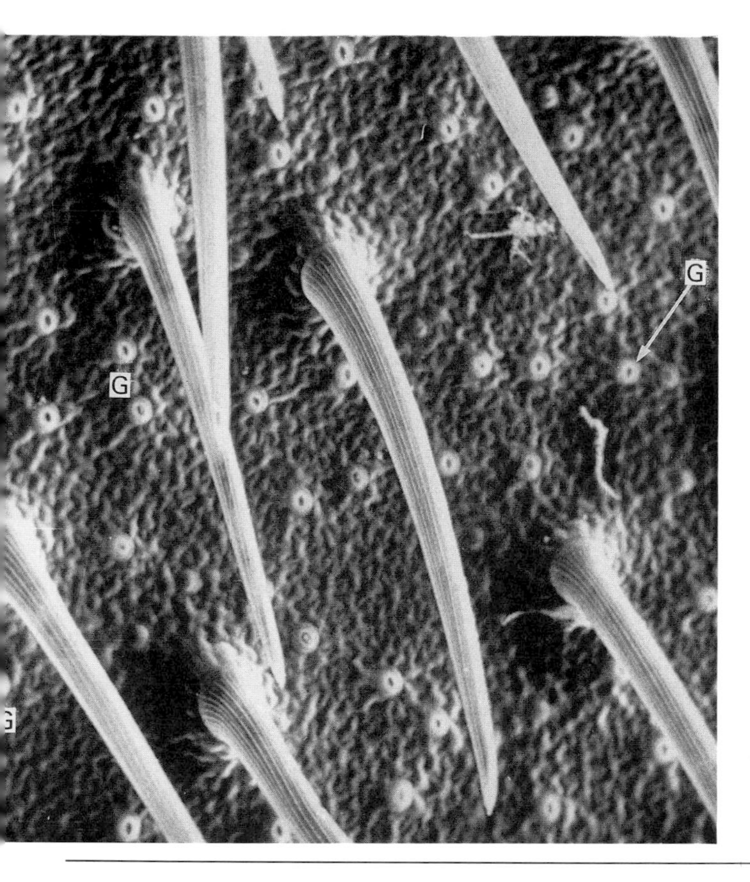

Fig. 6.28, Left. Hollow, downward-pointing, ridged hairs of *Sarracenia purpurea*. G = nectar glands. Zone 1 of Fig. 4.9. SEM by D. Joel.

Fig. 6.29, Below. A transverse section (TEM) through the ridged hairs of Zone 2 (see Fig. 4.9*D*) of *Sarracenia purpurea*. Note that the hairs are hollow and still have living contents and that the ridges not only surround the hairs, but also extend along the buttresses supporting them and on the adjacent epidermal cells.

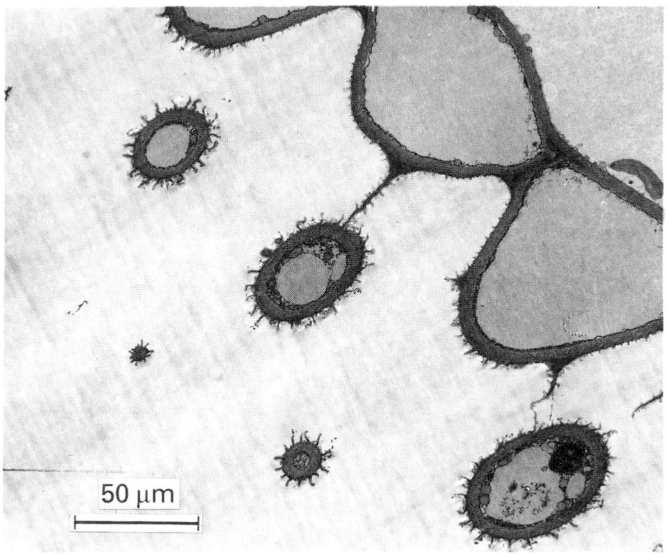

the venations and green between. The red coloration following the main veins may continue, but be less well marked, down into Zone 2. Zone 1's role is principally attractive, but the direction of the pointed hairs and the presence of mucilage tends to encourage the insect's movements down towards Zone 2.

The transition zone, as can be seen in Fig. 6.30, is relatively abrupt. Zone 2 (Fig. 6.31) continues downwards, virtually without change until it too

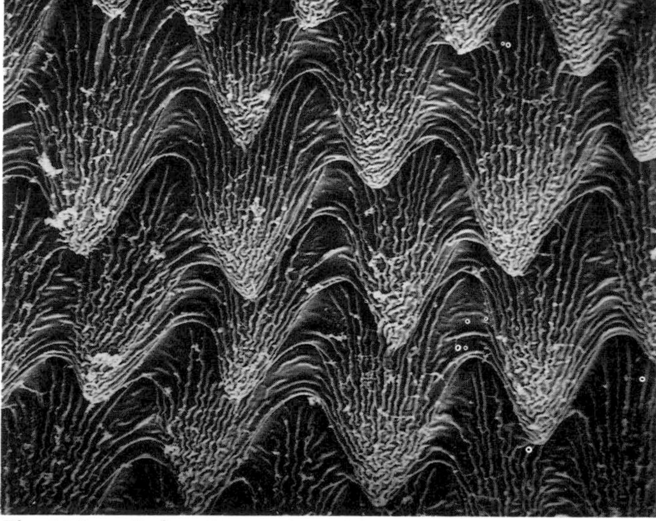

Fig. 6.30, Left. *Sarracenia purpurea*. Area of transition between Zones 1 and 2, Fig. 4.9*D*. (SEM by G. Wakley.)

Fig. 6.31, Above. *Sarracenia psittacina* The imbricate hairs of Zone 2; see Fig. 4.9D. (SEM by D. Kerr.)

ends abruptly, commonly about half-way down the pitcher at the digestive zone (Zone 3 of Fig. 4.9*D*). Zone 4, also an integral feature of the trap system, at least in juvenile traps, has been termed the 'eel-trap zone' by some authors (Fig. 4.11; see Heide-Jørgensen, 1986). Below this again is Zone 5, whose function is unclear; it is free of hairs and non-glandular.

The Trapping Mechanism of *Heliamphora*

Although, as suggested in Chapter 3, this genus may not be a very effective carnivorous plant it does, nevertheless, have a very complex trap mechanism. The conducting surface is most pronounced (Fig. 4.9*F*) and the hairs are relatively longer than in *Sarracenia*, more slender and distinctly flexible, the more so in the lower part of the conductive surface. The transition zone is shown in Figure 6.32. The hairs of Lloyd's retentive surface (Zone 4: Fig. 6.33) are short, very thick-walled, rigid and obviously designed to make escape difficult.

However, whereas the basal parts of pitchers, particularly those of *Sarracenia* and *Darlingtonia*, are noticeably robust and obviously designed to prevent the escape, through the walls of muscled or jawed insects, the walls of a *Heliamphora* pitcher are not so reinforced and collapse and crumple when flaccid.

The Trapping Mechanism of *Cephalotus*

The pitchers of *Cephalotus* fit tightly together

Fig. 6.32–6.35 Details of the peristome rim and nectaries of *Cephalotus follicularis*: Zone 2 in Fig. 4.9*B*.

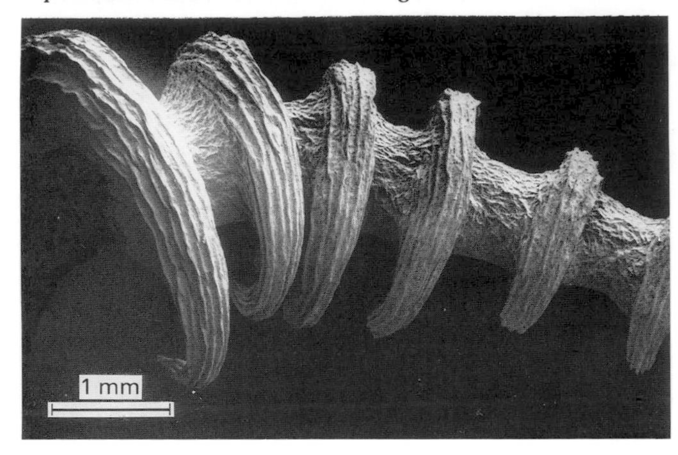

Fig. 6.32, Above. The 'teeth' of the peristome showing their marked ridging (SEM by S. Clarke).

Fig. 6.33, Below. The ridges and imbrications on the surface of the teeth with nectaries (N) at the tip just visible.

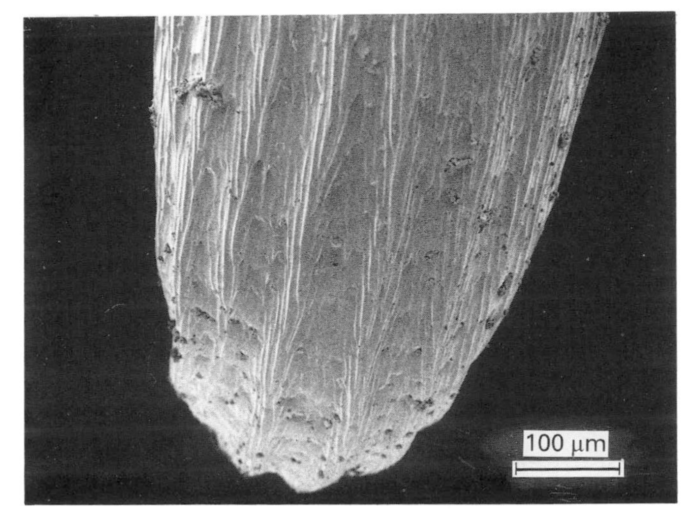

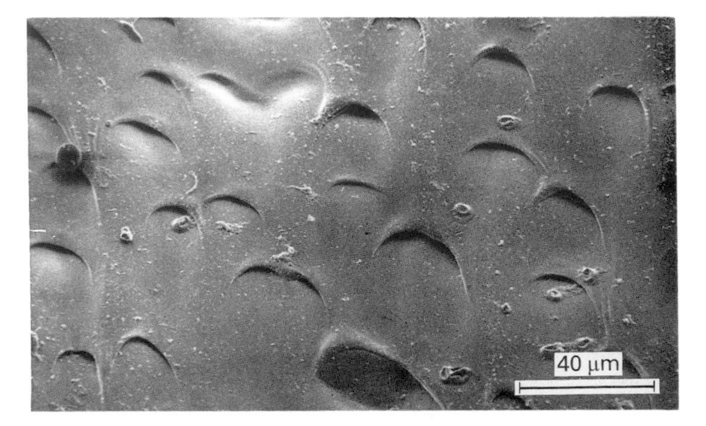

Fig. 6.34 Above. Surface detail of nectaries between the teeth (Liquid-nitrogen frozen SEM by S. Clarke.)

Fig. 6.35, Right. A transverse section, TEM,. through a nectary. Cu, cutinized wall; CuL, the cuticular lip visible in Fig. 6.34; Ec, endodermoid cell; R, reservoir cell; Cs, Casparian strip. (By kind permission of Heide-Jørgensen, 1981.)

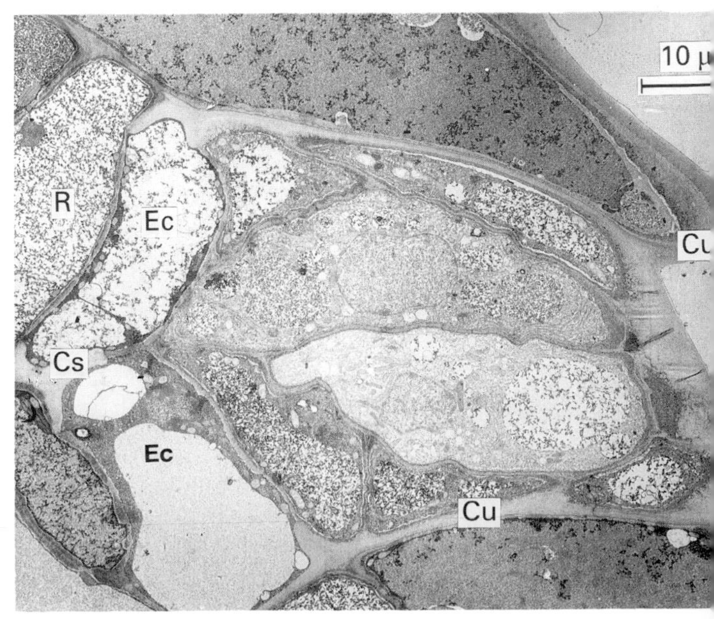

around the central flowering stem, often embedded in the soil litter, rather like a miniature *Nepenthes ampullaria* (see Plate 1 in Mabberley, 1985). The lids of the pitchers are said to alter their posture in response to changes in humidity, but whether this has any trapping function is unknown.

As we saw on page 62 there are usually about 24 teeth. Both the surface of the rim and the teeth are covered around the glands with sharp-pointed imbricate cells (Figs 6.35 and 6.37). Thus crawling insects (all authors from the earliest discoveries agree that ants are the principal prey) may be tempted by the glands to creep to the very tips of the sharp-pointed and imbricated teeth. Within the pitcher, inside the 'funnel' to use the terminology of Adams and Smith (1977), the surface is covered with very fine, downward-pointing hairs (Fig. 6.36). Below this region is the glandular patch (Zone 5 of Figs 4.9*B* and 6.37).

The Trap of *Brocchinia*

Brocchinia is so recent an addition to the association of carnivorous plants (see Ch. 1) that very little experimental work has been done on it other than on the details of absorption (see pages 19 and 204; Benzing *et al.*, 1980). The inner surface of each leaf is coated with a fine waxy powder that readily exfoliates (Fig. 6.43; Givnish *et al.*, 1984). *Componutus* ants, which are known as prey in these traps, were shown experimentally to be incapable of ascending this inner waxy surface (Givnish *et al.*, 1984). The detachable wax not only causes insects to fall into the trap (Fig. 4.9*C* and 6.43 and cf. 18.3) but, unlike *Nepenthes* (see below), entangles their legs and reduces their mobility.

The Trapping Mechanism of *Nepenthes*

The pitchers of the genus *Nepenthes* are very variable in varied size, often dimorphic (see Ch. 3 and Appendix 2) and even more in detailed architecture. The architecture of the peristome rim (Zone 1 of Fig. 4.9A) is particularly diverse and in extreme cases, as in *N. bicalcarata* and *N. villosa*, develop into cat-like fangs whose function is obscure but may serve to prevent robbing (see page 261). Zone 2, the slippery zone (Fig. 4.10) which most concerns us in this chapter, is variable in extent, sometimes almost non-existent in those species which form terrestrial pitchers (see Ch. 3). The fine structure of this zone appears to have no exact parallel in any other known plant (Juniper and Burras, 1962;

Fig. 6.36–6.38 Details of the 'funnel' and gland-patch of *Cephalotus follicularis*: Zones 3 and 5 in Fig. 4.9*B*. SEMs by S. Clarke.

Fig. 6.36 The imbricate, downward-pointing hairs of the funnel region (cf. *Sarracenia* and *Darlingtonia* Zone 4.)

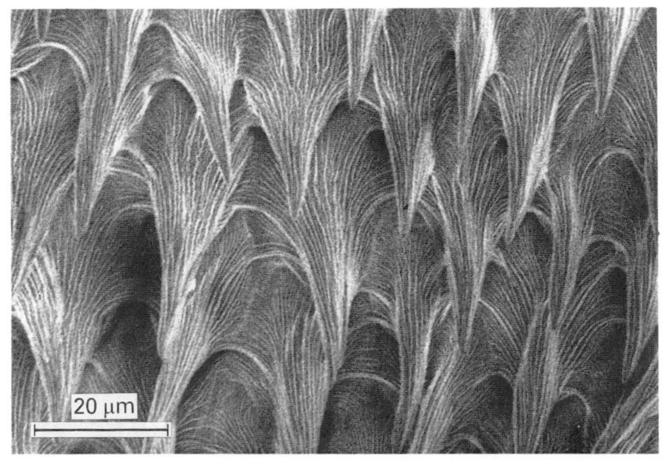

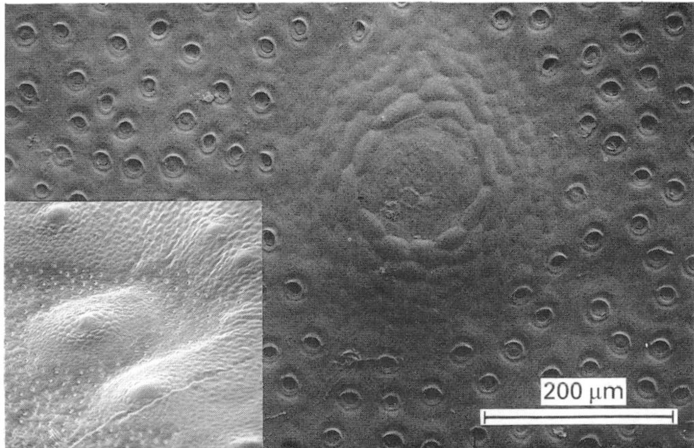

Fig. 6.37 Small inset bottom left. An area of the gland patch of Zone 5 of Fig. 4.9*B*, approximately 1 mm across. The whole micrograph shows the upper edge of the same gland patch at a higher magnification. One large multicellular digestive gland is in the centre and this gland is surrounded by the numerous single-celled 'small' glands, cf.Fig. 8.2A.

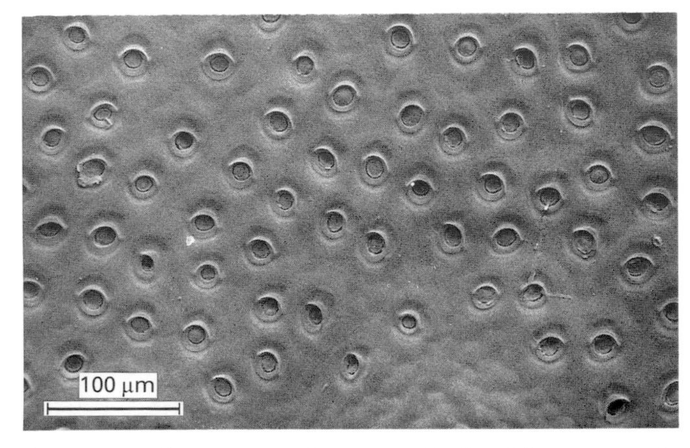

Fig. 6.38 Higher magnification detail of Fig. 6.37. These 'small' glands were once thought to be stomata, but the 'guard cells' of these symmetrical pores do not move and the pore is plugged.

Fig. 6.39 The intact, wax-scale-coated pitcher of a *Nepenthes* species (Zone 2 of Fig. 4.9*A*). This coat of scales is not readily detected under the scanning electron microscope (cf. Fig. 4.10) and is markedly thermolabile.

Fig. 6.40 Detail of the wax-scale-coated pitcher of *Nepenthes* species. (*A*) If the scale-like layer is removed with a camel-hair brush it reveals not the epidermal cells' cuticle but a more reticulate structure; the detached scales can be seen in Fig. 6.41 with an interpretation in Fig. 6.42. (Platinum-shadowed, carbon replica TEM).

(*B–D*) Wax crystals regrown on synthetic surfaces, e.g. millipore filter, from a chloroform or cyclohexane melt (SEM-liquid-nitrogen frozen specimen). (Unpublished work: B.E. Juniper, Tracy Scaysbrook and Sue Dossett, 1987.)

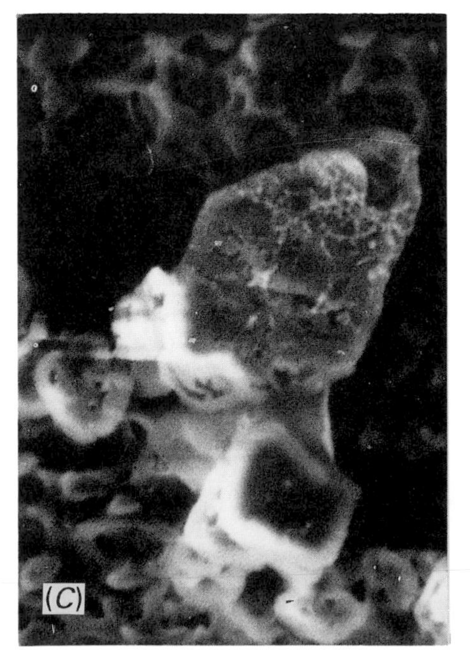

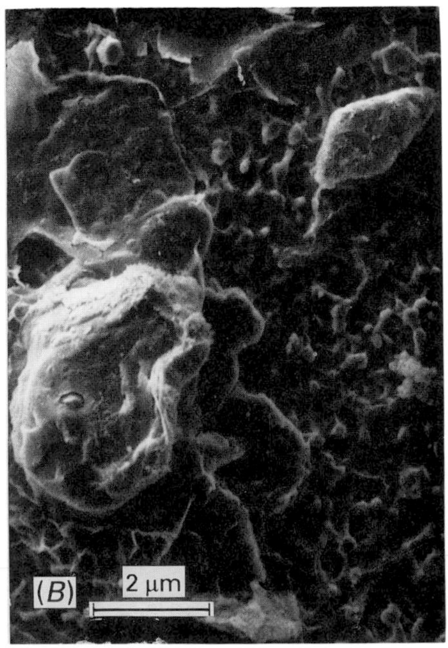

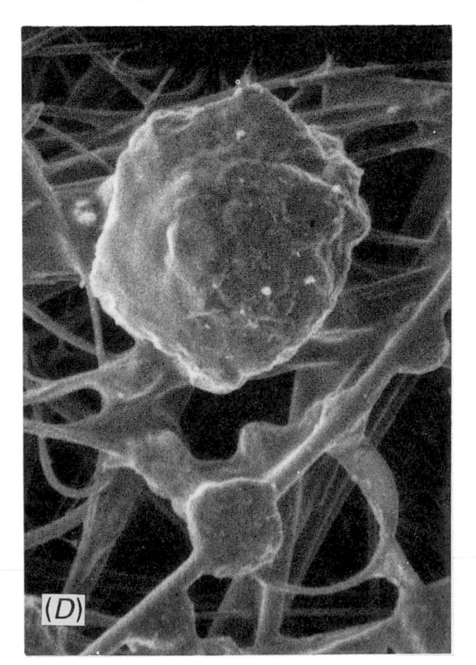

Fig. 6.41 TEM of platinum-shadowed carbon-replicas of the wax scales from a *Nepenthes* pitcher. Inset at higher magnification, one whole scale.

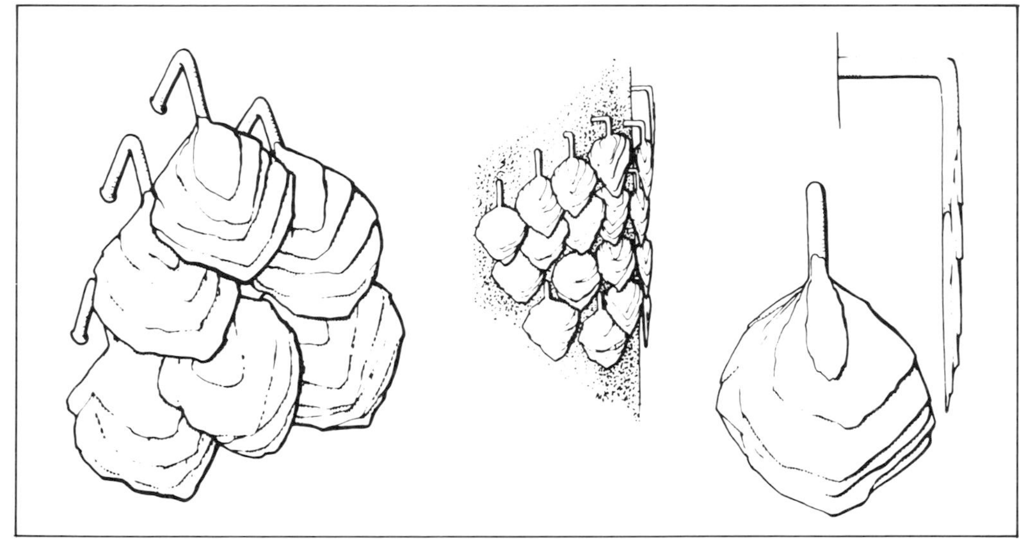

Fig. 6.42 Artistic interpretation of the structure of Fig. 6.41 *in situ*.

Martin and Juniper, 1970), although surfaces with similar functions are now known to be widespread (see page 299 and Fig. 18.3). Under the transmission electron microscope the intact surface is seen to comprise a layer of overlapping scales, each scale being irregular in shape, but about 1 μm across (Figs 6.39 and 6.40). When this surface is brushed away with a camel-hair brush there appears, not the true cuticle of the epidermal cells underneath, but another layer of wax, this time in the form of projections or ridges sticking out at right angles to the surface of the inside of the pitcher (Fig. 6.40). It is obvious from the ease with which the scales are detached from the supportive layer (Fig. 6.41) how the trap works.

The pad or the claw of the insect or the surface of a gastropod makes an attempt to adhere to this scale. The scale then pulls away readily from the underneath projection. Very large numbers of these scales (Fig. 6.42), and there must be several thousand *per cell*, will soon render the action of the claw of the insect or the surface of the gastropod foot ineffective. The behaviour of insects attempting to leave the traps is consistent. If they can climb at all on the walls of the pitcher they approach the peristome lip with great difficulty, but its smooth surface prevents them from scaling this barrier. They become very agitated and can be seen to wipe their pads against each other and against their mouthparts and abdomen as if in an effort to clean them. Finally, their adhesion fails altogether and they fall to the bottom of the pitcher.

Knoll (1914), without the help of electron microscopy, came very close to an explanation of the trap's success. He found that ants, for example, could climb comfortably on smooth polished glass or the plane surface of melted beeswax. But when he dusted the polished surface with talc or carbon the ant was unable to do so. He then removed the wax from the surface of *Nepenthes* pitchers with chloroform and found that ants could then escape. He dusted the solvent-cleaned *Nepenthes* surfaces with talc and found the trap efficiency restored. The surface of the leaf of *Iris pallida* behaves in a similar way. If the pads of insects that have been in contact with the *Nepenthes* surface are washed, the washings are found, under the electron microscope, to contain large numbers of wax plates of a curious structure which give a further clue as to the nature and function of the trap (Fig. 6.41 and inset).

Such a structure as shown in the inset of Figure 6.41 at first raises the problem of development. The most plausible interpretation to us is suggested in Figure 6.42. A surface of overlapping scales lies above projections and what appear to be ridges. We assume that the projections are first produced well before the pitcher lid opens. Surface replicas at this stage show the cuticle to be covered with small pegs. At some stage just before the lid opens, the production of a peg form shifts to production of a plate, and possibly under the weight of the growing projection the point of junction between peg and plate bends. This movement brings the plate parallel to the surface of the epidermal cell. Presumably the plate then goes on expanding by successive waves of wax secretion. There is some indication of waves of deposition of wax on a plate in the high magnification inset to Figure 6.41. The possible sequence of this development is shown in Figure 6.42. However, although the overall chemistry of this wax structure is known (Table 6.2) it at first seemed possible that the constituents or proportions of constituents change as the wax form develops. It now seems likely that, in a manner similar to other wax surfaces recrystallized from the melt under purely artificial conditions (Hallam, 1970; Jeffree *et al.*, 1975), *Nepenthes* wax-melts can be rebuilt on similar synthetic surfaces (Figs 6.40B–D).

No other crystalline wax is secreted by any other part of the plant, although areas secrete both sugars and proteolytic enzymes. Once formed, there does not seem to be any replacement mechanism if these scales are stripped off. Wax replacement does sometimes occur on other young plant surfaces (Martin and Juniper, 1970). Insects can negotiate the slippery zone of very old pitchers, suggesting that the bare, stripped epidermal cells themselves present no barrier.

The peculiar asymmetrical stomata on the surface of the slippery zone (Fig. 4.10) have been suggested by some earlier authors (see Lloyd, 1942, for discussion) to be a significant part of the trap. Although they cannot assist escape, since there is no stomatal lip to seize, their exact role is still a puzzle. These lunate cells are absent from the pitchers of *N. ampullaria* (Pant and Bhatnagar, 1977). No other significant interspecific differences have been seen, although a very wide range of surfaces from different species of pitchers has not yet been studied apart from a study of the variation in the slippery zone band's width (see Ch. 8).

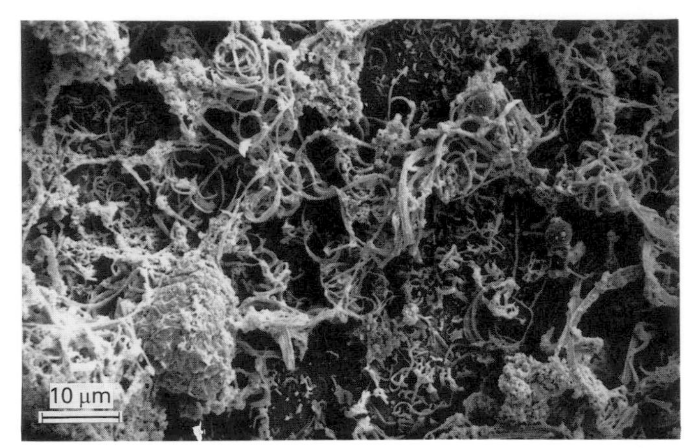

Fig. 6.43 Epicuticular wax on the adaxial surface of a leaf of *Brocchinia reducta*, cf. Figs. 6.39–42 (SEM by C. Merriman).

Table 6.2 Class composition of the epicuticular wax obtained from the inside of *Nepenthes* × *williamsii* pitchers[a] (source, Botanic Gardens, Oxford).

Preparative TLC fraction no.	Class	Percentage of total wax[b]
1	Alkanes (C_{19}–C_{35})	11.8
2	Alkyl esters (C_{34}–C_{54})	3.2
3	Alkanals (C_{20}–C_{34})[c]	28.2
4	1-Alkanols (C_{16}–C_{36})	26.9
5	Alkanoic acids (C_{16}–C_{36})	25.0
	Unidentified	4.9

[a] Epicuticular wax obtained from the surface of the inner 'iridescent layer' of the just-mature pitchers by swabbing with cotton wool soaked in chloroform. Yield c. 18 μg cm^{-2}. N.B. This wax layer was only partially removed by washing or immersion in the same solvent.

[b] Determined by preparative TLC and GC analysis using internal standards as described by Holloway *et al.* (1977).

[c] Estimated as the corresponding 1-alkanol TMSi ethers obtained after $NaBH_4$ reduction of the alkanals (aldehydes). Fraction gave strong positive reaction on TLC with the specific aldehyde detection reagent 4-amino-5-hydrazino-1,2,4-triazole-3-thiol (2% in M NaOH).

(Reproduced by kind permission of P. J. Holloway, Long Ashton Research Station, Ashton, Bristol).

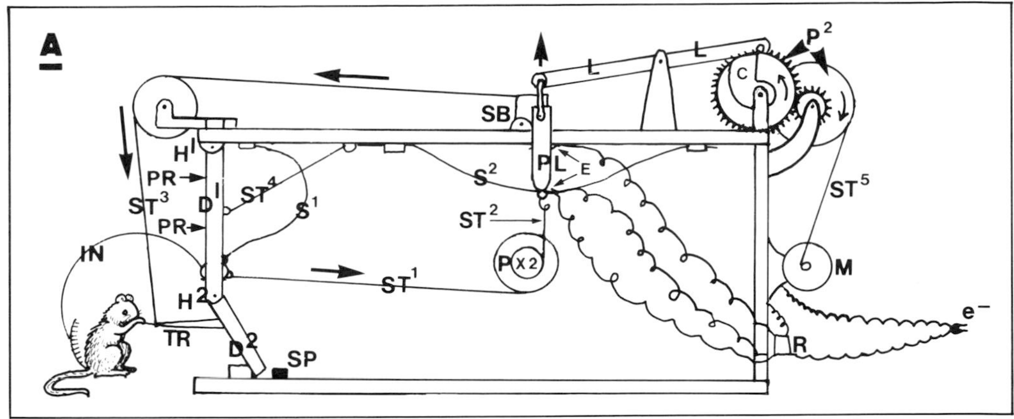

Fig. 6.44A An analogue of the *Utricularia* trap is provided by Lloyd (1942). The box-trap has two hinges (H^1 and H^2). Below the point H^2, the section D^2 swings independently from that of D^1 above. Pressure applied at PR does not allow the door to swing inwards but the rotation of D^2 on H^2, sufficient to clear the stop at SP, allows the door to swing inwards. An outward swing is prevented by the backstay ST^4. A handle, TR, on D^2, operated by the mouse, brings about the inward opening by pulling on the string ST^3, whereby the stop-block, SB, is pulled away from the top of the plunger (PL) allowing play to the spring (S^2). This spring then pulls on the string ST^2, actuating the double pulley Px2, one element smaller than the other in the ratio 1/3. The outer pulley then pulls on ST^1, rapidly opening the door. To the door front is attached the induction device IN, whose purpose, like the sudden inrush of water in a live trap, ensures the rapid entry of the mouse into the trap. At this moment an electric motor (M) starts, which is actuated by the closure of the contacts at E. The motor continues to rotate until the plunger, pushing the spring S^2 in its set position, allows the door to be pushed back into its set position by its spring S^1. When this is completed, the contact point on the plunger comes into contact with the contact point E below and the relay, R, then stops the motor. The power from the motor is applied to the plunger through the gear P^2 etc. ending in a cam, C, the whole being adjusted so that the cam comes into a position which allows the lever L to swing downwards when the door is actuated again by a second mouse. A digestion chamber may also be provided. (Redrawn from Lloyd, 1942, whose inspiration was drawn, in part, from Heath-Robinson.)

The Suction Traps

The *Utricularia* Trap

Darwin and other scientists of his generation believed that the prey itself pushed the door open and entered the trap, for the door opens only inwards and does not allow exit (Darwin, 1875; Meyers, 1982). Today, we know beyond doubt that the trap itself actively captures the prey. It is interesting to note that Darwin (1875) was not completely satisfied with his assumption because this did not explain the capture of tiny and weak creatures that are not strong enough to push the door open, e.g. nauplius of a crustacean and a tardigrade.

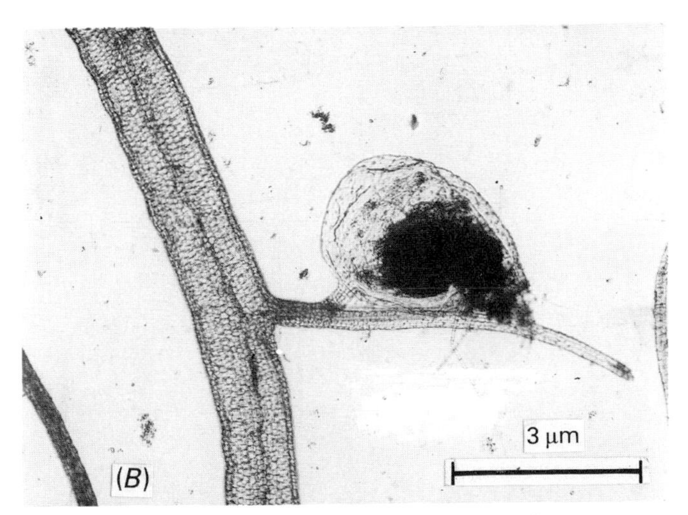

Fig. 6.44B Light micrograph of a bladder of *Utricularia neglecta* showing mud that has accumulated in the trap as a result of prey trapping (From Joel, 1982).

The 'Lloyd' analogy

We make no excuse for sharing both in Lloyd's admiration of the *Utricularia* trap nor for reproducing his proposed electromechanical model (Fig. 6.44*A*). Lloyd said at the time, 'A captious reader may find difficulty in accepting the analogy as complete'. We can only say that he would be right; but at least a purpose is served to indicate that the *Utricularia* is a complex bit of mechanism and offers, as yet, an intractable problem in evolution. Lloyd's analogy is totally inadequate, but probably not in the way he imagined it. We now know, through high-speed cinephotography (Oxford Scientific Films) that the recovery time of 1/33 of a second which he proposed is far too long; the actual speed must be shorter than 1/500 s, but the precise time is not yet known. With such a speed as a target Lloyd's simple electromechanical model would be far too slow and we would have to employ high-speed magnetic switches, linear actuators and proximity devices, all the gadgetry of an electronic age to produce a satisfactory analogue. Miniaturization would be a further problem. There is no complete natural analogue to this trap to our knowledge anywhere else in the plant kingdom, nor any satisfactory evolutionary path (see Chs 17–19). There are many trigger devices, often of impressive speed, such as the highly tensed style of *Sarothamnus* or the explosive seed dehiscence of *Ecballium* or *Impatiens*, but most of these are one-off events and none is able to carry out such rapid movements in opposite directions within fractions of a second.

Setting, firing and resetting of the *Utricularia* trap

When the trap of *Utricularia* is set, ready for trapping, it looks somewhat shrunken, because its lateral walls are drawn inward, often being concave (Fig. 6.45*A*). This shape is maintained through the negative hydrostatic pressure of the trap lumen (Brocher, 1911). When the trapdoor is stimulated by potential prey it opens and the trap walls, which are under tension, spring outward, forcibly sucking in water before the door rapidly slams shut. During this rapid movement, animals in the vicinity of the doorway are trapped by being swept (sometimes together with mud) into the bladder by the flow of water through the trapdoor (Fig. 6.44*B*) (Brocher, 1911, 1912; Meyers, 1982; Sydenham and Findlay, 1975).

At the end of this fast reaction, the trap looks swollen because its lateral walls are drawn outward, often becoming convex; the width of the trap through its centre is almost doubled during 'firing' (Sydenham and Findlay, 1973). The luminal volume increases by about 40% (Nold, 1934; Sydenham and Findlay, 1973; Sasago and Sibaoka, 1985a). It is the increase itself rather than the amount of the increase that is important in the process of trapping (Lloyd, 1942). This entire 'firing' process starting with stimulation and door opening and ending with closure of the door, takes about 30 ms (Lloyd, 1942; Sydenham and Findlay, 1973; Hill and Findlay, 1981). The firing of the *Utricularia* trap is considerably faster than the movement of other plant organs except, perhaps, for the movement of the column in some species of *Stylidium* (Findlay and Findlay, 1975 and 1981; Sydenham and Findlay, 1973; Hill and Findlay, 1981).

Because of the curved 'hinge-line' of the door, as the door opens, its plane is forced into a series of distortions which change successively. The changes can be seen in Figures 6.45*B* (front view) and 6.45C (side view). The extraordinary construction of the door with its semi-rigid outer layer and flexible buttressed inner layer as described by Lloyd (1935,

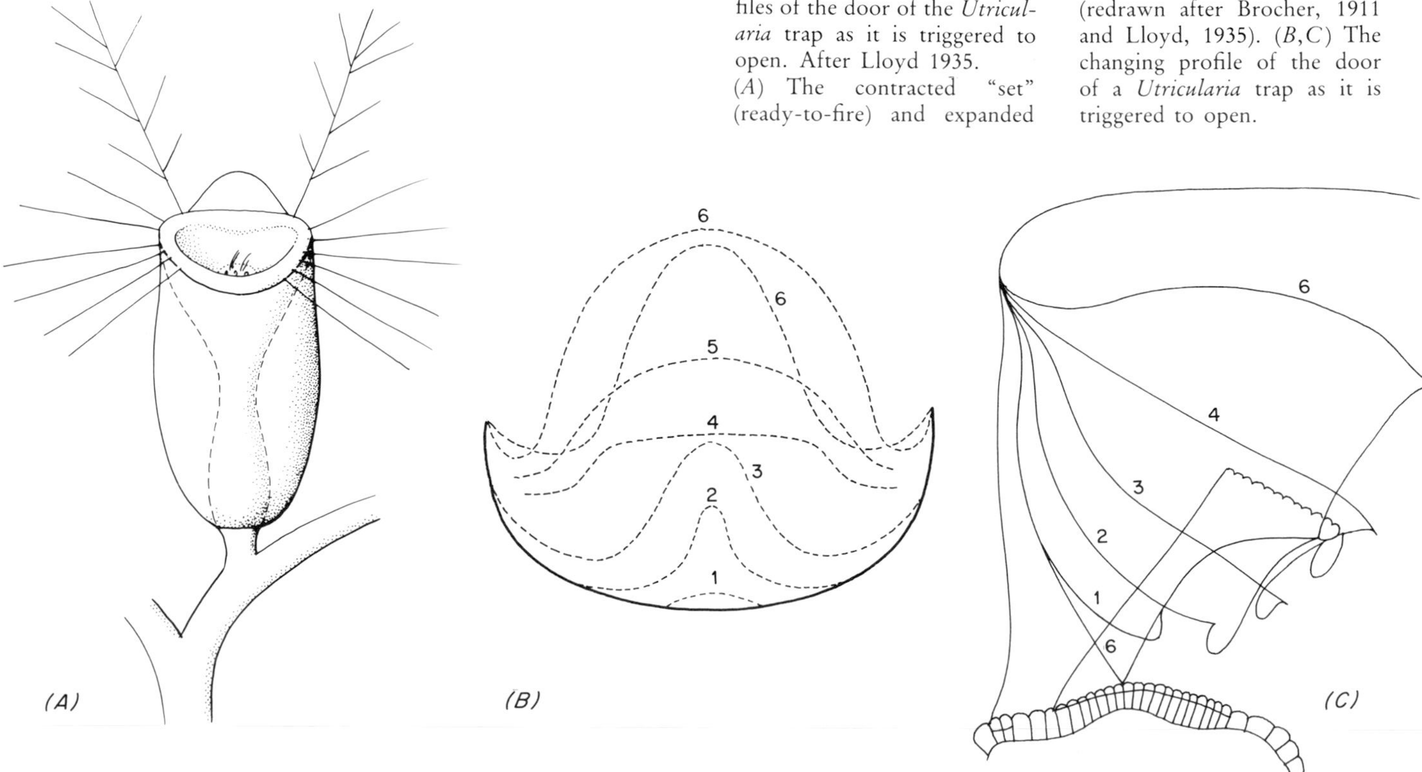

Fig. 6.45 The changing profiles of the door of the *Utricularia* trap as it is triggered to open. After Lloyd 1935. (*A*) The contracted "set" (ready-to-fire) and expanded "fired" profiles of a *Utricularia* (redrawn after Brocher, 1911 and Lloyd, 1935). (*B*,*C*) The changing profile of the door of a *Utricularia* trap as it is triggered to open.

his Fig. 22) is able to accommodate these very rapid, three-dimensional changes.

The trap restores its negative hydrostatic pressure a short while after firing. This is achieved by the removal of water from the trap lumen (see below) which is associated with an exponential decrease of the trap width until it reaches its compressed shape (Fig. 2 in Sasago and Sibaoka, 1985a). The trap is then able to capture again (Ekambaram, 1916; Sydenham and Findlay, 1975). Once set, the trap remains set for many days, regardless of changes in temperature (Sydenham and Findlay, 1975). The time required for the trap to reset, i.e. to re-attain the state of readiness, is species-dependent. It can occur in as little as 20 min, but usually the whole resetting period lasts 2–4 h. A single trap can be fired up to eight times at intervals of 40 min without appreciably upsetting the resetting rate (Sydenham and Findlay, 1975).

Negative internal pressure

The luminal hydrostatic pressure in fully set *Utricularia* traps was measured by Sydenham and Findlay (1973) and by Sasago and Sibaoka (1985a), both presenting similar rates of about −171 Pa with respect to the external solution, which is equivalent to 0.8 bar if the pressure of the external solution is assumed to be 1.0 bar. Upon firing, the hydrostatic pressure approaches that of the outside, but even then it is still slightly negative (Sydenham and Findlay, 1973), which seems to indicate either that the removal of water from the trap lumen is a continuous process which does not halt when the door opens, or that the door mechanically springs back to its closed position before pressure balance is achieved between the inside and outside of the trap. The structure of the door with its buttresses hints that the latter is likely to apply.

Water Removal from the Trap

Route of water out-flow

The glandular structures on both the outside and the inside of the trap are thought to be involved in the process of water extrusion from its lumen. Using tracer elements with both light and electron microscopy these glands were shown to be capable of passive absorption, due to their incomplete cuticular membrane (Fineran and Gilbertson, 1980; Fineran, 1985; and see also page 163). This view is consistent with anatomical and ultrastructural studies. Experimental studies of traps that were

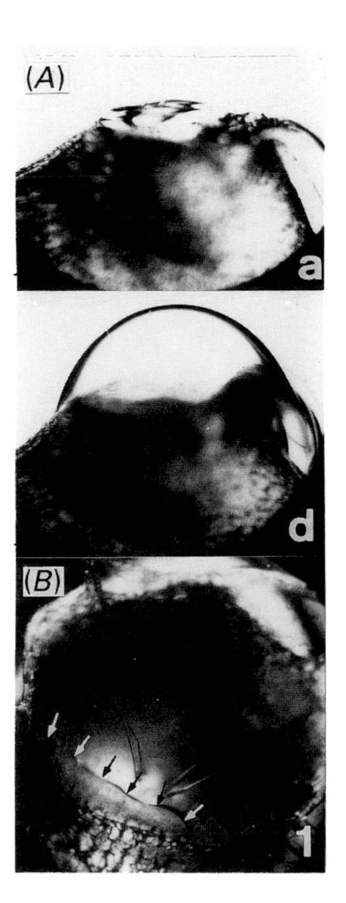

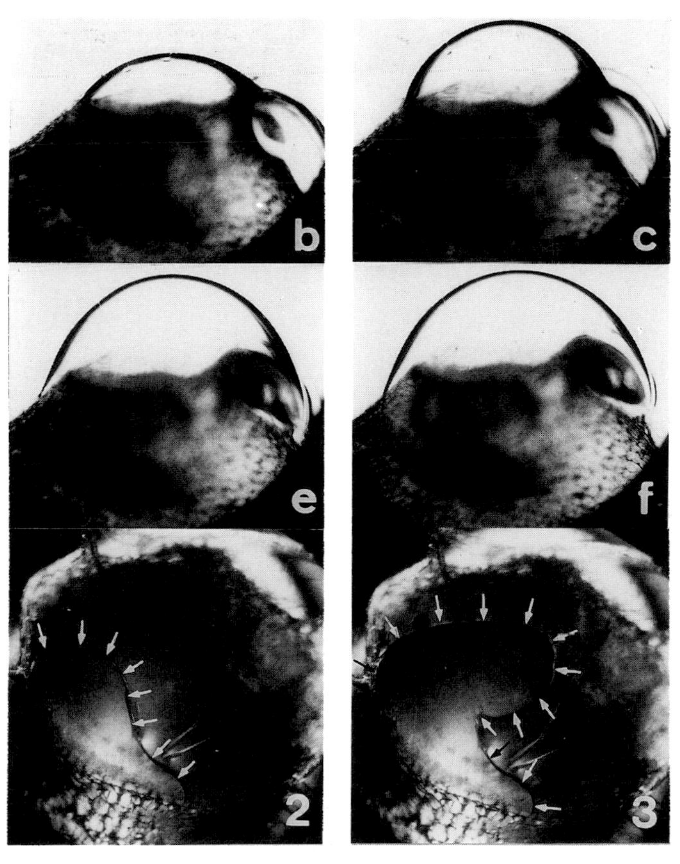

Fig. 6.46(*A*) Water outflow in the bladders of *Utricularia vulgaris*. The growth of exudate at the entrance of a bladder under paraffin oil. Lateral view: a, 0 min; b, 5 min; c, 10 min; d, 15 min; e, 20 min; f, 30 min. (*B*) Front view: micrographs 1–3 at 30 s intervals, black and white arrows delineate the margin of the excreted water. (By kind permission of Sasago and Sibaoka, 1985.)

immersed in paraffin oil have shown that water is discharged from the trap entrance (the mouth) (Sydenham and Findlay, 1975). Sasago and Sibaoka (1985a) observed that water appears first near the free edge of the door, and then gradually accumulates in front of the door (Fig. 6.46). Further, they could demonstrate that almost all reduction of the trap weight during the setting period, (0.6 μg in their system, using *U. vulgaris*) equals the amount of water which is exuded at the entrance of the bladder. They suggest that the site of excretion at the doorway is the specialized cells below the free edge of the door, i.e. the pavement epithelium (see also Fineran and Lee, 1975). Exudation was not observed elsewhere on the trap surface, and therefore they question the hypothesis that quadrifids provide the main absorptive site inside the trap. Instead they attribute this role to the bifids, which are densely developed in close proximity to the doorway.

Both arguments, the one regarding the site of

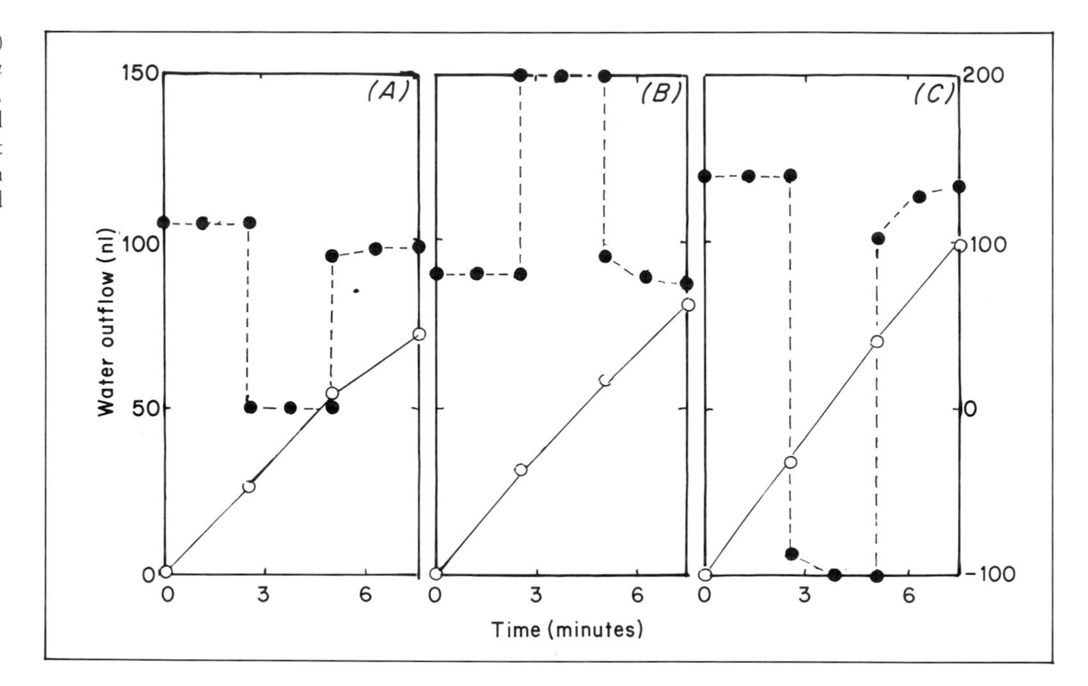

Fig. 6.47 Water outflow (○) and wall PD (●) of *Utricularia* bladders. Decrease (A), increase (B) or polarity reversal (C) of wall PD does not affect water outflow. (Redrawn, with permission from Sasago and Sibaoka, 1985.)

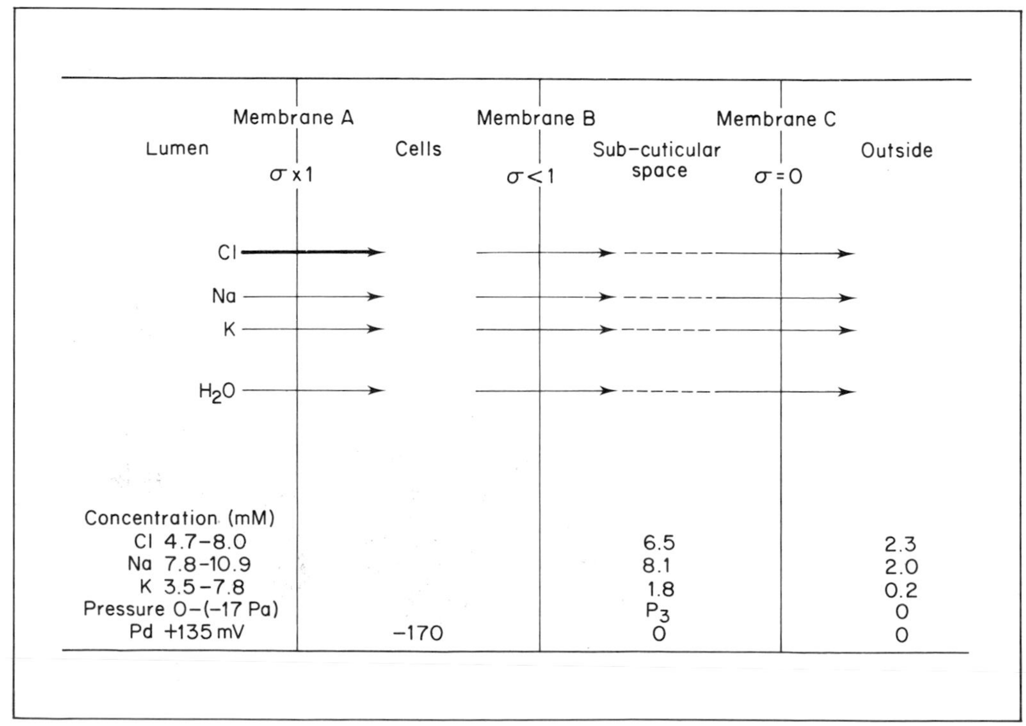

Fig. 6.48 A model of solute and water transport in the re-setting bladder of *Utricularia*. The heavy arrow indicates active transport and the light arrow passive flow. The dotted lines connecting the arrows indicate bulk flow of solution through the sub-cuticular space to the outside. The values of luminal ionic concentrations and pressure show the range between fired and reset bladders. The luminal Pd, apart from a relatively rapid change at the time of firing, is constant during the re-setting. The ionic concentrations shown for the sub-cuticular space are shown in Table 1 of Sydenham and Findlay (1975), and are values for the exudate collected from bladders under oil. For passive flow of water $P_2 > P_3 > 0$. (Redrawn with permission from Sydenham and Findlay, 1975.)

water removal from the lumen and the one regarding water excretion from the trap wall, were recently challenged by Fineran (1985). He suspects that the amount of exudate seen at the doorway represents only a portion of the water and ions expelled by the trap. Because of the concentration of glands at this site, the exudate is aggregated there and is therefore clearly apparent. Fineran suggests that the external glands on the trap surface (Fig. 6.14C), together with nearby glands on other parts of the plant, may also release a significant volume of water. But exudate released by these scattered glands over the outside of the trap might easily pass unnoticed because of the small volume at any one position (Fineran, 1985 and Figs. 6.49 and 6.50).

Factors affecting the rate of resetting

Temperature. The resetting rate is temperature dependent. The optimal temperature for trap resetting in *U. vulgaris* is between 24 and 34 °C. A

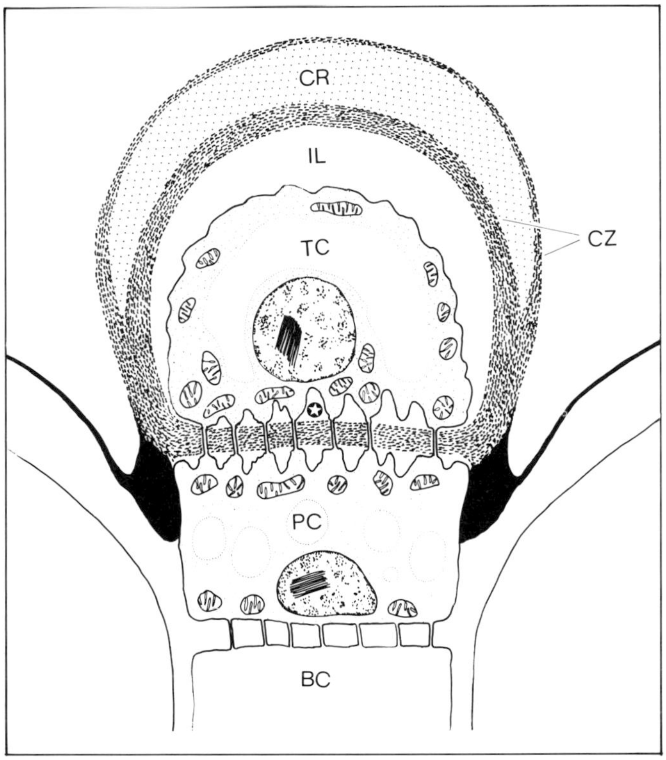

Fig. 6.49 Diagrammatic LS external gland of a *U. monanthos* leaf. The gland has the same basic organization as external glands found elsewhere on the plant. It differs in possessing a polysaccharide cap region (CR) differentiated within the cuticular zone (CZ) of the outer wall of the terminal cell (TC) and wall ingrowths (the star) on the inner transverse wall. The polysaccharide cap might function as a wick-like device by which water withdrawn from the sub-cuticular space might evaporate. The protoplast of the terminal cell possibly re-absorbs ions from the sub-cuticular space for recycling to the pedestal cell, to help maintain the standing osmotic gradient for expelling water from the gland. The nuclei of both the terminal and pedestal (PC) cells contain nuclear crystals. IL, thickened inner layer of terminal cell wall. (Redrawn, with kind permission, from Fineran, 1985; cf. Fig. 4.16C.)

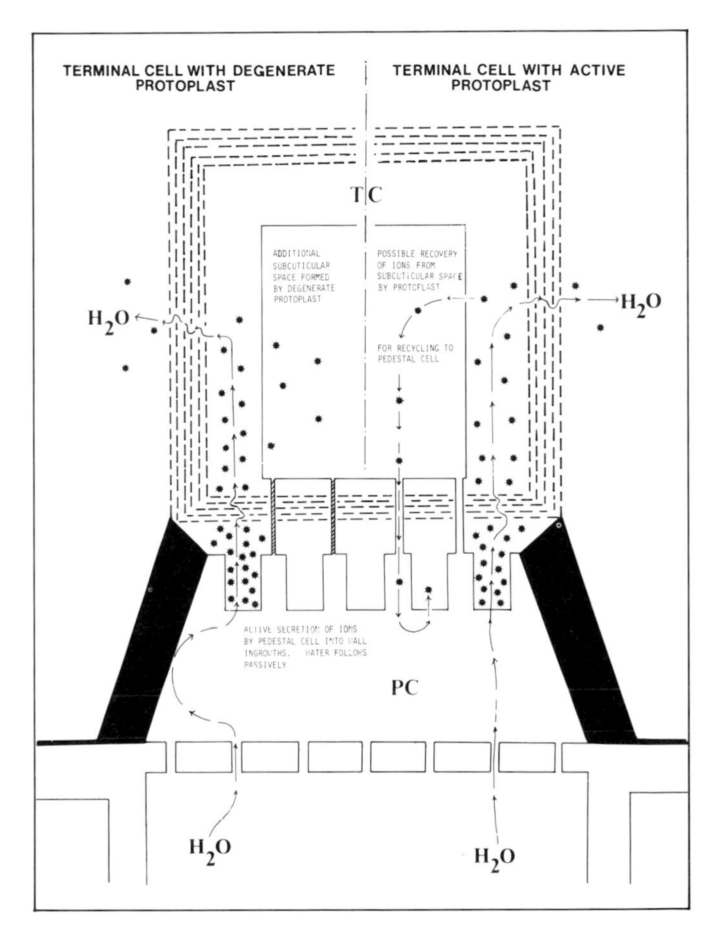

Fig. 6.50 Schematic diagram of the operation of external glands in water secretion in mature traps of *Utricularia*. At the left, the terminal cell is shown with a degenerate protoplast and the plasmodesmata occluded, with the lumen contributing to the sub-cuticular space used in the operation of a standing osmotic gradient between the pedestal cell (PC) and the cuticle of the terminal cell (TC). At right, the terminal is shown with a metabolically active protoplast. The pedestal cell actively secretes ions (*) into wall ingrowths developed on its outer transverse wall. Water transported from underlying tissues of the gland then flows passively from the pedestal cell along its potential gradient into the wall ingrowths. From there it follows an apoplastic pathway with ions in solution, into the sub-cuticular space and eventually leaves the gland through the permeable cuticle. The hydrostatic pressure developed within the sub-cuticular space may help to expel water through the cuticle by distending the pores. In glands where the protoplast of the terminal cell remains metabolically active (right) ions might be re-absorbed from the sub-cuticular space and recycled symplastically to the pedestal cell for maintaining the linked solute-water transport mechanism. (Redrawn from Fineran, 1985, with kind permission.)

temperature decrease to 6°C caused a rapid temporary 70% reduction in the resetting rate (Sydenham and Findlay, 1975; Sasago and Sibaoka, 1985b).

Metabolic inhibitors. The respiratory inhibitor, sodium azide, added to the external solution immediately after the trap had fired, reduced the rate of resetting to 60–65% of normal. After removal of the inhibitor, the resetting rate returned to near normal. Application into the trap lumen was more effective than only into the external bath. In the former case the restting rate was reduced to 20–40% of the normal value (Sydenham and Findlay, 1975; Sasago and Sibaoka, 1985b). Dinitrophenol (DNP), a commonly used uncoupler of respiration, similarly reduced the rate of transport of fluid from the lumen to the outside solution to 40–50% of the normal when applied to the trap lumen, and to 65% when given to the exterior.

Iodoacetamide, an inhibitor of the anaerobic stages of respiration in plant and animal cells had an effect on the rate of resetting only when applied to the trap interior. It reduced the resetting rate to a value approximately 25% of normal. Pentachlorophenol (PCP), potassium cyanide and monoiodoacetic acid were effective only when they were applied in the trap lumen. Sodium fluoride and dichlorophenyl dimethylurea (DCMU) had no effect on the water outflow (Sydenham and Findlay, 1975; Sasago and Sibaoka, 1985b).

pH. Changes of pH in the range of 3.4 to 9.0, when applied either in the lumen or outside, had no effect on the resetting rate.

Osmotic effects. When the fluid within the trap lumen is made increasingly hypertonic, the net transfer of water from the lumen to the outside decreases. Using polyethylene glycol (PEG 1000) Sydenham and Findlay (1975) increased the molar concentration of the trap lumen to slightly more than 15 mM, by which level the transport of water from the lumen is stopped. When the lumen was filled with 350 mM sucrose (in artificial pond water), the resetting was completely blocked. According to microscopic observations, incipient plasmolysis occurs in the arm cells of the internal glands at about this sucrose concentration (Sasago and Sibaoka, 1985b). It should, however, be remembered that sucrose is metabolizable, contrary to PEG, and that sucrose has been demonstrated to absorb readily from the lumen (Sydenham and Findlay, 1975), probably by the digestive glands. We have seen that water potential difference set up between the trap lumen and the transporting tissue affects water movement in resetting. On the other hand changes in external solute concentration do not affect water outflow (Sydenham and Findlay, 1975; see page 123).

Ionic composition and concentration. Salts added to the trap lumen reduced the rate of change of trap width in resetting. Sodium salts seemed to have a weaker effect on resetting as compared to calcium salts. Sydenham and Findlay (1975) showed that 1.5 mM Na_2SO_4 caused 60% reduction, whereas $CaSO_4$ in the same concentration led to a 74% reduction. Similarly, Sasago and Sibaoka (1985b) showed that incrementing of Ca^{++} or Mg^{++} in the lumen inhibits the outflow, but that of Na^+, K^+, Cl^- or NO_3^- shows no effect. The higher the salt concentration the stronger its effect on the resetting rate (Table 6.3). Further to their osmotic effect, these ions seem also directly to affect the transport process.

Wall potential difference. Increase and decrease in wall PD and its polarity reversal did not alter the rate of water outflow from the trap lumen (Sasago and Sibaoka, 1985b) (Figs 6.46 and 47). Clamping at various levels from 0 to 139 mV produced no observable change on the rate of resetting (Sydenham and Findlay, 1975).

Normally the potential difference between the vacuole of the trap wall cells and the outside at 20°C is around −160 mV, and the difference

Table 6.3 Effect of salt concentration in trap lumen on resetting rate of *Utricularia* traps. Values represent percentage reduction of the initial rate of change of trap width

	NaCl		$CaSO_4$		$NaNO_3$			$CaCl_2$	$MgSO_4$	KNO_3
mM	3	7	1.5	3	3	10	50	2.5	1	1
%	60±3	75±2	74±4	90±2	33±7	64±9	94±3	73±8	65±9	NS[b]
Author[a]	SF	SF	SF	SF	SS	SS	SS	SS	SS	SS

[a] SF = Sydenham and Findlay (1975): SS = Sasago and Sibaoka, (1985b).
[b] NS = not significant.

between the trap lumen and the wall cells is around 310 mV, so that the potential difference between lumen and outside comes to about 150 mV. These values do not change in the dark, but low temperature reduces them (Sydenham and Findlay, 1975). The positive potential in the lumen did not change with alteration of the pH of lumen fluid, and metabolic inhibitors like monoiodoacetic acid similarly had no effect (Sasago and Sibaoka, 1985b). On the other hand, 1 mM DNP decreased the wall potential difference by about 50 mV when applied to the trap lumen (not to the outside). The addition of sodium azide to the solution outside the trap caused changes in the wall potential (Sydenham and Findlay, 1975), but its application to the trap lumen showed almost no effect (Sasago and Sibaoka, 1985b).

The mechanism of water removal

The above mentioned experimental results showing that water removal from the trap lumen during resetting is independent of changes in PD (Fig. 6.47). Furthermore, estimates of the number of water molecules transported per ion from the lumen to the outside, eliminate electro-osmosis (Diannelidis and Umrath, 1953) as the possible mechanism of water removal during trap resetting (Sydenham and Findlay, 1975). Similarly, the observations that the resetting process is influenced by osmotic gradients set up by the addition of PEG and other solutes to the inner solution, but not to the outer, show that water outflow is not simply due to an overall osmotic gradient established across the trap wall by a one-step active solute transport between the lumen and the outside.

Based on a detailed study of the trap function, Sydenham and Findlay (1975) proposed a three-step model for the removal of water and solute from the trap lumen to the outside (Fig. 6.48) which is an extension of the model proposed by Curran and MacIntosh (1962) for transport in the rat ileum. A partially revised model was later suggested by Sasago and Sibaoka (1985b).

According to this model (Fig. 6.48), the movement of water from the trap lumen to the cells of the trap wall is a passive process which comes about in response to a local osmotic gradient. This osmotic gradient develops by active transport of Cl^- from the trap interior to the glandular cells on the inner side of the trap wall (see page 120), across the plasma-membrane designated in the model as membrane A. The active transport of chloride against the electrochemical gradient at membrane A is a respiration-dependent electrogenic process with sodium as the predominant accompanying ion. This flow of ions lowers the water potential in the glandular cells, which brings about an accompanying flow of water in the same direction, with an increase in the intracellular hydrostatic pressure (Sydenham and Findlay, 1975; Bentrup, 1979; Lüttge, 1983). This flow is regarded as the first step in the removal of water from the trap lumen. The water thus taken by the internal glandular cells is then transferred to the sites of water release on the outside of the trap.

In the second step, water is removed from the trap wall to the outside via the capital cells of the outer and middle zones of the pavement epithelium (Sasago and Sibaoka, 1985b), and presumably also via the external glands (Fineran, 1980; 1985). It is released into a sub-cuticular space in the outer side of the trap through membranes designated in this model as B. It has already been shown that when the external solute concentration is experimentally increased, leading to a drastic reduction in water potential around the trap, the rate of water outflow during resetting is not affected (see above). This lack of change seems to indicate that the process of water extrusion from the tissue to the outside is a bulk flow. The production of this bulk flow would require a hydrostatic pressure difference between the tissue and the outside. As we have seen above, it is assumed that such hydrostatic pressure develops in the cells of the trap wall as a result of the action of ion pumps (first step in the model). In order that water and solute would be able to move out across membrane B due to the intracellular hydrostatic pressure, the reflection coefficient of membrane B must be nearly zero. In other words, this membrane is assumed to have a higher permeability for chloride ions and also a greater hydraulic conductivity than membrane A. Thus there will be a passive flux of Cl^- and cations, leading a resulting flow of water across membrane B.

Though being the only acceptable model to date, a few points still need to be explained and further studied. The central idea of the model is that water removal is energetically driven in only one step, while the second step is passive. However, both published papers on the system (Sydenham and Findlay, 1975; Sasago and Sibaoka, 1985b) present significant but undiscussed results showing that, with the presence of sodium azide in the outside of the trap, as also with the presence of DNP, the rate of resetting is reduced. Further, it was shown that the addition of sodium azide to the solution outside the trap causes changes in the wall potential (see page 123). These results might indicate that ion pumps are also active in step 2. In this case, active transport of solutes would prevail across the plasma membrane of the glandular cells in the pavement epithelium and perhaps also in the external glands. These pumps would force chloride into the apparently closed sub-cuticular space under the velum and into the degenerated terminal cells of the external glands, which are covered with a discontinuous cuticle. In both sites a standing osmotic gradient, which enables the systems to release large amounts of water with a relatively small amount of solutes, can develop not only as a result of the existence of sub-cuticular spaces as suggested by Sydenham and Findlay (1975), but also (or instead) thanks to the presence of mucilage or mucilaginous wall ingrowths. These ingrowths might provide the systems with low solute diffusion coefficients (see Ch. 8).

Mucilage is secreted by terminal cells in the pavement epithelium. It accumulates in the cell walls and beneath the cuticle until this ruptures and allows the mucilage to stream out from the cell (Fineran, 1985). The velum may confine the secreted mucilage of all glands along the edge of the door in some *Utricularia* species, while in others the ruptured cuticle of each head cell holds its own mucilage separately. In both cases the mucilage covers the head cells enabling the development of standing osmotic gradients. In the external glands the outer wall of the terminal cell is characteristically thickened and consists of an outer open cuticular zone and a wider inner zone of unimpregnated wall (Fig. 6.49). This latter develops by the fusion of the meshwork of previously developed labyrinthine wall. The mucilaginous nature of the wall seems to permit the development of effective short distance osmotic gradients for the release of water from the glandular cell underneath, which has characteristics of both an endodermal cell and a transfer cell. Another problem arising in this system is concerned with the calculations of ion availability. If the reduction in trap luminal volume is approximately 40%, and if indeed the ionic concentration of the exudate is several times greater than the luminal, one would expect a drastic decrease of solute concentration in the lumen during resetting. Instead, luminal ionic concentration is found to increase significantly. This is another indication that ion pumps are also active on membrane B.

The relatively weaker and delayed response of the trap to the application of inhibitors to the outside of the trap can perhaps be explained by the fact that the active cells are 'hidden' either behind the velum or behind degenerated cells, and that the active substances need to penetrate against a flow of water.

Possible mechanisms of operation of the door

Two different theories exist today regarding the possible mechanism by which the *Utricularia* trap opens its door as the result of a touch-stimulus:

(i) The trigger hair acts as a handle pushing the door edge, which then bends along its central hinge. Due to the lower hydrostatic pressure inside the trap, water flows through the small slit between the raised door edge and the threshold, pushing the door open for a short while until the pressure difference no longer exists (Heide-Jørgensen, 1981). The role played by the trigger hairs, according to this theory, is purely mechanical.

(ii) The trap of *Utricularia* operates through an excitatory process. Such a process might be one in which the initial mechanical stimulus is transduced to an electrical signal which is propagated to other cells of the trapdoor. Some plasma-membrane modification causes a loss of turgor, which results in rapid bending of the door under the inwardly directed hydrostatic pressure gradient, thus allowing fluid into the lumen (Sydenham and Findlay, 1975).

Unlike the traps of *Dionaea* and *Aldrovanda* and other excitatory organs which can be shown to respond to external electrical stimuli (see page 104), there is no agreement on the existence of a similar response in *Utricularia*. Diannelidis and Umrath (1953) reported that they were able to 'fire' the bladders of *U. vulgaris* with pulses of electric current on either side of the trap. In contrast, Lou and Hsueh (1950) and Sydenham and Findlay (1975) failed to find any such response. The former reported that electric currents and shocks of various magnitude and frequency failed to induce a response in the trap. The latter group used a pulse of current between two plate-electrodes in the external solution, and reported that current pulses of up to 20 A did not 'fire' the traps. In addition, Sydenham and Findlay (1975) passed pulses of current of up to 3 s duration between the lumen and the outside solution through a microelectrode in the trap lumen. Even changes of 250 mV from its resting value of about 130 mV to +380 mV did not fire the bladder trap. Despite this controversy, one should stress that conspicuous changes in the potential difference between the trap lumen and the external solution were recorded as the result of 'firing' the trap (Sydenham and Findlay, 1975) as shown in Figure 6.48. The receptor potential (see below) is recognizable when the trap is 'fatigued' because only then is the threshold rate low enough to permit a detectable time gap between stimulation and firing (Fig. 6.51). This reproducible change in potential difference might be regarded as an electrical stimulus; possibly an action potential, propagated from the trap door to the trap wall. To test this assumption one should check whether this phenomenon is the result of mechanical stimulation of the trigger mechanism or whether it is the result of short-circuiting of the system by water leakage when the door opens. Sydenham and Findlay (1975) were of the opinion that there is some evidence supporting the excitatory step in the firing process. They showed that there is indication of a loss of turgor in the cells of the trapdoor which occurred as a result of the mechanical stimulation, and could explain this only by the electrical hypothesis, not by the mechanical. In addition they found that, during repetitive stimulation, the potential difference gradually drifted to a less positive value, which is consistent with excitatory processes, provided certain assumptions are made. For instance, when the amount of water lost by cells of the door decreases during repetitive stimulation, the loss of turgor by cells adjacent to the trigger hairs after the first stimulus would be sufficient to fire the bladder. Following each additional stimulus an increasing number of cells should lose turgor.

In old traps and in those that had been treated with sodium azide or low temperature, the lack of response to the normal intensity of mechanical stimulation is also consistent with a loss of excitability in cells of the door (Sydenham and Findlay, 1975).

Although these two theories are superficially contradictory, we suggest that both apply to at least some species of *Utricularia*. As mentioned on page 70, the traps of some species lack trigger hairs. These species, which might be regarded as primitive, are more likely to be operated by a purely mechanical mechanism. Other species might possess triggering systems based on the propagation of electrical signals by which similar bending movements are performed by the trapdoor. Electrical mediation should improve the amplification effect gained by the build-up of negative hydrostatic pressure in the trap, which converts the small stimulus exerted on the trigger hair into movement of the whole door. Electrical mediation should simplify this conversion by rapidly changing the permeability of the cells responsible for the door deflection.

Operation of the Trap of *Genlisea*

All available knowledge of the trap of *Genlisea* (see page 71) has been obtained by examining dried or preserved material rather than living plants. There is no description of the trap in action in a living, albeit artificial, environment; therefore any explanation of how the trap works is purely speculative.

Darwin (1875) thought that 'animals are captured in *Genlisea* by a contrivance resembling an "eel-

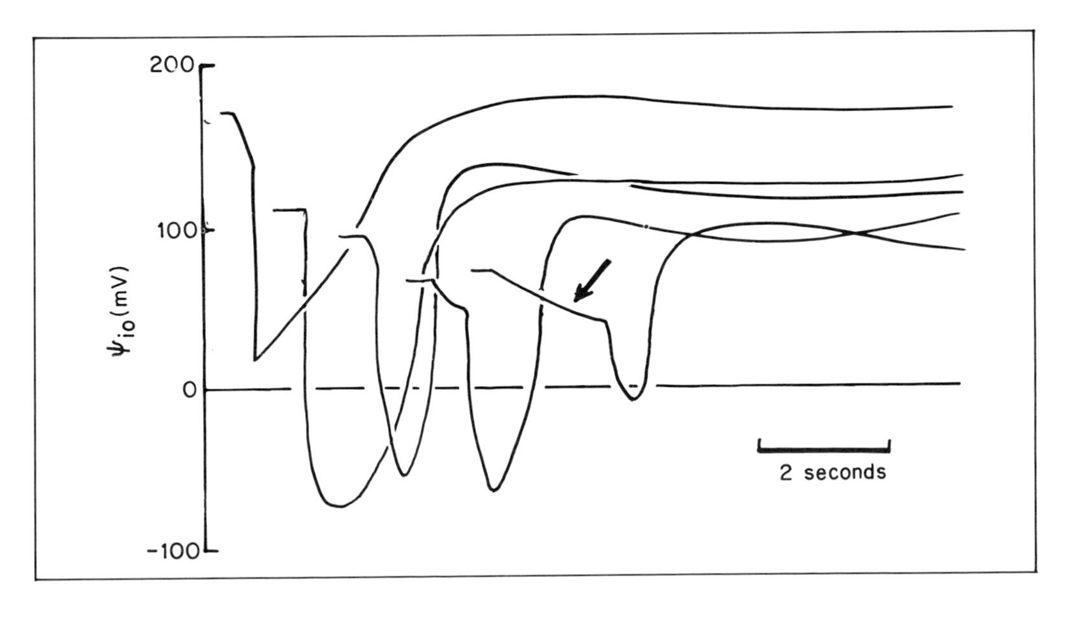

Fig. 6.51 A series of changes in the potential difference, following successive firings of one bladder at intervals of about 2.4 ks. Each consecutive curve has been displaced in time. A pronounced hyperpolarization is seen after each potential drop. The apparent lag-phase, indicated by the arrow, may indicate a receptor potential which is more obvious when the trap is 'tired'. (Redrawn, with permission, from Sydenham and Findlay, 1973.)

trap", though more complex'. This view was adopted in all descriptions of this plant until very recently. The mechanism *seems* quite obvious. As stated by Lloyd (1942) each section of the arms and tube is of the form of the entrance to an 'eel-trap', or 'lobster-pot', with its funnel extending into the next eel-trap below, the whole organ comprising a series of such traps. Darwin wrote:

> 'It is not known what induces small animals to enter, nevertheless it seems that the trap somehow inveigles them toward the interior. The sharp incurved hairs render their egress difficult, and as soon as they passed some way down the neck, it would be scarcely possible to return, owing to the ridges and to the downward pointing hairs. (Darwin, 1875).'

In this manner small animals will be trapped and conducted to the utricle. Lloyd (1942) observed that considerable signs of prey disintegration first occur only in the lower half of the tube, which means that the arms and upper half of the tube mainly serve as the 'eel-trap', while prey is digested in the lower parts of the trap.

We therefore see that the trap is considered, by these authors, to be passive, employing a trapping mechanism similar to that found in certain pitcher-plants, (see page 108 and Fig. 4.11) where detentive hairs permit only forward movements and block the way back, like a valve.

This theory seems satisfactory as long as only animals are found to be trapped in the system: animals can provide the system with the energy which is required for their own movement into the digestive pool. However, recent observations by Joel (unpublished) revealed that *Genlisea* traps also accumulate soil particles in addition to animal remains in their utricles and tubes. Considering the incapacity of soil particles for spontaneous movement, and taking into account that external water currents are unlikely to reach the narrow and closed space of the tube and utricle, we must assume that somehow the trap itself is also responsible for the prey movement along the arms and the tube. Further studies are therefore needed to elucidate the trapping mechanisms in *Genlisea*. The success in growing *Genlisea* under cultivation (Slack, 1986) is a step forward in this direction.

PART III

Nutrition and Digestion

CHAPTER 7

The Exploited Prey and its Nutritional Value

General Considerations

Darwin (1875) was the first to demonstrate beyond any reasonable doubt that certain plants could catch insects and were specifically adapted to that function; and that, if provided with insects and/or other animal proteins, showed enhanced growth.

However, as recently as 1974, Daubenmire was able to write: 'None of the above carnivorous plants [*Sarracenia*, *Nepenthes*, *Drosera*, *Dionaea*, and *Utricularia*] seems in the least dependent upon its animal prey for nitrogenous compounds; therefore it must be concluded that the carnivorous habit is *only an incidental feature of their nutrition*' [our italics].

Although, since Darwin, there has not been any serious doubt about the carnivorous habit of certain plants, what he failed to demonstrate, and what has remained a source of contention ever since, is whether or not insects are essential to the full development of a particular plant and, if so, which particular nutrients are required. There is a corollary to this question. If carnivorous plants have made available to them inorganic sources of nutrients, are they as successful as if supplied with a varied insect diet?

There can be no doubt that axenically grown cultures of, e.g. *Utricularia* species, can, if given the right light conditions, grow and flower without any insect supplement whatsoever (Pringsheim and Pringsheim, 1962; Dore Swamy and Mohan Ram, 1969, 1971). However, as Gibson (1983a) has pointed out, laboratory physiology studies which demonstrate that a single carnivorous plant can grow to maturity without catching a single insect, are irrelevant: such experiments do not consider nutrient availability of bog soils, or the importance of rapid growth and seed-set, to lead a successful fugitive existence. Nor can there be any measure of the competitive nature of most carnivorous plants' existence in the sheltered nature of a laboratory greenhouse or growth chamber.

It cannot be overemphasized that, as the brief review below will show, as Figure 7.1 graphically illustrates and as we shall discuss on page 144, there is no simple answer on the cost/benefit ratio of trap systems. In this, as in almost every other respect, the carnivorous plants lie along a spectrum from (almost) total indifference to almost total dependence for their full functioning and spread upon a broad arthropod diet. From this diversity has sprung some of the misunderstandings implicit in Daubenmire's statement.

The Insect as a Source of Food

An insect, in comparative terms, is a rich source of food. The mean energy content of insects is 22.8 J mg^{-1} dry weight, whereas that of land plants is 18.9 J mg^{-1} dry weight (Cummins and Wuycheck, 1971). These calculations do not, of course, take into account the fact that the chitin component of a trapped insect is, except to a very limited extent (see p. 202), discarded. The extent of the discard by a carnivorous plant must, of course, be variable, and is probably greater amongst the fly-paper types as opposed to the pitcher-plant types. Dixon *et al.* (1980) measured the nitrogen content of *Drosophila* carcasses on *Drosera erythrorhiza* leaves. One hundred *Drosophila* flies (total 2.68 mg nitrogen) were applied to ten plants over a two-week period. The carcass residues remaining two weeks after the end of the feeding period contained a total of 0.635 mg nitrogen. Thus 73.6% of the insect nitrogen was consumed either by the plant or by various microbial agencies. In a

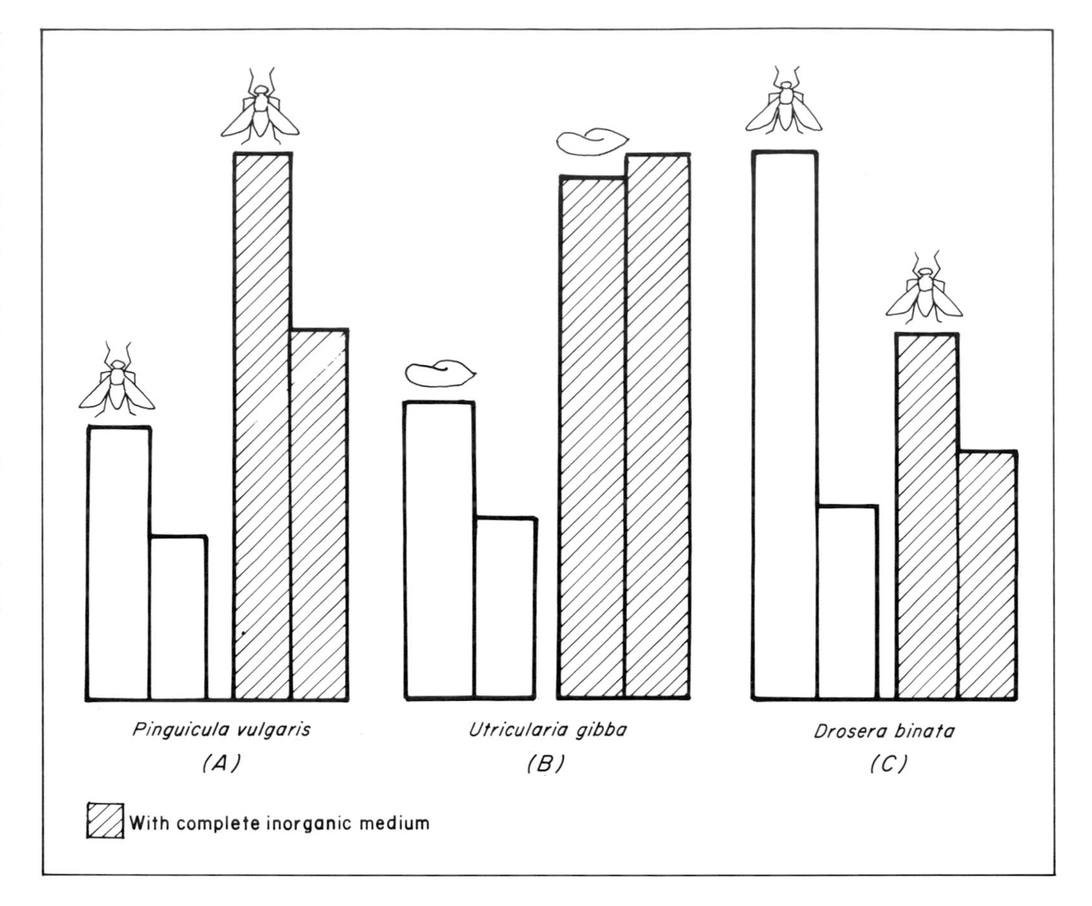

Fig. 7.1 A comparison of relative vegetative growth to illustrate the differences in response by different carnivorous plants to inorganic and/or organic nutrition.
(*A*) *P. vulgaris* from Aldenius *et al.* (1983) demonstrating the superiority of autotrophic nutrition over the heterotrophic state, but with the enhancement, overall, by small dipterans.
(*B*) *U. gibba* from Sorenson and Jackson (1968) again showing the superiority of an inorganic medium, but showing no enhancement of vegetative growth by feeding with *Daphnia*.
(*C*) *D. binata* from Chandler and Anderson (1976a) showing the markedly superior rates of growth by plants fed with *Drosophila melanogaster* alone as compared with those fed inorganically or with an insect supplement.

similar experiment, ^{15}N feeding studies indicated that 76.1% of the nitrogen was absorbed by the *Drosera* plants (Dixon *et al.*, 1980).

Reichle *et al.* (1969) have calculated that of six major nutrients required by plants the mean percentage of dry whole insect biomass is: nitrogen = 10.5, potassium = 3.18, calcium = 2.25 (excluding calcium-rich Diplopoda), phosphorus = 0.594, magnesium = 0.094 and iron = 0.024. Gibson (1983a) therefore calculates that a mature pitcher of *Sarracenia leucophylla* catches roughly 0.533 g (= 3.2 g × 16.6%) of these major nutrients during a season. These observations will be discussed in detail in the sections below.

The Utilization of the Insect as a Source of Food

Insects then are both a concentrated and diverse source of organic and inorganic material. To what extent can the equally different carnivorous plants take advantage of this cornucopia?

Insects as a carbon source

All carnivorous plants are chlorophyll-containing, photosynthesizing eutrophic organisms. Some, such as *Dionaea*, would seem to have devoted a significant fraction of their biomass to the trap system (see page 145). Others, such as *Catopsis*, the epiphytic bromeliad, have a marked yellow-green appearance suggesting either that the chlorophyll is not fully developed or is masked in some way. Nevertheless, there is no evidence that any carnivorous plant can grow heterotrophically using insects as the sole carbon source, even when the supply of prey is saturating. In fact the evidence from ecological tolerance (see Ch. 3) suggests most strongly, in almost every case, that carnivorous plants, even at the seed germination stage, are dependent upon uninterrupted sunlight, and shaded colonies of almost every species, deteriorate rapidly.

The experimental evidence for this belief is, however, limited. Chandler and Anderson (1976a) grew plants of *Drosera whittakeri*, both in the light and dark and in combinations of inorganic and insect feeding. The results are shown in Table 7.1. The growth of any dark-grown plant was poor, regardless of the feeding method.

As we shall see on page 131, in contrast with carbon nutrition, both the nitrogen uptake and the carry-over of nitrogen to subsequent generations is significant. In Dixon *et al.*'s (1980) observations,

Table 7.1 Effect of light on the growth of *Drosera whittakeri* in the absence and presence of insects

Treatment			Increase in dry weight per plant (mg)
Addition of salts to water culture	Insect application	Light regime	
Complete salt solution	nil	light	47.8 A
Complete salt solution	nil	dark	12.8 B
No additions	applied	light	40.8 A
No additions	applied	dark	9.2 B

Plants of *D. whittakeri* grown in aerated water cultures with or without inorganic salts. The initial dry weight of plants was 8.7 mg/plant. The plants were raised in the laboratory; light-regime under a bank of six fluorescent tubes containing two 'grow-lux' tubes (total light intensity: 8.5 watts/m^2); a dark regime under a box which decreased the light intensity to 1.5 watts/m^2 and allowed light mediated development without any significant photosynthesis. Values followed by the same letters are not significantly different ($P<0.05$).

Reproduced with permission from Dixon *et al.* (1980).

over 70% of the nitrogen was carried over from one generation of *D. erythrorhiza* to another but ^{14}C-labelled insects gave a very different picture. Only 47.1% of the applied label was recovered in the daughter and replacement tubers at a September harvest.

It is, also, regularly observed that most carnivorous plants, but not those adapted specifically to forest conditions (see page 26) fail to grow completely in low-light conditions. Green (1967) noticed that *Nepenthes* species in private collections, or botanic gardens in too great a shade, failed to develop pitchers. Failure of pitcher development due to lack of light is also suggested by Schmid-Hollinger (1979). On the other hand, Lüttge (1964c, 1983) fed a *Nepenthes* pitcher for 17.5 h with ^{14}C-alanine, and a large number of different compounds in various fractions of the pitcher wall tissue proved to be radioactively labelled. Lüttge however agrees that it is unlikely, except possibly in the case of *Utricularia* that we shall review below, that carnivorous plants can grow heterotrophically with animals as the sole carbon source. Certain species of *Utricularia*, it would seem, can make significant dry-weight increases if supplied with sucrose and acetate in the correct proportions in the growth media (Harder, 1970a, b).

On the other hand, *Utricularia* species such as *U. purpurea* in a natural environment only have the ability to use CO_2, since they are not known to be C_4 plants, and seem unable to use bicarbonate (Moeller, 1978). This inability to use other sources of inorganic carbon, unlike such other similar aquatic plants as *Elodea canadensis* and *Ceratophyllum demersum* may restrict *Utricularia* to acid waters.

Insects as sources of nitrogen, phosphorus, potassium and trace elements

As we shall see on page 138 onwards, the diet of most carnivorous plants is catholic in the extreme. For example, in one study by Erber (1979), of ten *Nepenthes* pitchers, he found the corpses of 1994 individual arthropods of 150 animal species. Wray and Brimley (1943), investigating *Sarracenia purpurea* pitchers, found victims belonging to 14 orders and 115 families of insects, not counting the mollusca. Given the range of the diet, it is virtually impossible to say at any one time what balance of nutrients a carnivorous plant may be receiving, but on the whole it would appear that there is seldom a shortage.

An insect as a source of nitrogen and phosphate

Chandler and Anderson (1976a) studied the nutrition of *Drosera whittakeri* and *D. binata* grown on sand cultures. They found that the application of insects (*Drosophila melanogaster*) (one insect per plant per week after the residue of the previous insect had been removed) enhanced growth of the plants in all but one soil condition. They noticed enhancement in the absence of:

(i) soil inorganic salts
(ii) omission or very low concentrations of inorganic nitrogen
(iii) absence of inorganic sulphate.

There was a failure of enhancement when insects were fed to *D. whittakeri* on a complete medium, but there was enhancement of *D. binata* on a complete inorganic medium.

We shall observe these specific different responses to insect feeding in several other examples below, emphasizing that few generalities can be made in carnivorous plant nutrition. Chandler and Anderson also noted that insects, in these two species, did not appear to be an important source of micro-elements. Application of insects to plants raised in the absence of Mo, Cu, Zn, B and Mn, caused no significant increase in growth (but see below). Watson *et al.* (1982) working with *D. erythrorhiza* in the Perth area of Western Australia found that an arthropod diet was sufficient to supply 100% of the observed nitrogen and phosphate pool. Arthropods contributed a negligible fraction of the potassium content of the plant, and here soil sources were predominant.

Likewise, Christensen (1976) fed eight plants of *Sarracenia flava* for five months under controlled conditions. Her experimental distribution was similar to that of Chandler and Anderson (1976a), i.e. low nutrients – no insects; low nutrients with insects; high nutrients – no insects; high nutrients with insects. After the five months the leaf tissue was analysed for N, P, Ca, Mg, and K. As expected, plants grown in nutrient-deficient environments had lower tissue concentrations of all the nutrients measured. The availability of insects resulted in significantly higher tissue concentrations of N and P, but had no effect on Ca, Mg, and K.

These data clearly indicate that, in low nitrogen and phosphorus environments, carnivory, at least in certain droseras and *Sarracenia flava*, may be particularly important. For potassium and certain other trace elements, the issue is less clear-cut.

A slightly different pattern of response to insect feeding was seen by Aldenius *et al.* (1983) in *Pinguicula vulgaris* (Fig. 7.1). In most cases, plants supplied with insects alone or in combination with fertilizers showed higher values for dry weight, number and length of leaves, and concentrations of nitrogen and phosphorus. The difference between this plant's response and that of *Drosera binata* above was that the amount of nitrogen absorbed when insects were added to fertilized plants was larger than the insects contained. Thus, *P. vulgaris* uses both nitrogen and phosphate from insects, but at the same time either these or some other components are used to enhance the uptake of major minerals by the roots. A comparison of these two rather different situations, and that of the rootless *Utricularia gibba*, is given in Figure 7.1 (and see Lüttge, 1985).

Confirming the above situation, but adding the information that nitrogen nutrition can be transferred to subsequent generations is the work by Dixon *et al.* (1980). They were able to show that if they fed ^{15}N-labelled *Drosophila* flies to leaf rosettes of *Drosera erythrorhiza* the result was an enrichment of leaves, stems, daughter and replacement tubers with ^{15}N. A transfer of 76% of the labelled nitrogen from the insects to the *Drosera* was recorded, and by the end of the growing season the new set of tubers had 70% of the applied ^{15}N. The ^{15}N was traced through two subsequent seasons of growth to study the extent of carry-over of insect-derived nitrogen between successive generations of tubers.

Seasonal changes in nitrogen of *D. erythrorhiza* in natural habitat suggested a high degree of transfer of nitrogen from the tubers to vegetative parts and vice versa. Tubers had 50–60% of their nitrogen in soluble form, principally as arginine. Stems and leaf rosettes had lower proportions of soluble nitrogen and less arginine. Arginine in the tubers became labelled with ^{15}N after feeding with ^{15}N-labelled *Drosophila*.

Arthropods caught by naturally growing clones of *D. erythrorhiza* included Collembola, Diptera, Hymenoptera, Hemiptera, Coleoptera and Arachnida. Assuming there was a 76% transfer of nitrogen from the fauna to *Drosera*, the season's catch of from 0.25 to 0.39 mg nitrogen per plant provided the equivalent of from 14 to 21% of the nitrogen transferred to the new season's tubers, and from 11 to 17% of the plant's net uptake of nitrogen during a growing season. A similar rapid transfer of nitrogen from a fed trap (in this case *D. rotundifolia*) was observed by Shibata and Komiya (1972, 1973). The leaves do not, in this instance, accumulate more than 10% of the applied nitrogen. Transfer is discontinuous, with two peaks in 24 h.

We should not, however, overemphasize the importance of nitrogen. Karlsson and Carlsson (1984) concluded that, in the nutrition of *Pinguicula vulgaris*, phosphate is the most important nutrient gained by carnivory. This distinction may, in part, explain why *Pinguicula* (see Fig. 7.1) seems less dependent upon the carnivorous habit than most other genera.

Insects as a source of trace elements

It is now well established that carnivory is of significant benefit to some genera when plants are grown in conditions of mineral deficiency. Sorenson and Jackson (1968) showed that plants of *Utricularia gibba*, grown in culture in a medium deficient in Mg and K, and then fed with *Paramecia*, increased their growth, as measured by the number and length of internodes and number of traps, in comparison with unfed controls in the same medium.

In addition to the nitrogen enhancement, Green *et al.* (1979) noticed that *Triphyophyllum* was able to accumulate K to significant levels in tissues, although the soils in which the plants grew were deficient in this mineral. Again, apart from the studies on nitrogen uptake of *Drosera whittakeri*, Chandler and Anderson (1976c) noted that plants in sulphur deficient medium showed a recovery in growth after being fed on insects.

On the other hand, plants of *Sarracenia flava*, grown under both field and controlled conditions by Christensen (1976), gave a somewhat different picture. In the nutrient deficient conditions, as we have seen above, nutrient contents of nitrogen and phosphate were significantly enhanced by insect feeding, whereas insects had no significant effect on the concentrations of calcium, magnesium or potassium.

The redistribution of insect-derived nutrient within the carnivorous plant

When radioactively-labelled *Daphnia* were fed to plants of *Drosera capensis* and *Aldrovanda vesiculosa* differences were observed in the redistribution of the labelled sugars and amino acids (Fabian-Galan and Salageanu, 1968). In the *Drosera* plants the animal nutrients were, in time, redistributed throughout the whole of the plant. In *Aldrovanda* plants, on the other hand, the derived nutrients were transported almost entirely from the mature traps to the growing point of the plant. Moreover, from young traps, the insect-derived nutrients were not redistributed at all (see page 209).

Alternative Nutrition: Non-Arthropod Trapping by the Glands of Carnivorous Plants

Pinguicula species, with their broad cauline trichomatous leaves, may benefit from the rain of pollen, seeds and leaf fragments, augmenting a conventional insect diet. We might expect some of the cauline *Drosera* species to benefit in the same way.

Harder and Zemlin (1968) grew axenic cultures of *Pinguicula* plants on inorganic agar medium with no added nitrogen or phosphorus. After eight weeks, 20 of the plants were fed four times, over a period of five weeks, with *Pinus* pollen. The fed plants grew larger, turned deeper green, and aged more slowly. On the fed plants, bud initials were visible after the second feeding. All the treated plants flowered before the last feeding, and developed a total of 127 flowers. None of the unfed plants flowered at all. Pollen grains were found to germinate on glands of *Pinguicula* (Joel, unpublished) and we might assume that the digestion of germinated pollen grains is much faster because the permeability of the pollen tube to digestive enzymes is greater than that of the pollen wall. There can be no doubt that certain carnivorous plants, both episodically and continuously, must be able to benefit, at least to some extent, from the aerial rain of pollen, spores, seeds and other random fragments. Austin (1875F, see Appendix 1) showed that *D. rotundifolia* is both capable of responding to and enzymically degrading plant as well as animal material.

Alternative Nutrition: Through the Roots

Carnivorous plants have, for the most part, atrophied root systems. Even those with substantial root systems, e.g. *Triphyophyllum* are known to use their traps as major sources of nutrients (see page 21). There is some evidence that nutrition through the roots may, in certain situations, actually be deleterious (see page 134).

For the most part the soils where carnivorous plants grow are extremely barren. Plummer (1963) studied the pitcher-plant habitats in the Georgia coastal plain. The overall level of nutrients on these 'Moist Pine Barrens' at any given time is quite low compared with other soils in the area. Yet the rather dense plant cover seems inconsistent with the apparent inability of the soil to support it. But *Sarracenia*, and the few associated plants, may decompose slowly and gradually release nutrients. There may be a rapid recycling of available soil minerals, e.g. potassium through weathering and leaching in the warm, moist winters and through fires. In addition, the soil water may bring into the ecosystem some soluble nutrients, small at any given time, but constituting a steady flow throughout the season. These habitats, according to Plummer, are particularly low in calcium, potassium, and magnesium. He suggests that the carnivorous plants may derive their greatest benefit from scavenged metal ions in their prey rather than nitrogen, as generally assumed. Nevertheless, as Figure 7.1 shows, a few terrestrial carnivorous plants such as *Pinguicula vulgaris* actually do better in most respects on an inorganic medium supplied through the roots than on a straight insectivorous diet. They do best of all on an inorganic diet supplemented by insects. The rootless *Utricularia* (Fig. 7.1) is intermediate in this respect; it does well, at least vegetatively, with an inorganic diet; a supplement of *Paramecia* does not improve its performance (Dore Swamy and Mohan Ram, 1969, 1971; Sorenson and Jackson, 1968). However, Pringsheim and Pringsheim (1967) grew *U. minor* and *U. ochroleuca* on a mineral medium, but only obtained satisfactory vegetative growth if the medium was supplemented with traces of peptone and beef extract. Their development was further enhanced by the addition of glucose and acetate.

Mineral nutrition: the negative evidence

Dionaea plants, transplanted into conventional clay-loam garden soil (Roberts and Oosting, 1958), show little vigour. The leaves do not develop traps and plants of only a third of the size of those left *in situ* developed flowers; none set seed. In addition, five months after transplantation, only two plants of the original twelve were still alive. A similar pattern was observed in *Dionaea* plants transplanted to greenhouse potting soil. The roots of the plants in the richer soils were atrophied, no new root growth took place after transplantation.

In further studies, Roberts and Oosting (1958) showed that all plants receiving mineral nutrients grew poorly, whilst all the controls, watered with distilled water, showed much more satisfactory growth. The inorganically fed plants steadily declined in weight and died after about three months. The plants fed on organic material, on the other hand, showed more vigorous vegetative growth. Roberts and Oosting do, however, qualify their tentative experimental findings with the statement that the poor growth of the experimental plants may have resulted from too high a concentration or the wrong proportions of inorganic nutrients.

Eleuterius and Jones (1969) studied bog soils in Mississippi and found no deficiencies in nitrogen, phosphorus or potassium. When they applied 6–6–12 (N.P.K.) fertilizer and ammonium nitrate, productivity in *Sarracenia alata* actually declined.

We should, however, be careful to qualify this apparent denigration of root-derived minerals. As Christensen (1976) showed in *Sarracenia flava*, a significant insect diet may be unable to provide this species with appreciable amounts of potassium, calcium, or magnesium.

The importance of soil pH in artificial nutrition studies

As well as the qualifications suggested by Christensen (1976) we should also be cautious of all experiments on plants grown, for example, on garden soil plots, or on field plots without close control of soil conditions. As Rychnovska-Soudkova (1953, 1954) has suggested, the calcifuge behaviour of *Drosera rotundifolia* is complex. Regardless of the level of certain ions in the soil solution, the pH of the soil is critical. Lime is not necessarily toxic to these plants in very acid solutions. At low concentrations of calcium, the soil pH may be permitted to rise to as high as 6.7 without limitation, but at high concentrations pH 3.0 may be the optimum. At higher concentrations, and at more alkaline pH levels, the plant will preferentially absorb cations, thus inducing the observed toxicity. An opposite effect is observed with artificial nitrogen feeding. She used both nitrate and ammonia salts in solution. To the acidic side of pH 4.0 there was marked stimulation of growth from the nitrate ion; to the alkaline side of that level from the ammonium ion.

The interaction of light and mineral nutrition

Roberts and Oosting (1958) concluded (see above) that insect material, digested and absorbed through the leaves, resulted in healthier, more vigorous plants than controls without insects. On the other hand, high rates of insect feeding resulted in either a failure of the development of floral initials or a failure of their production. But these greenhouse experiments were done under light intensities much lower than in the field. It is possible, they suggest, that the rate of feeding was too high for the balance of photosynthesis occurring. A critical carbohydrate-nitrogen balance in *Dionaea* is necessary for the complete development of floral initials.

These effects (i.e. an absence of flowering, if not a reduction in general vigour, even when soil nutrients are available), have been observed by several workers in the totally unrelated *Nepenthes*. Smythies (1963) wrote 'it is quite common to find pitcher plants growing vigorously without producing any pitchers and I believe it has been demonstrated in Singapore Botanic Gardens that a nitrogen-rich soil is one factor that will inhibit pitcher development'.

To expand this view, Green (1967) reported 'Where shade is too deep pitchers and tendrils do not develop on stem or rosette leaves . . . laminas become large and thin . . . cultivators of *Nepenthes* in private or botanic gardens are sometimes troubled by a scarcity of pitchers. This can be due, not only to too much or too little shade, but also to too rich a soil . . .'

***Darlingtonia*: the negative benefit of the roots**

As we saw in Chapter 3 *Darlingtonia* grows, almost exclusively, on metal-rich soils, or at least on seepages in which the water is heavily charged with metal ions. There is little experimental evidence to confirm the field observations, but the assumption is that *Darlingtonia* is indifferent to the mineral contribution made by its root system. It is frequently noted that *Darlingtonia* colonies grow semi-submerged in water that is very cold. In cultivation, *Darlingtonia* is both stunted in growth and reluctant to flower unless this growth condition is met.

Trap Nutrition vs. Root Nutrition: A Field Comparison

Gibson (1983a) has calculated from Reichle *et al.*'s (1969) data (see page 130) and from the field soil analyses of Roberts and Oosting (1958) that a mature *S. leucophylla* pitcher-plant gains approximately 66.3 times more mineral nutrients from insect resources than from the bog soil in which it grows. This figure he derives from the following calculations.

A flowering *S. leucophylla* pitcher-plant catches approximately 3.2 grams of dry insect biomass during a season (Gibson, 1983a, his Fig. 24). The spectrum of mineral nutrients in these taxa is not known, but, using the general chemical analyses of whole insects by Reichle *et al.* (1969), a mature pitcher-plant is calculated to absorb roughly 0.533 g (= 3.2 g × 16.6%) of these major nutrients during a season.

Roberts and Oosting (1958) analysed the top 10 cm of carnivorous plant bog soils in North Carolina. Chemical analyses of the coastal plain soils indicate very low levels of available nutrients. There was a complete lack of detectable calcium, manganese and nitrate, and very low amounts of ammonia (2 ppm), iron (1 ppm), magnesium (0 ppm), potassium (2 ppm), and phosphate (less than 2 ppm). Therefore, bog soil contains 6 ppm of the six major nutrients cited above for insects or 0.000006 g/g bog soil. The roots of a mature pitcher-plant occupy about 1000 cc of bog soil (Gibson, 1983a). In general, Gibson states, the root system of herbs in the southeastern bogs are poorly developed and restricted in area. The bulk density of the *Rains Series* pitcher plant bog soil is 1.34 g/cc. Therefore, 1000 cc of bog soil contains roughly 0.008 g of these major nutrients.

This calculation, as Gibson points out, does not take into account any renewal rate for mineral nutrients in bog soil, only that all nutrients are available to the plant. In fact, renewal rates vary greatly between these nutrients, but in general, they may be quite low in bogs, where nutrients become available mainly through the oxidation of organic material. High renewal rates would obviously reduce the relative contribution of insects in Gibson's calculations. These calculations do not consider the relative efficiencies of a plant at absorbing mineral nutrients through its leaf-traps from insect victims versus through its roots from bog soils, as such data, except for the limited comments made on page 133–134 above, do not exist.

We should note, too, that not all bog soils are as low in nutrients as is suggested by the Roberts and Oosting study. As we saw on page 134 (Eleuterius and Jones, 1969) some bog soils are, in N, P and K terms, comparatively rich.

Insect Digestion and the Promotion of Flowering

Common observation and horticultural study of carnivorous plants grown in greenhouses suggest that total insect starvation may inhibit flowering. *Utricularia longifolia* grown under greenhouse conditions, but in 25 cm plastic pots plunged into a peat bed, develop vigorously vegetatively, but never flower. Other specimens of the same plant, grown under identical conditions side by side with the above, but unrestricted in their growth through the peat bed, flower regularly (R. Gardner, Plant Sciences Oxford, pers. obsv. 1987).

Francis Darwin (1878) was probably the first to demonstrate, unequivocally, that the feeding of a carnivorous plant by insects promoted flowering. He grew plants of *Drosera rotundifolia*, obtained from the field, in two lots – the one excluded from catching insect prey, the other artificially fed. His results, the control expressed as 100 were as follows.

	Control	Artificially fed
Number of inflorescences	100	164.9
Number of capsules	100	194.4
Total number of seeds	100	122.7
Total weight of seeds	100	379.7

At the end of the above paper Darwin describes other work by Kellerman and von Raumer (also cited by Lloyd, 1942) which, in somewhat more sophisticated experiments, came to exactly the same conclusion.

Gibson (1983a) has calculated the number of days taken by artificially fed, control and starved

Sarracenia leucophylla plants to reach the same critical flowering size threshold, based on the daily growth rates of these plants. The number of new pitchers and phyllodia produced per plant per day is: 0.039, 0.023 and 0.016 for fed, control and starved plants respectively. These data are estimated by linear regression (see Gibson, 1983a, his Fig. 16). A pitcher-plant of this species has a high probability of flowering after it has produced its thirtieth pitcher. Given the same threshold, but different daily growth rates, Gibson calculates that fed, control and starved plants take 760, 1288 and 1898 days to produce their thirtieth pitcher. Since in northwest Florida these plants can only grow from mid-March to early December, or for about 250 days, fed, control and starved plants take 3.04, 5.15 and 7.59 years respectively to mature. These estimates are based on a number of assumptions, e.g. a constancy of trap size and sequence regardless of feeding rate, but they do indicate strongly the importance of insect diet, at least in the establishment of new colonies.

However, most of the above experiments were probably performed under optimal light conditions. As we have previously emphasized, and as Dore Swamy and Mohan Ram (1969, 1971) have demonstrated in *Utricularia inflexa*, not only the quantity but the balance of light is critical. *U. inflexa* may not require insects but it does require 8 h of light and 16 h of darkness for satisfactory flowering.

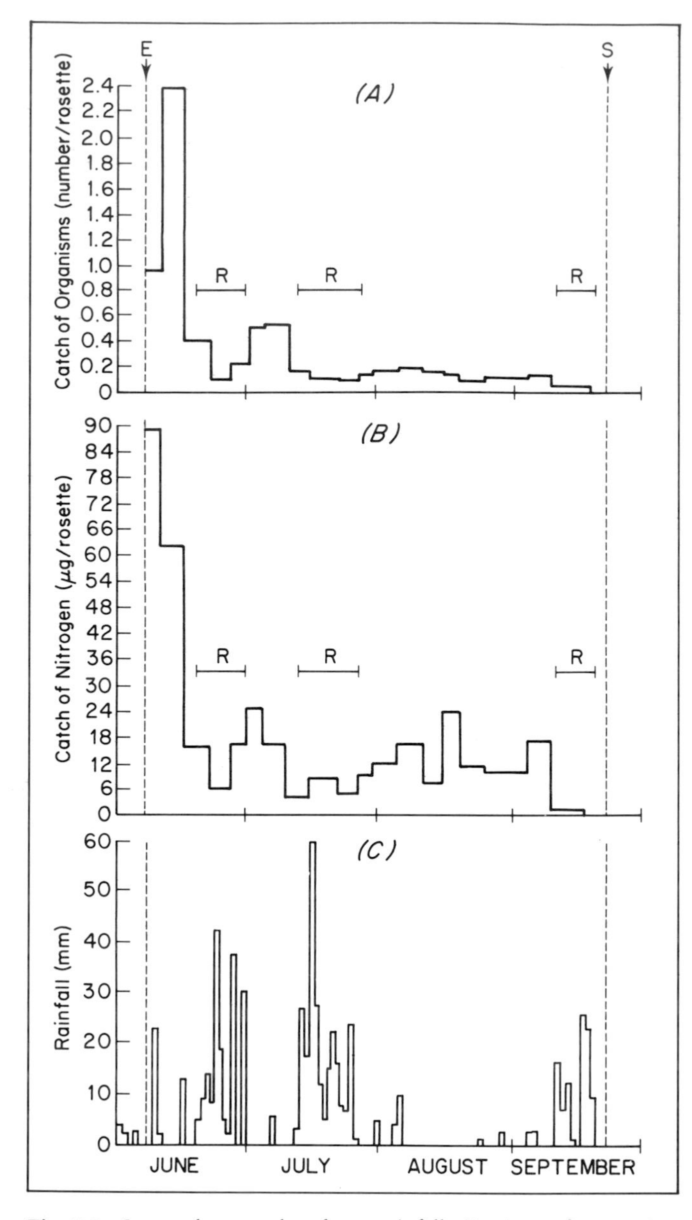

Fig. 7.2 Seasonal record of catch of fauna by three clones (157 plants) of *Drosera erythrorhiza* in 1978 at Bayswater, Perth, W. Australia. (*A*) numbers and (*B*), nitrogen content of plant catches are related to (C), the season's pattern of rainfall; E, time of expansion of rosette, S, time senescence of vegetative parts at end of growing season; R, period of heavy rainfall. (Reproduced, with permission from Dixon *et al.*, 1980.)

The Prey of Carnivorous Plants

A principle which cannot be emphasized too strongly is that the prey of any one species of carnivorous plant is not constant either in space or time. Individual records of a single trap on a single day, or even a population of traps over several days will tell us very little. In so far as they have been studied in any detail it is clear that carnivorous plants trap a very varied and everchanging diet throughout their 'open' season. The diet may often change from day to night, and night-trapping must be significant (Austin, 1875F, see Appendix 1). Green *et al.* (1979) point out (see page 138) that many species of the prey they identified from *Triphyophyllum* traps were nocturnal or crepuscular arthropods. We should also note, in *Triphyophyllum*, that the glandular leaves of the juvenile plants are produced in May and June, just before, and only before, the height of the rainy season. Sometimes they are observed in six-month-old seedlings, but sometimes they are not produced until the plant is in its second year.

The diet is certainly different between young and old traps of the same species on the same site. This is very clearly indicated by Dixon *et al.* (1980; Fig. 7.2), working on *Drosera erythrorhiza*.

However, again as a very general rule, individual

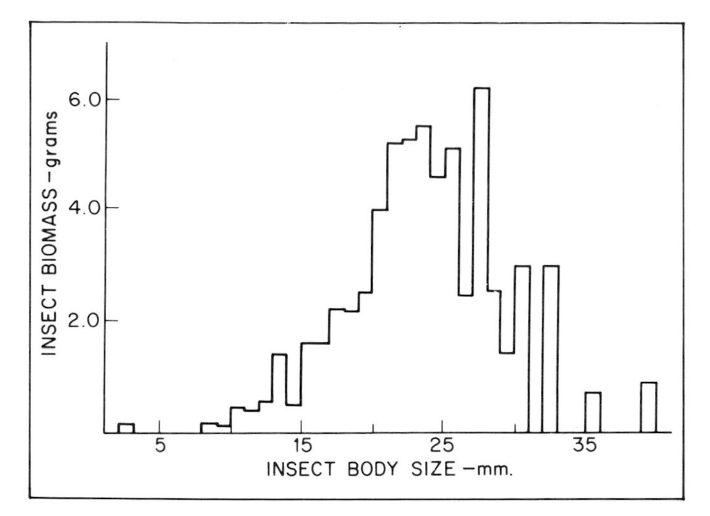

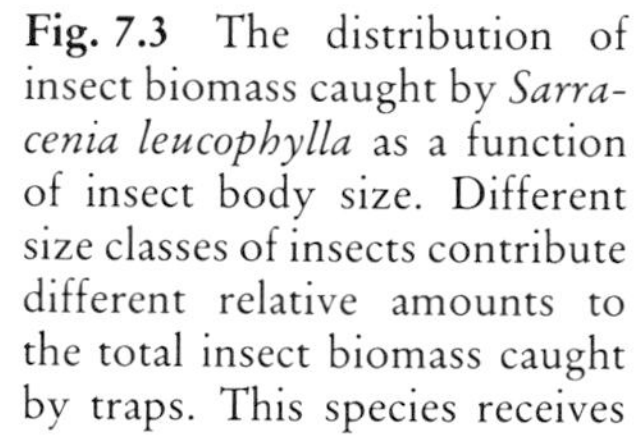
Fig. 7.3 The distribution of insect biomass caught by *Sarracenia leucophylla* as a function of insect body size. Different size classes of insects contribute different relative amounts to the total insect biomass caught by traps. This species receives most of its insect biomass from insects approximately 25 mm long (N = 892 insects in 50 pitchers). Larger insects are both rarer in the environment and rarely caught. (Reproduced by kind permission from Gibson, 1983a.)

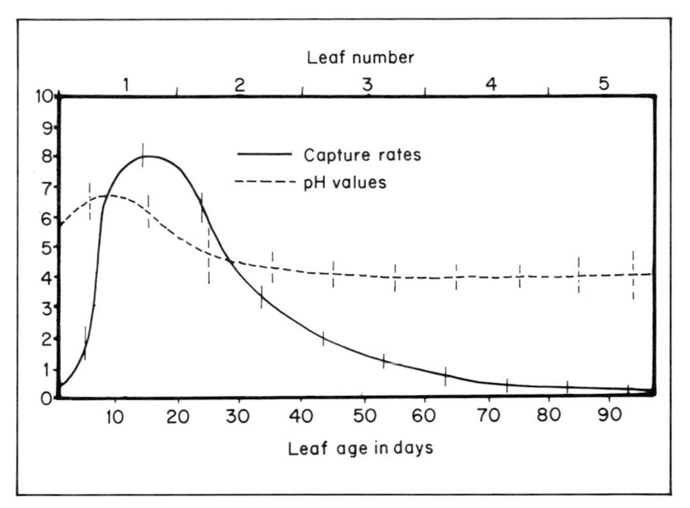

Fig. 7.4 Mean and standard error insect capture rates (ordinate) of *Sarracenia purpurea* pitchers at various ages (bottom abscissa) from greenhouse experiments (N = 350) compared to leaf chronology of a mid-season plant at the field study site with a 20-day interval between leaf production (top abscissa). Broken line curve: mean and standard error pH values (ordinate) of the pitcher fluid of various ages of greenhouse-reared pitcher plants (bottom abscissa) (N = 50) compared to leaf chronology of a mid-season plant at the field study site with a 20-day interval between leaf production (top abscissa). (Redrawn, with permission, from Fish and Hall, 1978.)

species of carnivorous plants will only catch insects within a restricted size range. This bias has been clearly shown by Gibson (1983) in *Sarracenia leucophylla* (Fig. 7.3) which although mechanically capable of trapping larger insects rarely does so.

Wolfe (1981) showed that plants of *Sarracenia purpurea* captured significantly more *Drosophila melanogaster* as prey with their newly opened leaves than did plants without new leaves. In fact, a pitcher-plant leaf catches most of its prey in the first 30 days of its life (Fig. 7.4). A new leaf is such a dominant feature of a whole pitcher-plant (see below and Frontispiece) that it may catch 82% of the total prey caught by the whole plant in a given time. Wolfe speculates that the attraction may be based either on the finite amount of nectar produced by a pitcher, or by a decreasing rate after thirty days or again by the progressive loss of some active component of the nectar.

There is also evidence (see page 140) that, since records began, the diet of specific carnivorous plants may have changed. To what extent this may be due to changes in the plant, changes in the agricultural practice in the area surrounding the plants, or gross environmental changes is not known. Lastly, there may be huge episodic 'crashes' of insects, such as have been recorded for populations of *Drosera* (see page 140) which can make nonsense, in the short term, of statistical analyses of prey numbers and types. Such 'catastrophes', however rare, may not be unimportant in, for example, the seed-set and colonization of further areas by the plants concerned.

The Seasonal Nature of the Trap and the Prey

We have seen that *Triphyophyllum* is the most striking of the part-time carnivorous plants. Yet this characteristic, although less well marked, is present in many other groups. Some species of *Sarracenia*, e.g. *S. oreophila*, *S. leucophylla* and *S. alata*, produce attenuated, flat, phyllodia, not traps, in the autumn. In *Dionaea* the trap area is much reduced in leaves formed late in the growing season. The petiolar wings become enlarged, thereby presumably increasing the photosynthetic capacity. In *Cephalotus follicularis* the traps are formed in early spring and ordinary foliage leaves are produced in the autumn.

Untutored as she was in the finer details of entomological taxonomy, Mrs Austin's long-term observations on *Darlingtonia* give powerful emphasis to this seasonal pattern. She records (Austin, 1875K, 1876G, see Appendix 1) the following.

May	Ants, spiders in the majority
June–early July	Beetles, moths, butterflies and small snails
August and September	Flies, grasshoppers (Orthoptera), 'Katydids' (a name given to several species of long-horned grasshoppers) 'yellowjackets' (Vespidae), honey and bumble bees
October	Moths and butterflies

She also notes that, although honey bees were common throughout the year (she kept bees), they were only caught in small numbers in the autumn. On 24 September 1875 she recorded (in Butterfly Valley, California): '. . . twelve honey-bees, four green and yellow spotted "worms" about an inch and a half long, and an eighth of an inch in diameter, three small snails, large hairy flies, ants, moths of various kinds, large red-colored grasshoppers, "katydids", yellowjackets, wasps, hornets, bumble bees, spiders, beetles, white larvae in great numbers and various smaller winged insects found in the tubes.' The 'white larvae' were almost certainly the larvae of dipteran commensals (see Ch. 14).

We can see a most marked pattern in the quantity of prey caught over a season from the work of Dixon *et al.* (1980). As we can see from Figure 7.2*A*, the newly expanded rosettes of *Drosera erythrorhiza* are the most effective, and periods of heavy rain (R) rapidly depress both the numbers and nitrogen content of the catch.

The Diversity of the Prey

The diet of almost any carnivorous plant taken over a season, is highly diverse. Erber (1979), for example, examined ten *Nepenthes* pitchers and in them found 1994 individual arthropods of 150 identified different species, belonging principally to the Diptera, Hymenoptera and mostly Formicidae, Collembola, Aphididae and Acarina.

Although only a part-time carnivore, *Triphyophyllum peltatum* has no doubt evolved so that the insect-trapping phase of its life cycle coincides not only with a quantity but with a diversity of insect prey. This is exemplified by the data from Green *et al.* (1979) (Table 7.2), which is, it should be noted, from a single set of collections by day from a limited area. But then few have been privileged to see, let alone study *Triphyophyllum*.

Table 7.2 Types of prey trapped on the glandular leaves of *Triphyophyllum*, (numbers are from a random sample of eight leaves collected from the wild from separate plants. Total number of carcasses identifiable = 165). (Reproduced with permision from Green *et al.*, 1979.)

Type of prey caught Class	Order (common name)	Numbers and (% of total by numbers)
Crustacea (subclass Malacostraca)	Isopoda	
Myriapoda	(millipedes)	Occasional[a]
	(centipedes)	Occasional[a]
Insecta	Orthoptera, sub-order Saltatoria (grasshoppers)	13 (7.9)
	Isoptera (termites)	25(15.2)
	Hemiptera, sub-order Heteroptera (bugs)	7 (4.2)
	Mecoptera	1 (0.6)
	Lepidoptera (mostly nocturnal moths)	10 (6.1)
	Coleoptera (beetles)	60(56.3)
	Hymenoptera (ants, small wasps)	25(15.1)
	Diptera (mostly mosquitoes)	20(12.1)
Arachnida	Araneida (all from family Araneae, hunting spiders)	4 (2.4)

[a] Trapped prey observed, but not part of the eight leaf sample.

From a completely different set of plants on a different continent, Dixon *et al.* (1980) collected 1090 arthropods in the four-month growing season, June to September (Table 7.3). This time Collembola (springtails) and Hymenoptera (the social insects, with ants predominant) formed the major part of the diet of *Drosera erythrorhiza*. But we can be sure that the species trapped varied over the season.

A confirmation of this pattern of trapping of *D. erythrorhiza*, was made by Watson *et al.* (1982), this time in a bushland site of the Spearwood Dunes System, also near Perth. The majority of the prey they observed was the collembolan *Hypogastrura vernalis*, whose presence in the Perth metropolitan area is a new record for Australia.

Table 7.3 Taxonomic breakdown and nitrogen content of the catch of arthropod fauna by leaf rosettes of three clones (157 plants) of *Drosera erythrorhiza* during the 1978 growing season (June–September) Bayswater, Perth, Western Australia (Reproduced with permission from Dixon *et al.*, 1980.)

Taxa captured	No. of individuals	Percentage total catch of organisms	Nitrogen capture	Percentage total catch of nitrogen
	Total season's capture			
Collembola				
Poduridae	5	71.0	17.27	33.2
Onychiuridae	1			
Isotomidae	16			
Sminthuridae	752			
Blattodea	1	0.1	0.12	0.2
Isoptera	1	0.1	0.31	0.6
Orthoptera				
Acrididae (nymph)	1	0.1	0.12	0.2
Hemiptera				
Homoptera				
Aphididae	24	2.6	3.14	6.0
Cicadellidae	4			
Heteroptera				
Miridae	4	1.3	1.57	3.0
Unidentified adult (a)	2			
(b)	6			
Unidentified nymph	2			
Coleoptera				
Curculionidae (larval)	6	1.4	1.57	3.0
Staphylinidae	1			
Unidentified adult	8			
Lepidoptera	1	0.1	0.02	0.04
Diptera				
Nematocera	189	18.8	9.42	18.1
Brachycera	10			
Unidentified larvae	6			
Hymenoptera				
Ichneumonidae	1	3.4	17.30	33.2
Chalcididae	4			
Formicidae	30			
Apoidae	1			
Unidentified larvae	1			
Arachnida				
Mites	10	1.1	1.60	3.0
Ticks	2			
Nematodes	1	0.1	0.16	0.3
Totals for season	1090		52.6	

The diversity of prey from *Sarracenia purpurea* introduced into the Irish Republic

Sarracenia species are well known for their catholic diet (Judd, 1959; Gibson, 1983a). Even when carnivorous plants are displaced from their natural habitats they may still be able to catch a highly diverse range of arthropods. C. Aldridge, (pers. comm., 1986) collected the prey from six *Sarracenia* pitchers naturalized at Bellacorich, Eire (see Ch. 3) on 5 July 1986. The contents of pitcher 5 are listed in Table 7.4, with identification kindly provided by M. Amphlett (pers. comm., 1987) of the Department of Zoology, Oxford. Pitcher 5 was not dissimilar to any of the other pitchers examined except that it contained the greatest number of different species. Mites (Acarina) in every case provided the largest number of individual victims.

Table 7.4 Contents of *Sarracenia* pitcher

Pterogota
Lepidoptera: 1 × adult, ? Tortricidae
Trichoptera (Caddis flies):
 1 × *Phryganea obsoleta*
 1 × other, ? *Limnephilus* sp.
Diptera, Nematocera:
 Tipulidae 1 × adult ? sp.
 Others × 25
 Other countless remains probably attributable to family c.50
Diptera-Brachycera and -Cyclorrhapha:
 1 × Muscid type
 1 × Syrphidae, possibly *Orthonerva* sp.
 1 × ? orange-bodied sp.
 2 × Lonchopteridae sp. ?
 5 × Dolichopodidae sp. × 2. Possibly *Dolichopus* and *Poecilobothrus*
 21 × ? family. Heleomyzids
Hymenoptera:
 "Parasitica" × 35 sp. ?
 Formicidae 7 × *Myrmica ruginodis*
Coleoptera:
 Chrysomelidae 2 × *Lochmaea capreae*
 ? 13 × sp. and family. (Phytophagous)

Other Orders
Acarina, Mites × 34
Araneae, Spiders. fam.
 Thomisidae (Crab spiders)
 Tibellus maritimus × 1
 Linyphiidae (money spiders)
 Gonatium rubens × 1
 Tapinopa longidens × 1
Opiliones, Harvest spiders:
 1 × *Oligolophus agrestis*
A total of 205 victims in this one trap

The Episodic Capture of Prey

All carnivorous plants, particularly those living near patches of open water or on the migration routes of insects, are likely to experience episodic 'crashes' of vast numbers of one or a few species. One such was observed on 4 August 1911 by Oliver (1944).

On an island in Sutton Broad, Norfolk, England, just under one hectare in extent, the level sward consisted almost entirely of *Drosera anglica*. At 1.30 p.m. that day each *Drosera* plant had from four to seven Cabbage White butterflies (*Pieris rapae*) adhering to it. An approximation suggested that six million Cabbage Whites had been caught, presumably shortly after daylight. No Cabbage Whites were seen still flying in the air. About 30% of the bodies of the *Pieris*, which presumably had been caught for no more than nine hours, were in an obvious state of solution. The swarm had probably come, by sea from the continent of Europe, early that morning and had moved inland about 8 km.

Changes in Prey Composition with Time

Williams (1980) records that *Dionaea muscipula*, in its native habitat in the Carolinas of eastern USA catches principally ants and spiders. The percentages he estimates are: ants 30%; spiders 39%; grasshoppers 10%; beetles 10%; flies and mosquitoes 1–4%. He states that subsequent investigations at different times of the year have confirmed this pattern.

However, Jones (1923) opened 50 different sprung traps in the Wilmington area of North Carolina and identified the following prey: Hymenoptera (wasps and ants) 10; Diptera 9; Arachnida 9; Coleoptera 9; Orthoptera 7; Hemiptera 4; Lepidoptera 2; and a much higher proportion of winged insects. The average length of the prey was 8.6 mm.

A possible explanation for these frequent discrepancies may be offered by Gibson (1983a). He observed that ants increase dramatically when the incidence of fires increases. He also offers, as more than a coincidence, that ant specialist plants (e.g. *Sarracenia minor*) also increase in numbers when fires are prevalent.

Prey-Partitioning by Closely Related Species in a Similar Habitat

We have seen with *Drosera* and *Sarracenia* (see pages 27 and 37) that closely related species may

Table 7.5 Composition of the catches of the *Drosera* spp.

Drosera species	*intermedia*				*rotundifolia*				*anglica*			
	Total		Partly digested		Total		Partly digested		Total		Partly digested	
Groups of Arthropods	no	%	no	%	no	%	no	%	no	%	no	%
Diptera	412	93.5	332	83	137	57.0	89	65	49	66.2	41	83
Nematocera	403		330		116		83		39		37	
Brachycera	9		2		21		6		10		4	
Coleoptera	4	0.9	2	50	38	15.8	16	42	16	21.6	5	32
Hymenoptera	9	2.0	3	33	23	9.6	8	32	2	2.7	—	—
Formicoidea	1		—		6		3		—		—	
Chalcidoidea	5		2		8		3		—		—	
Proctotrupoidea	2		1		4		—		—		—	
Braconidae	1[6]		—		2[7]		1		2[8]		—	
Ichneumonidae	—		—		2		—		—		—	
Cynipidae	—		—		1		1		—		—	
Homoptera	3	0.7	—	—	18	7.5	2	11	4	5.4	—	—
Cicadomorpha	1		—		4		—		—		—	
Aphididae	2[1]		—		8		2		1		—	
Heteroptera	1[2]	0.2	—	—	4[3]	1.7	—	—	1	1.4	—	—
Trichoptera	1[4]	0.2	—	—	—	—	—	—	—	—	—	—
Odonata	—	—	—	—	3[5]	1.3	—	—	—	—	—	—
Orthoptera	1	0.2	—	—	1	0.4	—	—	—	—	—	—
Thysanoptera	—	—	—	—	4	1.7	1	25	—	—	—	—
Collembola	7	1.6	—	—	1	0.4	—	—	—	—	—	—
Arachnida	2	0.5	1	50	7	2.9	2	29	2	2.7	—	—
Acarina (Oribatei)	1	0.2	—	—	4	1.7	—	—	—	—	—	—
TOTAL	441	100.0	338	77	240	100.0	118	50	74	100.0	46	62

(1) Wingless
(2) Macropterous form of *Hebrus ruficeps*
(3) Two brachypterous forms of *Hebrus ruficeps*
(4) Hydropsychidae
(5) ♂♂ of *Enallagma cyathigerum*
(6) ♀ of *Blacus* sp.
(7) ♂ of *Dacnusa* sp. and ♂ of *Opius* sp.
(8) ♂ of *Apanteles* sp. and ♂ of *Dacnusa* sp.
(Reproduced with permission from Achterberg, 1973.)

live side by side in a similar habitat. Two questions arise from the juxtapositions. Is there selective advantage in related species, but of different growth form, growing side by side? Do these side-by-side species partition the prey available to them in the habitat?

Prey partitioning by three *Drosera* species

Three closely related species of *Drosera* (see Fig. 3.3 and Table 7.5) are sometimes found in a specific zonation. Achterberg (1973) has studied the prey caught by these three species in June of 1973 in the eastern Netherlands. As can be seen from Figure 3.3 the most striking difference between the species is that the leaves of *D. intermedia*, often in the water or wetter areas of the habitat, are narrower and more or less erect. Those of *D. rotundifolia* are more horizontal, like a rosette. No flourishing specimen of *D. intermedia* was ever observed by Achterberg in the habitat of *D. rotundifolia*, but the reverse did occasionally occur.

As we can see from Table 7.5, *D. intermedia* caught significantly more Diptera/Nematocera than the two other species. This correlates with the wetter habitat of *D. intermedia*, from whence midges emerge and stay for a while. Beetles, aphids and arachnida, as one might expect on this hypothesis, were caught in smaller numbers. The catches of *D. rotundifolia* and *D. anglica* were, again as one would expect, similar, and comprised especially Diptera, Coleoptera, Homoptera and Arachnida. Thus the habitat of a specific plant rather than the orientation or size of the leaves, seems to be the most significant factor, in this case, in partitioning of the prey.

The catches of arthropoda in more open habitats of *D. intermedia*, into which *D. rotundifolia* had penetrated showed, as one would have expected on the 'habitat' hypothesis, a more diverse and numerous catch.

The negative association of similar species of *Sarracenia* and other species of carnivorous plants

McDaniel (1971) was probably the first clearly to observe and speculate upon the non-random juxtaposition of two or more carnivorous plant species. He noticed that, in localities where several species of *Sarracenia* occur together, as they do in much of the southeastern United States (see Fig. 19.6), they usually have different spatial arrangements. For example, populations of *S. alata*, *S. flava* or *S. leucophylla*, all tall species as can be seen from Fig. 7.5, often have plants of the much smaller *S. psittacina* intermixed. The striking difference in size and shape of the leaf orifice of the three larger species compared to *S. psittacina* suggests different environmental niches. *Sarracenia flava* and *S. leucophylla*, although they have the same general range over a wide area of Georgia and Florida, are rarely found together. *Sarracenia leucophylla* is more likely to occur in boggy areas, or at least areas that are moist during most of the year, while *S. flava* is more frequent in savannas, i.e. intermittently flooded areas. Where the two species do occur together, he often observed striking zonation of species. At one locality in Bay County, Florida, the two species occur at opposite ends of a small, low-pine flatwoods area, with the two coming into contact along a well defined line. At other localities, micro-habitat differences may account for spatial separation of two or more species. In the large hillside bogs from western Florida to Mississippi, *S. purpurea* usually occupies the edge of the bog or some local area within the bog margin, while the larger-leaved, taller species such as *S. leucophylla* and *S. alata* occur in the bog proper.

In part, the habitat selection observed here may be related to the partitioning of available prey (Gibson, 1983a). As Gibson observed, most carnivorous plants have a fairly restricted size range of prey (Fig. 7.3). We shall review Gibson's theories as to the extent to which this prey selection may permit closely related species to occupy the same habitat. Prey-partitioning may also have arisen from evolutionary pressure to prevent the formation of non-viable hybrids (see page 38).

Sarracenia begins flowering along the Gulf Coast in early to mid-March with *S. purpurea*, *S. alata*, *S. flava* and *S. leucophylla*, and ends with *S. psittacina*, *S. rubra* and *S. minor* in mid-May. Northward, each species progressively blooms later. *Sarracenia purpurea* at the northern extremity of its range, blooms in late July and early August. *Sarracenia rubra* in the mountains of North Carolina blooms

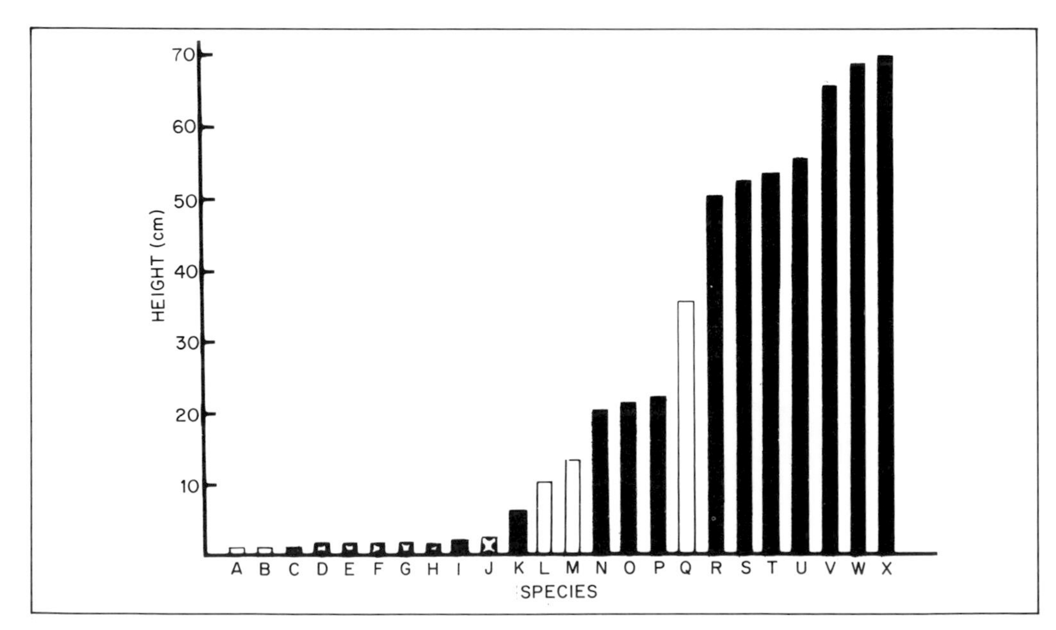

Fig. 7.5 Carnivorous plant species place their traps at different heights. Note ground forms of pitcher plants (I = *S. psittacina* and K = *S. purpurea*) and aerial forms of sundews (Q = *D. filiformis tracyi*). Black bars = *Sarracenia* species; white bars = *Drosera* species, stippled bars = *Pinguicula* species, and bar C = *Dionaea muscipula*. For mean maximum height distributions for mature plants (Gibson, unpublished data). Mean heights for the following species: A = *D. brevifolia*, B = *P. pumila*, C = *Dionaea muscipula*, D = *D. capillaris*, E = *P. lutea*, F = *P. caerulea*, G = *P. planifolia*, H = *P. ionantha*, I = *S. psittacina*, J = *P. primuliflora*, K = *S. purpurea*, L = *D. intermedia*, M = *D. filiformis* (short Florida form), N = *S. minor*, O = *S. alabamensis wherryi*, P = *S. rubra*, Q = *D. filiformis tracyi*, R = *S. oreophila*, S = *S. alabamensis alabamensis*, T = *S. jonesii*, U = *S. alata*, V = *S. flava*, W = *S. minor* (Okefenokee form), X = *S. leucophylla*. (Redrawn with permission from Gibson, 1983a.)

as late as mid-June. As we have seen in Chapter 3, Bell (1952) suggests that seasonal isolation in any given locality prevents extensive hybridization. Often species occurring in the same area have different flowering periods. For example, such species as *S. flava* and *S. psittacina*, frequently occurring together and not known to hybridize, usually flower at quite different times. Yet data from herbarium material and field observations suggest that seasonal isolation may under certain conditions be lacking.

We should not forget, as Rymal and Folkerts (1982) and Folkerts (1982) have pointed out, that the pollinators are important. The major pollinators of the large-flowered species (*S. alata*, *S. flava*, *S. leucophylla*, *S. oreophila* and *S. purpurea*) are the newly emerged queens of the genus *Bombus*. There is a range of species. Although there is a range of size, most of these queens are too large to enter the flowers of the smaller species, e.g. *S. minor*, *S. psittacina* and *S. rubra* complex. The smaller worker bees do not emerge until, generally, the main flowering season of *Sarracenia* is past.

Prey-partitioning between different, juxtaposed species of *Sarracenia*

Several workers have commented upon the apparent specialization of certain species of *Sarracenia* with respect to the type of prey that they usually capture. Fish (1976b) observed that *Sarracenia minor* may be particularly specialized to catch ants. According to Folkerts (1982) *S. purpurea*, on the other hand, may be adapted to catch a broad spectrum of prey comprising grasshoppers, crickets and snails and many insects. This catholic diet may be one of the features of its general success as a species and its broad geographical range (see Fig. 19.5).

Gibson (1983a) has taken this prey-specialization further. Figure 7.5 (from Gibson, 1983a) shows the heights of 24 species of carnivorous plants from the southeastern United States. Gibson argues that the most potent factor restricting the juxtaposition of similar-sized species is not the danger of possible hybridization, but competition for insect resources. Such competition, he argues as we shall see later, would be most intense between the tallest, most similar-sized pairs of species. He puts forward evidence that life-history parameters (growth, flowering and survival rates) of individual plants in the field strongly depend on the quantity of insect biomass captured and that natural levels of insect availability in the field limit these rates. Although

Fig. 7.6 Relative abundance of pairs of carnivorous plant species within bog systems. (*A*) For two tall similar-sized pitcher plant species, *S. leucophylla* and *S. flava*, there is a strong negative correlation in their relative abundances within bogs (r = .83, P < .01, *N* = 11 bogs). (B) In contrast, for two dissimilar-sized species, *D. filiformis tracyi* and *S. psittacina*, there is a strong positive correlation in their relative abundances within bogs (r = .81, P < .01, *N* = 12 bogs). (Redrawn with permission from Gibson, 1983a.)

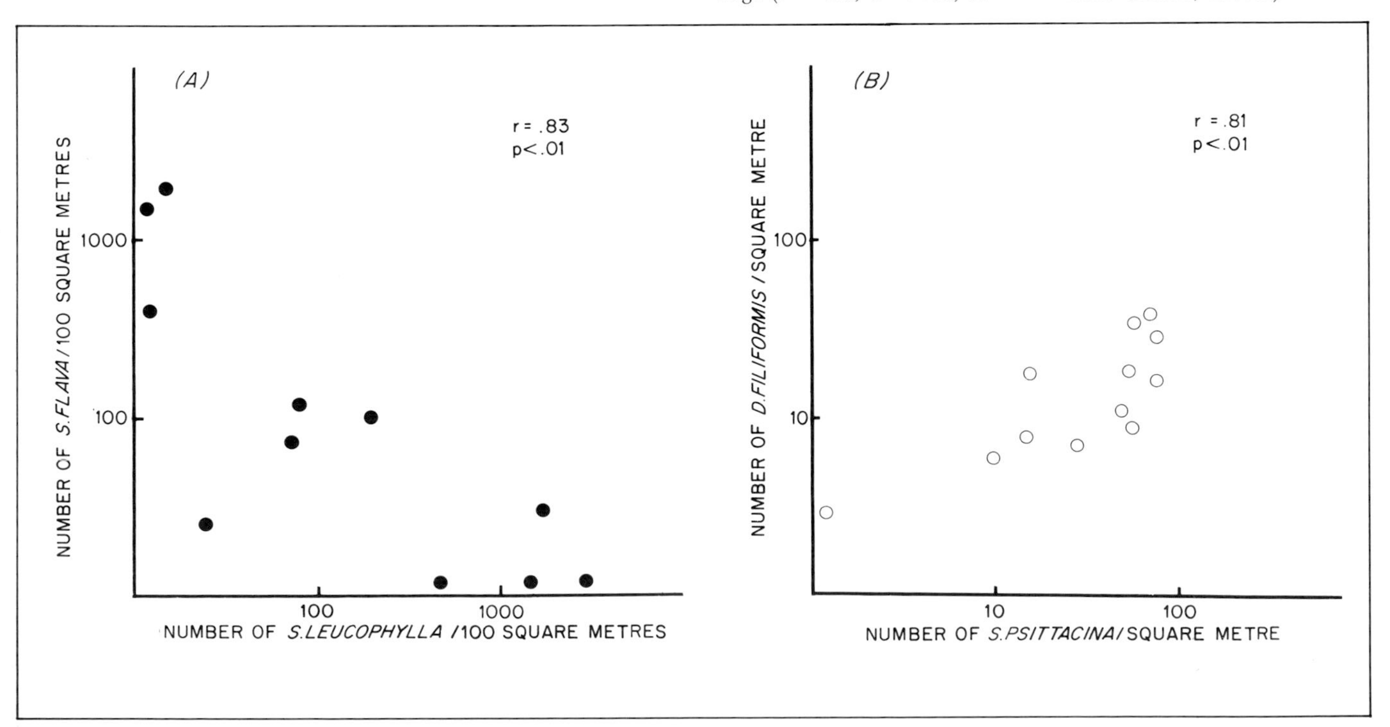

trap height may vary in a given habitat within a given species, e.g. *Sarracenia flava* (40–90 cm) this scatter is small compared to the range in mean adult heights between species, e.g. *S. psittacina* and *S. flava* being 20 cm versus 65 cm.

The essence of Gibson's hypothesis is summarized in Figure 7.6 (from Gibson, 1983a) in which the relative abundances of two similar-sized *Sarracenia* species are compared with *Drosera filiformis* and *S. psittacina*.

We have already argued that, although an insect diet may not be essential for the simple existence of a carnivorous plant, it shifts the balance decisively in favour of effective colonization and subsequent seed-set. According to Gibson, species composition and relative abundance are consequences of a process which operates at the levels of individuals: interspecific competitive interactions between individual plants for limited insect resources within, in this case, each bog. Species pairs (Fig. 7.6) which are most negatively associated and correlated in relative abundance are then those whose individuals compete most intensively for limited insect resources. Taller carnivorous plants, e.g. *S. flava* and *S. leucophylla*, will compete preferentially for larger flying insects. The replacement rates for such insects in acid bogs of low nutrient status is likely to be slower than those for smaller flying or crawling insects or other fauna. We should therefore expect these species (M to Z in Fig. 7.5) to show the most striking negative correlations and this is indeed the case as we can see.

It seems reasonable then, in the absence of more complete data on more species pairs, to accept the insect-competition hypothesis, but also to accept that the suppression of hybridity and individual habitat selection may, in certain situations, contribute to the observed phenomena.

Carnivorous Plants and Insect Control

Since the insect-trapping ability, and in particular the mosquito-larva trapping ability of *Utricularia* was recognized, there have been many suggestions made in the literature that such plants might be used as a form of biological control. In fact, as Gibson (1983a) and others have shown, and as we enumerate in Chapter 16, in numerical terms, carnivorous plants are inefficient trappers of insects. There is no doubt, though, that *Utricularia minor* when predating mosquito larvae for example, can significantly reduce both the numbers of eggs and mosquito larvae in plant-filled ponds as opposed to control, plantless ponds (Angerilli and Beirne, 1974, 1980). B.E.J. observed in the high Sierras of California, USA, that ponds in the open, well-stocked with *Utricularia vulgaris*, were virtually free of mosquitoes. In the shade of the forest canopy (see page 130) *U. vulgaris* was absent and the mosquitoes abundant and vicious.

Utricularia obtusa, of general distribution in the Caribbean, has been shown to capture *Schistosoma mansoni*, the causative agent of schistosomiasis, along with other water-borne organisms such as *Miracidia* and *Cercaria* (Gibson and Warren, 1970). However, it would seem that, given the problems outlined above and the impossibility of draining every patch of shaded water, the biological control of disease vectors in this way is limited.

The Carnivorous Habit: The Cost-Benefit Analysis

In 1958 Sir Ronald Fisher posed the very powerful question: '. . . what circumstances in the life history and environment would render profitable the diversion of a greater or lesser share of the available resource towards reproduction . . .?' This question stimulated a wealth of thought and experimentation. We might, in a more limited context, rephrase the question to focus, not on the reproduction of the species, but on the fraction of the resources that can be devoted to a trap mechanism.

Most carnivorous plants are poor competitors. Those few species that seem able to compete in conventional habitats, e.g. *Triphyophyllum*, are only part-time carnivorous plants and may, according to Green *et al.* (1979) be exploiting a brief insect abundance perhaps for some specific nutrient, such as potassium, rare in the surrounding soil.

Apart from *Triphyophyllum*, and perhaps *Byblis gigantea*, carnivorous plants (and we shall consider why there do not seem to be very many of them overall) are restricted mainly to sunny, nutrient-poor sites.

In Chapters 1 and 2 we documented both the reluctance of many investigators to accept the carnivorous habit and also the components of the carnivorous syndrome. As work has proceeded, many species and genera which seemed not to possess significant parts of the syndrome have now been shown to have alternative features. For example, the bladderworts (*Utricularia* and relatives) and *Pinguicula* species were thought not to have specific attractants. The former group is now suspected to have commensal arthropods as 'bait' and the latter is known to possess, if not necessarily be able to use, UV signalling on the trap itself (Joel *et al.*, 1985). (See pages 76 and 268).

Plants which were once dismissed as primitive or doubtful carnivores because they seemed not to secrete enzymes, e.g. *Sarracenia* and *Darlingtonia*, are now known to have a sophisticated mercenary army of dipteran larvae to aid the digestion. We must be very reluctant to use the word 'primitive' when in fact what we may be observing is a plant or group of plants, on the limit of its range, restricting the expression of certain segments of the carnivorous syndrome (Fig. 1.1) to relevant and environment-limited features.

Givnish *et al.* (1984) and Benzing (1987), as many have done, point out the rarity of carnivorous plants. They estimate in the angiosperms as a whole about 535 species out of a total of perhaps 250 000. Amongst the bromeliads, the only monocotyledonous family known so far, the relative position is worse: two or perhaps three only in the whole tribe, and among the estimated 15 000 epiphytes there may be 18 examples. Despite this, carnivorous plants occupy a very wide range of habitats and it is no longer true to say that they are restricted to sunny and moist habitats, at least in the growing season. *Drosophyllum*, for certain, and *Ibicella* (at least in part) are able to grow and perform in marginal desert conditions and, as we have seen (Ch. 3), there may be several devices for enduring drought as well as low nutrient conditions.

Carnivory, then, will evolve in a range of habitats if the benefits of the enhanced nutrition exceed the cost of the investment in carnivorous adaptations (Givnish *et al.*, 1984; Benzing, 1987). The currency is carbon; the prize, new habitats.

Dore Swamy and Mohan Ram (1969) and Gibson (1983a) have shown convincingly that many carnivorous plants can survive to maturity, and even in many cases set seed and disperse, in the complete absence of prey capture (see page 135). What carnivory does ensure, however, is more rapid maturity, greater seed-set and all the features of enhanced chances not only to maintain populations but also to establish through propagation new colonies of the same species. In support of this Benzing (1980, 1986; Benzing *et al.*, 1976) has shown that epiphytic bromeliads divert a significant fraction of their nitrogen and phosphorus preferentially to flowers and seeds, this floral redistribution suggesting that, at least these two nutrients, are paramount and the same may be true for carnivores (see page 132). Carnivory rarely seems to be relevant as a carbon source (see page 130) but we should note that in one species of *Pinguicula* and in *Darlingtonia* the choice of habitat may suggest that light is limiting (see Ch. 3).

As Givnish *et al.* (1984) point out, there are three potential energetic benefits associated with carnivory. First, carnivory may increase the plant's total rate of photosynthesis as a result of the enhanced mineral absorption by an increased rate of photosynthesis per unit leaf mass or an increase in the total leaf mass that can be supported. These suppositions are given credence by nitrogen-feeding experiments on at least some non-carnivorous plants (Figs 7.1 and 7.7). These studies show that an increase in soil nitrogen can enhance the levels of Rubisco (ribulose bisphosphate carboxylase) for CO_2 capture.

Secondly, carnivory may result in an increased level of nutrients in seeds or in actual seed production, and these assumptions seem to be borne out by Gibson's (1983a) work. We can add to this benefit the additional observations of Gibson's that successful carnivory significantly accelerates seed-set as well.

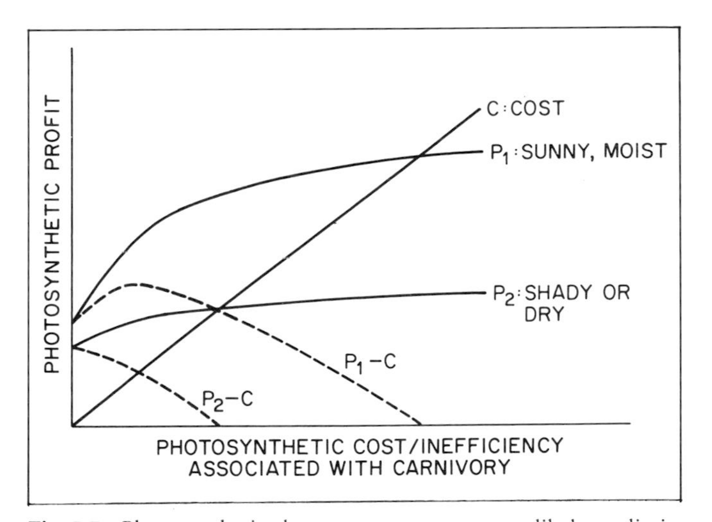

Fig. 7.7 Photosynthetic benefits and costs associated with different levels of investment in carnivorous adaptations in a nutrient-poor site, as a function of environmental conditions (see text). Enhancement of photosynthetic rate (*P*) resulting from added nutrients should be more rapid, and show less tendency to plateau, in a well lit and moist environment (P_1) than in sites where light or water are more likely to limit photosynthesis (P_2). Dashed lines = net difference between photosynthetic benefit and cost (*C*) of obtaining nutrients through carnivory. Carnivory should evolve whenever the benefit of a small investment in carnivory exceeds its own cost, i.e. when the net profit curve slopes upward near $C = 0$. (Redrawn with permission from Givnish *et al.* 1984)

Thirdly, it has been suggested that carnivory may partly replace autotrophy with heterotrophy as a source of chemical energy. This is dismissed both by Givnish *et al.* (1984) and the additional evidence given on page 130.

Givnish *et al.* (1984) synthesize their argument by considering a plant with a given biomass in leaves and roots (Fig. 7.7). As the amount of energy devoted to carnivory in terms of attractants, traps with reduced photosynthetic efficiency, digestive enzymes or consideration given to commensals (Joel and Gepstein, 1985; see Ch. 14) increases, there should be a corresponding increase in the amount of nutrient absorbed. As a result, the effective rate of photosynthesis per unit leaf mass, either in absolute terms or relative to the cost of accumulating nutrients and carbohydrates for new structures, should increase (Fig. 7.7). Furthermore, as the amount of energy devoted to carnivory and the resulting mineral gain continues to increase, the photosynthetic gain expected should tend to reach a plateau, as factors other than nutrients limit photosynthesis or the conversion of photosynthate into new plant structures. Just as we see that xerophytes, which may have extensive root systems, deploy a small photosynthetic area, so too (again commonly for reasons of desiccation), carnivores must limit the exposed areas of their traps with their absorbent and vulnerable surfaces.

The greatest benefits are going to be seen on extremely nutrient-poor sites under high light conditions and this seems borne out by the evidence presented on page 144–145. Hence the large numbers of *Drosera* found in parts of Australia (Fig. 19.7). As the nutrient availability increases, the rate at which new leaves can be produced will depend less on normally limiting materials like nitrogen and phosphorus, and more on the availability of carbon skeletons (Givnish *et al.*, 1984). There are hints, too, on special requirements for trace elements (see page 132).

Givnish *et al.* (1984) argue, from Figure 7.7, that carnivory should have its greatest impact in highly sterile habitats that are sunny and moist; photosynthetic gains (curve P1), resulting from added investments in carnivory, should rise quickly and plateau slowly. There is abundant evidence that there is a close relationship between maximum potential photosynthesis (Amax) and the nitrogen content of the leaves (Field and Mooney, 1986).

Whilst Givnish *et al.*'s hypothesis seems to hold good for many situations, they also argue that this type of analysis explains why carnivory is rare in epiphytes in general, and bromeliads in particular. The bromeliad is either on a shady perch or, if exposed, subject to limiting desiccation. We should note, however, that a far greater range of carnivorous plants than was previously expected seems to be able to shift from the terrestrial to the epiphytic mode, e.g. *Nepenthes* (see page 34), *Pinguicula* (see page 22) and *Utricularia* (see pages 22 and 43) and others are more drought resistant than was previously realized, e.g. *Drosophyllum* and *Ibicella*.

On the Givnish model, however, we can speculate why certain carnivorous plants, e.g. *Dionaea*, are both restricted in habitat and apparently shrinking in a changing world. *Dionaea* has heavy commitments to a highly sophisticated trap system. The trap system relies only to a marginal extent or not at all on commensal organisms for the digestion sequence. There are no drought-resistant devices, as in *Drosophyllum* (see page 32) or *Utricularia* (see page 44) nor are there any photosynthetic alternatives to be mustered in times of low light intensity, as with the phyllodes of *Sarracenia* (see page 137), the 'leaves' of certain terrestrial *Utricularia* species (see page 65) or the true leaves of *Cephalotus* (see Ch. 3). The genus *Drosera*, as we shall see in Chapter 19, would seem to have kept its options open; *Dionaea* would seem to have advanced too far down curve P_2–C (Fig. 7.7).

CHAPTER 8

The Digestive Glands and the Secretion of Digestive Fluids

Introduction

In the majority of plant species, essential nutrients are absorbed by the roots from the soil. While it is quite clear that carnivorous plants possess this capacity, although their roots are often poorly developed or soil may be absent, their adaptation to a carnivorous habit confers on them the advantage of being able to grow in soils or water almost totally lacking certain nutrients. The details of the extent to which prey-derived nutrients contribute to the success of these species has been discussed in Chapter 7. Now, we wish to consider those features of the traps to which we assign an important role in the production of the secretion, followed by a consideration of the current ideas about breakdown of the prey (Chapter 9) and the absorption of the products of digestion (Chapter 10). The features described in Part II enable the plant to attract and capture suitable prey. All these traps provide an environment in which degradation of the prey takes place and breakdown products are made available for incorporation into the tissues of the plant. This process is controlled by the activity of the digestive glands.

The Distribution of Digestive Glands

In all carnivorous plants, with the apparent exception of the doubtful *Heliamphora* (see pages 42, 60 and 64), a restricted area of the surface of their trapping organs possesses glandular structures that have been shown to be active in the digestive phase. These glands form a digestive apparatus. The early assumption that the secretion of a digestive fluid is carried out by the digestive glands was based on the observation that the digestion of prey takes place in the above mentioned digestive apparatus (Darwin, 1875). However, not until 1971 did Heslop-Harrison and Knox, using the gelatin substrate-film method, elegantly demonstrate that the sessile glands of *Pinguicula* serve as a source of hydrolases.

In this method a piece of a trapping organ (the leaf) carrying digestive glands is pressed against photographic film for a few hours in conditions promoting protease activity. When the film is then exposed and developed, areas devoid of silver grains are seen where proteolysis has degraded the emulsion, releasing the silver halide (Fig. 8.17*A*–*E*). This technique has subsequently given positive results in a number of other genera (Heslop-Harrison, 1975, 1976; Green *et al.*, 1979; Parkes, 1980; Vintéjoux, 1979; see Table 9.1). Labelling methods showing enzyme synthesis in the glandular cells and cytochemical methods demonstrating enzyme activity in the glandular cell walls support these observations (see page 173).

In *Pinguicula*, *Drosera* and *Drosophyllum*, the leaf surface has stalked mucilage glands in addition to the sessile digestive glands (Fig. 8.1 and Table 8.1). While the digestive gland may be distinct from those glands involved in the production of attracting and trapping agents (Chs. 5 and 6), this need not be so. For example, in *Drosera* and *Drosophyllum*, the glands involved in the production of trapping mucilage also secrete hydrolases and are pigmented, indicating a possible additional role in attraction (see Ch. 5). In *Utricularia*, the bifid and quadrifid glands concerned with water removal may produce proteases (see Ch. 9 and Figs 4.13*A*–4.15*B*). Absorption of digestive products seems to take place through the same glands, though, as we will propose later, not always by the same pathway (see page 210).

The structure/function aspects of secretory

activity have been most extensively examined in *Dionaea* (Lüttge, 1963; Scala *et al.*, 1968b; Schwab *et al.*, 1969; Robins and Juniper, 1980a,b,c,d,e; Joel *et al.*, 1983; Rea, 1982a,b), in *Drosophyllum* (Schnepf, 1963b,c; Joel and Juniper, 1982), in *Nepenthes* (Lüttge, 1965, 1971) and in *Pinguicula* (Heslop-Harrison and Knox, 1971; Heslop-Harrison, 1975, 1976; Heslop-Harrison and Heslop-Harrison, 1980, 1981; Vassilyev and Muravnik, 1988a,b). This evidence must be drawn upon heavily in attempting to establish the mode of action of other species. Parkes (1980) examined glands in many genera and provides much useful comparative information.

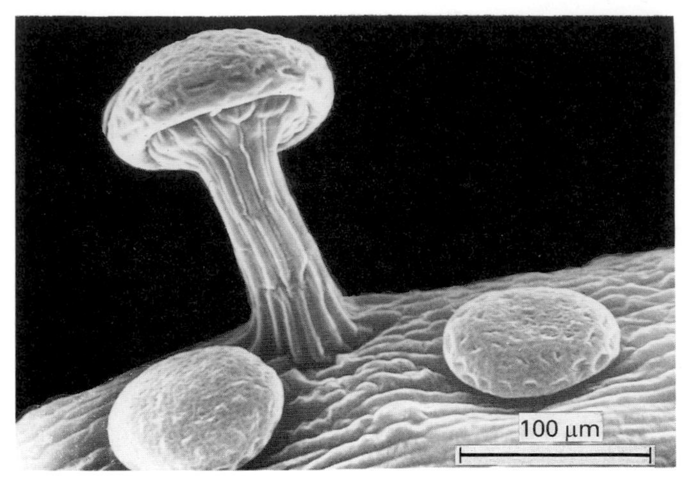

Fig. 8.1 Stalked and sessile glands of *Drosophyllum lusitanicum* (SEM by G. Wakley).

Table 8.1 Comparisons between the stalked glands of *Triphyophyllum*, *Drosophyllum* and certain *Drosera* species. ((?) denotes fact not established.)

	Triphyophyllum	*Drosophyllum*	*Drosera* spp.
A. Characteristics of the head			
Gland size	Two size classes, with some intergradation	Relatively uniform	Two size classes, lateral glands (tentacles) larger than the remainder
Gland shape	Hemispherical, asymmetrically placed stalk	Hemispherical, domed, stalk symmetrically placed	Inner heads ellipsoidal, symmetrical; lateral heads markedly asymmetrical in some species
Secretion droplet	Comparatively low viscosity; hydrolytic enzymes present	Comparatively low viscosity; hydrolytic enzymes present	Initially viscid; hydrolytic enzymes present
Secretory cells:			
(a) no. of layers, and cell no. at maximum width of gland in T.S.	2–4 layers; 100–200 cells	2 cell layers; 10–25 cells	2 cell layers; 25–40 cells
(b) surface characters	Cuticle thin; (?) gaps present	Cuticle thin; with gaps	Cuticle of variable thickness, often with surface particles; gaps over anticlinal walls
(c) shape of outer cells	Columnar in longitudinal section, tapering inwards; isodiametric in transverse section	Isodiametric or transversely flattened in longitudinal section, not tapering. Roughly isodiametric in transverse section	Columnar, tapering inwards in longitudinal section; upper cells of the head somewhat elongated, lower less so. Isodiametric in transverse section
(d) radial walls	Transfer-cell type, but without crenellations or platelike invaginations. Enzymes probably held in wall sites	Transfer-cell type, with crenellations and plate-like extensions. Enzymes held in wall sites	Transfer-cell type, with crenellations and plate-like extensions. Enzymes held in wall sites

	Triphyophyllum	*Drosophyllum*	*Drosera* spp.
Geometry of endodermoid cells, and cell no. in transverse section at maximum width	Domed; 50–70	Domed; 10–15	Bell-shaped; 15–25
Tissues within endodermoid layer	500–750 cells, with 8–12 groups of tracheids with annular, helical or pitted thickening, dilated at contact face with endodermoid layer. Parenchyma between tracheids, with amyloplasts. No distinct neck region	10–20 tracheids, with pitted or reticulate thickening, dilated at contact face with endodermoid layer. No distinct neck region (?)	A mass of 10–20 tracheids with helical thickening. Lower part forming a 'neck' with parenchymatous cells with numerous chloroplasts surrounding a central vascular strand
B. Characteristics of the stalk			
Branching	Sometimes branched	Never branched	Never branched
Response to prey	No movement in response to prey	No movement in response to prey	Lateral tentacles bend inwards in response to prey; other stalked glands do not move
Epidermis: circumferential cell number	30–40, with thick outer cuticle	10–20, with thick outer cuticle	10–15, with thick outer cuticle
Cortex: cell number in cross section	50–60	8–15	4–8
Vascular core	10–20 tracheids and associated phloem	5–10 tracheids and associated phloem	4–3 tracheids; no phloem

(Reproduced from Green *et al.*, 1979, with permission.)

Minimal Requirements of Digestive Glands

Digestive glands fulfil three main functions with respect to the assimilation of captured prey: they serve as receptors which sense the arrival of potential prey; they are the main source of the digestive fluid; they form the absorptive site for the products of digestion. A digestive gland is, therefore, a sensory, a secretory and an absorptive structure. Each function requires specific characteristics, resulting in a complex organ which is often difficult to analyse.

For the perception of chemical stimuli the minimal requirements are:

(i) a receptor site or sites;
(ii) minimal dilution of the stimulant on its way to the receptors;
(iii) minimal resistance to its movement;
(iv) shortest pathway to reach the receptors.

The minimal requirements for an effective absorptive organ are:

(i) a large surface area;
(ii) an active transport mechanism;
(iii) sufficient space for maintaining solute concentration gradients;
(iv) minimal diffusive resistance;
(v) proximity to a sink;
(vi) prevention of reverse flux;
(vii) selective permeability;
(viii) defence mechanisms;
(ix) control over the movement of water and solute.

The minimal requirements for an effective digestive secretory structure are:

(i) a large surface area;
(ii) active enzyme synthesis;
(iii) active secretion of enzymes;
(iv) rapid release of secreted substances;
(v) means to maintain concentration gradients of secreted substances;
(vi) minimal diffusive resistance;
(vii) sufficient supplies of water and certain ions, as well as amino acids for enzyme synthesis;
(viii) prevention of reverse flux.

To what extent, then, do the various digestive glands found in carnivorous plants fulfil these requirements? Are they not contradictory? Many features of the glands are still obscure; others not well understood. We shall try to summarize those structural features that are interpreted as having a role in their function and, in particular, indicate those features that they have in common, while highlighting the distinct features of individual functional mechanisms.

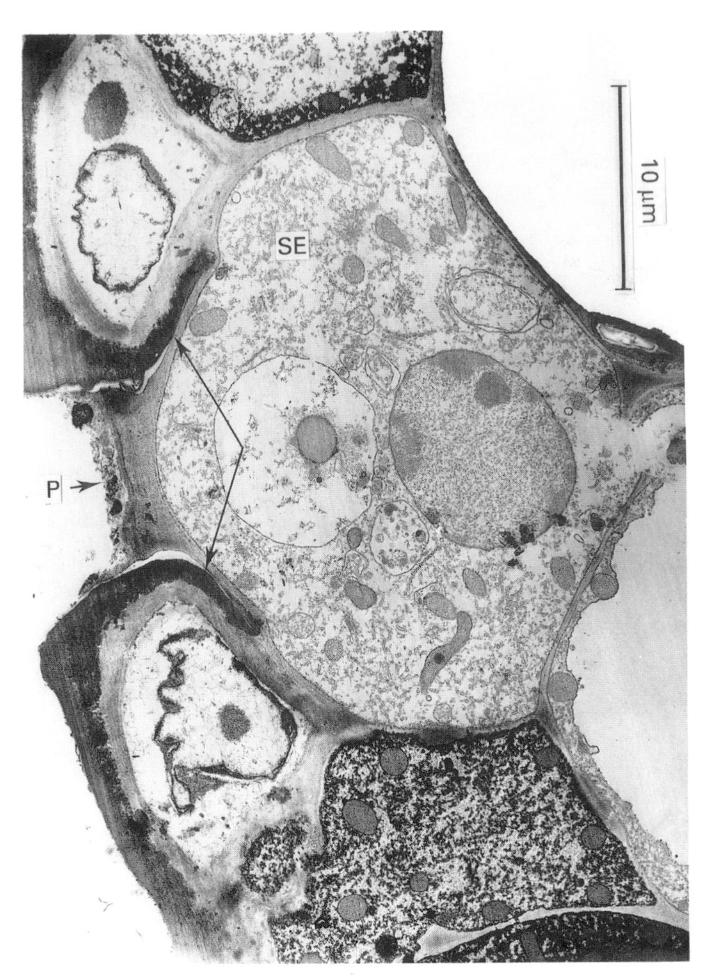

Fig. 8.2A Median TEM section through a 'small' gland of *Cephalotus follicularis* (cf. Fig. 6.37). These small glands were once thought to be stomata (see Lloyd, 1942), but the 'guard cells' do not move and the aperture is sealed by a thick, but non-cutinized 'plug' (P). Note the massive endocuticle (arrows) which almost encircles the non-functional 'guard cells' and various, as yet unidentified vacuolar structures in the subepidermal cell (SE). (Unpublished micrograph by kind permission of Drs D.M. Parkes and N.D. Hallam, Monash, Australia.)

The Architecture of the Digestive Glands

The digestive glands in carnivorous plants vary from a small organ containing a few cells in *Cephalotus* or *Utricularia* to the multicellular, multilayered glands of *Drosera*, *Drosophyllum*, *Triphyophyllum* and some *Nepenthes*. Despite this, they all show a common architecture with a group of glandular cells surmounting an endodermoid layer (Figs 4.1 and 8.5*A–D*) and, in most cases, one or more basal cells below the endodermoid layer. This structure is common to many other plant glands secreting hydrophilic compounds (Fahn, 1979). The degree of variation in complexity correlates well with the diversity of the roles in which any particular gland participates. Thus, all digestive glands appear to take part in absorption but the less complex participate in only one other role – that of hydrolytic fluid secretion. Those which are additionally involved in the trapping process by producing mucilage (see Ch. 6), as is the case in *Drosera*, are the most complex.

Types of Digestive Glands

Structurally, the digestive glands fall into three basic categories.

(i) embedded in the wall of the trap (*Cephalotus*, *Darlingtonia*, *Sarracenia* (Figs 6.36, 6.38, and 8.2*A*); these are termed the 'sunken' glands;

(ii) lying directly upon the epidermis, protruding by no more than the thickness of the gland; these are the so-called 'sessile' glands (*Aldrovanda*, *Brocchinia*, *Byblis*, *Catopsis*, *Dionaea*, *Drosophyllum*, *Genlisea*, *Nepenthes*, *Pinguicula*, *Triphyophyllum*, and *Utricularia* (Figs 4.7*C*, 4.13, 4.18, 4.19, 4.20, 8.1, 8.2*B* & *C*, 8.18 and 9.9). In some cases, as in *Brocchinia* and *Nepenthes* (Fig. 8.18*F*) and some species of *Pinguicula* (Fig. 8.7*C*), the glands are seated in an epidermal concavity and do not extend beyond it; thus, while in some cases appearing embedded, they actually lie on the epidermis, not within it;

(iii) raised on a stalk of varying complexity and length (*Byblis*, *Drosera*, *Drosophyllum*, *Ibicella*, *Pinguicula*, *Triphyophyllum* (Figs 4.2, 4.18, 4.19, 4.20, 8.1, 8.16*D* and 9.9); in *Drosera*, uniquely so far as we are aware (Figs 6.9 and 6.24) the stalks show nastic movements, therefore often regarded as 'tentacles'.

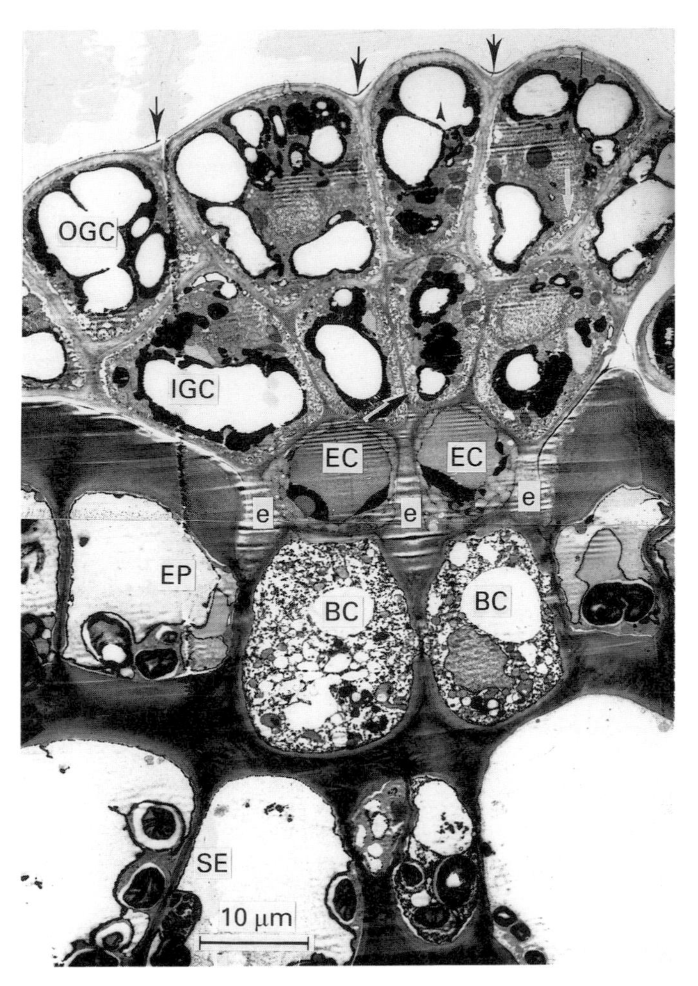

The digestive glands of all genera are composed of two main components: *glandular* and *endodermoid*. These components may, individually, comprise a single cell or multiple groups of cells. Below the endodermoid layer is often a large *basal* cell (Figs 8.2*A*,*B* and 8.3.*A*,*B*).

A thin but distinct *cuticle*, usually perforated, covers the gland and provides it with a permeable protective cover (see page 162 and Figs 8.9–10). The glandular head cells mostly consist of one or two layers, though in *Triphyophyllum* there are up to four and, in some species of *Nepenthes*, up to ten or more. These layers provide both the sites of synthesis of enzymes and mucilage and a large surface area for absorption. Each layer, as in the Droseraceae, may show individual features

Fig. 8.2B A median vertical TEM section through a gland of *Dionaea* (cf. Figs 4.3 and 6.17*A*) to show the general ultrastructure. OGC and IGC, outer and inner glandular cells; EP, epidermal cells; EC, endodermoid cells; BC, basal cells; SE, sub-epidermal cells. The endocuticle is marked with 'e's, and the well-developed transfer-cell protuberances of the OGC and IGC are marked with white arrows. Note too the vestigial cuticle which is slightly thicker over the anticlinal cell walls (black arrowheads).

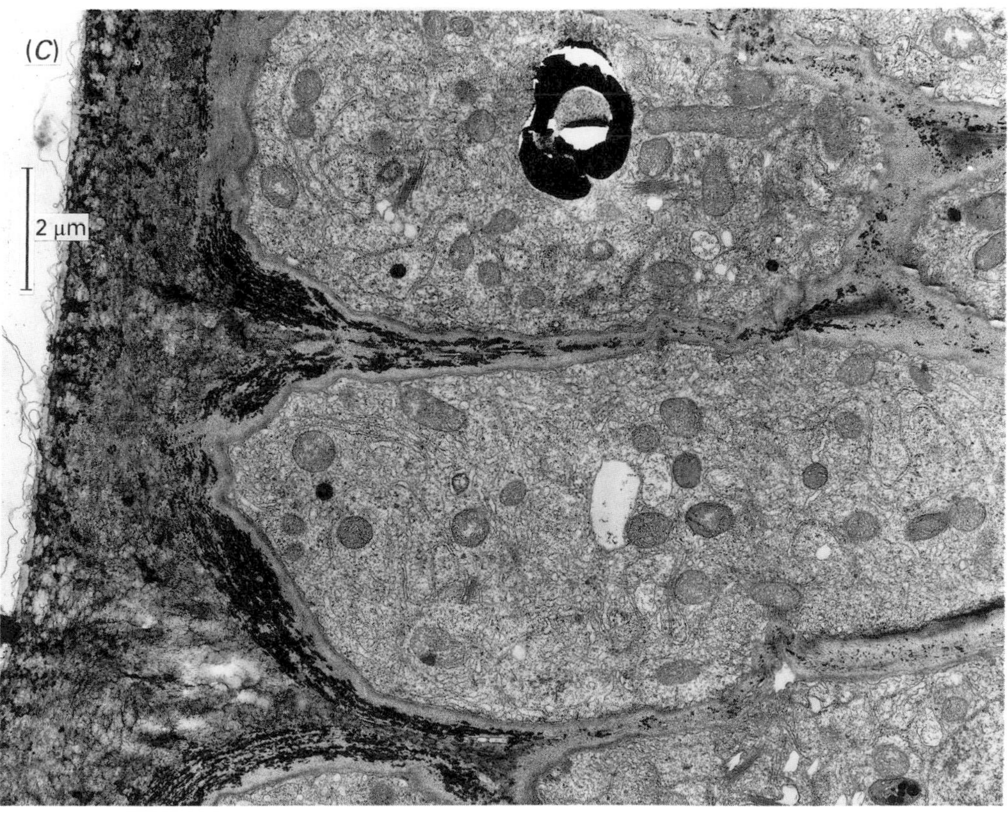

Fig. 8.2C *Nepenthes khasiana*: the ultrastructure of the digestive gland. The pitcher has just started to open and contains a significant amount of fluid. Note the very heavily thickened outer cell wall (left) which may be up to 2.5 μm in thickness. On its surface is a thin discontinuous cuticle; cf. Fig. 8.9*F*. The cuticle of the adjacent epidermal cells of the pitcher is both continuous and may be up to 2 μm thick. The outer wall of these secretory cells is heterogenous in structure and contains cuticular 'cystoliths' similar to those found in *Utricularia* (cf. Fig. 8.9*E*). The osmiophilic precipitate extends down into the upper part of the anticlinal cell walls. Both light and dark (phenolic rich?) vacuoles are present in the cells, along with numerous dictyosomes and ER. The ER, which is very well developed, appears to be predominantly cisternal in the centre of the walls and more markedly tubular towards the margins. (Unpublished micrograph by kind permission of Drs A.E. Vassilyev and Lyudmila E. Muravnik, Leningrad, USSR.)

indicating that their functions may be dissimilar (Schnepf, 1961a). Differences in the distribution of labyrinthine protuberances in these layers are especially noticeable and are interpreted as reflecting different aspects of glandular activity (see page 165; Robins and Juniper, 1980b,c,e; Figs 8.2*B*, 8.5 and 8.13). Each terminal cell of the divided glands of *Utricularia* (bifids and quadrifids) can similarly be thought of as a pair of functional cells because of the highly specialized regional separation of structure and function evident in each of them (Fineran and Lee, 1975). The roles of the different glands in *Utricularia* have recently been reviewed (Fineran, 1985 and see page 119–124). The glands and organs of the pitcher plants often show similar internal structural segregation (see Ch. 10, 13 and Fig. 14.1).

The *endodermoid layer*, as distinct from the functionally similar endodermis of higher plant roots, and stems of some water plants, mediates between the glandular cells and the underlying leaf tissues. This layer is usually composed of a single cell or a group of cells each having an impermeable wall region effectively blocking the apoplastic pathway between the head cells and the basal cells (see Table 8.1 and Figs 8.2*B*, 8.5*A–D*). The endodermoid cells of the pitcher epithelium of *Sarracenia* are completely encapsulated in a suberized wall layer, similar to the bundle-sheath cells of C_4 plants (Joel and Heide-Jørgensen, 1985; Joel, 1986; Fig. 14.1).

The sunken glands

Cephalotus and *Darlingtonia* have strikingly similar small, simple glands embedded in the epidermis of the lower region of the trap. These glands appear to be modified stomata (see Chs. 6 and 18), with one or two gland head cells occluding the cavity of the ancestral stoma (Figs 6.38, 8.2*A*). In *Darlingtonia* they are scattered at low density across Zone 3 (Fig. 4.9*E*), whereas in *Cephalotus* they are confined exclusively to the *glandular patch* (Figs. 4.9*B* and 8.2*A*) (Dakin, 1919; Parkes, 1980; Parkes and Hallam, 1984), a raised mass, densely packed with glands on either inner lateral wall of the pitcher. Lloyd (1942) does not mention such glands in *Darlingtonia* while those of *Cephalotus* he considers to be stomata. This is now proved not to be so but, although it is reasonable to assign a similar role to the glands of *Darlingtonia*, there is no evidence as to their function.

Cephalotus additionally contains much larger, domed, multicellular glands embedded in the surface which are clearly not modified stomata (Parkes, 1980; Parkes and Hallam, 1984). They are largest

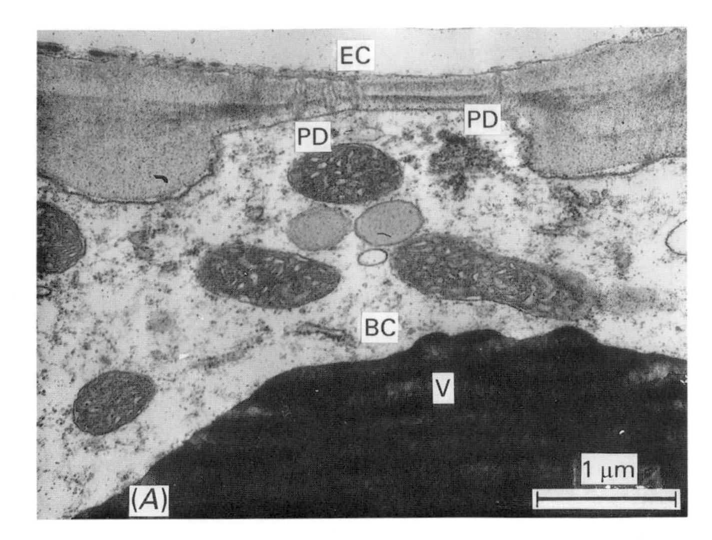

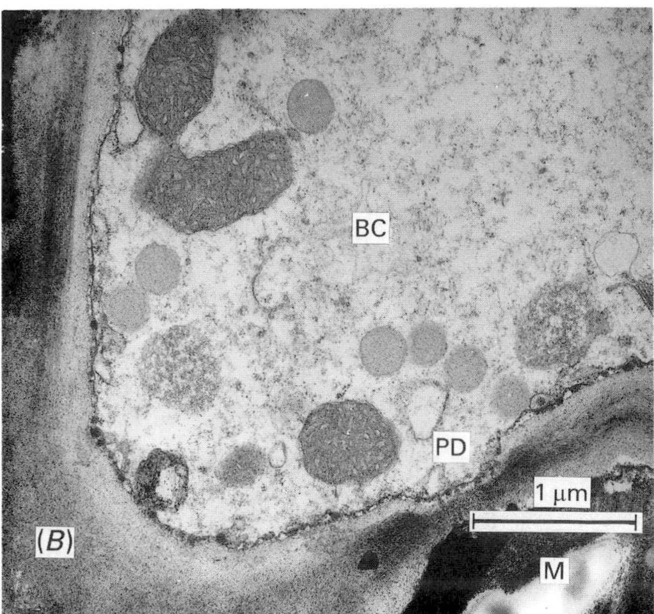

Fig. 8.3 Basal cells of the digestive gland of *Dionaea*, cf. Fig. 8.2B. BC, – basal cells. (*A*) Upper side of cell facing the endodermoid cell (EC); note the numerous plasmodesmata (PD) crossing the wall in a pit field. V, phenolic-rich vacuole. (*B*) Lower side of a basal cell facing the mesophyll (M); plasmodesmata in a pit-field also join these cells. (unpublished results by DMJ, 1983).

on the top edge of the glandular patch but spread, at decreasing size and density, right up to the pitcher wall to just below the cornice (Figs 4.9*B* and 6.37). These glands differ from most carnivorous plant organs in being buried deep within the wall tissue, the dome being composed of a layer of columnar head cells. Despite this, the cutinized endodermoid layer is still present and tracheid elements are closely juxtaposed to the base of this junction.

In *Sarracenia* two principal types of digestive glands develop in the trap:

(i) unit glands, which resemble the digestive glands of other carnivorous plants; these glands are sunken in the pitcher wall at zone 3;
(ii) the pitcher epithelium of Zones 4 and 5 (the bottom gland).

The pitcher epithelium of *Sarracenia* is regarded as a gland, because it comprises a glandular epidermis and an endodermoid hypodermis (Fig. 14.1). This gland is composed of an unlimited number of cells and extends over the whole surface of the bottom zone of the pitcher, instead of being divided into many unit glands, each having a limited number of cells and each restricted to one small portion of the glandular surface (Joel, 1986).

Some species of *Sarracenia* (*S. flava* and *S. oreophila*) are apparently devoid of the sunken unit glands in Zone 3, as also is *Heliamphora* (Juniper and Joel, unpublished). Parkes (1980) finds small glands in *S. psittacina* very similar to those of *Darlingtonia* and numerous digestive glands in *S. purpurea*.

The digestive glands of *Nepenthes khasiana* closely resemble the nectary glands of this species (Fig. 8.2*C*) being enclosed within a pit formed by partly overlapping flanges of epidermis (Parkes, 1980; Figs 4.10 and 8.18*F*). In the upper part of the digestive zone this flange overhangs more than half the gland but, by the bottom of the pitcher, the glands are almost completely uncovered. This feature is common to digestive glands throughout this genus. The gland is reminiscent of the structure of *Cephalotus* large glands, having a head of long columnar cells overlying one or more layers of secretory cells and an endodermoid layer. Tracheids are often present near endodermoid cells. The glands differ, however, in being free of the epidermis to the sides and this makes them resemble the sessile glands of *Drosophyllum* and *Dionaea* sunk in a pit. Parkes (1980) suggests that the overhanging ridge protects the gland from damage by prey trying to escape and prevents the use of the gland as a foothold. In other genera with sessile glands, this protective feature is not necessary and they are raised above the surface.

The sessile glands

The simplest sessile glands are those of *Byblis*, *Genlisea* and *Pinguicula*, which are very similar, all containing 4–8 head cells surmounting a single endodermoid cell and sometimes closely associated with tracheid elements (Fig. 6.1). Although the sessile glands of *Genlisea* (Fig. 4.18) have received little attention, and those of *Byblis* even less, their structural and topographical similarities to the thoroughly-studied sessile glands of *P. grandiflora* make it tempting to suggest a similar function and mechanism. In *Byblis* and *Pinguicula* they are partially sunken in a furrow or pit respectively, but in *Genlisea* the more swollen endodermoid cell causes the gland head to stand proud of the epidermis.

Sessile glands in *Drosera*, *Aldrovanda*, and *Dionaea* (Fig. 8.2*B*) show increasing complexity. In *Drosera* they are small, comprising two layers, each of 6–10 secretory cells arranged in various patterns, surmounting a pair of endodermal cells, much like in *Pinguicula*. *Aldrovanda* has 12 secretory cells in a single layer surmounting a large spherical core of four endodermoid cells, more like a complex form of *Genlisea*. In *Dionaea* (Fig. 8.2*B*) the sessile gland has two structurally distinct layers of secretory cells, about 40 in all, with two endodermoid cells. In some cases (e.g. *Pinguicula*; Heslop-Harrison, 1975) vascular connections are present in the cell layers immediately below the gland (see page 156).

The glands of *Brocchinia reducta* closely resemble the water-absorbing glands of related bromeliads (Benzing *et al.*, 1976). They do not extend beyond the epidermal concavity in which they are seated, nevertheless they are connected to the leaf tissue only by their basal cell. The gland comprises a series of single cells, one upon the other, which end distally with a multi-seriate head. Partially concentric in arrangement, this trichome lacks the upper tier disc cells which form the flat cap of typical tillansioid trichomes (Benzing, 1986).

The sessile gland of *Dionaea* (Fig. 8.2*B*) is probably the most complex true sessile digestive structure, as the sessile glands of both *Drosophyllum* and *Triphyophyllum* closely resemble the heads of the stalked glands but without the stalk.

The stalked glands

The stalked glands of *Pinguicula*, *Drosera*, *Drosophyllum* and *Triphyophyllum* are all engaged to some extent in digestive activity as well as in the production of mucilage. Those of *Byblis* and *Ibicella* may only perform the latter function. In all cases these glands have three distinct regions:

(i) a number of head cells arranged in 1–4 layers external to the endodermoid layer;
(ii) an endodermoid layer;
(iii) a clearly defined stalk.

The level of complexity of each of these regions varies greatly from *Pinguicula*, the simplest (Fig. 6.1), through *Byblis*, *Drosera* (Fig. 10.10) and *Drosophyllum* to the massive glands of *Triphyophyllum* (Fig. 4.19*A*). Regions (i) plus (ii) essentially form an equivalent to a sessile gland: region (iii) serves to raise the gland head above the surface of the leaf and, in the more complex glands, is composed of several further cell types, notably a head of spongy tracheids, xylem and phloem elements and one or more layers of specialized cortical cells.

The stalked glands of *P. grandiflora* have been particularly well described (Heslop-Harrison and Knox, 1971; Heslop-Harrison and Heslop-Harrison, 1981) and those of other species are essentially similar. The whole gland arises from sequential divisions of a single epidermal cell (Fenner, 1904), the cell remaining embedded in the epidermis becoming the basal reservoir cell which maintains contact with 4–8 cells of the epidermis through numerous plasmodesmata (Figs 8.3*B* and 8.4*A–C*). Unlike sessile glands, however, this cell does not appear to connect directly to the vascular system. A single elongated cell comprises the stalk, linking the reservoir and endodermoid cells, this being the main anatomical feature differentiating sessile glands from stalked glands of this type. The gland head cells are arranged in a single radial layer, typically composed of 8–32 cells, all in contact with the single endodermoid cell. The structure of stalked glands in *Byblis* is very similar (Lange, 1901; Fenner, 1904; Lloyd, 1942). The head normally contains 32 cells (though it can be as few as four) surmounting a single collar cell on a long, single stalk cell which has walls thickened in such a way that it twists on drying. *Byblis* differs from *Pinguicula* in not having a large basal cell but rather a number of smaller cells which appear to perform this role.

However, *Drosera* (Fenner, 1904; Ragetli *et al.*, 1972; Gilchrist, 1974), *Drosophyllum* (Fenner, 1904) and *Triphyophyllum* (Green *et al.*, 1979; Marburger, 1979) all show a markedly more complex development of the stalk, which is multi-cellular and multi-layered. In these glands the centre of the stalk contains vascular tissue running up its length and making contact with a mass of tracheids or tracheid-like dead cells which form the core of the gland head. This core is surrounded by a domed or bell-shaped endodermoid layer which, in the case of the massive glands of *Triphyophyllum*, may contain up to 750 cells (Table 8.1). The cells of the endodermis show typically impregnated walls and numerous plasmodesmata connecting them to the head cells, which form 2–4 layers of isodiametric or columnar cells with wall elaborations and a thin cuticle containing pores. While the endodermoid cells show characteristics typical of this layer in sessile glands, they differ in not having plasmodesmata connecting them to the underlying layer, consisting in these glands of dead tracheid cells. Nor, as in most sessile digestive glands, is a transfer cell anatomy found in them even though here both a symplastic and an apoplastic discontinuity exists. Only in the neck region where the stalk cells make direct contact with the endodermoid layers are plasmodesmata present (Fig. 10.8).

The stalk itself consists of at least three layers of cells. The simplest form is found in *Drosera* (Fig. 10.8) where a central single xylem element (rarely two or three) is surrounded by a distinct layer of inner stalk cells and an external layer of outer stalk cells. The inner stalk cells show a highly elongated structure, between 130 and 510 μm in *D. capensis* (Ragetli *et al.*, 1972), and have thin tapering transverse walls pierced by numerous plasmodesmata. While not evacuolate, they show many characteristics of phloem tubes, though are not vascular tissue. At the head end they make symplastic contact with the lower end of a group of transfer cells, the upper part of these being appressed to the core of tracheids. The cortical cells are shorter, 104–314 μm long and at the head end make symplastic contact both with the transfer cells and the endodermoid cells. The importance of these various interconnections for transport within the gland head is discussed in Chapter 10.

Drosophyllum shows a slightly different arrangement with the vascular core containing 2–3 xylem elements and some phloem elements and is sheathed by two layers of cortical cells (Fig. 6.4). *Triphyophyllum* is similar to this but with 10–12 groups of vascular elements (xylem plus phloem) and 2–3 layers of cortical cells.

The xylem elements in all these complex glands penetrate the tracheid core, though this feature is very much more marked in *Triphyophyllum* where the xylem has dilated terminations abutting directly onto the endodermoid cells. This genus is the only one to contain both tracheid and parenchymatous cells within the core, the latter containing numerous amyloplasts and forming sheaths around the xylem. Surprisingly, they do not appear to have transfer cell anatomy.

Triphyophyllum is also the only genus in which branched glands are found. *Drosera*, however, has small six-celled epidermal glands placed up the sides of the stalk (Fenner, 1904; Ragetli *et al.*, 1972; Fig. 6.1*B*). These cells contain large nuclei, a small fragmented vacuole and have a dense cytoplasm with numerous vesicles. The external wall is very thin compared with the cortical cells of the stalk and the inner surfaces have highly elaborated labyrinthine walls on all but the innermost side of the two cells abutting the outer stalk cell. Their function remains obscure, though it has been suggested that they act as touch sensors or in the release of volatile attractants.

The Digestive Gland in the Resting State

The Basal Cells

Basal, or reservoir, cells are found in a number of carnivorous plant glands underlying the endodermoid layer (see page 156). They are particularly prominent in *Dionaea* (Robins and Juniper, 1980a) (Fig. 8.3) and *Pinguicula* (Heslop-Harrison, 1975, 1976; Fig. 8.7C). In both these genera they have been shown to play a role in secretory activity (see page 182) and must therefore be considered to constitute part of the gland even though they lie outside the region of tissue bounded by the impregnated endodermoid walls.

In *Pinguicula* the reservoir cell is in contact with 4–8 cells which are arranged in a concentric manner around it (Heslop-Harrison and Heslop-Harrison, 1981). These surrounding cells are often termed

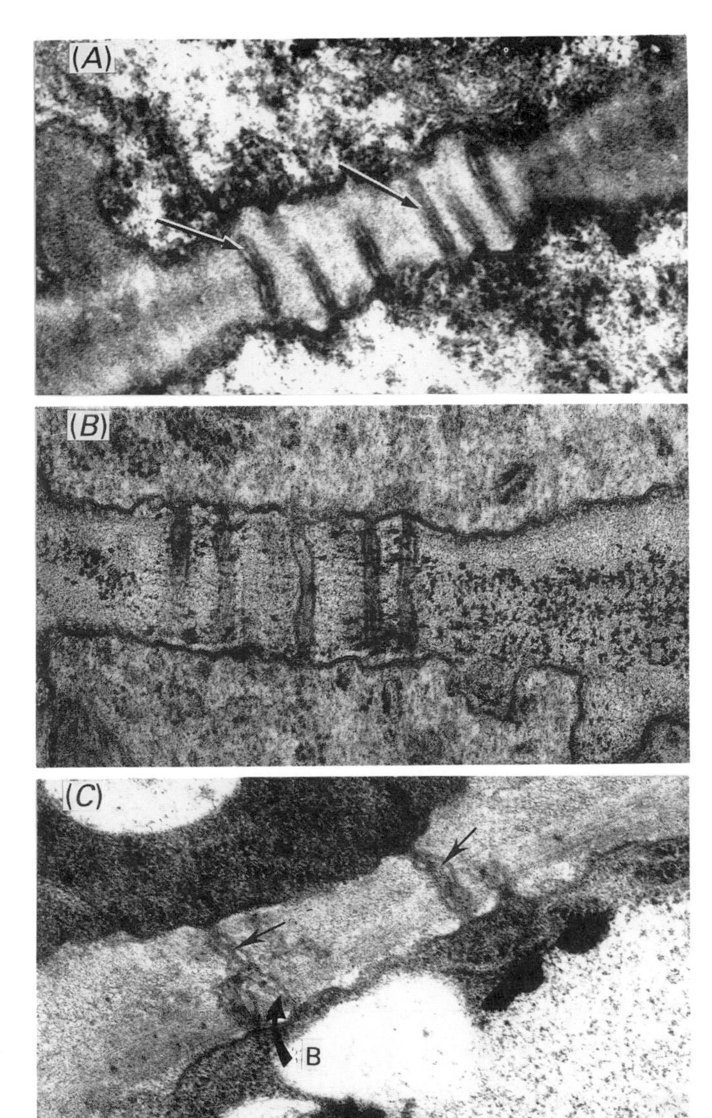

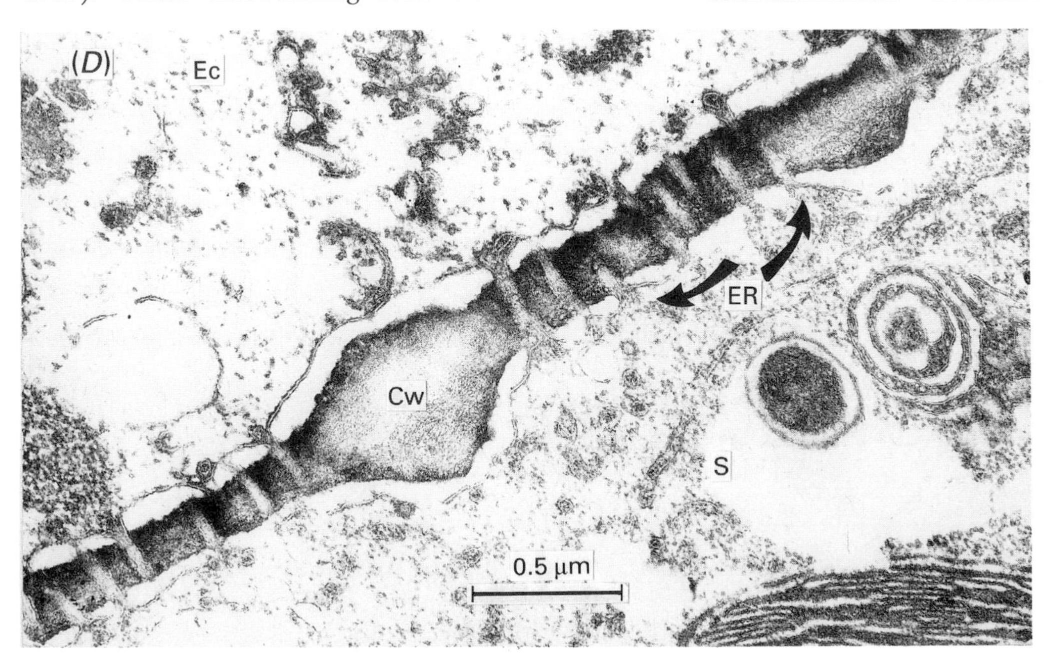

Fig. 8.4 Plasmodesmata between the secretory cells of the gland of *Dionaea* (*A*–*C*). (*A*) Between two of the inner gland cells; note the clear desmotubules (arrows); stimulated day 4. (*B*) Between an outer (upper) and inner (lower) gland cell; stimulated tissue, day 4. (C) Plasmodesmata between an outer (upper) and inner (lower) gland cell; note the desmotubules (arrows) and also the branching (B); stimulated tissue, day 2.

(*D*) *Drosera capensis*. A part of a wall (Cw) between endodermoid cells (Ec) and stalk cells (S). Note the abundant plasmodesmata in the wall and the association of ER (ER) with these plasmodesmata. (By permission from Gilchrist, 1974.)

'collecting cells'. Similar large basal cells are found in *Genlisea* (Heslop-Harrison, 1976) and *Utricularia* (Fineran and Lee, 1975). In *Cephalotus*, both the nectaries and the small digestive glands have one or two similar cells present while the large glands are surrounded by a ring of reservoir cells (Parkes, 1980; Parkes and Hallam, 1984).

All these cells have, more or less, been shown to have several common features:

(i) They are connected to the endodermoid cells by numerous plasmodesmata, often gathered in simple pitfields (Fig. 8.4*A–D*). In *Pinguicula* and *Dionaea* these plasmodesmata are typically straight but in *Utricularia* they are often compound, the single opening being to the endodermal side (Fineran and Lee, 1975). Pit-fields also connect the reservoir cells to each other and to adjacent epidermal and mesophyll cells.
(ii) The cells are highly vacuolated, their cytoplasm often confined to a very narrow region round the wall (Figs 8.3*A*,*B*).
(iii) In some instances notably *Pinguicula* (Heslop-Harrison, 1975) and the large glands of *Cephalotus* (Parkes, 1980) vascular connections are located immediately below the reservoir cell.

There are several variations on this theme. In the stalked glands of *Drosera* and *Drosophyllum* which show a similar arrangement, the spongy tracheid cell-mass forming the centre of the head of the gland is equivalent to the reservoir cells, being connected to both the endodermis and vascular tissue (Fenner, 1904; Gilchrist, 1974; Ragetli *et al.*, 1972).

In *Nepenthes*, basal cells are not apparent in any of the gland types (Parkes, 1980). Stern (1917) noted, however, that vascular connections are frequently seen just below the gland tissue.

The role of the basal cells appears to be in the accumulation and distribution of solutes during transport both into and out of the glands. This is entirely compatible with the numerous symplastic connections these cells make with the surrounding tissue (Figs 8.3*A*, *B* and see also Fig. 8.4*A*, *D*).

The basal cells appear to play a major role in the rapid initiation of chloride fluxes during the secretory activity (see pages 183–188). This feature may explain their absence from the continually secreting pitcher plants.

The Endodermoid Layer

Between the basal cells and the digestive cells lies the endodermoid layer, providing a zone of impregnated apoplast (Figs 8.5 and 8.6). The shape of the layer differs in the various digestive glands but is present in all cases examined. Just a single cell forms the endodermoid layer of all the Lentibulariaceae. In *Dionaea* and *Aldrovanda* there are two cells, while other genera have several or many cells often arranged in an hemispherical layer. In the pitcher epithelium of *Sarracenia* (Zone 4) the whole hypodermis serves as an endodermoid barrier between the glandular epidermis and the leaf mesophyll (see page 253).

The impermeable wall deposits of the endodermoid layer

A consistent feature of the endodermoid cells is the thickened, heavily impregnated radial walls (the 'endocuticle') which resembles the well known 'Casparian strips' of root endoderm. These heavy deposits of suberin or cutin look peg-like in longitudinal sections (Figs 8.5*A*, 8.6 and 8.7*C*) but are seen completely to surround each cell in tangential sections and have the plasmalemma intimately attached to them even in plasmolysed cells (Fig. 8.5*B*). Thus they form a 'tight junction' of the type found in the primary root endodermis and similar to that of many other external glands (Fahn, 1979). The endocuticle is continuous with the thin surface cuticle of the secretory head cells and with

Fig. 8.5 Details of the stalked gland of *Drosera capensis*. (*A*) Stalk (S), endodermoid (EC) and glandular cells (G) near the base of a *Drosera capensis* gland. The orientation is shown in the diagrammatic sketch (*B*), square 1 (SQ[1]). A Casparian strip (Cs) is visible between endodermoid cells. Chloroplasts (Ch) can be seen in the stalk cells and pigment granules (Pg) in the glandular cells. Note the vestigial cuticle (C) on the outside of the gland cell. (Reproduced by kind permission of Dr. A. J. Gilchrist.)

(*B*) Diagrammatic sketch of the head of a stalked gland of *Drosera capensis* to show the locations of the TEM micrographs in Figs 8.4*D*, 8.5*A* (SQ[1]), 8.5*C*, *D* (SQ[2]) and in 10.10 (SQ[3]). Abbreviations as in Fig. 4.1 and the micrographs above.

(*C*, *D*) Further detail of the endodermoid, glandular and stalk cells of *Drosera capensis*. Terminology as above. Transfer-cell type protuberances can be seen in the walls of both the stalk cells and the glandular cells (Tc). In this preparation, which is slightly plasmolysed, the plasma membrane (Pm) can be seen to have pulled away slightly from the unsuberised region of the anticlinal wall in *D*, but to be firmly attached to the suberised region (S) of the wall. Plasmodesmata (Pd) can be seen in the walls between the endodermoid and glandular cells. T = a young tracheid of the stalk. (Reproduced by kind permission of Dr. A. J. Gilchrist.)

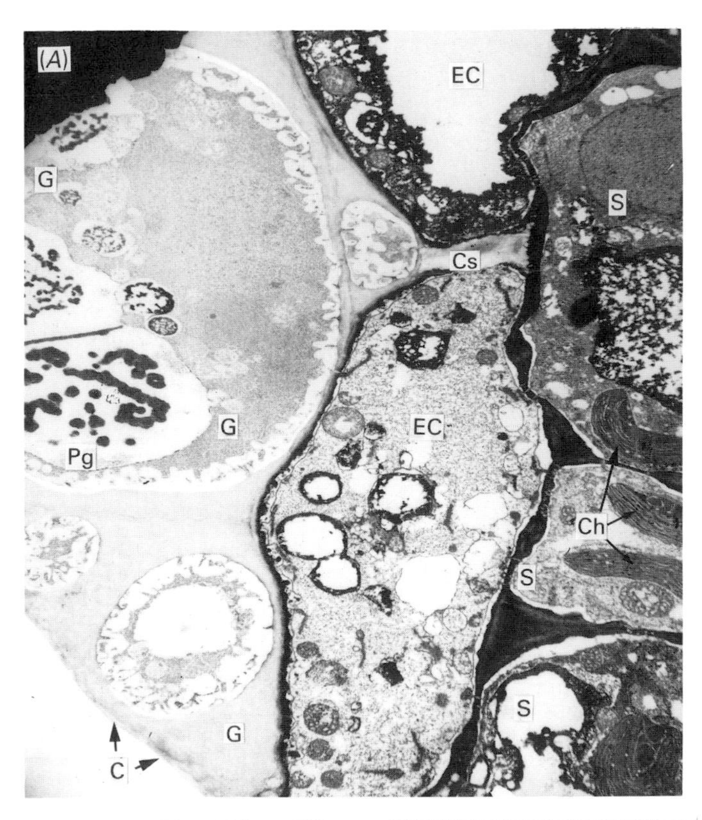
(A)
EC
G
S
Cs
G
EC
Pg
Ch
S
S
C
G

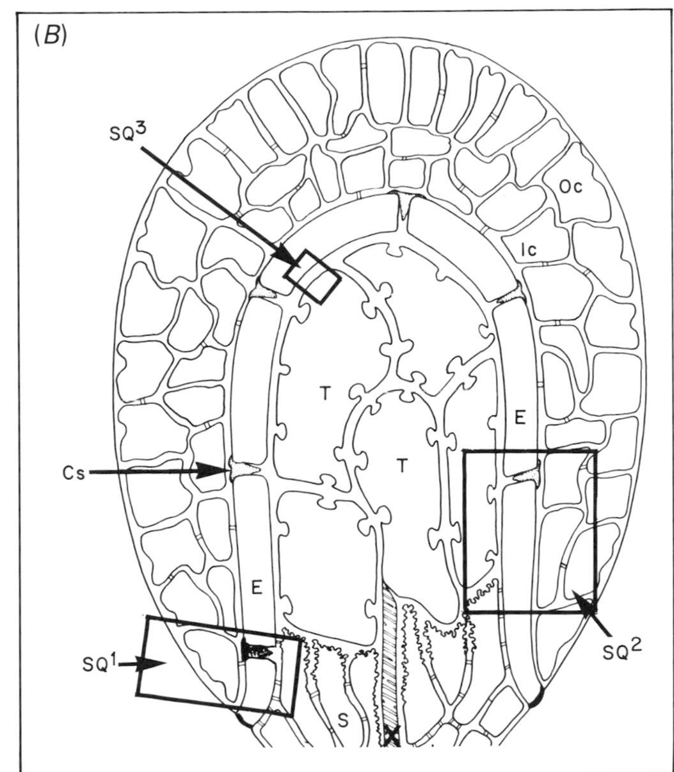
(B)
SQ3
Oc
Ic
T
T
E
Cs
E
SQ1
S
SQ2

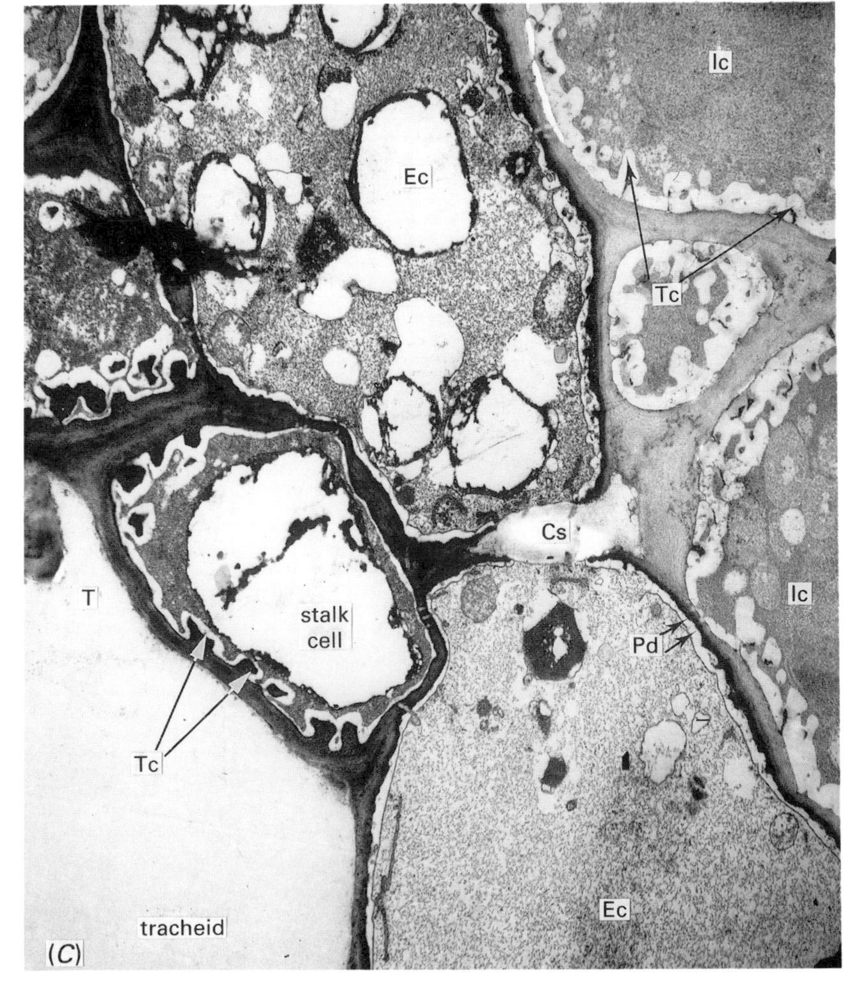
Ic
Ec
Tc
T
Cs
Ic
stalk cell
Pd
Tc
tracheid
Ec
(C)

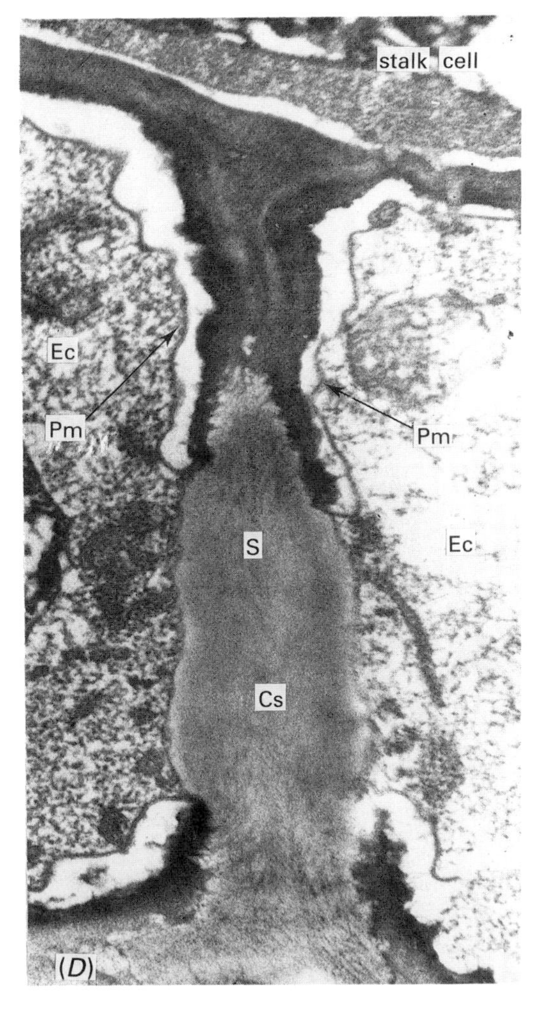
stalk cell
Ec
Pm
Pm
S
Ec
Cs
(D)

the thicker cuticle of the leaf epidermis (Fig. 8.6). The impermeability of the endodermoid apoplastic space to the passage of small molecules has been demonstrated in *Pinguicula* (Heslop-Harrison, 1976) and *Utricularia* (Fineran and Gilbertson, 1980) with electron-dense tracers. In *Drosera* stalked glands a similar impermeability was established by tracing the calcium which is bound to milk proteins, by X-ray microprobe analysis (Gilchrist, 1974; Juniper and Gilchrist, 1976; Juniper *et al.*, 1977) and in *Dionaea* by following Cl^- uptake by silver precipitation (Robins, 1978; Robins and Juniper, 1980*c*). Contact between the gland and the leaf is thus confined to the symplast, and numerous plasmodesmata are formed in the secretory/endodermoid and endodermoid/basal cell walls (Fig. 8.4*A–D*). These plasmodesmata may, as in *Dionaea*, be grouped together in pits (Robins and Juniper, 1980a).

In *Sarracenia* each endodermoid hypodermal cell of the pitcher epithelium is completely encapsulated in a continuous suberin sheath which is interrupted only by plasmodesmata (Joel, 1986 and Fig. 14.4). The secondary, suberized wall layer is adcrusted to the thin primary wall during maturation of the pitcher. Later, a thick tertiary wall is also added. The suberized wall layer shows typical lamellation. The permeability of this wall layer seems to be determined by the impregnating soluble lipids which constitute a physical barrier to apoplastic transport of water and solute (Joel and Heide-Jørgensen, 1985). The hypodermal cells closely resemble bundle-sheath cells of certain grasses not only in having a suberin wall layer, but also in showing silver hexamine-positive 'plugs' between the suberin layers of adjacent hypodermal cells (Fig. 14.1*A*). The impermeable nature of the suberin lamellae, together with that of the intercellular matter which is presumably phenolic, would not apparently permit any apoplastic transport across the hypodermis (Joel, 1986).

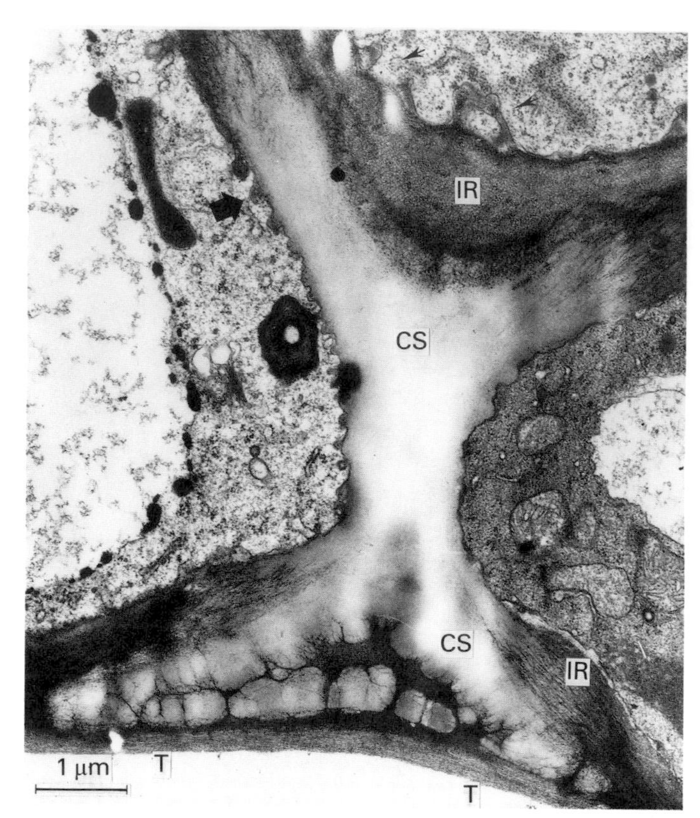

Fig. 8.6 Part of two endodermoid cells and an anticlinal wall in a 'mature' unstimulated digestive gland of *Drosophyllum lusitanicum*. Note the internal thickenings (IR) of the glandular cells. All plasmamembrane/wall interfaces have numerous protuberances or wall ingrowths (arrows). The impermeable cutinized or suberized wall region of the endodermoid layer (CS) is very conspicuous; cf. Figs 8.4*C*, *D* and 8.5*A*. T = tracheid. (Unpublished work by kind permission of Drs A.E. Vassilyev and Lyudmila E. Muravnik, Leningrad, USSR.)

The cytoplasm of endodermoid cells

The cytoplasm of the endodermoid cells varies considerably in the different glands. In general, although they have a common developmental origin with the secretory cell, their cytoplasm is less dense and not as rich in organelles. Typically, they possess a large vacuole, often containing various osmiophilic substances. Lipid bodies are particularly prominent in the endodermoid cells of *Dionaea* and *Pinguicula* (Heslop-Harrison and Heslop-Harrison, 1981; Robins and Juniper, 1980a), forming 25% of the cell volume in *Dionaea*. The nucleus of the endodermoid cells in *Pinguicula* contains massive protein bodies of a distinctly fibrillar organization (Schnepf, 1960b; Heslop-Harrison, 1975). Such protein inclusions are commonly found in many cells of *Utricularia* (Genéves and Vintéjoux, 1967; Vintéjoux, 1970; Thomas and Gouranton, 1979), particularly in the turions. Though these authors do not find them in the secretory cells, Honsell and Ghirardelli (1972) reported intranuclear proteins in gland cells of *U. vulgaris*. These latter authorities considered that the presence or absence of such deposits reflect the availability of nutrients to the plant examined.

It seems likely that, while having a common role in blocking apoplastic movement of solutes and controlling intercellular transport, other specific functions unrelated to transport are also performed by endodermoid cells in some glands. In *Drosera* tentacles, for instance, where action potentials are

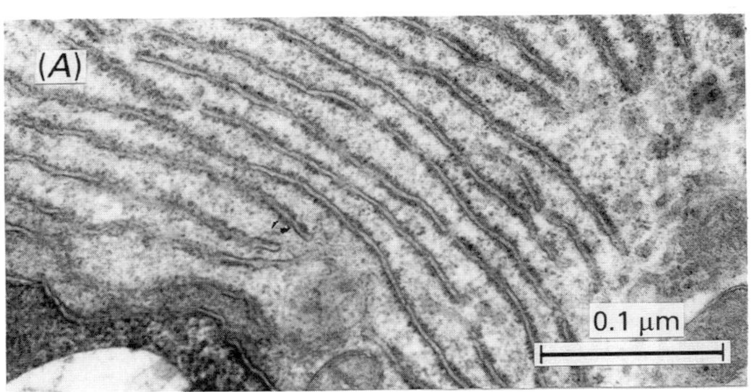

Fig. 8.7A TEM of stacks of parallel elements of rough ER in an epidermal cell of the glandular epithelium of *Sarracenia* (Zone 4) (from Joel, 1986).

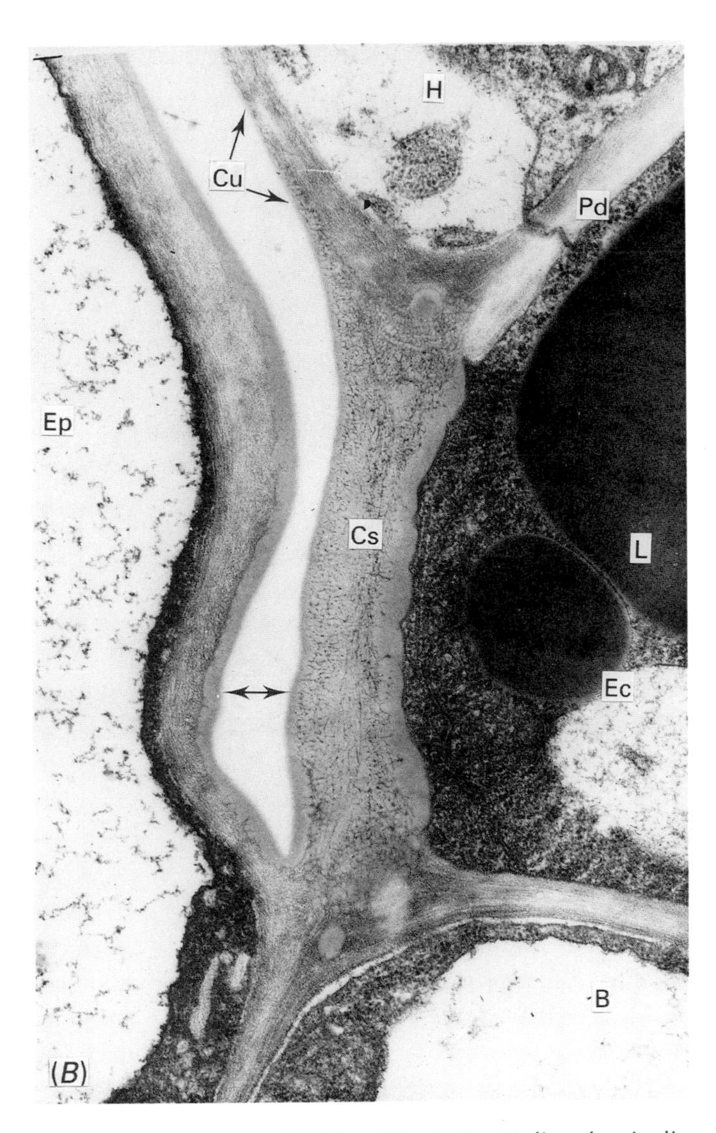

Fig. 8.7B TEM of part of a digestive gland of *Pinguicula grandiflora*, showing parts of all of the components of the gland – epidermal cell (Ep) left, head cell (H), endodermoid cell (Ec) and reservoir or basal cell (B). For the exact location see Fig. 8.7C (rectangle). Note the well-developed ER adjacent to the wall of the endodermoid cell. The Casparian strip (Cs) is contiguous with the thin and interrupted cuticle (Cu) of the head cell. Note also the difference in thickness of the cuticle of the endodermoid and epidermal cell (double-headed arrow). Note the plasmodesmata (Pd) joining the endodermoid and head cell and the prominent lipid bodies (L) in the former. (Reproduced by kind permission of Dr Y. Heslop-Harrison, redrawn from Heslop-Harrison and Heslop-Harrison, 1981.)

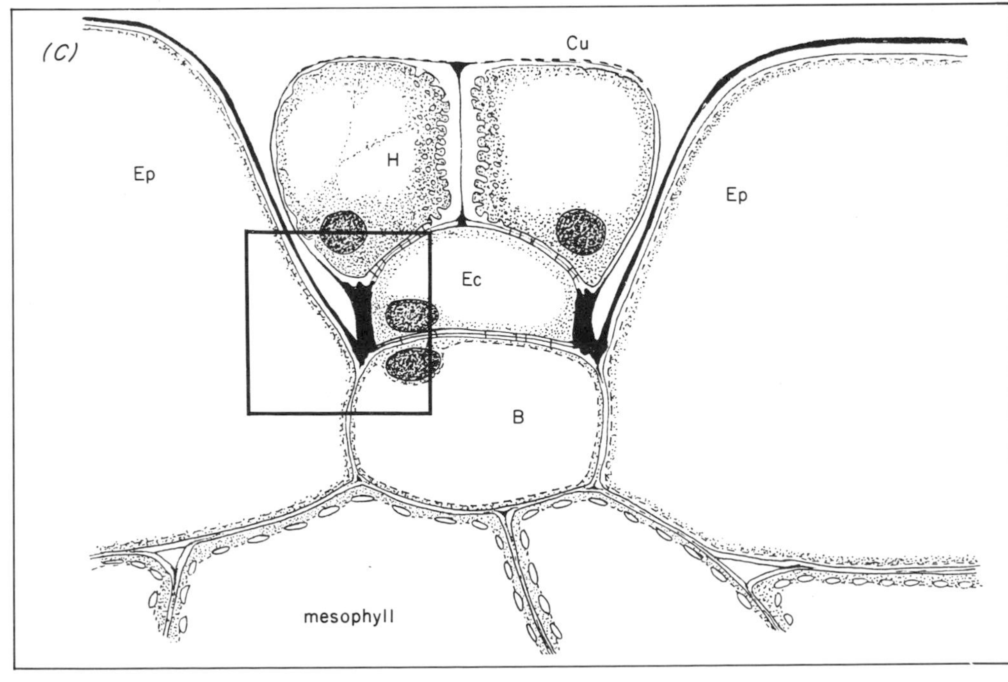

Fig. 8.7C Median longitudinal section of a sessile gland, from upper leaf surface of *Pinguicula*. Note the distribution of the transfer-cell type protuberances in the head cells (H) and the fragmented cuticle. Note also the contiguous Casparian strip of the endodermoid cells (Ec) with the cuticle of the epidermal cells (Ep). The positions of the nuclei shown are also consistent. The rectangular area delimited approximates to the TEM micrograph in Fig. 8.7B. (Unpublished drawing by Dr Y. Heslop-Harrison, Aberystwyth, UK).

generated as a result of a mechanical or chemical stimulation (see Fig. 6.16), the endodermoid layer might serve as the best site for receptor (or generator) potentials to occur as the endocuticle has high electrical resistance (Williams and Pickard, 1972a).

In the small sessile glands of *Drosera* the endodermoid cells show conspicuous labyrinthine projections. These wall elaborations develop on the upper tangential wall facing the neighbouring head cells (see Figs 10 and 11 in Ragetli *et al.*, 1972 and Fig. 35 in Heide-Jørgensen, 1987).

Similarly in the divided glands of *Utricularia* (bifids and quadrifids), the endodermoid cells are unique in having a highly developed labyrinthine wall, characteristic of typical transfer cells (see page 164). These labyrinthine walls, believed to be involved in the removal of water from the lumen when the trap is being set (see Fig. 6.49), show changes in structure related to functional activity within the gland (Fineran and Lee, 1975). Compatible with this role of a pumping cell is the very high density of mitochondria observed in the endodermoid cells of *U. monanthos* (Fineran and Lee, 1974). During trap activity, the degree of distension of these mitochondrial cristae increases. The contrast between the structure of these *Utricularia* glands and that of, for example, *Dionaea* glands emphasises the quite distinct mechanisms by which they act (see page 182).

The Head-Cells of the Gland

In general, the appearance of the cytoplasm reflects the state of differentiation of the cell and its degree of activity, as well as its phase in the digestive cycle. The dense cytoplasm, characteristic of glandular cells viewed in the light microscope, can partly be attributed to relatively large cellular inclusions. As in many other glands (Schnepf, 1969, 1974; Fahn, 1979), the cells of digestive organs have large nuclei with prominent nucleoli. These nuclei are generally spherical and often distal from the absorptive surface. Their nuclear envelope has many pores (c. 20 μm^{-2} in *Dionaea*; Robins, 1978), which are sometimes filled with a dark osmiophilic substance (Schnepf, 1960b; Scala *et al.*, 1968b). Nuclear crystals, such as those found in the non-glandular cells of *Pinguicula* (Thomas and Gouranton, 1979) and *Utricularia* (Genéves and Vintéjoux, 1967) are not seen in the secretory cells themselves. The mitochondria are numerous and contain well-developed cristae, consistent with the high metabolic activity in these cells. In several cases a correlation exists between the density (and arrangement) of the cristae and the state of activity of particular cells (Schnepf, 1960a, 1963c; Ciobanu and Tacina, 1973; Robins and Juniper, 1980a,b; Fig. 8.23).

The plastids

While active chloroplasts are found only in the glandular cells of the pitcher epithelium of *Sarracenia* (Joel and Gepstein, 1985; Joel, 1986; Joel and Heide-Jørgensen, 1985; see page 253) no photosynthetic plastid could be found in any digestive organ of other carnivorous plants. The plastids in the non-photosynthetic glands have various shapes and dimensions, typically with a dense finely grained stroma; they frequently contain plastoglobuli, probably composed of lipid and inner tubular structures arranged in groups reminiscent of prolamellar bodies. In mature glands, the plastids are devoid of starch, but young non-secreting glandular cells of *Drosophyllum* frequently contain starch grains which disappear later in the development of the gland (Schnepf, 1961a). In the secretory cells of *Pinguicula* the plastids are considerably enlarged with amoeboid or vermiform extensions, frequently ensheathed by ER elements (Vogel, 1960; Schnepf, 1961a; Heslop-Harrison, 1975). The ER is linked to the plastid envelope by fine fibrils and has ribosomes bound only to its outer face (Heslop-Harrison and Heslop-Harrison, 1981).

Plastids are quite common in many species, occupying 2–3% of the cellular volume in *Dionaea* (Robins and Juniper, 1980a). While usually clearly unable to photosynthesize (but see page 253) it is not improbable that they do perform a function, as yet undefined. Their similarity to 'secretoplasts' (Joel and Fahn, 1980) might indicate a role in the production of alluring volatile compounds, but this is purely speculative as no such production has been confirmed (see Ch. 5).

The endoplasmic reticulum

The endoplasmic reticulum is well established as the key site of protein synthesis and the selection of proteins for secretion is controlled at the point of synthesis on the ER membrane (Scheele *et al.*, 1978). One would therefore expect the specialized protein-secreting glandular cells of digestive organs to show high rates of ER activity and this is confirmed by the presence of a considerable amount of ER in the gland cells of all carnivorous plants (Scala *et al.*, 1968b; Heslop-Harrison and Knox, 1971; Vogel, 1960; Fineran and Lee, 1975; Ciobanu

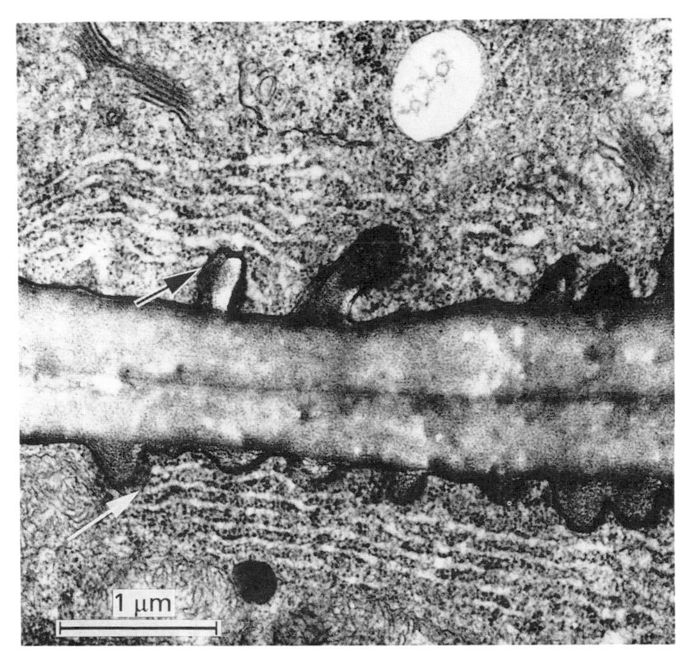

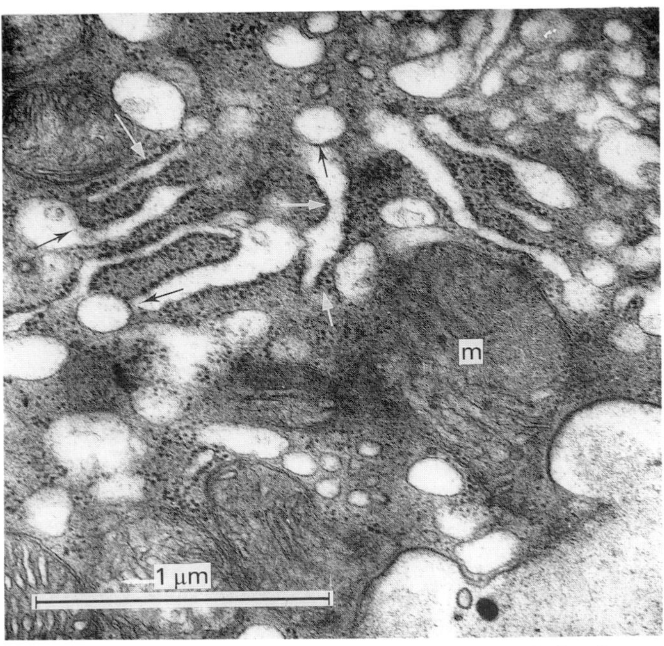

Fig. 8.8 Regions of two outer gland cells of *Dionaea*. (*A*) Parallel stacks of ER run adjacent to a cell wall region with marked labyrinthine outgrowths. Note the apparent points of fusion between the ER and the plasmamembrane (arrows); cf. Fig. 8.15. In the upper cell is a dictyosome apparently in the act of dividing; cf. Fig. 2 of Dexheimer (1978b) showing similar juxtapositions in *Drosera*. (*B*) Four days after stimulation. The rough ER (white arrows) is swollen and apparently releasing smooth vesicles (narrow black arrows) which may be fusing with the plasmamembrane. The mitochondria (m) show the typical massed cristae of stimulated cells (cf. Fig. 8.23).

and Tacina, 1973; Parkes, 1980; Panessa *et al.*, 1976; Barkhaus and Weiert, 1974; Robins, 1978; Robins and Juniper, 1980a,b,c; Heslop-Harrison, 1975; Joel and Heide-Jørgensen, 1985). In several genera, the ER is arranged in long parallel stacks lying around the periphery of the cell (Figs 8.2*C*, 8.7 and 8.8) and long profiles follow the edges of vacuoles. The concentration of ER in *Dionaea* secretory cells at the start of the digestive cycle is high (13 μm^2 μm^{-3} cell volume) and can be compared to that of pancreatic exocrine cells or tapetal cells (Robins and Juniper, 1980a). Only a few ribosomes are found irregularly spaced and widely distributed along the cisternal membrane (Fig. 8.7) although numerous ribosomes are also present in the cytoplasm. In some gland cells, such as *Cephalotus*, concentric whorls of ER are found (Parkes and Hallam, 1984, and cf. Fig. 6.20*B*).

Protein destined for secretion is apparently synthesized on the rER during the maturation of the gland and packaged in distended profiles of sER and small vacuoles which may be derived from the sER. These comprise the important intracellular store of protein ready to be secreted in response to stimulation (Robins and Juniper, 1980a,b,c). Developmental studies of *Pinguicula* have shown that certain vacuoles are derived exclusively from the enlargement of cisternae of the ER (Heslop-Harrison, 1975) and it is tempting to suggest that they are equivalent to those in *Dionaea*. ER is often closely associated with the plasmalemma (Fig. 8.8*A*) and is either smooth or bears ribosomes only on its side distal to the plasmalemma (Fineran and Lee, 1975; Robins, 1978; and cf. Fig. 5.3*E*). This apparently forms transient fusions to the plasmalemma (Heslop-Harrison, 1975; Robins, 1978; Robins and Juniper, 1980a,b,c). The rER in *Dionaea* may extend ribosome-free processes that pinch off and, in the manner of a dictyosome-derived vesicle, fuse with the plasmalemma (Fig. 8.8*B*; Robins and Juniper, 1980a). However, what appear to be tubular processes of rER may join directly to the plasma membrane (Fig. 8.8*C*; Juniper *et al.*, 1982a) and appear to transfer their contents to the apoplast without contributing any membrane components (Juniper *et al.*, 1982a). As these processes are observed mainly opposite those regions of cell wall where enzymes are also stored in the apoplast, we consider that the ER cisternae secrete their products directly into the extracellular space, loading labyrinthine walls and certain other wall compartments with enzymes (see page 174).

The vacuole

The vacuole of glandular cells is always large, occupying 50% of the cell volume in *Dionaea* (Robins and Juniper, 1980; Fig. 8.2*B*) and the whole of the middle and distal region of the arm of the bifid and quadrifid gland cells in *Utricularia* (Fineran and Lee, 1975). This cytoplasmic space is multi-functional and in *Dionaea* can be sub-divided into large and small osmiophilic vacuoles and clear vacuoles on the basis of their apparent function (Robins and Juniper, 1980a,b). In *Nepenthes* and *Cephalotus*, Parkes (1980) similarly finds clear, densely filled osmiophilic and granular vacuoles. Densely staining vacuoles always occur in pigmented gland cells such as *Dionaea* and the stalked glands of *Drosophyllum* and *Drosera*, but not in unpigmented glands such as the sessile glands of *Drosophyllum*, *Pinguicula* and all the glands in *Utricularia*. It seems likely that the osmiophilia is due to the presence of pigment which in the Droseraceae is anthocyanins derived from quercetin (*Dionaea*, *Drosera*) or luteolin (*Drosophyllum*) (see Ch. 11), and that the precipitation of these during fixation leads to the typical granular appearance.

The dictyosomes

All secretory cells contain active dictyosomes during their maturation and some activity is also visible in resting, unstimulated glands. Whether dictyosomes play a role in secreting digestive fluid in active glands has not been clearly established and the various possible contributions this organelle makes will be discussed in the relevant Sections (pages 167, 168 and 169).

The Cell Walls

The Cuticle

The cuticle, common to almost all plant leaves, presents a layer largely impermeable to hydrophilic substances. This property is inconvenient in glands and must be circumvented either by physically rupturing the cuticle during secretion, or by metabolically inhibiting the deposition of cutin or changing its permeability over all or part of the glandular surface. The former possibility, which is typical of certain nectaries and mucilage glands, e.g. the pavement epithelium of *Utricularia* (see Fig. 4.14), is clearly destructive and only suitable for 'one-off' glands (Juniper and Jeffree, 1983); the latter permits functional integrity over long periods. Furthermore, it makes possible bi-directional flow. All immature digestive glands are covered by a cuticle. Nevertheless, functional digestive glands are highly permeable. This apparent paradox is resolved in all cases by one of the following possible mechanisms.

In many non-digestive secretory glands a continuous cuticle is distended by the secretion itself (e.g. resin, nectar, salt) and, when incapable of further distension, breaks allowing the secretion to escape (Schnepf, 1969; Fahn, 1979). The only known possible example of such a mechanism in carnivorous plants is in the large glands of *Cephalotus*, where the cuticle is often seen in EM preparations lifting away from the underlying wall. Parkes (1980), however, suggests this to be an artefact resulting from tissue preparation. Even in the sessile glands of *Pinguicula*, which apparently perform a one-off function, this mechanism is not evident.

The rapid exchange of material across the outer boundary of the digestive glands indicates that the cuticle covering them must be highly permeable. As early as 1901, Haberlandt's careful observation of the cuticle of *Drosophyllum* in the light microscope suggested to him the presence of 'innumerable microscopic pores' (Haberlandt, 1906). Fenner (1904) similarly observed pores in *Byblis*. Lloyd (1942) doubted, however, that pores existed in the cuticle of digestive glands (except for *Byblis*) and attributed the rapid passage of solutes entirely to the 'very thin' nature of the cuticle.

At the beginning of the 1960s, when glands were examined for the first time under the electron microscope, Schnepf (1960a,b) and Vogel (1960) found large gaps in the cuticle of the stalked glands of both *Drosophyllum* (Fig. 8.11) and *Pinguicula*. Though at first rejecting the reliability of their own micrographs, it was later widely accepted that these cuticles are indeed discontinuous and Haberlandt was vindicated.

Unexpectedly, it later proved very difficult to detect openings in the cuticle under the SEM. Only in the head of *Drosera* tentacles (Williams and Pickard, 1969, 1974) was it possible to detect cuticular pores, about 30 nm in diameter, with the aid of the SEM (Fig. 8.10*A*). Using TEM (Fig. 8.10*B*) these pores were found to be wall-free openings in the 80 nm thick cuticularized layer (Ragetli *et al.*, 1972; Gilchrist, 1974; Heslop-Harrison, 1970; Joel and Juniper, 1982).

In contrast, the digestive glands of most other carnivorous plants develop a cuticle which is almost or completely devoid of this thickened layer. Instead, their cuticle comprises a cutinized wall-layer

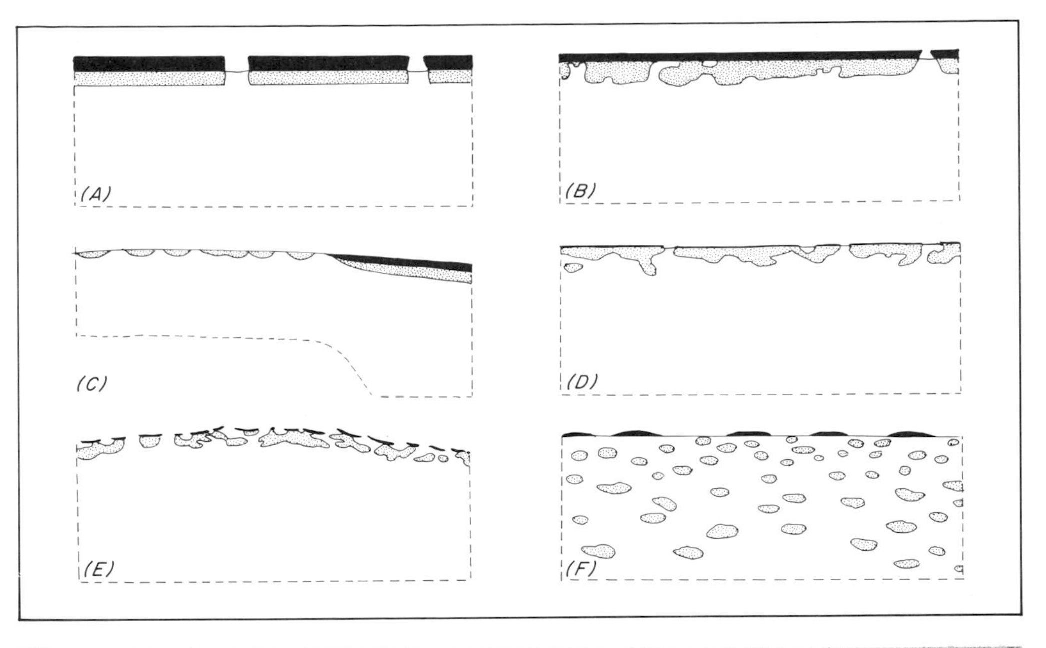

Fig. 8.9 Cuticular discontinuities in the different digestive organs of carnivorous plants. The cuticularized wall layer is shown as a solid black line and the cutinized wall as dotted. *A. Drosera*, showing real cuticular pores. *B. Pinguicula*. *C. Sarracenia*, with cuticularized layer only above anticlinal walls. *D. Drosophyllum*, showing typical cuticular gaps. *E. Utricularia*, showing the open structure of the cuticle. *F. Nepenthes*, with a very loose cuticle composed of 'cuticular droplets'.

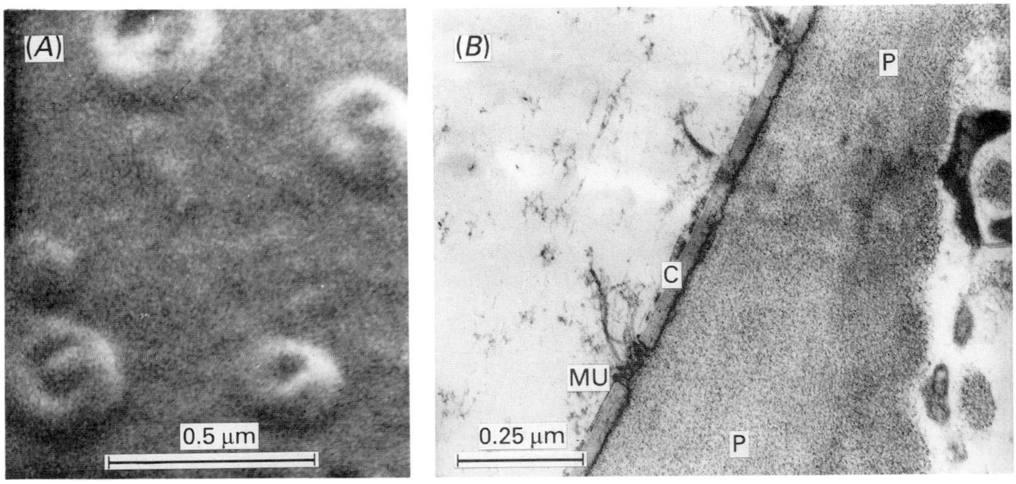

Fig. 8.10 Two views of the discontinuities in the *Drosera* stalked gland cuticle (cf. Fig. 8.9A). (*A*) In the SEM external view. (Reproduced from Williams and Pickard, 1974, their Fig. 4E, with permission.)
(*B*) In TEM section. Reproduced from Dexheimer (1976) with permission. MU, mucilage; C, cuticle; P, pectocellulosic wall.

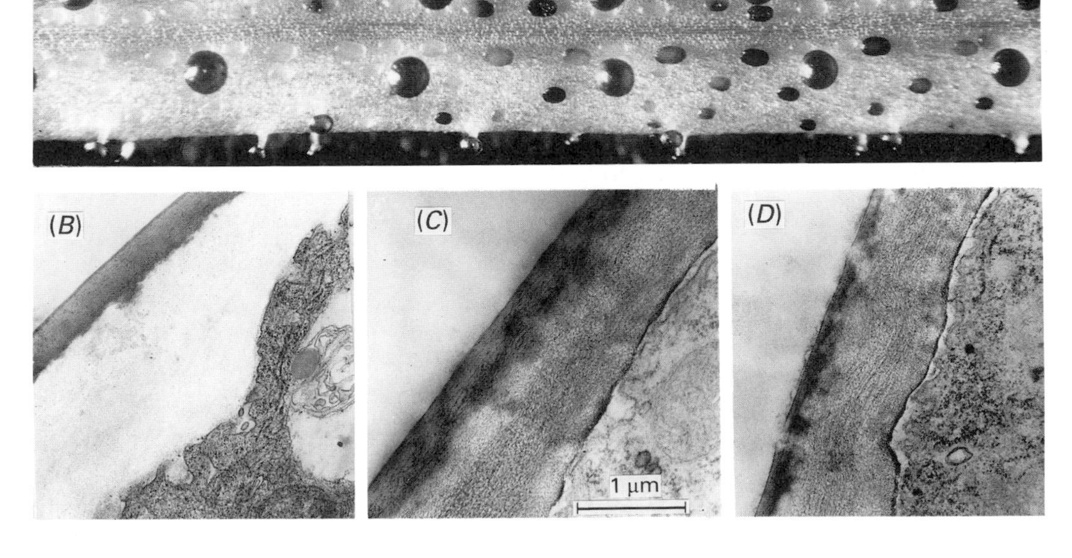

Fig. 8.11 Formation of cuticular gaps in *Drosophyllum lusitanicum*. (*A*) Light micrograph of a portion of a developing leaf which was immersed in Neutral Red solution for one minute showing young sessile digestive glands which did not absorb the dye (left), and fully developed sessile glands which absorbed the dye and stained red (right). (*B, C, D*) TEM illustrating the changes shown in the whole tissue above. (*B*) The outer wall of a neutral-red negative very young sessile gland with a continuous cuticle; (*C*) intermediate stage of development; (*D*) the outer wall and cuticle of a fully mature, neutral-red positive, digestive gland showing the well-developed cuticular gaps.

only, i.e. an outer pectocellulosic layer impregnated with cutin (Joel and Juniper, 1982). This can easily be seen in sections stained by the PA-TCH-SP method, when polysaccharide can be demonstrated up to the very surface of the outer wall. The cuticular discontinuities in these digestive glands are thus cutin-free wall regions, invisible in the SEM simply because they are filled with wall material. This type of a cuticular discontinuity is referred to by Joel and Juniper (1982) as a *cuticular gap*.

The development of cuticular gaps in *Drosophyllum* was followed by immersing a circinate developing leaf, which forms a crozier at its apex, in a 1% solution of neutral red. The glands become permeable to neutral red only when they reach maturity. A clear correlation was established between stainability and the presence of cuticular gaps (Fig. 8.11). The outer walls of the secretory cells become considerably thinner during the last stages of the gland development. This led us to suggest that the formation of cuticular gaps takes place at the final stage of maturation of the glands. The stretching of the outer wall probably separates cutin bodies from one another, creating the observed discontinuities (Joel and Juniper, 1982).

A similar phenomenon was observed in the pitcher epithelium of *Sarracenia* (Joel, 1986; Joel and Heide-Jørgensen, 1985), in the digestive glands of developing pitchers of *Nepenthes* and in the divided glands of young traps of *Utricularia* (Joel, unpublished). In *Nepenthes*, the cuticle of the young glands is composed of both cuticularized and cutinized layers (Figs 4.9*A*, 8.9*F* and 8.12). Micrographs taken of developing glands of *Nepenthes* show that the cuticularized layer ruptures and almost disappears as the glands develop, only small portions remaining attached to the outer wall of the mature gland (Figs 8.2*C* and 8.9*F*). At the same time the continuous cutinized wall layer changes by separating small cutin bodies from each other until it becomes very loose (Joel, unpublished) (Fig. 8.9*F*; see also Schnepf, 1966).

Various other forms of cuticular discontinuities have been described (Fig. 8.9). In *Utricularia*, the cuticle shows an 'open structure' (Fineran and Lee, 1975) while in *Pinguicula* many 'ill defined cuticular discontinuities' were found (Heslop-Harrison, 1975, 1976). *P. vulgaris* very rapidly absorbs neutral red (Desiré, 1946). In *Sarracenia* and perhaps also in *Darlingtonia*, an extremely thin cuticularized layer and a discontinuous cutinized layer cover the centres of the glandular pitcher epithelial cells (Joel and Heide-Jørgensen, 1985; see page 253–254). We believe that these differences in the cuticular discontinuities may be the result of cutin of different composition responding in various ways to similar stretching forces (Joel and Juniper, 1982). The same functional ability is thus produced by a number of distinct mechanisms.

Dionaea, which appears to possess a similar mechanism for producing cuticular gaps, is unique in forming these gaps only when the glands are chemically stimulated (Joel *et al.*, 1983). This cuticular fenestration is further discussed on page 178.

The labyrinthine walls

The secretory cells of digestive glands of many carnivorous plants contain some degree of labyrinthine wall development. These walls, widely found in many other plant organs, are the important

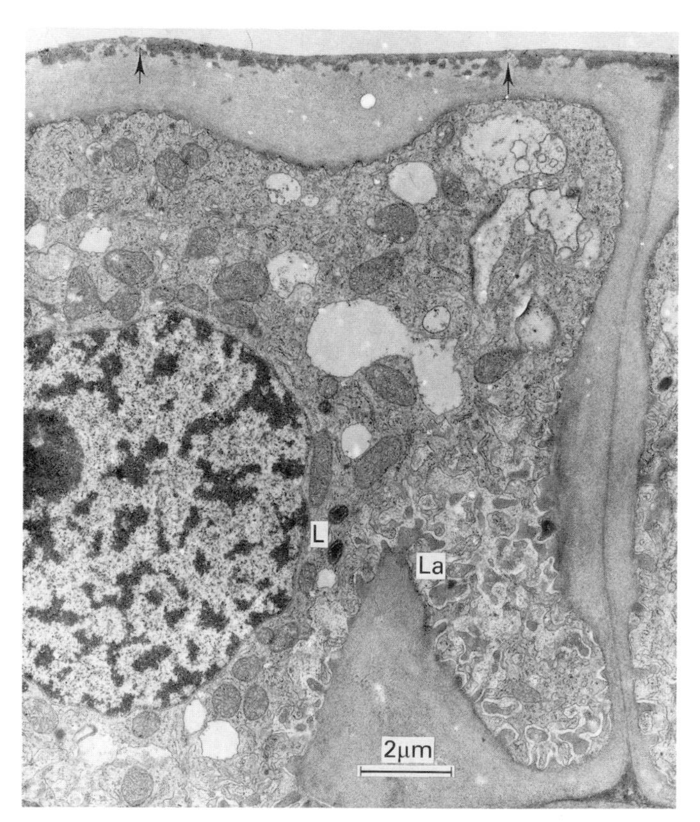

Fig. 8.12 TEM general view of a mature, unstimulated outer digestive gland cell of *Drosophyllum lusitanicum*. Two very small leucoplasts (L) with very dense stroma can be seen in the centre. There are numerous mitochondria and much of the ER seems to be in the tubular form (small arrows and cf. Fig. 6.8). The cell walls are prominently thickened with marked ribs above, and particularly below. The lower rib also has a well-developed labyrinthine region (La). The vacuoles are small and the cuticle (cf. Fig. 8.9*D*) had very small discontinuities (large arrows). (Unpublished work by kind permission of Drs A.E. Vassilyev and Lyudmila E. Muravnik, Leningrad, USSR.)

feature of 'transfer cells' – i.e. cells specialized in the short-distance transport of solutes (Gunning and Pate, 1974). In digestive glands, these walls seem to fulfil important roles in both secretion and absorption (see page 183 and Ch. 10).

The labyrinthine wall comprises numerous wall protuberances of irregular shape, each continuous with the bulk of the wall on its inner side and together forming an extensive labyrinthine surface lined by the plasmalemma (Figs 8.12 and 8.13*A–E*). The wall ingrowths give a positive reaction with PAS reagent and usually stain less densely than the primary cell wall but with a darker, fibrillar core (Figs 8.12 and 8.13*B*) indicating a chemical distinction between the two structures (Scala *et al.*, 1968b; Robins and Juniper, 1980a). The electron opacity of core material increases following PTA staining (Heslop-Harrison and Heslop-Harrison,

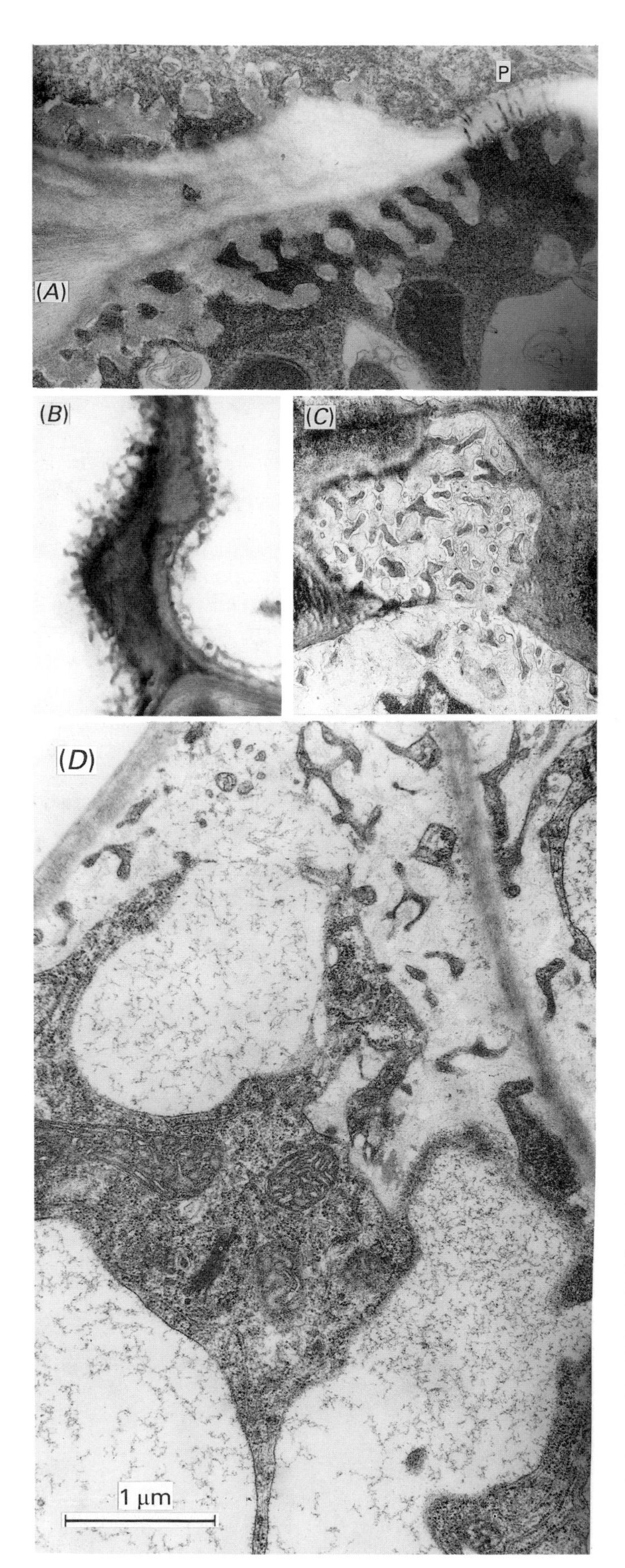

Fig. 8.13 Labyrinthine walls in various carnivorous plant species. (*A*) *Drosophyllum*, wall protuberances in the outer (above) and inner gland cells of a sessile gland. The pit field (right) with plasmodesmata (P) seems virtually devoid of protuberances. (*B*) High-voltage TEM of a thick (c. 1 μm) section of *Dionaea* showing the labyrinthine wall layer on both sides of a wall between the inner secretory cells. (*C*) Tangential section of an outer glandular cell of a stalked gland of *Drosophyllum*, showing interconnections between adjacent protuberances. (*D*) Wall ingrowths on the anticlinal and outer tangential walls of a gland of *Pinguicula* showing the ingrowths beginning to surround the vacuoles. (Unpublished micrograph by kind permission of Dr Y. Heslop-Harrison.) (*E*) *Pinguicula* gland cell showing the protuberances on the wall surfaces in particular the anticlinal ribs or partial septa descending from the outer wall; cf. Fig. 8.12. Note also the very thin true cuticle. (Unpublished micrograph by kind permission of Dr J. Dexheimer, France.)

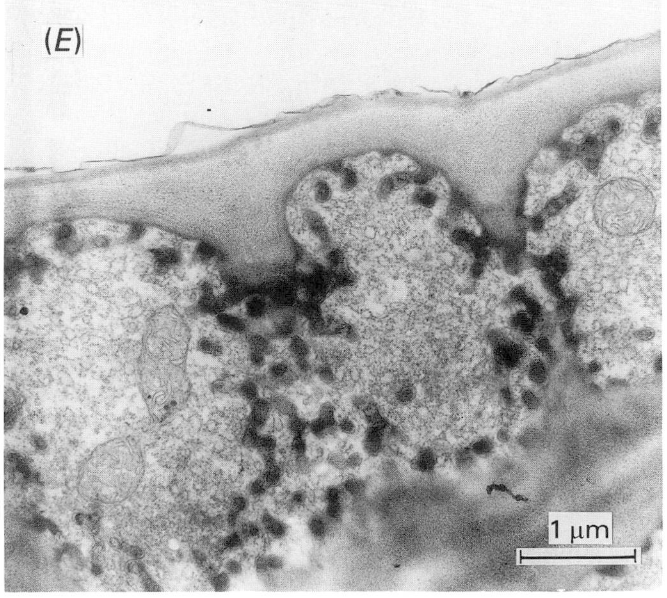

1981). A clear interfacial zone is usually observed between the plasmalemma and the wall ingrowth, similar to that found in transfer cells (Gunning and Pate, 1974). Browning and Gunning (1977a,b,c), however, using freeze substitution and freeze fracture techniques, have shown that, in well-preserved tissue which can be considered to be a faithful reflection of the *in vivo* state, the interfacial zone is lacking and the wall ingrowths appear more compact than in conventional preparations. This view should be widely adopted for the labyrinthine walls of digestive glands.

The asymmetry of the labyrinthine walls

Labyrinthine walls tend to show a clearly asymmetrical distribution within the gland (Figs 8.5*B*; Tc and 8.12). In *Dionaea*, *Drosera*, *Drosophyllum*, *Nepenthes* and *Pinguicula* a gradient of wall ingrowths is apparent along the radial walls of the secretory cells (Fig. 8.7*C*). These walls are often highly elaborated at the boundary with the endodermoid cells, decreasing in the extent of development across the secretory layer until on the external walls there are only a few small ingrowths or, as often happens, none at all. This polarity has been studied in *Dionaea* (Robins, 1978; Robins and Juniper, 1980a,e) and is discussed further on page 184.

Glandular head-cells without labyrinthine walls

Not all glands show labyrinthine walls in their head-cells. The digestive structures of *Cephalotus*, *Sarracenia* and *Utricularia* are all devoid of such structures. However, the available information concerning them does not permit us unequivocally to state that they are all functioning as normal digestive glands. In the divided glands (quadrifid and bifid) of *Utricularia*, the walls of the terminal glandular cells are thickened rather than labyrinthine (Fineran and Lee, 1975). It is the endodermoid cell which here, in total contrast to all the other digestive glands, has an elaborate labyrinthine structure. The importance of this is discussed elsewhere (see pages 160 and 211).

Modification of the labyrinthine wall during secretion

It has been suggested (Schwab *et al.*, 1969) that in *Dionaea* the degree of elaboration of the wall surface changes with the state of activity of the gland. While no change in the surface area of the ingrowths occurs, the volume of the labyrinthine space increases (Robins and Juniper, 1980a) and it is likely that secretory activity is influencing the internal structure of the wall. The chemical reactivity of the wall components alters markedly in the pH 4–3 range (Gunning and Pate, 1974) and alterations of pH are highly effective at inducing changes in the volume of wall ingrowths (Lauchli *et al.*, 1974). One may therefore assume that the active transport of specific ions across the plasmalemma of the glandular cells during secretion and absorption could also cause significant changes in the texture and dimensions of the wall, as observed in *U. monanthos* (Fineran, 1985).

The amplification factor

The architecture of these ingrowths has been quantitatively studied in *Dionaea* (Robins, 1978; Robins and Juniper, 1980a). The total number of ingrowths determines the extent to which the perimeter of the cell is enlarged. The amplification factor (i.e. the ratio of length of middle lamella to plasmalemma over unit length of wall), which indicates how much the length of plasmelemma is increased as the result of the presence of wall elaborations, depends both on their distribution and geometry. Both the density and volume of ingrowths seem to contribute to determining the amplification factor in outer glandular cells. In inner glandular cells, where a higher ingrowth density exists, the surface area and volume of the ingrowths are independent of the length of the cell perimeter. One can easily show that additional volume of the wall does not necessarily increase the amplification factor but does offer an increased apoplastic space for the accumulation of enzymes (see page 168).

In *Drosophyllum* and in some *Drosera* species, the glandular cells have large wall ingrowths in the form of septa, dividing the cells into small interconnected compartments (Fig. 8.13*C*). These septa are themselves covered with labyrinthine wall ingrowths, further increasing the amplification factor and providing additional apoplastic space for the storage of enzymes and/or mucilage.

The wall-plug in the stomatal gland of Cephalotus

The outer wall of the small uni- or bi-cellular digestive glands of *Cephalotus* is also thickened, forming a plug extending into the space between the two 'guard' cells (Goebel, 1891; Dakin, 1919; Lloyd, 1942; Parkes, 1980; Fig. 8.2A). This structure appears again to form an important apoplastic space in which enzymes are stored (see page 177). It seems to be the only instance where the external wall of the gland forms this function, perhaps rather surprisingly until one considers the mechanism employed to wash digestive fluid out of the gland (see page 182).

Dictyosomes in the formation of the labyrinthine walls
Two cell compartments, dictyosomes and lomasomes, are presumed to be involved in the formation of wall ingrowths. Labyrinthine walls are laid down very late in the development of the glands (Heslop-Harrison and Knox, 1971; Fineran and Lee, 1975; Robins, 1978; Robins and Juniper, 1980a; Outenreath, 1980). Some of the vesicles derived from dictyosomes in *Utricularia*, *Pinguicula*, *Drosophyllum* and *Dionaea* have a granular or fibrillar core resembling the core material of the labyrinthine wall (Fig. 8.14 and Schnepf, 1963b; Vintéjoux and Prevost, 1976; Vintéjoux, 1979; Robins and Juniper, 1980a) and are interpreted as supplying material to build the wall protuberances. Large dictyosomal

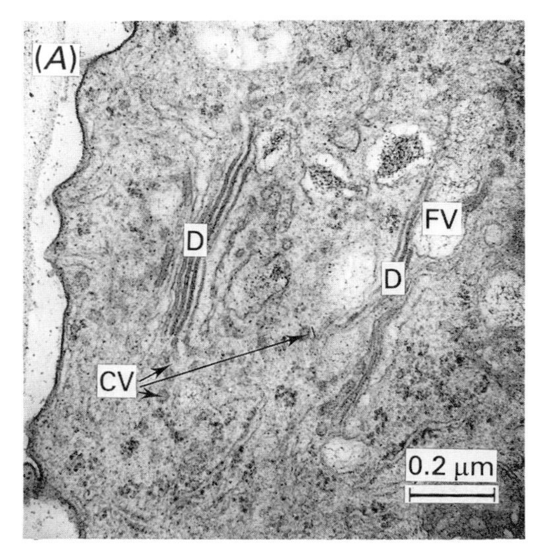

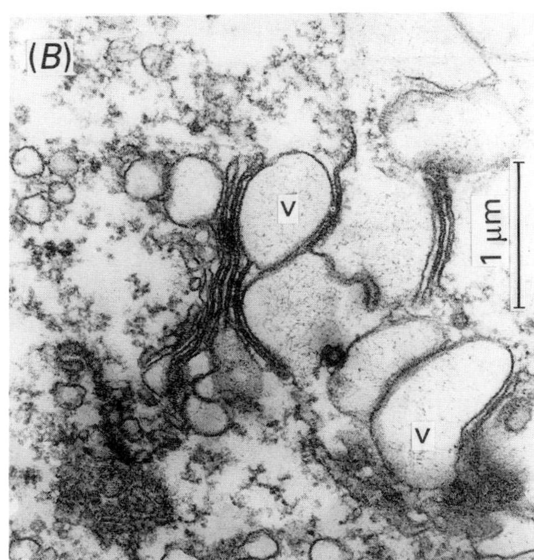

Fig. 8.14A The outer glandular cell of a digestive gland of *Dionaea* showing dictyosomes (D) producing both small 'coated' (CV) and large vesicles containing wall material (FV). **Fig. 8.14B** Detail of an outer glandular cell of *Drosera rotundifolia* during the early phase of secretion. The dictyosome (centre), one of the highly active Golgi complex at this phase, shows vigorous vesicle production (V). Some of the vesicles can be seen to contain fibrous polysaccharide material. (Unpublished micrograph by kind permission of Dr. J. Dexheimer.)

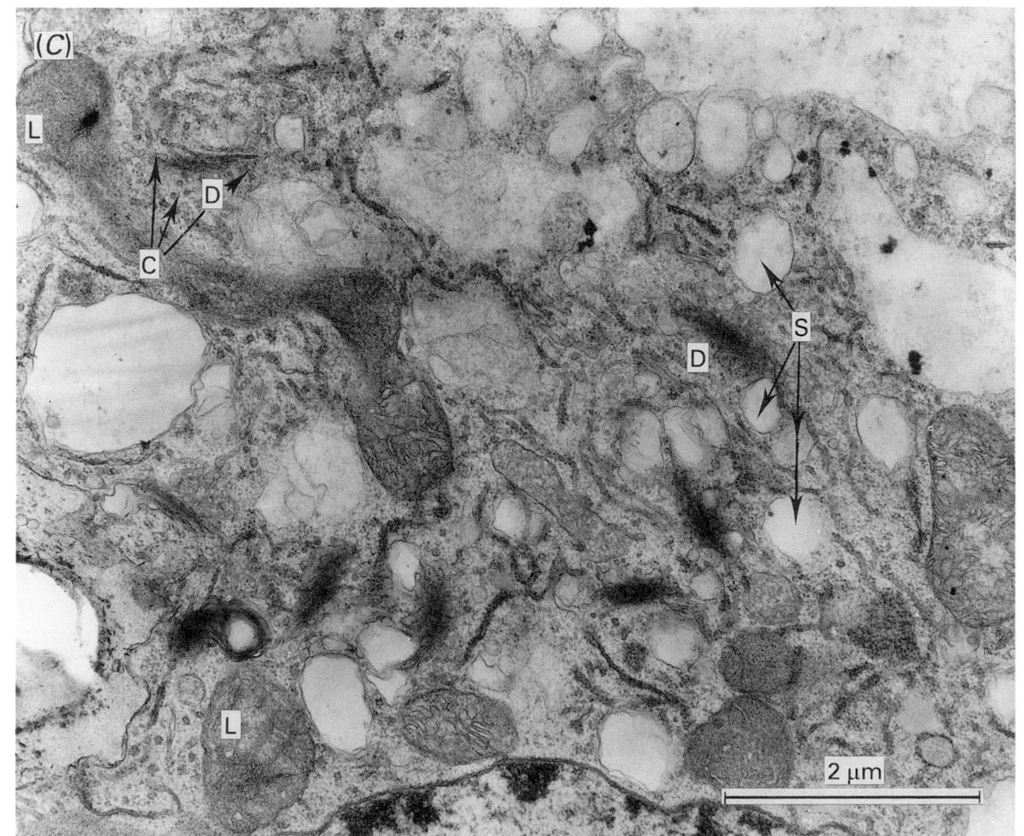

Fig. 8.14C Part of a mature mucilage gland of *Drosophyllum* before stimulation. Clearly visible are numerous dictyosomes (D) apparently producing both large smooth (S-arrows) and smaller coated (C-arrows) vesicles. Most of the ER appears to be in the tubular form. Two leucoplasts (L) can also be seen, similar to those of the secretory cells of the digestive glands. Like those of the digestive glands these leucoplasts do not seem to be associated with sheathing or tubular ER. (Unpublished micrograph by kind permission of Drs A.E. Vassilyev and Lyudmila E. Muravnik, Leningrad, USSR.)

vesicles only appear in the glands of *Pinguicula* during the interval of wall building (Heslop-Harrison and Heslop-Harrison, 1981) and in *Dionaea* dictyosomes of all degrees of activity occur throughout the digestive cycle, showing no correlation with secretory activity (Robins, 1978). The appearance of these vesicles differs from those found in *Drosophyllum* and *Drosera* which were shown to be involved with mucilage production (Schnepf, 1960a, 1961a,b and see Ch. 6) and which tend to be more uniform and flocculent (Fig. 8.14*B–C*). Dictyosomes are clearly involved in mucilage production (see Ch. 6). Dexheimer (1978a and personal communication) has suggested that in *Drosera* the protuberances represent, at least in part, the remains of mucilage vesicles.

The dual role of dictyosomes

Dictyosomes perform two major roles in the digestive glands of carnivorous plants. On the one hand they are the site of synthesis for the trapping mucilage and form its export vesicles (Robins, 1978; see also page 32); on the other, they synthesize and secrete wall material for the construction of the labyrinthine walls. Dictyosomes may also act as a source of suitable membrane to be supplied when new ER and vacuoles are formed.

Although the extent to which dictyosomes are involved in *Dionaea* with the secretion of hydrolytic proteins is controversial (Scala *et al.*, 1968; Robins, 1978; Robins and Juniper, 1980a,b,c; Henry and Steer, 1985) and will be considered on page 170, interconnections between ER and dictyosomes have been detected by staining with ZnI/OsO_4 (Fig. 8.15*A–D* and Juniper *et al.*, 1982). These connections are seen at all stages of the secretory cycle, although the impermeability of the unstimulated gland presents some problems. The presence of such interconnections supports the idea that dictyosomes are providing membranes for ER and ER-derived vacuoles. Alternatively, the dictyosome-derived vesicles apparently involved with labyrinthine wall synthesis may contain, in addition to cell wall polysaccharides, wall-synthesizing enzymes, made on the ER. Dictyosomes also do contain β-glycerophosphatase and acid phosphatase activity early in the secretory cycle (Henry and Steer, 1985).

Lomasomes in the formation of the labyrinthine walls

The role of lomasomes in the formation of the labyrinthine walls is not clear. It has been suggested for other plant tissue that lomasomes are involved with the movement of cell-wall synthesizing enzymes across the plasma membrane (Marchant and Robards, 1968). Using autoradiographic labelling, Cox and Juniper (1973) have further shown that lomasomes have an important role in the synthesis of non-cellulosic cell walls and they are implicated in the thickening of fibre walls (Lawton *et al.*, 1979). Accordingly, the lomasomes, which occur quite commonly in secretory cells of *Dionaea* closely associated with regions of growing labyrinthine walls, may indeed be involved with the synthesis of these structures (Robins, 1978). Surprisingly, in view of their possible function, lomasomes have only been reported in *Dionaea* and not in the genera such as *Cephalotus* and *Darlingtonia* which have especially thickened cell walls.

The Storage and Release of the Digestive Fluid

General Considerations

The sub-cellular localization of hydrolytic enzymes has only been extensively studied in *Pinguicula*, *Nepenthes* and *Dionaea*. Some cytochemistry has been done in other genera but they have received much less detailed attention. Inevitably, our discussion of secretory mechanisms must centre round these genera, with evidence for similarities and differences in other systems being brought in whenever it is available.

The Mode of Fluid Secretion

Any insect entrapped by a pitcher-plant normally falls into a well-prepared pool of digestive fluid (see page 170, Ch. 9 and Fig. 9.1). This fluid is spontaneously secreted before the full maturation of the trap and while the pitcher is still closed. The products of the digestion of each single prey are diluted in the pitcher fluid and are apparently absorbed by all digestive glands (see Ch. 10). In certain pitchers, the addition of food induces an additional secretion of fluid by the digestive glands. Feeding with beef broth produced a threefold increase in the volume of liquid in *Darlingtonia* pitchers and an 18-fold increase in closed pitchers of *Sarracenia flava* (Hepburn *et al.*, 1927). Early studies of *Nepenthes* revealed that abundant secretion is provoked by chemical stimulation (Clautriau, 1899; de Zeeuw, 1934). Surprisingly, the

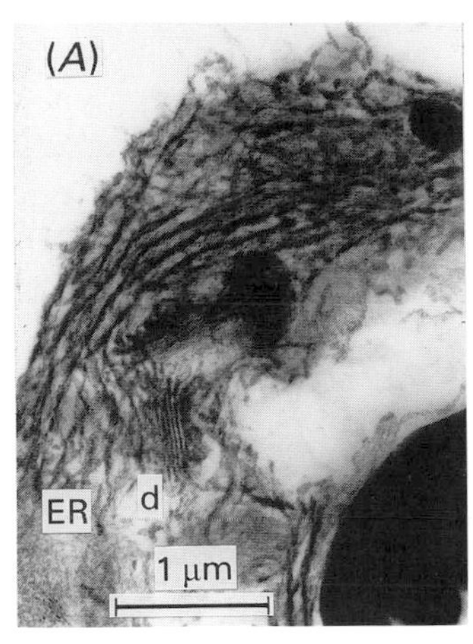

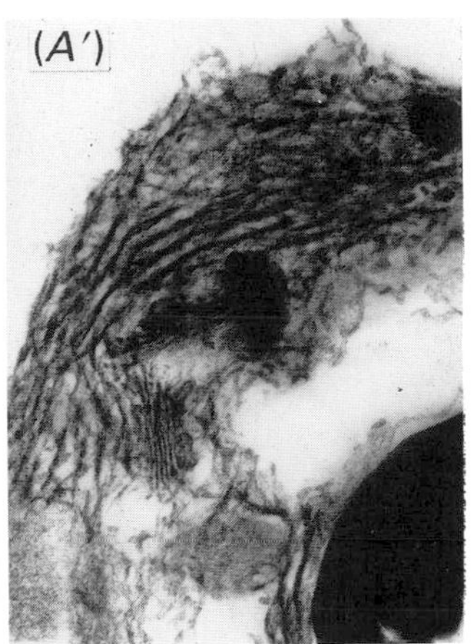

sterile injection of protein into *Nepenthes* pitchers did not cause any stimulation of enzyme secretion (Lüttge, 1964a; but see page 172). The acidity of the fluid in *Nepenthes* is, however, often affected, showing a drop in pH of several units suggesting that the proton pumps may be activated by stimulation (see Ch. 9 and Fig. 9.2).

The Digestive Pool

Three main strategies have been employed by carnivorous plants to digest the entrapped prey (Joel, 1986). These all involve a pool of fluid which

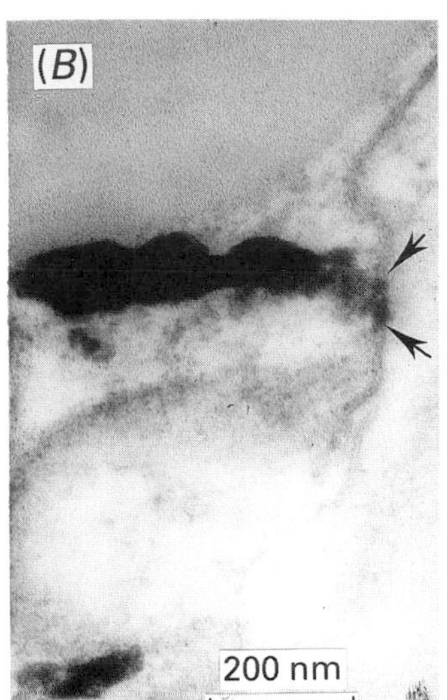

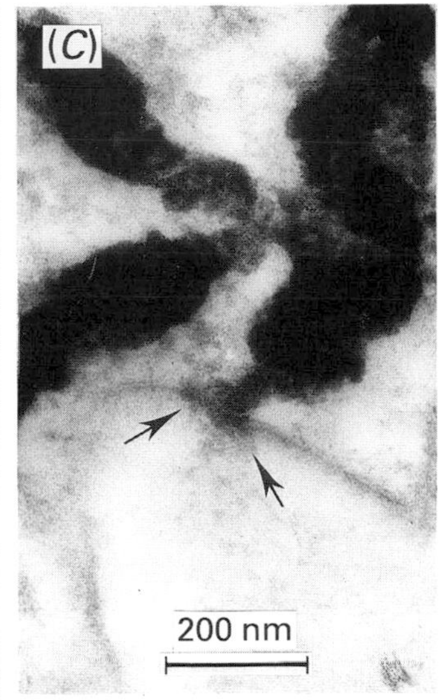

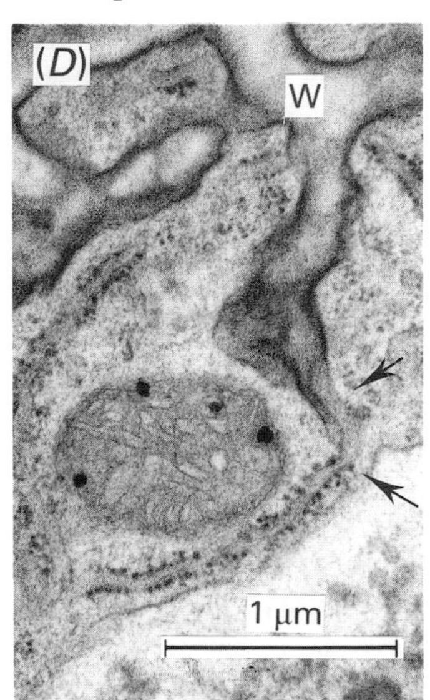

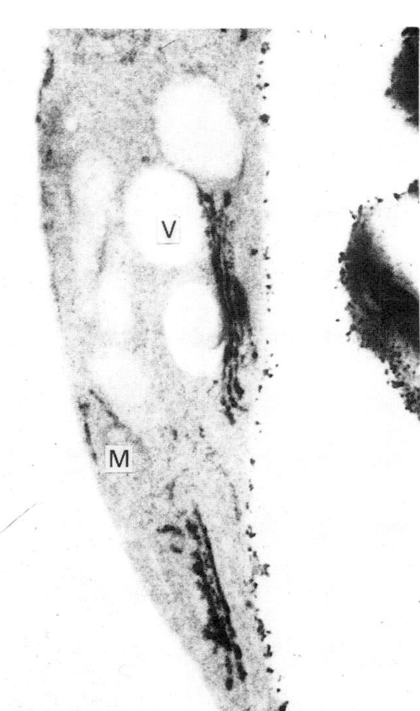

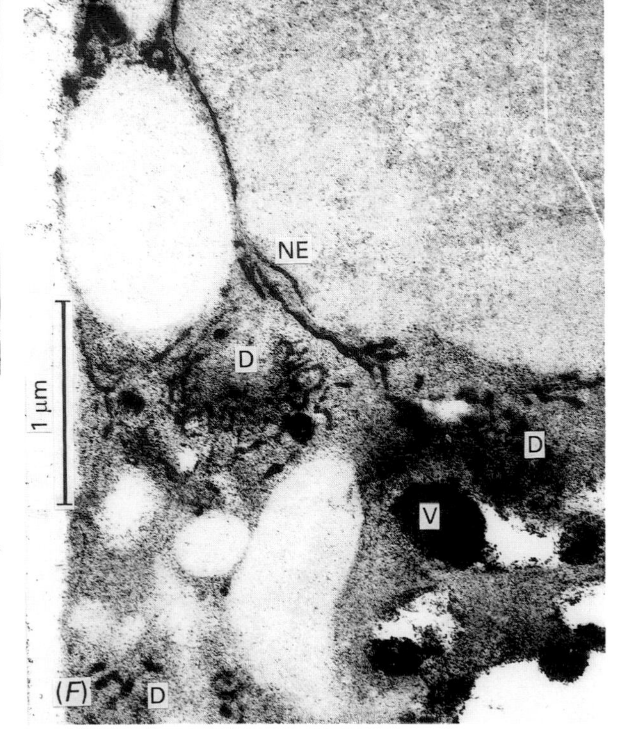

Fig. 8.15 (*A–D*) The fine structure of a *Dionaea* gland cell visualized by the ZIO-impregnation method for TEM. (*A*, *A*[1]) Stereo-pair of a dictyosome (D) enmeshed in a mass of tubular ER (ER). 1 μm thick section. (*B*, *C*). Connections between ZIO-impregnated ER and the plasma membrane of the gland cell (arrows). (*D*) Connection between the plasma membrane and the ER (arrows) in a conventionally fixed gland cell; W = cell wall ingrowth.
(*E*, *F*). The localization of acid phosphatase in the digestive glands of *Dionaea*, using the lead nitrate precipitation method. (*E*) 48 h after stimulation. Lead deposits occur throughout the cisternae of the two dictyosomes, with deposits also in the ER and tonoplast membrane. No deposits occur in secretory vesicles (V) or mitochondria (M). (*F*) Also 48 h after stimulation. Tangential section through dictyosome cisternae (D) showing the fenestrated network filled with lead. These deposits also occur in the ER, nuclear envelope (NE) and, in this case, the secretory vesicles (V). (*E* & *F* reproduced by kind permission of Henry and Steer, 1985.)

forms a direct apoplastic bridge between the glands and the prey:

(i) *continuous digestive activity*: All victims entrapped during the life of the trap are digested in the same digestive pool. This is employed by all pitcher plants (see below).
(ii) *digestive cycle*: each single prey is separately digested by the whole digestive surface forming a temporary digestive pool. This is employed by the snap traps of *Dionaea* and *Aldrovanda* (see below).
(iii) *limited digestive pools*: each single prey is surrounded by a small digestive pool with viscous fluid. The size of the pool corresponds to the size of the digested animal. This is employed by the adhesive traps (see page 90 and Fig. 6.7 and below).

The nature of the digestive fluid differs in the various plants, and is well correlated both with the behaviour of the glands and the nature of the trapping mechanism (see also Table 9.1).

The digestive pool in the pitcher-plants

The pitcher-plants hold permanent pools of a digestive fluid with low surface tension. Quantitative measurement of the liquor from closed pitchers of *Sarracenia flava* at 25°C showed a surface tension of 66.4 dynes cm^{-1}, which is considerably lower than that of water (Hepburn *et al.*, 1927). This depression is probably due to certain organic acids (P. A. Rea, pers. comm.) and allows the digestive fluids quickly to wet the normally water-resistant surface of most insects (see page 313). The pool may be large and usually all digestive glands are involved in digestive activity, or are at least in continuous contact with the fluid.

The digestive pool of snap traps

The traps of *Dionaea* and *Aldrovanda* do not hold digestive fluid prior to receiving a clear stimulus that food is available. When such a stimulus is perceived, a temporary digestive cavity is formed by closure of the trap and, as in the pitchers of *Nepenthes*, *Sarracenia* and *Cephalotus*, all the digestive glands become involved in the digestion of each prey (see page 184). In contrast to the pitcher-plants, however, which use the same pool of digestive fluid for the digestion of many victims (the digestive fluid is fully absorbed only when the trap senesces), in *Dionaea* and *Aldrovanda* the whole volume of fluid is absorbed by the time the digestion of each prey is completed. A new pool is formed only when new prey is captured. The difference between the activity of the traps of *Dionaea* and *Aldrovanda* and the activity of pitchers can therefore be regarded as a difference between a clear digestive cycle and continuous digestive activity.

The digestive pool in mucilage traps

In the mucilage traps there is no permanent digestive pool and, in contrast to the pitcher-plants and snap traps, only a relatively small number of glands are involved in the digestion of each prey. The stalked glands secrete a mucilaginous fluid which, in addition to its role in digestion, serves as a trapping mucilage (Fangschleim). This combined function of the secreted substance is an advantage where a real digestive cavity is absent. The viscous nature of the digestive fluid, which is due to the presence of polysaccharides (see page 89), enables the relatively small amounts of secreted fluid to form effective small pools which adhere to the prey (Fig. 8.16*A–D*). The mucilage smothers the insects' tracheae, suffocating it and in addition conferring rapid digestion and absorption through the maximal surface area of the prey. Functioning as an ion exchanger, the mucilage also provides a more effective transporting medium by coming into intimate contact with many parts of the prey. The absorption of nutrients from each single prey can therefore rapidly be completed in a relatively short time and by a relatively small number of glands. Movements of the leaf and gland of *Drosera* increase the active digestive area, without significantly changing the volume of the digestive pool itself.

Many mucilage traps also exhibit functional sessile digestive glands in addition to the stalked glands. Darwin (1875) wrote:

> 'When an insect alights on a leaf of *Drosophyllum*, the drops of trapping mucilage adhere to its wings, feet, or body, and are drawn from the gland. The insect then crawls onward and other drops adhere to it so that at last, bathed by the viscid secretion, it sinks down and dies, resting on the small sessile glands with which the surface of the leaf is thickly covered.'

This very precise description clearly indicates that the trapping mucilage of *Drosophyllum* which is secreted by the stalked glands mediates between the digestive glands and the prey. The digestive fluid secreted by the sessile digestive glands combines with the mucilage wrapping the prey and continuously supplies enzymes when stimulated. The glands in

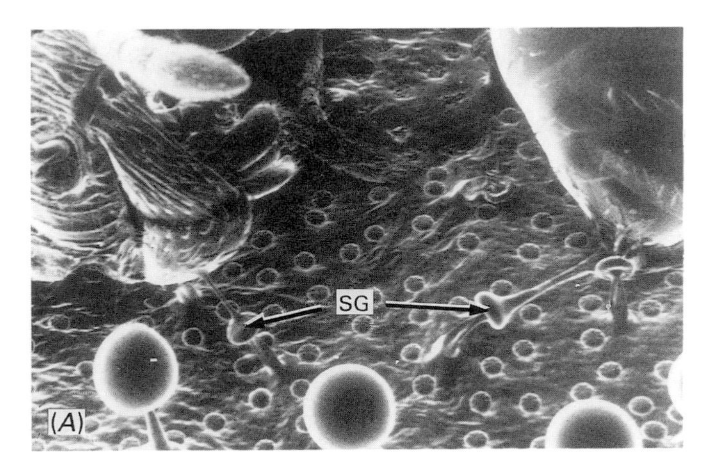

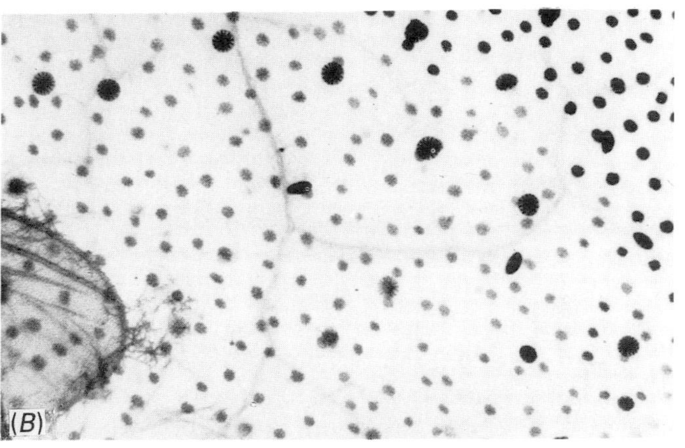

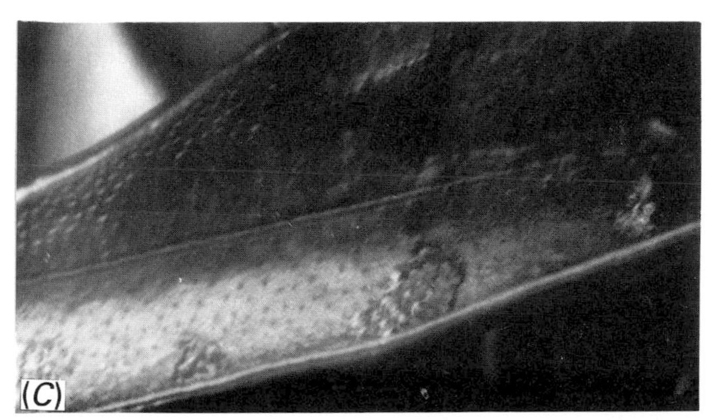

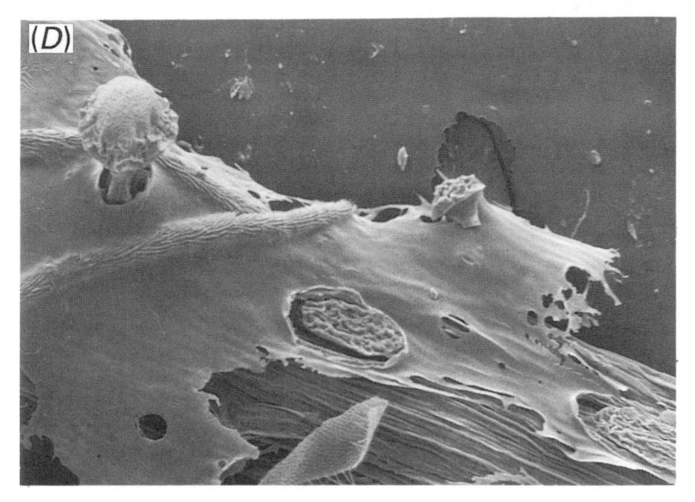

Fig. 8.16 The glandular surface of *Pinguicula grandiflora*. (*A*) SEM of a portion of a leaf bearing a captured ant. At the lower margin of the micrograph there stalked glands remain with there mucilage heads intact and undrained. The mucilage from several other stalked glands (SG) has been drained away by the ant's struggles and persists as 'cables'. cf. Figs. 6.1A & 18.1.
(*B*) Whole mount of a portion of a leaf bearing a captured fly of which a wing margin can be seen on the left. The whole leaf has been treated with α-napthyl acid phosphate – pararosanolin for acid phosphatase. The secretion pool from the fly is roughly demarcated by the dotted line. Adjacent to the fly the stimulated glands have discharged and show reduced activity. Beyond the secretion pool i.e. top-right, the glands still show intense activity. Reproduced by kind permission of Dr Y. Heslop-Harrison from Heslop-Harrison and Knox (1971).
(*C*) The leaf surface of *Pinguicula ionantha* showing small digestive pools. Photographed with a UV transmitting filter.
(*D*) *Drosophyllum* leaf surface showing the trapping mucilage spread over the glands and insect's feet. cf. *Byblis* in Fig. 9.9.

these genera are probably able to operate a number of times.

Pinguicula, which preys on numerous small insects, is somewhat different. Within a few hours of entrapping an insect a small pool of digestive fluid will form around it, the extent of the pool being only sufficient to cope with each insect (Figs 6.7 and 8.16*A*). Glands that become incorporated into the pool are triggered to release their total load of hydrolases in a 'one-off' secretory action. Once fired, these glands will not synthesize more hydrolytic activity (see page 182) (Heslop-Harrison and Knox, 1971; Heslop-Harrison, 1975, 1976; Heslop-Harrison and Heslop-Harrison, 1981).

The digestive fluid of *Pinguicula* appears to contain wetting agents, similar to the pitcher liquor (Heslop-Harrison and Knox, 1971). Copious amounts of this low viscosity digestive fluid are secreted by the sessile glands; the mucilage secreted by the stalked glands is mainly involved in trapping and seems to play only a minor role in digestion. The development of motile leaf margins which curl inwards, produces a (temporary) real digestive pool in which prey is either completely bathed, or in which at least there is a large contact area (Fig. 6.7).

Similarity between digestive mucilages and root-cap slime

The function of the digestive mucilage is closely comparable to that of root-cap slime. Both prevent abrasive damage and provide a mucigel across which the plant can establish an intimate contact with a source of nutrients: a prey in the case of the digestive gland, and soil particles in the case of the root. As with the root slime, the digestive mucilage may play an important role in the uptake of ions by adsorption-exchange between prey and mucilage. The cell walls of both root and digestive gland tissues largely consist of pectic substances through which, again, close adsorption-exchange with the external medium is possible.

The Perception of Stimuli

A digestive cycle often proceeds in response to stimuli provided by trapped prey. Experimentally it was established that a series of chemical compounds can induce such stimulation (see page 195). Traps are clearly capable of distinguishing 'useful' from 'non-useful' matter, and secrete digestive fluid only when the former is present (Darwin, 1875). Certain enzymes are, however, secreted with the trapping mucilage of *Drosera*, *Drosophyllum* (Dexheimer, 1978b) and *Pinguicula* (Heslop-Harrison and Knox, 1971; Heslop-Harrison, 1976).

Stimuli and nitrogenous compounds

Darwin (1875) suggested that the effective stimulants are mainly nitrogenous compounds and this has since been confirmed. It appears that ammonium ions or free ionized primary amino groups are necessary for a nitrogenous compound to be active and that nitrogen-containing compounds which do not meet this specification are not active (e.g. ovalbumin) or only slightly active (e.g. urea) (Robins, 1976; Lichtner and Williams, 1977). In addition, sodium ions are apparently capable of eliciting secretion (Lichtner and Williams, 1977), although the hydrolytic activity of the fluid so induced is not known.

Factors affecting the reception of the stimulus

Two factors might affect the selectivity of the gland's response to potential stimulants. First, the specific sensitivity of special receptors, as yet unknown but which are likely to lie in the glandular layer of the digestive organs. Secondly, the selective permeability of the cell wall and cuticle to potential stimulants.

For a substance to be an effective stimulant it must penetrate the cuticle of the digestive gland. The difference in the structure and in the composition of the cuticle of the various carnivorous plants (see page 162 and Fig. 8.9) might therefore determine certain differences in their response to similar stimuli. Large polar molecules are not likely to cross a continuous cuticle but might freely diffuse through cuticular discontinuities. Accordingly, egg-albumen and casein were successfully used by Schnepf (1963a,b,c) for eliciting the digestive activity in the sessile glands of *Drosophyllum* which show many gaps in their cuticle (Fig. 8.11). The digestive glands of *Dionaea*, on the other hand, which show a continuous cuticle when unstimulated (Joel *et al.*, 1983) do not respond to the presence of large protein molecules (Lichtner and Williams, 1977; Lüttge, 1964b).

It is ostensibly paradoxical that the cuticle is impermeable to exogenous solutes, when it is continuous. Yet, at the same time, elicitation of both cuticular gap formation and the secretory activity is nonetheless dependent on exposure of the glands to exogenous solutes (Robins, 1976; Lichtner and Williams, 1977; Rea, 1982a,b). In contrast to the secretory enzymes, which can only traverse the cuticle in its open phase, the stimulants must do so before the formation of cuticular gaps. The rate of uptake of small-molecular-weight nitrogenous elicitors is low before the onset of secretion (Fig. 10.13; Rea and Whatley, 1983). Nevertheless, the extremely low concentrations which gain access to the gland interior are sufficient to elicit gap formation and enzyme secretion (Joel *et al.*, 1983).

Digestive enzymes which are stored in the wall of the secretory cells might act to release small active residues from non-stimulating proteins once these have penetrated the gland through cuticular discontinuities. These residues will stimulate the glands as soon as they reach the receptors. Natural prey contains numerous small nitrogenous molecules (e.g. amino acids, amines) which will act as stimulants directly.

Chemical stimulants, penetrating the gland by diffusion, will be diluted in the apoplast before reaching the receptors of the glandular cells. This may be one factor contributing to the poor development of labyrinthine walls on the outer face of the head cells of almost all digestive glands. Ingrowths in this region would dilute stimuli reaching the plasmalemma, reducing the sensitivity of the glands to stimuli. The thin wall thereby enhances the sensitivity of the glands to even very weak stimuli.

Mechanical stimulation

There is some evidence that, in addition to chemical stimulation, mechanical stimulation of digestive glands is also possible in certain species. Repeated mechanical irritation of the trigger hairs in a closed trap of *Dionaea* without any chemical stimulation produced a viscous acidic secretion (Macfarlane, 1892; Lichtner and Williams, 1977). There is, however, apparently no enzyme release following

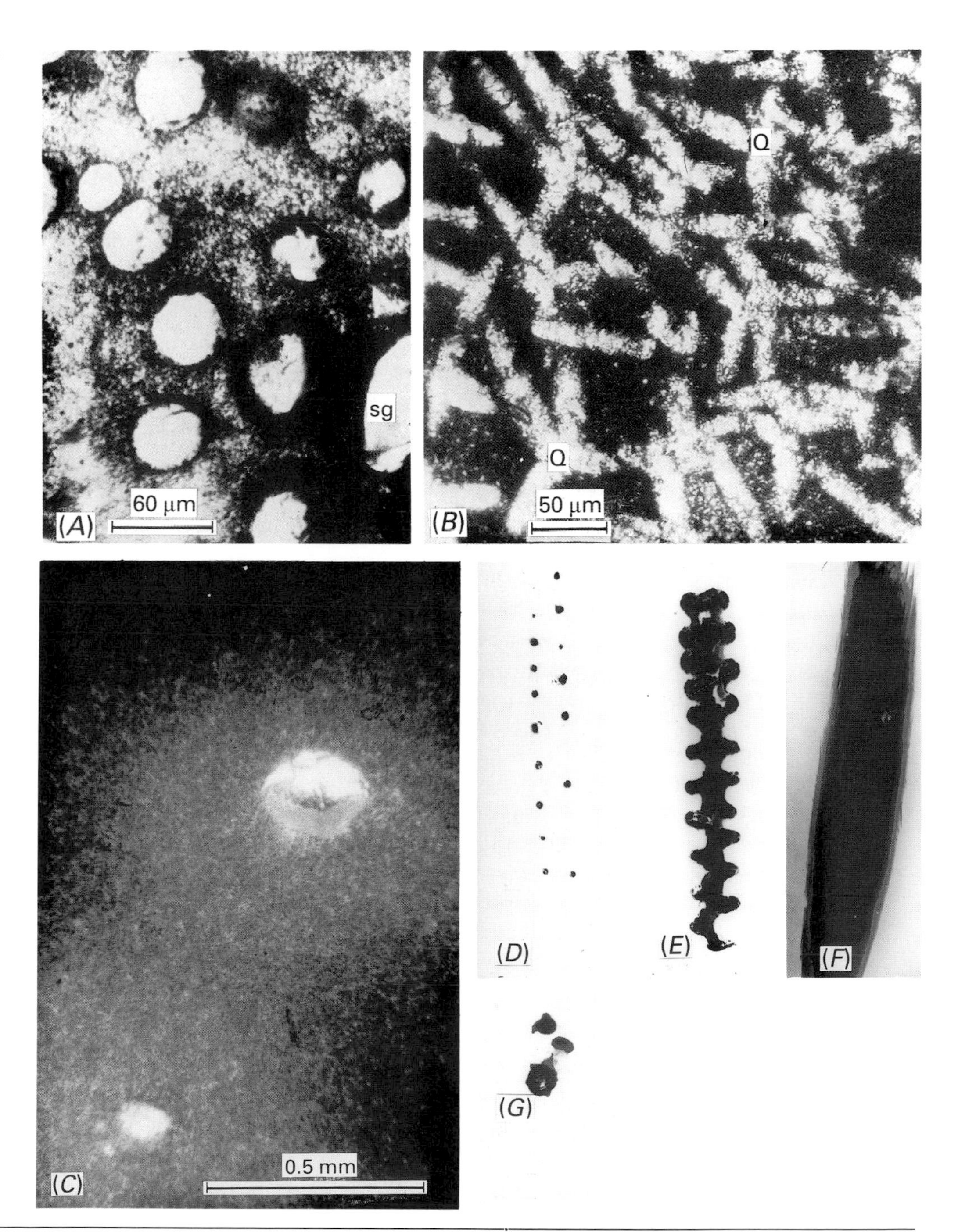

Figs. 8.17*A–G* The release of protease activity from the glandular surfaces of several carnivorous plant species as demonstrated by the digestion of photographic film.
(*A*) *Pinguicula grandiflora* – upper leaf surface, sessile glands with a stalked gland at the right (sg). Reproduced by kind permission of Dr. Y. Heslop-Harrison from Heslop-Harrison & Knox (1971).
(*B*) *Utricularia neglecta*. Bifid and quadrifid (Q) glands. Reproduced by kind permission of Dr. C. Vintéjoux from Vintéjoux (1979).
(*C*) *Nepenthes maxima superba*. Protease digestion by the the pitcher digestive glands of the gelatin layers in a colour film. There is an area of complete digestion directly over the gland and a surrounding 'halo' of slight digestion by enzyme which has leaked out during incubation. Unpublished micrograph kindly supplied by Dr. N.D. Hallam, Monash, Australia.
(*D–F*) *Drosophyllum*. Three stages in the digestion of a gelatine coating of a colour film by the protease secretion from the digestive glands. (*D*) after 30 mins incubation: (*E*) 90 mins. (*F*) 180 mins. Unpublished photomicrographs by kind permission of Dr. Y. Heslop-Harrison.
(*G*) *Triphyophyllum peltatum*. Protease digestion of colour film over three gland heads, after 30 mins. digestion. Unpublished micrograph reproduced by kind permission of Dr. Y. Heslop-Harrison. cf. Fig. 4.18.

such stimulation, indicating that fluid and hydrolase release need not be under precisely the same control. In this, *Dionaea* is similar to *Nepenthes* (see pages 106 and 188).

Some digestive organs respond rapidly to stimulation; others are comparatively slow. In *Pinguicula* digestive glands react rapidly. The first evidence of esterase activity in the leaf-surface fluid was found 1 h after stimulation and an intense activity was present after 2 h (Heslop-Harrison and Knox, 1971; Heslop-Harrison and Heslop-Harrison, 1981). In *Dionaea* on the other hand, though responses to stimulation can be detected within 15 min (Rea *et al.*, 1983) the first evidence of proteolytic activity in the trap of this plant occurs after about 20 h (Robins, 1978; Joel *et al.*, 1983).

The Origins of the Digestive Fluids

As can be seen from Table 9.2, it was established some time ago that digestive glands in many genera secrete proteases. The intracellular sources of these secreted enzymes will be dealt with next.

In 1926, Quintanilha suggested that the vacuome is the 'seat of the elaboration of ferments' and that it 'certainly has an important role in the process of digestion and secretion.' This idea had apparently become firmly rooted by the time the first examinations of the glands of carnivorous plants in the electron microscope were performed.

Vogel (1960) assumed that the vacuoles of *Pinguicula* glands contain the proteolytic enzymes identified from the secretion. His electron micrographs showed small vacuoles together with many mitochondria near the nucleus of the glandular cells. This lead him to suggest that the vacuoles are built near the nucleus and fuse together near the outer surface of the gland to form larger vacuoles from which enzymes are released on stimulation.

Similarly, from light and electron microscope observations on *Dionaea* (Scala *et al.*, 1968b; 1969; Schwab *et al.*, 1969), it was concluded that hydrolases activity was contained in the vacuoles as zymogens. These authors observed that the vacuolar 'protein bodies' seen in the light microscope nearly disappear when the gland cells are stimulated and release digestive enzymes but return to their initial staining intensity by the end of the digestive process.

While subsequent work has partially confirmed these earlier studies (see pages 175 and 177) the situation is seen to be rather more complex, particularly in *Dionaea*.

Synthesis and storage in *Dionaea*

Where, then, are the digestive enzymes synthesized, packaged and stored? Conclusive evidence that the secretion was an intrinsic property of the glandular compartment of the digestive gland in *Dionaea* was obtained by Robins and Juniper using radiolabelling techniques (Robins, 1978; Robins and Juniper, 1980b,c). Whole plants were fed ^{3}H-leucine every day for 2–3 weeks and the sub-cellular distribution of labelled molecules was examined in new leaves, which themselves were not fed but had grown during the feeding period. Glands from immature traps and freshly matured traps showed certain significant key differences (Table 8.2). Not surprisingly, the rER of immature glands showed an elevated level of labelling which fell away in mature glands, suggesting a higher rate of protein synthesis during development. Silver grains radiolabels were also frequent over the sER but this level remained high on maturation. Small osmiophilic vacuoles in mature traps showed much more labelling than in immature traps but the large vacuole showed no such effect and was not labelled more than might be expected by random. Thus it would appear that, during the late phase of development, protein destined for secretion is synthesized on the rER which either loses its ribosomes and becomes smooth or buds regions of sER (Fig. 8.8*B*), presumably due to a high concentration of protein in the cisternal space. Alternatively, smooth-membraned vesicles may be formed by dictyosomes. Juniper *et al.* (1982a) showed rER/dictyosomal connections in unstimulated *Dionaea* glands (Fig. 8.15*A–D*) while Henry and Steer (1985) found phosphatases in the cisternae at the mature face of dictyosomes during the early stage of secretion (Fig. 8.15*E*,*F*). Thus, a role for dictyosomes in the formation of hydrolase-containing packages in maturing glands is probable, though they appear to have no major role in the release of the secretion (see page 188). The smooth vesicles derived from sER or dictyosomes may then fuse to form the small osmiophilic vacuoles. Furthermore, profiles of rER are seen in mature unstimulated glands fused to the plasmalemma or, more frequently, very closely juxtaposed (Figs 8.7*A* and 8.8).

Very similar observations have been made in a number of other glands exporting proteinaceous mucilages. In *Nomaphila stricta*, Kristen (1975) was unable to detect hypersecretory dictyosomes and considered them to participate only in the release of cell wall material. Instead, he observed rER fusion to the plasmalemma and ER-derived vesicles and considered these to be the mechanisms of protein secretion (Figs 8.8*A–C* and 8.15). Direct release from the rER by fusion was also observed in *Isoetes lacustris* ligules (Kristen *et al.*, 1982) and by ER-derived vesicles in the young stigma of *Aptenia cordifolia* (Kristen, 1977).

Sites of storage in other genera

That synthesis on ER leads to major vacuolar and apoplastic stores of hydrolytic enzymes is amply borne out by light microscope cytochemical observations in a number of genera (Fig. 8.18 and Table 9.1). Acid phosphatase, esterase and leucine aminopeptidase have been widely found in the vacuoles (Heslop-Harrison, 1975; Parkes, 1980) and these enzymes are also plentiful in the labyrinthine walls of *Pinguicula* (Heslop-Harrison and Knox, 1971; Heslop-Harrison and Heslop-Harrison, 1981 and Figs 8.7*C* and 8.19). Several phosphatases are found in *Drosera capensis* walls associated with the mucilage (Dexheimer, 1978b). *Dionaea*, in contrast,

Table 8.2 Distribution of autoradiographic silver grains within the gland cells of *Dionaea muscipula* before stimulation as seen under TEM. Mature plants were fed daily with stimulating solution containing bactopeptone and L-[4,5 ^{3}H] leucine. After 14 d traps were prepared for autoradiography and examined under the TEM.

Silver grain distribution	Cytoplasmic compartment[a]															
	1° CW	LW	V_1	V_2	MVB	N	M	LD	P	D	DV	sER	rER	TER	GP	Total
Outer secretory cell																
Mature																
Observed % grains present	14.4	9.5	25.5	17.1	0	1.5	3.4	0	0.8	2.3	1.9	12.9	1.9	—	8.4	—
Observed total grains present	38	25	67	45	0	5	9	0	2	6	5	34	5	—	22	263
Calculated total grains	32	5	109	12	<1	26	10	0	5	4	1	—	—	1	58	263
Immature																
Observed % grains present	12.9	—	32.5	7.1	2.8	6.0	3.4	0.3	0.9	0.9	1.4	15.2	4.7	—	4.8	—
Observed total grains present	91	—	229	50	20	42	24	2	6	6	10	107	33	—	84	704
Inner secretory cell																
Mature																
Observed % grains present	8.1	9.4	20.8	18.8	0	6.7	8.7	0	0	0	2.0	16.1	0.7	—	8.7	—
Observed total grains present	12	14	31	28	0	10	13	0	0	0	3	24	1	—	13	149
Calculated total grains	13	5	23	24	0	19	10	<1	3	<1	<1	—	—	7	46	149
Immature																
Observed % grains present	13.1	—	28.8	9.3	3.1	10.0	3.1	0.2	1.7	0.7	0.5	13.6	4.1	—	11.9	—
Observed total grains present	55	—	121	39	13	42	13	1	7	3	2	57	17	—	50	420

[a] Abbreviations: 1° CW, primary cell wall; LW, labyrinthine cell wall; V_1, osmiophilic vacuoles greater than 2 μm in diameter; V_2, osmiophilic vacuoles less than 2 μm in diameter; MVB, multi-vesicular body; N, nucleus; M, mitochondrion; LD, lipid droplet; P, plastid; D, dictyosome; DV, dictyosomal vesicle; s-, r- and TER, smooth, rough and total endoplasmic reticulum; GP, ground-plasm.

χ^2-tests: Outer secretory cell: mature observed v. calculated $\chi^2 = 64$, $P < 0.001$; mature observed v. immature observed $\chi^2 = 58$, $P < 0.001$; Outer secretory cell: mature observed v. inner mature observed $\chi^2 = 23$, $P < 0.05$; Inner secretory cell: mature observed v. calculated $\chi^2 = 25$, $P < 0.01$; mature observed v. immature observed $\chi^2 = 40$, $P < 0.001$.

Reproduced from Robins and Juniper (1980b) with permission.

lacks these enzymes in the labyrinthine walls of unstimulated gland cells (Fig. 8.19*A* and Robins, 1978; Robins and Juniper, 1980b; Parkes, 1980). Nor do these walls stain with mercuric bromophenol blue (Fig. 8.19*B* and Robins, 1978), indicating them to be relatively free of protein. Yet acid phosphatase occurs in the primary wall of *Dionaea* (Fig. 8.19*A*) and ^{3}H-leucine accumulates in both parts of the wall (Robins and Juniper, 1980b). While it is difficult to resolve these observations it is possible that *Dionaea* differs from *Pinguicula* in only having a very minor store in the walls of unstimulated gland cells (perhaps showing much variation from one gland to another and with the age of the trap), enzymes being loaded in the walls largely as the result of stimulation.

This view is consistent with the results showing that the differentiation of *Dionaea* digestive glands is not complete until the gland is stimulated (Joel *et al.*, 1983). It is also consistent with recent observations made in *Drosera rotundifolia* by McEwan *et al.* (1988). These authors found that in unstimulated stalked glands, acid phosphatase activity (4-nitrophenyl phosphate) could only be detected in a few glands and, when present, was only associated with vacuoles and wall regions of outer head cells. Following stimulation, enzyme is much more frequently detected and is additionally

Fig. 8.18*A* *Pinguicula grandiflora*. Neighbouring sessile and stalked glands from an unstimulated leaf, RNase reaction. Strong ribonuclease activity is associated with the spongy anticlinal walls, but that in the sessile gland (left) is even greater. Reaction product is also present in the inner surface of the periclinal walls of the head cells of the sessile gland. Neither gland shows cytoplasmic activity, but the nuclei of the stalked gland show some activity. (Reproduced by kind permission of Dr Y. Heslop-Harrison from Heslop-Harrison and Heslop-Harrison, 1971.)

Fig. 8.18*B*, *C* *Pinguicula caudata*. Leucine aminopeptidase localisation showing the reaction product in the anticlinal walls of the sessile glands (*B*), but the absence of such reaction product in the stalked gland (*C*). (Unpublished micrographs by kind permission of Dr D.M. Parkes, Monash, Australia.)

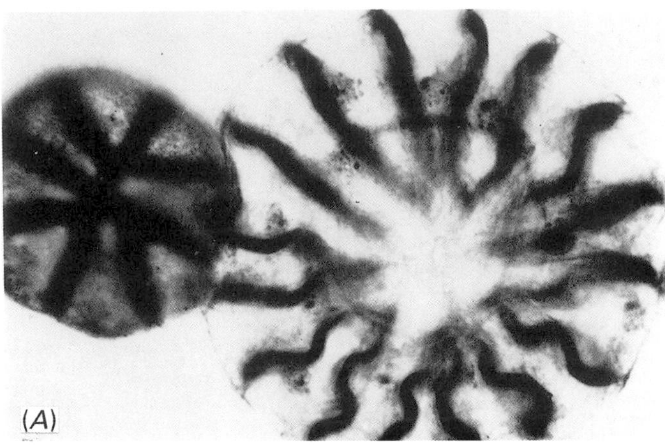

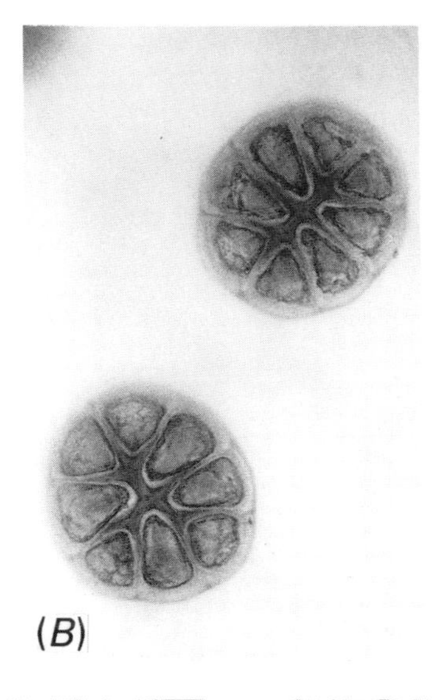

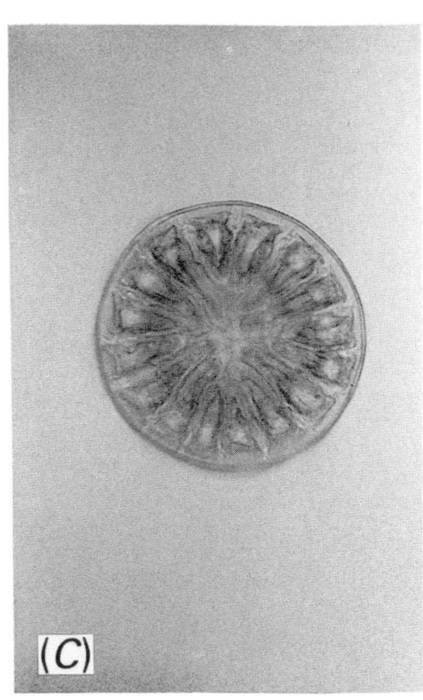

Fig. 8.18*D*, *E* Acid phosphatase and ATP-ase localization in the head cells of the stalked glands of *Drosera capensis*. Note how the reaction product is located in the rough ER and the primary cell wall region, but not in the dictyosomes (D), or their associated vesicles, nor in the labyrinthine region (*La*) of the cell wall. Reaction product is seen in the dictyosomes only in the stationary or resting phase of the gland cell's cycle. (Both micrographs reproduced by kind permission of Dr J. Dexheimer from Dexheimer, 1978c.)

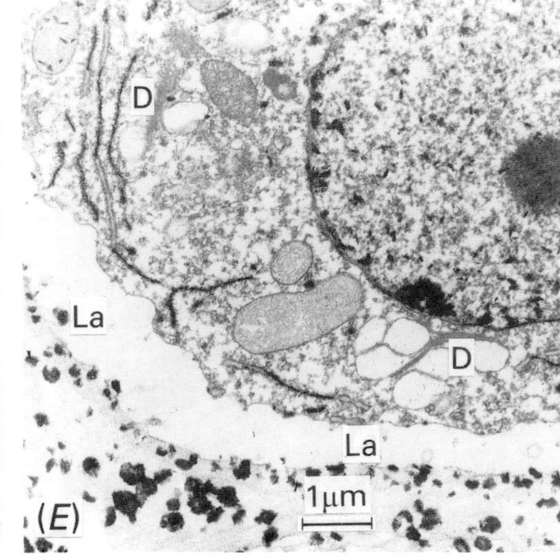

Fig. 8.18*F* *Nepenthes maxima superba*. Photomicrograph of a median section through a digestive gland showing vacuolar leucine amino peptidase activity in the head cells. (Photomicrograph by kind permission of Dr D.M. Parkes, M.Sc. thesis, Monash, Australia.)

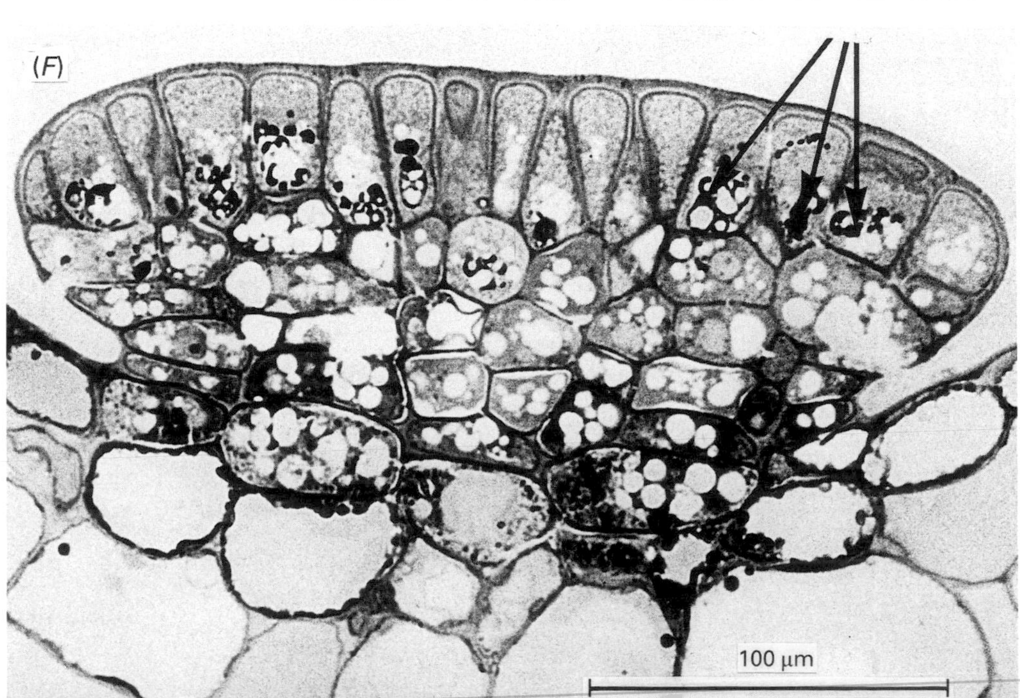

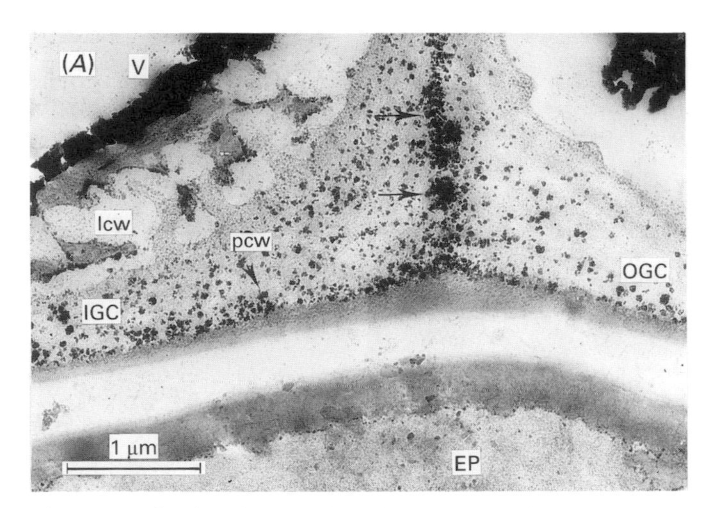

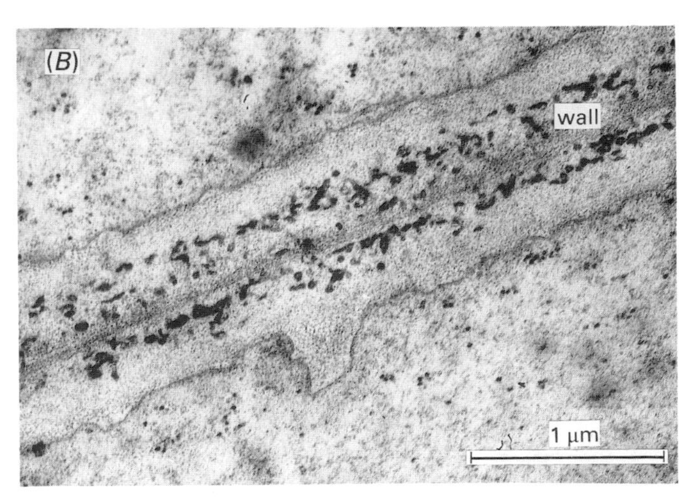

Fig. 8.19 The distribution of proteins within the walls of head cells in digestive glands.
(*A*) Detail of the boundary between an outer glandular cell (OGC) and an inner glandular cell (IGC) from an unstimulated *Dionaea* gland incubated for B-glycerophosphatase activity. Heavy deposits of the reaction product can be seen in the primary cell walls (pcw and arrowheads) and along the plasmamembrane, but are markedly absent from the labyrinthine projections of the cell wall (lcw) and the vacuole (V). Note also the absence of reaction from the epidermal cell (EP). (Unpublished micrograph R.J.R.)
(*B*) *Nepenthes rufescens*. A region of the wall between two columnar gland head cells stained for acid phosphatase by the Gomoryi lead salt method. Note the reaction product in the irregular cavities or channels in the wall material. (Photomicrograph by kind permission of Dr D.M. Parkes; M.Sc. thesis, Monash, Australia.)

found in the nuclear membranes, the endoplasmic reticulum and dictyosomal cisternae and vesicles. Both layers of head cells show activity and there is some polarization towards the outer edges of the outer head cells and towards the gland apex. Maximal reaction product was observed after 96 h stimulation, the peak of extracellular activity also (Clancy and Coffey, 1977). Thus, as in *Dionaea*, the evidence is that stimulation of the gland involves the initial release of vacuolar enzymes followed by a phase of *de novo* synthesis. Dictyosomes in *Drosera* are actively synthesizing and secreting polysaccharide mucilage and it appears that acid phosphatase is also being secreted by this mechanism. Whether a direct ER pathway comparable to that of *Dionaea* is also active requires further investigation. The examination of sessile glands may help to answer this question.

Ribonuclease and glucose-6-phosphatase have also been found in labyrinthine walls of *Drosera*, *N. khasiana* and the *small* but, interestingly, not the *large*, glands of *Cephalotus* (Parkes, 1980). In the small glands of *Cephalotus*, the space between the digestive cells and the 'guard' cells is filled by a wall-plug which is continuous with the outer wall. This plug contains enzymes (Parkes, 1980), and shows surface protease activity (Joel, unpublished).

The localization of protease

No protease has been localized in any wall or subcellular compartment as no suitable cytochemical test is available at present. The substrate-film method, however, shows them to be released from the glands (Fig. 8.17 and Table 9.1) and it seems reasonable to assume that they are held within the same compartments as the other hydrolases.

Secretory activity in *Pinguicula*

A very close correlation has been established in *Pinguicula* between the loss of histochemical reactivity and the onset of secretory activity (Heslop-Harrison and Heslop-Harrison, 1981). Unfortunately, however, all the cytochemical tests available are relatively non-specific and the enzyme activities examined are widespread in plant tissues. None of this work, therefore, unequivocally shows that these enzymes are destined for secretion, but it is hard to envisage that this is not their role.

The identification of the origin of the secretion in *Dionaea*

The autoradiographic study of Robins and Juniper (1980b) in *Dionaea* is also subject to question. These authors did show, however, that radiolabel, detected autoradiographically in the gland cells, appeared in the secretion. In this study, ^{3}H-leucine was fed to some of the leaves on a plant. After a short period other leaves were stimulated using bactopeptone and it is these, previously unfed, lobes that were used for autoradiography. The secretory fluid collected from these traps contained considerable radiolabel in its hydrolytic proteins (Robins, 1978),

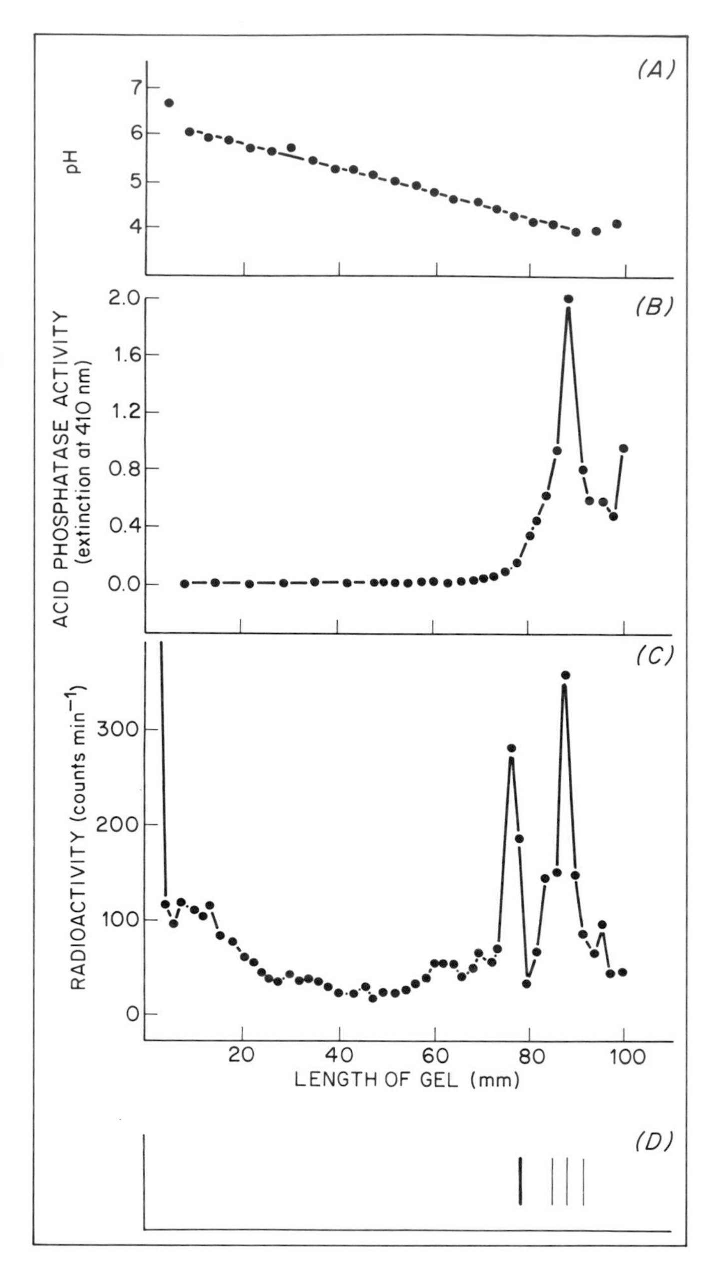

Fig. 8.20 Isoelectric focusing profile of secretion collected from thirteen *Dionaea* traps stimulated with 4% (w/v) bacto-peptone solution containing 41.6 kBq ml^{-1} L-(U-^{14}C) leucine. 1.2 ml crude secretion was mixed with 120 mg sucrose to give a 10% (w/v) sucrose solution. 400 μl of this was placed on to each of 3 gel tubes, and 400 μl 10% (w/v) sucrose in water placed on to a fourth gel tube. Isoelectric focusing was conducted at 350 V for 160 min followed by 700 V for 40 min. Gels were then extracted and treated as described in the methods. (a) pH gradient; (b) acid phosphatase activity; (c) radioactivity; (d) total protein. (Redrawn from Robins and Juniper, 1980b.)

showing that these had come from within the gland cells and must, at least partially, have been the proteins detected autoradiographically.

Furthermore, Robins and Juniper (1980b) demonstrated for the first time in carnivorous plants that *de novo* enzyme synthesis takes place in the stimulated gland of *Dionaea*. Plants fed ^{3}H-leucine through the leaves were shown to have incorporated a considerable activity into protein removed from the same traps 24 h later. Exhaustive analysis by dialysis, ion-exchange chromatography and isoelectric focusing showed the label to be incorporated into secreted proteins including one identified as acid phosphatase (Fig. 8.20). The incorporation was significantly depressed by cycloheximide, as was the activity of acid phosphatase and protease in the secretion. Deuterium oxide, another potent inhibitor of protein synthesis, had a similar effect.

Thus, it is now established that certain carnivorous plants make their own digestive proteins and in some types store them in sub-cellular compartments, releasing them following stimulation by a specific group of secretogogues.

Cellular Events following Stimulation

Changes in cuticular permeability

As already discussed, a prerequisite for the release of digestive fluid is a highly permeable cuticle. This permeability is achieved by cuticular discontinuities, which normally develop late in the differentiation of the glands (see page 162). In a few carnivorous plants, however, this happens only as the result of stimulation.

The formation of cuticular discontinuities in Dionaea

The digestive gland of *Dionaea* is the only digestive organ which has been clearly demonstrated to show a stimulation-mediated formation of cuticular discontinuities (Joel *et al.*, 1983). Shortly after the chemical stimulation of the digestive glands of *Dionaea*, discontinuities (40–80 nm wide) form in the cuticle outer walls of the head cells of the gland complex (Fig. 8.21*C*). When these discontinuities form there is an increase in the capacity of the glands to absorb vital stains (Figs 8.21*A*,*B*). Similarly the capacity of the trap lobes to absorb detectable amounts of ^{14}C-glutamic acid or $^{35}SO_4^{--}$ is dependent on prior exposure of the glands to chemical stimulation by the prey for several hours before the application of the radioactive tracer (Lüttge, 1963). (Fig. 8.24 *C–E*).

Two equally tenable models were proposed by Joel *et al.* (1983) for the formation of cuticular gaps in *Dionaea*: a stimulation-dependent increase in

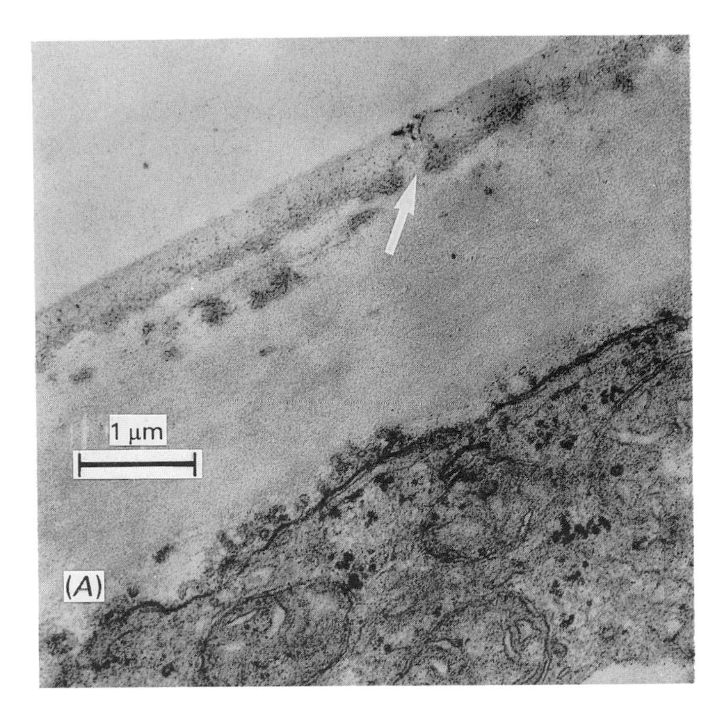

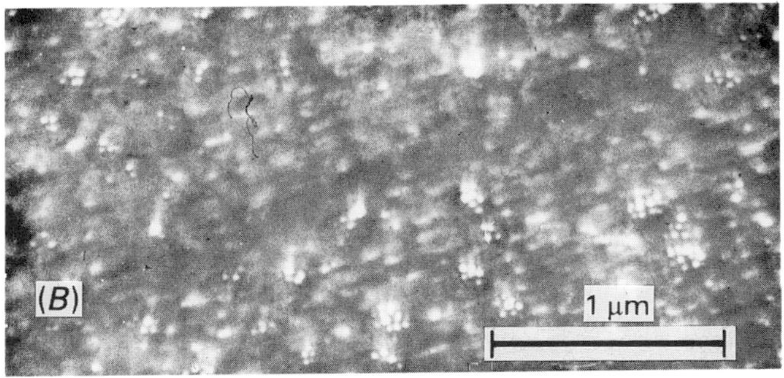

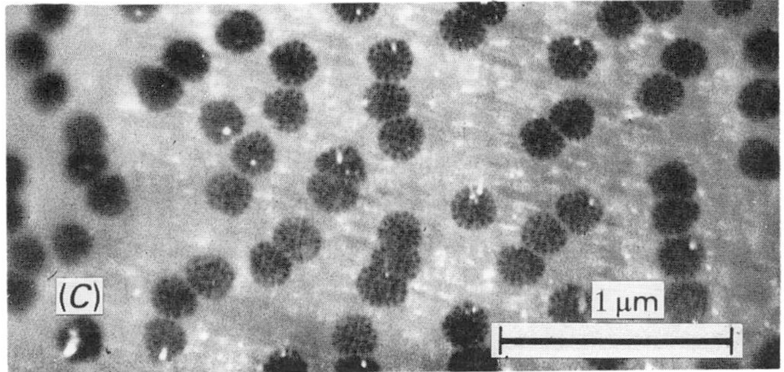

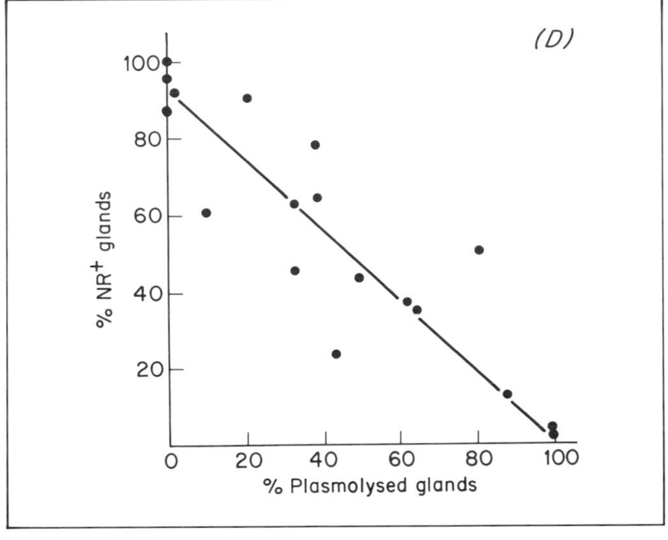

Fig. 8.21 Formation of cuticular gaps in the digestive glands of *Dionaea*. (*A*) Cuticular gap (arrow) of a stimulated gland as seen under TEM. (*B*, *C*) Whole digestive glands seen from above following treatment with neutral red for 2 min. (*B*) Unstimulated gland showing no neutral red absorption. (*C*) After exposure of the trap to thiourea for 7h, showing appreciable uptake of the dye. (*D*) A scatter diagram of the percentage of neutral-red staining *Dionaea* glands (% NR^+ GLANDS) versus the percentage of glands containing plasmolysed head cells (% PLASMOLYSED GLANDS). Each point represents the measurements taken from sets of 3–4 lobular disks treated with different concentrations of PEG_{100} for 10 h. The line of best fit was calculated by linear regression (least squares method) and the confidence limits on the gradient were calculated from the difference between the total sum of the squares and the sum of the squares for the regression. The mean difference was tested as $t_{(15)}NR^+ = -0.973P + 93.78$ and the limits on the gradient are +/−0.087, +/−0.121 and +/−0.167 (from Joel *et al.*, 1983).

head cell turgor, and/or an increase in the plasticity of the walls bounding the head cells.

(i) *Turgor-mediated expansion*. The response of the glands to external solute potential of more than −1 bar was found to be similar to that seen upon chemical stimulation of the gland *in situ*. Plasmolysis inhibits cuticular gap formation, while the generation of a positive turgor in the head cells causes the formation of cuticular permeability (Fig. 8.21*D*). These results indicate that turgor is involved in the formation of cuticular gaps.

(ii) *Changes in cell wall plasticity*. It is known that the digestive fluid is highly acidic, and the digestive glands extrude protons soon after the onset of stimulation (see pages 184, 194 and 223). It is therefore possible that acid-induced growth (Rayle and Cleland, 1977) is involved in the formation of cuticular discontinuities in *Dionaea*, acidification of the walls causing an increase in wall plasticity and accelerated expansion. If acidification does not cause a similar increase in the plasticity of the cutinized wall layer, cuticular fracture would result. Since it is known that chloride accumulates in the glandular head cells within 15 min of the start of stimulation (Rea *et al.*, 1983: see page 185), it would be anticipated that solute accumulation in the gland causes a concomitant movement of water, resulting in an increase in gland cell turgor. The subsequent active excretion of protons into the apoplast (Rea, 1983a: see page 197) suggests that acidification of the cell walls of the gland occurs within minutes of stimulation.

Thus, we can see evidence that both turgor pressure and acidification of the wall occur in response to chemical stimulation of *Dionaea* glands. Both models are consistent with the available evidence. The formation of cuticular gaps in *Dionaea* therefore seems to be a stimulus-induced process in which wall loosening due to modification results in the turgor-driven separation of cuticular particles from one another, forming cuticular gaps in a manner similar to that described for *Drosophyllum* (Joel and Juniper, 1982; Joel *et al.*, 1983 see page 164). The stimulus-induced secretion of chloride in *Dionaea* is a very early event and follows a comparable time course to the appearance of cuticular gaps (Joel *et al.*, 1983).

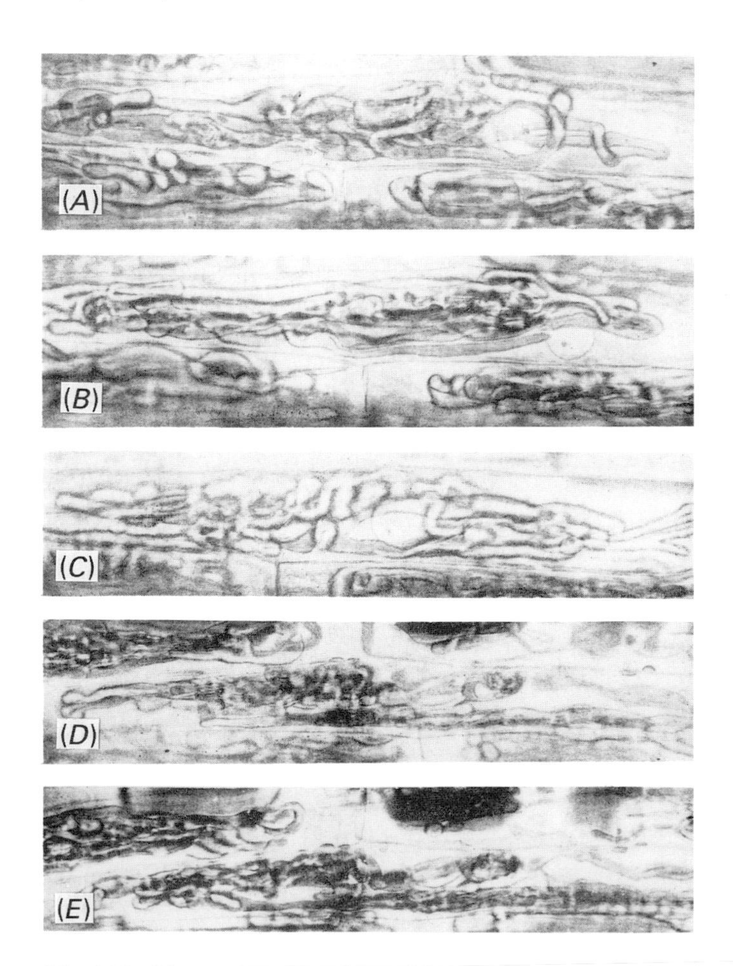

Fig. 8.22 The process of aggregation in the gland cells of *Drosera capensis* (reproduced from Åkerman, 1917).

Changes in the cuticle in Sarracenia *and* Darlingtonia

Batalin (1880) was of the opinion that in *Sarracenia flava*, which contains almost no pitcher liquor when unstimulated, a stimulus excreted by an adhering insect on the epidermal cells of the absorptive region causes the cuticle to detach from the cell wall at the stimulated cells. Similarly, he believed the cuticle to be absent in *Darlingtonia* from the entire surface of the absorptive cells and only remnants of cuticle to be left between adjoining cells. A recent electron microscopic examination of *Sarracenia* pitchers did not lead to support for Batalin's description. It seems that he was misled by using non-specific and insensitive staining techniques. Schimper (1882a,b) and Fenner (1904) found that the cuticle is provided with fine pores. This seems more consistent with the EM study carried out by Joel and Heide-Jørgensen (1985), revealing the presence of an extremely thin and discontinuous cuticle over the centres of the epidermal cells, and thicker cuticular 'rims' above the anticlinical walls (Fig. 8.9C). The resolving power of the light microscope is often incapable of distinguishing the thin cuticle over the centres of the cells, thus giving the impression that these wall regions are devoid of any cuticles.

'Aggregation'

The early studies of digestive glands were concerned with the structural changes in the glandular cells as seen in light microscopy of living material. The most conspicuous change in these cells is the start of cyclosis of the protoplasm and the phenomenon described by Darwin '. . . some hours after the gland [of *Drosera rotundifolia*] has been excited . . . the cells, instead of being filled with homogenous purple fluid contain variously shaped masses of purple matter . . .' (Darwin, 1875, his p. 33). He believed that these masses are composed of aggregated protoplasmic matter and termed the process 'aggregation'. Later it was established that these cellular masses are vacuolar and not protoplasmic. Various authors have studied the 'aggregation' process (Fig. 8.22) and found that it combines two separate phenomena, namely the 'swelling' of the protoplasm, termed 'true aggregation', and the precipitation in the vacuole, termed 'granulation' (Åkerman, 1917). At first the peripheral protoplasm is thin and displays slight rotation. This movement soon changes into a real circulation whereupon the cytoplasm thickens. 'Folds' of the protoplasm cross the large central vacuoles in various directions, breaking it into many small vacuoles. New vacuoles

Table 8.3 The structure of mitochondria in the outer secretory cells of *Dionaea muscipula* at different stages of the secretory cycle

	Days after stimulation			
	0	2	4	6
Volume of cell occupied (%)	3.6 ± 0.8	9.0 ± 1.2	7.3 ± 0.3	4.9 ± 0.7
Length of perimeter (μm)	2.9 ± 0.1	2.9 ± 0.2	2.9 ± 0.3	2.8 ± 0.2
Length of cristae (μm)	6.7 ± 0.6	10.2 ± 1.3	9.5 ± 2.0	8.4 ± 1.0
Area within perimeter membrane (μm^2)	1.2 ± 0.1	1.2 ± 0.2	0.9 ± 0.1	1.1 ± 0.1
Area within cristae membrane (μm^2)	0.3 ± 0.02	0.4 ± 0.06	0.4 ± 0.07	0.4 ± 0.05
Length of cristae membrane per unit cross-sectional area ($\mu m\ \mu m^{-2}$)	5.8 ± 0.5	9.0 ± 0.4	9.8 ± 0.6	7.7 ± 0.6
Area of cristae per unit volume ($\mu m\ \mu m^{-3}$)	49.2 ± 12.7	68.9 ± 6.7	82.4 ± 12.6	62.8 ± 9.7
Number of mitochondria analyzed	12	9	12	11

Reproduced from Robins and Juniper (1980a) with permission.

which did not contain a pigment are also formed in addition to the previously existing vacuoles (Schimper, 1882a,b; de Vries, 1886), so that the stimulated, active gland shows abundant small vacuoles rapidly moving in the cells in a complex pattern of currents. The reduction in the volume of the vacuole causes an increase in the concentration of vacuolar pigments, leading to granulation. To what extent have more recent studies shown that these phenomena are actually connected with digestion?

'Aggregation' and the vacuoles

According to electron microscopic studies, the vacuolar space of the secretory cells is divided into several types, including small osmiophilic vacuoles which contain enzymes and various solutes. In *Dionaea* large vacuoles rapidly sub-divide following stimulation and by 2 days their volume has decreased by 50% (Robins and Juniper, 1980a). As this division takes place before the onset of hydrolytic secretion (Joel *et al.*, 1983) we are led to suggest that it is related to the redistribution and secretion of chloride ions. Disintegration of the vacuole into smaller units will render possible a greater flux of vacuolar enzymes and ions across the tonoplast.

'Aggregation' and cyclosis

Cyclosis, which is associated with the disaggregation of the vacuoles, and which lasts throughout the digestive activity of the glands, is likely to play a key role in symplastic transport associated with both secretion and absorption.

'Aggregation' and cell organelles

Following stimulation, the mitochondria undergo changes indicative of increased respiratory activity.

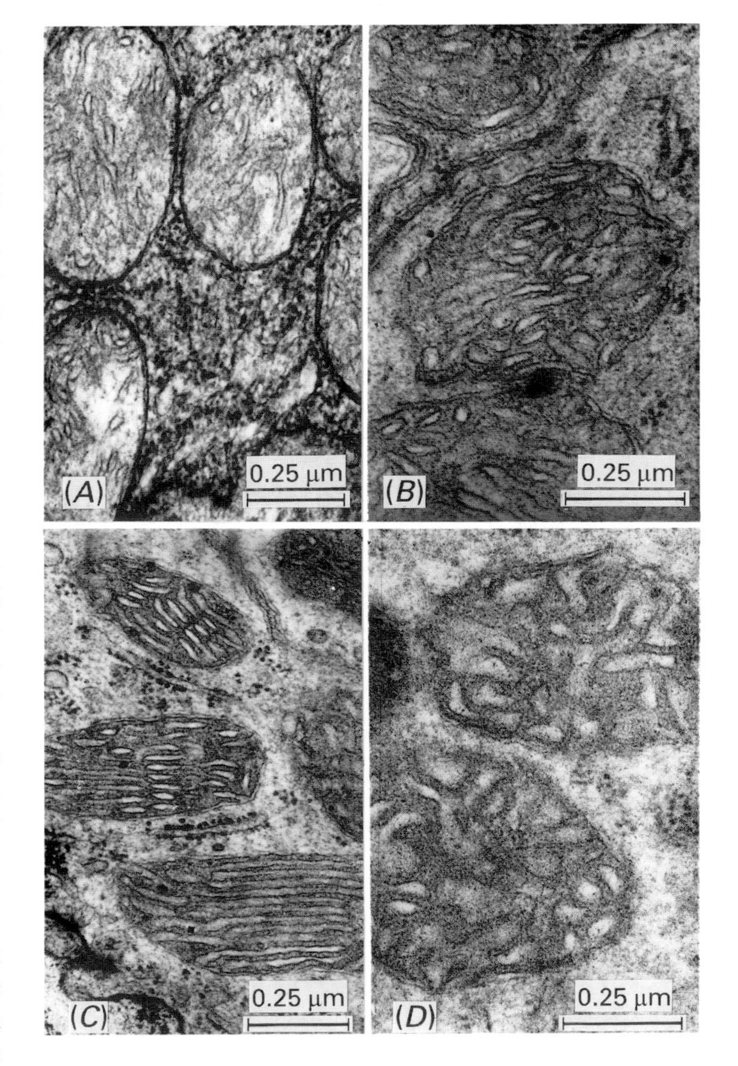

Fig. 8.23 Mitochondria from the outer secretory cells of the gland of *Dionaea* at different stages of the secretory cycle. (*A*) unstimulated tissue, day 0; (*B*) stimulated tissue, day 2; (*C*) stimulated tissue, day 4; (*D*) stimulated tissue, day 6. All tissue from the same preparation; CR = cristae of the mitochondria; cf. Table 8.3.

The surface area of the cristae increases compared to the overall inner volume of the organelle and they become densely packed (Schnepf, 1963b; Robins and Juniper, 1980a). These changes reach a peak during the height of the digestive cycle (Table 8.3), when an immediate supply of energy is crucial for maintaining high rates of active transport across the plasmamembrane. The mitochondria of stimulated *Dionaea* and *Drosophyllum* glands closely resemble mitochondria of active animal tissues such as those of the heart and liver (Robins, 1978). In *Dionaea*, the mitochondria frequently become extended (Fig. 8.23), greatly increasing their surface-to-volume ratio, which is probably essential to obtain a sufficiently rapid exchange of metabolites with the cytoplasm to maintain a high rate of ATP synthesis. Even so, Jaffé (1973) found a decrease in ATP levels in *Dionaea* glands following stimulation.

'Aggregation' and the nucleus

Early light microscopical studies revealed that, as the result of feeding, changes also occur in the nucleus of glandular cells. These changes were described for *Drosera* (Huie, 1897; Konopka and Ziegenspeck, 1929), *Drosophyllum* (Schnepf, 1963c), *Nepenthes* (Faber, 1912) and *Utricularia* (Kruck, 1931). Schnepf (1963c) in his electron microscopical studies of *Drosophyllum*, describes these changes in detail. About 30 min after feeding the glands with albumin, the chromatin agglutinates and concentrates around the periphery of the nucleus. Neither the size and shape of the nucleoli nor the 'Grundsubstanz' is affected. However, highly contrasted granules of 10–20 nm appear which are not seen in the resting glands. They disappear after 2–3 h. The granules are larger and more numerous in the sessile glands which are the major site of enzyme secretion. In addition, it seems that the number of nuclear pores increases when the digestive glands of *Drosophyllum* become active (Schnepf, 1963c).

It is interesting to note that similar changes were not seen in other electron microscopic studies of carnivorous plants. This may be the result of observations made at longer time intervals which might miss the rapid changes in the first hours after stimulation. It is very likely that the changes in the nucleus as described by Schnepf (1963c) are concerned with the digestive activity of the glandular cells. The high density of nuclear pores (c. 20 μm^{-2}) and the occurrence of condensed chromatin and a more convoluted nuclear envelope in the stimulated *Dionaea* glandular cells (Robins, 1978) may similarly indicate a more active state, borne out by the demonstration of *de novo* enzyme synthesis in stimulated glands of this genus (Robins and Juniper, 1980b).

The Mechanisms of the Release of the Digestive Fluid

A detailed mechanistic analysis can currently only be presented for two glands: *Pinguicula* and *Dionaea*. These have in common a stimulus-dependent release and appear to use a similar basic technique to move ions, water and digestive enzymes out of the gland head into the digestive pool.

Secretion in the One-off *Pinguicula* Gland

Maturation of the headcells

During the later stages of maturation of the digestive glands of *Pinguicula*, extensive polysomal activity is seen along the plasmamembrane and profiles of rER are found running very close to the labyrinthine protrusions of the head cells (Fig. 8.7*C*, 18.13 *D–E*: Heslop-Harrison and Heslop-Harrison, 1981; Vassilyev and Muravnik, 1988a) even forming parallel stacks (Morgan and Arnott, 1971).

The Heslop-Harrisons suggest, therefore, that protein is synthesized close to the cell wall and passes directly across the plasmamembrane into the expanding apoplast. Vassilyev and Muravnik (1988a) were of the opinion that in *P. vulgaris* digestive enzymes are transferred directly from the ER into labyrinthine walls through the continuity of ER membranes with the plasmamembranes. At the same time microtubules appear to control the directed transport of exocytotic vesicles into certain regions of the cell surface and their subsequent fusion to the plasmamembrane.

Heslop-Harrison and Heslop-Harrison (1981) suggested that during the final maturation of the sessile digestive glands of *Pinguicula grandiflora*, the membrane systems of the secretory head cells are degraded and the cytoplasmic contents undergo a kind of autophagic process, where the wall ingrowths become confluent with the vacuolar system (Fig. 8.24), ultimately forming a single enzyme-containing compartment. According to this theory the digestive glands of *Pinguicula* are 'pre-set' to be activated very rapidly after stimulation by captured prey, and each sessile gland is only able to function once (Heslop-Harrison and Heslop-Harrison, 1981). Contrary to this 'eccrine' hypothesis Vassilyev and Muravnik (1988a,b) showed that the head cells of *P. vulgaris* remain highly active during the whole of the secretion-absorption cycle, and suggested that the head cells are simultaneously performing both secretion and absorption activities (c.f. *Dionaea*, p. 214 and Fig. 8.27).

The mechanism of fluid movement

Heslop-Harrison and Heslop-Harrison (1980) found that the initial event associated with the onset of secretion in *Pinguicula* is the rapid movement of chloride ions across the gland and suggested that this ion movement induces the flux of water through the gland. The counter ions were not determined but, from studies of *Nepenthes* and *Dionaea* (see page 193) are likely to be H^+ and K^+. The gland complex of *Pinguicula* consists of three compartments, each being loaded with chloride ions in a different phase of activity. In the unstimulated gland the reservoir cell (see Fig. 6.1) has an extremely high concentration of chloride ions, while the level in the endodermoid and secretory head cells is low. The capture of prey initiates a rapid movement of chloride ions from the reservoir cell into the endodermoid cell and head cells respectively, with a consequent lowering of water potential in the latter. Presumably the sub-cuticular space is suitable for a local osmotic gradient to be established by pumping ions out of the cells. A flow of water is accordingly initiated from the reservoir cell into the endodermoid cell and head cell and finally through the discontinuities in the cuticle onto the leaf surface (Heslop-Harrison and Heslop-Harrison, 1980). The pumps controlling the movement of chloride ions are presumably located in the plasmalemma of the head cells. It is not known which ions are actively excreted.

The flush of acidic water thus created will carry the package of hydrolytic activity with it. The gland then seems to reverse this process and pump fluid and metabolites back into the leaf (Heslop-Harrison, 1975; see Ch. 10).

Certain aspects of the structural changes in *Drosera* as the sessile gland head cells reach maturation may indicate a similar mechanism (Heslop-Harrison, 1976) but this has yet to be elucidated.

Fig. 8.24 The digestive glands of *Pinguicula* at different stages of dissolution.

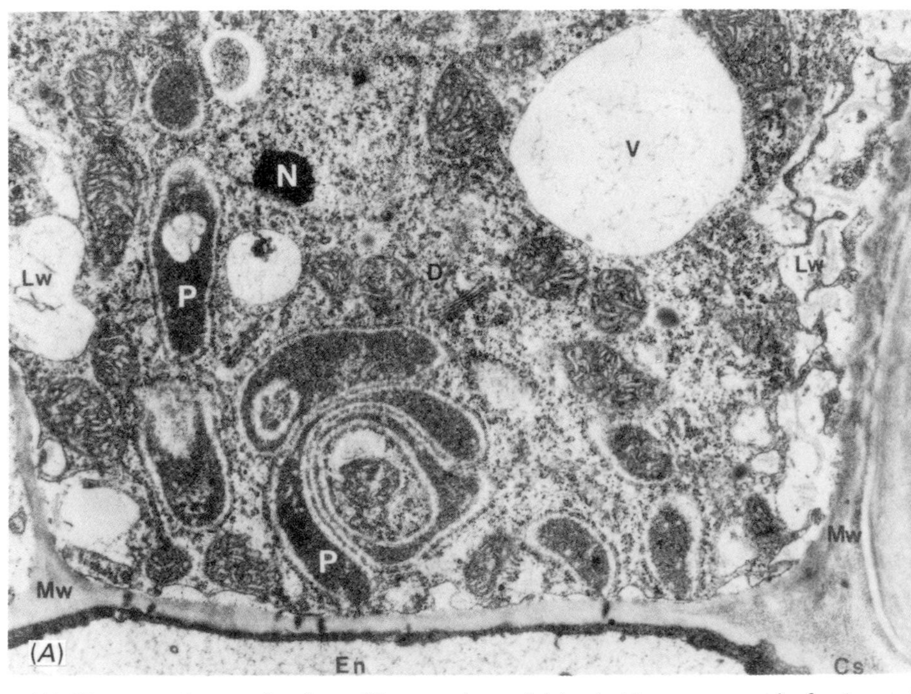

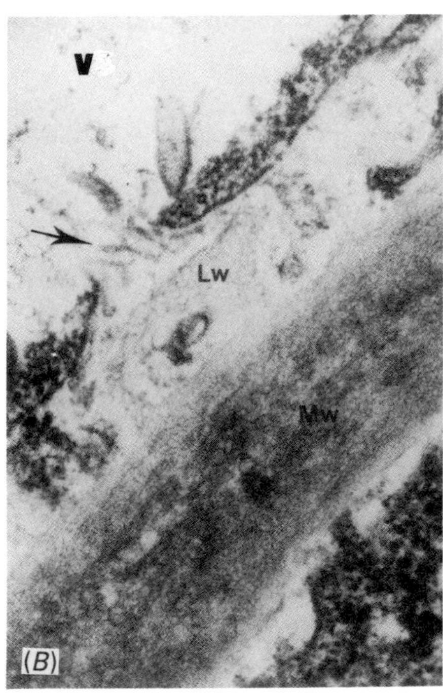

(*A*) Electron micrograph of gland head cells of *P. grandiflora*, prematuration stage, standard uranyl acetate-lead citrate staining. Corner of a head cell with a strip of an endodermoid cell at the base. The cuticle of the neighbouring epidermal cell can be seen to the right. Note the very complex convoluted form of the plastids (P), stretches of labyrinthine ingrowth (Lw) and the impregnation of the Casparian strip (Cs); N, nucleolus; Mw, microfibrillar wall; V, vacuole. Note also the numerous mitochondria with their very rich development of cristae. (For location see Fig. 2 of Heslop-Harrison and Heslop-Harrison, 1981.)

(*B*) Glutaraldehyde/osmium tetroxide fixation and staining with localization of acid phosphatase, by napthol AS-BI phosphate-hexazotized pararosanolin method, in a portion of gland head cell at a late stage of dissolution. The vacuoles (V), plasma-membrane and labyrinthine wall (lw) are becoming confluent. The tonoplast has disappeared and only stretches of the plasma-membrane remain. Acid phosphatase reaction product can be seen in the wall (Mw). (Reproduced by kind permission of Dr Y. Heslop-Harrison from Heslop-Harrison and Heslop-Harrison (1981).)

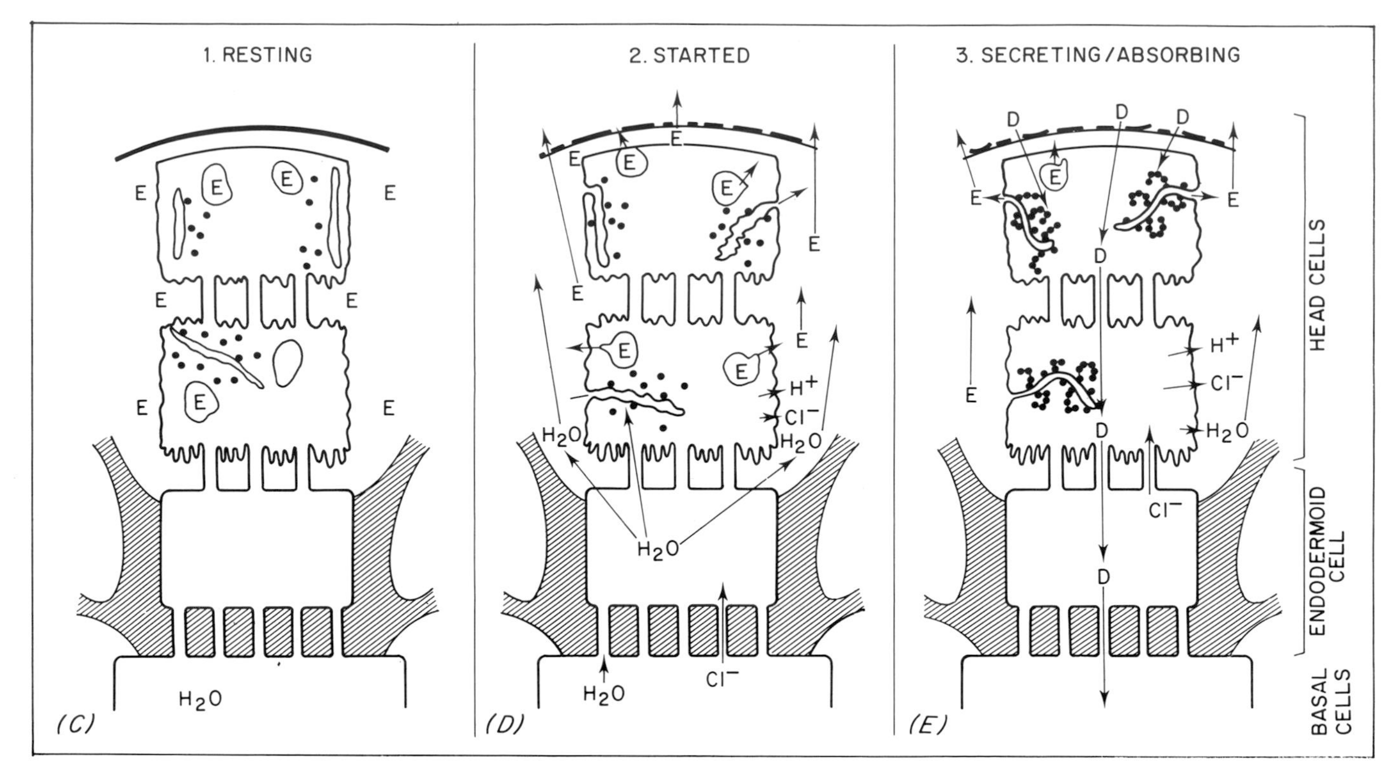

Fig. 8.24 ***C–E*** The stages of digestive activity in a gland from *Dionaea muscipula* showing a single file of cells, from top to bottom, the outer and inner head cells, the stalk cell (with surrounding endodermis) and the basal cell. (*C*) shows the unstimulated mature state, with hydrolytic enzymes (E) within vacuoles, distended profiles of the endoplasmic reticulum (er), some of which may be fused to the plasmalemma, and the cell wall. No cuticular gaps are present in the outer cutinised layer. Numerous free ribosomes (r) are present. (*D*) indicates the state shortly after stimulation has occurred. Cuticular gaps have formed. Chloride ions (Cl^-) and protons (H^+ are being rapidly excreted from the cells drawing both a flush of water and more ions from the leaf via the basal cell. This carries with it a flush of hydrolytic enzymes from the store in the cell walls which is replenished from stores in both the vacuoles and smooth endoplasmic reticulum. Some secretion via dictyosomal vesicles may also be involved at this stage. (*E*) indicates the steady state of a fully active digestive gland. *De novo* protein synthesis by polysomes (pr) maintains a supply of enzymes, many of these being released directly to the extracellular space. The products of digestion (D) are absorbed rapidly by the head cells and move through the gland in the symplast (see Ch. 10). Increasingly as the cycle comes to an end, small vesicles form intracellularly from the endoplasmic reticulum, either directly or via dictyosomal connections, and intracellular stores are reformed (not shown).

Secretion in the Cyclic *Dionaea* Gland

The mechanisms of fluid release

In *Dionaea* (Fig. 8.24*C*, *D* & *E*) it is clear that the active excretion of protons coupled with chloride ions similarly plays an important role in driving water from the digestive glands, as suggested by Robins (1978) and confirmed experimentally by Rea *et al.* (1983). However, the pattern of chloride ion accumulation and discharge in *Dionaea* differs considerably from that described for *Pinguicula*. The unstimulated *Dionaea* gland does not show any chloride ion accumulation but, as the result of stimulation two 'waves' of chloride ion secretion have been distinguished (Rea *et al.*, 1983; Fig. 8.25). The first 'wave' corresponds to the preparatory stage of the digestive cycle and is possibly involved in the formation of cuticular gaps (see page 178), whereas the second 'wave' corresponds to the phase of hydrolytic activity (Fig. 8.25). The endodermoid cells are loaded with chloride ions in the transition period between the two 'waves' and an internal gradient of chloride, consisting of a high chloride content in the endodermoid cells and of smaller amounts in the secretory cells, is formed in the digestive glands and maintained throughout the digestive activity. It seems that the solute pumps are localized in the plasma membrane of the head cells of *Dionaea*, excreting the chloride which is drawn symplastically from the basal cells through the endodermoid cells into the secretory head cells. Chloride ions have been shown to pass only through plasmodesmata between the endodermoid cells and the inner secretory cells (Fig. 10.7 and Robins, 1978; Robins and Juniper, 1980e).

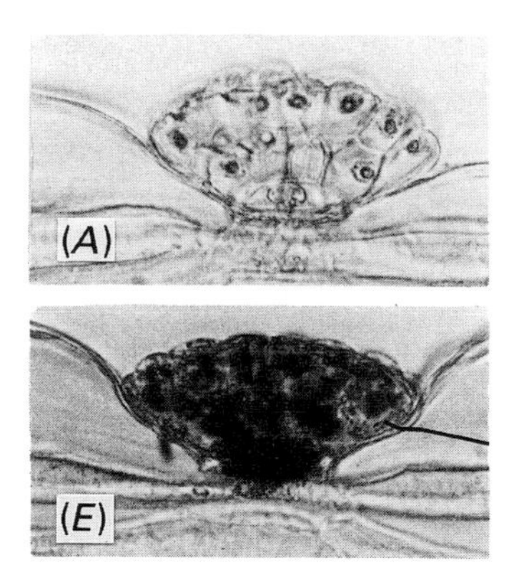

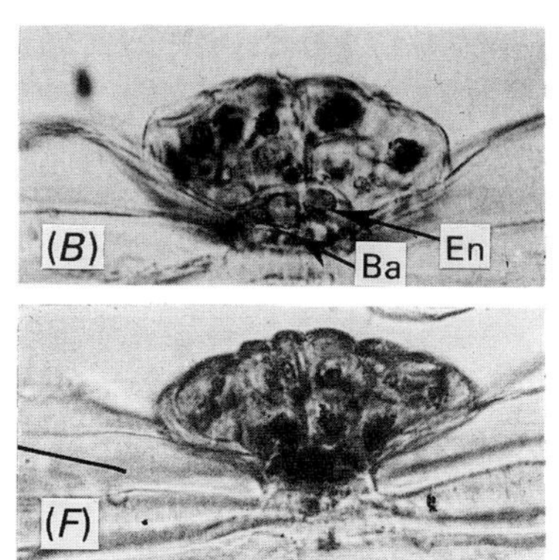

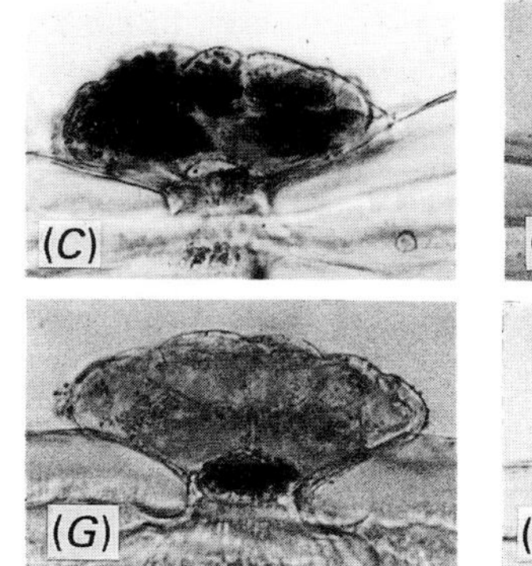

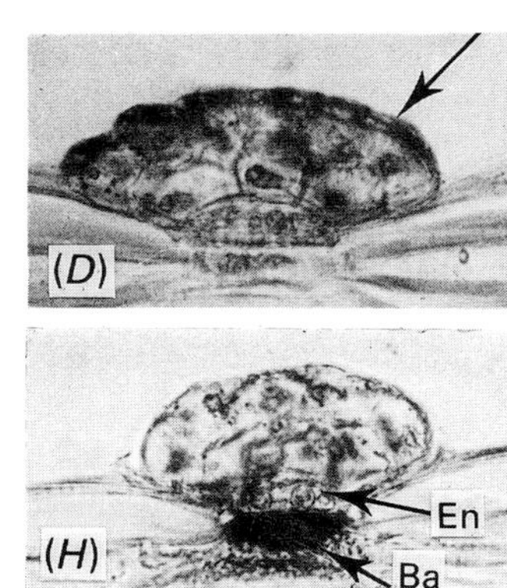

Fig. 8.25 The distribution of Cl^- in the digestive glands of *Dionaea* after different periods of continuous stimulation with 30 mM thiourea, as seen with the AgCl precipitation technique, showing two 'waves' of chloride release; preparatory (*B–D*) and secretory (*F–G*). (*A*) unstimulated gland, no chloride accumulation; (*B*) 15 min, Cl^- accumulating in the head cells; (C) 1.5 h, the staining reaction of the head cells approaches maximum. (*D*) 6.5 h, the head cells have lost most of their deposits, and at the same time chloride appears in the outer walls (arrow); (*E*) 10 h, the transition period between the preparatory and secretory phases; the endodermoid cells are heavily stained, together with the head cells; (*F*) 24 h, the secretory phase; while the secretory cells release some of their accumulated Cl^-, the endodermoid cells remain loaded. (*G*) Same as (*F*) but *in vitro* with minimized reabsorption of Cl^-, indicating that the Cl^- in the endodermoid cells is of endogenous origin. (*H*) 72 h followed by 10 days without stimulation; resting phase between successive digestive cycles; only the basal cells with heavy silver deposits, indicating chloride accumulation (After Rea *et al.*, 1983.) (*J*) The time courses of digestive fluid (O) and ^{36}Cl (O) secretion by the traps of intact plants and petiolar explants respectively. The volume of fluid secreted was estimated as the mean volume (mm^3 per trap) collected each day from 25 traps inoculated with 10 mol m^{-3} thiourea. The secretion of $^{36}Cl^-$ was estimated as the mean +/− s.d. of four experiments. (Reproduced by kind permission of Dr P.A. Rea from Rea *et al.*, 1983.)

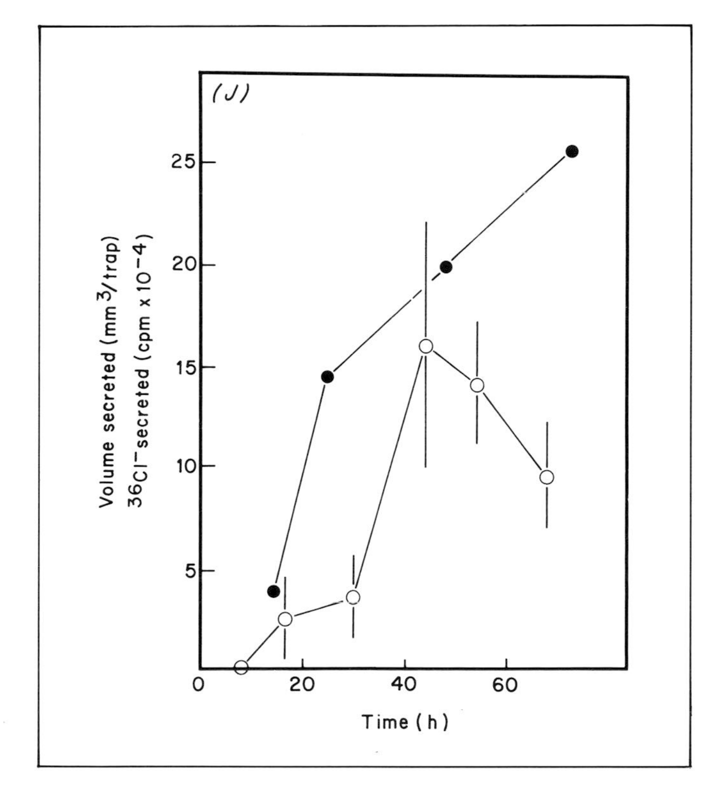

Table 8.4 Effect of small nitrogenous molecules on the release of cell products by the digestive glands of *Dionaea muscipula*

Stimulus given	Cumulative mean[a] protein secretion		Cumulative mean[a] fluid secretion			
	Maximum value attained (M_v) (µg)	M_p as %of control	Maximum value attained (M_v) (µl)	M_v as %of control	M_p/M_v ratio (R)	R as %of control
Control	89.6	100	73.5	100	1.22	100
Uric acid	56.1	63	78.5	107	0.72	51
Ammonia	38.6	44	88.8	121	0.44	36
Glutamine	18.5	20	68.7	94	0.24	20
Urea	8.0	9	19.6	27	0.41	34
Phenylalanine	—	—	11.5	16	—	—

[a] 'Cumulative mean' defined as the mean daily secretion per secreting lobe summated over the total experimental period. For experimental details see Fig. 8.26.
Reproduced from Juniper *et al.* (1977) with permission.

The distribution of labyrinthine walls in *Dionaea* is markedly different to that of simple salt-secreting systems, such as the *Frankenia grandifolia* salt glands (Campbell and Thomson, 1976), in which the labyrinthine walls are at the outer-most side of the gland. This difference is partly explained on page 172.

The secretion of chloride is linked to active proton extrusion (Rea, 1983a). This creates a standing osmotic gradient in the mainly labyrinthine apoplast of the secretory cells which is believed to drive water out of the cells at a steady rate. Secretion is linear over a number of days (Fig. 8.26; Robins and Juniper, 1980c), probably due to the limiting volume of the closed trap. The importance of impregnating the walls of the endodermoid layer is now clear. Without this apoplastic block, ions and water secreted from the gland head cells would just flow back into the leaf and neither a gradient nor a flush of fluid could be established. How counter-current absorptive flow takes place is discussed in Chapter 10 and Fig. 8.27.

The relationship of fluid and protein secretion

Although fluid and protein release are intimately linked, Robins (1976) has shown that in *Dionaea* the secretion of protein is not solely dependent on the quantity of water released by the digestive glands. When traps were stimulated with different nitrogenous secretogogues the ratio of fluid:protein varied considerably indicating that the volume of secretion and its protein content may be controlled by semi-independent mechanisms (Table 8.4). In addition, the secretion of a low-pH enzyme-free fluid was recorded when the trigger hairs were subjected to repeated mechanical irritation (Macfarlane, 1892; Lichtner and Williams, 1977). As this stimulus is known to generate electrical impulses (see Ch. 6), which are probably propagated by the movement of ions across membranes (Ijiima and Sibaoka, 1985; Hodick and Sievers, 1986), it looks likely that fluid release alone can be induced by stimulating ionic movements. Furthermore, the walls of the secretory cells of many carnivorous

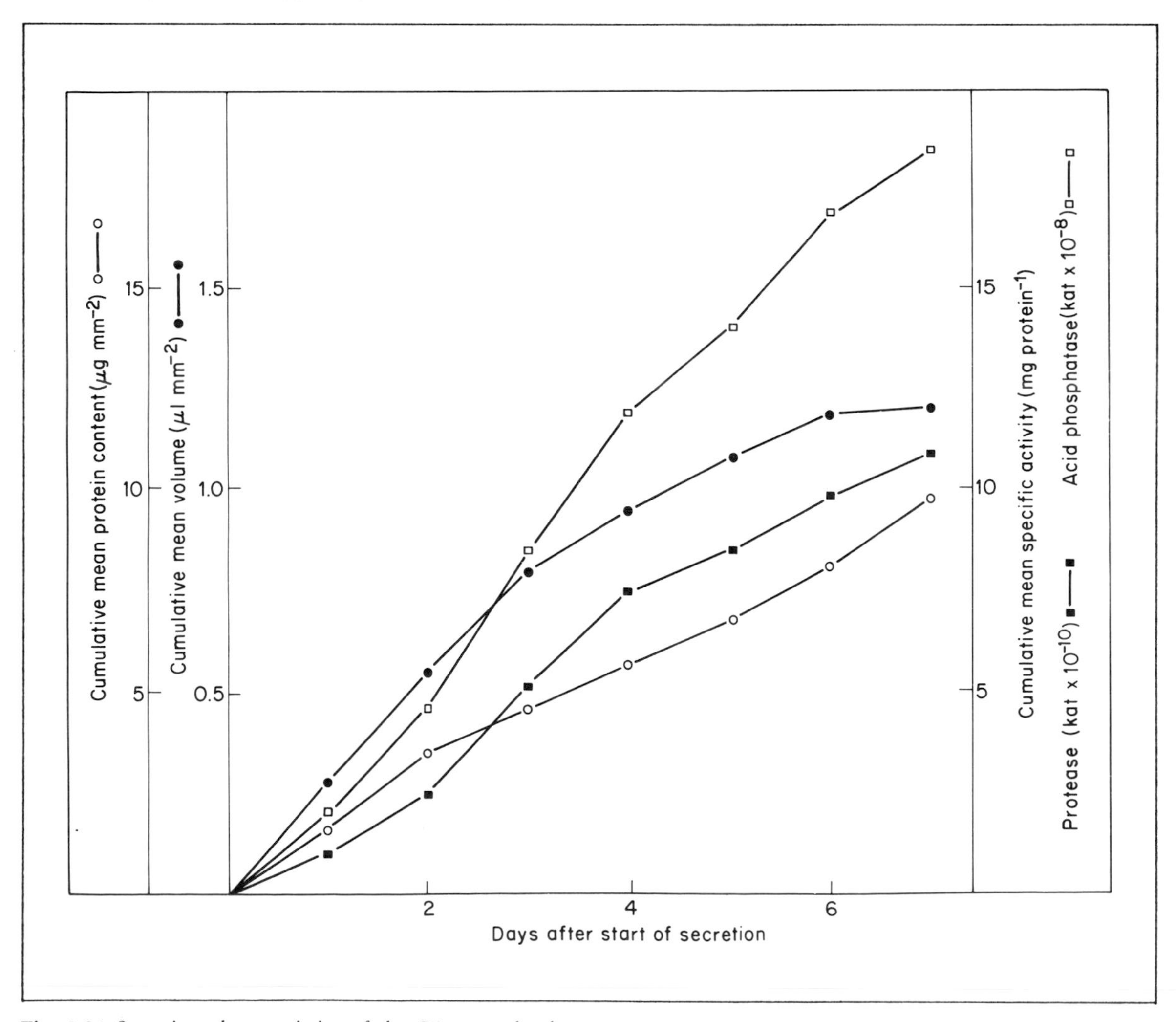

Fig. 8.26 Secretion characteristics of the *Dionaea* gland.

plants are progressively loaded with enzymes during the last phase of the development of the gland without any major release of water.

Recycling in the *Dionaea* gland

The ability of the glands of *Dionaea* to function several times makes this a particularly complex secretory system. As already discussed (see page 174), enzyme stores are established during maturation so that in the unstimulated gland hydrolases are present in vacuoles and, to some extent, in the walls but in addition are found in sER profiles. Furthermore, in both unstimulated and stimulated glands, profiles of rER are seen fused to the plasmalemma (Figs 8.8 and 8.15) sufficiently frequently to conclude that this is an important means of discharge. By following, autoradiographically, the redistribution of ^{3}H-leucine-labelled protein after stimulation, Robins and Juniper (1980c) were able to plot the probable sequence of events in the active gland (Fig. 8.24, 8.26 and 8.27). Within two days almost all the label is lost from ER and small osmiophilic vacuoles, while simultaneously the grain density over the apoplast rises. After two days the level in both ER and small vacuoles steadily increases, suggesting that *de novo* synthesis is replenishing these stores. EM images taken at this stage show that numerous polysomes have formed on the ER membranes, supporting the autoradiographic results. After 6 days the grain-count over the large vacuoles also decreases, suggesting that further reserves of hydrolases or stored amino acids (as proteins) are being mobilized. Robins and Juniper (1980b, c) found that the dictyosomes did not appear involved at all in this process, showing neither any stimulus-induced ultrastructural changes nor any significant redistribution of silver grains. Henry and Steer (1985), however, considered there to be stimulus-induced changes in dictyosomes early in the cycle, but unfortunately, did not examine these later stages in which they appear to play no role.

Secretion in the *Cephalotus* Glands

In the trap of *Cephalotus* (Fig. 4.9*B*), the secretion of fluid and protein are not only under separate control mechanisms, but take place in different secretory structures. While the small uni- and bicellular glands both showed a positive histochemical reaction for various hydrolytic enzymes (Table 9.1) in their outer thickened walls, vacuoles and ER, it was impossible to demonstrate any hydrolase in the walls and cytoplasm of the large digestive glands

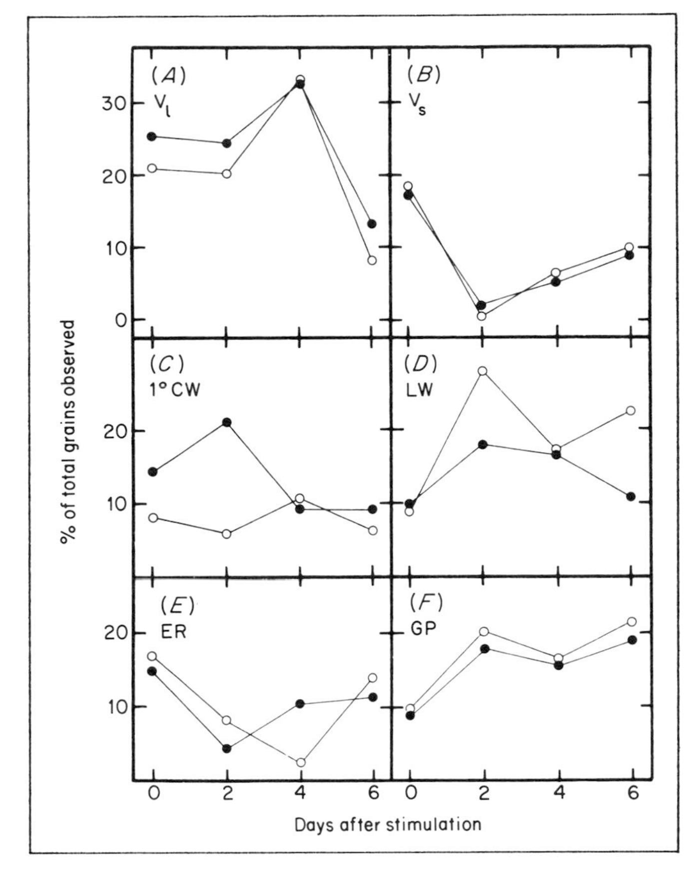

Fig. 8.27 The dynamic redistribution of radio-label as indicated by silver grains in electron micrograph autoradiographs within the different sub-cellular compartments of the secretory cells following stimulation with 4% (w/v) bactopeptone solution: (*A*) large osmophilic vacuole; (*B*) small osmophilic vacuole; (*C*) primary cell wall; (*D*) labyrinthine cell wall; (*E*) endoplasmic reticulum; (*F*) ground plasm; ●, outer secretory cells; ○, inner secretory cells. (Redrawn from Robins and Juniper, 1980c.)

(Parkes, 1980). On the other hand, acid phosphatase was shown to be active in the vicinity of the plasma membrane and in the vacuoles of the secretory cells in the large glands (Parkes, 1980), showing a distribution similar to that of certain water-, nectar- and salt-secreting glands. Using the substrate-film technique, D.M.J. (unpublished) found that only the small glands secrete proteolytic activity, the large glands apparently secreting a non-proteolytic fluid. It appears, therefore, that *Cephalotus* has developed two distinct secretory systems, one located in the large glands and responsible for the release of fluid, the other in the small glands and responsible for the supply of hydrolytic enzymes. In the other genera, however, economy of effort has meant that both processes occur in a single glandular type. Possibly, therefore, *Cephalotus* represents an intermediate stage in the development of pitchers, falling between *Heliamphora* where no hydrolytic activity is known and the well-developed secretory activity of *Nepenthes*.

Secretion in Other Glands and the General Role of Chloride

Accumulation of chloride ions is found in the digestive glands of *Drosophyllum*, *Drosera*, *Utricularia*, *Nepenthes* and *Sarracenia* (Joel, unpublished) and chloride secretion seems to be a widespread phenomenon in the carnivorous plants. Chloride ions in the digestive fluid were first believed to be necessary for digestion (Morissey, 1955), as when protons are the counter ion, the digestive fluid becomes acid. While this is still true, studies of both *Pinguicula* and *Dionaea* clearly show that chloride is also playing a key role in driving fluid out of the digestive glands (see pages 183 and 184), achieving two important results by a single process. Probably, as carnivorous plants evolved they adapted the ubiquitous ion pumps to provide a water flow and, as this provides an acidic environment, this pressure caused the selection of hydrolases of suitable pH optimum for maximal activity at the relevant pH (see Ch. 9).

Although in *Pinguicula* a fluid flush appears sufficient to discharge the secretory cells of their enzymes, it is not clear whether the enzymes in the labyrinthine wall are free or bound to cell wall polysaccharides. If, as seems likely, the wall enzymes in carnivorous plants are bound to the wall matrix, then protons, which are typical of almost all digestive fluids of carnivorous plants, might serve as the key factor releasing the wall enzymes in a manner suggested by Rayle and Cleland (1978) for enzymes responsible for the extension of cell walls. Enzyme release should then be dependent on the activity of the ion pumps responsible for the release of water. Indeed a clear correlation was found between the ion content of the pitcher fluid in *Nepenthes* and its enzymatic activity (Lüttge, 1966b). However, these results are not sufficient to support the above assumption and further experimental work must be conducted before final conclusions can be drawn.

Unfortunately, no study of the mechanism of discharge of digestive fluid in a continuously producing *Nepenthes* gland is yet available to be compared with these. It might be expected that many similarities will be seen with the active phase of *Dionaea*, there being little need for major stores of hydrolases to be established. In fact, however, the acid phosphatase content of glands in *N. rafflesiana* varies considerably (Heslop-Harrison, 1976), suggesting that either the glands go through waves of activity or, less likely, that they also perform as a one-off process. *Drosera* and *Drosophyllum* mature glands also show some degree of disintegration, possibly indicative of a similar mode of action. Whether *Dionaea* is in fact unique – other species tending to act more like *Pinguicula* – still remains an intriguing question. An investigation of these aspects in *Aldrovanda* will be particularly valuable in this context.

The General Control of Secretion: Conclusions

What controls the stimulus-induced fusion of secretory ER vesicles or vacuoles to the plasmalemma in *Dionaea* remains a mystery. Although the drugs cytochalasin B and colchicine both decrease the levels of enzyme activity in the secretory fluid, they do not decrease the overall protein content (Robins and Juniper, 1980c). Hence, it appears that neither microfilaments nor microtubules are involved, despite their commonly being associated with such events. The lack of involvement of these sub-cellular skeletal structures is perhaps not surprising as most of the secretory stores are closely juxtaposed to the plasmalemma and, at least in the initial event, no long-range transport is required. Cycloheximide has a similar effect. It looks like all three drugs are disrupting complete synthesis of active enzymes, causing incomplete or inactivated proteins to be secreted. In addition these drugs are known to decrease the ease with which membranes fuse, which may lead to some autolysis of the enzymes between stimulation and secretion.

Similarly, nothing is known of the mechanism by which cells respond to the presence of prey and the control of the proton pump and hydrolase release activities. How, for example, are pre-loaded *Pinguicula* glands triggered? Only those glands that come into contact with the pool of secretion discharge their contents, so the stimulus is directly perceived by the gland. Do *Nepenthes* respond to prey or at least monitor the status of the pitcher fluid? The long periods over which they are active favours their doing so, yet no evidence is available on this either. Much remains to be done.

CHAPTER 9

The Hydrolytic Environment and Processes of Digestion

Introduction

Is the enzymic activity involved in digesting captured prey of intrinsic or extrinsic origin? The controversy has continued for many years (Lloyd, 1942). As with nutrition (see Ch. 7), there is no comprehensive answer to this question for all the carnivorous plants.

Only recently has it been demonstrated definitively that the hydrolytic activity present in the digestive fluid of *Dionaea muscipula* is produced by the plants (Robins and Juniper, 1980b). Cytochemical evidence, particularly from *Pinguicula* (Heslop-Harrison and Knox, 1971; Heslop-Harrison and Heslop-Harrison, 1981), *Utricularia* (Vintéjoux, 1974) and *Drosera* (Palczewska, 1966), showed various sub-cellular enzyme activities present in the gland cells of activated traps (Table 9.1). These enzyme indicators are lost on activation but, as the stains used are not specific for enzymes identified in the secretion, these experiments are still not unequivocal.

Hydrolytic activity, particularly against proteins, has been established for the traps of all carnivorous plants (Tables 9.2 and 9.3). But, although decay of captured prey occurs, it has been difficult to assess the relative importance to digestion of enzymes of plant origin and those that may be produced by bacteria present in the traps. No controlled analysis of secretion produced under sterile conditions has been done except for the Sarraceniaceae (Hepburn *et al.*, 1927) and *Nepenthes* (Lüttge, 1964b) from which sterile secretion may be collected before the pitchers are fully mature and open to the atmosphere. Once open, however, the continuously digesting 'open' systems of *Nepenthes*, *Sarracenia*, *Cephalotus* and *Heliamphora* contain a bacterial flora which,

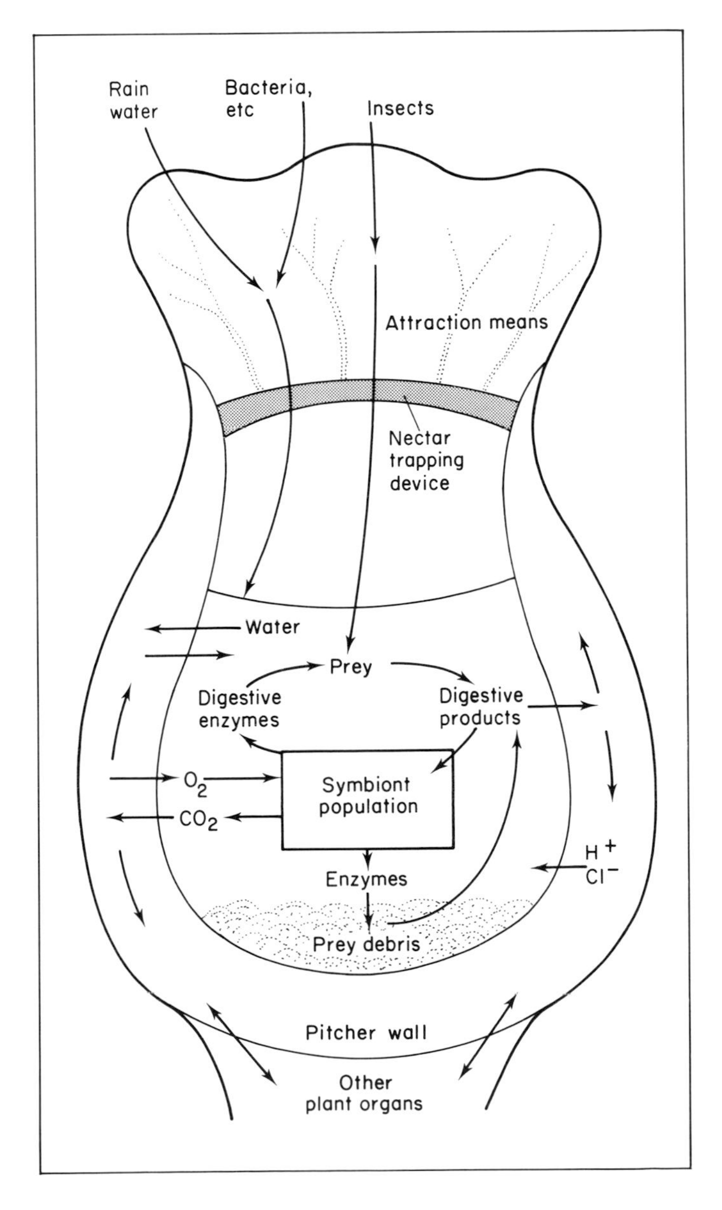

Fig. 9.1 The micro-environment of a generalized 'pitcher'.

Table 9.1 Enzymes localized cytochemically in the digestive glands.

Species	Gland type	Enzyme															References
		esterase	acid phosphatase	β-glycerophosphatase (neutral)	glucose-6-phosphatase	adenosine triphosphatase	adenosine diphosphatase	adenosine monophosphatase	inosine diphosphatase	thiamine pyrophosphatase	amylase	protease	leucine aminopeptidase	carbonic anhydrase	peroxidase	ribonuclease	
Cephalotaceae																	
Cephalotus follicularis	large	−	+		−	−						−	−				1,16
	small	+	−		+	+						+	+				1,16
Dioncophyllaceae																	
Triphyophyllum peltatum	stalked	++	+									++					14
Droseraceae																	
Dionaea muscipula		+	+	+		+						++	+				1,2,3,4,8
Drosera capensis	stalked	++	+	+	+	+	+	+	+	+		+			+		1,4,5,6,7
rotundifolia	stalked	+	++														9,17
Drosophyllum lusitanicum	stalked	+	+									++	+		++		1,4
	sessile	+	+									+	+				1
Lentibulariaceae																	
Genlisea africana		+	+														4
Pinguicula caudata	stalked	+	+								+	+				+	4
	sessile	+	+		+								+				1,10
grandiflora	stalked	+	+								+	+			−	+	4,10
	sessile	++	++								−	++			−	++	4,10
ionantha	sessile	+	+														13
lusitanica	stalked	+	+								+	+				+	10
	sessile	+	+								+	+				+	10
moranensis	sessile	+	+														13
vulgaris	stalked	+	+								+	+				+	10
	sessile	+	+								+	+				+	10
Urticularia aurea		+	+									+					1
neglecta	quadrifid		+									+					11,12
sp.		+	+														4
Nepenthaceae																	
Nepenthes khasiana		+	+		+	+							+				1
rafflesiana		+	++												+		4
rufescens			+														1
maxima superba		+	+		+							+	+			+	1
sp.														+			15
Sarraceniaceae																	
Sarracenia purpurea		+	+									+	+				1

Enzyme activities were detected variously by histochemical or substrate-film methods. + = detected; ++ = strongly detected; — = absent; space = not tested.
References: 1, Parkes (1980); 2, Robins and Juniper (1980b); 3, Henry and Steer (1985); 4, Heslop-Harrison (1975); 5, Dexheimer (1978a); 6, Dexheimer (1978b); 7, Dexheimer (1979); 8, Robins (1978); 9, Palczewska (1966); 10, Heslop-Harrison and Knox (1971); 11, Vintéjoux (1973c); 12, Vintéjoux (1974); 13, Heslop-Harrison and Heslop-Harrison (1981); 14, Green *et al.* (1979); 15, Morrissey (1964); 16, Joel (unpublished); 17, McEwan *et al.* (1988).

Table 9.2 Authorities providing evidence for protease secretion in carnivorous plants, 1875–1974. Analyses based on gland secretion or whole leaf extracts, usually using standard biochemical techniques.

Genus	Authorities
Nepenthes	Hooker, 1875; Tait, 1875; Gorup and Will, 1876; Vines, 1877, 1897, 1898, 1901; Goebel, 1891, 1893; Clautriau, 1900; Abderhalden and Terruchi, 1906; Hepburn, 1918; Hepburn *et al.*, 1919; Stern and Stern, 1932; Zeeuw, 1934; Kandler and Schmiderer, 1952; Lüttge, 1964; Steckelburg *et al.*, 1967; Nakayama and Amagase, 1968; Jentsch, 1970, 1972; Amagase, 1972; Amagase *et al.*, 1969, 1972.
Sarracenia	Mellichamp, 1875; Zipperer, 1885, 1887; Robinson, 1908; Hepburn *et al.*, 1920, 1927.
Darlingtonia	Hepburn *et al.*, 1920.
Cephalotus	Dakin, 1919.
Drosera	Darwin, 1875; Rees and Will, 1875; Tate, 1875; Robinson, 1909; White, 1911; Dernby, 1917a; Okahara, 1930; Holter and Linderstrøm-Lang, 1933; Whitaker, 1946, 1949; Fabian-Galan and Sălăgeanu, 1968; Amagase, 1972; Amagase *et al.*, 1972; Takahashi *et al.*, 1974.
Drosophyllum	Darwin, 1975; Goebel, 1891–93; Meyer and Dewévre, 1894; Quintanilha, 1926.
Dionaea	Darwin, 1875; Lüttge, 1964b; Scala *et al.*, 1969.
Aldrovanda	Fermi and Buscaglione, 1899.
Pinguicula	Darwin, 1875; Goebel, 1891–93; Dernby, 1917b; Heslop-Harrison and Knox, 1971.
Utricularia	Luetzelburg, 1910; Adowa, 1924; Hada, 1930.
Byblis	Bruce, 1905.

(From Heslop-Harrison, 1976 with permission.)

Table 9.3 Enzymes (other than proteases) detected in the trapping organs of carnivorous plants, 1875–1974.

	Enzymes detected (with authority)								
	Transferases	Hydrolases							
Genus	Ribonuclease EC2.7.7.16	Lipase EC3.1.1.3	Esterase EC3.1.1(1,2,4)	Phosphatase (non-specific) EC3.1.3(1,2)	Amylase EC3.2.1.1	Maltase (α-glucosidase)	Invertase (β-fructo-furanosidase) EC3.2.1.36	Urease EC3.5.1.5	Chitinase EC3.2.1.14
Nepenthes	+(M)	−(R)	−(R)	+(M)	−(G, V, C, R)				+?(A)
Sarracenia		+?(H)	−(H,R)		+(R, Z) −(H)	−(H)	+(R, H)	−(H)	
Darlingtonia		−(H)			+?(H)	−(H)	+?(H)	−(H)	
Cephalotus					−(Dk)				
Drosera		−(D)			−(D)				+?(A)–(D)
Dionaea				+(S)					
Drosophyllum					−(MD)				
Pinguicula	+(HH)		+(HH)	+(HH)	+(HH)				

+, indicates enzyme found to be present; −, indicates enzyme tested and found to be absent; +?, indicates doubtful presence, or possible contamination by micro-organisms. Data based mainly on detection in whole leaf extracts or pitcher fluid using standard biochemical techniques. Letters in brackets indicate authorities: A (Amagase, 1972); C (Clautriau, 1900); D (Darwin, 1875); Dk (Dakin, 1918); G (Gorup and Will, 1876); H (Hepburn *et al.*, 1927); HH (Heslop-Harrison and Knox, 1971); M (Matthews, 1960); MD (Meyer and Dewèvre, 1894); S (Scala *et al.*, 1969); R (Robinson, 1908); V (Vines, 1877, 1897, 1898, 1901); Z (Zipperer, 1885).
(Reproduced from Heslop-Harrison, 1976, with permission.)

potentially, may contribute greatly to degradation (Fig. 9.1). Indeed, in *Byblis*, *Heliamphora* and in some *Sarracenia* species it is doubtful whether any enzymes are secreted by the plant at all (Hepburn *et al.*, 1927). In the genera stimulated by the presence of prey, e.g *Dionaea*, *Drosera*, *Drosophyllum* and *Pinguicula*, which provide no hydrolytic environment until prey is secured, microbial contamination is likely to be much less important. A number of genera can now be propagated under sterile conditions (Pringsheim and Pringsheim, 1962; Small and Hendrikz, 1974; Chandler and Anderson,

1976b; Beebe, 1980; Minocha, 1985) and therefore the systems in which to perform these important studies are now available. Unfortunately, very little enzymology using them has yet been done (see pages 197 onwards).

Before discussing the hydrolytic activity of the secretion, the hydrolytic environment must be considered. This is followed by an analysis of the enzymes of the fluid and other enzymes isolated from carnivorous plants which play a role in the assimilation of nutrients but have not been found in the secretion. Finally, the role of exogenous factors in digestion is to be discussed.

The Hydrolytic Environment

Pitchers (Continuous Digestion)

In all carnivorous plants, digestion occurs in a wet acidic solution produced by the surfaces of the trap. Although Linnaeus believed that the fluid within *Nepenthes* pitchers was rain-water (he never accepted the carnivorous nature of these genera), as early as 1832/33 Treviranus described organs in *N. distillatoria* which he considered secretory and in 1849 Voelcker pointed out that unopened pitchers contain a clear, colourless acid solution which becomes yellow and cloudy in open pitchers. He analysed the fluid of unopened pitchers of an unidentified *Nepenthes* and found K^+, Na^+, Ca^{++}, Mg^{++} and Cl^- present in a dry matter of 0.5–1% of the fluid. Of this, about 25% was lost on combustion and 50% was KCl. In contrast to Turner's findings (see Voelcker, 1849) he was unable to detect oxalic acid and in addition showed the absence of tartaric, acetic, formic, racemic, sulphuric and phosphoric acids in both open and closed pitchers. He did, however, detect malic and citric acids as about 39% of the dry matter of unopened pitchers.

The ionic content and pH of the pitcher fluid of *Nepenthes*

The ionic composition found in closed pitchers of *N. henryana* Hort. confirms Voelcker's work. Nemček *et al.* (1966) showed there to be 19.5 mM Cl^-, 9.7 mM K^+, 0.7 mM Na^+, 4.8 mM Ca^{++} and 1.0 mM Mg^{++} present. Unfortunately, they did not test for organic anions. Morrissey (1955) did, however, and like Voelcker (1849), failed to detect any of the common organic acids in fluid of freshly opened pitchers from 15 species of *Nepenthes*. She showed this fluid to contain 17–25 mM Cl^- (see

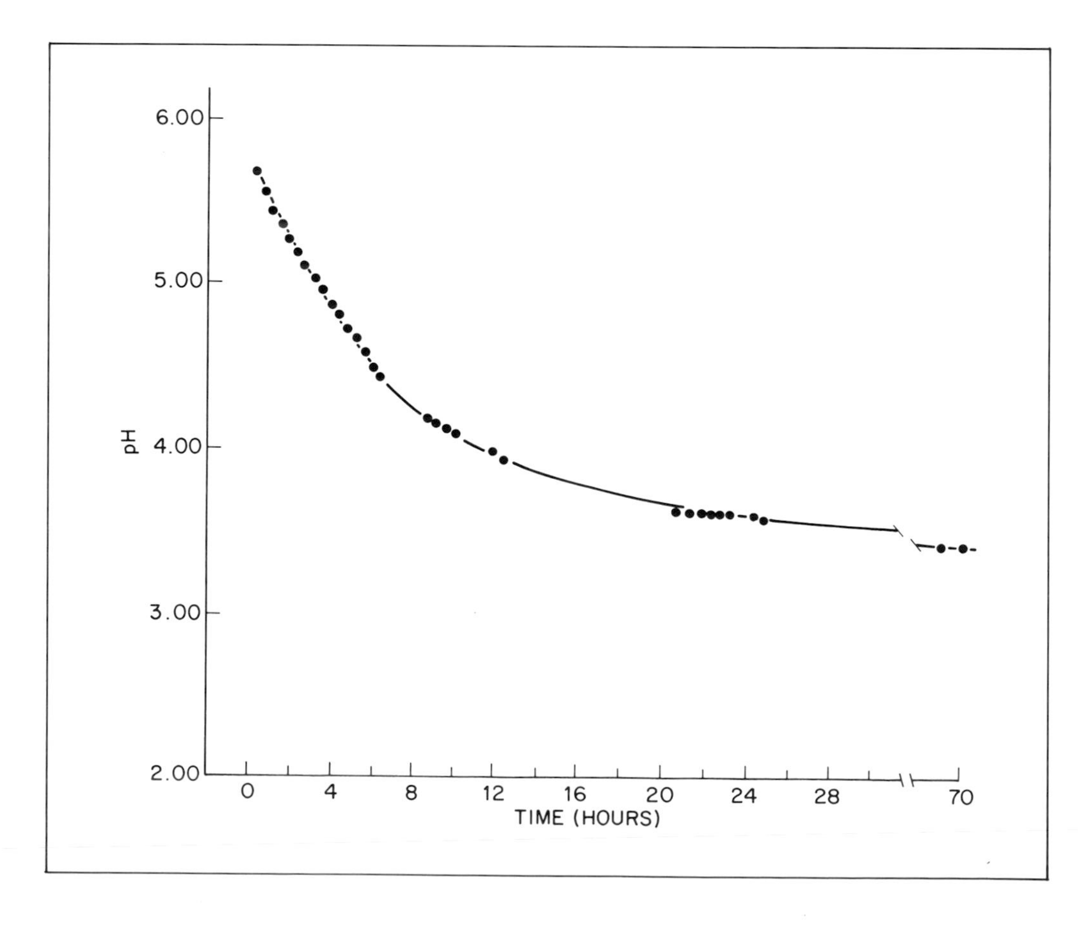

Fig. 9.2 The effect of adding 0.5 ml of sterile bactopeptone to an unopened pitcher of *Nepenthes* × *benecti*. The total volume of fluid within the pitcher was measured at the end of the experiment and found to be 1.8 ml. The pitcher was kept attached to the parent plant in the Botanic Garden, Oxford during the whole of the experiment. (B.E.J. unpubl., 1981.)

below) and claimed unopened pitchers to contain 'very little' else, in contrast also to the findings of Nemček *et al.* (1966).

Morrissey (1955) showed the pH of pitchers to decrease following opening from 5.5 (range 4.2–6.6) to pH 3.3 (range 2.3–4.6). This shift is rapid and can be induced with nitrogenous solutions (Fig. 9.2). *N. gracilis* and *N. albomarginata* show similar values (Tökés *et al.*, 1974), while the fluid in *N. mixta* Hort. (Nakayama and Amagase, 1968) and *N. chelsonii* Hort. (Heslop-Harrison, 1975) show greater changes, from pH 7.0 to 2.5 on opening. *N. macfarlaneii* seems unusual in having a low pH (3.2) in unopened pitchers, which only decreases 0.8–1.0 pH units on opening (Tökés *et al.*, 1974).

The mechanism of acidification of *Nepenthes*

Morrissey (1955) suggested that H^+ was excreted accompanied by Cl^-, resulting in a large increase in Cl^- concentration in open pitchers. Lüttge (1966a) showed that Cl^- secretion into the pitcher lumen is an active process dependent on metabolic activity, being inhibited by arsenate and cyanide. Secretion will take place against a high concentration gradient even though *in vivo* the cellular and fluid concentrations are similar. Lüttge also demonstrated that Cl^- transport takes place in intact pitchers (Lüttge, 1966a) and the glands are the site of Cl^- flux, though Cl^- does not accumulate in the gland cells (Lüttge, 1966b, 1967). The uptake of Cl^- at the glands, however, appears to be passive (Lüttge, 1964a, 1967) (but see page 211).

In an attempt to elucidate the mechanism of ion secretion in *Nepenthes*, Nemček *et al.* (1966) measured the membrane potentials of *N. henryana* pitcher cells using micro-electrodes and compared them to the calculated equilibrium potentials at the pitcher–liquid interphase according to the Nernst equation. Rather surprisingly, they found that about 0.6 mM of the Na^+ in the pitcher was firmly bound to non-diffusible material as compared to 0.4 mM K^+, 1.5 mM Cl^- and negligible amounts of Ca^{++} and Mg^{++}. They corrected for this in their calculations.

Repeated measurements in ten plants gave a mean membrane potential of 35 mV, the digestive fluid being positive with respect to the epidermal cells. This is close to the calculated value for K^+ but differs greatly from those for Na^+ (132.6 mV) and Cl^- (−80.3 mV), from which these authors suggest that independent active pumps for Na^+ and Cl^- are operating on the inner surface of the pitcher, driving these ions from the lumen into the cells, the K^+ flux being passive. The potential differences and short-circuit currents across the walls of the pitcher when the lumen wall and external wall were bathed in various media were also examined. With Ringer solution in both chambers, a secretory potential of about 50 mV and a short-circuit clamp current of 15 mA cm^{-2} were established, the inner surface of the pitcher being positive. Both of these were rapidly dissipated by metabolic inhibitors such as 2,4-dinitrophenol (0.4 mM), $HgCl_2$ (0.4 mM), NaF (10 mM) or $CuSO_4$ (0.4 mM), indicating a role of active transport in maintaining the ionic gradients. With Na^+ or Cl^- solutions the short-circuit current was interpreted as compatible with net ion transport against the electrochemical gradient.

The close correlation of the Donnan equilibrium and the measured membrane potential for K^+ implies that the distribution of K^+ depends on a passive equilibrium. But, in the presence of K^+, the short-circuit current and membrane potential are reversed, indicating an active transport of K^+ into the pitcher lumen. Nemček *et al.* (1966) conclude that either K^+ must be actively transported into the pitcher fluid by a pump of low capacity at the inner lumen surface and that this membrane has a high passive permeability to K^+, or that the K^+ pump is distinct from the inner surface of the pitcher. While they are unable to distinguish between these possibilities, Lüttge (1971) points out that if the K^+ pump were distal, loading K^+ into the mesophyll cells, this would be identical to the system in *Atriplex* salt glands. Unfortunately, however, Nemček *et al.* (1966) seem to have misinterpreted their data. Hill and Hill (1976) point out that for Na^+, for example, a positive secretory potential must be linked to a negative clamp current and that in fact Na^+ and Cl^- are therefore being pumped into the lumen while K^+ is being absorbed. Lüttge (1971), however, suggests that, if the K^+ pump is electrogenic, passive Cl^- fluxes can be associated with it. This model is compatible with his data that Cl^- influx is sensitive to metabolic inhibitors (Lüttge 1964a, 1966a). Overall, because of the operation of a number of different pumps at different sites within the tissue, it would appear that Nernst-analysis is insufficient to solve this problem (Lüttge, 1971).

Since the period of investigation in the 1960s, a number of metabolic effectors of ion pumps have been identified. Lüttge has more recently returned to this problem, probing the system with FC, ABA and 4PA. Using discs of *N. rafflesiana* (*hookeriana*), Jung and Lüttge (1980) showed FC significantly to inhibit Cl^- excretion after 20 h at levels above 1 μM while excretion and uptake were much less affected. Conversely, ABA increased Cl^- excretion, again with little effect on accumulation or uptake. K^+ transport in both directions was stimulated by ABA. 4PA, after 15 h, inhibited Cl^- efflux by about 80% at 2 mM while accumulation was less markedly affected (Jung *et al.*, 1980).

What, then, can be deduced from these results? The 4PA treatment, which causes aberrant protein synthesis, indicates an important role for proteins in the Cl^- transport process and that these proteins turn over quite rapidly. The differential effect on excretion and uptake may indicate the presence of separate pumps for these two processes. Are H^+-pumps involved? The stimulation of K^+ excretion by ABA is compatible with this but, if Cl^- were following H^+, FC-stimulation and ABA-inhibition of H^+ pumping should increase and decrease Cl^- efflux respectively, the opposite to these observations. It is unlikely that these effectors are acting directly on a Cl^- pump, leading Jung *et al.* (1980) to propose that an anion-OH^- counter-transport operates at the plasmalemma of the gland cells (Fig. 9.3). Thus, Cl^- actively pumped out of the cells is transported back in exchange for OH^- ions, the movement of OH^-/H^+ being affected by FC. K^+ transport inversely follows H^+ transport. Thus, the metabolic inhibition of Cl^- excretion observed earlier is also explained.

Nemček *et al.* (1966) only examined the situation in unopened pitchers which they found to be pH 7.0 ± 0.1. The shift in pH on opening will result from a stimulation of Cl^- excretion in Lüttge's model (Fig. 9.3). Morrissey (1960, 1963) has shown this shift to be due to a plant-based trigger and not simply to exposure to air. She extracted a compound from the gland-bearing part of the pitcher which will mimic acteyl-choline and histamine in bio-assays. The fall in pH can also be achieved by injecting histamine into unopened pitchers at concentrations as low as 10 ng/ml. This phenomenon needs further investigation but it is likely that the shift of pH is again brought about by altered activity of the ion-pumps in the lumen wall. The demonstration by Morrissey (1964) that carbonic anhydrase activity may be extracted from the glandular region of the walls shows that controlled H^+ production can take place, as in mammalian gastric acid secretion. The enzyme was histochemically localized using cobalt sulphide deposition, to the gland cells and could not be detected in non-glandular regions (Morrissey, 1964). Heslop-Harrison (1975) showed that acid phosphatase and esterase can only be detected in the lumen walls from the onset of the drop in pH. It is tempting to suggest that the secretion of these enzymes is linked to the same trigger, particularly as Lüttge (1966b) showed that the proteolytic activity of pitchers was correlated with their Cl^- content (and hence pH).

Other pitchers

No such detailed analysis of the control of the digestive environment exists for any other pitchers.

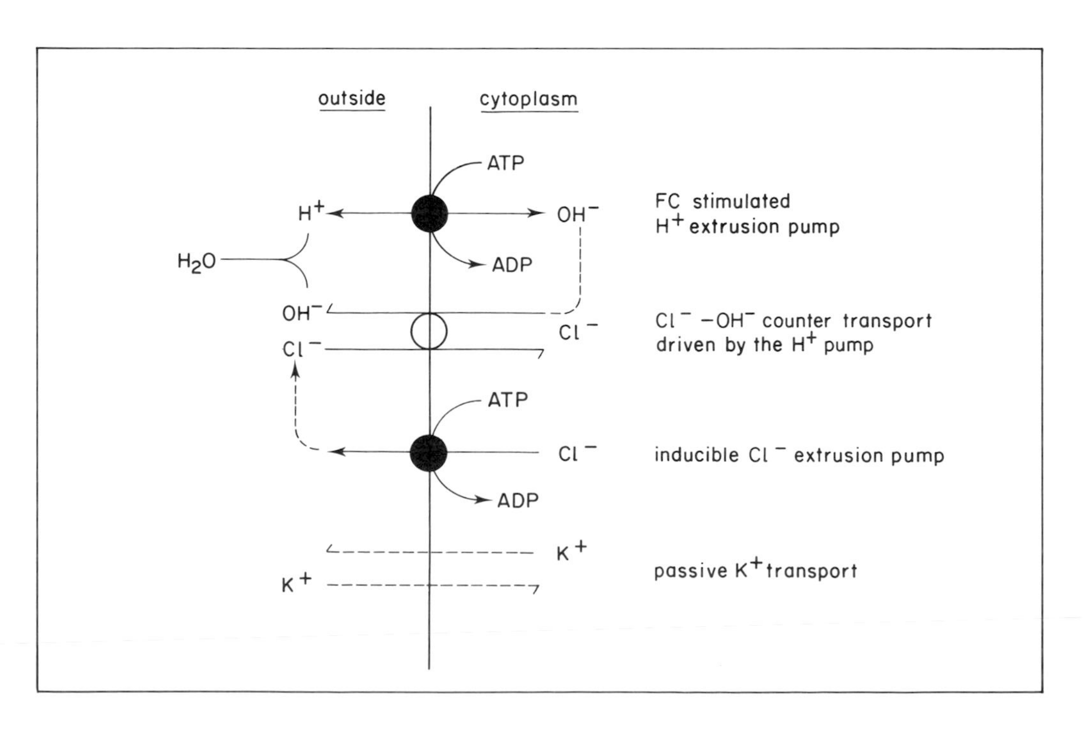

Fig. 9.3 Hypothetical scheme to explain the inhibition by fusicoccin (FC) of Cl^- excretion by *Limonium* glands. (Redrawn and adapted, with permission, from Jung and Lüttge, 1980.)

The pH of the pitchers of numerous *Sarracenia* plants has been assayed and this is summarized in Table 9.4. Wherry (1929) concluded that, apart perhaps from *S. minor* and *S. flava*, this genus could not control the pH within the pitchers, citing the general shift towards neutral or alkaline conditions in older traps. This shift towards neutrality takes place, however, in fluid isolated from young traps without any prey (Plummer and Jackson, 1963) yet acid conditions are established and maintained over long periods (90 days) in fluid within intact traps (Fish and Hall, 1978). These authors show that the pH of *S. purpurea* pitchers rose slightly on opening but after about 20 days decreased, attaining a steady pH of about 3.8 after 40 days (Fig. 7.4). This profile explains the variability seen in the results of 'single' observations by Wherry (1929) and indicates that *Sarracenia* may be able to control the pH of its pitchers, in contrast to Wherry's conclusion (Wherry, 1929). The mechanism awaits elucidation.

Table 9.4 The ranges of pH determined in pitchers of *Sarracenia* species, data pooled from various authors, localities and ages of pitcher.

Sarracenia species	Open/closed	pH (2 3 4 5 6 7 8 9)	Reference
S. alata [*sledgeii*]	o&c	—	1
	o&c	—	2
S. flava	c	—	1
	o	—	1
	o	—	2,4
S. flava × *purpurea* [*tolliana* Hort.]	o	—	3
S. leucophylla	c	—	1
[*drummondii*]	o	—	1
	o&c	—	2
S. minor	o&c	—	1
S. psittacina[a]	o	—	1
S. purpurea	o&c	—	1
S. purpurea × Hort.			
young	o	—	3
old	o	—	3
S. rubra[a]	o	—	1
S. rubra ssp. *jonesii* [*jonesii*]	o&c	—	2

[a] Pitchers normally dry; pH as determined after adding distilled water.
References: 1, Wherry (1929); 2, Hepburn *et al.* (1927); 3, Atkins (1922); 4, Plummer and Jackson (1963).
(Reproduced from Heslop-Harrison, 1976, with permission.)

Utricularia

Digestion and the assimilation of prey occurs in the traps of *Utricularia*. During the phase immediately following the trapping of prey, water is rapidly removed from the traps (see Ch. 6) but little is known of the environment within the trap following this process. While proteases are probably released by the glands (Vintéjoux, 1973b, c; Fineran, 1985) it is not known whether the plant controls the pH within the trap. In view of evidence from pitchers (see page 193) and other systems (see page 196), it is likely that it does but confirmation is required.

Droseraceae and *Pinguicula* (Non-continuous Digestion)

The pH of the traps

Early studies of Darwin (1875) and Balfour (1875) showed the secreted fluid to be acidic and, at least in *Drosera* and *Dionaea*, to contain formic, butyric and propionic acids and, significantly, Cl^-. The fluid present in stimulated traps is typically pH 2.5–3.5 (Heslop-Harrison, 1975; Rea, 1982a, b). The surfaces of the mucilaginous traps are acidic in the unstimulated state. Depending on the species, *Drosera* shows pHs between 3.0 and 5.0 and, again depending on the species, the pH may not (Whittaker, 1949) or may (Heslop-Harrison, 1975) decrease on stimulation by up to 2 pH units. *Drosophyllum* secretion is acidic (pH 2.5–3.0) (Heslop-Harrison, 1975) while *Pinguicula* is only mildly acidic (pH 5.0), dropping to pH 3.0 following stimulation (Heslop-Harrison and Knox, 1971).

The ionic composition of these mucilages is markedly similar to that of *Nepenthes* pitcher fluid, although present at lower concentrations (Heinrich, 1984). The divalent cations Mg^{++} and Ca^{++} tend to be more prevalent than K^+ and Na^+, but the major anion is again Cl^-. Unfortunately fluid from actively secreting traps has not been analysed in detail and no distinction has been made between the secretions of the stalked and sessile glands.

The control of fluid release in *Dionaea*

The importance of Cl^- in the mechanism of fluid release in both *Pinguicula* and *Dionaea* has already been discussed (see page 184–185). Recently, the mechanism by which fluid release is stimulated has been investigated (Rea, 1982b, 1983a, b).

That stimulation is needed for fluid release in *Dionaea* was demonstrated by Darwin (1875) and further investigated by Robins (1976) and Rea (1982b). Following Darwin's lead, Robins (1976) showed that a range of small-molecular-weight elicitors caused the release of a proteolytic solution. The potency of various elicitors varied, uric acid and ammonia being particularly good (Fig. 9.4). Furthermore, the fluid:protein ratio seemed partially dependent on the elicitor (Table 8.4). Rea (1982b) greatly extended the range of known active elicitors and showed that, in all cases, the release of proteolytic activity was linked to the release of Cl^- (cf. *Nepenthes*; Lüttge, 1966a). The osmolarity of secreted fluid is typically 10–14 mosM and Cl^-, present at 5–10 mEq, constitutes at least 90% of the total anion contribution to the osmolarity and 70–100% of the anionic electrical balance. The principle cation is protons (1–2 mEq) while K^+, Na^+, Mg^{++} and Ca^{++} are all present at less than 1 mM. In contrast to earlier reports, Rea (1982a,b) using GC-MS, did not detect short chain carboxylic acids such as formic, though several long-chain fatty acids and some neutral compounds (glycerol, inositol, glucose) were present. The mineral ion composition of the fluid varied considerably with the elicitor molecule used.

Does the secretion involve the active extrusion of chloride anions or protons? Rea (1983a) used trap lobes stimulated with ammonium or bactopeptone to investigate this problem. First, he showed that most of the measured efflux from stimulated traps was from the adaxial (glandular) surface of the lobes. Secondly, he demonstrated that the rate of H^+ efflux from lobes was dependent on the time for which traps had been previously exposed to an elicitor. Repeated stimulation every 8 h caused a progressively greater rate of H^+ efflux up to a maximum at about 55 h, after which the rate declined (Fig. 9.5). Other elicitors gave a similar response with a 7- to 11-fold increase, while a non-stimulatory solution of calcium sulphate only caused a two-fold increase. Thirdly, he showed the effect of elicitors to be rapid, increased H^+ efflux being detectable within 5–10 min of exposure of bactopeptone-stimulated traps to ammonia. Interestingly, urea, shown by Robins (1976) to be a poor elicitor, actually decreased H^+ efflux if used to replace the bactopeptone solution. The effect of ammonia was shown likely to be due to protogenic ammonia assimilation (Rea, 1983a). Fourthly, Rea (1983a) showed that H^+ efflux is markedly affected by FC, 2,4-dichlorophenoxyacetic acid, diethylstilbestrol and ABA, the former two causing a marked

Fig. 9.4 Cumulative time-course of the volume of secretion produced in response to different stimuli: plotted per secreting lobe from the first day after each lobe started secreting. The volume of stimulant solution administered each day varied from 50 to 20 μl. The stimulants applied were as follows: ●, 2% (w/v) bactopeptone, (15): ○, live *Calliphora* sp. (approximately 60 mg) sealed in a dialysis membrane, (8); ▲, 6.8 mM L-glutamine, (11); △, 16.7 mM-urea, (11); ■, 6.0 mM-uric acid, (7); □, 58–57 mM–ammonia, (7); ◆, 6.1 mM L-phenylalanine, (8). The number of traps is given in parentheses. (Reproduced from Robins and Juniper, 1980c, with permission.)

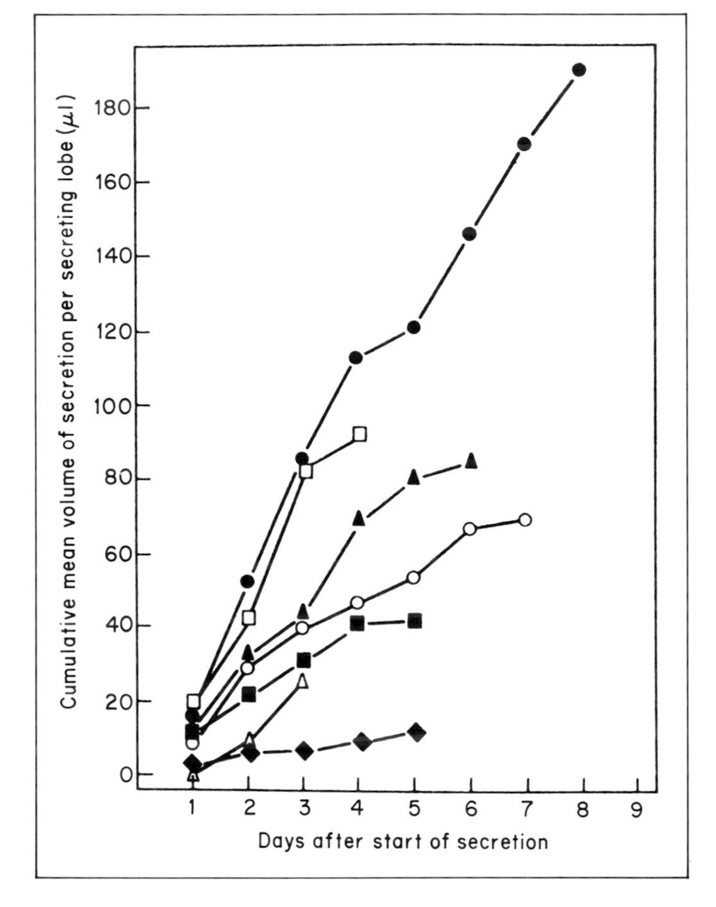

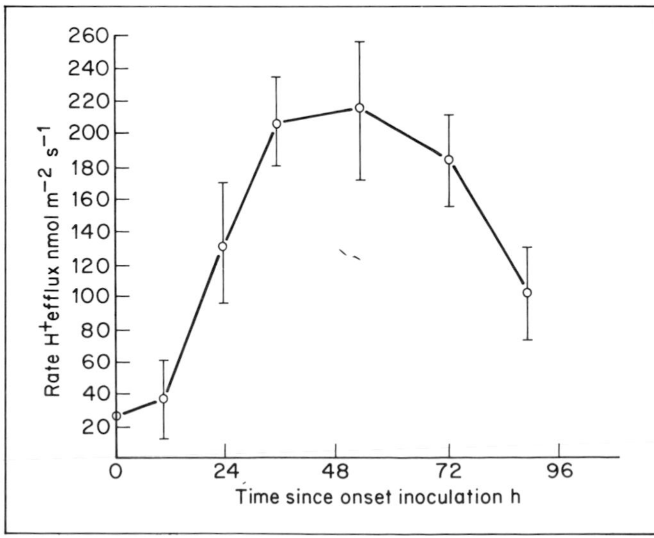

Fig. 9.5 Influence of repeated inoculation every 8 h with 20 kg m^{-3} bactopeptone on the steady-state rate of H^+ efflux from the trap lobes. Each value is the mean +/− SD for n = 3–5 traps. (Reproduced from Rea, 1983a, with permission.)

stimulation (enhanced by K^+), the latter two being inhibitory. In view of the known action of these effectors of ion flux, he concludes that in *Dionaea* pH-control involves direct H^+ extrusion via an H^+–K^+ ATPase on the plasmalemma, the Cl^- moving passively, acting to maintain electrical neutrality. This is in marked contrast to *Nepenthes* (see page 193).

Rea recognized that the rapid rate of H^+ efflux from the glands would cause the intracellular pH to become alkaline unless a mechanism to prevent this were also activated on stimulation. In most higher plants, malate metabolism is central to the control of pH-stasis and malate accumulation is apparently ubiquitously linked to H^+ extrusion. By $^{14}CO_2$-labelling the organic anion fraction of excised discs of *Dionaea* lobes, Rea (1983b) demonstrated that *de novo* malate synthesis is closely linked to elicitor-enhanced H^+ extrusion. In contrast to the protons, the malate is retained within the tissues. The rate of incorporation of $^{14}CO_2$ into malate shows a dependence on the period of prior exposure of tissues to bactopeptone, increasing three-fold after 48 h elicitation. The effectors FC, 2,4-dichlorophenoxyacetic acid and K^+, which stimulate the rate of H^+ efflux from the lobes (Rea, 1983a) caused a 1.5 to 2.5-fold enhancement in the rate of malate synthesis (Rea, 1983b). In contrast, protogenic NH_4^+ predictably decreased the rate. These findings led to the conclusion that malate synthesis is intimately linked to H^+ extrusion in this system, providing the necessary acidification of the cytoplasm to counteract H^+ extrusion and that this pH-controlling mechanism is tightly regulated so as to respond rapidly to changes in the secretory activity of the glands.

Although the mechanism of trap closure in the related *Aldrovanda* has now been quite well studied (see Ch. 6 and Fig. 6.10), little is known of its secretory mechanism. Darwin reported Cohn's observation 'that when rather large crustaceans are caught between the closing lobes [of *Aldrovanda*] they are pressed so hard whilst making their escape that they often void their sausage-shaped masses of excrement, which were found within most of the leaves' (Darwin, 1875: cf. *Dionaea* in Ch. 6). Although this led him to try urea, it would seem likely that *Aldrovanda* will also respond to uric acid, NH_4^+ and so on, like *Dionaea*, and probably show similar features of H^+-regulated pH control.

The capture of prey seems to stimulate secretion in *Triphyophyllum peltatum* (Green *et al.*, 1979); the details of stimulation in this genus have not been examined.

Enzyme Activities in the Secretion

The nutritive value to the plants of digesting prey has been discussed in Chapter 7. In this section we consider the enzymic activities found in the secretion. The principal requirements of carnivorous plants – nitrogen, phosphorous and sulphur – are largely held in the prey in organic forms. While some of these can be absorbed directly, others must be released from complex molecules. Other requirements, notably potassium, are already available and can be absorbed directly as can the relatively small amounts of free phosphate and nitrate present.

In order to break complex molecules down into their simpler, absorbable components, most carnivorous plants secrete hydrolytic enzymes into the environments just described. The package of enzymes released is known to contain several types of activity with different functions. The best studied are the proteases, enzymes that cut up proteins to release free amino acids. These act both within (endopeptidases) and at the ends of (exopeptidases) peptide chains. Esterases, in particular phosphatases, are important as, by releasing phosphate from molecules, they render both the free phosphate and the dephosphorylated product available for absorption. Chitinase will degrade the hard exoskeleton to produce free N-acetyl sugars; ribonucleases will degrade the nucleic acids and lipases break down the phospholipids and triglycerides of the cell membranes. Thus, the package of enzymes is designed to degrade all parts of the prey though in some cases, notably chitin, degradation is not complete and chitinase might aid access of other activities into the carcass (Fig. 9.8).

Proteases

Historically the proteolytic activities have received the major attention. A number of early works were reported in which some attempts were made to characterize these activities in secretion from several carnivorous species. Inevitably such work largely consisted of crude experiments in which meat or gelatin was left for a few days and examined for signs of degradation, such as the production of peptones. Table 9.2 indicates some of the more important of these studies. Vines (1877) was the first to extract an active principal from the walls of *Nepenthes* pitchers while several workers showed the importance of an acid medium to obtain activity (Gorup and Will, 1876; Zeeuw, 1934). These results are all discussed in Lloyd (1942).

Endopeptidases in *Nepenthes*

Since 1942, a number of enzymological studies have been conducted on carnivorous genera. In particular, several species of *Nepenthes* have been examined. The most detailed work is that of a Japanese group who used a mixed secretion of *N. mixta* Hort. and *N. maxima* to examine the proteolytic specificity of the digestive enzymes (Nakayama and Amagase, 1968; Amagase *et al.*, 1969, 1972; Amagase, 1972). The secretion was concentrated by dialysis and separation attempted by both gel filtration on Sephadex G200 and ion-exchange chromatography on DEAE-Sephadex A50. Neither method resolved the activity into more than one component and the rather broad peak obtained from the ion-exchange column was used for the majority of the subsequent work. The use of what they called 'partially purified' material was considered justified by the difficulty of obtaining sufficient digestive fluid, and they point out that the specific activity of the protease had been greatly increased. The active material they call 'nepenthesin'. Amagase (1972) showed, however, that this preparation was not pure, containing at least 2 bands on a zymogram. The fact that this group was working with a mixed protease preparation might explain the discrepancy between their results and those of Tökés *et al.* (1974), who question the validity of Amagase's work.

Steckelberg *et al.* (1967) and Lobareva *et al.* (1973) both claimed to obtain electrophoretically pure protease by a single run on Ectola-cellulose but Jentsch (1970, 1972) felt these simple procedures did not render a pure enzyme. He considered the dilute nature of the enzymes in the pitcher fluid and the presence of carbohydrate to be serious problems and studied 20 different species of *Nepenthes* in an effort to resolve them. Secretion was concentrated by lyophylization and low-molecular-weight contaminants removed by gel filtration on Sephadex 50. The high-molecular-weight fractions were desalted on Sephadex G-75 and the protease-positive fractions applied sequentially to two columns of DEAE-Sephadex A50 before a final desalting. Only after this extensive procedure was the protease substantially free of non-proteolytic enzymes and carbohydrate. The preparation is stable and can be stored for several years.

Tökés *et al.* (1974), in contrast to Jentsch (1972) found two distinct activities in *N. macfarlaneii* separable by a single stage Sephadex G-75 filtration. The major one was at 59 000 ± 4000 daltons and the minor at about 21 000. Both enzymes are present in sterile, unopened pitchers. β-Mercaptoethanol does not alter the elution ptatern, indicating that the larger molecule is not just a disulphide-bonded trimeric form of the smaller. The larger enzyme, which they referred to as nepenthesin and used for the specificity studies, retains apparent homogeneity on DEAE-Sephadex and Ectola-Cellulose. Despite the work of Jentsch (1970, 1972) they took this to indicate the presence of a single enzyme and failed to examine seriously the homogeneity of their preparation. Thus, although they had removed one minor component, others may still have remained.

A major problem in the determination of proteolytic specificity is the choice of substrate. Amagase *et al.* (1969, 1972) used tryptones produced by the tryptic digest of bacteriophage T-4 or rabbit liver cytochromes or a peptic digest of lysozyme. But, as Tökés *et al.* (1974) point out, the use of short peptides presents a very limited range of amino acid combinations and their results, using whole horse-heart cytochromes, indicate that this choice has strongly influenced the apparent cleavage pattern. The Japanese workers obtained the fragmentation pattern shown in Figure 9.6 and from this concluded that nepenthesin has a unique hydrolytic action primarily directed towards peptide bonds to either side of aspartic acid residues. Figure 9.6 also shows the positions where pepsin hydrolyses those peptides. It is surprising that these workers used a tryptic digest, as there is evidence from the 1930s (Stern and Stern, 1932; Zeeuw, 1934) that *Nepenthes* juice itself has tryptic action. Thus, to present it with a tryptic digest may restrict considerably the number of bonds presented for proteolytic action. However, at the pH of the pitchers trypsin (pH optimum c. 8.0) is not active and the proteolytic activity with this pH optimum shown by these authors is unlikely to be functional *in vivo*. Mostly, the activity is described as 'peptic' by earlier workers, due to the acidic pH optima attained (Lloyd, 1942).

The digestion of horse heart cytochrome c by nepenthesin produced six major peptides and the cleavage pattern shown in Figure 9.7 (Tökés *et al.*, 1974). While some of the hydrolases are similar to those of Amagase *et al.* (1969), there is a further group comparable to the cleavage points for pepsin (Tökés *et al.*, 1974). They were, however, unable to quantify these.

Fingerprint patterns from the 21 000 dalton protease were very different, indicative of an exopeptidase hydrolysing the peptide products of nepenthesin to give free amino acids. Lack of material has, unfortunately, prevented any further

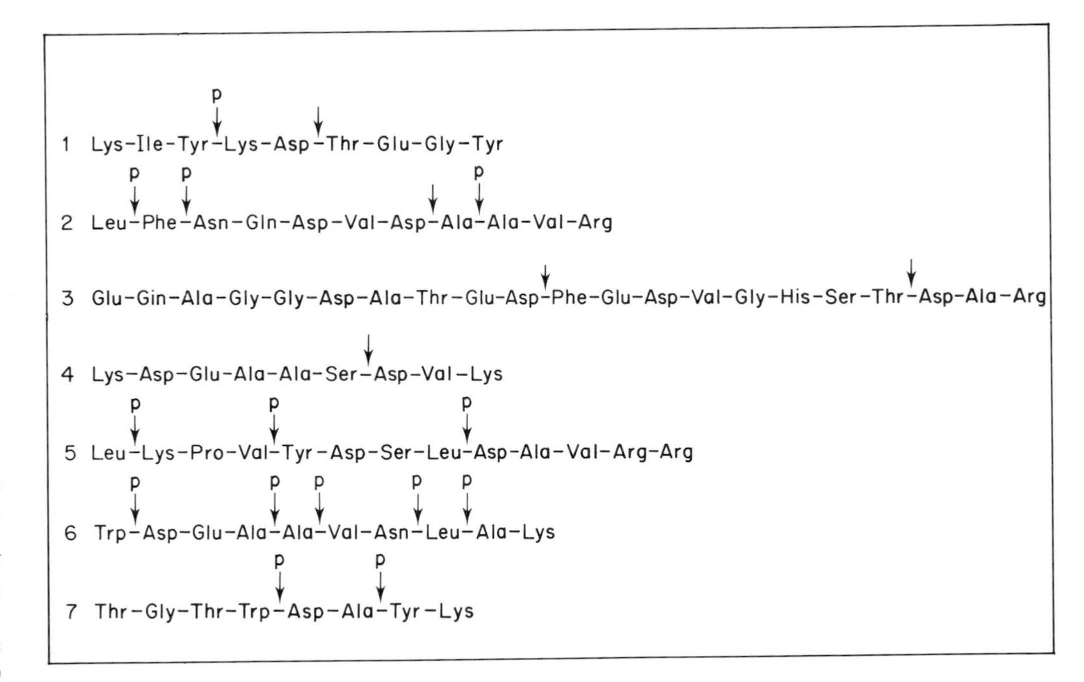

Fig. 9.6 Yields of peptide fragments obtained by nepenthesin digestion of various kinds of peptides. The peptides 2,5,6,7, were obtained by the trypsin digestion of the lysozyme of bacteriophaze T4; and peptides 3,4 by the chymotryptic and peptide 1 by the peptic digest of this protein. The plain arrows indicate the sites of splitting with nepenthesin, while arrows with p indicate those with pepsin. (Redrawn with permission from Amagase *et al.*, 1986.)

Fig. 9.7 Purified peptides from nepenthesin-digested horse heart cytochrome c. Arrows indicate the sites of hydrolysis. Numbers refer to the amino acid sequence position. (Reproduced from Tökés *et al.*, 1974, with permission.)

A: -Thr-Val-Glu-Lys-Gly-Gly (19 … 24)
B: -Leu-Phe-Gly-Arg-Lys-Thr (35 … 40)
C: -Lys-Glu-Glu-Thr- (60 … 63)
D: -Glu-Asp-Leu-Ile Ala-Tyr (92 … 97)
E: -Ile-Ala-Tyr-Leu-Lys-Lys-Ala- (95 … 101)
F: -Lys-Ala-Thr-Asn-GluCOOH (100 … 104)

work on this enzyme. However, it may be that the removal of this from the major component, nepenthesin, further helps to explain the differences between these results and those of Amagase *et al.* (1969, 1972).

The similarity of nepenthesin and pepsin is confirmed by inhibitor studies. Lobareva *et al.* (1972) found the potent carboxyl-directed-inhibitor N-diazoacetyl-N′-2′4′dinitrophenyl-ethylenediamine in the presence of copper ions to inhibit completely nepenthesin and Takahashi *et al.* (1974) found a 90% inhibition of proteolysis by the crude secretion of *N. ampullaria* with diazoacetyl-DL-norleucine methyl ester, which is active against a wide range of acid proteases related to pepsin.

The strongest evidence for the similarity of nepenthesin to pepsin was obtained using pepstatin, a pentapeptide inhibitor showing very high specificity to pepsin. Complete inhibition of nepenthesin by pepstatin at 0.004 mM is reported by Takahashi *et al.* (1974) and Tökés *et al.* (1974) obtained 80% inhibition at 0.05 mM, a concentration inhibiting pepsin by 99%. Thus, an aspartic or glutamic acid residue may play a key role in the catalytic action of nepenthesin, but this has yet to be confirmed.

DIPFP, which inhibits serine proteases by binding to the active site serine residue, did not inhibit nepenthesin at 0.1 mM, (Tökés *et al.*, 1974), trypsin being inhibited by 97% at the same concentration. Alkylating agents such as 0.5 mM iodoacetate (Tökés *et al.*, 1974) or 0.01 mM p-CMB (Nakayama and Amagase, 1968) similarly had negligible effects against nepenthesin when 92% inhibition of papain was achieved by alkylation of the active site cysteine residue. No metal ion is apparently involved as 20 mM EDTA only causes a 7% drop in activity while acetylation with acetic anhydride has a similar negligible effect, making it unlikely that an amide group is involved (Tökés *et al.*, 1974).

Most workers agree on a single pH optimum for *Nepenthes* secretion around pH 2.2, very similar to pepsin. Stern and Stern (1932) also report a peak of activity in the range of pH 6–8 but Nakayama and Amagase (1968) failed to find this in their partially purified extract using casein as substrate and Tökés *et al.* (1974) find the activities of both their enzymes to be optimal at pH 2.2. This neutral optimum is probably due to a bacterial protease. Stern and Stern (1932) did not detect bacterial contamination in their extracts but Lüttge (1964b), who found pH optima at pH 2.2 and pH 6–8 for *N. khasiana* and *N.dormanniana* Hort. in unpurified juice from open pitchers, showed that if sterile material is taken from unopened pitchers then only the acidic optimum is found. This shows that the acid protease is most probably intrinsic to *Nepenthes*, while the neutral activity is not.

The enzymes appear to be quite stable, showing little loss of activity at any pH above 3.0 at 40°C. At 50°C, however, thermostability shows considerable pH-dependence, being maximal at pH 5.0 and at 60°C well over 50% of the activity is lost even at this pH (Nakayama and Amagase, 1968).

Exopeptidases in *Nepenthes*

In addition to these endopeptides, Lüttge (1964b) showed the exopeptidase leucine aminopeptidase to be present in mature pitchers of *N. khasiana* and *N. dormanniana* Hort. although he was unable to detect glycyl-glycine dipeptidase, glycyl-L-leucine dipeptidase or prolinase. He points out that the plant origin of these is not proven and that leucine aminopeptidase is not apparently present in either sterile or unopened traps of *N. khasiana*. Parkes (1980) showed leucine aminopeptidase by histochemical staining to be present in the transverse wall of digestive glands of *Nepenthes*, making it probable that this is a plant-derived activity (but see pages 206–207).

Peptidases in other pitchers and *Utricularia*

Beyond demonstrations such as that of Hepburn *et al.* (1919, 1927) that some *Sarracenia* fluids have proteolytic activity, there is little further knowledge of these activities in pitchers of Sarraceniaceae or Cephalotaceae (Lloyd, 1942). Similar peptidolytic activities are, however, likely to be present. A hint of support for this comes from Parkes (1980), who has shown leucine aminopeptidase in the glands of *S. purpurea*, *Cephalotus* and of *Utricularia*.

Proteases in Droseraceae

The specificity of proteases from *Drosera peltata* leaves was examined by Amagase (1972); unfortunately, as extracts rather than secretion were used, contamination by intracellular enzymes will have occurred. After chromatography on Sephadex G-75 followed by Sephadex G-200 the preparation still contained at least two proteases, as shown by a zymogram, although Amagase refers to this as pure! The cleavage pattern was examined as with *Nepenthes* and again the activity was found to split peptide bonds with good yield to the carboxyl side of aspartic acid. Some cleavage to the carboxyl side of lysine also occurred and it seems likely that this is a separate activity. Certainly, the observed removal of a terminal alanine could be due to a carboxypeptidase A-type activity, as present in *Dionaea* (Robins and Juniper, 1980d). While some of the cleavage is similar to 'nepenthesin' a number of the peptide fragments differ and the pattern bears little relationship to that expected from pepsin. One of the substrate peptides has a free N-terminal leucine and as this remains intact it appears that no leucine aminopeptidase is present. Lüttge (1964b) was unable to detect this enzyme in *D. binata* secretion using a specific substrate, though Parkes (1980) located it in the head cells of both stalked and sessile glands of *Drosophyllum*.

When grown in axenic culture both *D. filiformis* secretion (Whittaker, 1949) and *D. binata* extract (Chandler and Anderson, 1976b) exhibit a single pH optimum at pH 2.2 and 2.6 respectively. The *D. peltata* extract of Amagase (1972) also has a single optimum at pH 2.2, but the secretion of *D. rotundifolia* (Clancy and Coffey, 1977) shows an irregular peak, around pH 3.0, suggesting that more than one enzyme is active. Extracts made from field-grown material of *D. whittakeri* and *D. peltata* ssp. *auriculata* show numerous pH optima depending on the length of time after feeding, with 2.6 and 4.0 being predominant (Chandler and Anderson, 1976b). These authors put this down to microbial contamination, showing bacterial colonies cultured from the leaves to produce protease with some of these optima. Using leaf extracts may also have clouded the issue. As with *Dionaea* (Lüttge, 1964b, Robins and Juniper, 1980d), the nature of the macromolecular substrate used may affect the result obtained. Whittaker (1942) used haemoglobin and Amagase (1972) casein, a substrate found to be better than bovine serum albumin by Clancy and Coffey (1977), while Chandler and Anderson (1976b) used gelatin, a poor substrate, offering limited cleavage potential.

The thermostability of the enzymes from *D. peltata* is, like *Nepenthes*, quite good, showing little degradation at 40°C at any pH above 3. At 50°C considerable loss of activity occurs except when heated at pH 4–6 and by 60°C about 75% activity is lost even at pH 5.0 (Amagase, 1972).

Only two inhibitors have been exhibited to *D. peltata* extract, 0.01 mM p-CMB and 0.1 mM DIPFP, but neither had any effect (Amagase, 1972). Further work on the nature of the active site is needed to see how closely the enzyme resembles that from other genera.

In contrast, the proteases from *Dionaea muscipula* have not been tested for their cleavage pattern but have been more thoroughly examined in the crude secretion for their substrate selectivity and sensitivity to inhibitors (Robins and Juniper, 1980d). A wide range of activities is present. Crude secretion was shown by an electrophoretic zymogram to contain at least four proteases active against casein, a substrate readily hydrolysed by *Dionaea* secretion (Lüttge, 1964b). Once again the nature of the substrate seems important. Scala *et al.* (1969) used a dye-fibrin complex to assay for general proteolytic activity and obtained a single sharp pH optimum at pH 5.5. With casein, however, two major peaks are obtained at pH 4.0 and pH 5.2 (Robins and Juniper, 1980d) suggesting that Scala *et al.* (1969) only assayed a single protease.

The nature of the major protease is not tryptic or chymotryptic. Scala *et al.* (1969) obtained no activity against specific substrates for these enzymes while Robins and Juniper (1980d) confirmed this, also showing that PMSF did not inhibit activity in crude secretion. Thus, the presence of a serine protease seems unlikely and the pH optima are also too low for this. Scala *et al.* (1969) assumed that the protease must be like papain. This was not born out by Robins and Juniper (1980d). About 25% inhibition of activity is obtained by alkylation with iodoacetamide, suggesting that a thiol may be involved in activity, but no physiologically significant activity was obtained using specific substrates for papain or bromelain. Acetylation with acetic anhydride causes a 40% loss of activity suggesting a very important role of an amide group in a major protease. In contrast to *Nepenthes*, pepstatin did not inhibit activity, showing this not to be pepsin-like and involving a carboxyl group. The less acidic pH optima in *D. muscipula* also suggest that the major proteolytic activity in this genus differs greatly from that of *Nepenthes* and, perhaps surprisingly, from that of *Drosera* also.

The peptide hydrolases, leucine amino peptidase, glycyl-glycine dipeptidase, glycyl-leucine dipeptidase and prolidase, were shown in the secretion by Lüttge (1964b) and we have added carboxypeptidase A to his list on the basis of activity with N-carbobenzoxy-L-alanine-4-nitrophenyl ester and of strong inhibition by EDTA (Robins and Juniper, 1980d). These are clearly important for reducing the products of endopeptolytic cleavage to a suitable size for absorption (but see pages 206–207).

Protease in *Triphyophyllum*

Green *et al.* (1979) identified abundant protease activity in the secretion of *T. peltatum* by the substrate film technique. The nature of these awaits examination.

Esterase and Anhydrase Activities

Phosphatase activity has been tested against a range of substrates with secretion from both *Drosera rotundifolia* (Clancy and Coffey, 1977) and *Dionaea muscipula* (Scala *et al.*, 1969; Robins and Juniper, 1980d). The results of this work are summarized in Table 9.5.

In *D. rotundifolia* the general phosphatase pH/activity profile shows two maxima at pH 2.5 and 4.0., suggesting at least two major enzymes. With most substrates the activities are fairly comparable, but the pH 4.0 peak is rather more active against nucleoside tri- and di-phosphates, indicating that this may be an anhydrase.

Table 9.5 Phosphatase activity of secretion from *Drosera rotundifolia* and *Dionaea muscipula* against a range of substrates

	Activity[a]		
	D. rotundifolia		*D. muscipula*
Substrate	pH 2.5	pH 4.0	pH 4.5
Adenosine 5′-triphosphate	2.17	5.49	1.0
Cytidine 5′-triphosphate	2.30	4.60	
Uridine 5′-triphosphate	1.76	4.18	
Thymidine 5′-triphosphate			0.3
Adenosine 5′-diphosphate	1.60	4.09	0
Thymidine 5′-diphosphate			0
Adenosine 5′-monophosphate	0	0	0
Adenosine 2′-monophosphate	0.06	0.31	
Uridine 5′-monophosphate	0	0.11	
D-Fructose 1,6-diphosphate	0.39	0.84	0
D-Fructose 6-phosphate			0
D-Glucose 6-phosphate	0.21	0.12	0
D-Glucose 1-phosphate	0.08	0	
D(-)3-Phosphoglyceric acid	0.30	0.48	
α-Glycerophosphate			0
β-Glycerophosphate	0.09	0.13	0(+)
4-Nitrophenyl phosphate	8.06	12.88	350
Phenyl phosphate			0
Tripolyphosphate	1.60	3.14	
Hexametaphosphate	0	3.81	
Phytin	0	0	
Reference	1	1	2

[a] Activity expressed as enzyme units/gram fresh weight for *Drosera* and enzyme units/mg protein for *Dionaea*. Space = not determined. (+) indicates that activity is detected at a different pH.
References: 1, Clancy and Coffey (1977); 2, Scala *et al* (1969).

Dionaea muscipula, in contrast, only shows a single pH optimum at pH 5.0 but three activities of differing molecular weight on polyacrylamide electrophoresis (Robins and Juniper, 1980d). Because 4-nitrophenyl phosphate was the only substrate tested by the authors, it is not possible to assign roles to these bands. Scala *et al.* (1969), however, found several activities but unfortunately do not report their results in a quantitative manner. Very low activities of acid phosphatase were found in the secretion of *Nepenthes* and in some instances it appeared to be absent (Matthews, 1960). At the most, its activity is below 0.1% that of the ribonuclease present (see page 203) and, if 4-nitrophenyl phosphate was used to assay the activity, might represent the action of some other esterase.

While esterase or anhydrase activities have not been demonstrated in the secretions of any other genus there is good histochemical evidence for their occurrence in secretory cells (Table 9.1). Thus, in *Pinguicula grandiflora* the intracellular stores of these enzymes are clearly lost during the secretory process during which the gland-head contents are washed into the digestive pool (Heslop-Harrison and Knox, 1971; Heslop-Harrison, 1976; Heslop-Harrison and Heslop-Harrison, 1981). The tentacle-head cells of *Triphyophyllum peltatum* contain esterase (Green *et al.*, 1979) and those of *Drosera rotundifolia* contain a 4-nitrophenol-sensitive esterase (Palczewska, 1966), while those of *D. capensis* contain glucose-6-phosphatase (Dexheimer, 1978c, 1979).

Parkes (1980) surveyed several enzymes in different glands. He found acid phosphatase in the walls of glands of *Sarracenia purpurea* (cf. *Pinguicula* and *Dionaea*) and glucose-6-phosphatase and ATPase in the cells. *Cephalotus* only showed acid phosphatase in the vacuoles of the large gland cells (see Ch. 8), though other esterases were present in the small glands. *Nepenthes* similarly only had acid phosphatase intracellularly, though here in the cytoplasm, but glucose-6-phosphatase and esterase occurred in the walls. Acid phosphatase and esterase also occur in *Utricularia* glands.

The variation observed in these histochemical localizations may be partially due to problems with penetration of the substrate (Henry and Steer, 1985). Thus, the absence of an activity is of little value, only positive results being of use. What evidence there is, however, indicates that these activities, like the proteases, are probably found in the secretion of all carnivorous plants.

Other Hydrolase Activities

Chitinase

Chitin, the hard amino-sugar component of insect cuticle contains the monomer N-acetyl glucosamine and is thus a potential source of nitrogen for carnivorous plants. Its degradation might also increase the ability of other enzymes to gain access to the soft body tissues within. Chitinase may be present in conventional higher plants (Boller *et al.*, 1983). Chitinase was first detected in a carnivorous plant in secretion from *Nepenthes mixta* and *N. maxima* pitchers by Amagase *et al.* (1972) using both ant exoskeleton and colloidal chitin on substrate, although their results are only qualitative. They also found chitinase in their extract of *D. peltata* and Chandler and Anderson (1976b) confirmed their findings using extracts of *D. binata* and *D. whittakeri*. The only quantitative assessment of chitinase is that of Robins and Juniper (1980d) in *Dionaea muscipula* where the enzyme is found in the secretion at a specific activity of 1.64 pkat/mg protein. This is sufficient to digest most of the chitin in an exoskeleton during a secretory cycle of 6–10 days. Nevertheless, in all carnivorous traps the exoskeleton of the digested prey does not seem to be significantly affected by the digestive process (Fig. 9.8).

Fig. 9.8 An old leaf of *Drosophyllum* carrying the exoskeleton remains of digested prey. (SEM by G. Wakley.)

The origin of this activity, however, remains uncertain. In contrast to Amagase *et al.* (1972), who used fluid from sterile unopened pitchers, Chandler and Anderson (1976b) could not detect any chitinase in extracts from axenically grown *D. binata* and suggest it may be of microbial origin. In *Dionaea*, microbial contamination is low but, apart from with *Nepenthes*, the possibility cannot be excluded that the enzyme is microbial and not an intrinsic component of the plant secretion.

Nucleases

Ribonuclease activity found in *N. gracilis*, *N. ampullaria* and *N. rafflesiana* secretion shows activity over a wide pH range (2–9), optimal at pH 5.0 (Matthews, 1960). Filtered, but unconcentrated, fluid from open, wild grown, pitchers was incubated at pH 5.0 and 37°C with TMV RNA, and almost complete quantitative degradation to the nucleoside 2′,3′-cyclic phosphates occurred within a few hours. Further degradation of adenosine 2′,3′-cyclic phosphate and guanine 2′,3′-cyclic phosphate occurs more slowly, yielding the 3′-monophosphates. The pyrimidines resist further attack. Activity is lost during storage over a few months at 4°C but more slowly if frozen. Both activities are stable at 70°C for 5 min but are 95% destroyed at 80°C. Ribonuclease occurs in the transverse walls of the glands of *N. khasiana* and *Pinguicula* (Parkes, 1980).

High deoxyribonuclease activity was found in *Dionaea muscipula* secretion (Scala *et al.*, 1969). These authors also found nucleoside triphosphatase activity but, surprisingly, did not detect any nucleoside diphosphatases.

Lipases

Sterile secretion from *N. macfarlaneii* was tested by Tökés *et al.* (1974) for activity against glycerol trioleate, glycerol tripalmitate and lecithin. While glycerol tripalmitate was not degraded, glycerol trioleate was broken down to oleic acid and dioleate at both pH 6.0 and pH 2.6. This may suggest some selectivity for unsaturated fatty acids, but this has not been tested. Lecithin was only degraded to diglycerides at neutral pH, indicating phosphatidase D activity, but at pH 2.2 was totally degraded to fatty acids.

Amylases

While Tökés *et al.* (1974) were able to detect amylase activity in the sterile secretion of *N. macfarlaneii* at pH 6.0 and 37°C, no detectable activity occurred at either pH 4.0 or 2.0, the range of pitcher acidities. This suggests that amylase may be present in the secretion but not play an important role in digestion. It may become active as digestion proceeds, as the pH of the pitchers tends towards neutral when they become choked with prey.

In *Dionaea muscipula* amylase activity is also very low (Scala *et al.*, 1969) and only detectable during the first few days of secretion. This makes it unlikely that the enzyme is of microbial origin.

Amylase has been detected histochemically in *Pinguicula* stalked glands (Heslop-Harrison, 1975).

General points

The stability observed in many of these enzymes is as expected for extracellular activities, in which a prolonged lifetime in an environment subject to less control than that in the cells is required. Furthermore, as a wide range of substrates might be predicted in the prey, so a wide range of enzyme activities is observed. The plants show considerable economy of effort by having in the secretion a relatively small number of enzymes each showing a broad substrate specificity. This is particularly true in the esterases and anhydrases, where the acid phosphatases show activity with many different substrates.

Intracellular Activities Involved in the Assimilation of Nutrients

Sulphate Assimilation

The leaf tissue of *D. whittakeri* contains both adenosine triphosphate sulphurylase (EC.2.7.7.4) and adenosine diphosphate sulphurylase (EC.2.7.7.5) at specific activities of 84 and 51 pkat/mg protein respectively (Burnell and Anderson, 1973).

These enzymes are widely found in plant tissue and play a key role in sulphate assimilation, the former activating sulphate ions with ATP. Clearly, utilization of sulphate absorbed from digested prey will be an important nutritional requirement for this and many other carnivorous plants.

Nitrate-Assimilating Enzymes

It has long been known that *Drosera rotundifolia* will grow in a purely inorganic nutrient solution with N supplied to the roots either as NO_3^- or NH_4^+ (Rychnovska-Soudkova, 1953, 1954; see Ch. 7). The effectiveness with which this occurs shows considerable pH dependence, NO_3^- being assimilated best at low pH, NH_4^+ from alkaline solutions. The enzyme nitrate reductase is essential for the assimilation of NO_3^- and has been shown present in *D. whittakeri* leaves (Chandler and Anderson, 1976a) and leaf and root extracts of *D. aliciae* plants grown in sterile culture (Small and Hendrikz, 1974; Small *et al.*, 1977).

In *D. aliciae*, nitrate reductase is induced in response to NO_3^- in the growth medium, only traces of activity being detected in plants grown with only NH_4^+ (Small and Hendrikz, 1974). Plants took more than 48 h to adapt fully from NH_4^+ to NO_3^-, suggesting that the switch involves *de novo* synthesis, as is well-known for nitrate-induced reductase formation. Small *et al.* (1977) complete the picture of inorganic nitrogen assimilation in *D. aliciae* by showing that plants grown on NO_3^- contain, in addition to nitrate reductase, nitrite reductase, glutamate dehydrogenase, glutamine synthetase, glutamate synthase and a peroxidase capable of reducing nitrate. As might be expected NH_4^+ or cheese-grown plants lack the former two enzymes. The identity of the enzymes was checked by inhibition of the various activities with KCN (nitrate and nitrite reductases), NaN_3 (nitrate-reducing peroxidase), azaserine (glutamate synthase) or p-CMB (glutamate dehydrogenase and glutamine synthetase). It is tentatively suggested that assimilation from NH_4^+ and cheese takes place via a peroxidase-nitrate reducing enzyme and glutamate dehydrogenase while the glutamine synthetase/glutamate synthase system is most active with NO_3^- as N-source, in agreement with previous evidence. Insect-derived nitrogen was found to decrease the amount of nitrate reductase activity in *D. whittakeri* (Chandler and Anderson, 1976a), suggesting that prey-derived organic nitrogen represents a more important source of nutrients than the inorganic nitrate present either in the prey or in the soil.

The Role of Micro- and Macro-Organisms in Digestion

Both a microfauna and, in the case of pitchers, a macrofauna (see Ch. 14) are known to be associated with the actively digesting traps of carnivorous plants. The argument over the extent to which these contribute to the degradation of prey has lasted many years (Lloyd, 1942) and only recently have the means become available to resolve it satisfactorily. It can, however, now be categorically stated that micro-organisms are not the sole source of hydrolytic activity in the fluid on the basis of several lines of evidence. Firstly, proteolytic enzymes are present in sterile liquid from unopened *Sarracenia* and *Nepenthes* pitchers. Secondly, the cytochemical and radiolabelling evidence clearly establishes that digestive glands release hydrolases. Thirdly, axenic cultures are able to digest prey in the complete absence of any micro- or macro-fauna. These aseptic cultures, which are available for a range of carnivorous plants (see page 189), offer an excellent opportunity for studies to define the importance of external sources of digestive activity within the traps.

Nevertheless, digestion by associated organisms must be taking place. In the Bromeliaceae, with the exception of the few identified carnivorous species (see Chs 1 and 19), the plants are entirely dependent on their macro- and micro-fauna for the degradation of the tank-held detritus. Lindquist (1986, pers. comm.) observed that non-carnivorous bromeliads, grown in the greenhouse at the Botany Department of Wisconsin-Madison, harboured purple non-sulphur bacteria (*Rhodopseudomonas* and *Rhodomicrobium*) as well as purple sulphur bacteria. To what extent, then, do carnivorous plants benefit (or suffer) as a result of such activity?

Pitcher Traps (Continuous Digestion)

The phytotelm is particularly highly developed in the *Nepenthes* pitchers (Beaver, 1983) from which a wide range of micro-organisms (Thienemann, 1932, 1934), protozoa (Oye, 1921; Ghosh, 1928), rotifers, nematodes and oligochaetes (Menzel, 1922; Menzel and Micoletzky, 1928; Ghosh, 1928; Thienemann, 1932) have been recorded. Insect larvae also abound (Beaver, 1983: Ch. 14; and Figs 14*A* and B). *Sarracenia* pitchers similarly contain micro-organisms (Hepburn *et al.*, 1927; Hegner, 1926b; Plummer and Jackson, 1963; Lindquist, 1975), mosquito and chironomid larvae (see Ch. 14). In *Sarracenia*, mosquito larvae are known to consume protozoans (Addicot, 1974) and probably also bacteria (Fish and Hall, 1978), thereby decreasing the microbial population. The same seems true for *Nepenthes* (Beaver, 1983).

It would seem that microbial populations, while present, may not be extensive in the earlier, most active, period when traps are recently opened. Okahara (1933) found 3×10^6 to 3×10^7 bacteria/ml in *N. mirabilis* pitchers seven days after opening.

Beaver (1983) reports micro-organisms to be abundant only in old pitchers already devoid of phytotelms. Putrefaction is only found in old pitchers (Beaver, 1983; Fish and Hall, 1978; Plummer and Jackson, 1963), in particular those which have captured an excess of prey. Considering the warm, nutrient-rich environment provided, it might be expected that microbial growth should rapidly escalate, despite predation, and dominate the utilization of the nutrient supply. Yet this only seems to happen in old pitchers. Thus, for example, Albrecht (1974) observed that in pitchers of *Sarracenia purpurea* in Wisconsin, USA during the late summer the numbers of bacteria may rise to 10^8 to 10^9/ml fluid. These bacteria are probably adventitious, having ridden in on prey and grown rapidly in the temporarily elevated temperatures and pH (about 7.0 as recorded) of late summer. These arrivals may have displaced the more 'exotic' species that, in lower numbers (10^4 to 10^5), are the usual inhabitants of lower temperature and lower pH (c. pH 4.0) pitchers. Some of these *Sarracenia* bacteria of both the bog pools and the pitchers were initially identified as *Rhodopseudomonas acidophila* (Lindquist, 1975), but are now known to be *R. palustris*. In addition Lindquist detected *Escherichia coli* at levels between 430/ml and 2300/ml in a few pitchers suggesting indirect fecal contamination, possibly by flies. Also, the pitcher of *S. purpurea* is a reliable source of *Chromobacterium violaceum*, *Caulobacter* and a still-unidentified *Acinetobacter*-like pink bacterium (J. Lindquist, pers. comm., 1986) *Chromobacterium violaceum* is normally more commonly associated with tropical habitats.

But bacterial contamination of normal, fully active young pitchers is generally low. How then are these low levels normally maintained? There is no evidence for anti-microbial activity within the pitcher fluid (see Ch. 12) but the key to the inhibition of microbial growth probably lies in the ability of the plant to maintian an acidic pH (see page 193). In general, bacteria grow poorly in acidic conditions and pitchers typically are maintained at pH 2.5–4.5. Plummer and Jackson (1963) showed that liquid removed from a pitcher slowly became more alkaline, reaching neutrality in 22 days. This process was accelerated by infecting the liquor with bacteria. Thus, it appears that microbial activity only increases in old leaves in which pH-control has ceased or in cases of a super-abundance of prey capture when the plant is unable to maintain acidic conditions. In this latter case, putrefaction is likely to initiate premature necrosis.

Bacteria isolated from *Sarracenia* show proteolytic and chitinolytic activity against a range of substrates (Hepburn *et al.*, 1927; Plummer and Jackson, 1963; Lindquist, 1975), though Hepburn *et al.* (1927) considered the activity 'slow' compared with that from *Sarracenia* itself. Similarly, isolates from *Nepenthes mirabilis* were active against a range of proteins (Okahara, 1933). Significantly, however, these assays were all conducted at pH 6–8. Okahara (1933) demonstrated that at pH 3.3, that of the pitcher fluid, none of the bacteria was able to degrade the substrates used, though fungal isolates were still active. Hence it would appear that the acidity of pitchers maintained by the plant helps to limit bacterial degradative activity as well as growth.

Dakin (1919) also found that fibrin digestion by fluid from *Cephalotus* pitchers only occurred in acidic conditions despite bacteria being present. Parkes (1980) showed a *Pseudomonas* from *Cephalotus* to secreted extracellular protease but only tested their activity at a weakly acidic pH.

Utricularia

As discussed earlier (see page 195), the contribution of *Utricularia* to digestion within the trap is poorly characterized. Like the pitchers, *Utricularia* provides a micro-environment which is colonized by small organisms (see Ch. 14), some of which are apparently able to live within the traps (Hegner, 1926a; Botta, 1976). Paramecia captured by traps die on average in 75 min and, washing the trap interior prior to capture did little to extend this period (Hegner, 1926a). Old traps, or heat-killed traps (100°C for 2 min), however, failed to kill paramecia, indicating that digestive activity was involved.

Proteolytic activity occurs in the traps (Luetzelberg, 1910; Adowa, 1924; Kiesel, 1924) but Kiesel (1924) considered this entirely due to microbial activity. This cannot be so, however, as axenic cultures of several *Utricularia* species have been shown capable of obtaining nutrition from captured prey (Harder, 1963, 1970; Pringsheim and Pringsheim, 1962), while Harder (1970) showed daphnia feeding to improve the growth of such cultures. Thus, the role of micro-organisms in *Utricularia* nutrition remains unresolved but they are clearly not the only source of digestive power.

Droseraceae and *Pinguicula* (Non-Continuous Digestion)

The trap surfaces of *Drosera*, *Dionaea* and *Pinguicula* are not free of micro-organisms even in the unstimulated state. Indeed, *Drosera* and *Pinguicula* have rather interesting bacterial floras which have been used for many years to make 'ropy milk' (Thomas and McQuillan, 1953; Nilsson, 1950; see Ch. 13). No analysis of secreted proteases is available from axenic culture for any of these genera as, unfortunately, Chandler and Anderson (1976b) used leaf extracts rather than secretion. These authors did find, however, that axenic cultures showed a single pH optimum at pH 2.6, while field-grown plants showed optima in the pH 4–6 range and around pH 8.0. Bacteria from *D. whittakeri* leaf surfaces showed optimal proteolytic activity around pH 3.5–4 in contrast to the findings of Okahara (1933) who only found bacterial protease activity in *D. rotundifolia* at pH 5–6. Lüttge (1981) points out that in *Dionaea* the range of exopeptidases present (see page 201) is likely to be due to microbial activity. Chitinase, identified in *Dionaea* secretion (Robins and Juniper, 1980d) and *D. binata* extracts (Chandler and Anderson, 1976b) may also be microbial as it is absent from axenic cultures of the latter species.

In adhesive traps as in the pitcher there is no evidence for anti-microbial activity in the secretion (see Ch. 12) and it may be that in the short-lived digestive pools in these species microbial contamination is simply not a problem. The acidic pH will again hinder rapid bacterial growth, though it appears that bacteria in some *Drosera* have active acid proteases (Chandler and Anderson, 1976b). Indeed, these authors have argued that the microbial flora is beneficial to the plant, insect-enhanced growth of *D. whittakeri* being inhibited by antibiotics and axenic *D. binata* failing to show any insect-enhanced growth. Hence, what little evidence there is indicates that a small microbial population is beneficial. In *Dionaea*, excessive bacterial contamination which might develop when a very large prey is trapped is detrimental, leading to rapid necrosis of the lobe (R.J.R., unpublished observations). In *Byblis* fungal activity is very common. As no direct evidence has yet been found for the secretion of enzymes by this genus, it must be assumed that the fungi provide the main digestive apparatus of *Byblis* (but see Fig. 9.9).

Competition for Nutrients

Even if, as may be the case in some instances, proteolytic activity is predominantly derived from the plant, this does not prevent microbial associates benefiting from the digestive process, to the detriment of the plant (cf. also the cost-benefit analysis of a carnivorous plant discussed on page 144).

A potential strategy that the plants may adopt,

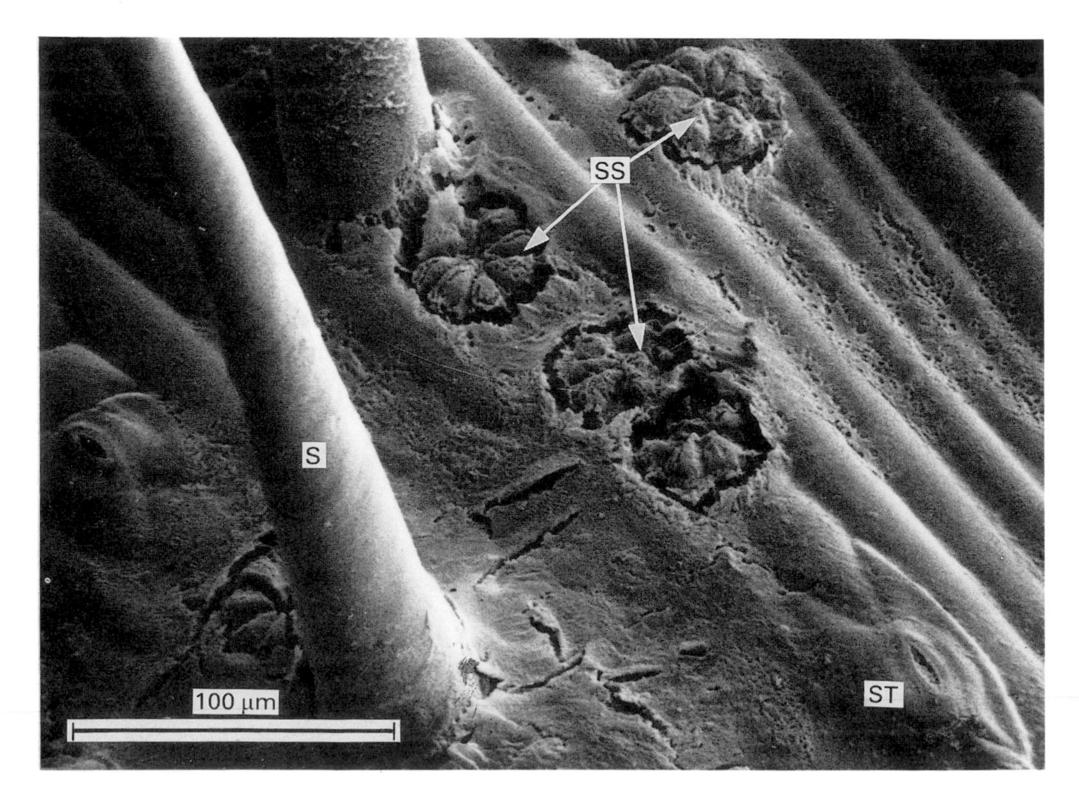

Fig. 9.9 The surface of *Byblis liniflora* smeared, after insect disturbance, with its own mucilage. Dried and cracked mucilage is clearly visible at the base of the stalked gland (S) and the sessile glands (SS) have apparently digested their way through this layer (arrows). The raised stomata (ST) are clearly visible, apparently unaffected by the spread of mucilage; cf. Fig. 8.16D. (SEM by C. Merriman.)

which could confer an advantage, would be only partially to degrade the proteins in the extracellular environment, to absorb the majority of the digestion products as small peptides and to degrade these to amino acids intracellularly. Is there any evidence for this?

Firstly, secreted fluid is surprisingly deficient in exopeptidase activities and in *Nepenthes* these are notably absent from unopened pitchers (Lüttge, 1964b), the only sterile source analysed. Hence, those found in other digestive fluids such as *Dionaea*, could be microbial in origin (Lüttge, 1964b).

Secondly, it has been demonstrated that peptides are absorbed. Plummer and Keithley (1964) showed by autoradiography that alanyl-methionine, alanyl-leucine and alanyl-asparagine were all absorbed by *Sarracenia flava* leaves. Significantly, none of these substrates were hydrolysed by the pitcher fluid. Simola (1978) showed that a range of dipeptides could act as sole N-source for *Drosera rotundifolia* axenic cultures, while Juniper and Gilchrist (1976) have argued that larger peptides may be absorbed by *D. capensis* stalked glands. The speed of absorption of ^{14}C-label from protein by *Pinguicula grandiflora* (Heslop-Harrison and Knox, 1971) might also indicate peptide absorption. Free amino acids are also readily absorbed (see Ch. 10).

Thirdly, hydrolytic activities capable of releasing free amino acids have been demonstrated cytochemically within gland tissue (Table 9.1), notably leucine aminopeptidase. Plummer and Keithley (1964) found that the dipeptides they fed were degraded to the free amino acids once within the tissue. Hence, while no direct evidence is available, certain facts are compatible with such a strategy.

A further strategy adopted by carnivorous plants is that absorption is very rapid. In *P. grandiflora*, products of digestion are seen within the leaf after 2 h (Heslop-Harrison and Knox, 1971) and ^{3}H-asparagine fed to *D. capensis* tentacles is detected in the head cells after 2 min (Juniper *et al.*, 1977). In only 24 h, 80% of the radioactivity in a fly was absorbed by *Drosera* (Ashley and Gennaro, 1971), though this fell to only 12% in excised leaves, indicating an important role in redistribution within the plant. The vascular connections within the *Drosera* glands aid this process.

It may be, therefore, that by adopting strategies involving a low pH environment, minimal degradation in the digestive fluid and the rapid absorption of products, the carnivorous plants can compete favourably with microbial associates for prey-derived nutrition. That they do benefit has already been discussed (see page 145).

CHAPTER 10

The Absorption of the Products of Digestion

Introduction

The extent to which carnivorous plants benefit from captured prey (see Ch. 7), the structural features of the digestive apparatus (see Ch. 8) and the hydrolysis of the constituents of prey (see Ch. 9) have now all been considered. In this chapter we turn to the final process in the acquisition of nutrients by carnivory, that of absorbing the products of hydrolytic activity and other prey-derived nutrients.

The early assumption (e.g. Darwin, 1875) that the absorption of products is carried out by the digestive glands was based on a number of observations:

(i) 'aggregation' (see page 181) typically started in the glandular cells when plants were fed but not in neighbouring epidermal cells;
(ii) vital stains were seen to penetrate the glands rapidly, but not to pass directly into the epidermal cells;
(iii) the cuticle of glandular cells was clearly much less well formed than in neighbouring epidermal cells and was often apparently absent or porous (see page 162).

These observations have been entirely substantiated by recent physiological and high-resolution anatomical studies. The transport of many substances has been shown to require energy-consuming processes (Arisz, 1953) and in *Dionaea* small molecules are found to be more readily taken up if charged (Rea and Whatley, 1983). Uptake apparently involves specific membrane-located transport systems (Arisz, 1953) and there is only the slightest evidence in any species for pinocytotic- or phagocytotic-type processes (Gilchrist and Juniper, 1974; Honsell *et al.*, 1975). The pathway of absorption in a number of species has been examined with tracer techniques. Nevertheless, many key features are not understood and, as previously, a balanced analysis cannot yet be presented.

Evidence for Absorption from the Traps

That carnivorous plants absorb the products of digestion is implicit in the nutritive effect of feeding traps with prey (see Ch. 7). Using gravimetric determinations, Hepburn *et al.* (1927) demonstrated the disappearance of a range of nitrogenous compounds (ammonium salts, acetamide, urea, asparagine and peptone) introduced into the pitchers of *Sarracenia alata* (*sledgei*), *S. flava*, *S. leucophylla* (*drummondii*) and *S. purpurea*. They found water and solute not to be absorbed at the same rate and that absorption took place over a length of time. By feeding lithium ions they were able to demonstrate that absorption took place by detecting this mineral in the pitcher tissues, from which it is normally absent.

Firm evidence for the absorption of organic materials was, however, only obtained following the advent of radioactive tracers. Using autoradiography, Lüttge (1963) was the first to apply this technique, feeding ^{14}C-labelled glutamic acid to leaves of *Dionaea*. He showed convincingly that label applied to a small spot on the adaxial lobe surface spread throughout the plant. Radiolabelling with L-leucine has also been used to show absorption in *Dionaea* (Robins and Juniper, 1980b). Label was again found in parts of the plant not exposed to an exogenous supply.

Lüttge also demonstrated ^{14}C-L-alanine to be

absorbed and assimilated by pitchers of both *Nepenthes dormanniana* (Lüttge, 1964c) and *N. khasiana* (Lüttge, 1965). $^{32}PO_4^{3-}$ and $^{35}SO_4^{--}$ were similarly absorbed and assimilated (Lüttge, 1965). Plummer and Kethley (1964) performed similar experiments on *Sarracenia*. They collected *S. flava* from a range of habitats in Georgia, USA, and plugged pitchers to prevent contaminants entering. Treatment solutions were syringed into the pitchers. Radioactive zinc, iodine, cobalt, sulphur, calcium and phosphorus were all absorbed; though cobalt and iodine showed little redistribution. In the cases of ^{32}P and ^{35}S, label administered in pre-labelled ants showed good redistribution throughout the plant and was detected in the apex of young growing leaves. Similarly, ^{32}P fed to *Utricularia inflata* via ostrocods was distributed throughout the plant (Lollar *et al.*, 1971). Uptake of label was diminished by 80% if the traps were removed prior to feeding. In *Drosera binata*, $^{35}SO_4^{--}$ was shown to be incorporated into sulphur amino acids following feeding either with pre-labelled *Drosophila* or free ions (Chandler and Anderson, 1976c).

The uptake of unlabelled amino acids by *S. flava*, when fed individually or in mixtures, has been studied (Plummer and Kethley, 1964). Within two days all but arginine, cystine and threonine had been completely absorbed, the amino acids appearing in other leaves, roots and the rhizome. A number of dipeptides were also reported to be absorbed by the traps of some species.

Drosera capensis and *Aldrovanda vesiculosa* absorbed ^{14}C-labelled sugars and amino acids from *Daphnia* grown on these substrates (Fabian-Galan and Salageanu, 1968). In both cases activity was seen to spread from the point of application towards the growing centre of the plant (Fig. 10.1). A similar effect was observed in *Pinguicula grandiflora* fed ^{14}C-protein (Heslop-Harrison and Knox, 1971). Using ^{15}N as a tracer, Dixon *et al.* (1980) were able to demonstrate the absorption and long-term distribution of nitrogen in *D. erythrorhiza* plants (Fig. 10.2) and Shibata and Komiya (1972, 1973) showed the transfer of nitrogen from protein to plants of *D. rotundifolia*. These studies have demonstrated that a wide range of both mineral and organic substrates are readily absorbed.

Macromolecular constituents of the digest are largely, but not necessarily totally, degraded before absorption (see Ch. 9). This would explain the 2 h delay when ^{14}C-protein was fed to *P. grandiflora* (Heslop-Harrison and Knox, 1971) compared to the 2 min required to detect ^{3}H-asparagine in *D. capensis* tentacles (Juniper *et al.*, 1977) or the few minutes to detect ^{14}C-glutamic acid in *Dionaea* (Lüttge, 1963). On the other hand, Ca^{++} is thought to be co-transported with proteins (Juniper and Gilchrist, 1976) and ferritin can be located in structures well within the gland head (Panessa *et al.*, 1976; Juniper *et al.*, 1977). The extent to which these proteins were degraded before entering the gland was not examined, but they served to examine the pathway of absorption in these genera.

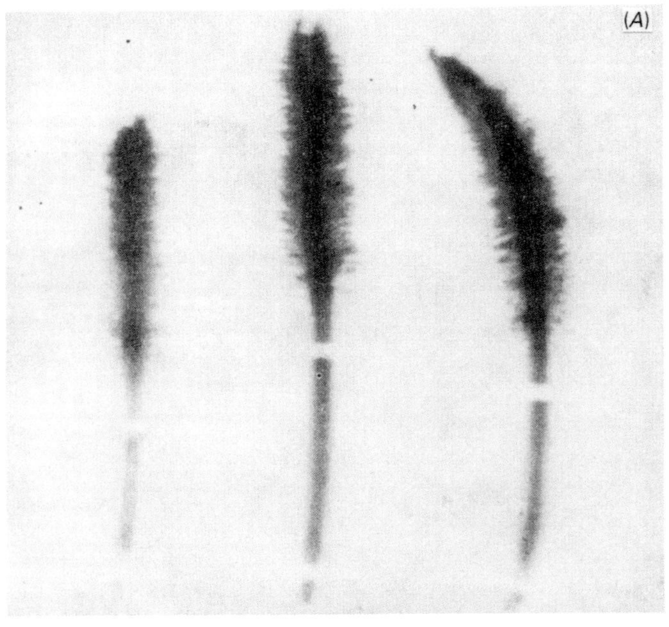

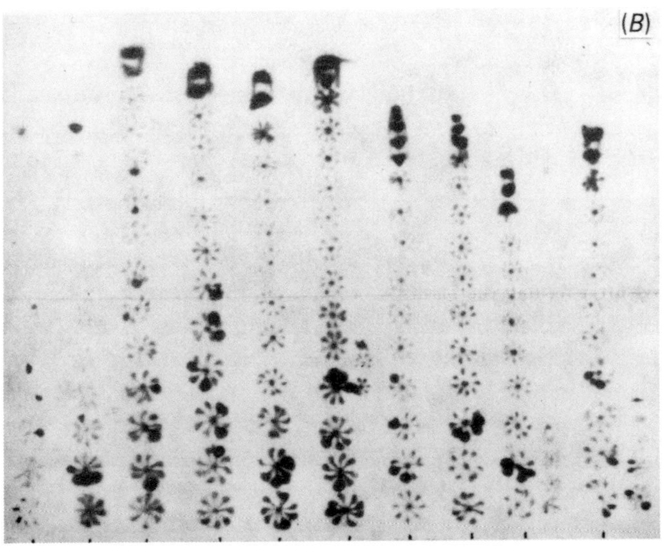

Fig. 10.1 Whole-tissue micrographs of complete leaves or stems of carnivorous plants fed with radioactive *Daphnia* incubated on labelled sugars. (*A*) Whole *Drosera capensis* leaves after 5, 6 and 8 days. (*B*) Distal stems of *Aldrovanda vesiculosa* of approximately 12 whorls (see Fig. 4.6), – 8 days after feeding. (Reproduced, with permission, from Fabian-Galan and Salageanu, 1968.)

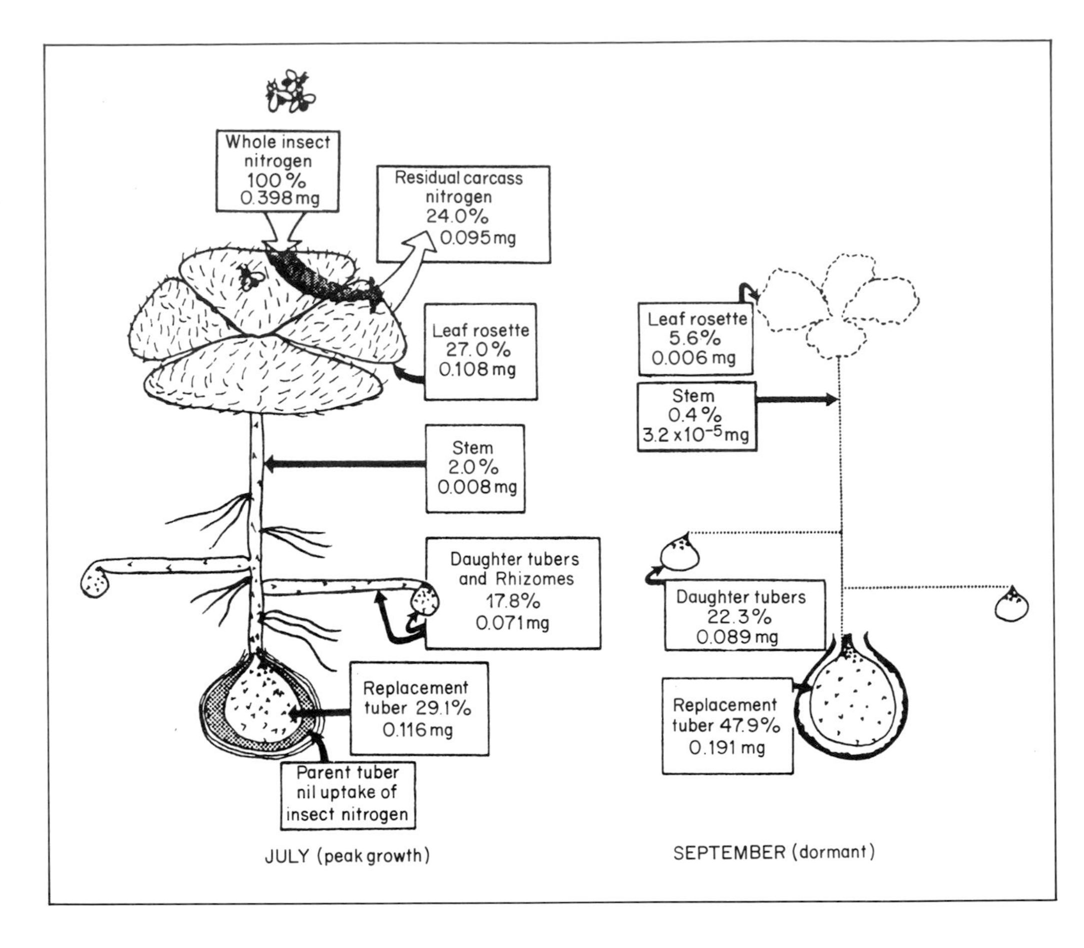

Fig. 10.2 The distribution of nitrogen fed as ^{15}N-labelled *Drosophila* to the leaf rosettes of *Drosera erythrorhiza*. July harvest four weeks after the end of the feeding period; September harvest at the end of the growing season. Amounts of nitrogen are given as mg per plant and as percentages of the total insect nitrogen applied. (Redrawn with permission from Dixon *et al.*, 1980.)

The Pathways and Mechanisms of Absorption

This section presents evidence that absorption takes place through the glands of the trap and, where such evidence is available, how it occurs. Structural evidence, presented in Chapter 8 showed these to be the organs from which fluid was released. In those systems studied in any detail (*Nepenthes*, *Pinguicula*, *Drosera* and *Dionaea*) they are also shown to be the sites of absorption. Two basic models for absorption have been proposed; one for *Pinguicula*, and one for *Dionaea*. The former involves absorption at the endodermal layer during a phase subsequent to secretion; the latter absorption in the gland-head cells while secretion is still taking place. The rationale for these two possibilities will be considered and their relevance to other, less exactingly studied genera discussed.

Pitchers (Continuous Digestion)

The site of absorption

Using micro-autoradiography, Lüttge (1965) demonstrated that $^{35}SO_4^{--}$ only entered the tissues of *Nepenthes khasiana* through the roofed glands at the bottom of the pitcher. Hence these glands appear to be performing three functions:

(i) the secretion of Cl^-;
(ii) the secretion of hydrolases;
(iii) the absorption of a wide range of products.

This multiple physiological activity, common to other glands of carnivorous plants (see Chs 6 and 8), makes them rather different to most, more specific glands (Lüttge, 1971). Little further detail of the pathway of absorption is known for *Nepenthes*. (see page 162 onwards).

In *Sarracenia purpurea*, however, iron and uranium uptake has been followed by X-ray microanalysis (Panessa *et al.*, 1976). Ferritin, fed to isolated digestive surface of sterile pitchers, was shown to be absorbed specifically by the glands. The maximum X-ray signal was reached in the digestive glands within 45 min but, curiously, the signal in the abaxial external nectary glands continued to accumulate up to 90 min. Ferritin feeding appeared to increase both ER and small vesicles within 60 min of feeding. Some of the latter were associated with the plasmalemma, possibly indicating phagocytosis or blebbing similar to that observed by Gilchrist and Juniper (1974) in ferritin-fed *Drosera capensis* glands. By 60 min the ferritin core could be visualized throughout the cytoplasm of the gland cells. By 150 min the gland cells were densely labelled and ferritin could be detected in adjoining mesophyll cells and vascular tissue and was already accumulating in the nectary cells. X-ray microanalysis confirmed these particles to be ferritin. Uptake and transport was completely inhibited by boiling tissue or pre-treating with KCN, indicating a requirement for metabolic activity. Uranium showed very similar uptake characteristics.

High-resolution examination showed ferritin to be located in the plasmodesmata between the glandular cells and the endodermal cell but to be excluded from the adjacent cell wall. Thus, it appears that inward transport in these glands utilizes a symplastic route. As pointed out by Joel (1986), however, Panessa *et al.* (1976) examined glands in Zone 3 (see Fig. 4.9*D*), which may not play a major role in absorption as they are not always submerged, in contrast to the Zone 4 glandular epithelium. Similar experiments need to be performed, therefore, on other areas of *Sarracenia* as well as *Nepenthes*.

The mechanism of absorption

The uptake by *N. khasiana* of alanine, PO_4^{3-} and SO_4^{2-} shows different rates, the order being alanine $> PO_4^{3-} > SO_4^{2-}$ when an equimolar solution is exhibited in the range 1–10 mM (Lüttge, 1965). The rate of uptake of PO_4^{3-} with increasing concentration followed a Michaelis-Menten relationship, but no saturation was observed over the physiological range and was, if anything, slightly stimulated by the metabolic inhibitor KCN (Lüttge, 1965). In contrast, Cl^- absorption in both *N. khasiana* and *N. rafflesiana* (*hookeriana*) is apparently passive yet shows sensitivity to metabolic inhibitors, though to a much lesser extent that Cl^- excretion (Lüttge, 1966a; Jung *et al.*, 1980). Further work is clearly needed to define the complex processes involved in mineral-ion fluxes in *Nepenthes*.

There are no data available on the mechanism of uptake of amino acids in pitcher-plants. It should be noted, however, that the distribution of labyrinthine walls in *N. khasiana* (Parkes, 1980) resembles *Dionaea*, indicating that perhaps a similar pathway is involved (see page 214).

Utricularia

The pathway of absorption

Although Lollar *et al.* (1971) showed *U. inflata* to absorb $^{32}PO_4^{3-}$, no radiotracer study has been done in *Utricularia* using organic solutes and therefore it has yet to be demonstrated that the traps are important in the absorption of these nutrients. High-resolution inorganic tracer studies have, however, been performed. In the first instance, the arms of these glands are shown to be the only permeable sites within the trap. Secondly, the swollen apoplast of the gland-head cells of *U. monanthos* is, like that of *P. grandiflora*, freely permeable to lanthanum and uranyl salts, which are excluded from the totally cutinized endodermoid cell wall and the cutinized regions of the 'stalk' (Fineran and Gilbertson, 1980). The tracers readily pass, however, into the transverse wall of the endodermoid cell, penetrating the

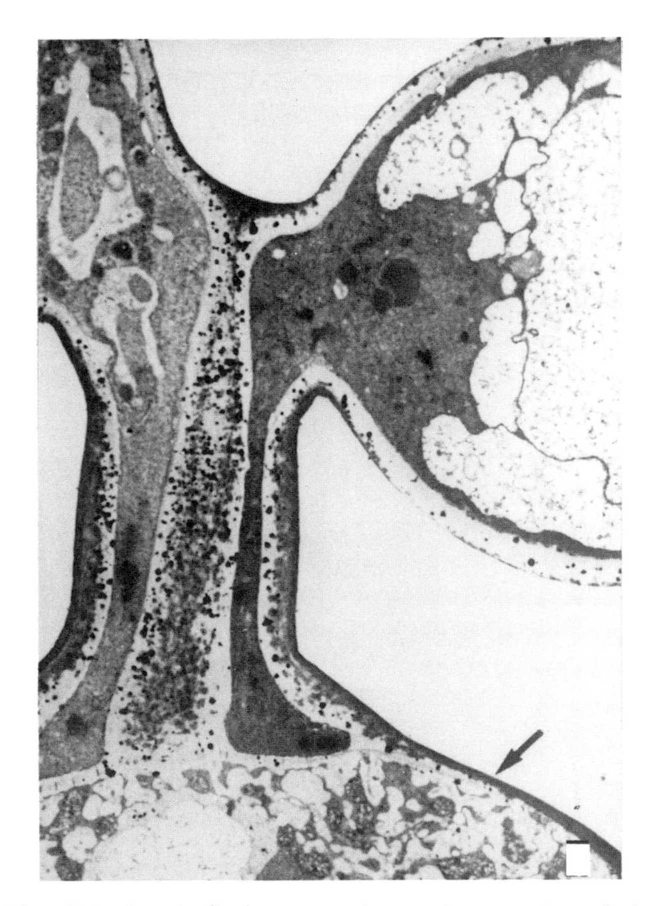

Fig. 10.3 Longitudinal section through a quadrifid from a trap incubated in 1% $La(NO_3)_3$ for 1 day. The tracer extends from the arms (*top right and left*) through the stalk and has just entered the wall ingrowths of the pedestal cell. The La^3 has failed to penetrate the continuous impregnation of the lateral wall (arrow) and is also excluded from the outer cuticularized parts of the stalk. (Reproduced by kind permission of Drs. B. Fineran and J.M. Gilbertson, Christchurch, New Zealand).

extensive labyrinthine wall, where they accumulate, unable to cross the endodermoid layer (Fig. 10.3). Such cytoplasm as persists in these cells is largely occupied by numerous mitochondria at a much greater density than in the head cells (Fineran and Lee, 1974, 1975). Both these cells and their labyrinthine walls show structural changes compatible with different levels of metabolic activity (Fineran and Lee, 1975). In one state the mitochondria have narrow cristae, while the wall ingrowths are distended; in the other the cristae are swollen and the wall ingrowths constricted (cf. *Dionaea*; see page 181). Fineran (1985) interprets these as indicating low and high activity states respectively, associated with the rapid removal of water from the traps.

As argued by Fineran and Lee (1975), it thus seems probable that the absorption of solutes in *Utricularia* takes place at the endodermoid labyrinthine surface. The extensive convolution of the plasmalemma at this boundary indicates an ability to establish a considerable rate of flux. Symplastic transport out of the endodermal cell to the basal cell may take place through the large, compound plasmodesmata present at about 7 μm^{-2} (Fineran and Lee, 1975). Numerous simple plasmodesmata (35 μm^{-2}) also occur in the outer transverse endodermoid wall. These may be associated with water movement through the gland, which may use both apoplastic and symplastic pathways.

The mechanism of absorption in *Utricularia*

Nothing is known, however, about the mode of absorption of solutes. After capturing prey the bladders are rapidly reset (see page 119) and further prey often caught long before the first is degraded or even dead. Degradation is slow and possibly dependent primarily on microbial activity (see page 205). Two models are thus possible invoking either a slow, continual uptake or a 'burst' of uptake linked to the resetting of the bladder. In both cases transport systems must be located in the plasmalemma of the endodermal cell and the apoplast of the head-cells will thus present a considerable diffusion-limited barrier. If continual uptake occurs this will be partially overcome by high-affinity systems creating a concentration gradient. There are advantages, however, in linking absorption to the phase of water resorption. Firstly, the rapid flow of fluid through the head-cells (resetting takes about 30 min [Fineran, 1985; Sasago and Sibaoka, 1985a]) will decrease the long diffusion barrier by invoking mass flow; secondly, metabolic energy will be conserved as the solute transport systems could involve ion-cotransport. Further work is required to elucidate this problem. It should be noted, however, that water removal is an active process, being inhibited in *U. vulgaris* by various respiratory metabolic inhibitors (Sasago and Sibaoka, 1985a, b, see page 119).

Pinguicula and Droseraceae (Non-Continuous Digestion)

The site of absorption in *Pinguicula*

Although we have speculated that absorption is rapid in *Utricularia*, in *Pinguicula* this is known to be the case (Heslop-Harrison, 1975, 1976), taking about the same time as required to produce the pool of secretion. As noted by Darwin (1875) the pool is completely resorbed, in contrast to *Utricularia*. Absorption is apparently already occurring in 'fired' glands, however, while the pool of fluid is still expanding. Thus, working with *P. grandiflora*, Heslop-Harrison and Knox (1971) demonstrated autoradiographically that the sessile glands absorb ^{14}C-labelled amino acids within 2 h of stimulation. By 12 h, these have been distributed throughout the leaf along the lines of the vascular tissue (Fig. 10.4).

During the release of digestive fluid the protoplasm of the gland-head cells becomes considerably degraded (see page 183), making it clear that these cells are unlikely to play any active role in uptake. There is some confusion in the literature, as in earlier works (Heslop-Harrison, 1975, 1976) the extent of this degradation was not apparently appreciated (cf. Heslop-Harrison and Heslop-Harrison, 1981). Thus, Heslop-Harrison (1975, 1976) reports that lanthanum ions added to the digest pool temporarily accumulate in the spongy wall space of the head cells but they would fill the entire head-cell space were the plasmalemma no longer intact. What is clear is that the lanthanum, as in *Utricularia*, does not penetrate either the lateral wall or the protoplast of the endodermoid layer. Thus the plasmalemma of the endodermoid cell has remained intact, providing a boundary at which active uptake of solutes may occur, again allowing control over their movement from the apoplast to the symplast. The way that digestion products concentrate in the region around the gland indicates this (Fig. 10.5).

The mechanism of absorption in *Pinguicula*

Because total resorption of water and, apparently, solutes takes place, Heslop-Harrison (1976) has

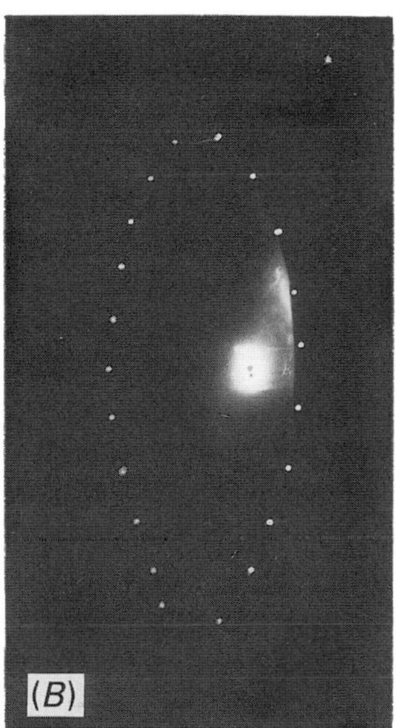

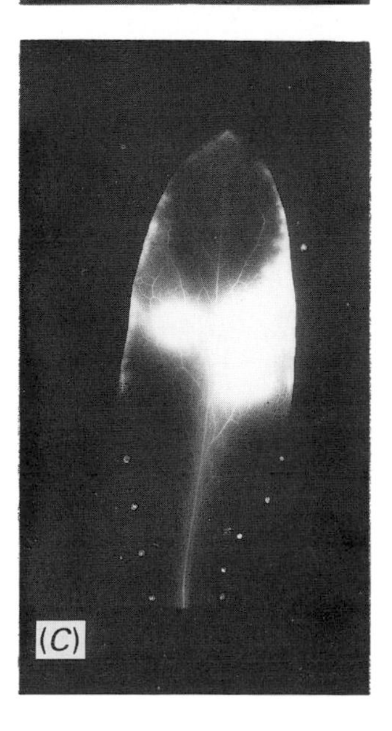

Fig. 10.4 Micro-autoradiographs of leaves of *Pinguicula grandiflora* excised 2 h, 4 h and 24 h after feeding with ^{14}C-labelled protein. A needle-point load of protein was placed to the right of the mid-rib in each case except that in the load of 'C' another small particle contaminated the area to the left. The blade is outlined, in each case, by the white chain-dot line. The emulsion is saturated over the central area of intense radioactivity. The radial spread of this area is due to the increase in the size of the secretion pool, which is partly drained during drying for autoradiography, and partly by the movement within the leaf. The early transport to the margin may be seen in *A* and *B*. (Unpublished autoradiographs by kind permission of Dr Y. Heslop-Harrison, with reference to Heslop-Harrison and Knox, 1971.)

proposed a model invoking mass flow in response to water tensions induced in the xylem. For such a model to work, the protoplast of the endodermoid cell must become progressively more permeable with total loss of membrane-mediated control and the removal of any osmotic barrier. This model, however, seems too simplistic. Firstly, the rate of resorption is not too great to be accounted for by active processes: it is comparable to the rate of secretion which is controlled by ion pumps (see page 184) and is slower than the resetting of *Utricularia* bladders, where water is actively removed from the bladder (see Ch. 6.). Secondly, the endodermoid cells continue to provide a barrier to the passage of lanthanum nitrate (Heslop-Harrison, 1975, 1976) indicating that the membranes remain intact. Thirdly, if the integrity of the endodermoid layer is disrupted, there is a total loss of control over water movements within the leaf, making it difficult to see how other glands could subsequently be activated.

It would appear that the endodermoid cell plays a key role in absorption. Probably the polarity of the ion pumps used to establish water flow and the release of digestive fluid is reversed, causing water and solutes to flow back into the leaf. The large number of mitochondria in these cells apparently

Fig. 10.5 Micro-autoradiographs of the head cells of a sessile gland of *Pinguicula grandiflora*. (*A*) The glands have absorbed ^{14}C-labelled protein applied to the leaf surface. The distribution of silver grains suggests that digestion products have been taken into the head cells. (*B*) As *A*, but showing a concentration of radioactivity in a ring defining the annular pit in which the sessile gland is sunken (cf. Figs 4.2 and 6.1*A*). This concentration may also be seen in *A* above, although the focal point is too high to resolve it perfectly. (Both autoradiographs by kind permission of Dr Y. Heslop-Harrison from Heslop-Harrison and Knox, 1971.)

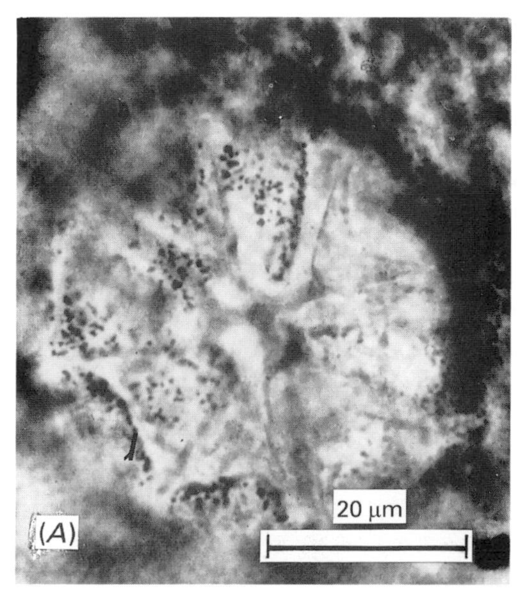

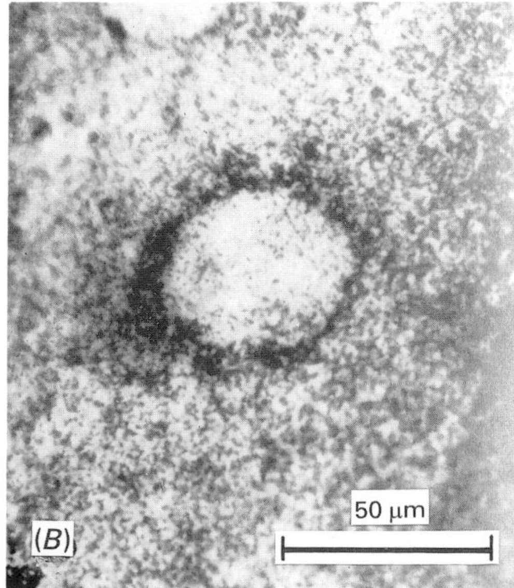

remain coupled during resorption, implying that the cells are using a lot of metabolic energy. This energy turnover is closely analogous to *Utricularia* and it is, perhaps, surprising that inwardly directed labyrinthine walls are not found here also. The tracheids always found just below *Pinguicula* glands provide a route for the rapid removal of resorbed materials from the vicinity of the gland.

It thus appears that *Pinguicula* is a little more sophisticated than at first thought and is, in fact, more like the other well-studied genera than realized. Nothing is known, however, about the metabolic control of absorption.

The site of absorption in the Droseraceae

Studies such as those of Darwin (1876), Gardiner (1885), Vries (1886) and Åkerman (1917) on the induction of 'aggregation' in the stalked glands of various species of *Drosera* clearly established their ability to absorb solutes. More recently, ^{3}H-asparagine was shown to be absorbed rapidly by the head of *D. capensis* glands and to pass progressively down the stalk (Juniper *et al.*, 1977), while the absorption and transport of a range of substances in these organs has been examined in detail (Oudman, 1936; Arisz and Oudman, 1937; Arisz, 1941, 1942a, b, 1944a, b, 1953). Thus, without doubt, these glands are the absorptive surface of *Drosera*.

The absorption of solutes by the lobes of *Dionaea* has also been recognized for some time. Darwin (1875), when studying the stimulation of secretion in the trap, realized that the absorption of a nitrogenous secretogogue was the primary event of the digestive cycle. It has now been shown that the glands are not fully matured until that stimulation occurs (see page 162). Lüttge, having shown autoradiographically that the lobes of *Dionaea* readily absorb both Ca^{++} and glutamic acid (Lüttge, 1963), went on to show by micro-autoradiography that the glands were, as previously indicated, the sites of absorption (Lüttge, 1965). There has been only minimal investigation of uptake in *Drosophyllum* and none in *Triphyophyllum*. In the former genus, Schnepf (1963c) demonstrated that the glands are the site of absorption. The close structural similarity of the glands of these genera to *Drosera* and *Dionaea* makes it probable that their absorptive processes are similar, despite *Triphyophyllum* belonging to a different (though closely related) family (see Fig. 17.3).

The pathway of absorption in *Dionaea*

The demonstration that ^{14}C-leucine fed to traps is rapidly detected in the hydrolases being secreted by the same trap, clearly shows that absorption and secretion are taking place simultaneously (Robins and Juniper, 1980b). *Dionaea* thus differs fundamentally from *Pinguicula*. As the secretory activity of the glands involves the flow of fluid containing hydrolytic enzymes outwards through the apoplast of the gland-head cells (see page 184) it is difficult to conceive an apoplastic counter-flow of digestion products. Even if this were envisaged, the ultrastructural anatomy of the glands is not compatible with a mechanism like that in the Lentibulariaceae. The endodermoid cells are notable in lacking labyrinthine walls, in contrast to *Utricularia*, yet the ratio of the outer surface of the gland to the endodermoid outer transverse wall is about 10:1 (Fig. 10.6; Robins and Juniper, 1980e) and two layers of head-cells would present an even greater barrier to diffusion.

On the basis of measurements of the surface areas at different boundaries within the gland, Robins and Juniper (1980e) proposed that absorption takes place primarily at the outer surface of the gland, solutes being transported into the symplast of the outer layer of head-cells and traversing the gland through the plasmodesmatal pathway. Such a mechanism is compatible with the known structural and functional features of the gland and is consistent with studies carried out in *Drosera*, which trace calcium uptake by X-ray analysis (see page 216). Particularly, it allows for bi-directional flow. The greatest absorptive surface is found in the outer head-cells (Fig. 10.6*A*) and these are connected with the inner head-cells by numerous plasmodesmata. The density of these symplastic connections increases further in the transverse wall between the inner head-cells and the endodermoid cells and their dimensions ensure that there is adequate symplastic continuity to ensure a steady flux across the gland (Fig. 10.6*B*). The endodermoid cells are connected to the basal cell by massive 'pit-like' structures (Fig. 8.3) which also occur in the basal/epidermal cell walls (Fig. 8.3) providing onward routes for the digestion products into the leaf. Curiously, *Dionaea* glands have no tracheid connections like *Pinguicula*, despite their greater complexity. As these connections occur in the other Droseraceae, and a direct evolutionary line is strongly indicated (see Ch. 19 and Fig. 19.1), this absence is hard to understand.

The problem with the use of lanthanum and uranium salts to trace pathways of uptake is that these salts fail to show a symplastic route, due to

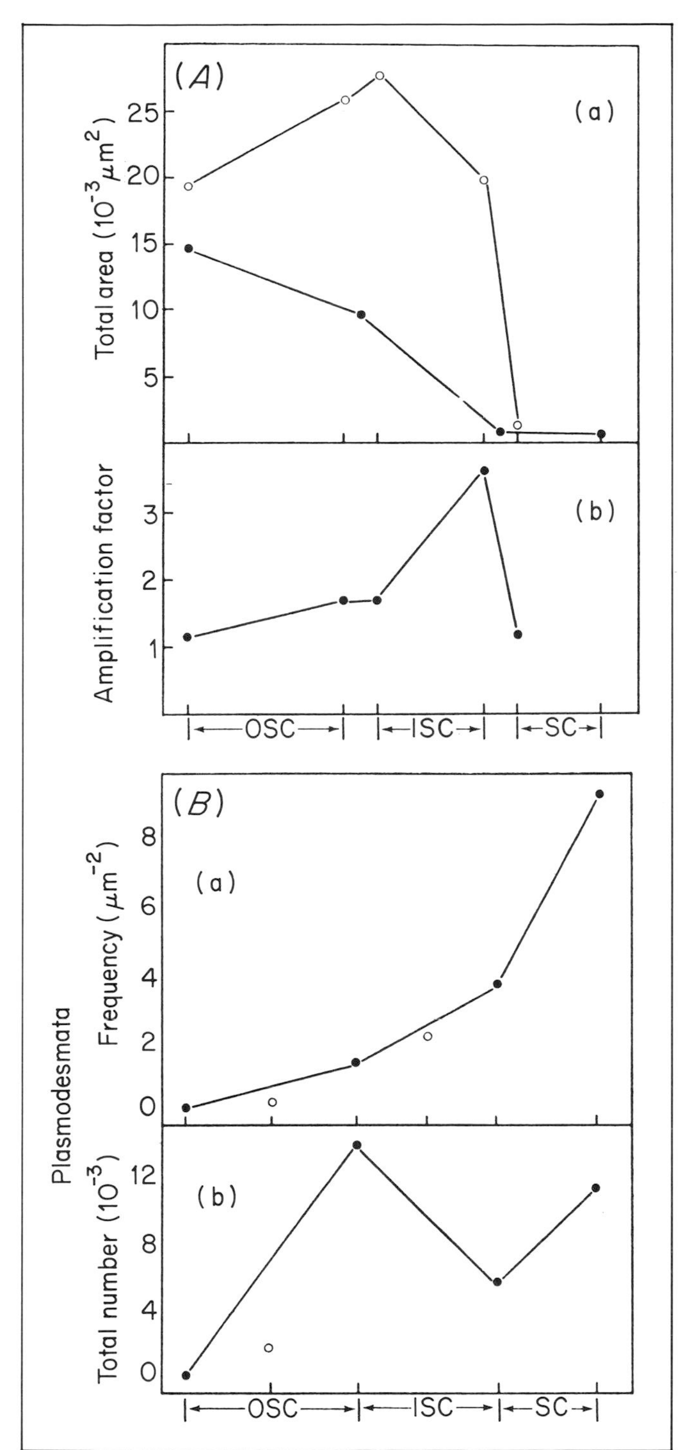

Fig. 10.6 (*A*) (a) = area of each of the boundaries within the gland. ○, area of boundary; ●, area of labyrinthine wall; (b) = relationship between the amplification factors at each boundary within the gland. OSC, outer secretory cell; ISC, inner secretory cell; SC, stalk cell.

(*B*) Relationship between the plasmodesmata at each boundary of the gland; (a) = number of plasmodesmata per μm^2. (b) = total number of plasmodesmata in that boundary. OSC, outer secretory cell; ISC, inner secretory cell; SC, stalk cell. Open symbols show the frequency or number in the boundaries between outer secretory or inner secretory cells. (Redrawn from Robins and Juniper, 1980e.) Gland from *D. muscipula*.

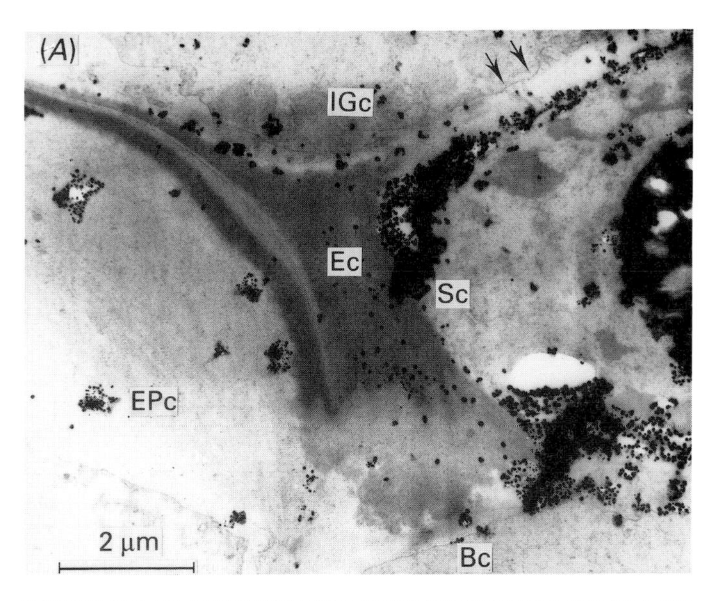

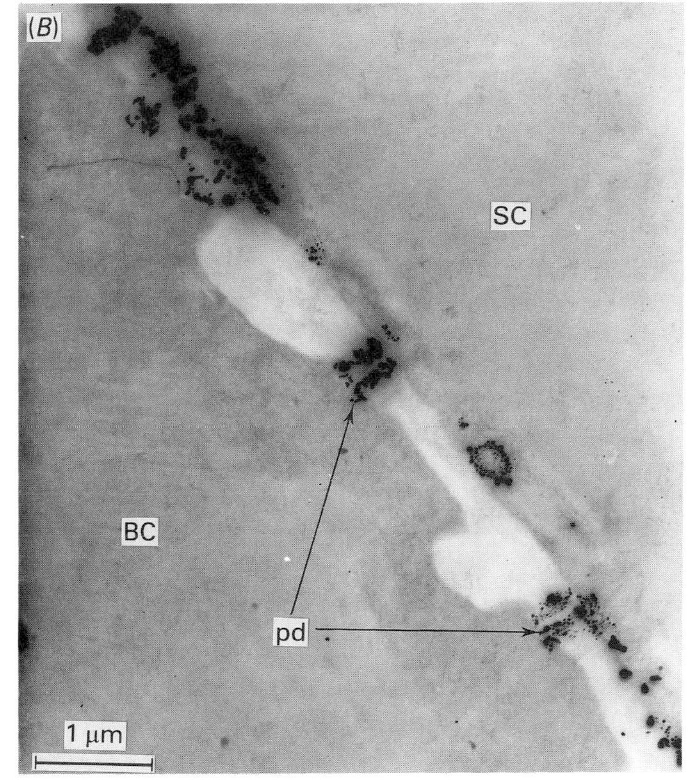

Fig. 10.7A Detail of a cell wall between an epidermal cell (EPc) and a stalk cell (Sc) of a sessile gland of *Dionaea*, three days after stimulation as determined by silver precipitation. The pathway of chloride movement is shown by the dark glandular deposits. The region adjacent to the endocuticle (Ec) shows heavy deposits within the cytoplasm of the stalk cell abutting both the epidermal cell and the basal cell (Bc) cell walls. The deposits are almost completely excluded from the thickened and impregnated wall region (Ec). Plasmodesmata between stalk cell and inner glandular cell (IGc) arrowed. (Reproduced with permission from Robins and Juniper, 1980e.)

Fig. 10.7B The cell wall between a stalk cell (SC) and a basal cell (BC) of a *Dionaea* gland three days after feeding with silver nitrate. Note AgCl deposits in plasmodesmata (pd). (TEM by R.J. Robins, D Phil thesis.)

their inability to enter the protoplast. In an attempt to show the symplastic pathway, stimulated traps were fed Cl^- for various lengths of time and then fixed in the presence of silver nitrate (Robins, 1978; Robins and Juniper, 1980e). This reaction precipitates electron-dense deposits of silver chloride in those regions where Cl^- ions have penetrated. Once again the endocuticle was found to exclude the tracer (Fig. 10.7*A*), confirming the block that this presents to apoplastic flow. Instead, dense deposits were located in the walls of the endodermoid cells where plasmodesmata were located. These were particularly prominent in the pits of the endodermoid/basal cell wall and largely absent in other parts of that cell wall (Fig. 10.7*B*). This supports the suggestion that the plasmodesmata are providing the absorptive pathway through the glands. Further support for this proposal is found in the distribution and activity of mitochondria within the gland. Endodermoid cells contain far fewer mitochondria than head cells and do not show the stimulus-induced coupling seen in the latter (Robins, 1978; Robins and Juniper, 1980a). This situation is in complete contrast to the densely packed, coupled mitochondria in *Utricularia* endodermoid cells (Fineran and Lee, 1974, 1975), once again showing a marked difference between *Dionaea* and the Lentibulariaceae.

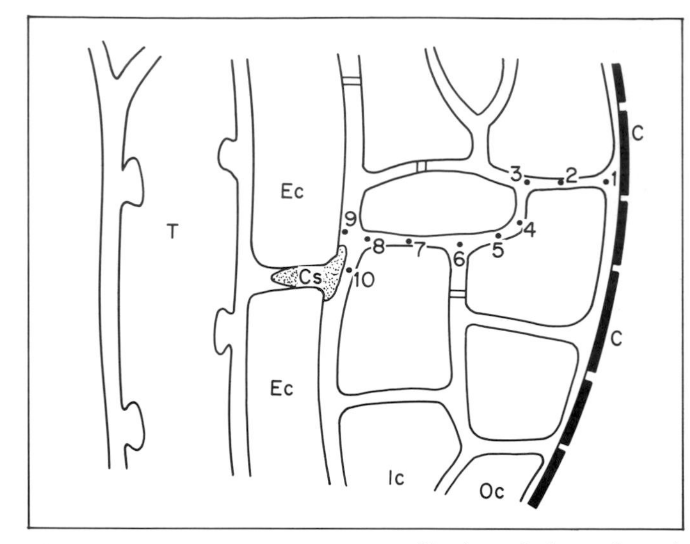

Fig. 10.8 Diagrammatic representation of a part of a *Drosera capensis* gland showing sites where the apoplast (the region external to the plasma membrane) was analysed for calcium in the EMMA-4 X-ray analysing electron microscope. The Ca solution penetrates readily through the perforated cuticle (C) and along the walls between the outer (Oc) and inner (Ic) gland cells until apparently blocked by the barrier of the Casparian strip (Cs) between the endodermoid cells (Ec). T = tracheid. (Redrawn from Juniper *et al.*, 1977.)

Relative peaks for Ca at the sites numbered above

Position	1	2	3	4	5	6	7	8	9	10
Relative peak value	145	145	150	92	55	55	45	35	15	85

The pathway of absorption in *Drosera*

The increased structural complexity of the stalked gland of *Drosera capensis* (see Ch. 6 and Figs 10.8–10.10) need not necessarily suggest a more complex pathway for the initial absorption of solutes than those already discussed. The two (or in places, three) layers of head-cells responsible for mucilage and protein secretion, overlying an endodermoid layer, creates a longer diffusion pathway but with apoplastic and symplastic continuity. *Drosera* dif-

Table 10.1 Mean counts per minute for calcium at different locations within the stalked gland of *Drosera capensis* as a function of time

<table>
<tr><th rowspan="2">Position within gland (see Figs. 4.1 and 10.8)</th><th colspan="6">Duration of feeding with milk (min)</th></tr>
<tr><th>0</th><th>1</th><th>2</th><th>15</th><th>60</th><th>120[a]</th></tr>
<tr><td></td><td colspan="6">Counts per minute (equivalent) on EMMA-4</td></tr>
<tr><td>Outer gland cell—cytoplasm</td><td>0.1 (7)</td><td>6.2 (6)</td><td>11.0 (5)</td><td>8.2 (5)</td><td>27.3 (10)</td><td></td></tr>
<tr><td>Inner gland cell—cytoplasm</td><td>0.4 (6)</td><td>2.2 (5)</td><td>5.3 (4)</td><td>10.7 (6)</td><td>25.6 (10)</td><td>0.2 (10)</td></tr>
<tr><td>Outer gland cell—cell wall</td><td rowspan="2">0.1 (9)</td><td>4.0 (4)</td><td>11.8 (5)</td><td>1.9 (5)</td><td>31.5 (10)</td><td>0.6 (10)</td></tr>
<tr><td>Inner gland cell—cell wall</td><td>3.1 (4)</td><td>4.7 (5)</td><td>4.0 (7)</td><td>15.4 (10)</td><td rowspan="2">0.0 (3)</td></tr>
<tr><td>Endodermoid cell—cytoplasm</td><td>0.3 (4)</td><td>6.8 (6)</td><td>4.0 (6)</td><td>9.3 (10)</td><td>7.6 (10)</td></tr>
<tr><td>Tracheid wall</td><td>0.3 (4)</td><td>−0.1 (3)</td><td>—</td><td>0.8 (3)</td><td>7.8 (5)</td><td>1.8 (6)</td></tr>
<tr><td>Transfer cell wall</td><td>0.9 (4)</td><td>—</td><td>—</td><td>—</td><td>12.6 (5)</td><td>8.8 (3)</td></tr>
<tr><td>Stalk cells</td><td>0.0</td><td>0.0</td><td>—</td><td>11.7 (11)</td><td>−0.3 (5)</td><td>—</td></tr>
</table>

[a] 120 min feeding with milk then 'chased' for 22 h before fixation. Figures in parentheses indicate number of analyses presented. (Redrawn from Juniper *et al.*, 1977.)

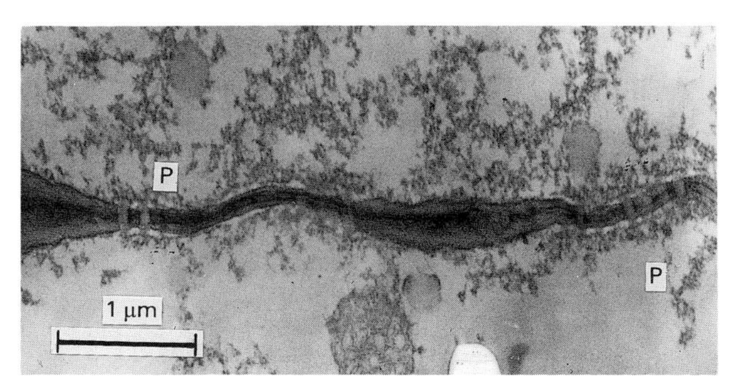

Fig. 10.9, Above. Parts of two stalk cells of the outer edge of the gland/stalk boundary of *Drosera capensis*. See (S) at the base of Fig. 8.4*D*. The cross wall between these two stalk cells is penetrated by abundant plasmodesmata (P) clustered into pit fields. (Unpublished micrograph reproduced by kind permission of Dr A.J. Gilchrist; see also Gilchrist, 1974.)

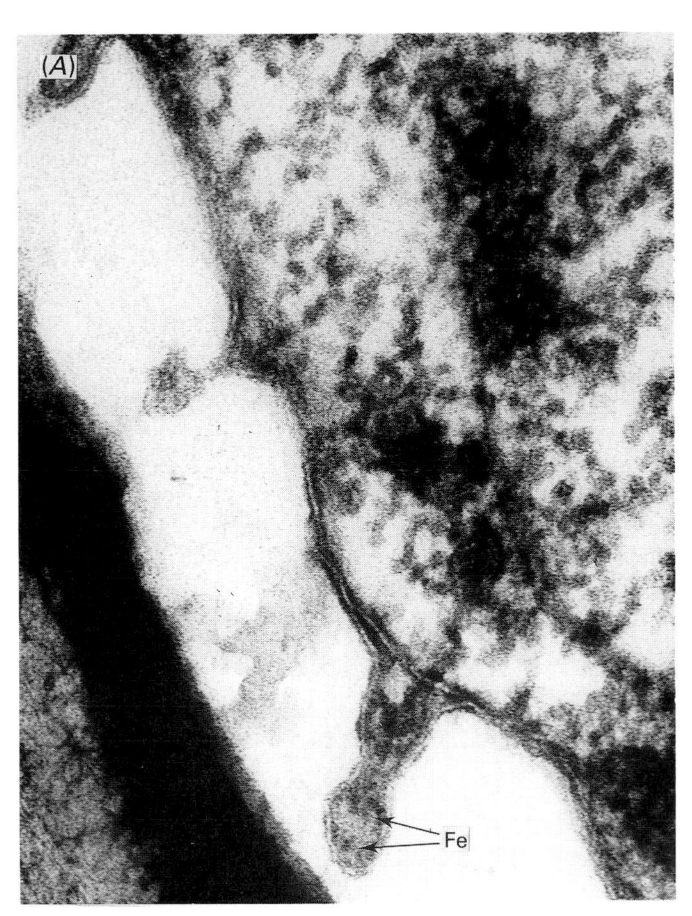

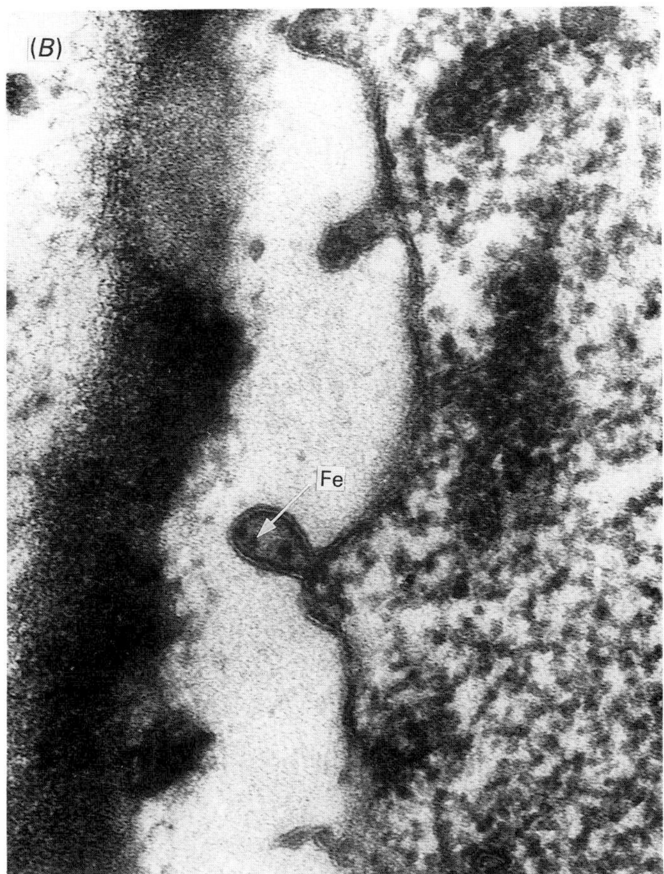

Fig. 10.10, Right. Evaginations of the inner plasma membrane of an endodermoid cell of *Drosera capensis* after feeding. These 'blebs' form in the space between the inner glandular cells and the central tracheid mass (SQ' in Fig. 8.5*B*). These evaginations are believed to represent the sites of exocytosis of proteins from the living endodermoid cells to the tracheids, and thence down the stalk cells to the leaf. (*A*) In this micrograph there appears to be a septum across the base of one evagination, but this may be due to the projection having a thickness only slightly greater than the thickness of the section. Ferritin particles (Fe) may be present in the evaginations. Parts of the plasmamembrane appear to stain heavily with phosphotungstic acid, which may be an indication of enhanced membrane activity. TEM, conventional thin section with PTA stain. (Juniper *et al.*, 1977 and by kind permission of Dr A.J. Gilchrist.)

fers, however, from those systems so far considered, in neither showing degradation of the head-cells (cf. *Pinguicula*) nor a rapid phase of uptake (cf. *Utricularia*). Furthermore, beyond the endodermoid layer are several structures unique to this group of stalked glands. The passage of solutes between these cells and then down the stalk adds a further level of complexity to be considered.

By feeding milk to the glands it has proved possible to follow the movement of calcium through the head and into the stalk at the ultrastructural level (Gilchrist, 1974; Juniper and Gilchrist, 1976; Juniper *et al.*, 1977). The calcium appears to be co-transported with partially degraded proteins and is retained when material is prepared for electron microscopy. Its passage can thus be followed by X-ray microprobe analysis (Table 10.1 and Fig. 10.8).

Calcium enters the gland very rapidly and within 1 min is detectable within the head-cells and the endodermal cells. By 15 min it is detectable throughout the gland and head-cells and into the stalk-cells. During the first few minutes there is little difference between the concentrations in the apoplast and symplast of the outer or inner layers of head-cells, although nearer to the exterior the level is greater. After 15 min, however, by which time a steady-state has probably been achieved, there is a markedly greater concentration of Ca^{++} within the symplast than in the apoplast of these cells. Furthermore, the content of the endodermoid cells is similar to the head-cell's symplast. It appears, therefore, that the major part of the solute is taken into the head-cells and transported from them symplastically to the endodermoid layer. A scan across the apoplast after 2 min supports this (Fig. 10.8). The Ca^{++} content decreases nearly five-fold from the outer wall to the head-cell/endodermoid boundary, where a dramatic increase occurs due to the blockage of further passage by the endocuticle. If the main site for absorption was at the endodermoid external transverse wall this dramatic increase would not occur. Also, the apoplast should retain far more Ca^{++} than the symplast. Hence, it appears that absorption takes place in the head cells, primarily into the outermost layer. After 60 min feeding the whole apoplast and symplast exterior to the endodermoid layer is saturated with calcium.

The ultrastructural similarity of the head-cells of *D. capensis* to those of *Dionaea* adds further support to this proposal. The labyrinthine walls of the heads-cells show a similar polarity and the outer transverse endodermoid walls lack elaborations. Plasmodesmata are fairly frequent between the outer and inner layers of head-cells although notably absent from the outermost part of the anticlinal wall of the outer head cells (Ragetli *et al.*, 1972). Between the inner layer of head-cells and the endodermoid cells they occur in groups, rather like in *Dionaea*, with unbranched wide pores (Ragetli *et al.*, 1972; Gilchrist, 1974; Williams and Pickard, 1974). They do not appear to be so numerous as in *Dionaea* but the bell-shaped structure of these glands results in a much lower ratio between the surface area of the outer gland head-cells and the outer transverse wall of the endodermoid layer. Thus each endodermoid cell is, effectively, linked to a smaller surface area than in the more hemispherical gland of *Dionaea*. The numerous cuticular pores already described (see page 163) ensure that there is no block to absorption at the surface of the gland. It has been suggested that absorption only occurs at the exposed ends of the endodermoid 'bell' (Ec in Fig. 4.1; Williams and Pickard, 1969, 1974). While this would simplify the absorptive pathway considerably it has been reasonably argued that this is not the case (Juniper and Gilchrist, 1976).

The translocation of absorbed products in *Drosera* tentacles

So far then, the pathway is apparently the same as that of *Dionaea* and may be rather general for glands coping with a bi-directional flow of this nature. The added complication in the *Drosera* glands is their tri-functionality, as they are probably secreting both proteins and carbohydrates as they absorb. The complexity of the internal structural features are interpreted as allowing symplastic bi-directional flow using different layers of cells.

The entire core of the gland head is filled with a spongy mass of tracheid cells. Juxtaposed to these at the open end of the 'bell', Gilchrist (1974) describes a layer of stalk cells which have extensive labyrinthine walls facing the spongy tracheids (TCW in Fig. 4.1). This layer is penetrated by the xylem element which forms the core of the stalk. In the stalk itself the xylem is surrounded by two quite distinct layers of cells (Fig. 6.4). All of these cells show features of importance to transport activities.

The central tracheid mass shows little internal structural detail in mature glands. When immature material is examined, however, these cells are found still to be very like parenchyma, with protoplasts closely resembling those of mature inner stalk cells (Ragetli *et al.*, 1972). These conducting cells are very elongated (130–570 μm), often tapered at one end, and are extensively vacuolated, with an extremely unstructured cytoplasm containing numerous chloroplasts and mitochondria. Down the entire length of the stalk the transverse walls between these cells contain large accumulations of plasmodesmata in simple pits (Fig. 10.9). No plasmodesmata are found, however, between these inner stalk cells and the tracheid mass, despite their apparently common developmental origin. Instead, this junction contains an elaborated labyrinthine wall, making it a true transfer cell (Gunning and Pate, 1974). These transfer cells have plentiful plasmodesmatal connections with the basal-end of the endodermoid layer as well as between each other and with the inner layer of stalk cells. The outermost layer of stalk cells are more rectangular and less than half the length (60–310 μm) of the inner sheath. A large vacuole, chloroplasts and a structural cytoplasm suggests that these are typical epidermoid cells (Ragetli *et al.*, 1972). Again, the transverse walls contain large numbers of plasmodesmata and in the wall between the epidermal and endodermal cells massive concentrations of these channels are found in simple pits. Scattered groups of plasmodesmata also occur in the longitudinal walls between inner and outer stalk cells, although at a lower frequency than in the transverse walls.

There are three obvious potential pathways for solutes to be translocated from the endodermoid cells out of the gland along the stalks:

(i) symplastic movement through the endodermoid cells into either or both layers of stalk cells;
(ii) secretion by the endodermoid cells into the spongy tracheid mass and reabsorption by the transfer cells into the inner layer of stalk cells, with subsequent symplastic movement primarily through this layer.;
(iii) secretion by the endodermoid cells into the spongy tracheid mass, reabsorption by the transfer cells followed by symplastic movement through the xylem.

Translocation does not appear to involve transport through the xylem. Increasing the osmotic flow of water towards the glands does not influence transport (Helder, 1967) and tracer Ca^{++} was never detected in this tissue (Gilchrist, 1974; Gilchrist and Juniper, 1976). Hence, translocation appears to be symplastic. Studies observing aggregation in tentacles of *D. capensis* fed with caffeine showed transport to take place through the longitudinal cells of the stalk. Aggregation spreads rapidly throughout each cell but there is a marked time-lapse until it appears in the adjacent cell (Kok, 1933), indicating a limitation to flow at the plasmodesmata. It should be borne in mind, however, that caffeine is toxic to the cells and may not be transported in a typical manner (see page 223). Working with asparagine, phosphate and K^{+}, Arisz (1953) showed that almost all the material absorbed by the glands could be recovered from the leaf blade.

When the EM was used to examine stalk tissue, excellent support for this pathway was obtained. Numerous plasmodesmata were observed in the transverse walls of the endodermoid cells, arranged in simple pits (Ragetli *et al.*, 1972; Gilchrist, 1974) and, as noted already, excellent symplastic continuity was found to exist at the endodermoid/stalk and stalk/stalk cell boundaries (Fig. 10.9). Hence, it appeared initially that the first pathway (i) was probably the preferred route for translocation.

When calcium movement is traced beyond the endodermoid, however, a complex picture emerges which, unfortunately, still fails to resolve the problem. What is clear is that such a proposal is too simple, as passage of absorbed substances into the tracheid mass is found to occur. After 10–15 min feeding the membrane at the innermost plasmalemma of the endodermoid cells starts to show evaginations outwards, producing small blebs 50–100 nm across (Gilchrist and Juniper, 1974). These blebs do not originate from the fusion of a vesicle to the membrane and stain strongly with phosphotungstic acid (Fig. 10.10*A*). They contain high concentrations of calcium or, when fed with ferritin, the iron core of this protein (Fig. 10.10*B*; Juniper *et al.*, 1977). It appears that the blebs do not rupture, but actually pinch off and disintegrate in the space between the endodermoidal plasmalemma and the tracheid wall. Although at 15 min the tracheid mass contains negligible calcium its content after 60 min is comparable to that of the endodermoid cells. Unfortunately, the transfer cells were not analysed at 15 min but by 60 min they too contain a high concentration of calcium, especially in the labyrinthine region. As these cells are directly connected to the inner sheath of stalk cells, it is logical to expect further basipetal flow to take place along this potential pathway. Ragetli *et al.* (1972) point out that the extreme length of these cells, their numerous transversely located plasmodesmata and the appearance of the protoplast are all features reminiscent of developing sieve tubes. The tapering at the transverse junctions supports this view. Yet when analysed for calcium after 15 min the concentration in the cytoplasm of the inner stalk cells is only slightly raised above background (Table 10.1). Some calcium is, however, detected in the 'pit fields' between the inner stalk cells. Very much more is located in the cytoplasm and transverse walls of the epidermal stalk cells (Table 10.1).

On the basis of this rather scanty evidence, it is difficult to reach any conclusion about transport beyond the head of the gland. Structural evidence supports a primary role for the inner stalk cells in basipetal movement along the stalk. The distribution of calcium at first suggests flow through the epidermis but could be interpreted as showing rapid flux through the phloem-like inner stalk cells. If the flux down the stalk exceeded the rate of loading by the endodermoid cells then a low concentration in the conducting tissue would be expected. Why, then, is so much found in the epidermal cells of the stalk?

Further experiments using soluble tracers are required to solve this intriguing problem. It does appear probable that the inner layer of stalk cells is the primary route for basipetal transport. The large core of spongy tracheid cells filling the head may act as an internal reservoir, being rapidly loaded with material to be transported. The ability to absorb nutrients quickly is important when the plant is competing with a microbial fauna for the available solutes (see Ch. 9). A 'storage area' within the head-space could confer a distinct advantage. The epidermal cells are then left free to provide symplastic flow towards the gland head of precursors for the synthesis of mucilage and hydrolytic proteins, and the xylem to provide mass flow of water and, perhaps, inorganic ions.

The mechanism of absorption in *Drosera*

Between 1936 and 1953, Oudman and Arisz, working in the Groningen Plant Physiology Laboratory, published a series of papers reporting their detailed investigations into the mechanism by which a wide range of substances are absorbed and transported by *D. capensis* leaves. They were well aware of the problems inherent in the system they were using which consisted of whole leaves with agar pads applied to the tentacles and a strap to prevent leaf curling (Arisz and Oudman, 1937; Arisz, 1953). In some experiments granulation or aggregation were used as indicators of uptake (see page 180) but in most cases the analysis of the nitrogen, phosphorus or ionic content of the tissue was employed, leading to quantitative results.

The uptake of nitrogenous substrates

Some substances are clearly transported by diffusion. Thus, Kok (1933), Oudman (1936) and Arisz and Oudman (1937) all showed that the rate of uptake of caffeine was dependent on the square root of time, indicating a purely diffusive movement. Caffeine uptake is readily followed by granulation (Kok, 1933).

When investigating asparagine, however, a different picture emerged. With this amino acid a constant rate of transport was maintained for at least 72h (Arisz, 1944a). The rate of uptake was quite low in the first 4 h period but thereafter increased (Arisz and Oudman, 1937). The increase was associated with aggregation within the tentacle cells and pre-aggregated tissue did not show a lag prior to the maximal velocity. Plasmolysis did not affect asparagine uptake, indicating a cytoplasmic route through the cells, compatible with the observed enhanced cyclosis of the cytoplasm associated with aggregation (Åkerman, 1917). Furthermore, caffeine, which inhibits or eliminates aggregation, decreased the rate of uptake of asparagine (Arisz and Oudman, 1937).

Caffeine and asparagine also show marked differences in the relationship between external concentration and uptake (Oudman, 1936; Arisz, 1953). The rate of caffeine absorption increases steadily with concentration, showing no saturation even at 50 mM (Fig. 10.11). The amino acid, in contrast, shows saturation kinetics and substrate inhibition at high concentrations. Glycine behaves similarly. The data is insufficient for detailed kinetic analysis but is not fitted by a simple Michaelis-Menten hyperbolic relationship. Both glycine and asparagine are, nevertheless, accumulated against a concentration gradient, accumulation ratios (internal concentration:external concentration) at low concentration (0.2 mM) being determined at 135 (Arisz, 1944a) and 37 (Arisz, 1942a) respectively. That for caffeine never exceeded unity (Arisz, 1942b). Leucine and phenylalanine are also accumulated against a concentration gradient (Arisz, 1942b).

From this evidence it was possible to conclude that the absorption of amino acids involved active transport. This was confirmed by showing that anaerobic conditions (Arisz, 1942b) severely inhibited the absorption of several amino acids. Caffeine, in contrast, was much less affected, the small decrease observed presumably representing a requirement for metabolic energy in the granulation process and in the transfer of caffeine between the vacuoles of adjacent cells. Urea, methyl- and thiourea absorption also showed partial inhibition by anaerobic conditions.

As Arisz (1953) points out, anaerobic conditions pertained throughout the whole experimental tissue. He therefore investigated the influence of inhibitors of oxidative phosphorylation, which could be applied only to the tentacles. Examining asparagine absorption, KCN, NaN_3, arsenite, DNP and iodoacetate were all found to cause inhibition in a concentration-dependent manner, 50% inhibition occurring at 10^{-4} to 10^{-6} M depending on the conditions and the inhibitor used. Caffeine uptake, however, was inhibited only 50% by 3×10^{-3} M KCN. A number of metabolic poisons such as eosin and fluorescein were also inhibitors of amino-acid uptake (Arisz, 1953).

The influence of temperature on uptake of these compounds varied, again in parallel with the degree to which they are actively absorbed. Thus, for example, the rate of uptake of caffeine increased 2.6-fold between 5 and 25°C whereas that for

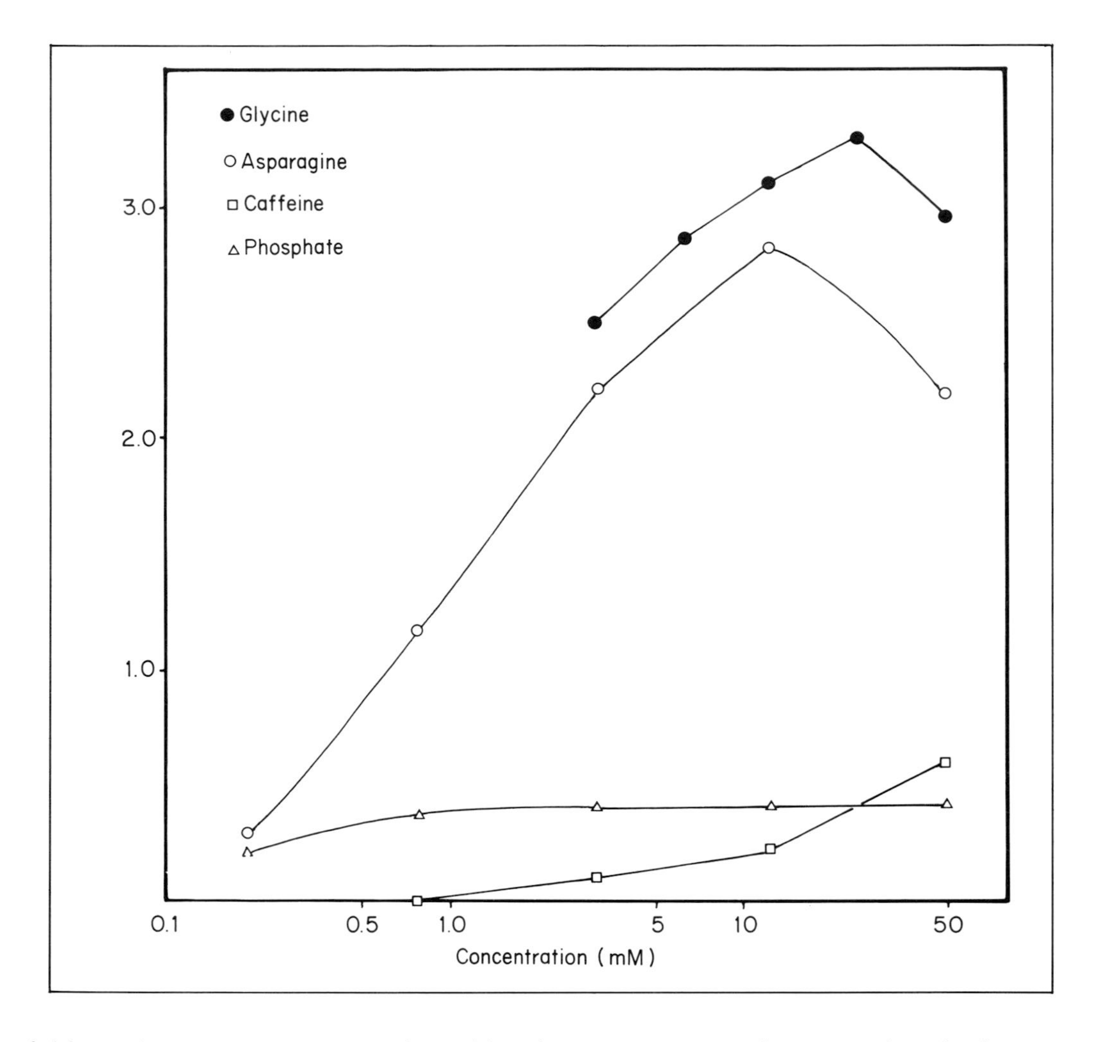

Fig. 10.11 The effect of concentration on the rates of absorption of various substances by the stalked glands of *Drosera capensis*. (Data taken from Arisz, 1953).

asparagine increased 5.3-fold (Arisz, 1953). Urea uptake, despite being less affected by anaerobiosis than amino-acid uptake, showed a more marked temperature-dependence, being increased 9.5-fold over this range.

The uptake of inorganic ions

In a similar series of experiments, it became apparent that phosphate is also absorbed by active transport (Arisz, 1942b, 1944a, 1953). The process again shows concentration dependence (Fig. 10.11), oxygen-dependence and inhibition by inhibitors of oxidative phosphorylation. In addition, phoridzin, known to influence cellular phosphate metabolism, was antagonistic. The results were comparable for both sodium and potassium phosphates. Potassium nitrate or chloride, however, showed no uptake (Arisz, 1944a), although potassium chloride in combination with calcium sulphate inhibited phosphate uptake (Arisz, 1953).

With ammonium salts, however, different effects were observed. The uptake of ammonia from NH_4Cl showed all the features of active transport although the chloride was apparently not absorbed. In contrast, ammonia uptake from the carbonate appeared passive, even though the rate of uptake (3.2 mmol/h) was greater than for other substances (Arisz, 1953).

The uptake of other substances

Sucrose appeared not to be absorbed, although examining this substrate presented problems with osmotic instability caused by the sugar (Arisz, 1944a). However, it had little effect on phosphate absorption (Arisz, 1953). No other substances appear to have been examined.

Competition for uptake

Arisz (1944b, 1953) clearly demonstrated that the simultaneous absorption of two or more substances can cause interactions to occur. These are of two types:

(i) *Indirect*: caffeine and ammonium carbohydrate both decrease the rate of absorption of actively transported substrates. Thus, equimolar caffeine decreased the rates of absorption of glycine and asparagine to 4.9% and 12.4% respectively. The effect on phosphate, however, was much less marked, even causing a slight stimulation of absorption from potassium phosphate. Only at an eightfold molar excess (12.5 mM vs. 100 mM) was substantial inhibition achieved with either caffeine or ammonium carbonate. Coupled to the observed effects of toxins such as phoridzin, this indicates quite distinct properties of the amino acid and phosphate transport processes beyond their common requirement for active metabolism.

Probably phosphate is sequestered by the vacuoles, while the amino acids move through the cytoplasm. Thus, at high caffeine levels the granulation induced may be precipitating phosphate within the vacuoles. Ammonium carbonate is probably acting quite differently to inhibit phosphate uptake. The absorption of ammonia causes alkalination of the cytoplasm, affecting proton-gradient-dependent transport process. In this instance, the effect is apparently to decrease net phosphate flux into the vacuole, hence net accumulation.

In contrast, amino-acid absorption is affected by the antagonistic effect of caffeine on aggregation and cyclosis.

(ii) *Direct*: substrates of a similar type compete for a single uptake mechanism, while dissimilar actively-transported substrates do not (Arisz, 1944b). Thus, as shown in Table 10.2, at saturating concentrations (50 mM) glycine and alanine inhibit each other's absorption (expt. III). At low levels (expt. II), an additive uptake is observed. A similar pattern is seen with combinations of glycine with asparagine (Arisz, 1944b). These findings indicate that a general amino acid permease is active in this tissue. Unfortunately, the effect of ammonium carbonate on amino acid transport was not examined. Rea (1984) has postulated for *Dionaea* that amino acid absorption involves co-transport with protons (see page 224 and Fig. 10.12). Ammonia-dependent alkalination of the cytoplasm would inhibit this process due to the ammonia forming ammonium ions by the sequestration of protons.

Table 10.2 Absorption of various substances alone or in combination by tentacles of *Drosera capensis*.

Experiment	Substrates present (mM)				Increases in tissue content after[a]	
	glycine	alanine	asparagine	KH_2PO_4	24h	48h
I	0.8	—	—	—	0.29	—
	—	0.8	—	—	0.32	—
	0.8	0.8	—	—	0.44	—
II	1.6	—	—	—	—	0.45
	—	1.6	—	—	—	0.38
	0.8	0.8	—	—	—	0.47
III	50	—	—	—	0.91	1.16
	—	50	—	—	0.94	1.27
	50	50	—	—	0.99	1.10
IV	50	—	—	—	0.79	—
	—	—	—	50	0.48	—
	50	—	—	50	0.69(N) 0.57(P)	—
V	—	—	50	—	1.21	—
	—	—	—	50	0.39	—
	—	—	50	50	1.18(N) 0.31(P)	—

[a]Nitrogenous materials as mg N/g fresh wt; phosphate as mg P_2O_5/g fresh weight.
(All data from Arisz, 1944b and 1953.)

Differential processes in gland-heads and tentacles

All the experiments discussed so far have made no distinction between the two processes absorption by the head-cells and transport down the tentacles. While the two tissues can be physically separated, this causes injury, loss of osmotic turgor and exposure of the xylem elements making it difficult to perform meaningful experiments. Recognizing these points, Arisz (1953) nevertheless was able to show with reasonable clarity that decapitated tentacles absorb phosphate as rapidly as whole glands and that transport of phosphate to the leaf involves active processes, being inhibited by anaerobiosis, KCN, azide and arsenite. With asparagine, the rate of transport in decapitated tentacles was 70% that of whole glands at 50 mM, but only 20% at 10 mM, indicating that a higher contribution by diffusion occurs than in whole glands. Nevertheless, at 10 mM, transport was inhibited by anaerobiosis and azide, indicating an active component. The transport of these substances has also been shown to be polar (Kok, 1933). Caffeine, however, shows only diffusion-controlled transport through decapitated tentacles and is not polar (Kok, 1933).

Hence, it would appear that uptake by the head-cells and translocation through the tentacles involve the same mechanisms, the nature of which is dependent on the substrate. The data presently available are complex, showing non-linear double-reciprocal plots, indicative of a combination of processes occurring. A marked difference is seen between phosphate and amino acid absorption in that only the latter shows substrate inhibition. Further investigations into the mechanism of amino acid uptake by *Drosera* are clearly needed, in particular using radiotracers and currently available

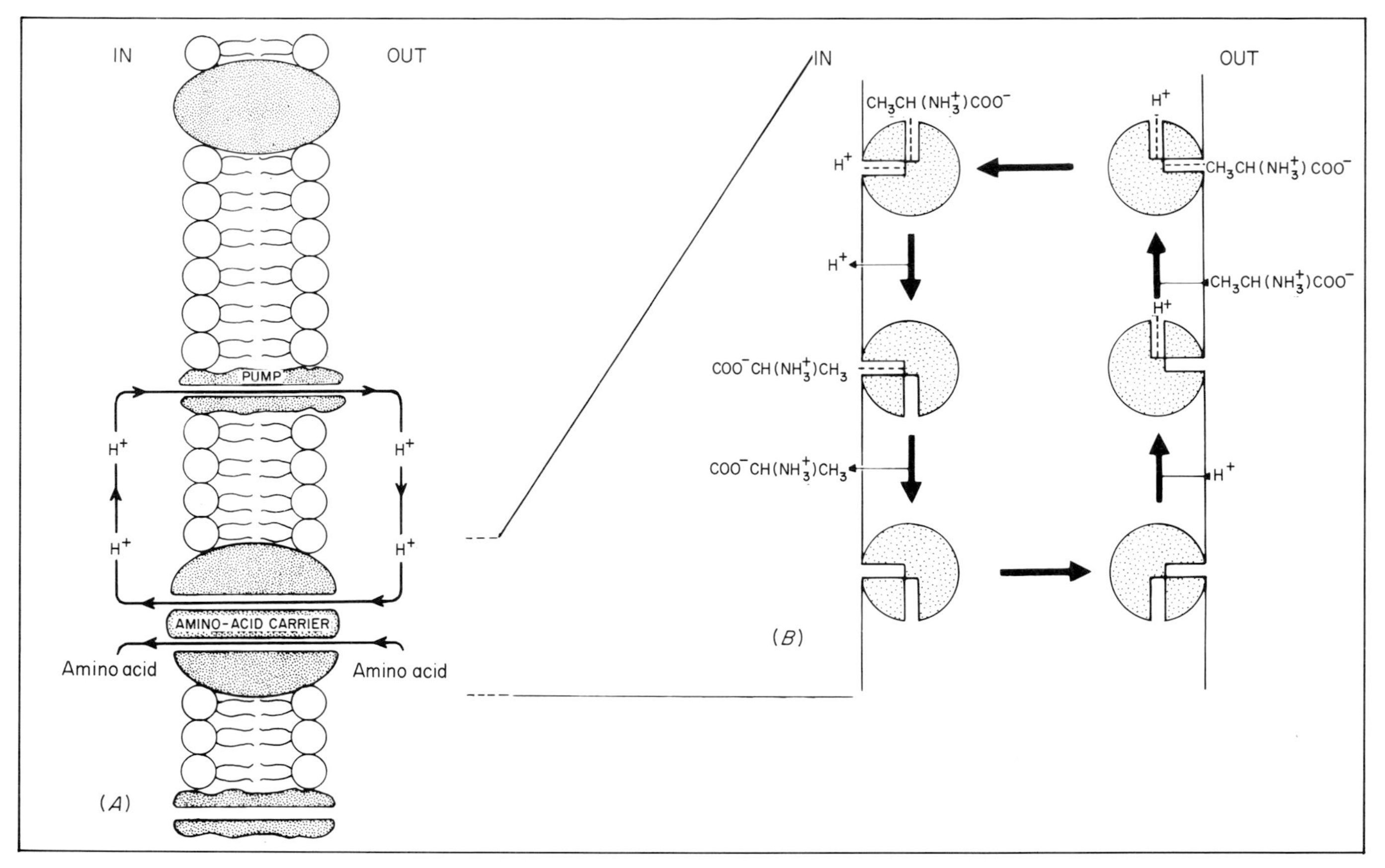

Fig. 10.12 A provisional model for the H^+-dependent transport of D-alanine into the trap lobes of *Dionaea*. H^+ pumping establishes a pH of 3 units within the trap cavity and thereby creates a large H^+ electrochemical potential difference between the absorptive tissue and trap cavity down which H^+ ions and D-alanine may be transported. (Modified and redrawn with the kind help of P.A. Rea from Rea, 1984, and Lüttge, 1985.)

specific inhibitors. (Helder, 1967, quotes work by Boasson, Van Reatte and Zweep using radiotracers but this appears never to have been published). A start has been made using such techniques in *Dionaea* and the results obtained will now be discussed.

The mechanism of absorption in *Dionaea*

Lüttge (1964) demonstrated that the absorption of various solutes from within *Dionaea* traps leads to enhanced oxygen consumption. Feeding with peptones leads to a 2.5-fold increase in respiration over the control while glutamic acid, fructose and glucose all caused about a two-fold stimulation. This strongly indicates that active transport is also required in *Dionaea* to absorb these molecules. Surprisingly, leucine and aspartic acid do not stimulate respiration, indicating that they are possibly absorbed by non-active processes.

The uptake of amino acids by *Dionaea* traps has subsequently been studied in detail (Rea, 1982a; Rea and Whatley, 1983; Rea, 1984). In initial experiments, Rea (1982a) demonstrated that the rates of amino acid uptake by the digestive glands are not related to their metabolism in the plant cells. This was shown by comparing the uptake of D-alanine with that of L-aspartate. D-alanine, the least readily metabolized amino acid in *Dionaea* traps, showed the highest rate of uptake, while L-aspartate, the most readily metabolized, showed a lower uptake rate (Rea, 1982a).

The uptake system in *Dionaea* traps stimulated for 10–20 h, approximates to Michaelis-Menten kinetics and appears to involve carrier-mediated transport. It shows clear saturation with external substrate concentrations, yields high-affinity constants and is pH-dependent. It is also temperature dependent, with a Q_{10} of 2.5 which is indicative of a process that is dependent upon metabolic energy. In contrast, the rate of uptake in unstimulated traps shows neither saturation kinetics nor pH-dependence, indicative of passive diffusion (Rea and

Whatley, 1983; Rea, 1982a). Furthermore, the velocity was at only 15–20% of stimulated traps, indicative of a requirement for the induction of transport mechanisms subsequent to stimulation. The time course of induction (Fig. 10.13) seems similar to that for the formation of cuticular gaps (see pages 162 onwards).

The effect of pH on the active uptake is dependent on whether the amino acid concerned is neutral, acidic or basic, although the uptake of all three is facilitated by low pHs around pH 4.0 (Rea, 1982a). Because of the virtual absence of substrate protonation over the pH range examined (decreasing the pH by 2 units causes the zwitterionic activity to decrease by less than 1%; Rea, 1984), the facilitating effect of low pHs on D-alanine uptake into *Dionaea* glands (Fig. 10.14) is assumed to result from an increase in the affinity of the carrier system for alanine due to carrier protonation (Rea and Whatley, 1983).

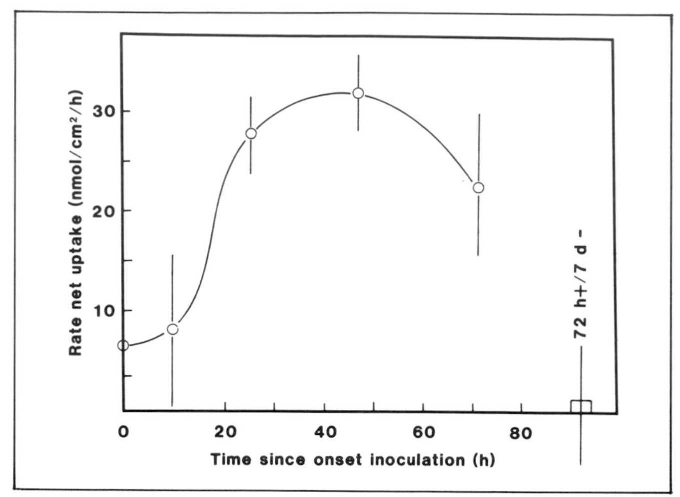

Fig. 10.13 The dependence of the rate of D-alanine uptake by lobular disks from *Dionaea* traps on prior inoculation with 20 kg m^{-3} bactopeptone. The traps, while still attached to the plant, were inoculated with bactopeptone every 12h and at the times indicated were excised for the measurement of uptake. Uptake was measured at pH 3.8 and 30°C from 5 mol m^{-3} solutions of ^{14}C-labelled D-alanine. The bars above and below each point delimit the 95% confidence limits for the rate of net uptake computed from five experiments by linear regression. 72h+/7d− denotes traps that have undergone a 72-h inoculation period followed by a 7-day period without inoculation. (Reproduced, with permission, from Rea and Whatley, 1983.)

Models for amino acid uptake

Based on the above evidence a tentative model as shown in Figure 10.12 has been put forward (Rea, 1982a). In this model, the carrier binds a proton and subsequently a substrate at the outer membrane surface to form a positively charged ternary complex of carrier, substrate and H^+. At the inner membrane surface, the substrate and H^+ dissociate from the complex and the carrier is allowed to recycle. Directionality is imposed on the transport system by the net positive charge of the ternary carrier complex, by the normal negative value of the membrane potential and the normally inwardly directed H^+ chemical potential difference (Rea, 1982a).

The possible participation of an H^+ pump in this system is supported by investigations, *in vitro*, of the dynamics and effector sensitivity of H^+ efflux from the digestive surface of the *Dionaea* traps (Rea, 1983a).

On the basis of this model, accelerated H^+ efflux would enhance amino acid uptake. Additionally, enhanced H^+ efflux would increase the rate at which the H^+ ions circulate in the system, so diminishing the otherwise pH-perturbing consequences of H^+ uptake. The co-transport of amino acids with H^+ ions necessitates a mechanism for the disposal or extrusion of the absorbed H^+ ions. Uptake without simultaneous H^+ efflux would result in a membrane depolarization and alkalinization of the fluid contained in the trap cavity with the eventual loss of the capacity for mediated uptake (Rea, 1984).

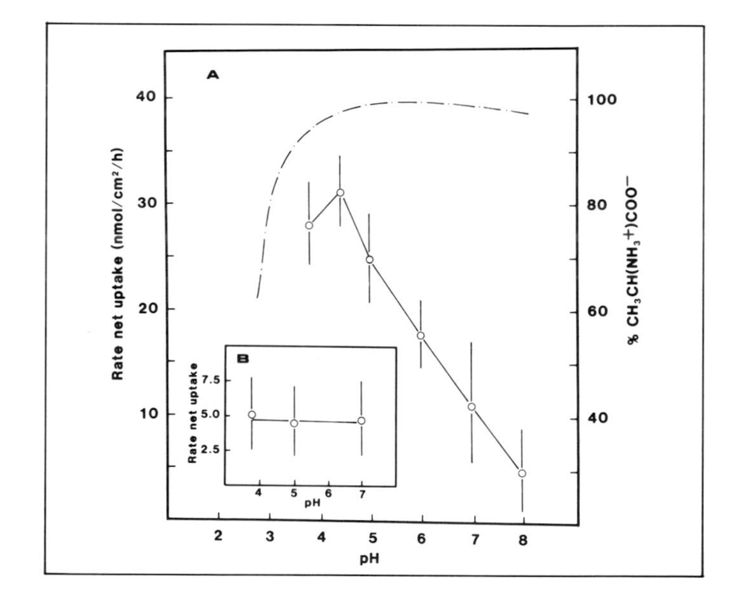

Fig. 10.14 The effect of the pH of the incubation medium on the rate of D-alanine uptake via the adaxial surfaces of lobular disks from *Dionaea* traps. (*A*) Uptake into disks from traps that have undergone a 48 h period of inoculation with 20 kg m^{-3}. —— = rate of net uptake (nmol cm^{-2} h^{-1}); –·–·– = zwitterion as % total D-alanine concentration calculated from the Henderson-Hasselbach equation, pH = pKa + $\log_{10}$ (conjugate base/(acid)). (*B*) Uptake into disks from uninoculated traps. The rate of uptake from five mol m^{-3} solutions of D-[^{14}C]alanine was measured at 30°C. (Reproduced by kind permission of Rea and Whatley, 1983.)

Additional support for this model comes from the effects of compounds that influence H^+ transport on D-alanine uptake (Rea, 1984). FC and 2,4-D stimulate, and diethylstibestrol inhibits H^+ efflux (Rea, 1983b). These compounds stimulate and retard D-alanine uptake in parallel to their effect on H^+ efflux.

While this model fits the pH-dependent uptake of D-alanine, it cannot entirely account for the uptake behaviour of L-aspartate and L-histidine. Rea (1982a) suggests that in contrast to the effect on D-alanine uptake, the principal effect of low pHs in facilitating the uptake of L-aspartate and L-histidine is by way of substrate protonation. In this way, the aspartate anion is converted to its zwitterion, for which it is proposed the carrier has a higher affinity. At the inner membrane surface the proton dissociates from the complex, where the pH is nearly neutral, and aspartate ions are rapidly released from the carrier. For histidine, it is proposed that the cationic form of the substrate binds to the carrier. In all cases the direction of net transport will be determined by the ratio of the intracellular and extracellular substrate and proton activities (Rea, 1982a).

The role of pH in amino acid uptake
Rea's conclusion that low extracellular pHs facilitate the uptake of all three classes of amino acids into the digestive glands is of obvious functional significance with regard to the absorption of the products of digestion from the trap lumen, during the digestion cycle. Since the digestive fluid secreted by *Dionaea* has a mean pH of 3 (Rea, 1982b), and secretion-elicitor-enhanced H^+ effluxes can be demonstrated in trap explants (Rea, 1983a), the implication is that the acid secretion directly facilitates amino acid uptake from the trap cavity (Rea and Whatley, 1983).

The implication of absorption studies for the evolution of carnivorous plants

It has generally been assumed that the primary reason for the acidity of the digestive fluid is to promote the proteolysis by the secretory proteases which possess acidic pH optima (see page 192). However, *a priori*, the secretion of alkaline proteases could also be justified by this argument. Amino acid absorption, on the other hand, is facilitated directly by proton-pumps that lower the external pH as a secondary effect (Rea, 1984). It seems obvious that, in the consequent acidic medium, only proteases active at acidic pHs can be effective. From an evolutionary point of view, it is perhaps more logical to assume that the development of the absorptive mechanism, based on a proton-carrier system, preceded the development of the enzyme-secreting system than *vice versa*. This can explain why the Bromeliaceae show many species with absorptive properties but very few carnivorous species (Benzing, 1980, 1986). Similarly, in *Sarracenia*, the secretion of hydrolases is dubious in some species, while it is very doubtful in *Heliamphora*. All these examples absorb products. Coincidentally, the low pHs may also assist in suppressing microbial contamination (see page 205).

The Metabolic Fate of Absorbed Nutrients

As discussed earlier (see page 210), absorbed nutrients are rapidly redistributed throughout the plant. In many cases they are not retained for any period in the chemical state in which they are absorbed but undergo rapid metabolism. Thus, for example, Plummer and Keithley (1964) found that when peptides were fed to pitchers of *Sarracenia flava*, only the amino acids of which they were composed were detectable in the tissue.

Using ^{15}N-fed *Drosophila* as a dipteran prey, Dixon *et al.* (1980) followed the absorbed nitrogen in *Drosera erythrorhiza* plants. At peak vegetative growth levels, 25% of the nitrogen was transferred to the leaves and 70% of the nitrogen supplied by the insects was accumulated in daughter and replacement tubers (see Ch. 7). The compound arginine was singled out as being of special significance in the storage of nitrogen in the tubers, asparagine was second in abundance in the stems, and it and glutamine, aspartic acid and alanine were the major amino acids of the leaf rosette.

Rea (1982a), using ^{14}C-labelled amino acids, showed that in *Dionaea* a large proportion of the absorbed matter is converted into other compounds. Furthermore, he found that most of the absorbed nitrogen that enters the free amino acid pools of the trap is accumulated as amides. In fact, all four classes of the metabolic nitrogen matter tested (NH_4^+, urea, glycine and xanthine) caused an elevation of the pool size of glutamine in the trap lobes. The other amino acids underwent only small changes. Glutamine and aspargine therefore appear to be the principal free amino acid repositories for the absorbed nitrogen in *Dionaea* (Rea, 1982a).

Glutamine accumulation in *Dionaea* is light-dependent while the accumulation of asparagine is slightly stimulated by darkness. If the traps are transferred from the dark to the light during the course of feeding, the glutamine pool undergoes a rapid and large increase in size. On the other hand, illumination causes a transient decrease in the rate of asparagine accumulation. The light-dependence of glutamine accumulation was related by Rea (1982a) to a requirement for photosynthetic reducing-equivalents for the synthesis of its immediate precursor, glutamine. This is the required substrate for glutamine synthase, known to be active in ammonia assimilation in at least one carnivorous plant, *Drosera aliciae* (Small and Hendrikz, 1974; Small *et al.*, 1977).

The relationship between the levels of glutamine and asparagine seems to indicate a precursor–product relationship (Rea, 1982a). Asparagine seems to be synthesized by a mechanism involving the transamidation of aspartate by glutamine. This process is light-independent and does not require any photosynthetic reducing equivalent (Rea, 1982a). The results of short-term $^{14}CO_2$ incorporation experiments (Rea, 1982a) show that asparagine is an ultimate product of CO_2-fixation in *Dionaea*, and that prior exposure to nitrogenous elicitors is a necessary condition for rapid incorporation into this amide. The kinetics of radio-incorporation indicate that glutamate, aspartate and the non-amide, neutral amino acids are the primary products of ^{14}C-incorporation, while asparagine is secondary (Rea, 1982a).

Of the two amides, asparagine is the more long lived and subject to less metabolic turnover. Asparagine, therefore, appears to be the primary storage form for free organic nitrogen in the traps of *Dionaea*. In contrast to *Drosera*, the dicarboxylic, non-amide, neutral and basic amino acids do not make a significant contribution to the storage of the nitrogen assimilated as a result of artificial feeding in *Dionaea* (Rea, 1982a).

PART IV

Phytochemical Aspects

CHAPTER 11

Phytochemicals

Introduction

A number of members of the carnivorous genera, and closely related genera e.g. *Roridula* have featured in phytochemical investigations. These considerations have been in terms of:

(i) those extracting and identifying compounds for their own sake;
(ii) chemo-taxonomic analysis;
(iii) the isolation of potential pharmaceutical compounds.

In addition some members of the Droseraceae possess a novel pathway for the biosynthesis of naphthoquinones.

These studies have rarely been related to the plants' carnivorous habits. In this chapter we have therefore catalogued the compounds according to their chemical nature. Inevitably this has led to a long list of compounds (largely tabulated) about which, at present, little more can be said. A major problem, from the comparative viewpoint, is that different extraction methods may lead to a false impression of the differences between genera and, needless to say, the analyses are far from systematic. In the two subsequent chapters we discuss the roles of some of these compounds in carnivory (Ch. 12) and the herbal and medicinal uses of these plants (Ch. 13).

It should be borne in mind that, although evolved for catching insects as a nutrient source, this group of plants is equally susceptible to phytophagous insects as other, non-carnivorous species. Thus, they are just as in need of an armoury of protectant chemicals. Some insects are able to feed on carnivorous plants and are discussed in Chapter 14.

The Compounds

Minerals and Aliphatic Organic Acids

As carnivorous plants undergo all the usual metabolic activity of plants they will contain a normal complement of minerals and organic acids. Because, however, their growth is limited by the nutrient supply it is likely that free metabolites may be at a minimal level. While the elemental composition of some species is given in Table 11.1, it is clear that the mineral content may vary greatly not only with the site from which the plants are obtained but also with the season. Pate and Dixon (1978) analysed the seasonal changes in key nutrients for *Drosera erythrorhiza*, and found these to vary four- or five-fold not only with the season but also with the part of the plant analysed. The importance of minerals in nutrition has been discussed in Chapter 7. Other minerals, notably chloride, may play key roles in the mechanism by which the traps operate (see Ch. 6) or in the composition of the hydrolytic environment (see Ch. 9).

Formic, propionic, butyric, malic and citric acids have been isolated from both stem and unripe seeds of *Drosera rotundifolia* (Sablitschka, 1923) and ascorbic acid detected in the leaves and secretion of *D. intermedia* (Weber, 1940). Acrylic acid has been reported in the leaves of *Sarracenia purpurea* (Björklund, 1864). Although Turner (in Lindley, 1846) claimed to detect oxalic acid in the fluid of *Nepenthes*, Voelckler (1849) contested this, considering the crystals to be of potassium chloride. He also found malic and citric acid in the fluid (Voelckler, 1849).

Table 11.1 Elemental composition of various carnivorous plants

Species	Part of plant	Dry weight (%)	Element (mg/g dry weight)																							References
			Al	B	Ba	Ca	Ce	Cl	Co	Cu	Fe	Hg	K	Li	Mg	Mn	N	Na	Ni	P	Pb	Si	S	Sr	Ti	Zn	
Bromeliaceae																											
Catopsis berteroniana	leaf blade					4							7.2		1.2		11.4	5.5		0.9							18
Droseraceae													P														
Aldrovanda vesiculosa	lobe					P		P					P		P			P									24
Dionaea muscipula	lobe																										1,21
Drosera erythrorhiza	rhizome					0.7							12		1.1		10	6		1.7						0.06	11
intermedia	whole	5.8	++	+	+	++	+			+	++				++	++				+	+	++		+	+	+	2
rotundifolia	whole	7.2	++	+	+	1	+			+	++		10		2.8	++	12	2.2		0.8	+	++		+	+	+	2,3,20
Drosophyllum lusitanicum	leaf							P																			1
Lentibulariaceae																											
Pinguicula ionantha	leaf							P																			4
Utricularia benjaminiana	leaf															1											15
foliosa	whole	4.1				5							22		2		24	17		1		18					5
inflata	whole			0.008																							6
vulgaris	whole									0.01	1.8	0.0005	20		8.7	1	24.4	10.1		2.1	0.03		0.6	0.06		0.2	12,13,19
sp.	whole		0.53			8.8			0.004	0.008	4.1	0.08							0.002		+	0.6		0.01		0.08	16,22,23
Nepenthaceae																											
Nepenthes 'henryana' Hort.	pitcher	9.3				16.9		26					26		1.4			5.2									14
Sarraceniaceae																											
Darlingtonia californica	whole																P										7
Sarracenia flava	leaf					1.4							8	–	1.9		8.8			0.6		P					7,8,9
minor	—					P							P									P					9
purpurea	leaf					P		P			P		4.7	–	P		11	P		0.9		P					7,10,17

Values are given as mid-range; P = presence shown; ++ = major part of ash; + = minor part of ash; – = absent.

References: 1, Rea *et al.* (1983); 2, Denoel (1949a); 3, Moussli (1930); 4, Heslop-Harrison and Heslop-Harrison (1980); 5, Howard-Williams and Junk (1977); 6, Boyd and Walley (1972); 7, Hepburn *et al.* (1927); 8, Christensen (1976); 9, Porcher (1849); 10, Bjørklund (1864); 11, Pate and Dixon (1978); 12, Kovacs (1978); 13, Panin and Crishin (1975); 14, Nemçek *et al.* (1966); 15, Roels *et al.* (1968); 16, Ryabtsev (1978); 17, Small (1972); 18, Benzing and Renfrow (1974); 19, Cajander and Ihantola (1984); 20, Slater (1981); 21, Buchen and Schröder (1986); 22, Bosserman (1985); 23, Newman and McIntosh (1981); 24, Iijima and Sibaoka (1983).

Aliphatic Hydrocarbons, Fatty Acids, Alcohols and Sugars

During an examination of the lipid geochemistry of a coastal bog in Mississippi, Sever *et al.* (1972) determined the *n*-alkane series removed by chloroform extraction of whole plants of *Drosera filiformis*, *D. rotundifolia* and *Sarracenia psittacina*.

All three species contain *n*-alkanes of chain lengths between C17 and C33. *D. filiformis* proved most interesting with C23 (20%), C25 (23.5%) and the rare, even-length C28 (33%) dominating, C18, C20, C32 and C33 being absent. In contrast *D. rotundifolia* possessed the entire range except C32 with C27 (12.9%), C29 (14.9%), C31 (11.9%) and the unusually short C17 (20%) as the major constituents. C17 is rare in higher plants, although a common constituent of blue-green algae. Traces of C14 were also found. In *S. psittacina* a more even distribution over the complete range was found, C29 (15.9%) being the commonest single component with significant amounts of C28 (9.9%), C30 (9.3%) and C31 (11.4%) also present. In addition, this plant contains another series, probably either *iso*- or *anteiso*-alkanes but these are present in much smaller quantities and have yet to be further characterized.

By steam distillation and ether extraction the essential oils were prepared from *S. flava* leaves (Miles *et al.*, 1975). In contrast to *S. psittacina*, C17 (4.6% total oil) proved the major *n*-alkane, with C11 (0.2%), C13 (0.2%), C15 (0.6%) and C16 (1.8%) also detected. This suggests that C17 is about 60% of the *n*-alkane in *S. flava* compared to 2.2% in *S. psittacina*. While this may partially reflect different extraction procedures it does suggest that the more volatile, short-chain components may be concentrated in the leaves, the longer chained *n*-alkanes being more prevalent in the roots.

The variety of aliphatic *n*-fatty acids extracted by Sever *et al.* (1972) is smaller than for the *n*-alkanes. In *D. filiformis* 16:0 (38.8%), 20:0 (9.9%), 22:0 (17.3%) and 24:0 (13.8%) were the principal fatty acids with small amounts of 12:0, 14:0, 18:0, 18:1 and 26:0 present. The major fatty acids occurring in *D. rotundifolia* are 16:0 (23.7%), 22:0 (9.1%), 24:0 (22.4%), 26:0 (15.4%) and 28:0 (10.6%) with 18:0, 18:1, 20:0, 23:0, 25:0 and 27:0 also present. Thus in the droseras the only unsaturated fatty acid is 18:1, and this only in trace amounts. In complete contrast, 18:1 forms 60% of the fatty acids of *S. psittacina* with 16:1 (29.4%) as the other major component and 14:0, 16:0 and 18:0 in small quantities. Thus, the fatty acids show much more inter-generic variation than the *n*-alkanes.

3-Hexene-1-ol is the only aliphatic alcohol identified from *S. flava* (Miles *et al.*, 1975), but ethanolic extraction of the leaves of *Nepenthes albomarginata* yielded C24, C26, C28, C30 and C32 straight-chain alcohols (Wan *et al.*, 1972).

As part of their study of changes in *Utricularia* turions during break of dormancy, Simola *et al.* (1985) investigated the glycolipids of *U. vulgaris* turions. In this species, the turions float and can withstand freezing to −8°C but not lower (Winston and Gorham, 1979). Low temperature cultivation is known to affect the level of saturation of lipids in plants. The dominant glycolipids belonged to the classes of mono-(MGDG) or di-galactosyldiacylglycerides (DGDG) or to cerebrosides, the lattermost class being little studied in plants. No marked changes were found in the relative proportions of the three classes at different developmental stages. All classes contained acyl side chains ranging from 14:0 to 24:0, the most common being palmitic (16:0), stearic (18:0), linoleic (18:2) and linolenic (18:3), though the latter was markedly low in the cerebrosides. High 18:2 was characteristic of the MGDG (27%) and DGDG (40%) fractions of resting turions but within six days of sprouting it dropped to only 8 and 24% respectively while, conversely, 18:3 rose markedly from 17 to 63% and 20 to 45% respectively. In contrast, there was little change in the fatty acid composition of the cerebrosides, which was dominated by 16:0 (24–21% over six days) and 18:0 (21–18%). The levels established by six days were fairly stable over subsequent growth with the exception of 18:3 which tended to decrease to an intermediate level.

Lea (1976) tentatively identified lysophosphatidic acid in extracts of *Dionaea*. The content of stimulated traps was increased relative to unstimulated ones.

The reducing sugar content (glucose equivalents) of *Drosera rotundifolia* and *D. intermedia* were, on average, 5.21 and 8.21% dry weight respectively (Denoël, 1949a).

General Terpenes, Iridoids, Phytosterols and Aliphatic Carbonyl Compounds

General Terpenes

The essential oils of *Sarracenia flava* leaves contain the monoterpenes, humulene and β-pinene, and the sesquiterpene, caryophyllene (Miles *et al.*, 1975), as well as the triterpenes α-amyrin, betulin, betulinic acid, betulinaldehyde and lupeol (Bhattacharyya *et al.*, 1976; Miles and Kokpol, 1976; Miles *et al.*, 1974). The roots of *S. purpurea* contain both α- (5–10%) and β-amyrin (< 5%) and lupeol (< 5%) while the leaves contain a similar spectrum, but lacking lupeol, at lower levels (Hooper and Chandler, 1984). α-Amyrin (39%) and β-amyrin (61%) were the only major free triterpenes found in the pitcher tissue of *Nepenthes albomarginata*, whereas in the esterified fraction only 1% was β-amyrin and α-amyrin was present in trace quantities (Wan *et al.*, 1972). Instead, cycloartenol (68%) and 24-methylene cycloartenol (38%) constituted this fraction.

Iridoids

Iridoids are widely distributed and have been found in several unrelated genera of carnivorous plants. Many of these compounds have repellent properties and are of some chemotaxonomic value (Bate-Smith, 1984). Thus, interestingly, iridoids occur in Lentibulariaceae, Martyniaceae, Sarraceniaceae and Roridulaceae, in the latter genus as alkaloids (see page 293 and Ch. 17).

Wieffering (1966) reported catalpol from leaves of *Pinguicula vulgaris*. Recently, Marco (1985) and Damtoft *et al.* (1985) have separately identified globularin from whole plants while the former author also isolated globularicism, scutellarioside-II and 1-*O*-*p*-cumaroyl-β-D-glucopyranoside and the latter authors also isolated the novel compound 10-(*Z*)-cinnamoyl catalpol. *Utricularia australis* was also shown by Damtoft *et al.* (1985) to contain aucubin, gardoside and 6-deoxycatalpol (a new compound), the foremost having previously been found in *Utricularia* sp. (probably *U. vulgaris*) by Wieffering (1966). In *Martynia louisiana* leaves and stem, ajugol and mioporoside occur, while the unripe fruits contain catalpol (Sasaki *et al.*, 1978). These types of iridoids are particularly associated with the Scrophulariaceae and related families, indicating that the Lentibulariaceae probably is correctly placed in the Scrophulariales and that the Martyniaceae may have some affinities (see page 284 and Fig. 1.2).

Sarracenia, in contrast contains simple seco-iridoids (Dahlgren *et al.*, 1981) and the unusual enol diacetyl, tricyclic monoterpene sarracenin was isolated from roots of *S. flava*, Miles *et al.* (1976). The unusual structure of sarracenin and the initial evidence for its action as an antitumour agent (see page 246) stimulated considerable chemical interest. Miles *et al.* (1976) described its crystalline structure and subsequently two groups successfully achieved its chemical synthesis (Whitesell *et al.*, 1977, 1981; Tietze *et al.*, 1982).

Phytosterols

Nepenthes albomarginata and *N. sanguinea* pitchers, extracted with ethanol, have yielded a number of phytosterols (Wan *et al.*, 1972). The 4-desmethyl sterols cholesterol, campesterol, isofucosterol, sitosterol and stigmasterol were all found both free and esterified with sitosterol as the major component in both the free (52%) and esterified (67%) fractions. The *N. sanguinea* extract proved hard to characterize but campesterol, sitosterol and stigmasterol were all apparently present, a mixed glycoside being purified containing them in the ratio 5:85:10. β-Sitosterol is present in the roots of *Sarracenia flava* (Miles and Kokpol, 1976) and is the major identified phytosterol in both roots (> 20%) and tops (10–20%) of *S. purpurea* (Hooper and Chandler, 1984). Campesterol also occurs throughout this plant (tops 5–10%; roots, < 5%) while stigmasterol has only been isolated from the pitchers (< 5%) (Palmer *et al.*, 1978; Hooper and Chandler, 1984). The leaves contain a number of other unidentified, non-saponifiable lipids (Hooper and Chandler, 1984).

No free 4α-methyl sterols were detected in *N. albomarginata* but citrostadienol (31%), cyclolencalenol (24%) and obtusifoliol (45%) were all found esterified (Wan *et al.*, 1972).

Palmer *et al.* (1978) injected the radiolabelled precursor 5α-lanost-24-ene-3β,9α-diol [2-$^3H_2$] into the cavity of immature pitchers of *Sarracenia purpurea*. After ten days no radiolabel could be detected in stigmasterol, in contrast to the metabolism in algae, but significant incorporation was achieved with 28-isofucosterol-[7-$^3H_2$]. The poor incorporation may indicate a limited ability for the immature pitchers to absorb these compounds.

Carbonyl compounds

The essential oil of *Sarracenia flava* contains small amounts of *n*-nonanal (0.2% total oil), 2,4-dimethyl-2-4-heptadienal (0.1%) and methylcyclohexanone (0.1%) (Miles *et al.*, 1975).

Carotenoids

The carotenoids of a number of carnivorous genera have been examined as a possible taxonomic character by Neamtu and Bodea (1972). They extracted leaves and flowers with acetone/ether/methanol (6:2:2): the results are given in Table 11.2. Surpris-

Table 11.2 Carotenoids in leaves and flowers of various carnivorous plants

Species	Part of Plant	Carotenoids (mg/100 g dry wt)									
		α-ionines			β-ionines						
		α-carotene	lutein	lutein-ester	β-carotene	cryptoxanthin	cryptoxanthin-ester	violaxanthin	violaxanthin-ester	neoxanthin	zeaxanthin-ester
Droseraceae											
Aldrovanda vesiculosa	leaf	0.2	1.3		2.4	0.8		0.2		0.04	
	flower	0		1.2	1.4		tr		0.1		0
Drosera rotundifolia	leaf	0	1.0		2.1	0.7		0.1		tr	
	flower	tr		1.3	1.4		0.2		0.3		0
Lentibulariaceae											
Pinguicula alpina	leaf	tr	1.3		2.3	0.9		0.3		0.1	
	flower	0		0.8	1.0		0		0.1		0
Utricularia vulgaris	leaf	0.2	1.2		2.1	0.7		0.2		0.1	
	flower	0.8		1.2	1.5		0.2		0.3		0.4
Nepenthaceae											
Nepenthes mixta	leaf	0	0.8		2.4	0.3		tr		0	
	flower	0		0.5	0.7		0		tr		0

Reference: Neamtu and Bodea (1972).
tr = trace.

ingly, they found that carotenoids of the β-ionene type dominate in both parts. The variation between the genera is considerable, *Nepenthes mixta* flowers contain comparatively little carotenoid, while the Lentibulariaceae contain a wide range. β-Carotene and lutein are present in all the genera examined, but the distribution of minor constituents is more specific. Apart from α- and β-carotene, the floral carotenoids are all esterified, while those in the leaves are not.

Miscellaneous

Fresh leaves and stems of *Martynia louisiana* contain roseoside, an abscisic acid analogue-glycoside (Sasaki *et al.*, 1978).

Aromatics

Phenolic compounds

A number of these were identified in the essential oil of fresh leaves of *Sarracenia flava* by Miles *et al.* (1975), who found *p*-tolualdehyde (7.3% total oil), 1,1,3,-trimethyl-*p*-3-phenylindan (2.6%), phenylacetaldehyde(1.7%),*o*-xylene(1.4%), acetylnaphthelene (0.7%), *m*-xylene (0.1%), 1-methyl-2-ethylbenzene(0.3%),1-methyl-2-ethylbenzene(0.3%),and 1-methyl-4-ethylbenzene(0.1%).

Benzoic acid was reported in the stems and unripe seeds of *Drosera rotundifolia* (Sablitshka, 1923) and as the principal acid in the leaves of *Pinguicula vulgaris* (Loew and Aso, 1907). Subsequent analysis, however, showed this to be *trans*-cinnamic acid (Christen, 1961), a conclusion confirmed in both *P. vulgaris* and *P. alpina* by Bauquis and Mirimanoff (1970) who showed about 0.08% fresh weight to be *trans*-cinnamic acid but that this was degraded to benzoic acid on desiccation above 45°C. They found the acid both free and esterified and also found small amounts of cinnamaldehyde.

In the *S. flava* extract the most common phenolics were carvacrol (4.1%) and 2-phenylethanol (3.1%). *o*-cresol (1.0%), p-methoxyphenol (0.2%), benzyl alcohol (0.1%), methylengenol (0.3%) and hydroxyphenylbenzene (0.1%) were also detected (Miles *et al.*, 1975).

Table 11.3 Flavonoids in various parts of carnivorous plants and related genera and species

		Flavones, Flavanones and Flavonols																	
Species	Part of plant	apigenin	diosmetin	eriodictyol	6-hydroxyluteolin	8-hydroxyquercetin[a]	hypolaetin[b]	isocutallarin[c]	kaempferol	luteolin	3-methylquercetin	myricetin	quercetin	scutellarin[d]	apigenin-6-C-β-D-glucoside[e]	apigenin-6-C-β-D-glucosyl-7-O-glucoside[f]	apigenin-7-O-glucuronide	cyanidin-3-glucoside	cyanidin-glycosides
Cephalotaceae																			
Cephalotus follicularis	leaf			+					+/−	+		+	+						
Droseraceae																			
Dionaea muscipula	leaf								−			+	++					++	
Drosera anglica	whole																		P
auriculata	leaf glands																		P
binata	leaf/glands								−			+	++						P
capensis	leaf								tr			+	++						
intermedia	whole																		P
longifolia	leaf								−			+	++						
pygmaea	leaf glands																		
ramentacea	—												P						
rotundifolia	leaf, whole								P/−			+	++						P
spathulata	leaf								−			tr(?)	++						
Drosophyllum lusitanicum	leaf								−	+		−	−						
Lentibulariaceae																			
Pinguicula vulgaris	leaf	++			+		+	+		+				+					
Utricularia dichotoma	flower																		
vulgaris	leaf	+	+		++					++									
Martyniaceae																			
Martynia annua	flower	P								P							P		
proboscidata	leaf														P	P			
Nepenthaceae																			
Nepenthes 'chelsonii' Hort.	leaf									tr		−	++						
distillatoria	leaf									tr		−	++						
'morganiana' Hort.	leaf									tr		−	++						
rafflesiana	leaf									tr		−	++						
Sarraceniaceae																			
Darlingtonia californica	leaf									−		−	++						
Sarracenia alata	leaf																		
flava	leaf					tr				tr		−	++						
leucophylla	leaf									tr	tr	−	++						
minor (variolaris)	leaf									tr	tr	−	++						
oreophila	leaf																		
psittacina	leaf									tr		−	++						
purpurea	leaf									tr		−	++						
rubra	leaf																		
rubra spp. *alabamensis**	leaf																		
rubra spp. *jonesii**	leaf																		

P = presence shown; ++ = major component; + = minor component; tr = trace. *Schlauer (1986) lists only eight full species with six sub-species.
References: 1, Jay and Lebreton (1972); 2, Di Gregorio and Di Palma (1961); 3, Benz and Lindberg (1968); 4, Paris and Delaveau (1959); 5, Benz and Lindberg (1970); 6, Bienenfeld and Katzlmeier (1966); 7, Paris and Dennis (1957); 8, Jay and Gonnet (1973); 9, Jay and Gonnet (1974); 10, Romeo *et al.* (1977); 11, Nair and Gunasegaran (1982); 12, Pagani (1982); 13, Gascoigne *et al.* (1948); 14, Nicholls *et al.* (1985); 15, Ayuga *et al.* (1985).
Synonyms:

[a] gossypetine
[b] 8-hydroxyluteolin
[c] 8-hydroxyapigenin
[d] 6-hydroxyapigenin
[e] isovitexin
[f] saponarin
[g] hyperoside

Anthocyanins																					Leucoanthocyanins			References
eriodictyol-7-O-glucoside	isorhamnetin-3-galactoside	isorhamnetin-3-gluco-galactoside	kaempferol-3-glucoside	kaempferol-3-gluco-galactoside	kaempferol-3-O-rutinoside	luteolin-7-O-glucoside	luteolin-7-O-glucuronide	malvidin-glycosides	myricitin-3-O-glucoside	myricitin-3-O-rutinoside	pelargonidin-glycoside	quercetin-3-arabinoside	quercetin-3-digalactoside	quercetin-3-galactoside[g]	quercetin-3-O-gluco-galactoside	quercetin-3-O-glucoside	quercetin-3-glycoside	quercetin-3-O-rhamnoside	quercetin-3-O-rutinoside	unidentified (number)	leucocyanidin	leucodelphinidin	leucopelargonidin	
+			−		+	+			+	+						++		++	+		−	−	−	1,14
																					+	−	++	1,2
											P													3
																								13
								P													+	+	+	1,13
																					+	+	+	1
											P													3
																					+	+	+	1
											P													13
																								4
											P		P	P						P	+	+	+	1,5,6,7,15
											P										+	+	+	1,13
																					+	++	tr	1
																								8,9
								P																13
																								9
							P																	11
																								12
																					++	−	−	1
																					++	−	−	1
																					++	−	−	1
																					++	−	−	1
																					++	−	−	1
	−	tr	tr	−								+	−	+	tr		tr			2				10
	−	+	tr	−								++	−	+	++		−			1	++	−	−	1,10
	−	+	+	−								++	−	+	++		−			0	++	−	−	1,10
	tr	+	−	−								+	−	+	++		−			0	++	−	−	1,10
	−	tr	−	tr								+	−	++	+		tr			3	++	−	−	10
	−	tr	tr	−								++	tr	+	+		tr			0	++	−	−	1,10
	++	+/−	tr/−	tr/−								++/−	tr	++	+/−		tr			2				1,10
	+/−	++/−	tr	−								++	+−	++	++/−		tr			2				10
	−	+	tr	tr								++	+	++	−		tr			2				10
	+	tr	tr	−								++	−	++	++		tr			1				10

During their investigations of naphthoquinone synthesis in *Drosophyllum lusitanicum*, Durand and Zenk (1974b) showed that the vital intermediate homogentisic acid was present in small quantities.

Tannins were reported by Sartory *et al.* (1947) in *Drosera rotundifolia* and Sablitschka (1923) identified gallic acid amongst the tannins in stems and unripe seeds of the same species. Ellagic acid (an oxy-dimer of gallic acid) was subsequently found in the leaves of all the Droseraceae and *Cephalotus follicularis*. But it was not found in the Sarraceniaceae or Nepenthaceae (Jay and Lebreton, 1972) (see Table 11.3 for an enumeration of the species studied). This confirmed Shephard's findings (Porcher, 1849) that neither gallic or tannic acids could be detected in the rhizomes of *S. minor* and *S. flava*. Björklund (1864), however, detected tannic acid in the leaves and, probably, caffeic acid in the rhizomes of *S. purpurea*.

Sasaki *et al.* (1978) isolated two phenylpropanoid glycosides, martynoside, a novel compound, and acetoside from leaf and stem tissue of *Martynia louisiana*. These are very closely related compounds, the former being the O-methylated form of the latter. Martynoside also occurs in the leaves of *M. proboscidata* (Pagani, 1982).

Other benzene derivatives

Benzothiazole (0.3%) and 3,5-dimethylbenzoate (0.3%) are constituents of the essential oils of *Sarracenia flava* (Miles *et al.*, 1975).

Flavones and Anthocyanins

The distribution of these colouring compounds in the leaves of carnivorous plants is better studied than any other chemotaxonomical character. Representatives of five families have been considered, particularly by Jay and Lebreton (1972) while Romeo *et al.* (1977) have examined all ten species of *Sarracenia*. The results are given in Table 11.3.

In the Droseraceae, the leaves frequently contain a red pigment probably, as in *Dionaea muscipula*, located in the vacuole. The pigment of *D. muscipula* is cyanidin-3-glucoside, a very common anthocyanin (Di Gregorio and Di Palma, 1966). Cyanidin-glycoside has been found in *Drosera anglica*, *D. intermedia* (Bendz and Lindberg, 1968) and *D. rotundifolia* (Bendz and Lindberg, 1970) and quercetin, the aglycone of cyanidin and virtually ubiquitous in plants, found in nearly all the species examined (Table 11.3). Only in *Drosophyllum lusitanicum* is it absent, this monotypic genus instead contains luteolin, the reduced form of quercetin. Thus the Droseraceae all seem to derive their coloration from a closely related group of flavonoids.

Quercetin again appears as the major flavonoid in the Sarraceniaceae and the 3-galactoside and gluco-galactoside esters are the major anthocyanins. In *Nepenthes* quercetin is also the major flavone, probably again present in esterified form, although the glycoside residue has not been identified. Only in the Lentibulariaceae is there a very different complement of flavonoids, with apigenin, 6-hydroxy-luteotin and luteotin as the major constituents.

This group does not appear to provide a satisfactory basis for chemotaxonomy, a conclusion reached by Schnell (1978) when examining the pigments of *Sarracenia* flowers. With one or two exceptions, all the species contained all the anthocyanins he separated by tlc. Even red-flowered species contained yellow pigments; overall colour resulting from quantitative rather than qualitative differences. This mixture helps to explain the variation in flower colour observed between different races of the same species. In contrast, however, Kondo (1972) was able to conclude that *Utricularia cornuta* and *U. juncea* were probably separate species on the basis of their qualitatively distinct petal-pigment patterns.

Quinones

Benzoquinones

The quinol-glucoside, cornoside, has been identified in extracts of unripe fruits of *Martynia louisiana* (Sasaki *et al.*, 1978).

Naphthoquinones

Either ramentaceone or the widespread plumbagin has been found in the leaves and shoots of all the Droseraceae examined (Table 11.4). The content is very variable (Denoël, 1949a, b). Ramentaceone was first isolated from *D. ramentacea* (Paris and Delaveau, 1959), which is also the only known source of the dimer biramentaceone (Krishnamoorthy and Thompson, 1969). Several other naphthoquinones are known only from *Drosera* including a chlorinated naphthoquinone, at present unique to higher plants, isolated from *D. anglica* and *D. intermedia* (Bendz and Linberg, 1968). Droserone was first isolated from *D. whittakeri* by Rennie (1887) and characterized by Asano and Hase (1943a, b). The structure was confirmed by Thomson (1949). It also occurs as the 5-methyl ester in *Diospyros* (Thomson, 1971) but the other identified free

Table 11.4 Naphthoquinones in various parts of carnivorous plants

Species	Part of plant	Aglycones													Metabolites		Glycosides				References
		biramentaceone	3-chloroplumbagin	droserone[a]	hydroxydroserone	2-methylnaphtharazin	nepenthone-A	nepenthone-B	nepenthone-C	nepenthone-D	nepenthone-E	plumbagin[b]	ramentaceone[c]	unidentified	diomuscinone	diomuscipulone	droserone-glucoside	plumbagin-glucoside	ramentaceone-glucoside	rossoliside[d]	
Byblidaceae																					
Byblis gigantea	leaf											–	–								1
Dioncophyllaceae																					
Triphyophyllum peltatum	leaf											+									22
Droseraceae																					
Aldrovanda vesiculosa	leaf											++	–								1
Dionaea muscipula	leaf											++	–		P	P					1,19
Drosera aliciae	leaf, shoot											–	++								1,2
anglica	whole		tr			tr						++	tr	tr							4,5
binata	leaf, shoot											++	–	P							1,2,18
binata (dichotoma)	leaf, shoot											++	–								1,2
burkeana	leaf, shoot											–	++								1,2
burmanni	whole											–	–	P							8
capensis	leaf, shoot											+	++				–	–	–	–	1,2,16
capillaris	shoot											++									2
cistiflora	leaf											+	++								1
cuneifolia	leaf											–	++								1
erythrorhiza	shoot, root				P							++									2,11
filiformis var. *tracyi*	leaf											–	++								1
gigantea	root, flower, seed-head				P							++									11,12
hamiltonii	leaf											–	++								1
indica	leaf											++	–								1
intermedia	whole, shoot		tr			tr						++	tr	tr						P(?)	2,4,5,14,20
longifolia	leaf											++	–								1
madagascariensis (congoliana)	leaf											++									13
microphylla	shoot											++									2
peltata (lunata)	leaf			P								++	–								1,9
peltata ssp. *auriculata*	leaf glands											++	–	P							1,2,6
ramentacea	leaf	+				tr						+/–	++								1,3,17
rotundifolia	whole, shoot	–				–						++	+	tr			P			P	2,5,7,14,20
spathulata	leaf, shoot											–	++								1,2
stolonifera	shoot, root											++									2
trinervia	leaf											–	++								1
whittakeri	leaf, shoot, root			P	P							++	–	tr							1,2,9,10,11
zonaria	root				P																11
Drosophyllum lusitanicum	leaf											++	–					–	–		1,2
Nepenthaceae																					
Nepenthes rafflesiana	root			++	+	+	tr	+	+	+		++									21

P = presence shown; ++ = major component; + = minor component; tr = trace – = not detected.
Synonyms: [a]3-hydroxyplumbagin [b]2-methyljuglone [c]7-methyljuglone [d]1,4,5-trihydroxy-7-methylnaphthalene-glycoside

References: 1, Zenk *et al.* (1969); 2, Durand and Zenk (1974a); 3, Krishnamoorthy and Thomson (1969); 4, Benz and Linberg (1968); 5, Benz and Linberg (1970); 6, Paris and Delaveau (1959); 7, Vinkenborg *et al.* (1969); 8, Paris and Denis (1957); 9, Asano and Hase (1943a,b); 10, MacBeth *et al.* (1935a,b); 11, Russell (1958); 12, Russell (1959); 13, Bouquet (1970); 14, Sampara-Rumantir (1971); 15, Vinkenborg *et al.* (1970); 16, Durand and Zenk (1976); 17, Luckner *et al.* (1969); 18, Thomson (1971); 19, Miyoshi *et al.* (1984); 20, Denoël (1949a,b); 21, Cannon *et al.* (1980); 22, Bruneton *et al.* (1976).

naphthoquinones and the esterified trihydroxy-methyl-naphthoquinone of rossoliside (Vinkenberg *et al.*, 1970; Sampara-Rumantir, 1971) are all unique to *Drosera*. Luckner and Luckner (1970) have confirmed the identity of the four naphthoquinones from *D. ramentacea* by mass spectroscopy and NMR. The absence of both plumbagin and ramentaceone from *Byblis gigantea* confirms that, despite morphological similarities (see Chs 1 and 17), the families are not taxonomically closely related (Zenk *et al.*, 1969). In contrast, the presence of plumbagin in *Triphyophyllum peltatum* confirms the close association of Droseraceae and Dioncophyllaceae (Bruneton *et al.*, 1976; Fig. 1.2).

The major naphthoquinones of the Droseraceae have been found to be synthesized by the novel homogentisate ring-clearage pathway (Durand and Zenk, 1974a, b, 1976), previously only known from fungi. A survey of the family showed *Drosophyllum lusitanicum* to be the most active in plumbagin synthesis (Durand and Zenk, 1974a) and this was used in the initial experiments. Plants were fed a range of specifically ^{14}C-labelled precursors, known to be constituents of various synthetic pathways for naphthoquinones. Only with DL-[β-^{14}C]-tyrosine was significant incorporation achieved. By controlled chemical degradation of labelled plumbagin they showed the ^{14}C in carbon atoms 3, 6, 8, 10 and 11, compatible only with synthesis by condensation of acetate molecules derived from the side-chain of tyrosine. This was confirmed using other acetogenic precursors (Durand and Zenk, 1974a, b, 1976). The work was repeated with sterile grown plants and cell-suspension cultures (Durand and Zenk, 1974b) to rule out the possibility of fungal contamination. Additionally, they have shown the presence of the complete catabolic sequence of enzymes required to form acetate from the side-chain of tyrosine to be present in the cultured tissue (Durand and Zenk, 1974b) and that homogentisic acid is produced in the whole plants (Durand and Zenk, 1974a, b). The acetate generated is incorporated into these naphthoquinones by the acetate-polymalonate pathway (Durand and Zenk, 1976).

In *Plumbago europaea*, plumbagin cannot be labelled from tyrosine (Durand and Zenk, 1971). Possibly, in the Droseraceae the homogentisate ring-cleavage pathway is an adaptation to the scarcity of nitrogen in the environment. Following the removal of nitrogen from amino acids for other synthetic purposes the degradation of the ensuing carbon skeleton will lead to excess acetate being available. The readiness of *D. lusitanicum* to degrade amino acids supports this view.

Subsequently, this pathway was shown to occur in a number of *Drosera* species and *Dionaea muscipula* (Durand and Zenk, 1976). In the same paper, using *Drosera capensis*, they showed that ramentaceone (7-methyljuglone) is also produced by the homogentisate ring-clearage pathway. It will be most interesting to discover whether plumbagin in *Triphyophyllum peltatum* is also formed by this route.

Recently, Miyoshi *et al.* (1984) isolated two structures, diomuscinone and diomuscipulone, from *Dionaea muscipula*. They are very probably closely related biosynthetically to plumbagin, though not themselves naphthoquinones, but rather appear to be due to further metabolism. Evidence as to this relationship is lacking and, as yet, compounds of these types have not been found in other Droseraceae.

Naphthoquinones also occur in the Nepenthaceae. Cannon *et al.* (1980) investigated the roots of *Nepenthes rafflesiana* and isolated eight naphthoquinones, five of which were new and were named nepenthones A to E. Interestingly, the other three (Table 11.4) all occur in the Droseraceae. Unfortunately, the leaves, and in particular the pitchers, have not been examined.

Cyanogens

Greshoff (1909) found traces of hydrocyanic acid in *Drosera intermedia* and *D. rotundifolia*. Cyanogenic derivatives have also been reported in the Australian *D. peltata* and considered the cause of death to a number of sheep, while *D. binata*, *D. gigantea*, *D. spathulata* and *D. whittakeri* in addition are all suspected of having poisoned livestock by HCN production (Hurst, 1942). Nahrstedt (1980) has proved, however, that the detection of HCN in various *Drosera* as well as *Dionaea muscipula* and *Drosophyllum lusitanicum* was spurious, having been performed using the picrate paper test. He shows that plumbagin gives a false positive in this test and that the presence of this naphthoquinone correlates closely to the apparent positive results.

It is perfectly possible, however, that the poisonous nature of these plants could be due to these naphthoquinones, such as those well characterized from *D. whittakeri*. Reports of toxic effects of *Drosera* ingestion are not new: both Gerard (1633) and Hill (1756) considered red rot, a serious liver condition in sheep, to be due to the animals eating *Drosera* plants.

In Chapter 14 we consider a number of insects able to feed on the leaves of *Drosera*. Not only are these predators able to cope with or circumvent the mucilaginous secretions but also, it appears, deal with these poisons.

Nitrogenous Compounds

Amino acids

Only the common amino acids were found by Romeo *et al.* (1977) when examining all ten species of *Sarracenia*. The major α-amino acids were glutamate, aspartate, alanine, serine, valine and arginine while γ-amino butyrate was also detected. All these components were found in rather low concentrations.

Amines

Histamine has been detected in the leaf tissues of a number of genera (Werle, 1955). In both *Nepenthes* and *Drosera* the level appears to be higher in those parts of the leaf associated with the traps, though the level is variable. In these cases, as well as *Sarracenia* and *Pinguicula*, the concentration is in the range 2–13 μg/g fresh weight. Acetylcholine-like compounds were also detected in *Nepenthes* (Morrissey, 1963).

Villanueva *et al.* (1985) have shown that a number of amines and polyamines change in levels when turions of *Utricularia intermedia* break dormancy. Polyamines are probably involved in the control of gene expression and therefore are likely to be involved in controlling dormancy in this vegetative overwintering state. Dormant turions contain cadaverine (0.01 μmol/g dry weight), putrescine (0.1), spermidine (1.0) and spermine (0.07) but after the break cadaverine disappears and significant amounts of homospermidine (0.17) and norspermidine (0.02) are formed. Unfortunately, the significance of these changes is not known.

Alkaloids

Alkaloids, while not unknown, are relatively uncommon amongst the carnivorous plants. In view of their requirement for nitrogen in the molecules it is perhaps not surprising that these plants, living in nitrogen-limited environments, use other types of compounds as protective agents. Porcher (1849) was unable to detect morphine, nicotine or quinine in either *S. flava* or *S. minor* although Shepard (in Porcher, 1849) reported a new alkaloid, possibly related to cinchonine. *S. purpurea* plants yielded veratrine (Hétet, 1879), which possibly was Shepard's alkaloid. Björklund (1864) isolated coniine from roots of *S. purpurea* but not leaves, though Lambert (1902) subsequently identified coniine as a volatile base produced by fresh leaves of this species. Romeo *et al.* (1977) could not, however, isolate any alkoloid from all 10 species of *Sarracenia* but Mody *et al.* (1976), using large amounts of *S. flava* leaves (4.5 kg), showed that the unknown (1.9% total oil) $C_8H_{17}N$ extracted by Miles *et al.* (1975) is again coniine. The other unknown $C_5H_{11}NO$ (0.5%) may also be an alkaloid but awaits identification. Recent work has not been able to confirm the presence of veratine in *Sarracenia*. The variability in these reports may indicate seasonal and/or regional differences in alkaloid production, possibly related to carnivorous activity. *Roridula* also contains simple, unidentified alkaloids, probably iridoid in origin (Dahlgren *et al.*, 1981).

Pinguicula vulgaris does not appear to contain any alkaloids (Christen, 1961): nor does *Nepenthes rafflesiana* (Cannon *et al.*, 1980).

The best-studied alkaloids are those of the tropical liana *Triphyophyllum peltatum*. This plant, along with the closely related lianes of the genus *Ancistrocladus* are the only known sources of the naphthyl-isoquinoline alkaloids (Bringmann, 1986). This small group of compounds are made by a unique polyketide pathway from acetate, the nitrogen apparently being derived from free ammonia, or, more likely, pyridoxylamine (Bringmann and Jansen, 1986). Coniine, the alkaloid of *Sarracenia*, is also made by amino-group transfer to a diketide precursor. In both genera the use of amino acids, commonly the precursor of alkaloids, is avoided. It should be noted, however, that *T. peltatum* is in the Dioncophyllaceae and much more closely related to the Droseraceae and Nepenthaceae (see Fig. 17.2).The presence of plumbagin in *T. peltatum* strengthens this association (see Table 11.4).

The first alkaloids isolated from *T. peltatum* were named triphyophilline and triphopeltine (Bruneton *et al.*, 1976) and subsequently five more were described (Lavault *et al.*, 1977; Lavault and Bruneton, 1978; Bringmann, 1986). They are all structurally very similar, show a high degree of substitution and are present at about 0.3% of the dry matter. They are not confined to this genus in the family, having also been isolated from *Dionchophyllum tholonii* (Lavault and Bruneton, 1978) but no investigation of the Droseraceae for such products has been performed. *T. peltatum* differs, however, in showing seasonal carnivory; the glandular tissue only being borne on juvenile leaves (Green *et al.*, 1979). It may be that this plant is not, normally, nitrogen-limited and hence has no limitation on alkaloid production. The alkaloids may also show seasonal variation though this has not been examined. Bruneton *et al.* (1976) did, however, appear to extract gland-bearing tissue.

Bringmann and colleagues have successfully synthesized a number of naphthyl-isoquinoline alkaloids using, at least in part, biomimetic reactions (Bringmann, 1985a, b, 1986; Bringmann and Jansen, 1984, 1986).

Porphyrins

Maier (1973c) showed that both chlorophyll *a* and *b* are present in *Utricularia vulgaris* but presented no details of the distribution of pigment within the plant. He found that the concentration of chlorophyll *a*, but significantly not that of *b*, to rise to a peak in June, showing an annual cycle.

Nepenthes gracilis was found to have several features indicating that it can fix CO_2 by the mechanism known as crassulacean acid metabolism, including a high chlorophyll *a*/*b* ratio (Avadhani and Higgs, 1977) see Fig. 1.2.

CHAPTER 12

The Role of Phytochemicals in Carnivory

Introduction

The range of known phytochemicals is vast and products of staggering complexity have been isolated and identified. As seen from Chapter 11, carnivorous plants contain representatives of most groups of products, although the alkaloid content is limited. While many of the compounds, such as carotenoids and anthocyanins in flowers, are playing a conventional role, others are apparently located in the trapping organs and some may be assigned a role in carnivory. These form the subject of this chapter.

Attractants

Visual Stimuli

The morphological features we believe to be involved in attracting prey and the types of prey caught were discussed in Chapters 5 and 7. The bright-red anthocyanins present in the tentacle-heads of many Droseraceae may act as a visual attractant, since many *Drosera* species and *Drosophyllum lusitanicum* catch aerial prey. Judd (1969) found 47% of the prey of *D. rotundifolia* in the Byron Bog (Canada) to be flies; similarly, 57% of the prey for this species in the Netherlands were also dipterans (Achterberg, 1973). Yet, in Australia, 99% of the prey of the similarly rosette-forming *D. erythrorhiza* comprised collembola (Watson *et al.*, 1982). Data on the degree to which different types of prey are captured are clearly very dependent on season and environment (see Ch. 7), thus it is difficult to make these comparisons. In contrast, in other species such as *Dionaea muscipula*, where, on the whole, terrestrial prey is caught, the coloration is unlikely to play this role. Furthermore in *Byblis*, which takes similar prey to *D. lusitanicum*, there is no red coloration present; nor is it yet clear that such insects can distinguish colours at the red end of the spectrum.

Flavonoids are present in all the species examined, but only in the Droseraceae and Sarraceniaceae have anthocyanins been reported. The visual guidelines of *Sarracenia* and *Nepenthes* pitchers (Joel *et al.*, 1985) are almost certainly composed of the quercetin and quercetin-esters found in these genera (see Table 11.3). Quercetin and other flavonoids are known to form guidelines in petals and to protect plant tissues by acting as a UV absorbent (Luckner, 1984).

The role of the various naphthoquinones has not yet been determined. However, many of them have been identified in the leaves and, as they also strongly absorb UV radiation, may too be involved in producing the UV patterns.

Olfactory Stimuli

Evidence is now plentiful that a wide range of volatile wind-borne attractants is produced by certain flowers (Yeo, 1972). The close parallels between many floral and carnivorous features makes it probable that similar attractants may be released by the traps. While no analysis of the air around a plant has been attempted, Miles *et al.* (1975) have shown the presence of a number of suitable aliphatic and aromatic compounds in the leaves of *Sarracenia flava*. These include aliphatic alcohols and esters, lower terpenes, phenolic derivatives, hydroaromatics and at least one sulphur-containing compound, benzothiazole. Similar compounds are produced in some orchids (Dodson *et al.*, 1969). As no bioassay has been performed on any of these, any proposed role in luring prey into the pitchers remains speculative.

It has been found that, following suitable training, the bluntnose minnow can distinguish between various aquatic plants, including *Utricularia vulgaris*, on the basis of the water-soluble compounds released from the fronds (Walker and Hasler, 1949). On the basis of sessile rotifer colonization patterns, Wallace (1980) also considers that the traps of these plants may be detected by their chemical exudates (see Ch. 14). Perhaps *Utricularia* also uses attractants to draw prey towards the bladders. Unfortunately, no attempt has been made to identify the compounds involved but the fish and rotifers could apparently detect them at extreme dilution (see page 268).

Paralysing and Wetting Agents

Lambert (1902) noted the 'mouse-like' smell of freshly opened pitchers of *Sarracenia purpurea* and, although unable to isolate the causal agent, suggested it to be coniine. This may have been Hétet's unidentified amine (Hétet, 1879) and it has more recently been confirmed in the volatile fraction from leaves of *S. flava* (Mody *et al.*, 1976). If present in the alluring nectar it would effectively serve to prevent prey escaping, as it has a powerful insect-paralysing activity, with 100 ng causing total paralysis of small ants within 30 s (Mody *et al.*, 1976). This is the only evidence of a carnivorous plant possessing the means to 'stun' prey. The presence of this volatile alkaloid in the atmosphere within the pitcher, or in the nectary-secretion of the neck area, may explain the apparent disorientation and loss of locomotive control observed among insects caught in the trap (but see Austin, 1875K, 1877A and Appendix 1). Pollinators of various flowers may become drugged or intoxicated from drinking the nectar (Dressler, 1968). Such intoxication or poisoning might also be induced by naphthoquinones or phenolics in the secretion. Histamine might serve to irritate, and hence disorientate, insects or may play a purely protective role.

It has frequently been suggested that pitcher-plants, particularly *Nepenthes*, secrete a wetting agent into the pitcher-fluid but there is no evidence to support this. To be effective it would need to be present at high concentration and while a number of the compounds identified in *S. flava* could fulfil this role none is at a high enough concentration to make it very likely. But see p. 62 and Appendix 1, 1877A.

Antibiotic Agents

The controversy over the role of microbes in the digestive activity of carnivorous plants has been discussed in Chapter 9 but there is no evidence that the traps contain any active chemical principle to control microbial activity within the digestive fluid. Harder (1967) was unable to inhibit microbial growth in a culture medium by adding the secretion of *Drosera binata*, *D. burkeana*, *D. capensis*, *D. pygmaea*, *D. rotundifolia*, *D. spathulata*, *Pinguicula lusitanica*, *P. vulgaris*, *Darlingtonia californica* or *Dionaea muscipula*. It would appear, therefore, that in *Drosera* and *Pinguicula* the reported benzoic acid (see Ch. 11) though not released from the tissues does inhibit microbial growth. Luetzelberg (1910), however, claimed to find benzoic acid in the fluid of the traps of *Utricularia vulgaris* and also that the role of this was to inhibit microbial activity.

In the pitchers, the acidic conditions will themselves be quite effective at controlling microbial growth, particularly as the pH is kept acidic by an active proton pump (see Chs. 6 and 9). In addition, the low pH, typical of digestive fluids (see Ch. 9), may be important for the antibiotic activity of released compounds.

The secretions of *Dionaea muscipula* and probably also of some of the droseras contain a considerable amount of non-protein UV-absorbing material (Scala *et al.*, 1969; Robins, 1978). Some of this may prove to have an antibiotic action. Although not specifically identified in the secretion, quercetin, widely distributed as various glycosides in the carnivorous plants, only has anti-bacterial action at acidic pH (Karel and Roach, 1951). The free solubility of the glycosides in water makes this an interesting candidate for this role, especially in the pitchers. However, the glycosides are inactive, but the aglycone could be released as a result of microbial glycosidase activity. Having looked at more neutral conditions, Harder (1967) would have failed to detect this activity.

Extracts of *Sarracenia purpurea* have also been shown to have antimicrobial activity (Schnell *et al.*, 1949), an activity possibly associated with the flavonoids present.

Plumbagin, found in a number of Droseraceae (Table 11.4) is known to be toxic to certain microorganisms (see Ch. 13). Its high concentration in *D. erythrorhiza* tubers may help protect the plant against soil microorganisms (Russell, 1959: see Ch. 3). Durand and Zenk (1974) suggest that, as it is relatively water soluble, it may be discharged into the secretion, decreasing microbial competition for nutrients. The presence of such compounds in the secretion has not been examined but, although, disagreeing with Harder's conclusions (Harder, 1967), it does account for the large amount of this and related naphthoquinones (Table 11.4) in the leaves. It is equally likely, however, that they play an important role in protecting the plants. Juglone, which is closely related to plumbagin and ramentaceone is known to be liberated by some plants, to oxidize and form a compound that inhibits the growth of other plants (Bell, 1980). Another known action of juglone is to repel predators by reacting with free amino groups in proteins and inactivating them (Luckner, 1984). Both these actions will tend to act as favourable survival strategies, decreasing phytic competition and protecting against predation. In some droseras, a very few insect species are known to feed on the mucilage, leaf tissues and even captured prey (see Ch. 14). These insects include the moth caterpillar *Trichoptilus* (see page 267). Lepidopteran larvae are known to be able to sequester noxious compounds from their host plants and, while there is no evidence that this occurs in the case of predators on *Drosera*, such an advantageous survival strategy would be a strong evolutionary pressure off-setting the disadvantages of feeding on such a difficult host plant.

Stimulatory Agents

Traps of *Dionaea muscipula* show rapid movements similar in rate to mammalian muscle (see page 97). Lea (1976) investigated this system for muscle-contracting material. By chromatography, he isolated a compound causing contractions in isolated frog muscle which was not antagonized by atropine and was destroyed by acid phosphatase. From these properties he concluded that the active substance was lysophosphatidic acid. He showed this activity to increase about three-fold within 15 s of trigger-hair stimulation and then to decay over about 24 h to the basal state. He hypothesized that the action of phospholipase D may lead rapidly to elevated levels of lysophosphatidic acid, which is known to affect membrane permeability, and that this causes changes in membrane potential that are transmitted to the motor response centre. The activation of phospholipase D presumably results from trigger-hair flexing and in this context the redistribution of Ca^{++} during the flexure is interesting (Hodick and Sievers, 1986) as Ca^{++} plays an important role in membrane fluidity by forming ionic links to phospholipid head groups on adjacent molecules.

CHAPTER 13

Herbal and Medicinal Uses of Carnivorous Plants

Introduction

The exploitation of plants by humans for curative reasons is an ancient one, involving many plant species. Those of a carnivorous nature are no exception. Indeed, their curious structures have evoked interest amongst many peoples, resulting in a wide range of herbal and even mystical uses. Properties such as the 'perpetual dew' of *Drosera* (see page 13) were considered magical and we present a light, and by no means comprehensive review of such interests. Two properties in particular, anti-spasmodic activity in *Drosera* and anti-leukaemic activity in *Sarracenia*, have been the subjects of more detailed scientific investigation. The *Triphyophyllum* alkaloids may also have these activities but the active components occur at very low levels. Their recent chemical synthesis (see page 240) may enable suitable tests to be performed.

Herbal and Mystical Uses

A number of 'virtues' were assigned to both *Drosera* and *Pinguicula*. Gerard (1633) describes how, in Yorkshire, the juices of *Pinguicula* [*vulgaris*] were applied to the udders of cattle when these had become chapped, wounded or were bitten by snakes. By 1756 these juices, mixed with hog-lard were also used to cure wounds and sores in people (Hill, 1756). When bruised in white wine the leaves provided a remedy for dropsy and, taken as a syrup, act as a purge and diuretic (Hill, 1756). This lattermost action may be due to the extraction of quercetin and other hydroxylated flavones, known to act as diuretics. Garsault (1767) reports similar uses in France. Unlike *Drosera*, the use of *Pinguicula* appears to die out and it does not feature in modern herbals (Grieve, 1959).

Magical powers were also ascribed to *Pinguicula*. Grigson (1958) tells how, in the Scottish islands, butterwort was picked to protect against witches. Milk or butter from cattle which had eaten of the plant was supposed to protect newborn babies and their mothers.

The leaves of *Pinguicula*, as they were once in Britain, are still used by the Lapps to curdle and thicken milk, producing *Toetmoelk* or *Filmjölk*. In this action it resembles rennet and, while this could be due to the hydrolytic enzymes present, there appears to be a number of micro-organisms, particularly associated with the mucilage, that cause this effect when isolated (Thomas and McQuillan, 1953). The leaves of *Drosera* may similarly be used.

The perpetual 'dew' of the *Rosa Solis* (*Drosera*) makes this plant particularly powerful as a herbal remedy. Although William Turner (1568) reported it held good for 'consumptions and swouning and faintness of the harte' he was sceptical as to its effectiveness. Culpepper (1813), however, considered an extract with *aqua vitae* and spices good 'in qualms and passions of the Heart'. Small quantities of certain flavones can act as cardiac stimulants, so there may be some substance to these beliefs. When distilled with wine it is strengthening and nourishing, while 'cattle of the female kinde are stirred up to lust by eating even of a small quantitie' (Gerard, 1633). This belief in its aphrodisiacal power was perpetuated and Grigson (1958) details the development of a number of liqueurs in France, Italy and England based on this plant. In some places in Europe it is known as 'Youthwort' (Grigson, 1958) and Grieve (1959) reports its use in America as a tincture against old age and arteriosclerosis.

Juices squeezed from the fresh leaves are also used against warts and corns (Grieve, 1959) but Hill (1756) considered this a highly dangerous practice, causing both internal and external damage and when leaves are applied to the skin they act as an escharotic, causing painful inflammations. The secretions of the tentacles, however, appear to have been valued for the treatment of freckles and sunburn (Grigson, 1958).

The major role assigned to *Drosera* in continental Europe is as an agent against bronchial infections, in particular whooping-cough, and this has been shown to have some scientific foundation (see below and Fig. 2.1). Neither Gerard (1633) nor Hill (1756) mention such a use in Britain but Dujardin-Neametz and Egasse (1889) discuss this use in France, describing the alcoholic extract as anti-spasmodic. Grieve (1959) lists it as active against whooping-cough, phthisis, chronic bronchitis and asthma. França (1925) notes that *Drosophyllum* is used in Southern Portugal to combat conjunctivitis. Again, this may be due to an anti-microbial element (see page 246).

Belief in the power of *Drosera* continues: Frenzer (1980) has found extract of *Drosera* to be an ingredient in certain sweets made in Vienna.

North American Indians apparently used *Sarracenia* to treat smallpox; the juices were supposed both to save life and prevent unsightly pitting. In modern herbal usage it is considered valuable as a tonic, laxative or diuretic and in treating complaints of the liver, kidney, uterus and stomach, notably dyspepsia (Grieve, 1959). As with *Drosera*, the extract of *Sarracenia* can prove harmful. Björklund (1864) extracted a volatile amine from *S. purpurea* which, when 5 mg was fed to a mouse caused indolence and frequent urination without the animal drinking. It died after 12 h and a post-mortem showed damage to the heart, blood leakage into the lungs and brain and putrefication of the intestine. The toxicity is not typical of an alkaloid and presumably represents the combined effect of coniine, amines and the volatiles later identified by Miles *et al.* (1975) from *S. flava*. The possible role of phytosterols and triterpenes in herbal applications of *S. purpurea* is being investigated (Hooper and Chandler, 1984).

Nepenthes are used by Malays (who call them, amongst other names, 'baboon's bamboo jug') in several ways (Burkill, 1966). Powdered roots are used for treating stomach aches and dysentery, common problems in Malaysia (Ridley, 1906). These effects may well be due to the naphthoquinones present in this tissue (see page 238). An infusion of the stem is also used for treating fevers and the plant is used as an emetic. The pitchers are used in ceremonies of exorcism (Burkill, 1966).

Antispasmodic Agents

The demonstration by Paris and Quevauviller (1947) that the injection of an ethanolic extract of *Drosera* (species unidentified) into guinea-pigs would protect them against nasally administered histamine- and acetylcholine-induced bronchospasm put the previous herbal beliefs on a sound scientific basis. This finding was confirmed by Gordonoff (1951) using isolated organs, while Bézanger-Beauquesne (1954) and Paris and Delaveau (1959) showed a similar protective action of tincture from *D. ramentacea* using intestine isolated from rabbits and guinea-pigs. Ramanmanjay and Boiteau (1968) repeated the work on whole guinea-pigs under more controlled conditions, checking the susceptibility of each animal first. They found that *D. ramentacea* ethanolic extract would protect the animal for at least 48 h when administered at 100 mg per animal.

Simultaneous interest in the nature of the active ingredient led to the suggestion that it was plumbagin (Bézanger-Beauquesne, 1954), used widely from various sources to treat bronchial, gastric and epidermal problems (Thomson, 1971; Wurm *et al.*, 1984). Paris and Denis (1957) isolated plumbagin in various alcoholic extracts and by steam distillation from *D. burmanni*, *D. intermedia*, *D. ramentacea* and *D. rotundifolia*. They found fresh material contained far more naphthoquinone than commercial extracts and that *D. ramentacea* was the best source. It appears, however, that the amount of naphthoquinone present and the ease of its extraction is very variable (Bézanger-Beauquesne, 1954; Bézanger-Beauquesne and Perrin, 1972). Plumbagin and several other naphthoquinones have now been shown in a number of species (Table 11.4) but there is no firm evidence that this is the active ingredient. All these experiments have used simple alcoholic or chloroform extracts of whole plants or leaves without any extensive purification taking place. Stellfeld (1959) summarized the confusion over the formula of the active ingredient. Krahl (1956) reported that alcoholic extracts of *D. rotundifolia* contained a carboxy-hydroxy-naphthoquinone, which apparently had a similar anti-peristaltic action as the tincture from *D. ramentacea*. Yet Thomson and Wilkie (1961) confused the picture by showing this to be an impure preparation of the flavone quercetin and not a naphthoquinone.

The uncertainty of the identification of plumbagin as the active ingredient is compounded by the work of Christen (1961) with *Pinguicula vulgaris*. The use of this plant in herbal remedies also prompted an evaluation of its usefulness as an anti-spasmodic agent in modern therapy. Applying standard systematic extraction procedures to a mascerate of whole plants he obtained a fraction containing the anti-spasmodic agent. This was proved by chromatography, infra-red spectral analysis and elemental composition to be pure *trans*-cinnamic acid. A further active fraction contained *trans*-cinnamic ethyl ester. Both fractions were shown to be active against acetylcholine-induced spasms in isolated rat intestine.

Is, then, the action of tincture of *Drosera* also due to this acid? Pure plumbagin can be obtained from *Drosera* (Thomson, 1971; Vanhaelen, 1970) but does not appear to have been used in pharmaceutical tests. Thus, the anti-spasmodic activity of plumbagin is yet to be proven, although *Drosera* extract is used in preparations for the treatment of chronic bronchitis and has been shown active against the associated microorganisms (Wurm *et al.*, 1984). It should be noted, however, that pure plumbagin is quite toxic to mammals (LD_{50} = 20 mg/kg i.p.) and that the toxicity is against the central nervous system (Wurm *et al.*, 1984). This toxicity is compatible with a potential anti-spasmodic activity at lower doses.

The toxicity of these compounds is not, however, simple. Wagner *et al.* (1986) found extracts of *Dionaea* and *D. ramentacea* to show an immunosuppressive cytotoxicity at high concentrations (1–0.01 mg/ml) but the reverse immunostimulatory effect at low levels. This reversal is similar to the effect of the anti-leukaemic drug vincristine. Pure plumbagin behaved similarly, indicating this to be the active principle in the extracts.

The growing importance of *Drosera* extracts in pharmaceutical preparations has led to recent descriptions of suitable analytical procedures for standardizing their naphthoquinone content (LeClercq and Angenot, 1984; Bonnet *et al.*, 1984).

Antimicrobial Agents

While it is possible that the anti-spasmodic action discussed above might be due to a direct effect of the extract on bacteria, this does not seem to be the case. Pure plumbagin is active against various *Staphylococcus* and this effect is found in extracts of several *Drosera* species (Denoël, 1949a; Karel and Roach, 1951). Tincture of *Drosera ramentacea* was, however, comparatively ineffective against *Hemophilus pertussis*, the cause of whooping-cough (Bézanger-Beauquesne, 1954). The action of the tincture against spasms in isolated tissues and those artificially induced rules out the likelihood of any direct effect on the bacteria. Quercetin, found in many carnivorous plants (Table 11.3) is also active against a wide range of bacteria (Karel and Roach, 1951) and, as discussed earlier, has been found as a contaminant in plumbagin preparations (Thomson and Wilkie, 1961).

Pure plumbagin does, however, have an antimicrobial activity which may be enhanced by the preparation of analogues methylated at the 3-position (Wurm *et al.*, 1984). Pharmaceutically, these analogues have the added attraction of being much less toxic to mammals. Didry *et al.* (1986) tested pure plumbagin against a range of pathogenic bacteria and showed this compound to inhibit both Gram-positive and Gram-negative cocci and bacilli. Furthermore, it showed a synergistic effect with several other antimicrobial agents. Alcoholic extracts of *Drosera longifolia* from Hungary, from which plumbagin was apparently absent (but cf. Table 11.4) were inactive against bacterial growth; in contrast, extracts of *D. peltata* from India showed activity against most bacteria in parallel with the plumbagin content. Broadly, these results and those of Denoël (1949a) are in agreement though some discrepancies, such as the action against *Escherichia coli*, should be noted.

Antitumour Agents

Following up the use of extracts from the roots of *Sarracenia flava* in various folk remedies, a number of species were screened for antitumour agents against lymphocytic leukaemia (Kokpol, 1974). The crude hexane extracts of *S. alata*, *S. flava*, *S. leucophylla*, *S. psittacina* and *S. rubra* were inactive (T/C < 125) and of the ethanol extracts only *S. flava* showed slight activity (T/C = 130). Further systematic extraction (Miles *et al.*, 1974) yielded a methanol-insoluble beige solid possessing considerable activity against human epidermoid carcinoma of the nasopharynx (KB), with an ED_{50} = 27μg ml^{-1}.

A methanol-soluble oil also showed some anti-KB activity but neither fraction was effective against lymphocytic leukaemia or Walker carcinoma. The beige solid was identified as largely betulinic acid but pure samples of this gave negative tests with KB cells. By further extraction of this fraction Miles and Kokpol (1976) concentrated betulin and lupeol in a benzene fraction, significantly active against lymphocytic leukemia at only 50 mg kg^{-1}. Betulin, a previously known anti-tumour agent, was also identified in another, chloroform-soluble fraction.

Subsequently Bhattacharyya *et al.* (1976) found that an extract from the leaves gave much greater anti-leukaemic activity and isolated betulin, betulinaldehyde and a new monoterpene sarracenin (Miles *et al.*, 1976) from this preparation. Sarracenin proved ineffective against lymphocytic leukaemia and, as the betulin was only present in trace quantities, they were able to show the active ingredient to be betulinaldehyde. This was confirmed with synthetic material.

The care taken in proving that betulinaldehyde is the active principle is in great contrast to the easy assumptions made about the anti-spasmodic agent discussed above.

PART V

Exploitation and Mutualism

CHAPTER 14

The Associated Fauna (Inquilines*) and Flora of Carnivorous Plants: The carnivorous plant as a phytotelm

Introduction

Catesby (1754), as we saw in Chapter 2, noticed that insects other than natural prey take up semi-permanent residence in or on carnivorous plants. He wrote, in the second volume of the Natural History of Carolina (1754): 'The hollow of these leaves [of *Sarracenia*], as well as of the other kind, always retain some water; and seem to serve as an asylum or secure retreat for numerous insects, from Frogs and other animals, which feed on them.' On another continent, just a few years earlier, that observant traveller Rumphius (1747) on looking into a *Nepenthes* pitcher wrote: '. . . in the open chamber [of the pitcher] grubs and insects of different sorts creep about; these die there except for a kind of small plump "larva" which is sometimes found there and remains alive.' [Our translation from the Latin of the 'Herbarium Amboinense'].

Riley (1873 and 1874) identified at least two of the 'insects which brave the dangers of *Sarracenia*' as the moth *Xanthoptera* and the flesh-fly *Sarcophaga sarraceniae* (see page 257). Some of these associations are no more than casual, providing a temporary refuge from a desiccating climate or predators, but others are of a more permanent nature. Some fall within the meaning of the word 'phytotelm' (see below), e.g. the 'pitcher' plants, but others, such as the aquatic species of *Utricularia* lie outside the normal definition. There are yet other associations, usually a looser kind, between commensal or predatory insects and what are generally regarded as the 'fly-paper type' of carnivorous plants.

The Concept of the Phytotelm

A *phytotelm* (Greek: phytos = plant; telm = pool or pond; plural, phytotelmata) is defined as the small aquatic ecosystem formed by a part of a terrestrial plant, which may be a modified leaf, leaf axil, flower, stem, hole and depression, open fruit or even fallen leaf (Frank and Lounibos, 1983). The potential size range of phytotelmata is vast, from protozoa which live in the 300 μl drop held in the concavity of a *Sphagnum* leaf (Corbet, 1973) to the complex communities in the 2 litres of a *Nepenthes rajah* pitcher. The food chains within such structures are obviously long and complex (Kitching and Pimm, 1986).

Frank and Lounibos (1983) list over 29 higher plant families comprising 1500 species with such phenomena, and the catalogue must, inevitably, be incomplete.

For convenience, we shall here broaden the term phytotelmata to include not only the obvious 'pitcher-type' carnivorous plants, but also the bladder-trap-forming *Utricularia* species, which are often richly occupied both inside and out. The 'fly-paper' types are dealt with separately, and the 'snap-trap' and 'lobster-pot' types are, so far as we know, unoccupied.

One of the principal advantages in the study of the phytotelm of the pitcher-plants is the limitation

*Inquiline = "An animal which lives in the . . . abode of another" (S.O.E.D.) frequently used in the early history of the study of living inhabitants of pitchers; not now considered to be a correct descriptive term, and not used in this text.

of the habitat. Most aquatic communities can only be studied by a team of workers. A *Sarracenia* or *Nepenthes* pitcher is a small and relatively simple ecosystem (Paterson, 1971) and this closed community with its subtly balanced co-evolution between plant and insect (see below) has attracted workers for over 70 years (see Fish and Beaver, 1978, for a bibliography of the immense literature). It should not surprise us that so many insects have become obligate inhabitants of 'pitchers' when we appreciate that these carnivorous plants enjoy a catholic diet (Ch. 7). In one study by Erber (1979) ten *Nepenthes* pitchers, for example, contained 1994 individual arthropods of 150 different species. *Sarracenia purpurea* pitchers from North Carolina, USA, examined by Wray and Brimley (1943), revealed victims of 115 families belonging to 14 orders of insects, not counting the mollusca.

Phytotelmata, in the broadest sense, are of immense economic importance as breeding sites for disease-carrying mosquitoes and other biting insects (Lounibos *et al.*, 1985; Laird, 1988). However, although there are, as we shall see, many species of mosquito that inhabit carnivorous plant pitchers, they rarely bite humans (Goins, 1977; Beaver, 1979b) and do not seem to be of medical significance in that respect. It seems likely that the majority of the insect species found in these phytotelmata feed on nectar, decaying fruits or carrion and are not haematophagous.

The pitcher as a phytotelm: general considerations

As we saw in Chapters 8 to 10, prey insects falling into pitchers are rapidly digested. Early accounts of the living fauna found in pitcher-plants assumed that those animals living in the fluid must secrete an 'anti-enzyme' or be armoured in some sophisticated way to avoid their own digestion. There is no evidence that such special defence mechanisms exist. As Beaver (1979b) has pointed out, animals which do not drown in the fluid can survive for long periods, as we shall see on page 268, and the water-flea *Daphnia*, which is a common prey of *Utricularia*, may survive in the traps for many hours before dying. The boundary between being a victim and an inhabitant is obviously a fine one. Beaver reared the mosquito *Aedes albopictus*, which is not a normal inhabitant of *Nepenthes* pitchers, from egg to adult in freshly opened pitchers of *N. albomarginata* in which no dilution of the fluid was allowed. Development was as fast in this medium as in a conventional fluid and there was no increase in mortality. Aquatic insects, such as corixid bugs, survive for weeks in fresh pitcher fluid and seem to die, not from enzymic action but from starvation.

Animals and Plants that Exploit *Sarracenia* Species

The level of study of *Sarracenia* has, as one might expect from its geographical location, reached a higher level of sophistication than that in other genera, to include detailed modelling of pitcher temperatures and water loss (Kingsolver, 1979). The study of the predators and commensals of most of the other carnivorous plants is, with a few exceptions such as certain *Nepenthes* species, little more than a list of species. It is worth noting, too, that the commensal insects of the genera *Sarracenia* and *Darlingtonia* often lie within the same genera, bearing out the fairly recent divergence from a common, more primitive *Heliamphora*-like stock, as suggested by Macfarlane (1893), Thanikaimoni and Vasanthy (1972), De Buhr (1977a) and Maguire (1978) (see also Fig. 19.5). Species of the mosquito genus *Wyeomyia*, which is ubiquitous in the Sarraceniaceae, are also found in the totally unrelated bromeliad carnivorous plant, *Catopsis berteroniana* (Frank and O'Meara, 1984). On the other hand, species of *Nepenthes* seem to have, as we shall see on page 262, a very diverse commensal fauna with few species or even genera in common, suggesting both a recently expanding carnivorous genus and a very different mode of dispersal and colonization by insects.

Special Features of the *Sarracenia* Pitcher

The *Sarracenia* pitcher is, as Kingsolver (1981) showed, a highly evolved phytotelm, which only infrequently dries out in a given season. The pitcher shape decreases desiccation by minimizing the adverse effects of an infrequent or irregular rainfall pattern. In addition to its enhanced humidity, the cavity within a pitcher enjoys a lower light intensity and is somewhat less variable in temperature than an open pool. Moreover, in spite of heavy colonization by one or more species of respiring and feeding dipteran larvae, it can maintain the oxygen tension of the pitcher liquid at, or close to, the atmospheric level.

Cameron *et al.* (1977) were probably the first to reveal that, despite high oxygen consumption by symbiotic organisms, dissolved oxygen is kept at a high level. (The percipient speculations of MacBride of S. Carolina, 1818, have been described on page 15). Cameron *et al.* concluded that this came about by diffusion from the leaf tissue. Bradshaw (1983) suggested that the leaves of *S. purpurea* removed CO_2 from the pitcher fluid, particularly in the presence of light, and Joel and Gepstein (1985) by showing that a photosynthetic apparatus is present in the epidermal cells of the inner face of the pitcher of *Sarracenia purpurea* add weight to this argument (Figs 14.1 and 14.2 and see page 160). Chloroplasts are uncommon in non-stomatal epidermal cells, although known to be present in aquatic plants. *Sarracenia* is not aquatic, but does have more or less permanent water in contact with the chloroplast-containing epidermal cells. However, in this case the photosynthetic apparatus is separated from the mesophyll of the leaf by a non-photosynthetic endodermis-like hypodermis (Joel and Heide-Jørgensen, 1985; Joel, 1986 and see page 156) and predominantly receives illumination

Fig. 14.1 The pitcher epithelium of *Sarracenia*.
(*A*) Hypodermal cells with suberin wall layer (s). Silver deposits (Ag) are evident between neighbouring hypodermal cells in the sealed apoplastic gap between the cells: e, epidermal cells; h, hypodermal cells; p, plastids.

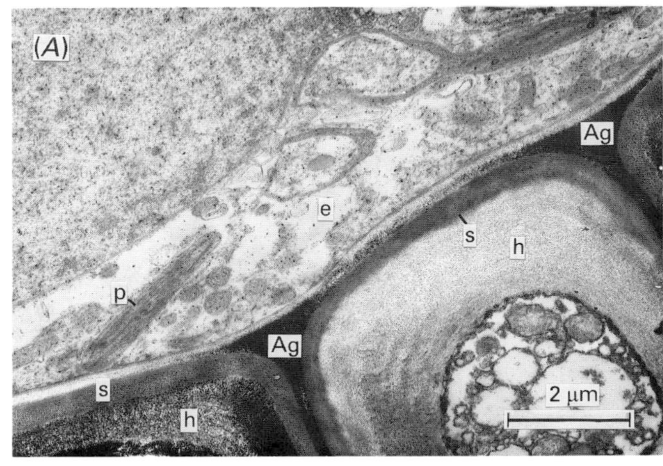

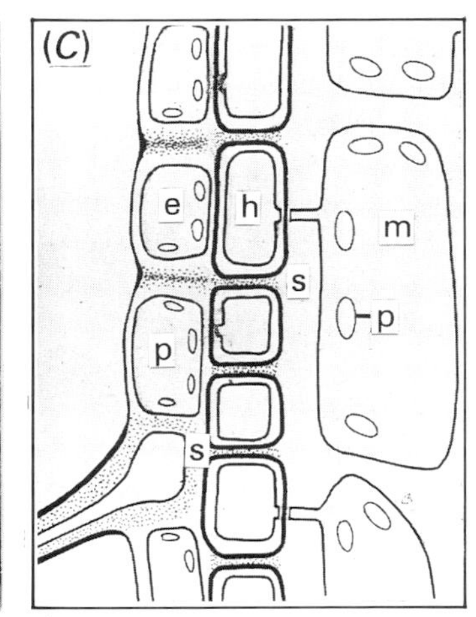

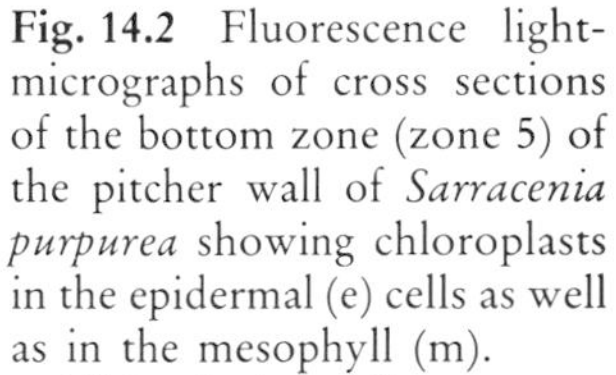
(*B*) Plasmodesmata (pd) crossing the suberin (s) wall layer in a pit field of a hypodermal (h) cell of *S. psittacina*: m, mesophyll cell. Suberin lamellae are clearly visible in the suberin wall layer. (Unpublished micrograph by kind permission of Dr H.S. Heide-Jørgensen.)

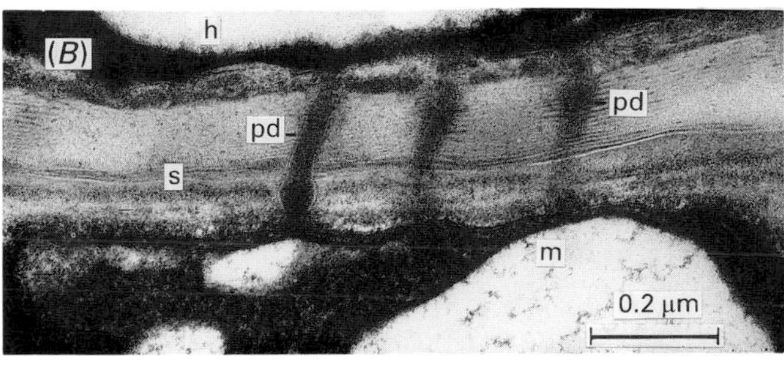

(C) Schematic drawing of a pitcher epithelium: e, glandular epidermis; h, endodermoid hypodermis; m, mesophyll; s, suberin wall layer; p, chloroplast. (Redrawn from Joel, 1986.)

Fig. 14.2 Fluorescence light-micrographs of cross sections of the bottom zone (zone 5) of the pitcher wall of *Sarracenia purpurea* showing chloroplasts in the epidermal (e) cells as well as in the mesophyll (m).
(*A*) Intrinsic red fluorescence of an untreated section, which corresponds to photochemical activity of the chloroplasts. No chloroplasts are seen in the hypodermis (h).
(*B*) FITC immunofluorescent reaction of the chloroplasts with a specific antibody against RuBP carboxylase, indicating possible capability of these chloroplasts to fix CO_2 photosynthetically. (Reproduced with permission from Joel and Gepstein, 1985.)

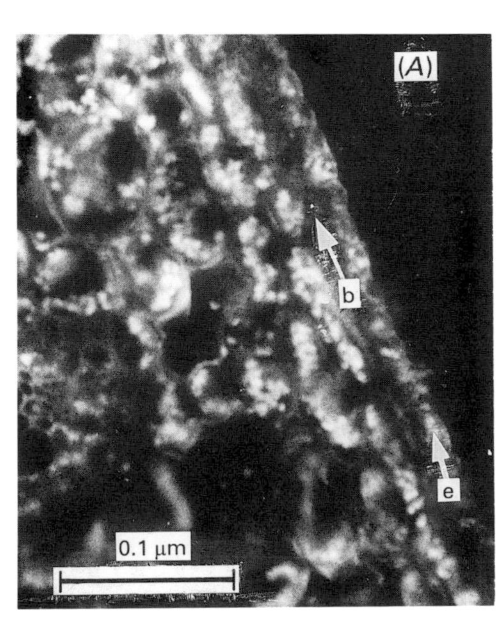

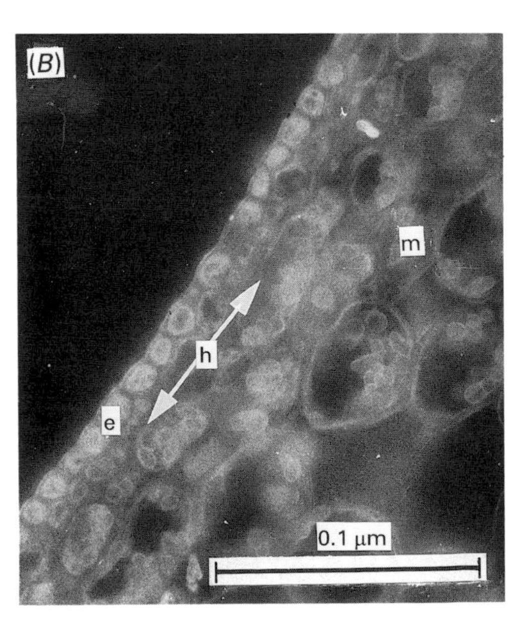

which has already passed through the main photosynthetic tissue.

The removal of CO_2 from the pitcher solution noticed by Bradshaw (1983) and the leakage of O_2 into the fluid is consistent with Joel and Gepstein's observations (1985) of functioning chloroplasts (albeit smaller than those in the mesophyll cells) and with the existence of a perforated cuticle which bounds the pitcher on the inner side (Joel and Heide-Jørgensen, 1985). It is obviously to the advantage of the plant to keep the chamber aerobic at all times in this way the plant will be able to scavenge more CO_2/g.

The stability and favourable nature of the habitat conferred by the various factors listed above may encourage many organisms to choose to live some or most of their lives in the confined world of a *Sarracenia* pitcher. These organisms include bacteria, algae, protozoans, rotifers, small crustaceans, occasionally such entrepreneurial species as the paper wasp, *Polistes* (Bernon, 1968) and, most particularly, a confirmed obligate association by at least 14 insects (Folkerts, 1982) of which the best known and studied are three dipterans, *Wyeomia smithii*, a culicid (Fig. 14.3); *Blaesoxipha fletcheri*, a sarcophagid, and *Metriocnemus knabi*, a chironomid midge.

Wyeomyia smithii (Culicidae)

Swales (1969, 1972, 1975) and Buffington (1970) were probably the first fully to appreciate the complex interrelationships between the above three dipterans of which the mosquito *Wyeomyia smithii* has been by far the most intensively studied.

Smith, as early as 1902, noticed that larvae of *W. smithii* could pass Canadian winters frozen solid in the ice cores of *Sarracenia purpurea*. He later placed some over-wintering larvae in a jar in his laboratory on 22 January and observed that metamorphosis did not take place until March of the same year.

These pitcher-plant mosquitoes, of which *Wyeomyia* is one of the best known in North America, range from the Gulf Coast to Labrador and northern Manitoba. We now know that *W. smithii* is not restricted to *S. purpurea* but at least in the southern part of its range inhabits *S. flava*, hybrids between *S. purpurea* and *S. flava*, *S. leucophylla*, *S. rubra* and *S. alata* (Bradshaw, 1983). However only the leaves of *S. purpurea* commonly persist in winter. Leaves on *S. purpurea*, at least in New England, USA (Fish and Hall, 1978), are produced at about 15-day intervals in the summer months. Each lasts only about 12 months, and there usually seem to

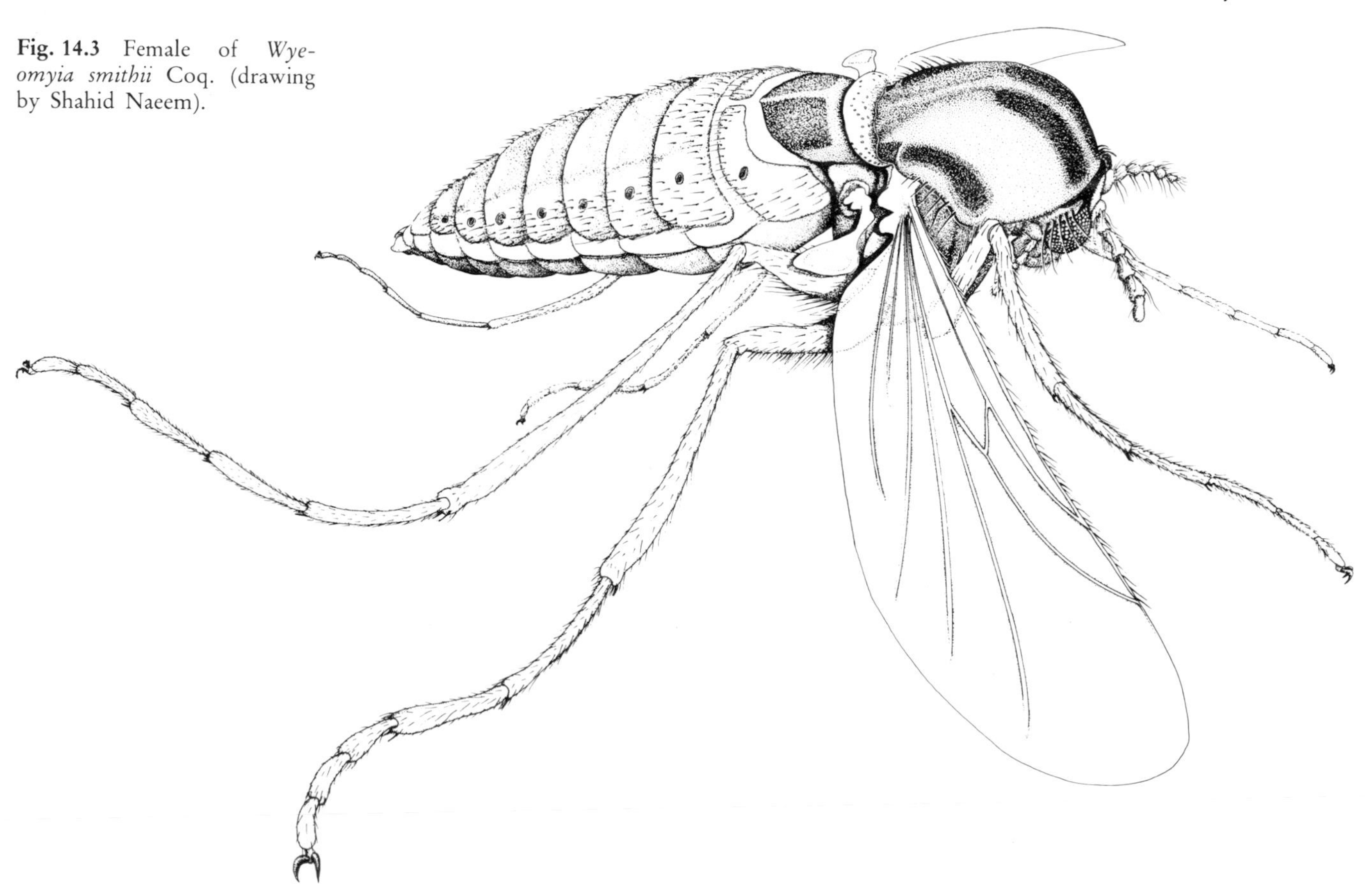

Fig. 14.3 Female of *Wyeomyia smithii* Coq. (drawing by Shahid Naeem).

be about five or six operative leaves per mature plant (see Fish and Hall, their Fig. 1). The situation in *Darlingtonia*, which probably lacks *Wyeomyia* but which possesses *Metriocnemus* (see page 257) in abundance, is very probably similar (page 258). Thus, *W. smithii* must vacate over-wintering leaves and colonize new leaves before the growing season ends and only one generation is produced in each leaf. This regular transference of phytotelm is in contrast to most mosquitoes which colonize ground pools of unpredictable duration and characteristics (Fish, 1983).

So far as is known (Istock *et al.*, 1983) *W. smithii* females oviposit only in the water-filled leaves of *Sarracenia* species. There have been infrequent reports of *W. smithii* being found in bog pools, but this may be because *Sarracenia* pitchers occasionally break or spill (Pennak, 1953). Istock *et al.* (1983) point out, that this obligate behaviour has not yet come to be a fixed trait of the mosquito, as is shown by a combination of artificial selection experiments and oviposition site preference experiments (cf. *Nepenthes pervillei* p. 264). The genus *Wyeomyia* is not restricted to *Sarracenia*. Other species of this genus inhabit the 'tanks' or phytotelmata of non-carnivorous bromeliads (Bradshaw, 1983).

W. smithii larvae are mobile, mainly grazing upon bacteria, a range of protozoans (Addicott, 1974) and small particulate material in what Bradshaw (1983) calls the plankton zone of the pitcher, or on the surface of the detritus on the bottom. As we might expect, they are in turn preyed upon by a polymorphic inclusion-body-forming Baculovirus (Hall and Fish, 1974). In their phytotelm home they may have to wait for a more powerful insect carnivore such as the *Metriocnemus* larvae to break down an insect victim, or for an insect-protozoan-bacterial food chain to develop.

Addicott (1974) used the relatively simple model of *W. smithii* as a predator to test the hypothesis that in the presence of predation more species will co-exist in the prey community than in its absence. He found that there was considerable variability in the density of predators in both time and space and that the protozoan community responded to these differences. However, contrary to the hypothesis, species number decreased with increasing predator density. This result probably relates to the low level of interaction between species of protozoan and the difficulty with which new species can enter the microhabitat. In this, the pitcher is probably a rather atypical environment.

The maintenance of larval dormancy in *Wyeomyia*

Over a Canadian winter the *W. smithii* larvae in the robust pitchers of *S. purpurea* maintain themselves in what was at first believed to be adult diapause, but is now known to be a larval dormancy. The maintenance of this larval dormancy is a function of photoperiod and independent of temperature. Daylengths greater than 14.5 h break dormancy, while periods of fewer than 14.5 h maintain dormancy for at least four months (Smith and Brust, 1971; Evans and Brust, 1972; Bradshaw, 1976; Bradshaw and Lounibos, 1972, 1977).

Bradshaw (1980a) and Bradshaw and Phillips (1980) showed that both in the dawn and in the dusk, diapausing larvae are photoperiodically most sensitive to blue light (390–450 nm) with a subsidiary response in the blue-green and green (480–540 nm) regions of the spectrum. The saturation curves for the response to blue light in the dusk have steeper slopes than those for the responses to blue-green and green light, suggesting that two distinct pigments or pigment complexes underlie photoperiodic response in *W. smithii*. The photic environment of *W. smithii* during twilight is rich in yellow-green light, but sufficient light is available at 390–540 nm to trigger photoperiodic response early and to sustain response until late in the evening. Comparison of action spectra with spectra of available light indicates that the zenith angles of the sun which would result in 50% response are 95°48′ and 94°52′ in the dawn and dusk respectively. Using these zenith angles to give approximate daylengths in nature provides a reasonable prediction of development in the field.

Smith (1902) also believed that *W. smithii* females, although obviously well-equipped for biting did not do so. It now appears that blood feeding only occurs erratically in the southern part of the range (Bradshaw, 1980b) and, in all parts of its range, *W. smithii* can complete the first ovarian cycle without biting. Subsequently, a blood meal may be needed by southern but not northern females (but see Goins, 1977). Lounibos *et al.* (1982) believe that the lack of blood-feeding in northern populations may have come about by periods of density-independent larval growth, i.e. a more generous diet.

Evolutionary adaptations in *Wyeomyia*

As one would expect over such a huge geographical range and climatic spectrum from near tropics to near arctic, the mosquitoes, and particularly their larvae, show a corresponding spectrum of evolutionary adaptation (Bradshaw and Lounibos, 1977). As we have seen, from the Gulf Coast northwards to Quebec and Saskatchewan (Burgess and Rempel, 1971) the larvae over-winter, in dormancy in the persistent leaves of *S. purpurea*. The Gulf Coast larvae in the south go into a shallow dormancy in the fourth instar. The depth of this dormancy increases as one follows northwards the distribution of the *Sarracenia*. At higher latitudes, the larvae diapause in the third larval instar (Smith and Brust, 1971) and the depth of dormancy increases with latitude and altitude, the cycle controlled by the photoperiodic shift (Bradshaw and Lounibos, 1977). Moreover clear anatomical differences can be seen between the southern races of this cline and the more northerly forms. Bradshaw and Lounibos (1977) and Lounibos and Bradshaw (1975) speculate that this evolution of synchronous behaviour has taken place from south to north and that *Wyeomyia haynei*, which some workers (e.g. Dodge, 1947) recognize as a different species, because of differences in detail (Goins, 1977), is in fact no more than a geographic sub-species of the major group *W. smithii*.

Istock *et al.* (1975, 1976a, b) have studied the genetic control of this evolution and have shown how artificial selection can promote or suppress the tendency to enter a warm-season diapause. Istock and his co-workers have also shown how the *W. smithii* population experiences an alternation of density-dependent and density-independent selection within the same season. Density-dependence occurs in the spring and late autumn. The first pulse of pupae comes in the spring. Density-independence holds good over two to three months of the summer and early autumn. During this period the population biomass is well below the carrying capacity of the pitchers.

Moeur and Istock (1982) have shown that some populations of *W. smithii*, in western New York State, have polymorphic chromosomes including a total of ten paracentric inversions, some features showing a change of frequency in successive seasons. Parallel with this, Istock (1978) has shown that diapause and development time are genetically variable and genetically correlated in local, northern and unpredictable environments.

Most eggs are laid in new pitchers, with some chemical stimulus, apparently specific to new pitchers (Mogi and Mokry, 1980) playing a dominant role. The resultant larvae then mature as the leaf ages. Istock *et al.* (1983) have shown that this stimulus is a water-soluble chemical moiety from the plant tissue that serves as an adult attractant, but a complete identification is not yet available. The ovipositing females apparently do not use visual clues, despite the presence of the recently identified UV patterns on *Sarracenia* pitchers (Joel *et al.*, 1985 and see Ch. 5). Mogi and Mokry (1980) also showed that larger pitchers contained more eggs than smaller ones. The female lays small batches of eggs at any one time, not randomly but in what these authors call a 'skip oviposition' strategy. This strategy obviously has selective advantage in spreading viable eggs over a larger number of potential breeding sites. The female may deposit only a fraction of her available eggs at any one time, or the follicles may develop asynchronously. Evidence for either of these hypotheses is not yet available. It has also been noticed that the diet significantly influences the onset of oocyte production by those females that hatch (Lang, 1978). Moeur and Istock (1980) found that there was a marked difference in the time to oogenesis between richly fed and sparsely fed females – 26.9 days as opposed to 31.6 days – but that the precocity of the richly fed females resulted in only a small and not significant difference (92.3 v. 81.9 eggs) in the number of eggs laid. The addition of sucrose to the adult diet increases both adult longevity and fecundity, but no such changes were noticed with a supplement of egg albumen.

The geographical range of *W. smithii* overlaps that of *Metriocnemus knabi* (see page 257). Thus *W. smithii* is dominant at the southern end of the range and *M. knabi* to the north, with a commensurate adjustment for high altitude locations. *M. knabi* is apparently slightly more tolerant of extremely low temperatures, and the degree of intra- and interspecific tolerance has been studied by Bradshaw (1983).

Blaesoxipha fletcheri (Sarcophagidae: Sarcophaginae)

B. fletcheri, the second-most prominent insect inhabitant of *Sarracenia* pitchers, exists in its larval state at the water surface in the pitcher. It is a typical cyclorraphous maggot with enlarged posterior spiracles which serve to prevent its submergence. The larvae, which are laid viviparously by the females sitting on the rim (see page 265), hang from the water surface and feed on dead, trapped animals, but only on those that remain floating on the surface (Forsyth and Robertson, 1975). Although many eggs are laid, the reproductive strategy appears to ensure that only a few, but very large, aggressive larvae are produced. *B. fletcheri* seems only to be found in pitchers that are at or very near their peak insect-catching ability (Fish and Hall, 1978). Fish (1976b) estimates that *Blaesoxipha* in *Sarracenia minor* may consume up to 50% of the captured prey.

Metriocnemus knabi (Chironomidae)

The larvae of *M. knabi*, a small chironomid midge species which cannot float or swim (Fish and Hall, 1978), live in the mixed detritus at the bottom of *Sarracenia* pitchers where, in spite of the concentration of living and dead insects, the oxygen concentration is still relatively high (Cameron *et al.*, 1977). *Metriocnemus knabi* scavenges for particulate material at the bottom and bores into drowned insects. A large insect lifted from a pitcher is, to quote Bradshaw (1983), a 'veritable Medusa of writhing *M. knabi* larvae.'

Whereas *W. smithii* can apparently survive in small bog pools (Pennak, 1953, but see page 255), *Metriocnemus knabi* appears to be an obligate commensal of *S. purpurea* (Paterson, 1971). In the New Brunswick area of Canada studied by Paterson, the fluid in the pitchers freezes solid for four months of the year. Even under these extreme conditions there was less mortality than that for larvae of *Wyeomyia*. Like *Wyeomyia*, *Metriocnemus* owes its survival in part to its control of dormancy by the photoperiod. This external control system was, in fact, discovered earlier than the system in *Wyeomyia*, by Paris and Jenner in 1959.

Niche Segregation of the Pitcher Commensals

The leaves of *S. purpurea* age at different rates, and ability to capture prey and the quantities of dead insects in different states of breakdown will vary greatly. Thus, although *Wyeomyia*, *Metriocnemus* and *Blaesoxipha* often occupy the same pitcher, the relative abundance of each insect will depend upon the state of decomposition of the prey, and only one generation of each species will be present at any one time (Fish and Hall, 1978). Each insect feeds on a different stage in the decomposition of the prey; *B. fletcheri* on the just dead and floating prey, *W. smithii* filter-feeds as we have seen on the particulate matter (Buffington, 1970), and *M. knabi* attacks whole sunken corpses. Thus a minimal interference in space and time is achieved.

The genus *Metriocnemus*, in this example *M. edwardsi*, is also found in the distantly related plant genus *Darlingtonia* (see page 259).

Other Fauna of *Sarracenia* Pitchers

The three insects discussed above are the best studied of the commensals of *Sarracenia*, but other fellow travellers abound (Swales, 1969, 1972). For example, when the quantity of prey insects in *S. flava* is abundant, the dipteran *Sarcophaga sarraceniae* can often be found (Plummer and Kethley, 1964). Other dipterans include *Dohrniphora cornuta* (a phorid) (Jones, 1918) and *Bradysis macfarlanei* (a sciarid) (Jones, 1920). In addition, the aphid *Macrosiphum jeanae* appears to live exclusively in the pitchers of *Sarracenia purpurea*. This new species (Robinson, 1972) has so far only been found in Manitoba but it may also live elsewhere. Both males and oviperae were found in the plants in September, suggesting that the entire life history is spent in the pitcher. It appears to live and breed in second-year pitchers only. If the pitchers are full, the colonies are found above the fluid line. If the pitchers are empty, because holes have been drilled near their bases, the colonies are found throughout the pitcher. The holes may be the work of the common paper wasp *Polistes fuscatus pallipes*, a nest of which has been found in a *S. purpurea* pitcher (Bernon, 1969). It may be, however, that another insect makes the holes and both aphid and wasp exploit the subsequent sheltered habitat.

A new species of alder fly, *Sialis joppa* has recently been found by Mather (1981) in *Sarracenia purpurea* pitchers in New Jersey, Delaware and Pennsylvania, and the mites *Anoetus gibsoni* and the new species *A. hughesi* are seasonal inhabitants of the same species of pitcher-plant (Hunter and Hunter, 1964; Mather and Catts, 1980; cf. *Nepenthes* pitchers, pages 262 onwards).

An insect of the more southern *Sarracenia* species is the phytoseiid mite *Macroseius biscutatus*. It is found from Florida through Georgia to North Carolina, and current information indicates that it is restricted to *S. minor*. Its diet appears to be almost entirely confined to nematodes, mainly of the genera *Panagrellus* and *Panagrolaemus* (Muma and Denmark, 1967). *Histiostoma* mites have been found in *S. purpurea* in southwestern Ontario (Judd, 1959).

Nematodes of the genus *Panagrodontus* were found in up to 50% of the pitchers of *S. alata* (*sledgei*) in March/April, but dropped to 3% in June (Goss *et al.*, 1964). Curiously these could only be cultured when culture medium was also inoculated with a bacterium and a *Mucor* fungus isolated from the pitcher fluid. No evidence was found for the nematodes burrowing into the plant from the soil; infection apparently needing an arthropod vector.

As we might expect, protozoa able to withstand the enzymic digestion, living and multiplying and probably obligate, are found in the pitchers of *S. purpurea* (Hegner, 1926b). They also occur in the *Nepenthes* pitchers and the bladders of *Utricularia* (see page 268).

Non-commensal, but probably obligate phytophagous inhabitants of *Sarracenia* pitchers are species of the noctuid moth *Exyra. E. rolandiana* is found in *S. purpurea* from Canada to the Gulf States (Judd, 1959; Brower and Brower, 1970), whereas the more catholic, but local *E. semicrocea* is found in *S. rubra*, *S. minor*, *S. drummondii*, *S. alata (sledgei)* and *S. psittacina* (Jones, 1904, 1907 and 1921). One or a few eggs are laid in each pitcher, of which only one survives to occupy the chamber. The larva then closes the peristome mouth with a web and feeds, thus protected, on the inner wall of the pitcher without penetrating to the outer surface. The necrotic, but hardened and totally enclosed pitcher, may provide a home over the winter for the pupating insect. The rate of infestation is usually not high: Jones examined 275 pitchers in a locality of *S. psittacina* and found only three to be plugged and to contain healthy moth larvae. On the other hand, in a population of *S. minor* in Florida, Fish (1976b) found that *E. semicrocea* had rendered two-thirds of the pitchers non-functional. Dead leaves have also been found to be occupied by colonies of the ant *Dolichoderus pustulatus* (Kannowski, 1967).

Jones (1908) and Brower and Brower (1970) also found a consistent infestation of *Sarracenia* roots by the root borer *Papaipema appassionata*. However, its obligate status is not known, Brower and Brower (1970) and Hilton (1982) also consistently found *Endothenia daeckeana* in the seed capsules. Bacteria and other micro-organisms are very widespread in the pitchers of *Sarracenia* (Plummer and Jackson, 1963), as they are in the pitchers of the closely related *Darlingtonia*. Their roles are considered under digestion of prey (see Chs 8 and 9).

We have assumed, in this account, that the presence of the commensal fauna is an integral part of the *Sarracenia* life cycle. This supposition is supported by the apparent co-evolution of the pitcher inner epidermis (see page 253 and Fig. 9.1) to enhance the oxygen tension within. It would be useful, therefore, to study, in the same fashion, the naturalized populations of *Sarracenia* in Ireland (see page 37) and Sweden. Do they possess the conventional commensal flora? First investigations suggest not. If not, have other entrepreneurial species taken over, or is an unoccupied pitcher still a viable community?

Moreover, as Bradshaw and Creelman (1984) have pointed out, from time to time *Sarracenia purpurea* may be swamped with a superabundance of prey, or may suffer lower light intensities and perhaps as a result of which the normal dipteran commensals may die. The water becomes foetid and takes on a red or wine colour which, they suggest, is due to *Rhodopseudomonas palustris* and possibly other bacteria. These anaerobes appear capable of continuing the degradation. Their role is further discussed under degradation of the prey (see pages 204 onwards).

Fungi in the Pitchers of *Sarracenia*

Mycelium-forming fungi, not further identified but apparently in a symbiotic condition, were seen by Barckhaus and Weinert (1974) in the absorbing zone of pitchers of *Sarracenia purpurea*. The hyphae were actually growing in the thickness of the wall, but they were not investigated further and their significance is not understood.

The Fauna of *Darlingtonia* Pitchers

Probably because of its distribution (see Ch. 3), the close relative of the sarracenias, *Darlingtonia californica*, has received far less attention from entomologists. Rebecca Austin (see Appendix 1, and Jepson, 1951), that determined and all-weather Californian botanist of the nineteenth century, was probably the first to record the existence of dipterous larvae feeding on the captured prey. Her description of the assault of the larvae on recently captured prey was just as dramatic as that of Bradshaw (see page 257). She wrote (in May 1875): 'Upon opening a tube that contained a fly in a perfect state, there were so many of these worms wound about it, that it had the appearance of a wet ball of cotton larger than a pea.' Often, she noticed, prey so recently caught that they were still floating on the surface of the liquid, were already the host to a thick felt of white 'worms' below the water surface. She was not able to make species identifications of these larvae (she did not even possess a hand lens for her field work until 1875) and we now know that there are several commensals. In his review, Fashing (1981) lists the chironomid midge *Metriocnemus edwardsi* (cf. *Sarracenia*), which is probably the most common inhabitant, and also the phorid dipterans *Megaselia ovestes* and *Botanobia darlingtoniae*, the histiostomatid acarine *Anoetus* spp. (cf. *Sarracenia* again), the web-forming lynyphiid spider *Eperigone trilobata* (cf. the noctuid moths in *Sarracenia*), the euploid acarine *Protereunetes* sp., which lives under the nectar roll, the eriophyd acarine *Leipothrix*, and the tarsonemid acarine *Tarsonemus* sp., which feeds on the tissue of the *Darlingtonia* above the water line. More recently, Fashing and O'Connor (1984) have added a new genus of histiostomatid acarine, *Sarraceniopus* to the list, one which seems exclusive to *D. californica*. Of all these, by far the most significant seems to be *Metriocnemus edwardsi*. The larvae of this chironomid midge were found in 80% of the pitchers examined by Jones (1916), (who also found the phorid *Botanobia*), and 100% of the pitchers examined in Butterfly Valley by Naeem (1985). Every pitcher examined by the senior author on two visits in 1985 (see page 39) was found to contain larvae which were tentatively identified as *Metriocnemus*. These larvae are found in the upper quarter of the captured insects in a given pitcher, eating the more freshly captured prey. Some of these obligate larvae over-winter as Austin (1876b) was the first to observe, but in comparison to *Sarracenia* little work on this phenomenon has been done.

Austin (1875–77 and Appendix 1), apart from her pioneering observations of the dipteran larvae in the *Darlingtonia* pitchers also noticed that many pitchers had spider's webs formed within them. Generally, at least half of the tubes she examined had webs, with either adult spiders or visible egg clusters present, and in some colonies virtually all the tubes were parasitized. She was not able to make identifications, and some of her spiders may have been the web-forming lyniphiid, *Eperigone*, noted above. However, her observations do suggest several species of web-forming arthropods, none of which had any difficulty in climbing about in, or leaving, the pitcher tubes.

Comparisons have been made between the arthropods of different species but of the same genera inhabiting, for example, *Sarracenia* and *Darlingtonia*. The *Anoetus* species observed by Hughes and Jackson (1958) and Fashing (1981) in *Darlingtonia* differs only in small morphological detail from the sarracenian *A. gibsoni* (Nesbit) and *A. hughesi* (Hunter) (see page 257) but, as one would expect, more markedly from the *A. guentheri* taken from *Nepenthes* species (see page 264). All of these mites are filter feeders and disperse phoretically on insects.

All of the inhabitants of *Darlingtonia* listed above are egg-laying arthropods which, in that they have not yet been found in other habitats, appear to be obligate inhabitants of the pitcher.

Although, as we have pointed out, the egg-laying obligates of *Darlingtonia* are less well studied than those in most species of *Sarracenia*, it should not surprise us if they turn out to be fewer in number and in diversity. Jones (1916) noted that *D. californica* seems 'almost destitute of insect associates other than the victims of the trap when compared to the species of *Sarracenia*.' While, with the benefit of more observations, this now seems to be something of an exaggeration, it is true that the water surfaces of most *Sarracenia* pitchers are more exposed than that of the deeply hooded *D. californica*. It would seem to present no small problem to a potential ovipositing female dipteran, for example, to fly into the enclosed chamber and 'bomb' her eggs on to the exposed water surface. But here we must note some more unique observations by Rebecca Austin which appear to question many of the assumptions made about oviposition by dipterans in the pitchers of the Sarraceniaceae.

The Fauna of Unopened Pitchers of *Darlingtonia Californica*

Rebecca Austin spent some six years, from about 1873 to 1879, in a very remote part of northern California, studying the flora of the area and *Darlingtonia* in particular. She had no access to the conventional sources of scientific literature and only a very erratic contact with academics in the east. Even had she been fully conversant with the latest literature (see page 17), there were no more than tentative observations on a few phytotelmic larvae. She had, moreover, only the sketchiest of botanical and no entomological training. We must, therefore, accept her observations as totally unbiased, if unsophisticated data.

She states categorically (Appendix 1: 1875B, E, F,, G; 1876D, E) that she found larvae ('worms' *sic*), of what she assumed to be more than one species, in the *unopened* pitchers. She notes (1875F) that in one particular colony of plants in a cool, shaded canyon, and one in which the plants were over 90 cm high, there was no such infestation. She frequently found larvae in the unopened pitchers of plants in full sun and in particular in those with orange or red pigments developing in the young pitchers (1875F). She also notes (19 June 1875), having made observations almost daily up to this date, that this was the first day of the year in which she found larvae in unopened pitchers. On 1 July of the next year (1876G) she notes:

> 'Examined many new leaves in reference to larvae in leaves having their orifices closed. They are in nearly all. How many days, before the opening of the orifice, I am unable to determine. I think from the size of the leaves gathered for noting, and those left standing, of similar sizes, that they are in the tubes from four to six days, before the hoods expand.'

How then did she think to explain this observation. She writes (1875E):

> '. . . six new leaves examined; they had the white larvae; on the lower part of most of the new leaves are small dots, as if pierced by some insect. During the month of June, found larvae in new tubes with closed hoods, – that is, in the first two leaves; since then they have not been so numerous, – I mean in the second and third pairs of leaves.'

She therefore appears to believe that, at some point from mid-June on, but perhaps varying with the colony and only in those colonies growing in full sun, ovipositing insects, and probably also mites, of more than one species penetrated the young and still-extending walls of pitcher tubes to lay their eggs. No such observations appear to have been made of this phenomenon, if it is correct, in any other species of the whole Sarraceniaceae family. Long vigils, both day and night, at substantial colonies of long-established Sarraceniaceae in undisturbed habitats are still required.

The Fauna of *Darlingtonia* Flowers

Although it is not usual to think of a flower as a phytotelm—a domatium would be a better description—Austin (1875–1977) noticed, as we shall see on page 271, many of the flowers had well established spiders' webs. She noted (10 May 1875):

> 'Seven spiders and nine webs in a dozen flowers, two of them large spiders. It is very common at this date to find flowers pretty well lined on the inside with these webs; they are attached to the stamens, carpels and stigmas, and a great many of them have webs attached to the stem and sepals, – filling the arch made by the drooping flower. May 11th. Examined fifty flowers; two of the newly opened ones contained no webs; all the others did. In this fifty, I found *thirty-five spiders*, in the corolla, or letting themselves down from the flower by webs. These spiders were from the size of a pin's head to that of a pea, more small ones than large.'

Despite these infestations, *Darlingtonia*, as we shall see (page 271), sets copious quantities of viable seed.

The Commensal Fauna of *Heliamphora*

Few have seen *Heliamphora* in its native habitat, and even fewer have been entomologists, so commensal information is rare. A very limited report is by Brewer-Carias (1972) cited by Mazrimas (1975), who saw *Heliamphora* species growing on the tepui (see page 42), the isolate mesa-type mountains of the southern border of Venezuela on the boundary with Brazil. In the base of the *Heliamphora* pitchers, Brewer-Carias saw a semi-liquid mass of decayed insects. Swimming in the water, in the mass, were numbers of what he assumed to be white nematodes. Brewer-Carias suggests a commensal relationship between these possible nematodes, although they might have been midge larvae, and the *Heliamphora*. More recently Zavortink (1985) has identified a new sub-genus of *Wyeomyia* which he calls *Zinzala* in the pitchers of *Heliamphora* in the Gran Sabana region of Venezuela. This sub-genus comprises two new species *Wyeomyia* (*Zinzala*) *zinzala* and *W.*(*Z*) *fishi*; the former was found in *Heliamphora nutans* and the latter in *H. heterodoxa*.

The Fauna and Flora of Non-Carnivorous and Carnivorous Bromeliads

We have seen that it is now accepted (see page 19) that some terrestrial bromeliads are carnivorous plants. Some of them, particularly the non–carnivorous epiphytes, are outstanding in their commensals and it is relevant to comment on them briefly. Large specimens of the non-carnivorous Jamaican bromeliad, *Aechmea paniculigera* and *Hohenbergia* spp. may contain about 2 litres of water. Laessle (1961) showed that these tank bromeliads (along with species of *Vriesia*) contained up to seven species of algae, as well as bacteria, protozoa, gastrotrichs, rotifers, flatworms, oligochaetes, copepods, ostracods, fresh-water crabs, numerous insect larvae of helodid and carabid beetles, chironomids, *Bezzia*, mosquito larvae and tadpoles of the tree frog *Hyla brunnea*. As we saw on page 252, another species of *Wyeomyia* is a common inhabitant of these tanks. Another genus of mosquito, *Rhynchomyia frontosa* has also been found in *Brocchinia reducta* (Zavortink, 1986). Even a carnivorous plant, *Utricularia humboldtii* (Maguire, 1967; Brewer-Carias, 1973) grows in the tanks of the high altitude, pan-tepuis, non-carnivorous species of *Brocchinia*, *B. tatei* and *B. micrantha*.

Animals and Plants that Exploit *Nepenthes* Species – General Considerations

As Fish and Beaver (1978) so pertinently write in their bibliographic study, an abundance of water-impounding plants or plant parts (phytotelmata) in malarial areas renders chemical control useless. In such circumstances some method of biological control must be found. Their works (and bibliography) apply not only to the southern United States and central America, where *Sarracenia* and bromeliad species abound, but also to the countries bordering the Indian Ocean where *Nepenthes* is found.

The Insect Fauna and Flora of *Nepenthes* Species: General

Nepenthes pitchers, especially the creeping species like *N. ampullaria* (see page 34) in which the pitchers most commonly are found half-buried in the litter of the forest floor, have the richest fauna of any southeastern Asian phytotelm (Beaver, 1979b; Kitching and Scholfield, 1986). Amongst the inhabitants in the liquid are fungi and slime moulds, protozoans of at least seven species (Hegner, 1926b), desmids, diatoms, rotatoria, oligochaete worms, crustaceans, the larvae of various flies and mosquitoes and even the occasional tadpole (Beaver, 1979a, b).

Above the fluid level are commonly found several species of spiders. One of these spiders, *Misumenops nepenthicola*, which is described in detail later (see page 265) commonly forms a web half-way down the pitcher tube, and robs the plant of its potential prey by intercepting insects as they plunge down the unclimbable wall (see pages 113 onwards).

M. Jebb (1986, pers. comm.) noticed that many pitchers of *N. mirabilis* in New Guinea were chewed through, below the fluid level, by unknown phytophagous insects. Other spiders, not of the genus *Misumenops* and not normal inhabitants of *Nepenthes* pitchers, would form webs in the upper part of the pitcher, predate the insects falling into the trap and make their escape, if necessary, through the chewed holes. Such a minor degree of damage renders the whole pitcher system useless, and Jebb speculates that the exterior glands of many pitcher-plant species may serve to encourage ants which might defend the pitcher from such random damage.

In Sabah, the small primate, *Tarsius spectrum*, visits the pitchers of various *Nepenthes* species, perches on the margin and feeds off the trapped insects (Shetler, 1974b). It has been suggested that the spines on the petiole and the curious over-arching 'claws' above the operculum of *Nepenthes bicalcarata* (Brown, 1880) are devices to prevent robbing of this kind (Burbidge, 1880). *N. bicalcarata*, is known also to be an obligate 'ant-plant'. In this species, the ant domatium is a swollen hollow petiole, which develops fully even in the absence of the ant association (Burbidge, 1880: Beccari, 1886). Could it be that this is an evolutionary response, unique within the genus *Nepenthes*, to defend this species against the predations of *Tarsius*?

The Obligate or Commensal Fauna of *Nepenthes* Pitchers

The associated insect fauna of *Nepenthes* species is vast and has a long history of study: for an early review see Dover (1928) and, for a comprehensive list, Beaver (1985). Thienemann (1932) divided the fauna inhabiting *Nepenthes* pitchers into three classes: *nepenthebiont* species which normally live only in the pitchers and are peculiar to them; *nepenthephil* species, which grow in large numbers in the pitchers but are also regularly found in other habitats; and *nepenthexene* species which are not found regularly in the pitchers, and occur therein only erratically or in small numbers. Such a group are the unnamed spiders mentioned above. According to Beaver (1979b) the great majority of the species studied (79%) are in the nepenthebiont class.

The ecology and morphology of *Nepenthes* species and their effects upon the food webs

Some differences between the fauna of different

species of *Nepenthes*, probably first noted by Barr and Chellapah (1963) are due, at least in part, to the different structure of the pitchers (Beaver, 1983). Barr and Chellapah noticed that the mosquito fauna, for example, of *N. gracilis* is rather different from that of *N. ampullaria*, whereas, at least for mosquitoes, the latter resembles *N. rafflesiana*: part may be structure, part habitat. For example, the glandular enzyme-secreting tissue is confined to the lower one-third to half of the pitcher in, amongst others, *N. albomarginata* and *N. gracilis*, but extends to the top of the pitcher in *N. ampullaria*. The pitcher of the latter species has an inturned rim, absent in *N. albomarginata* and *N. gracilis*, but the lid partially covers the opening in these two species, whereas in *N. ampullaria* the lid is turned back and does not stop rain filling the phytotelm (Beaver, 1979b).

Other differences may relate to the habitats in which the different species occur (see Ch. 3) and their paths of evolution. As Beaver (1983, 1985) points out, although the *Nepenthes* genus is widespread, the numbers of any one type of pitcher in a given patch are small and the number within that patch available to be colonized are, by reason of age, smaller still. We should, therefore, not be surprised at the differences between the inhabitants of different species of *Nepenthes*, which are much more marked than those observed between the once virtually contiguous species of *Sarracenia* (see page 252).

The evolution of a food web

The food webs of the nepenthebiont species of *Nepenthes* are extremely diverse (Figs 14.5*A* & *B*). Kitching and Schofield (1986) argue that the variety of webs supports the notion that those *Nepenthes* pitchers that are close to the presumed centre of diversity of the group in Borneo have the most complex webs, both in terms of numbers of species at each level and the number of levels. At the other end of the scale we would expect species on remote small islands such as *N. pervillei* on certain islands of the Seychelles to have the simplest webs.

The initiation of a food web

The pitcher of a *Nepenthes* is sterile when it opens (cf. *Darlingtonia*, page 260), but must rapidly become colonized by bacteria and protozoa. Each pitcher may remain viable for up to a year (Beaver, 1985), but as Southwood *et al.* (1974) and Beaver (1985) have pointed out, an individual pitcher is a temporary habitat and the phytotelm never reaches stability. However, a patch of *Nepenthes* under normal conditions will be continuously supplying new and uncolonized chambers.

Most of the phytotelm species (82%) have aquatic larvae and most are not known to breed in other habitats (nepenthebiont species). Only one medically important mosquito, the widespread *Aedes albopictus* has been reported. This is a nepenthexene (casual occupant) species and the pitchers only form a minute fraction of its breeding sites. Barr and Chellapah (1963), for example, list it only as an occasional accidental in old pitchers of *N. ampullaria*. Most mosquito species (71%), like *A. albopictus*, eat insects falling into the pitcher for food either directly or indirectly (by feeding on saprophagic microorganisms).

We have already seen (page 261) that nepenthexene spiders may, from time to time, occupy damaged pitchers. Undamaged phytotelm communities, depending upon the species of *Nepenthes* and the locality, may play host to up to five species of predator, two parasitoids and one herbivore. Two of the predators (a spider and the larvae of a mycetophilid fly) catch insects entering or leaving the pitcher; the others have aquatic larvae. The larvae of four species in the families Culicidae, Mycetophilidae, Sarcophagidae and Calliphoridae and the spider only occur as single individuals in the pitchers. They are relatively large and limited amounts of food are available in a single pitcher. Most of the nepenthebiont species seem adapted to an irregular food supply and their larvae can apparently survive long periods without food. They may, however, still be able to graze on bacteria. A few other species consume an occasional surplus of undigested food in the pitchers.

The food webs of *N. albomarginata* and *N. ampullaria*

It will be apparent by now, as Beaver (1983) has clearly demonstrated from his work in Penang, that the relationships of all these very different animals in a confined habitat under tropical conditions are complex; probably more so than the better studied, but season-limited relationships in *Sarracenia* pitchers (see page 252 and Fig. 14.4). Figures 14.5*A*,*B* represent this situation diagrammatically. The differences may be explained on the basis that in *N. albomarginata* and in *N. gracilis*, which is similar, the glandular tissue only occupies the lower half of the pitcher, i.e. there is room for *Misumenops* and others on the upper walls. In *N. ampullaria*,

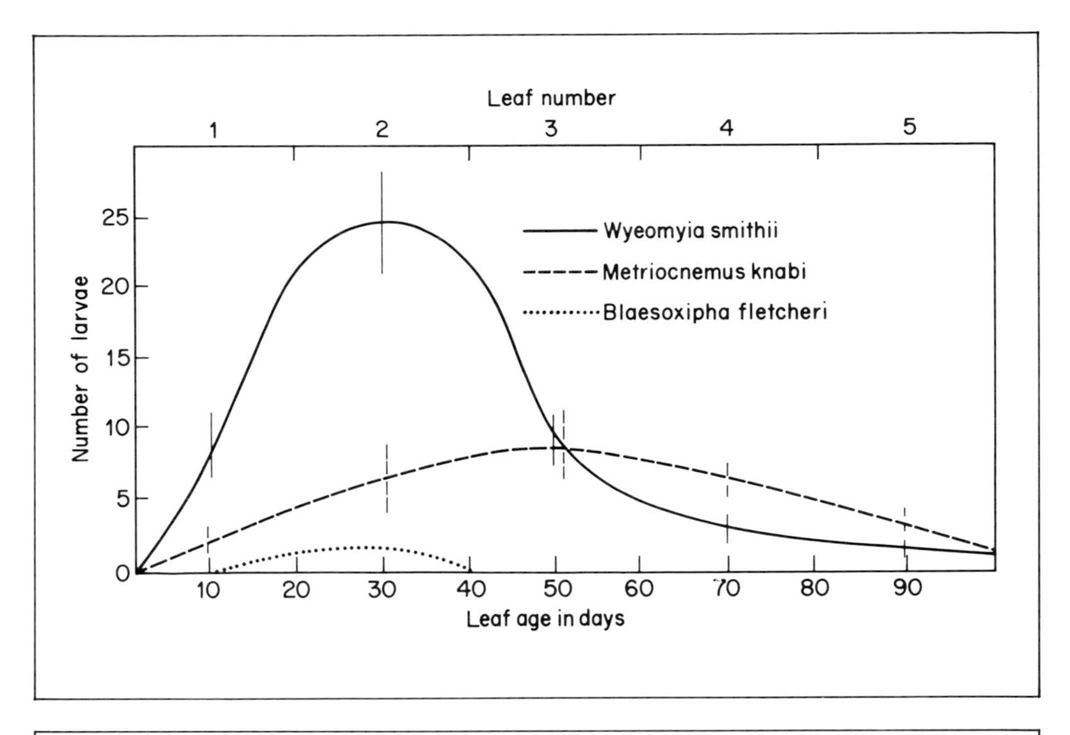

Fig. 14.4 Mean number and standard error of dipterous larvae (ordinate) inhabiting the pitcher fluid of 10 mid-season plants of *Sarracenia purpurea* (top abscissa) at the field study site compared to the range in age in days for each leaf based on a 20-day interval between leaf production (bottom abscissa); cf. Fig. 7.2. (Redrawn with permission from Fish and Hall, 1978.)

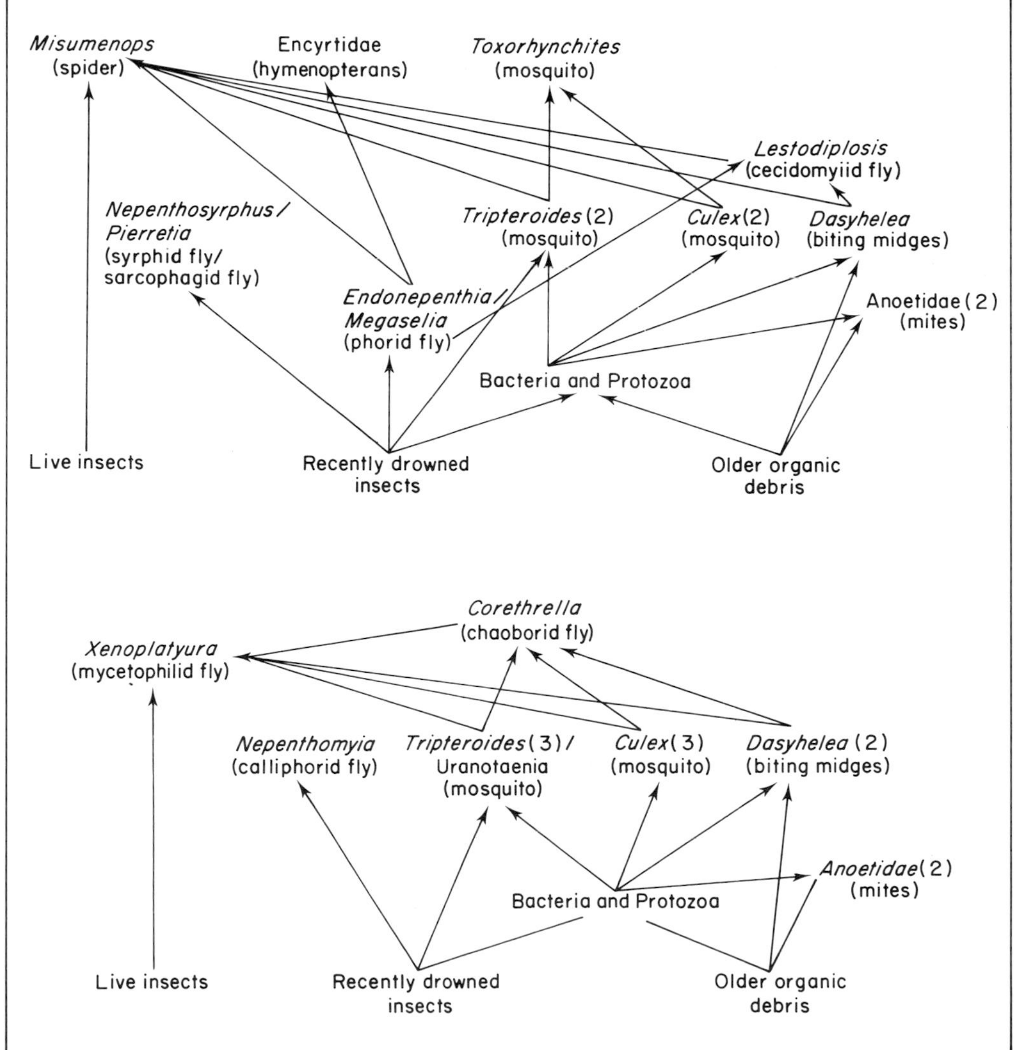

Fig. 14.5 Variations in *Nepenthes* food webs. (*A*) *Nepenthes albomarginata* and (*B*) *Nepenthes ampullaria*, in West Malaysia. (Both redrawn, with permission, from Beaver, 1985 by D. Thompson, Tucson, USA.)

on the other hand, the glandular area reaches to the top of the pitcher and the only resting or pupation site is under the peristome. These pitchers regularly fill with rain. It should be noted, too, that not every single species is present in each pitcher.

Beaver (1979b) gives an account of the biology of up to 25 species of insects and three species of arachnids living in the pitchers of *N. albomarginata*, *N. ampullaria* and *N. gracilis* in Penang. The fauna is normally dominated by Culicidae (ten species), but eight other families of Diptera are represented, as well as Hymenoptera, Lepidoptera, Araneae and Acari. Many, like the mite genus *Anoetus* (the nomenclature is complex and the genera *Zwickia* and *Creutzeria* are often included under this heading), have representatives in *Sarracenia* and *Darlingtonia* as well as *Nepenthes*, as already discussed. The mite commonly found in *Nepenthes* species is *A. guentheri* (Oudemans, 1915, 1924); Hirst (1928) has described *Zwickia (Anoetus) nepenthesiana* from *N. ampullaria* and Oudemans (1915, 1924) has found *Creutzeria tobaica* in *N. reinwardtiana* (*tobaica*).

Other Nepenthebiont Species

The geographical spread of the genus *Nepenthes* is, as we saw on page 33, vast, from northeastern Australia, north to mainland China, west to the borders of India, and south through Sri Lanka to the islands of the Seychelles and Madagascar in the Indian Ocean (see Fig. 3.7). Over this huge and diverse region are scattered at least 80 species of *Nepenthes*. For the most part, the list of insects which take advantage of all these different pitcher habitats, is far from complete, although many are new species.

A sarcophagid fly, *Pierretia urceola* was found by Shinonaga and Beaver (1979) in the pitchers of *N. albomarginata* and *N. gracilis* in western Malaysia. When fully grown, the larva crawls out of the pitcher, overcoming in some way as yet unknown the various trap devices, and moves into the soil to pupate. As we saw on page 262, this pattern of behaviour may be a consequence of the lack of an incurved rim in *N. albomarginata* and *N. gracilis* and to the fly's apparent restriction to these species. Both this and the phorid *Megaselia* (*Endonepenthia*) *schuitemakeri* are found in *N. albomarginata* and *N. gracilis*, while in *N. ampullaria* the ecological equivalent of *P. urceola* seems to be the calliphorid *Nepenthomyia malayana*. Both these species occur singly as larvae in the pitchers as a result of aggressive behaviour (Beaver, 1979a, b). In this detail they seem to resemble *Metriocnemus knabi* in *Sarracenia*. Also in *N. ampullaria* in western Malaysia occurs the new genus and species *Nepenthomyia malayana*, found by Kurahashi and Beaver (1979). Amongst the phorids the genus *Megaselia* was first found in *N. gracilis* by Lever (1956); in the genus *Endonepenthia* there are five species, all of which are adapted to live in *Nepenthes* pitchers (Beaver, 1979b). The larvae and pupae are found in the pitchers and the adults sitting on the wall. In this respect they seem to be unique in that they can move around on the waxy surface (see pages 113 and 261) without falling into the fluid.

Three species of mosquito have been reported from *N. mirabilis* in Queensland, Australia (Marks, 1971): *Tripteroides* (*Rachisoura*) *brevirhynchus*, *T. subobscurus* and *T. filipes*. These species are also known from New Guinea, but neither this study nor the works cited by Marks (1971) stated whether more than one species was found in a single pitcher. Two other species of *Tripteroides*, *T. aranoides* and *T. vicina*, along with the mosquito *Megarhinus metallicus* were found in the pitchers of *N. sanguinea* in the Cameron Highlands of Malaya (Lever, 1950).

A new species of dipteran, *Xenoplatyura beaveri* (Mycetophiloidea, Keroplatidae) was found, again in *N. ampullaria*, on the island of Penang (Malaysia) by Matile (1979). *Dasyhelea* sp. biting midges live and breed in a range of *Nepenthes* species over a broad expanse of southeastern Asia (Wirth and Beaver, 1979), and Disney (1981a, b) has reported two new species of the phorid *Megaselia*, *M. corkerae* in *N. mirabilis* from Hong Kong and *M. dengi* from *N. distillatoria* in Sri Lanka.

From Papua, New Guinea comes a new culicid dipteran, *Toxorhynchites nepenthicola* (Steffan and Evenhuis, 1982). Another new mosquito, *Aedes brevitibia* has been found in Brunei and Borneo (Ramalingam and Ramakrishnan, 1971).

From far away across the other side of the Indian Ocean, in *N. pervillei* comes yet another new anoetid mite of the genus *Creutzeria*, now named *C. seychellensis* (Nesbitt, 1979). Also in the Seychelles pitcher plants are the larvae of several new species of the mosquito genus *Uranotaenia* (Mattingley and Brown, 1955; Lambrecht, 1971a, b) and this culicid is also found on the same side of the Indian Ocean from *N. madagascariensis* in Malagasy (Grjebine, 1979). Field experiments in the Seychelles have shown that artificial pitchers filled with water and hung among the growing *N. pervillei* are ignored by the egg-laying females of *U. nepenthes*. However, artificial pitchers filled with fluid taken from opened pitchers of nearby *N. pervillei* were readily accepted as breeding grounds by the egg-laying females (Lambrecht, 1971a, b). Yet artificial pitchers filled with water, and in fact any receptacle in the area containing water, was colonized by other mosquitoes, e.g. *Aedes albopictus*. Larvae of species of mosquito other than *U. nepenthes*, although they could survive for several days in the pitcher fluid of *N. pervillei*, never seemed to reach pupation stage.

Misumenops nepenthicola and other Spiders

The habits of *Misumenops nepenthicola* were first described by Pocock in 1898. This thomisid spider ranges from Malaysia through Borneo to Sumatra. It normally occupies the pitchers of *N. gracilis* in the Singapore district where Bristowe (1939) found one spider to roughly every five pitchers. Its preference for *N. gracilis*, which was noted by Fage (1928), is probably in part accounted for as Fage suggests, by the internal architecture of this species. We have already seen how important this internal architecture may be in the distribution of the commensal fauna (see page 262). In *N. ampullaria*, which is often a close neighbour of *N. gracilis* but in which *Misumenops* is not found, the glandular layer extends to the peristome, so that the spider's eggs would be in immediate contact with the secretion. In *N. gracilis*, where the wax-coated slippery zone (Figs 6.39–6.42) occupies at least the top third of the pitcher, *Misumenops* appears to operate with impunity. Fage also suggests differences in the pattern of enzymes secreted by different pitchers, but there seems no evidence for this. In northwest Borneo, Reiskind (1978) found *Misumenops* in *N. rafflesiana* and in Sumatra it has also been observed in the pitchers of *N. gymnamphora*, which is architecturally similar to *N. gracilis*, in Sumatra (Sandved and Prance, 1985).

The dark maroon-brown colour of the spider looks, to the human eye, very similar to the pigmented spots on the pitcher wall. This camouflage may protect it from predation. Its behaviour has been described by both Bristowe and Reiskind and their accounts do not differ in any significant detail. A fly, attracted by the fluid or nectar-secreting glands around the peristome approaches the pitcher mouth. The spider may seize it directly or when it lands in the fluid (klepto-parasitism). If found to be distasteful, it may throw the insect down into the fluid to drown (mutualism). It is not known whether the spider can prey on the numerous dipteran species, mainly mosquitoes, which breed in the liquid of the pitcher.

If the spider is disturbed it slides down on a thread (fixed climbing rope) below the surface of the digestive fluid and remains there for several minutes before returning to its perch. The fixed threads suggest that it has not developed feet capable of dealing with the detachable wax scales (see page 113). Unlike most other thomisids, its body is covered by a thick chitinous armour. The exposed and unarmoured central surface of the abdomen around the tracheal spiracle is protected with fine hairs. When submerged, these hairs entrap a bubble of air. Except for wiping its mouth parts as it emerges from the digestive fluid the spider shows no sign of distress.

Less well known spider inhabitants are a second thomisid, *Thomisus nepenthiphilus*, from Sumatra, which behaves in an almost identical way, and *Theridion decaryi*, from Madagascar, which actually spins a web within the pitcher mouth (Fage, 1930).

The web-forming mycetophilid *Xenoplatyurus*

The ecological niche of *M. nepenthicola* is occupied in the pitchers of *N. ampullaria* by the larvae of the mycetophilid, *Xenoplatyurus beaveri*. This larva spins a web just below the peristome of the pitcher, often completely blocking the opening. Insects are trapped and killed by the highly acidic sticky droplets secreted by the larva onto the web (Beaver, 1979). Unlike *M. nepenthicola*, no gastronomic rejects are delivered to the pitcher fluid, but commensal insects emerging from the pitcher may be caught.

The Establishment of the Phytotelmic Fauna of 'Pitcher-Plants'

As we have seen, the phytotelmic fauna of pitcher-plants, in the broadest sense, is vast. Yet except for those brought in phoretically scarcely any report describes the actual deposition, by the female of the species, into the pitcher chamber. Even that devoted observer of *Darlingtonia*, Rebecca Austin (1875–1917) never observed any event which she could correlate with the presence of the various larvae in the pitcher fluid. The only report in the literature dates from 1818. Macbride, studying *Sarracenia flava* in South Carolina wrote:

'. . . a large fly caught my attention: it passed rapidly from one tube to another, delaying scarcely a moment at the faux of each, until it found, as it should seem, one suitable to its purpose; then hanging its posterior extremity over the margin, it ejected on the internal surface of the tube a larva with a black head, which immediately proceeded downwards by a brisk vermicular motion. This viviparous musca was more than double the size of the common house-fly, had a reddish head, and the body hairy, streaked greyish. I had often noticed it before among the *S. adunca* (= *S. minor*), but could never ascertain its object; the hoods probably obstructing my view.'

From this excellent description the dipteran was almost certainly a species of *Blaesoxipha* (G.C. McGavin, pers. comm., 1987; D. Fish, pers. comm., 1987). D. Fish, (unpubl. obsv., 1987) has also noticed *Wyeomyia* mosquitoes ovipositing in *S. purpurea*.

The well-developed hood of *Darlingtonia* would make it difficult for most egg-laying dipterans to achieve a successful infestation. *Metriocnemus* females may have learnt to deposit their eggs through the still soft walls of young *Darlingtonia* pitchers (see Austin, 1875E, page 260 and G.C. McGavin, pers. comm., 1987). Since the females of Metriocnemus normally predate grasshoppers, the transition from insect cuticle to plant cuticle is not a major step. Moreover, if *Metriocnemus* larvae are the first invaders, their presence, as the pitchers open, may account for the more or less complete absence of other commensals.

Fungi in the Pitchers of *Nepenthes*

Pant and Bhatnagar (1977) found microthyriaceous fungi in the pitchers of *N. khasiana*, and other fungi of unknown affinity in *N. gracilis* and *N. rafflesiana*. They also cite the observations of other workers on the identification of fungi in *Nepenthes*. But so far there seems to be no pattern in the occurrence of these fungi. Their significance, like that of the mycelium found in the walls of *Sarracenia* pitchers (see page 258), is not understood.

Fauna Associated with Non-pitcher Carnivorous Plants

Commensals and predators of carnivorous plants are not restricted to 'pitcher plants' although those found in other types of traps do not strictly come under our heading of insects in a phytotelm. Lloyd (1942) records the capsid bugs commonly found hunting on *Byblis gigantea* as well as on many of the Australian species of *Drosera*. A new species of an ant-like dicyphinid, *Setocoris bybliphilus* (China and Carvalho, 1951), also scavenges on *Byblis gigantea*. Better recorded are the species, again novel, of the taxonomically difficult *Cyrtopeltis* (capsid bugs) which appear to associate particularly with just a few species of *Drosera*, commonly *D. pallida* and *D. erythrorhiza*, of western Australia (China, 1953; Dixon and Pate, 1980; Watson *et al.*, 1982; Southwood, 1986).

The capsid bugs are principally phytophagous, but many are known to be at least partially carnivorous. As China (1953) points out, it must have been relatively easy for a phytophagous capsid to extend its diet from sucking plant juices to piercing small insects such as aphids that might be nearby and which themselves contain plant juices. This switch from herbivory to carnivory is now known to be widespread (Mattson, 1980). The most prolific source of such insects would be sticky or pubescent plants and most capsid bugs are found associated with plants such as tobacco, *Epilobium hirsutum* in Europe and *Cassia cathartica* in Brazil, all of which have hairy or sticky surfaces.

The capsids on *Drosera* species, *Cyrtopeltis droserae* and *C. russelli*, generally avoid the glandular leaves but, if accidentally caught, appear to be able to free themselves and remove the mucilage. All the various species of *Cyrtopeltis* have robust bristles which may be an adaptation to prevent the insect becoming completely entangled. The bristles appear readily to break off, without harm, if glued to the glands, but do not regrow.

Russell (1953) observed *Cyrtopeltis* on *Drosera* species and noticed that the insects are normally found on the underside of the leaf and have cryptic red markings perfectly matching the glands of the plant (cf. *Misumenops* page 265). These predators were able to move on the glandular zone and one adult was seen, under a lens, to place a tarsus directly on a gland and later move away without difficulty. A mucilage-coated specimen of *Cyrtopeltis*, covered as a result of being alarmed, was observed to clean itself systematically, like a house-fly, by the use of its first pair of legs. Russell noticed that, as a *Cyrtopeltis* moved over a glandular area, rarely were more than two legs placed at one time on the glands. The other four were positioned on the leaf or pedicels. Nevertheless, he observed adults with legs missing, which he surmised was a sacrificial mechanism for an irretrievably glued limb.

The scorpion fly, *Harpabittacus australis*, feeds on insects caught by *D. pallida*. Russell (1953) observed the fly to use its long legs to grasp the stem and petiole while keeping its body clear of the leaf glands. This casual commensalism must be quite dangerous to the fly. A number of groups of herbivorous insects are known to predate *Drosera* plants directly. Caterpillars of the 'plume moth', *Trichoptilus parvulus*, were found feeding on *D. rotundifolia* by Chapman in Britain (1906), while in 1965 they were behaving similarly in Florida (Eisner and Shepherd, 1965). The caterpillars, which chew both the leaves and the glands, are usually hidden in the surrounding *Sphagnum* moss by day, only emerging at night to feed. The younger larvae appear to prefer just the stalked glands, whereas the older larvae will also eat the leaf blade along with any captured insects (Lucas, 1976). Gunawardana (pers. comm., 1987) also saw the plume moth *Platyptilia*, probably *P. brachymorpha* on *D. indica* plants (see below), and Fletcher (1908) saw another plume moth, *Trichoptilus paludicola* on *D. burmanni*, also in Sri Lanka. A different strategy is adopted by caterpillars of the noctuid moth *Episilea monochromatea*. These predators avoid being trapped by crawling up the abaxial surface of the petiole and attacking only the lower surface (Hooker, 1919).

In a similar fashion, in Sri Lanka, Gunawardana (1986) noticed that a species of ant of the genus *Technomyrmex* preyed on the insects caught by *D. indica*. The ants were apparently unable to move over the glandular surface like the dicyphid bugs we shall describe below. These ants moved over the aglandular abaxial surface of the *Drosera* leaf, reaching up to pull away insects or parts of insects caught on the marginal trichomes above.

Adults of both a coleopteran, *Epitrix australis* (Bryant, 1953) and another dicyphid hemipteran *Dicyphus errans* live on *Drosera* species, the former on *D. pallida* in Western Australia, the latter on *D. binata* in Queensland. With the aid of ciné film, SEM and a mechanical/mathematical analysis, Southwood (1986) has studied the behaviour of one of the dicyphid plant bugs which live on *Drosera binata* in Queensland, Australia. The pretarsal structure, long tibiae and narrow body allow the femur to be raised clear of sticky trichomes when the legs are moved; all suggest a close adaptation to a surface hostile to insects (see p. 243).

Here again there may have been parallel evolutionary trends, this time the opposite of China's (1953) speculation, from insect-eating to plant-eating, perhaps mediated again by the carnivorous habits of *Drosera* species (Pinner, 1967).

In their behaviour adult dicyphids have adopted a similar strategy to adult capsids. However, adult aphids (*Acyrthosiphon pelargonii borealis*) were observed feeding on *D. anglica* in Scotland, adopting the policy described for *E. monochromatea* of feeding only on the underside of the leaf (Wood-Baker, 1972).

Direct Feeding on the Mucilage

Before leaving the commensal inhabitants of 'fly-paper' traps it is worth mentioning the otherwise unique observations of Kerner (1878) on the inhabitants of the mucilage of *Pinguicula* species. He noticed that on *P. alpina* and *P. vulgaris* in the Innsbrück area (Austria) that diatoms, including the genus *Epithemia*, grew and apparently mutliplied in the mucilage secreted by the glandular trichomes on the adaxial surface of the leaves. The genus *Epithemia* is more commonly found in the mucilage on the fronds of red algae. Kerner speculated that the 'flinty shells' of the diatomaceae might protect them from the enzymes contained in the mucilage. We now know that not only are these siliceous frustules porous, but also that far less well-protected species of animals and plants live and reproduce in these enzyme solutions.

The External and Internal Fauna and Flora of *Utricularia* Species

The External Commensals

All the surfaces of aquatic plants, as studied so far (McGaha, 1952), are generally very heavily colonized by insects and in this respect it is interesting that *Utricularia* is not heavily infested. The exterior surfaces of the bladders of *Utricularia* plants are, however, commonly colonized by other organisms (Frost, 1976). Sessile rotifers in particular are very frequent (Vos-Kelk and Davids, 1977; Wallace, 1978, 1980). This relationship might be termed commensal (but see below) because rotifers, although they colonize, do not feed on the plant or its prey and do not appear to affect the plant directly in any way. A rare sessile rotifer, *Collotheca trilobata*, was first discovered by Koste (1970) on *Utricularia*. Then came a series of observations by Wallace on the rotifer *Ptygura*. The sessile rotifer

P. beauchampi is restricted to the trapdoor areas of the large traps of *U. vulgaris* but, surprisingly, four other co-occurring congeneric species, *U. gibba*, *U. inflata* var. *minor*, *U. intermedia* and *U. purpurea*, were found not to be colonized (Wallace, 1978).

Another rotifer, *Floscularia conifera*, also has a high preference for the trapdoor areas of *U. vulgaris*; in one sampling, 65% of the colonies of this particular rotifer were found on the traps of this species (Wallace, 1980). However, 17 other species of more catholic sessile rotifers, including other species of the genus *Ptygura*, occur on *U. gibba* and *U. purpurea* as well as *U. vulgaris* (Wallace, 1977a, b). It is possible that yet another interrelationship is to be found here. Wallace is of the opinion that the mucilage glands of *Utricularia* secrete some attractive agent which encourages rotifers to colonize the trap entrance. (For a detailed description of these glands see Fineran, 1985.) Perhaps these rotifer colonies have sought the perimeter of the trap mouth as an area relatively free from grazing predators. However, these colonies do in their turn attract grazing insects which accidentally touch the trigger hairs when seeking for the rotifers and other 'epiphytes.' Could they be 'mercenary bait' organisms?

Apart from rotifers, desmids are abundant on *Utricularia* surfaces (Woelkerling, 1976; Gough and Woelkerling, 1976). Seventy-one species of desmids, 44 of them new to Argentina, were found on the surface of *U. foliosa* (De Diaz, 1981) and freshwater sponges were also found on several species of *Utricularia* (Frost, 1976).

The Inhabitants of the Traps of *Utricularia*

The 'duck-weed' *Wolffia* is commonly found inside the traps of *Utricularia* (Roberts, 1972). Ten per cent of the traps Roberts examined were colonized by *W. columbiana* and *W. punctata*. These minute flowering plants were presumably captured by small local water turbulences; they are not necessarily digested, but grow and apparently photosynthesize inside the traps.

Fifty-one taxa of algae were recorded from inside the traps of *U. oligosperma* in Corrientes province in the Argentine. Eighteen of them were new species and the colonies included many Cyanophyta (Mosto, 1979). Also in the Argentine, Botta (1976) examined the traps of *U. obtusa*, *U. platensis* and *U. foliosa*. In them she found large numbers of species of protozoans, rotifers and nematodes and plants which included Cyanophyta again, Chrysophyta, Euglenophyta and Chlorophyta. Many of the ciliates and *Euglena* not only grew well but reproduced. Schumacher (1960) found 25 taxa of desmids in the bladders of *Utricularia* in Georgia, USA. These observations were made on dried material but there was every reason to believe that these species survived and may even have reproduced in the bladders. The most common were *Cosmarium* ssp. and *Xanthidium* ssp., *Staurastrum ophiura* and *Triploceras gracile*.

Most protozoans captured by species of *Utricularia* are digested (Sorenson and Jackson, 1968). However, Hegner (1926a) noticed that, although most paramecia are killed within 75 minutes (cf. *Nepenthes*, page 261, and *Sarracenia*, page 258, pitchers where they survive) the euglenoids *Heteronema acus* and *Phacus longicaudus* were captured but not killed and, like captured *Euglena* species, multiplied.

The Carnivorous Phytotelm – Conclusions

We have seen that the pitcher-forming carnivorous plants are rich in phytotelmic species, while the non-pitcher species have a small, but specific association. The richness and diversity of behaviour should not surprise us since the carnivorous phytotelm possesses the clear advantage over, say, the epiphytic bromeliad or ferns in being set up to attract a broad spectrum of insects in the first place, whereas the conventional epiphyte is more concerned with marginal water conservation and liquid-nutrient scavenging. An eclectic insect-attracting mechanism opens the possiblities both for those attracted to feed in the first instance upon the accumulating prey and, secondly, on those prey able to surmount the killing mechanisms to achieve a relatively protected obligate life style. What perhaps should surprise us more is that certain insect groups, e.g. aquatic insects and the nematodes, although virtually ubiquitous, have not apparently managed to any significant extent, to penetrate into this exclusive habitat. Nematodes are mentioned by Goss *et al.* (1964) where they are seen to occupy *Sarracenia* for a brief period. They are also mentioned by Swales (1972), Fish and Hall (1978) and may perhaps have been seen by Brewer-Carias (see Mazrimas, 1975) in *Heliamphora*. A new species of nematode, *Paralongidorus droseri*, has been recorded in large numbers in the soil surrounding *Drosera burmanni* in West Bengal (Sukul, 1971). This association may turn out to be more than fortuitous, but it is not a phytotelm in the normally accepted sense. Most significantly, nematodes are unrecorded from the richest microfauna, that of *Nepenthes*. Yet these habitats are consistent in space, frequency, time and environmental characteristics. Possibly the chironomids and other dipteran larvae, so plentiful in the pitchers (Johannsen, 1932; Thienemann, 1934), have successfully overcome the digestive action of the pitcher fluid and totally dominate this niche.

Are Obligate Insects Essential to the Carnivorous Plant?

Much careful work and many observations suggest that a wide range of arthropods are obligate inhabitants of carnivorous phytotelms. Other work (see page 252) suggests that at least some of these phytotelms have co-evolved to modify the environment for some of the dipteran commensals. However, we should be cautious in accepting that any of these apparently close relationships, which suggest a dependence of the arthropod on the plant host, also imply a dependence by the plant on the larval digestive ability.

Although it is generally believed that most *Sarracenia* species and almost certainly *Darlingtonia*, for example, lack their own battery of secreted enzymes they are both grown successfully in captivity, rarely if ever with commensals. Moreover, *S. purpurea*, which is now widely naturalized throughout the world (see page 37), and in some cases invading and endangering the native vegetation, appears, in the few cases examined, to have no macroscopic commensal fauna (P.J. Foss, pers. comm., 1986; O., Almborn, pers. comm., 1986; C., Aldridge, pers. comm., 1986). The only living inhabitants of one of these alien populations in the Republic of Ireland were the alga *Ulothrix* and another unidentified unicellular alga.

Do Obligate Insects Pose a Threat to the Future of Carnivorous Plants?

It should be pointed out, as Rymal and Folkerts (1982) have concluded, that although almost every species of carnivorous plants is a phytotelm of one sort or another and often the clientele is numbered in dozens of species, none of these, insectivorous or phytophagous, offers any threat to species populations. Carnivorous plants of many species all over the world are suffering loss of habitat and a few, e.g. *Dionaea muscipula*, *Sarracenia alabamensis*, *S. oreophila*, *S. jonesii*, several *Nepenthes* and *Triphyophyllum*, are endangered through agriculture, forestry, urbanization and voracious collectors, not insects.

CHAPTER 15

Pollination and Reproductive Strategies

Introduction

Few workers have even made the briefest of comments on the pollination of carnivorous plants, preferring to concentrate on the insects that generally form their prey (Ch. 7). It is generally considered, although with incomplete evidence, that all carnivorous plants are insect pollinated, but, if true, this syndrome inevitably leads to a paradox discussed on page 273 below. In terms of pollination, only four species seem to have been studied in any detail and the observations are, it must be admitted, in every case incomplete and often paradoxical.

The Pollination and Reproductive Syndromes of the Droseraceae

Two completely different reproductive syndromes have developed in the Droseraceae. Both *Drosera* and *Dionaea* produce relatively small flowers. These flowers, which are white, violet or red, are pollinated by small insects that might also serve as prey. In some cases, pollination can therefore be limited by competition beween flower and trap for the same insects (see Ch. 16). But their adaptive devices of producing numerous dusty seeds per flower permits some seed dispersal, even when pollinator visits are rare. This syndrome is well known in other plant groups, e.g. orchids, certain parasitic plants and epiphytes (Van der Pijl, 1982).

Drosophyllum, which also belongs to the Droseraceae and is closely related to both *Drosera* and *Dionaea*, produces larger flowers of a conspicuous yellow colour which contrast strongly with the trap leaves. *Drosophyllum* seems to show no competition between its traps and flowers. The traps capture small insects, whereas the flower is pollinated by large insects. This seeming lack of competition is consistent with the fact that each *Drosophyllum* flower forms a capsule containing a small number of largish seeds. One can therefore see that *Drosophyllum* exhibits a reproductive strategy which is completely different from the strategy employed by its relatives in the same family. This distinction suggests an early divergence of this monospecific genus (see Ch. 19. and Fig. 19.1) from a common pre-*Drosera* stock.

An additional characteristic of both *Dionaea* and *Drosera*, and one which helps these plants to overcome possible reduced rate of effective pollination, is their ability to reproduce vegetatively. This characteristic has not developed in *Drosophyllum* which, as we have seen (page 31), is dependent exclusively on the distribution of seed for fresh colonization.

The Pollination of *Sarracenia* Species

Although much more intensively studied for other reasons, even less is known of the pollination syndromes of *Sarracenia* species.

Although the unique flower structure of *Sarracenia* is indicative of a highly evolved mechanism significant in the pollination process, there has been no extensive study so far (Rymal and Folkerts, 1982; O'Neil, 1983). The major pollinators of the large-flowered species (*S. alata*, *S. flava*, *S. leucophylla*, *S. oreophila*, *S. purpurea*) seem to be newly emerged queens of the genus *Bombus*. Along the Gulf coast, common pollinators are *Bombus bimaculatus*, *B. fraternus* and *B. impatiens*; further north, *B. pennsylvanicus* becomes more important.

Although there is some variation in size, queens of these species are generally too large to enter the flowers of the small-flowered species (*S. minor*, *S. psittacina*, *S. rubra* complex) (Folkerts, 1982). Worker bees of the above species are small enough to enter small *Sarracenia* flowers, but are not present in numbers until the *Sarracenia* flowering peak is past. Jones (1908) saw small bees of the genera *Augochlora* (Halictidae) and *Osmia* (Melittidae) visiting the flowers of *S. minor*.

A major factor affecting pollination success, especially with bumblebees, is the patch size of the plant species. When patch size is small, bees are forced to visit several species in order to secure sufficient pollen and nectar. This mixture must decrease pollination efficiency for any species involved. This restriction may account for the limited reproductive success of *S. alabamensis*, *S. jonesi* and *S. oreophila*, forms for which few or no large populations remain. Small patches of *S. flava* and *S. leucophylla* along the Gulf coast, which often represent populations decimated by competition resulting from habitat changes caused by drainage and/or fire suppression, seem seldom to receive much attention from pollinators (Rymal and Folkerts, 1982).

A critical comparative study of pollination processes in *Sarracenia* is needed to determine how they relate to reproductive success and hybridization.

The Pollination of *Darlingtonia Californica*

The peculiar flowers of *Darlingtonia*, protected from robbing arthropods by their peduncular bracts (Frontispiece) have five green-yellow sepals, 4–7.5 cm long, covering the shorter petals in the pendant flower. The five petals, maroon to purple in colour, converge at their tips. Near the tips, the edges of the petals are rolled inward so that the adjoining petals form five symmetrical openings into the stamens and pistil. In seed, the bracteate scape straightens and the obovoid capsule, 2.5–5 cm long, is held in a vertical position.

Both Hooker (1875) and Macfarlane (1893) were struck by the resemblance of *Darlingtonia* flowers to the traps:

> '. . . a remarkable analogy between the arrangement and colouring of the parts of the leaf and the flower . . . It is hence conceivable that this marvellous plant lures insects to its flowers for one object, and feeds them while it uses them to fertilize itself, and that, this accomplished, some of its benefactors are therefore lured to its pitchers for the sake of feeding itself.'

Intriguing speculation, but what is the pollination mechanism of *Darlingtonia?*

Rebecca Austin (1875–77; Appendix 1) spent six years in close proximity to several colonies of *Darlingtonia* in the Quincy area of northern California. Her principal area of study was Butterfly Valley (see page 39). Over many years of observation, often on a day-by-day basis and even, as she describes, taking her sewing into the woods and sitting by the plants, she saw just *two* flying insects enter flowers. One was a greenfly and the other another unidentified dipteran. No bees, butterflies or moths, although abundant in the area, were ever seen to enter the flowers. What she did find, however, was that virtually every flower, for example 48 out of 50 examined on 11 May 1875 (1875D), were occupied by the webs and/or spiders of several arachnid species. On one occasion, 27 April 1875, she found a number of small insects in the flowers amongst the webs and from her description these were probably *Thysanoptoa-thrips*. But she never saw these particular insects moving from flower to flower. In fact, in her own words, '. . . flying insects . . . all appear to shun the flowers. The butterflies flit over them but none light on them, while they do on alders and young fern leaves, among which these flowers are . . .'. She was convinced that the flowers were pollinated by spiders: 'To me it is as plain that it is accomplished by spiders, as the circles of root-growth show the age of the plant' (Austin, 1875D). *Darlingtonia* sets large quantities of viable seed, about 3.5 mm long with projections at one end (Dwyer, 1983), but whether self- or cross-pollinated and if the latter, how, must for the moment remain in doubt.

The Pollination of *Byblis gigantea*

Byblis gigantea is peculiar in several respects (page 42). It is also anomalous, in that its flowers are positioned amongst the traps unlike every other carnivorous plant (page 273). The only other plant with similar characteristics is *Roridula* (Obermeyer, 1970) not generally considered now to be a true carnivore (see page 293), and here the pollinating insect is capable of walking over the plant without being trapped.

DeBuhr (1973), whose work on *Byblis* is reviewed on page 42, also makes two other observations on the pollination characteristics of *Byblis* which deserve further study. First, there appears to exist floral mimicry between *B. gigantea* and a lily, *Thysanotus multiflorus*. Both species flower at the same time and have a similar floral structure, but whether the carnivore fails to reward the pollinating insect is not revealed. Such a convergence would appear to suggest a long evolutionary association. Secondly, it is also suggested that *B. gigantea* belongs to that diverse group of angiosperms which are buzz-pollinated, i.e. the pollen is released through pores at the tips of the anthers, the bees using their flight musculature to create the right pitch and vibrate the pollen free from the anthers. Generally buzz-pollinated flowers do not reward bees with nectar and for general reviews of this widespread phenomenon see Buchmann (1983, 1985).

The Pollination of *Nepenthes*

The small greenish or claret-coloured flowers of *Nepenthes* are borne in clusters at the tips of stems opposite the leaves. There may be as few as 15–20 or as many as several hundred flowers in one inflorescence, and the unisexual flowers, unlike other carnivorous plants, are found on different plants. Danser (1928) is doubtful about the few monoecious observations. The pollination ecology of the many species has not been well studied, but Shetler (1974b) notes that, in the wild, the flowering period is mainly from March to September. The inflorescences, at least in cultivation, attract clouds of insects and the male flowers give off a foetid smell which may be an insect attractant. Holttum (1954) notes that, in the Malayan species, the sepals in both kinds of flowers are covered with small glands producing a sweet nectar. He supposes that these are to attract small, pollinating insects, but records that he, unlike Shetler, has rarely seen insects in the wild on these flowers and speculates that they might be pollinated by night-flying insects.

After pollination the seeds mature in six to eight weeks. The problem of separation of pollinator and prey, because *Nepenthes* is one of the few carnivorous plants in which some species place their flowers near to the traps (see 273), has not been considered at all.

We have already seen from *Darlingtonia* (see p.271) how difficult pollination studies are. The climbing habit of most *Nepenthes* only adds to the problem.

Seed Formation, Dispersal and Germination in Carnivorous Plants

As with pollination, very little is known of any features of the seed cycle peculiar to carnivorous plants. The general horticultural literature and guides to growers give the impression that the seed of such plants is invariably dust-like and short-lived. There is no doubt that this is generally true, as with *Pinguicula* and *Nepenthes* (Slack, 1986). For example Ah-Lan and Prakash (1973) note '. . . the seeds (of *Nepenthes gracilis*) are viable only for a week after the fruit dehisces . . . germination takes place in about a month.' *Dionaea* seeds are about 3 mm long and, rare amongst carnivorous plants, smooth and shiny (Dwyer, 1983). Germination tests on field-ripened seeds on *D. muscipula* by Roberts and Oosting (1958) showed them to be 88% viable. After 100 days viability dropped to about 2% and after 140 days no germination occurred at all. The seeds germinated equally well in darkness or light (but see *Nepenthes* below). On the other hand, Slack noted that the seed of *N. khasiana* may last up to a year. As one might expect, more striking exceptions to the conventional turn up amongst the carnivorous plants growing in very extreme conditions. Dixon and Pate (1980) studying the tuberous *Drosera erythrorhiza* (see page 27) noted that the minute, ornamented seeds failed to germinate after one to two years, even under a variety of test conditions. Germination, although a very rare event, occurred in nature only in seeds older than three years. *Utricularia* seed, too, has a considerable longevity. From one *U. vulgaris* collection 80% germinated over a period of 11–12 months (Jane and Russell-Wells, 1935).

Distribution and Germination

Most carnivorous plant seed is, as we have seen, small, often dust-like, and conspicuously ornamented (Dwyer, 1983), but few carnivorous plants are obligate epiphytes, which would be one explanation for dust-like characteristics.

About 10 000 seeds are produced from one inflorescence of *N. gracilis* (Green, 1967). The seeds are light, winged (Dwyer, 1983), with each of the two filamentous wings hollow so that the seeds are buoyant, and may be dispersed by wind over long distances. They can float and there is evidence of short-range water dispersal. Light is essential for germination and enhanced light shortens germination time (Green, 1967).

In *Darlingtonia*, too, which invariably grows above moving water (see page 38), the seeds are small, light and papery and there is evidence of successful water dispersal; the same is true of *Utricularia* (Jane and Russell-Wells, 1935).

The only known large-seeded carnivorous plant is *Triphyophyllum*, an exception in many other ways as well. The seeds are papery discs, up to 10 cm across, distributed at the level of the forest canopy and presumably intended for long-distance transport. But apart from this morphological detail, nothing else seems to be known. According to Green *et al.* (1979) germination occurs at the peak of the rainy season (end of July) while the seeds are submerged or resting on the surface of waterlogged leaf litter on the forest floor. It is noteworthy that, in cultivation in Sierra Leone, the germinating seeds and seedlings of the species tolerate less variation in temperature than those of related Dioncophyllaceae in the same area. This selectivity must reduce its chances of success.

The Prey/Pollination Paradox of the Carnivorous Plant

All sexually reproducing carnivorous plants exist in a paradoxical situation. They are dependent upon insect prey for successful penetration into nutrient-poor habitats, yet they are, with no known exception, insect pollinated. Bees and hover-flies are occasionally found in pitcher traps. How then do they normally manage to separate their functions? Practically all traps use every device exploited by the insect-pollinated flower: colour, scent, nectar and UV guides (see Ch. 5). It is possible that subtle band-distinctions are achieved by trap and flower, but these features of the syndrome have not been studied. In fact, this whole facet of the carnivorous plant's growth cycle seems scarcely to have been considered before. Some solutions are obvious.

Firstly the carnivorous plant may separate the catching zone (trap) and the pollination zone (flowers) into different habitats. In *Utricularia* (page 43), *Genlisea* and *Aldrovanda* species the traps are buried deep in mud or water and the flowers in the air (Figs 4.12*B* and 4.17*A*). Secondly, in those plants which are totally terrestrial, the flowering stalk is often extended high into the air to attempt a physical, as opposed to a habitat separation of prey and pollinator (DeBuhr, 1973). These elongated peduncles or pedicels have caused pictorial problems for artists from Ferdinand Bauer (1801); (see Mabberley, 1985), with *Cephalotus*, Edwards (1804) and La Billiardière (1806) with *Genlisea* to Airy-Shaw (1951) with *Triphyophyllum*. The significance of these elongated structures seems to have been missed. To give a few examples, *Cephalotus* pitchers project about 2 cm above the soil surface. A typical flower (see Bauer, above) is raised 60 cm above the top of the pitcher. The newly discovered carnivorous plant, *Catopsis berteroniana*, unlike almost every other bromeliad, raises its flowering stalk up to 90 cm above the imbricate leaves (Rickett, 1967). In neither case is such a separation explicable by surrounding vegetation as almost every carnivorous plant grows in the open, whether terrestrial or as an epiphyte, like *Catopsis* above.

Nepenthes species often do carry their flowers amongst the pitchers, but the general habit is so lax and diffuse amongst the branches and leaves of other plants over which they scramble that a problem of misdirection of pollinator or prey may not arise (Kaul, 1982). *Byblis* does open its flowers amongst functioning traps, but is anomalous in many other respects as well (pages 41–43).

Thirdly, as in *Drosophyllum*, the trap system is only capable of dealing with tiny flies, midges, gnats or similar dipterans, whereas the pollination is by larger more muscular hymenoptera. The opposite seems to be the case in *Nepenthes* species, but it is admitted that the pollinating mechanism is not well understood. It is interesting that *Roridula*, which is not now generally considered to be a carnivorous plant but just a defensive insect-killer (see page 293), presents its flowers close amongst the sticky leaves (see Fig. 29 in Obermeyer, 1970). Pollinating insects must take their chance with the rest!

The peduncles or pedicels of carnivorous plants are often not only elongated, but also coated either with glandular hairs, as in *Drosera* species, or with detachable waxy scales, as in *Dionaea* and *Sarracenia* (see page 299). The glandular peduncles of *Drosera regia* are even known to trap small insects (Brokenbro, 1981).

A fourth method would seem to be the temporal separation of the major flowering period from the principal trapping period. This seems to be the case, for example, with most *Sarracenia* species which flower in the spring before new leaves appear (Shetler, 1974a, p. 253, but see page 276). Also Roberts and Oosting (1958) noticed in *Dionaea* that, with the appearance of the flowering axis, leaf initiation usually ceases and no new leaves are produced until after flowering. Diurnal separations would also seem to be possible, in either direction, but have not yet been detected. An avoidance method may be seen, as is observed with some Australian species of *Drosera*, in a move into a clonal, rhizomatous or tuberous habit, leaving little or no dependence on the seed habit. *Darlingtonia* too is known to spread rhizomatously (Austin, 1875a; Appendix 1). Too little though is known of the whole seasonal cycles of most carnivorous plants to attempt to do more than outline the problem.

CHAPTER 16

Mimicry or Mutualism?

Introduction

The carnivorous pitcher-plants have, as discussed in Chapter 5, developed visual and olfactory signals by which insects are attracted into the traps where they are digested and absorbed (see Part III). Williamson (1982) wrote that the pitcher-plants 'often bear flower-like appendages' and 'share the feature of *capturing insects who have innate floral preference, or provision experience with flower models, but little or no experience with the mimic*' (i.e. the trap).

This chapter explores the extent to which carnivorous plants deceive insects by mimicking other organs involved in plant/insect relationships, in which the relationship is beneficial, or at least not detrimental, to the insect.

The carnivorous traps were classified under the heading of 'aggressive mimicry' (Wickler, 1968; Wiens, 1978) and more recently as 'aggressive semi-abstract homotypy' (Pasteur, 1982) because in their case 'the model is virtual but nevertheless looks like a being that we know and can name'.

Currently there is little literature on the basic ecology and evolutionary biology of carnivorous plants in general, and of pitcher-plants in particular. Mimicry has never been studied in any carnivorous plant either from an ecological or an evolutionary point of view. The following discussion is taken from Joel (1988), who has recently analysed the available information on pitcher-plants and suggests a new approach to the interrelations between these plants and the insects on which they prey.

The Definition of Mimicry

Mimicry involves an organism (the mimic) simulating the signal properties of another organism (the model) which are perceived by a third living organism (the operator) as signals of interest, such that the mimic gains in fitness as a result of the operator identifying it as an example of the model (Vane-Wright, 1980). When reward is provided to the operator, it is always provided by the model but not always by the mimic. A learned image is developed and maintained through experience and reinforcement with the model and therefore mimicry depends on the frequencies of the operator's contact with the models and with the mimics (Matthews, 1977; Williamson, 1982). Frequent deception rapidly leads to selective pressures against the mimic. Hence, a mimetic system is effective only if the mimic is rare, either in space or in time (Joel, 1988).

Are Pitcher Plants Batesian Mimics?

The traps of pitcher-plants might be considered as mimics simulating visual signals as well as olfactory properties of nectar-producing flowers (see Ch. 5). Such signals, it could be argued, are perceived as attractive indicators by pollinators which are deceived, trapped and digested (Wickler, 1968; Wiens, 1978; Gibson, 1983a; Stowe, 1987). Since deceived insects are eliminated, insects experienced with the mimic will be rare and therefore avoidance will be negligible.

The data concerning the carnivorous pitcher plants of the Sarraceniaceae, Nepenthaceae and Cephalotaceae do not, however, fit the characteristics of Batesian mimicry (Joel, 1988):

(i) there is no evidence of any insect visiting a pitcher-plant by mistake;
(ii) no model is known for any one of the pitchers;
(iii) pitchers are not usually rare in their habitat;
(iv) a reward is provided to visiting insects by all pitcher-plants.

Interrelations between Insects and Pitcher Plants

Mutual relations are known to exist between pitcher-plants and their insect inhabitants. These interrelations were dealt with in Chapter 14. The following sections discuss factors which are involved in the interrelations between the pitchers and those organisms that visit their outer surfaces.

Shape and Colour of Pitchers

Patterns of shape and colour, including ultraviolet components (see Ch. 5), are believed to form guiding signals for certain insects in many carnivorous plants in general and in pitcher-plants in particular (Wiens, 1978; Joel *et al.*, 1985; Joel, 1986). Nevertheless, no behavioural data are available to examine whether these patterns are indeed deceptive.

Secretion of Nectar

Nectar-secretion by nectaries associated with the trap is common to all pitcher-plants, no matter whether they belong to the Nepenthaceae, Sarraceniaceae or Cephalotaceae (see Ch. 5). A pitcher does not act as a model of a flower: it cannot offer nectar to a 'deceived' insect, as successful deception leads to the demise of the operator. Floral nectar, in contrast, is regarded as a real energetic and nutritional reward provided to visiting pollinators (Ch. 15). Likewise, extra-floral nectaries in non-carnivorous plants also secrete nectar as a reward, supporting insects which seem to offer services to the plant such as defence or nutrition (see Bentley, 1977; Huxley, 1980, 1986).

The explanation given for the secretion of nectar by pitcher traps is that it serves, together with visual and olfactory cues, as an attractant, tempting insects to reach and remain at the trapping site until caught (Delpino, 1874; Macfarlane, 1893; Lloyd, 1942; Wickler, 1968; Heslop-Harrison, 1978; Williamson, 1982).

Many mimetic and deceptive systems utilize patterns of colour and shape (Dafni, 1984), but rarely are nectaries and nectar secretion involved. In those rare cases in which they are, nectar does not function as an attractant but is secreted inside traps where it maintains insects alive during their 'imprisonment' (Vogel, 1965; Dafni, 1984).

False Nectaries?

Certain flowers (e.g. *Parnassia*) mimic others by possessing false nectaries (pseudo-nectaries) advertising nectar scent and glistening without producing any real nectar (Daumann, 1960). If the pitcher-plants were real floral mimetic systems, one would expect them to provide only visual and olfactory cues and to have pseudo-nectaries, not, as is the case, nectar-secreting glands. No development of deceiving pseudo-nectaries is known in any of the carnivorous pitcher-plants (Joel, 1988). Such a system has developed in the orchid genus *Paphiopedilum*, of which the flowers resemble *Nepenthes* pitchers both in shape and colour as well as in their pseudo-nectaries which look like the real nectaries of *Nepenthes* (see page 36).

Pitcher-Plants and Flowers

No comparative study is available which compares insect pollinators and insect victims in the same habitat. Gibson (1983a), who studied the North American pitcher-plants, claimed that certain of these *Sarracenia* species specialized on insect taxa known regularly to visit non-carnivorous flowers blooming in the same habitat. However, two points contradict the possibility of mimicry in these cases. Firstly, there is often a temporal separation between the flowering period of associated flora and the trapping period of the pitchers. As Wiens (1978) pointed out, 'there is evidence that associated non-insectivorous species may have shifted their flowering periods, so that they do not compete with plants preying on their pollinators'. In other words, we cannot establish which plant has shifted its active periods. Secondly, pitcher-plants are generally most active in capturing insects (in terms of prey per pitcher) during their growth phase immediately after fire or drought (see Ch. 3 and below), when other flowering plants are far behind in forming their own community. Again, a temporal separation is indicated. This separation is, however, variable in the different species and in the different habitats. Nevertheless, when a population of a pitcher-plant densely covers a large area we may assume that certain insects hatching in this same area do not leave the boundaries of the pitcher-plant population and are therefore exposed only to the nectar provided by these plants.

Plant Community Structure and Mimicry

What then of the basic principles of mimicry in general, and of plant mimicry in particular, that deception is effective only when it is rare? A successful mimic builds either scattered, small populations or large populations which are only active for a very short period. Do pitcher-plant populations meet these criteria?

Size of the plant community

The populations of pitcher-plants are often extremely large and dense. In some cases, as in many species of *Sarracenia*, they may attain phenomenal densities under regimes of repeated disturbances from fires, drought and heavy grazing (Gibson, 1983a; Folkerts, 1982; and see Ch. 3).

Schnell (1976), recalling the heavily populated, multi-hectare stands of *Sarracenia* which were mentioned earlier in the century, describes some of the habitats which still show large populations of different species. For example, there are places in Georgia, USA, where one can see pitchers of *S. flava* filling a large savanna and 'melting to a vast golden blur when viewed from a distance' (Schnell, 1976). *S. purpurea* also forms extensive populations with plentiful pitchers in the *Sphagnum* bogs and heaths (Butler, 1985). This species can form dense, floating mats on water at the edge of the bog ponds and lakes and across acid streams (Schnell, 1976).

Many species require fire-frequencies of one per two years in order to maintain their dense populations (see Ch. 3). Fire removes plant cover and increases the percentage of bare ground in the bog, which is vital for seedling establishment. When the ecosystems mature, the carnivorous plants fail to compete with 'conventional' plants and therefore decrease dramatically in density with increasing density of non-carnivorous plants (Gibson, 1983).

Darlingtonia, like the *Sarracenia* species, also forms dense populations, as described for the Butterfly valley by Austin (1875–77; B.E. Juniper, pers. obsv., 1985).

Cephalotus follicularis again forms large clumps in its natural habitat (Mann, 1985). A great abundance of *Cephalotus* was observed in the Albany area, Australia, by L.T. Leschenault de la Tour, botanist on the expedition of Nicolas Baudin, who was an early visitor to this site in February–March 1803 (Willis, 1965).

Many species of *Nepenthes* also form large showy populations. Thousands of yellow pitchers of *N. madagascariensis*, for example, were described as shining in their habitat in Madagascar (Schmid-Hollinger, 1979). Thus, they are not only abundant but also very showy. *N. macfarlaneii* was described as *the most abundant plant species* in the moss forest of Ganung Ula Kali, one of the highest mountains in Tropical Malaysia (Shivas, 1983).

It would therefore appear that in all the pitcher plants the plant community structure is not compatible with a mimetic status (Joel, 1988).

Duration of trap activity

As far as is known, all pitcher-plants are perennial, their traps being active for whole seasons. According to Fish and Hall (1978), for example, each pitcher of a *Sarracenia purpurea* plant is active as a trap for about 1–2 months and a sequence of traps is formed by each plant throughout the season. In a similar manner, the pitchers of the different *Nepenthes* species persist for about 6–7 months, and the maximum length of life of a pitcher can be about one year. Austin, (1875C, Appendix 1) observed that individual *Darlingtonia* pitchers, having survived the winter, began trapping prey again about mid-April of the new season. Furthermore, once a pitcher-plant is established, it will provide a fairly regular and predictable supply of new pitchers over a number of years in a restricted area (Beaver, 1983). In this feature also, then, a mimetic status is not evident (Joel, 1988).

Trapping versus Pollination in Pitcher-Plants

As we saw in Chapter 15, many carnivorous plants of the families Droseraceae, Cephalotaceae, Bromeliaceae and Lentibulariaceae develop flowers that are separated in space from the leaf-traps by long flowering stalks. These scapes, we believe, enable a spacial separation to be achieved between pollinators and prey. Other plants, such as *Dionaea*, cease to produce traps when flowering (Roberts and Oosting, 1958), enabling a strategy of separation in time.

While *Cephalotus* and most *Nepenthes* species also separate their flowers from the traps by developing long flower stalks, a few carnivorous pitcher-plants develop flowers which are located between their traps and show similar colour. *Sarracenia flava*, for example, develops flowers with yellow petals and a light green pistil, together with greenish-yellow traps. In these plants nectar is provided by both the traps and the neighbouring flowers. Is there no 'danger' that the trap kills insects visiting its flowers, thus rendering selective pressures against its own pollinators? This dilemma has not been resolved (see page 273).

Nectar Consumption by Visitors which are not Trapped

Bünning (1947) observed an ant carrying a sugar crystal from the pitcher rim of *Nepenthes* and taking it down from the plant. A short while later a large number of similar ants came over to the same source of sugar. When this was consumed, the ants tried to find sugar in the inner side of the pitcher rim. Only then were many ants trapped (Erber, 1979). Macfarlane (1893) likewise described ants feeding on the extrafloral nectaries of *Nepenthes* with only rare events of trapping. Do the foliar nectaries then act as a guide-line, as suggested earlier (see Ch. 5)?

Austin (1875–77, Appendix 1) noticed that many flies alight on *Darlingtonia* pitchers, 'sip their fill and fly away', or 'enter the hood . . . often coming out'. She further noticed that 'of the few houseflies that ventured to taste the sweets of *Darlingtonia*, nearly all fed on the fold around the orifice, and flew away'. But of the few that did venture inside the hood, 'none made their escape'. Lloyd (1942) also '. . . on a warm sunshine day, watched flies in numbers busily sucking the nectar on the outer surface of the pitcher of *Sarracenia purpurea* and only some of them getting trapped'. With *Heliamphora* Slack (1986) followed a fly which lost its footing on the uncertain surface but 'was fortunate for it was just able to find its wings before entering the tube of the pitcher'.

Similar cases where insects were described to consume nectar, and leave the traps unharmed, are rare in the literature, which generally supports the view that nectar in pitcher-plants is a means of attraction. Such an assumption ignores the possibility that the nectar might be a reward. At the same time, observant growers of carnivorous plants, as well as many botanists who have had the opportunity of observing these plants in their natural habitats, often note that, while many insects visit these plants, a large proportion leave the trap unharmed. This is particularly obvious in the lower storey pitchers of *Nepenthes* (see M. Jebb, Appendix 2) and in the pitchers of *Cephalotus*, which are both much visited by ants. In these pitchers a 'two-way traffic' is generally observed (M. Jebb, and D. Mabberley, pers. comm., 1987), which is a clear indication of the rewarding nature of their nectar. Likewise, Gibson (pers. comm., 1986) also observed that many insects visit pitcher-plant traps to forage for nectar, then fly or crawl away, and that there is a strong differential escape-process of insects which varies with trap type (Gibson, in prep).

How successful, then, are pitchers at capturing these visitors? The average number of prey per pitcher of *N. mirabilis* was 140 in the lower pitchers, 134 of which were ants, and 41 in the upper pitchers, of which 33 were ants (see Appendix 2). These quantities of prey, found in mature traps, clearly indicate that the daily capture in pitcher-plants of this group is very low. Out of hundreds of ants visiting the pitcher every day only a few fall prey to the plant. Similarly, the mean number of insects captured daily in a single pitcher of *S. leucophylla* was 0.3–2.5, depending on pitcher density (Gibson, 1983a). Great numbers of insects were available in the vicinity of the plant and relatively high rates of visits per pitcher, yet the capture rate was low.

Analysis of data given by Gibson (1983a) show that, on the one hand, when the trap density of *S. flava* was ten per square metre, the average daily capture per pitcher was only 0.3. On the other hand, insect availability changes dramatically as a function of trap density, increasing to about three times when trap density changes from ten to 70 per square metre. One can assume (Joel 1988) that each insect that is attracted by the pitcher-plants visits many pitchers during its stay in the pitcher habitat and might also visit the same pitcher several times.

Real Batesian Mimicry in Pitcher Plants?

All the evidence presented to date contradicts the hypothesis that pitcher-plants mimic flowers. The only seemingly possible case of mimicry in this group of plants is not of traps resembling flowers, but of a trap resembling traps of another carnivorous genus (Joel, 1988). According to Givnish *et al.*, (1984) *Brocchinia reducta* occurs in the Gran Sabana of southeastern Venezuela in the same habitat as *Heliamphora heterodoxa*. *B. reducta* has bright yellow-green-coloured vertical leaves forming a conspicuous cylinder 15–35 cm high, closely resembling that of *H. heterodoxa*. In addition, it emits a sweet nectar-like odour (unique in the Bromeliaceae), again very like that of *H. heterodoxa*. In contrast, however, *H. heterodoxa* produces real fragrant nectar, whereas *B. reducta* does not produce any nectar at all (Givnish *et al.*, 1984). Is, therefore, this resemblance of *B. reducta* to *H. heterodoxa* in colour and scent, as well as in shape and size, a case of mimicry? It is very likely that insects which manage to gain some nectar from the pitcher lid (the 'spoon') of *Heliamphora* (Fig. 4.9F, Zone 1), without being trapped, learn that pitchers of this size, shape, colour and scent are rewarding. These insects might then be deceived by the similar attraction cues of *Brocchinia* traps. Analysis of the fluid contents in these bromeliad traps showed that more than 90% of the individuals were ants, and that the ant genera represented there are known to forage frequently at real nectaries (Givnish *et al.*, 1984). Similarly, *Heliamphora* pitchers are also known to attract ants.

Based on these data, one can suggest that *B. reducta* represents a case of Batesian mimicry. This seems to be the only Batesian mimicry known in the world of carnivorous plant traps. The model in this case is a rewarding trap, not a rewarding flower (Joel, 1988). A similar case might also exist with the orchid *Paphiopedilum*, the species *P. volontoneanum* and *P. dayanum* which were reported to grow in association with *Nepenthes*. As stated above, *Paphiopedilum* flowers closely resemble *Nepenthes* pitchers, and it is possible that these non-rewarding flowers with false nectaries mimic the rewarding pitchers of *Nepenthes* (see page 36 for reference).

Mutualism, not Mimicry

As the simple concept of mimicry does not appear to apply to the pitcher traps, is there any alternative explanation of the development of the complex interrelations between traps and insects? Clearly, carnivorous plants benefit by exploiting insects as a food (see Ch. 7), while some insects exploit carnivorous plants for much the same reasons (see Ch. 14). Do intermediate situations exist that are mutually beneficial?

Mutualism is defined as an interaction between species that is beneficial to both (Boucher *et al.*, 1982). The difference between mutualism and mimicry is therefore mainly in the bi-directional benefit of the former, while in mimicry benefit is gained only by one side in the interaction.

Mutualism between Pitchers and Insects

Joel (1988) suggested that pitcher-plants serve as important suppliers of nectar in certain habitats, mutualistically supporting insect communities which benefit from this nectar and sacrifice. In return, small portions of their community serve as prey. There may be a parallel here in the rotifers, which live on the trichomes of the trap door of *Utricularia* species (see page 267) and of which a few are sacrificed each time the trapdoor opens (Wallace, 1978).

This relationship is clearly mutually beneficial. The plants, which commonly live in nutrient-deficient habitats (see Ch. 3), benefit from the mineral and organic nutrients from digested prey. In return, the pitcher-plants can provide sufficient amounts of sugary nectar since they usually grow in *moist* and *sunny* habitats where water supply and photosynthetic energy are not limiting (Heslop-Harrison, 1978; Givnish *et al.*, 1984), enabling excess to be produced without serious cost to the plant.

Sacrifice of some members of an insect community which consume extra-floral nectar is not unique to this system. In a similar manner, members of ant communities associated with extra-floral nectaries of non-carnivorous plants are often lost – killed when attacking phytophagous invaders while 'guarding' the host plant (Janzen, 1985). Thus, in both carnivorous and non-carnivorous cases, the actual 'cost' paid by the insects community for extra-floral nectar is the same, although their services might be different.

Insects will only return to the same pitchers or to similar pitchers if they have gained some profit during their first visit. They rapidly learn that the visual and olfactory characteristics, typical of the pitchers in their vicinity, lead them to a reliable source of nectar. Those few insects which pay for the nectar with their lives cannot transfer their 'knowledge' of the possible danger because they die. In some cases, the trapped individuals are only the older ones. Selection against visits is not developed because the proportion of 'casualties' is limited (Joel 1988).

Interrelations with Ants

Such a mutualistic situation is most clearly established with ants. Some pitcher-plants develop structures apparently specifically adapted to encourage visitations by ants. The pitchers of *Cephalotus* and the lower pitchers of *Nepenthes* lie on the ground and are commonly visited by creeping insects, mainly ants. In contrast to the upper pitchers of

Nepenthes, the lower pitchers develop conspicuous 'wings' (see Ch. 3 and Appendix 2). Macfarlane (1893) wrote that 'the wing-like pitcher flap and areas between are more beset with alluring glands than the rest of the exterior, and along this . . . (the ants) pass till they come to the orifice'. Similar 'wings' are also well developed in *Cephalotus* where they are provided with hairs. These structures might be of some protective significance for the ants. In this respect, these pitchers resemble similar structures in ant plants, where ants prefer to move about under cover conferred by special hairs or spines (Huxley, 1978). In *Darlingtonia* a similar route is provided by the 'fish-tail' appendages which often lie on the surface of the ground, forming a ramp which leads arthropods directly to the pitcher mouth (Austin, 1875H; Lloyd, 1942; Frontispiece).

In *N. bicalcarata*, as we saw in Chapter 14, the relationship is apparently taken further and the tendrils connecting the pitchers with the leaf blades develop true ant *domatia*. The portion of the tendril opposite the pitcher bottom is swollen and hollow and ants commonly occupy it as a nest (Macfarlane, 1893; Lloyd, 1942; Huxley, 1986). On the same tendrils neighbouring the domatia, *N. bicalcarata* has large nectaries, the largest ever found in pitcher-plants (Macfarlane, 1893). These secretory regions are larger even than the nectaries of the pitcher peristome of the same species. It is hard not to argue that these structural features represent mutuality between insects and pitchers in this case (Joel, 1988).

Pitchers of *Sarracenia* are also sometimes occupied by nesting ants but, in contrast, Macfarlane (1893) and Kannowski (1967) both reported the nests to be in old, dry, non-functional leaves of active plants of *S. purpurea* and *S. minor* respectively. This association might be purely coincidental.

Interactions with Rotifers

As discussed in Chapter 14, rotifers are known to colonize the external surface round the entrance to *Utricularia* traps. Wallace (1978) considers that the plant secretes an attractive agent that encourages the rotifers to settle there. In addition, they may benefit by obtaining organic and inorganic nutrients from the plant. In return, the beds of twitching rotifers may attract grazing insects and crustaceans to the trap entrance, thus improving the capture rate of the traps. Again, a small proportion of the population of the mutualist partner is sacrificed to the overall benefit of the community.

PART VI

Evolution

CHAPTER 17

The Fossil Record, the Evolution of the Carnivorous Syndrome and the Phylogenetic Tree

The Fossil Record

The fossil record of the angiosperms suggests a hazy and probably unrecognizable origin somewhere in the early Cretaceous (Mabberley, 1984). Thereafter the quantity of angiosperm macrofossils is small (Müller, 1981) and those of carnivorous plants limited almost entirely to a handful of finds of fossil pollen (Wegmuller, 1972; Sohma, 1975) mainly confined to the genus *Drosera* (Navale and Misra, 1979; Dupont, 1986; Truswell and Marchant, 1986) or its near relative *Droseracidites*, both of which are easily recognized from their striking pollen morphology (Jones, 1964). *Aldrovanda* pollen (Kondratjev, 1973) has also been recognized and this last genus was obviously much more widespread and common in the European Pleistocene interglacials (Szafer, 1953; Kucowa, 1955; Kolesnikowa, 1961). Bromeliaceae pollen of the genus *Bromelia* is found in the Oligocene of Puerto Rico and what may possibly be *Dionaea* in the middle Miocene of North Borneo. (For a critical discussion of all of these records see Müller, 1981.)

The only macro-fossil exceptions to this skimpy pollen list seem to be the discovery of seeds of Dioncophyllaceae related to *Triphyophyllum* in the Eocene flora of Amur Oblast in Russia (Fedotov, 1982) and seed of *Aldrovanda praevesiculosa* from the early Tertiary of Thuringia (Kirchheimer, 1941).

The actual periods in which the carnivorous families evolved can only be guessed at, but we do know that the Saxifragales, within which and around which several of these families cluster (Fig. 1.2), were well established by the end of the Cretaceous (Friis and Skarby, 1981).

This fragmentary record, telling us nothing of evolutionary paths, serves only to demonstrate that the carnivorous plants of many genera and families were widespread from the beginning of the Tertiary period.

Occurrence of Carnivorous Taxa in the Flowering Plant Kingdom

It seems probable that the carnivorous syndrome (Ch. 1, Fig. 1.1 and Table 1.1) must have evolved several times in the period between the Upper Cretaceous and the present. This assumption is based on widely accepted taxonomic relationships between the various known carnivorous plants. To this end an evolutionary 'tree', such as we saw in Figure 1.2, may give us some idea of possible evolutionary paths concerning carnivory, although all evolutionary trees are the subject of dispute. Figure 1.2 is, we must emphasize, just *one* proposed tree of the angiosperms. Therein we indicated where carnivory was to be found. For reasons we shall explain later, two other striking, but distinct syndromes overlap, namely the C-4 photosynthetic sequence and 'ant-plants', (pages 240 and 261).

The first impression we chose to give in Figure 1.2 is that the carnivorous habit seems to have developed independently in the flowering plant kingdom at least six times. This may be more apparent than real.

Many carnivorous plant families are much closer to each other than Figure 1.2 might suggest. The relationships between the various carnivorous taxa have been dealt with using different taxonomic tools. While the structure of the reproductive system, upon which Figure 1.2 is principally based is the main criterion of plant classification, other criteria such as developmental and chemical features have also been employed, sometimes with puzzling results, as we shall illustrate below. The final conclusions are neither tidy nor comprehensive.

The Scrophulariales: The Zygomorphic Carnivorous Plants

The Scrophulariales, which contain the Lentibulariaceae and the Martyniaceae (see Table 1.1), have zygomorphic flowers and are generally regarded as relatively advanced, far away from the main cluster of carnivorous taxa. The Lentibulariaceae comprise the very different genera of *Pinguicula*, an apparently simple, 'fly-paper' trapper, and the very sophisticated free-floating rootless aquatic, *Utricularia* and its relatives. We shall return to this apparent anomaly later. All other groups of carnivorous plants, with the exception of the bromeliads in the monocots and with the inevitable boundary disputes (DeBuhr, 1975b; 1977a), have actinomorphic (radial) flowers and cluster more closely together; how closely is highly contentious.

The Actinomorphic Carnivorous Plant Families and their Possible Relationships

The Sarraceniales, in the Takhtajan scheme (Fig. 1.2) which are considered by some authorities to have a number of primitive features, are for this reason placed near the ancient group the Ranunculales. This order consists, per Takhtajan, of only the Sarraceniaceae. The Nepenthales is sometimes considered to be a unique order comprising only the Nepenthaceae. Sometimes the closely related Droseraceae are placed in this order and sometimes, as Figure 1.2 and Table 1.1 indicate, in the nearby Saxifragales along with the other carnivorous plant families, the Cephalotaceae and the Byblidaceae. In Takhtajan's scheme the Nepenthaceae, as can be seen, are set well apart from the Sarraceniaceae. However, several authors, in dispute with this separation, have discussed the relations between these two carnivorous orders, since both the Sarraceniaceae and the Nepenthaceae are the familiar, but not the only, 'pitcher-plants' of the layperson, while similar pitchers, as we have noted, have developed in *Cephalotus* (Saxifragaceae). In addition the suction traps of *Utricularia* have also been regarded as similar in ontogenetic terms (see page 56 and Fig. 4.8). Could all these 'epiascidiates' have had a common ancestry? The Takhtajan scheme, as we can see, sets the Nepenthaceae and the Sarraceniaceae well apart. In the opinion of some workers, however, ontogeny should carry some weight in a model of a phylogenetic sequence. Thus, Hooker (1859) noted that 'the resemblance between the pitcher of a seedling of *Nepenthes* and that of *Sarracenia purpurea* is very close, and leaves little doubt that the organ is strictly homologous in the two genera'. Similar early morphological studies (e.g. Markgraf, 1954, 1955), as well as the chemotaxonomic work by Jay and Lebreton (1972), indicate some affinity between the two families. However, the chemical differences appear to be too great to support a very close relationship between these two and the Droseraceae and Cephalotaceae, which other criteria seem to juxtapose.

Different carnivorous orders and families, although apparently separated from one another at least by some authorities, might nevertheless have some distant phylogenetic relationships. The Violales have some affinities to both the Nepenthales and the Sarraceniales (Marburger, 1979). Cronquist (1968, 1981) has assigned the Dioncophyllaceae to a broadly defined Violales wherein we find the monotypic genus *Triphyophyllum*, recently confirmed to be carnivorous (Green *et al.*, 1979). But others, such as Takhtajan (1980), would place the Dioncophyllaceae in the Theales and this dispute is suggested in Figure 1.2. The Saxifragales, which nowadays contain the Byblidaceae and Cephalotaceae, and also the near-carnivore *Roridula* which is discussed on pages 1, 3, 271 and 293, according to Takhtajan (1969) might have distant connections with the Nepenthales. Indeed at one time *Byblis*, now firmly in the Saxifragales, was placed in the Droseraceae by de Candolle (1873). As we saw with the Lentibulariaceae, the Saxifragales at first sight seem to comprise a group of disparate families (e.g. the Cephalotaceae with pitcher traps and the Byblidaceae with fly-paper traps). Yet a study of the wood anatomy of *Cephalotus* (Carlquist, 1981) shows that it is, beyond dispute, related to the Saxifragaceae and allied families.

Jensen *et al.* (1975) have used the presence of the monoterpinoid iridoid compounds as taxonomic markers. The Sarraceniaceae contain primitive secoiridoids and on this basis seem to be related to the Cornales, Ericales and, interestingly enough, the near carnivorous family the Dipsacales, which would seem, according to Figure 1.2, to be a long way away in the Asteridae.

Using this criterion alone, the Sarraceniaceae are not closely related to the Droseraceae and Nepenthaceae, but single-compound analysis of this type can obviously only have limited weight. The result is a muddled taxonomic picture.

The Juxtaposition of Different Trapping Systems

Carnivory, as such, presently carries little conventional taxonomic weight and it should not surprise us that, within each single taxonomic order, the trapping systems are not uniform. In the Scrophulariales, *Genlisea* and *Utricularia* have developed suction traps, whereas *Pinguicula* and *Ibicella* have apparently very 'primitive' adhesive traps. Nevertheless, floral evidence apart, the chemistry of the flavonoids (Jay and Gonnet, 1973, 1974; Jay and Lebreton, 1972) tells us that they are closely related and in the right area of a major phylogenetic tree. Other apparent anomalies abound. In the Saxifragales *Byblis* has developed adhesive traps, *Cephalotus*, pitchers; in the Nepenthaales, *Nepenthes* has developed pitchers, *Aldrovanda* and *Dionaea*, snap traps; and *Drosophyllum* and *Drosera*, adhesive traps. The Dioncophyllaceae have only the monotypic *Triphyophyllum*, which has developed adhesive traps on certain of its juvenile leaves. The bromeliads, far away in the monocotyledons, are simple pitchers, but with many features such as an epiascidiate pitcher-form in common with all other pitchers and a slippery wax zone in common with *Nepenthes* on the one hand and *Darlingtonia* on the other. We shall consider the scattered nature of these carnivorous features when we discuss both Croizat's integration on page 286 and the possible translocation of characters on page 288.

Early Attempts at the Integration of the Carnivorous Habit

All the distant relationships suggested above, transgressing traditional taxonomic groupings, might lead us to look again at the iconoclastic ideas of Markgraf (1954, 1955; Fig. 17.1) who synthesized all the then-known orders of carnivores, excluding the Scrophulariales of course, into a common scheme. We might even go the whole way with Schmid (1964) and consider the Dioncophyllaceae as the progenitor of three of the families of carnivores (Fig. 17.2) but we must bear in mind the isolation of the Scrophulariales and, more recently, that of the newly discovered carnivory in *Brocchinia* and *Catopsis* belonging to the distant monocotyledonous family Bromeliaceae. These latter are unique not only in being monocotyledons, but also in having carnivorous and non-carnivorous species in the

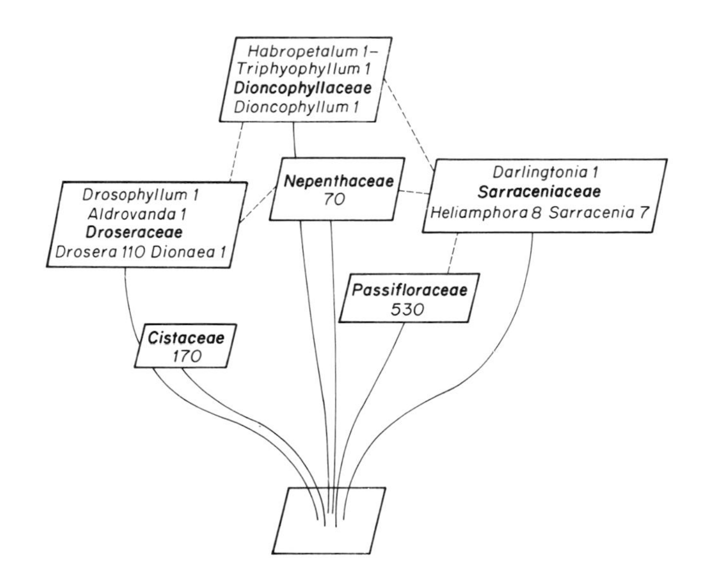

Fig. 17.1 Redrawn from Markgraf (1954) showing the possible relationships of four of the carnivorous plant families. The stronger evidence for direct lines of descent is indicated by solid lines; the weaker, reverse relationships, are suggested by the dotted lines. Revised species numbers from Schlauer (1986).

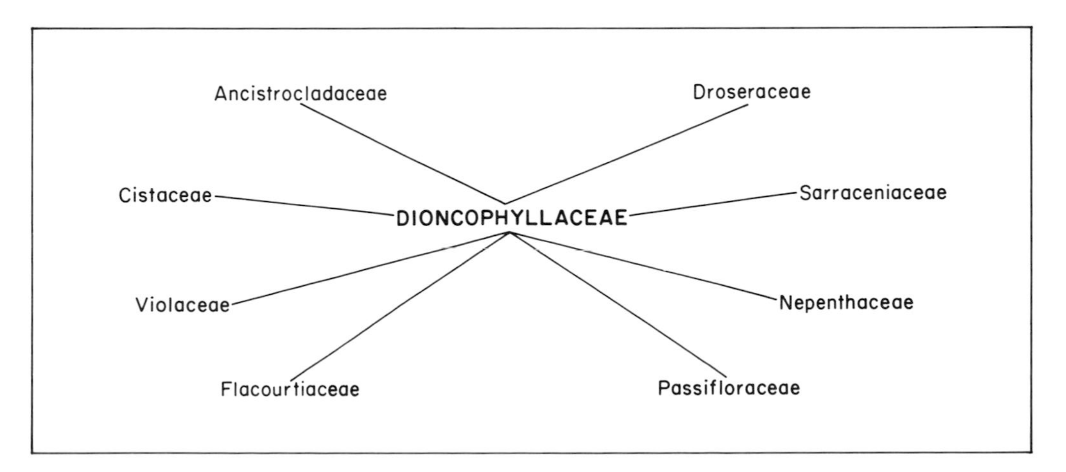

Fig. 17.2 The proposed centrifugal position of the Dioncophyllaceae. (Redrawn from Schmid, 1964.)

same genus. Thus, however one constructs an evolutionary tree or whatever criteria one uses, the polyphyletic nature of angiosperm carnivory, at least in part, seems in conventional taxonomic terms irrefutable. As already stated, we have in Figure 1.2 also indicated where other important strategems of the flowering plants lie, such as C-4 carbon fixation cycle (C-4) and Myrmecophily (Ant). They *appear* to be just as scattered (see also pages 240 and 261).

Croizat's Integration of all Carnivorous Taxa

Croizat (1961) (for review see Heads, in Craw and Gibbs, 1984) was not prepared to accept that all the taxa previously described arose independently and several times from other angiosperm families, as we have suggested. He argued that carnivory has a common root and arose as the result of differentiation of some unknown, but widespread, ancestor; an ancestor, he suggested, that was perhaps as old as angiospermy itself. What is Croizat's evidence for the co-sanguinity of the carnivorous plant assemblage?

The Interrelationships of the Carnivorous Plant Families According to Croizat

Marburger (1979) in the work that we described in Chapter 4 gave close attention to the stalked and sessile glands of *Triphyophyllum peltatum* (Dioncophyllaceae). It is argued that, unlike gross morphology, this is an aspect of anatomy least likely to show convergence if the groups in question were truly unrelated. Yet Marburger noticed that the structure of the glands was remarkably similar to those of *Drosophyllum lusitanicum* (Droseraceae) which (Fig. 17.3, Arrow 1) we have suggested, is far away (Fig. 1.2) in most conventional phylogenetic schemes.

The Lentibulariaceae are usually thought of as being totally removed from the major cluster of carnivorous taxa (Figs 1.2 and 17.3), e.g. having zygomorphic as opposed to actinomorphic flowers, as we have emphasized. However, if we look at evolution through the lens of peripheral and highly specialized carnivorous features, as in the *Drosophyllum* glands above, and do not concentrate upon the more conventional criteria, the Lentibulariaceae can be closely linked with other groups. It has some very diverse members. For example, the whorled *Utricularia tabulata* connects the family with the Hippuridaceae (Fig. 17.3, Arrow 2). Through the *U. 'avesicaria'* group the family is allied to the Podostemaceae (Fig. 17.3, Arrow 3). The pitchers and traps of the *U. 'dichotomamonanthos'* aggregate, along with *Genlisea* and the Cephalotaceae, help to create morphogenetic links with the Nepenthaceae. At this level of evolution, as Heads (1984) points out, the transition from floral zygomorphy to actinomorphy, of which so much is made in conventional phylogeny as if it were a major barrier, is easily achieved. He cites a progression through the genera *Cladopus – Dicraeia – Inversodicraea – Castelnavia – Jenmaniella – Dalzellia – Apinagia – Loncostephus – Tulasneantha*, all in the Podostemaceae. Such a sequence suggests that the zygomorphic flowers of the Lentibulariaceae are not necessarily remote from the actinomorphic flowers of the Droseraceae, Nepenthaceae and Cephalotaceae (Fig. 17.3, Arrow 5). The flowers of *Parnassia* connect the Droseraceae and Saxifragaceae (Arrow 6). Through the Sarraceniaceae, Nepenthaceae and Dioncophyllaceae (Arrow 7) the carnivorous plants can be seen to have connections with 20–30% of all angiosperm families.

We have already seen (Airy Shaw, 1951; Schmid, 1964) that there is evidence (Figs 17.2 and 17.3) for the affinities of the Dioncophyllaceae with the huge clusters of flowering plant genera around the Flacourtiaceae, Guttiferae and Droseraceae (Fig. 17.3 Arrows 1, 7 and 8) (Gottwald and Parameswaran, 1968). The Podostemaceae are, in vegetative terms, closely allied with the Hydrostachyaceae (Arrow 9). The evolution of the former from the latter can readily be explained on the basis of the reduction of the unisexual inflorescence to a lone ovary and the addition of a few stamens. Together, these two families can provide links for a vast range of angiosperms. With some apparent carpel 'fusion', i.e. intercalary growth, the Hippuridaceae can form a morphogenetic link between the Podostemaceae on the one hand, and the Haloragidaceae, Gunneraceae, Lythraceae and Onagraceae on the other (Arrow 10). The Lentibulariaceae and Droseraceae are bound together by the Byblidaceae (and *Aldrovanda*), which also serves to bring together *Cheiranthera*, (Pittosporaceae; see Fig. 17.3) the *actinomorphic* Pittosporaceae and the more or less *zygomorphic* Ochnaceae (Arrow 11).

Croizat's synthesis thus attempts to characterize carnivorous plants, and their allies (Figs 17.1, 17.2 and 17.3), not as scattered derivative groups, which we have suggested in Figure 1.1 and page 5, but as representing a level of evolution fundamental to and

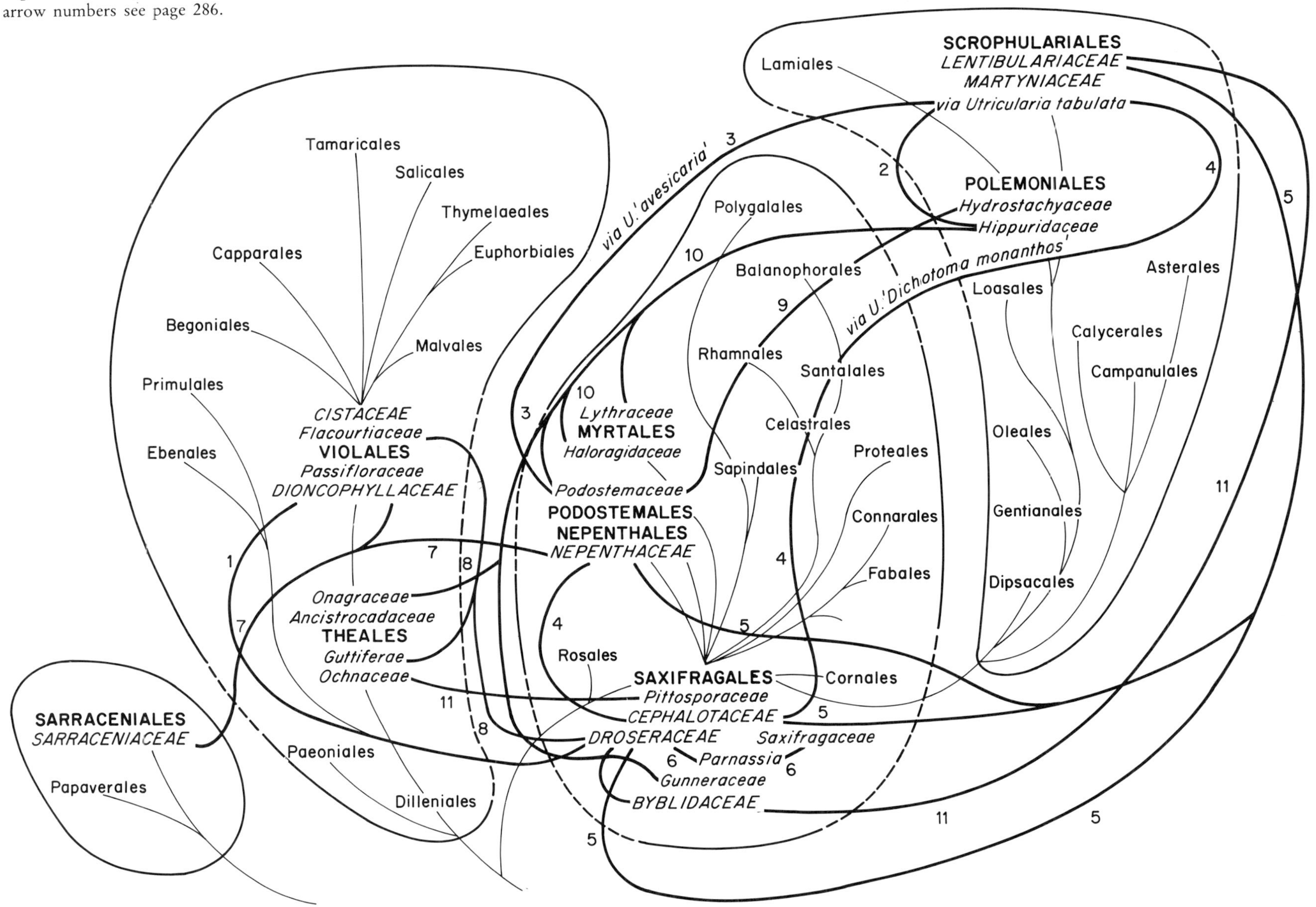

Fig. 17.3 A phylogenetic tree to show the unconventional relationships between all the dicotyledonous orders and families of carnivorous plants; cf. Fig. 1.2: for explanation of arrow numbers see page 286.

underlying all modern angiosperms. Whether a Croizat-type synthesis could bridge the chasm which now arises by the discovery of carnivorous genera in two sub-families of the monocotyledons (*Brocchinia* and *Catopsis*; see Figs 1.2 and 2.5) we shall now never learn from his pen. But if his general synthesis is correct, it might go a long way to explaining the distribution, not only of carnivory, but of many other syndromes such as C-4 at the enzyme level and ant-plants at the insect-interrelationship level. In the second half of this chapter we shall examine speculations which might point towards the origin of specific features of the carnivorous syndrome, setting aside for the moment how these features might have been synergized or when they might have arisen.

The Translocation of Characters in Plants

Croizat's speculations, whether we are prepared to give them credence or not, seem to imply that clusters of characters, or more likely factors controlling the explosion of characters, were tossed around, almost at random, somewhere in the early history of the angiosperms. Croizat (1961) was writing at a time when, although the rigidities of Linnean-type taxonomic thinking were long abandoned, fluidity between taxa which the very latest genetic research has revealed, was somewhat heretical. Do the latest ideas on the translocation of characters lend any support to Croizat's speculations? Is it possible to think, during the Golden Age of angiosperm evolution towards the end of the Cretaceous, that clusters of characters or more likely, as we have suggested above, factors controlling those characters, were randomly tossed around between taxa, just as the best bridge partners will, without total communication, muster their best card combinations for a tactical assault.

There are three principal ways in which such diversity could arise:

(i) hybridization;
(ii) the transfer of small clusters of DNA, via such systems as the Ti-plasmid of *Agrobacterium tumefaciens*.
(iii) the appearance of endogenous mutations on the lines of the 'pitcher' forms of *Ficus*, *Codiaeum* and *Taraxacum* (see page 296 below).

Hybridization does not seem to be a very likely source of fundamentally different syndromes, although it may be a potent source of new species, e.g. in *Sarracenia* (page 306), once a basic carnivorous type had arisen.

Plasmid-mediated vectors are feasible vehicles for the transfer of small clusters of genes, but it is highly unlikely that characters as complex as a pitcher or gland could be transferred in this way. What such a particle might do, however, is disturb the control of a sequence of events such as those, for example, leading to a complex leaf form (Fig. 17.4).

Endogenous mutants would seem to be the richest source of variability. If such internal defects were combined with sporadic breakdowns in control (Fig. 17.4) later and combined with hybridization, the further potential is almost limitless. With the passage of time and the development of ever more sophisticated incompatibility mechanisms and subtle insect–plant relationships, have angiosperm taxa become relatively more arthritic? If harder boundaries between groups have evolved, does this colour our thinking with respect to earlier and potentially more flexible evolutionary periods?

Examples of the Translocation of Whole Organs or Tissues in Carnivorous Plants

Characters so complex as a trap or digestive gland must inevitably comprise a large cluster of genes. It is a matter of common observation by all plant growers, from gardeners to research workers, that complete plant characters are readily multiplied. The carnivorous plants are no exception to these aberrations. As long ago as 1905, Leavitt observed a phenomenon that many growers of *Drosera* species have noticed, i.e. complete trap characters transposed to other parts of the plant (First step of Fig. 17.4). Carpels, for example, in *Drosera rotundifolia* may be transformed into complete and functional tentacular leaves. Whole flowers may revert to the tentacular leaf form. What is more, the mutant forms, as Leavitt observed, often breed true. Minton and Jeffreys (1972) noticed that in certain populations of *Dionaea muscipula*, both wild and cultivated, the floral-development control mechanism was obviously breaking down. There is nothing unusual about the normal flower of *Dionaea*, but in these populations some sepals were modified to form small traps. In other individuals the carpels, stamens and even the petals were modified. In the most perfect floral mutants, the

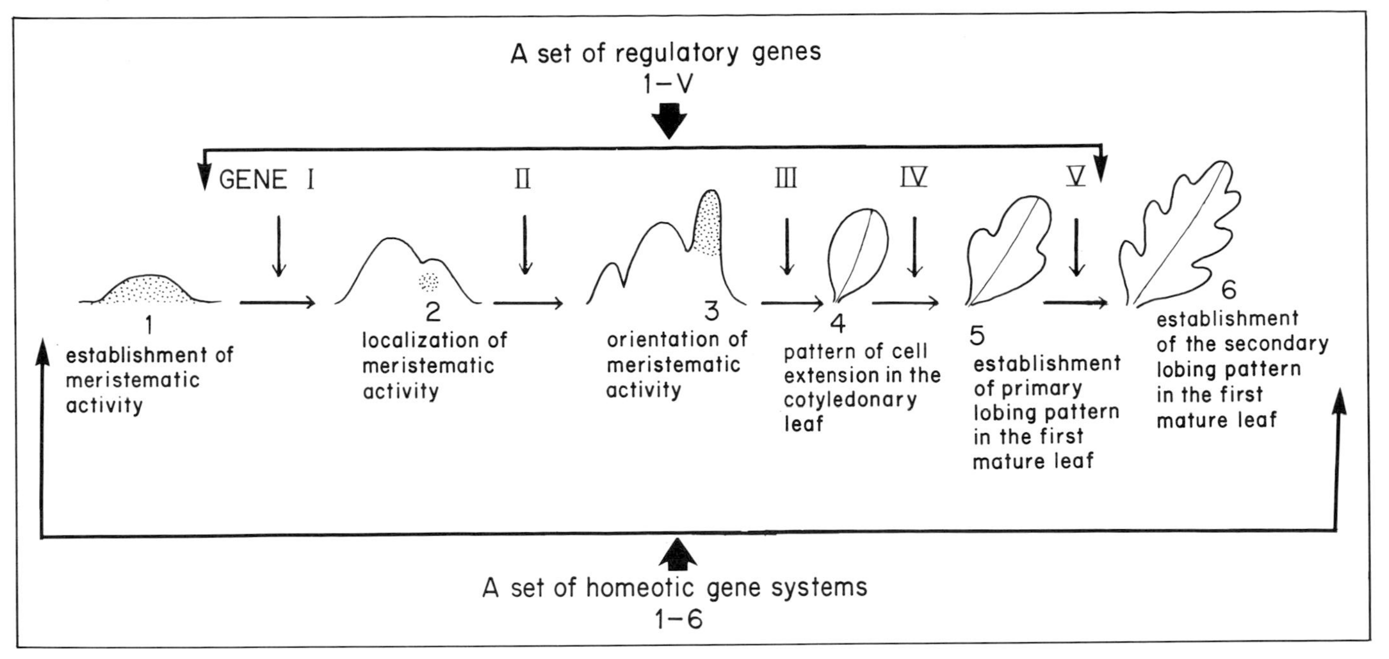

Fig. 17.4 A hypothetical interactive sequence of gene-controlled events leading to the development of a lobed pinnate leaf. Regulatory (switch) genes detect the completion of a homeotic sequence and permit the next linked gene system to begin to operate. The 'failure' of a regulatory gene to act will result in the repeat of the previous sequence or diversion to another homeotic sequence, e.g. 'doubling' of flowers, peloric flowers or fasciated stems. Such a system is completely speculative and no such sequence has been confirmed in a higher plant.

traps not only resembled those of small normal seedlings but could also be shown to have small but conventional electrical impulses. *Dionaea* lobes have been found with eight to thirteen trigger hairs (MacFarlane, 1892; cf. Fig. 19.1 and *Aldrovanda*).

We can hazard a guess that controller gene systems, on the lines of those suggested in Figure 17.4, are either failing to work or are switching *on* complete and complex sequences at the 'wrong' moment in a developmental succession.

Beebe (1980), attempting to establish *Dionaea* in tissue culture, cites, *inter alia*, observations on the spontaneous development of plantlets on the base of the petiole, on traps and on parts of the inflorescence. He found that these spontaneous aberrations could be enhanced by certain combinations of auxin and cytokinins. After treatment, shoot buds formed in the place of the traps at the apices of the leaves. Both spontaneously and artificially, *Dionaea* seems uniquely unstable.

Compartmented Gene Systems

The many defects of normal development listed above may possibly be ascribed to duplications of compartmented gene systems along the lines of blocks of partly autonomous genes which, it is proposed (Lawrence and Morata, 1976), control areas of development of an insect cuticle. Each semi-autonomous block of cells, i.e. each step in Figure 17.4, is a polyclone, switched on or off by regulator genes (homeotic gene systems; Garcia-Bellido, 1975). Regularly in the wild or in cultivation we see double or peloric flowers, fasciated stems, fused fruits or multiple meristems (Sättler, 1977). Some of these duplications may be due to regulator genes of the insect types switching on frequently to trigger another complete but irrelevant sequence of development. Many of these teratological or neotenous forms in plants like the 'hopeful monsters' of Goldschmidt (see below) and like those observed by Leavitt (1905), are inherited and may survive for a while, even if not selected as cultivars.

The Patio Ludens and 'Hopeful Monsters'

It is not necessary to envisage the development of many of the prominent features of carnivorous plants in a step-by-step aggregation of thousands of gene mutations. Nor is it difficult to see the accumulation of advantage that could be gained by a half-way carnivorous plant (Fig. 1.1 and page 288). As van Steenis (1969, 1977) has pointed out, saltatory evolution is surprisingly common

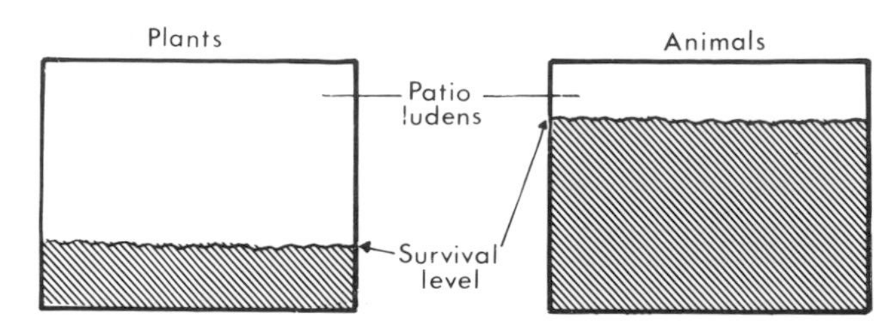

Fig. 17.5 A diagrammatic illustration of the 'Patio Ludens' of Van Steenis (1969).

particularly in tropical regions. Unlike animals, plants possess what he terms a large 'patio ludens' or 'room to play' (Fig. 17.5). The patio ludens, at least for advanced animals under the immense and somewhat uniform competition for survival, is narrow. Gross mutants and monsters in these groups are rare and ephemeral. In plants, even higher plants, the mosaic of selective advantage is uneven and thus there is greater diversity in the perfection of adaptive structures. There is ample licence, as we shall see later, for structural features which may arise, to quote Goldschmidt (1933, 1960), as 'hopeful monsters' – a useful phrase in spite of its teleological overtones. These hopeful monsters are, in the first instance, irrelevant to competition and adaptation (van Steenis, 1957) but may, if stored and switched on (Fig. 17.4) at the right moment, be a key in the development of a new strategy. The supposed flaw in the Goldschmidt theory, that such a 'monster' being by definition unique could have no sexual future, now loses its significance by the recent discoveries of 'transposons', small sections of mobile DNA that occasionally migrate across boundaries. If the monsters were generated by something like a viral infection, there opens the possibility that whole populations, rather than single individuals, might have arisen at any one time. Such populations might have an evolutionary future. Transfer of genes or gene complexes between species would allow much more rapid evolution than at present envisaged (Lewin, 1982). Moreover, the ability of certain carnivorous plants to switch to a vegetative form of reproduction would enhance the spread and establishment of such mutants (Nolan, 1978a). Such ideas of species flexibility are still speculative, but gaining ground (Erwin and Valentine, 1984; Syvanen, 1985). In Chapter 18 we shall investigate the many instances where features of the carnivorous syndrome may occur in non-carnivorous plants and how these features might have come together to form the complete and functional carnivorous plant.

CHAPTER 18

Features of the Carnivorous Syndrome in Non-Carnivorous Plants and General Aspects of their Possible Evolution

Non-Carnivorous Plants that Trap and Kill

Many higher plants are known to defend themselves against their herbivorous predators by the accumulations of certain toxic, noxious or abrasive substances. It is less widely appreciated just how many members of the angiosperms not only dissuade their would-be attackers by chemistry but may, in very sophisticated ways, trap and often kill them, thus removing them completely from a potentially herbivorous population.

Non-carnivorous Plants that Trap and Kill but do not Absorb

Many plants traps and actively kill, probably for a variety of evolutionary reasons, a wide range of insects, but then fail to *use* them in the sense understood in the carnivorous syndrome (see Fig. 1.1).

Although it can be of no selective advantage to the plant, insects are frequently found dead in or on the flowers of otherwise conventional entomophilous plants (Eisner and Aneshansley, 1983). The flowers of aroids, as we shall see in the analogies drawn with pitcher-plants, are often the graveyards of unsuccessful insects. *Sauromatum guttatum* is notorious in this respect. Another group of indiscriminate fly-killers is the genus *Stapelia*, and Marianne North (1980) noticed in her travels in South Africa that dipterans were often trapped and dying in the flowers of several of these species. The same phenomenon can sometimes be seen in greenhouse-grown stapelias (C.G. Vosa, pers. comm., 1986).

One species of *Passiflora* (*P. adenopoda*) has as part of its defence mechanism hard, sharp and curved trichomes on its leaf surfaces which can puncture the cuticle of the caterpillars of Heliconiine butterflies. The caterpillars die from total haemorrhage (Gilbert, 1971). *Passiflora* is not a carnivorous plant, but it is interesting to note how close the family is placed (Figs 17.1 and 17.2) by some authorities to true carnivorous groups. *Nicotiana* and *Solanum* are well known for their capacities in killing insects. Darwin noticed (1875) that *N. tabacum* was often covered with trapped insects. *Nicotiana* trichomes produce alkaloids that are, for example, toxic to green peach aphids which may become a serious pest to the plant. Aphids placed on a tobacco leaf surface at first become paralysed and then die (Levin, 1973). Several species of *Solanum*, *S. berthaultii*, *S. tarijense* and *S. polyadenium* but not, unfortunately for the farmer, cultivars of the edible potato *S. tuberosum*, bear four-lobed glandular hairs. These glands contain a very sophisticated binary defence system. When ruptured by aphid movements, they release two compounds which combine and glue the insect to the leaf

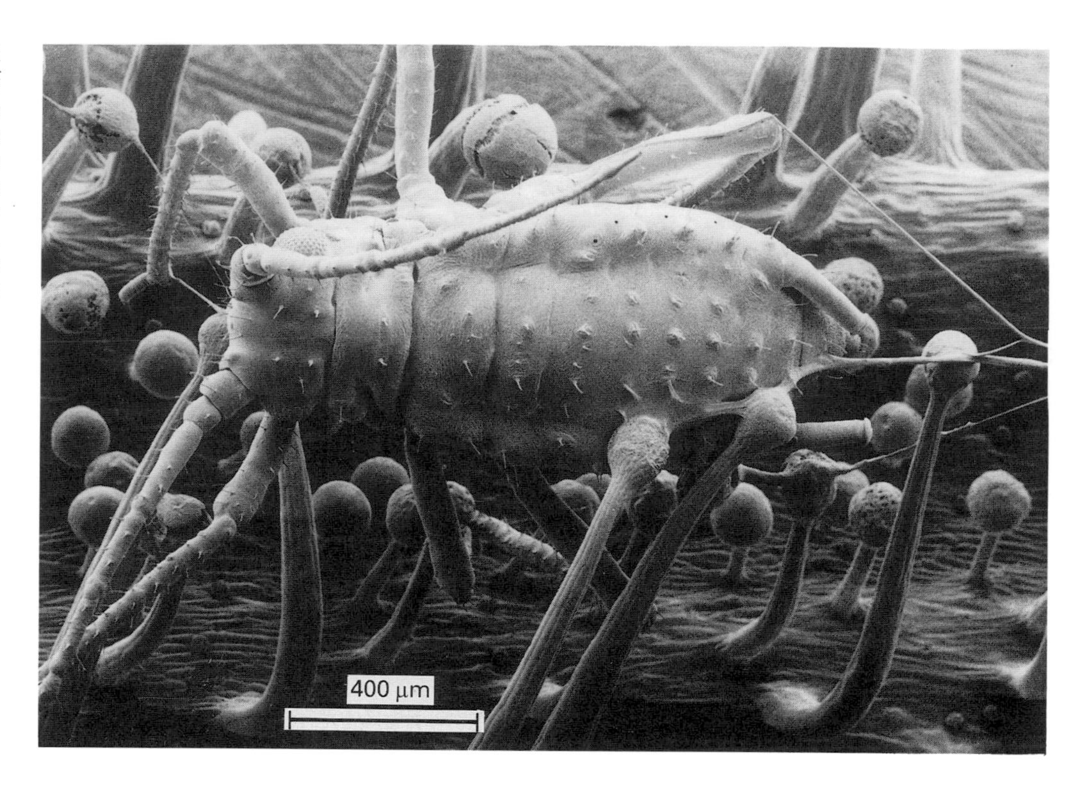

Fig. 18.1 An aphid trapped on the surface of a leaf of *Petunia* × *hybrida*. Note the multiple attachments of the sticky trichome tips to the insect's body and the mucilage strands (arrows) pulled out by its antennae, wings and legs. (SEM by kind permission of Dr C.E. Jeffree, Edinburgh, UK.)

(Gibson, 1974; Duffey, 1986; Gregory *et al.*, 1986; Fig. 18.1). Such a property is obviously of considerable value to the genus *Solanum* in resisting both aphid predation and, probably, the resultant spread of viral diseases, but is of no direct carnivorous benefit to the plant.

Another well-known trapper of insects is the 'catch-fly' of the northern European flora, *Silene*, which was given consideration by Darwin (1875) as a possible carnivore. Some of the other insect trappers he noted were *Primula viscosa*, *P. villosa*, *P. hirsuta*, *Saxifraga luteo-viridis*, *S. bulbifera*, *S.tridactylites*, *Saponaria viscosa*, *Silene viscosa*, *Cleome ornithopodioides*, *Sedum villosum* and *Sempervivum montanum*. The whole of *Picris echioides*, the well-named 'bristly oxtongue' of northern Europe is capable, with its miniature grappling hooks, of trapping a wide range of insects. Whalley (1986) saw damselflies (Odonata: *Ischnura elegans*), as well as moths and soldier beetles, caught on the leaves and stems. Eastop (1986) has reviewed a number of similar examples of non-carnivorous plants catching insects in this way. The advantage to *Picris echioides* of this behaviour is not clear, and it is certainly not carnivorous. Glandular calyx hairs of *Plumbago* (Rachmilevitz and Joel, 1976; Joel, 1978) and some species of *Stylosanthes* (see below; Sutherst and Wilson, 1986), again are not complete carnivorous plants. They not only catch insects on their flowering stems, but kill them as well.

The insect-killing genus *Stylosanthes*

Stylosanthes embraces about 25 species of tropical legumes of which several are planted in Australia as cattle fodder. Many of the species are covered with sticky, glandular trichomes and it has been noticed that, in areas where cattle ticks (Ixodidae) are prevalent, the ticks may be found immobilized and dead on the surfaces of these plants (Sutherst *et al.*, 1982).

These workers noticed that, on varieties of *S. scabra* and *S. viscosa*, but not on *S. harmata*, the climbing ticks were both glued down by the viscous secretion and apparently rapidly killed, shortly after trapping, by some exudate from the plant. The lethal exudate was effective only in the vapour phase. These *Stylosanthes* species appear to derive no direct benefit from this selective predation since the Ixodidae have no effect on plants. The results are, however, highly beneficial to the cattle ranchers. It should be emphasized, though, that these particular species are being grown far from their native habitat of South America, and the trapping and killing mechanisms may have some local defence role not so far investigated.

Seeds that trap insects

Many seeds have a mucilaginous coating which may enable them to maintain their water balance and aid germination (Harper, 1977). Other seeds such as those of *Capsella bursa-pastoris* and *Descurainia sophia* for example (Barber *et al.* (1974), Page and Barber (1975), Barber *et al.* 1976; Barber, (1977)) both attract and trap mosquito larvae. The ecological significance of this killing is not understood and a Kerner-type defence system (see below) does not seem to apply.

The near-carnivore *Roridula*

Well known and with a long history of investigation is the southern African genus *Roridula*. There are only two species and, although placed by some authorities in the Droseraceae, Carlquist (1976) has shown this to be incorrect, with the wood anatomy being closer to *Byblis*. The plant is so similar to *Drosera* that it was considered by Darwin (1875) and at least briefly by Lloyd (1934) to be a true carnivore. It bears tentacles, virtually identical to those of *Drosera*. It traps insects, but there are no absorbing glands and no use seems to be made by the plant itself of this potential source of nutrient. The dying insects are preyed upon by crab spiders which are not themselves caught (see page 266). It is possible that the insect-trapping here provides some indirect benefit to the *Roridula*, as Carlquist suggests, either through the commensal situation or by littering the ground at abscission with nitrogen-rich leaves: an indirect carnivory. It may, however, be simply defensive, as with *Solanum berthaultii*.

In many of the instances quoted above it is relatively easy to see the selective advantages of killing or impeding insects. But much insect killing occurs in floral systems where the death of the insect can be of no advantage to the plant concerned. As we shall see, there are even snap-trap analogues, in which the advantage of the insect killing is not apparent.

Crossing the Boundary between Defensive Trapping and Killing and True Carnivory

Many plants, and far more than we have listed on the previous page or beyond the outer boundary in Figure 1.1, carry out what we could call defensive killing. Could this defensive killing have provided one of the springboards towards the evolution of the whole syndrome? We should not forget that although, in most of the instances above, we may be able to provide 'reasons' for a particular evolutionary path the plant kingdom is rich in what appears to be non-adaptive evolution.

Flowers and their unbidden guests

Kerner (1878) in his intriguing but rarely read text *'Flowers and their unbidden Guests'*, suggests that the viscid radical leaves of *Pinguicula* species should not be regarded as a trapping system (the current view) but as a defensive mechanism to keep off non-pollinating, phytophagous creeping insects. He did not exclude the possibility of caught insects remaining adhered and being digested, and 'serving as welcome, if not very luxurious, food'. Recent evidence does, however, favour a carnivorous role now for *Pinguicula* (but see Fig. 7.1). It is therefore possible to speculate that it was not only selection towards carnivory that brought about the secretory trapping gland, the origin of which we shall examine in detail on page 294. A broad spectrum of higher plants, outside the conventionally accepted carnivores, and from many diverse families, as we have seen, trap and retain insects on their leaves. Many but not all of the examples we have quoted above, could be interpreted on a Kerner model.

The gland in transition from secretion to absorption

As we shall discuss on page 298 a major advantage might accrue to a gland-bearing, defensive-type leaf were it to become dished and able to hold a pool of fluid, i.e. a 'proto-pitcher'. If the glands could absorb as well, the advantage would have been even greater. Dishing of the leaf occurs in *Drosera* as shown by Darwin (1875), who also tested the absorptive ability of a whole range of plant genera (e.g. *Saxifraga*, *Pelargonium*, *Erica* and *Mirabilis*) to see whether their trichomes could absorb as well as secrete. It would seem that at least two species of *Saxifraga*, which is distantly related to *Drosera* as seen in Figure 17.3, could absorb matter from an infusion of raw meat and solutions of nitrate and carbonate of ammonia.

There are many other analogues of absorptive trichomes in non-carnivorous plants. In the bromeliads, all stalk cells of tank and atmospheric tillansioid trichomes absorb at least two amino acids from the solutions in contact with the leaf (Benzing *et al.*, 1976) and readily absorb inorganic ions (Benzing, 1980). In the 'ant-plant' *Hydnophytum formicarum* it has been shown by radioactive tracer experiments that animal tissue breakdown products are absorbed into the plant stem and translocated (Rickson, 1979). Whatever pressure or pressures may have brought about the evolution of the gland, and whether or not we finally conclude that it is mono- or polyphyletic, the evolution of the gland is a major consideration in the evolutionary syndrome.

The Evolution of the Gland: General Considerations

The gland is so salient a feature of the carnivorous syndrome that it is useful here to consider its possible paths of evolution. Glands are not confined to the angiosperms, although suggestions as to their role in more primitive plants are vestigial. In contemporary plants below the angiosperms, they are found in *Pteridium*, and several other ferns, some tropical like those on which Darwin (1877) speculated. It seems possible that under tropical conditions bracken (*Pteridium aquilinum*) may use the stem nectaries to attract carnivorous ants. The ants rewarded by the 'junk food' may, in their turn, defend the plant from phytophagous insects. In fossil plants glands are found in the primitive Carboniferous fern *Lyginopteris* (Stewart, 1983) and what possible role this capitate structure might have played is almost beyond speculation. We must remember though that the gland has many roles outside this syndrome and almost certainly is polyphyletic in its evolution (Jeffree, 1986).

The initial development of glands on a carnivorous plant surface is generally held to be associated with improving the efficiency of absorption. Bacteria, commonly followed by protozoa (see Ch. 9), are the usual initiators of insect decay: circumventing or supplementing their activity is apparently beneficial to the plant in improving its share of the breakdown products. But this is not the only possible evolutionary route. Kristen *et al.* (1982) have shown that in the lycopod *Isoetes* the ligule, which forms at the basal region of the microphylla, secretes a mucilaginous layer of which at least 22% is protein. They propose that this organ once had an extracellular lytic function, combined with absorption, analogous to that of the glands of carnivorous plants. Horner and Beltz (1975) have described the ligule in *Selaginella* and shown that it too may have a secretory and absorptive role.

The Possible Evolution of Glands from Stomata

Some other workers believe that some types of glands may have arisen from modified stomata. Even stomata may themselves have had a polyphyletic origin (Wilkinson, 1979). An example of possible stomatal-gland cell evolution can be seen in certain Sarraceniaceae. The head-cells visible at the surface in *Sarracenia* glands are similar to the two guard cells of a stoma closed together. In *Darlingtonia*, as Parkes (1980) has pointed out, the two cells are still raised slightly above the surface and occasionally possess an opening between them, whereas in *S. purpurea* the two cells are completely flattened and show the most modification from the possible ancestral guard cell.

Another example of the possible evolution of glands from stomata noted by Lloyd (1942) is in the small digestive gland of *Cephalotus*. In these glands the guard cells are still separate and become redundant. A glandular cell develops in the sub-stomatal cavity and occludes the apparently ancestral stomatal pore. Similar occluded stomata were described for *Hakaea* by Heide-Jørgensen (1978). This feature may suggest that, in the ancestors of this species, fluid was absorbed through the stomata. Certain carnivorous species have developed large, multi-cellular glands. The ontogeny of these from epidermal cells in, for example, *Dionaea*, shows no evidence for a stomatal origin (Franstadt, 1877). Immature tissue in the genera *Cephalotus* and *Nepenthes* have, in their large glands, glandular cells of epidermal origin and it is possible they are also stomatal in origin. They have, however, been modified to such an extent that there is now no evidence for this.

It is worth remembering that the evidence for the evolution of certain glands from stomata is strengthened by the fact that both stomatal guard cells and glands possess proton-pumps and may have some similarities in the first stages of ontogeny.

The Specialization of the Carnivorous Gland

In the 'fly-paper' species, after the initial development of the glands, there came a split either into specialization for mucilage or for enzyme secretion (but see Kerner's speculations of 1878; page 293). The exception is *Drosera*, in which the tentacular glands alone are responsible for both capture and digestion and have solved the problem of distance from the leaf surface by developing the ability to move towards the secretion pool (Chs. 4 & 6). However, on the *Drosera* leaf there are small, sessile glands, possibly functionless or involved in the transport of the digestive fluid. These simple glands may represent the precursors of the large, tentacular glands and may be homologous, at a primitive level, as Parkes (1980) suggests, to the sessile glands of *Drosophyllum*, another member of the Droseraceae. These small, sessile glands may have developed by the outgrowth of guard cells of ancestral stomata.

Thus, although the links are very tenuous in some examples, the digestive glands of many carnivorous species may have their origins in the stomata of ancestral leaf surfaces.

The similarities of digestive glands

Although at first glance there appears to be a wide variety of gland types in carnivorous plants, this is an anatomical rather than a functional variation. Glands show closer similarities than the trapping mechanisms, if their functionality is considered (see Chs. 4 & 8). This is because, regardless of the prey, the requirements for production of a secretion and transport of fluid are similar, so that digestive glands of different groups and nectar glands are similar at several levels.

The relationship between nectar glands and 'mucilage' glands

The question of which came first in the early evolutionary stages of the syndrome is also raised by the presence of carbohydrate-secreting glands which appear homologous to digestive glands (Joel, 1985). This phenomenon occurs in *Sarracenia*, *Nepenthes* and *Dionaea*, and the structural similarity is very marked. The functional specialization may have evolved recently, but as Parkes (1980) points out, it is difficult to hazard the nature of the glands before this separation in function. Possibly the glands developed initially under selective pressure to improve the absorption of bacterial breakdown products (but see Kerner, 1878 and page 293) and the secretion of nectar or enzymes was developed secondarily.

In *Darlingtonia*, which we consider to be a relatively primitive carnivore in some respects, there are nectar glands, but the glands of the digestive zone do not appear to secrete enzymes. Thus nectar secretion may have developed before enzyme secretion in a species which relies on a larval–protozoan–bacterial chain breakdown for digestion of prey. Presumably this dependency could have been the integral feature in the development of present-day highly adapted species. Also, nectar secretion may more readily develop than enzyme secretion because it was probably based initially on the exudation of phloem contents already present, needing only to become more concentrated. A mechanism was already present in floral nectaries.

On the other hand, nectar secretion could also have been a recent refinement in some groups, as is found in *Cephalotus*. In this species, mature juvenile pitchers which are produced by slender rhizomes do not have the recurved teeth and gland development that occurs around the rim of normal pitchers. Small as they are, they appear to function successfully in capturing and digesting correspondingly small prey (Lloyd, 1942) and may illustrate an intermediate stage in evolutionary development, with modification to the rim being a recent refinement. *Nepenthes* shows marked variation in the extent of rim development, yet all the pitchers are successful at catching prey. These then may represent some of the pressures that brought about, probably polyphyletically, diverse gland systems.

The Diverse Roles of Glands

We have suggested that more than one directional pressure may have given rise to a particular type of gland. The stalked glands of *Drosophyllum*, to take a single example, may have evolved at various times and various rates:

(i) as a lure;
(ii) as a trapping glue (*Fangschleim*);
(iii) as a viscous drowning mechanism;
(iv) as an enzyme medium; and
(v) as miniature hygrometers to absorb water vapour by night (see page 32).

They may also have evolved just once, but during evolution attained a very wide range of characters and functions which have enabled them to take advantage of using a single structure in multiple ways. Such possible conflicting or parallel pressures show how difficult the study of the evolution of a single character may be.

For a consideration of the evolution of the digestive mechanisms see pages 294 and 303.

The Evolution of the Pitcher Form: General Considerations

The pitcher form of developed flowers, often designed to retain or impede the escape of potential pollinators, is common. The Aristolochiaceae, Araceae, Nympheaceae and Annonaceae all contain striking examples. Many species of plants trap insects, at least temporarily, as part of pollination (see also page 301). Amongst the best known are certain aroids, including *Arum maculatum* and *Cryptocoryne griffithii*, some *Aristolochia* species and several members of the Annonaceae (Corner, 1964; Vogel, 1965; and see next page).

The Temporary Trapping of Insects for Pollination: Traps with Movement

In *Arum maculatum*, it is widely believed, insects are attracted by the heat and smell generated by the rapidly expanding and respiring spathe and are held as short-term prisoners. In its temporary trap *Arum* has many features of *Nepenthes*. Some aroids produce the foul smell of rotting meat and this feature is also used as an insect attractant by several genera in the Asclepiadaceae, most notably the genus *Stapelia*. In *Cryptocoryne griffithii*, which is an aquatic, the floral bract may elongate as a tube 40 cm long up to the water surface; a close 'mimic' of a pitcher trap. Into this tube tumble beetles, most of which subsequently escape bearing a pollen load.

One striking feature of two different pitcher-type carnivorous plants, namely *Darlingtonia californica* and *Sarracenia psittacina* is the presence of areolae (window-panes) at the hooded apex of the pitcher. These areolae are supposed (Ch. 5) to impede the escape of flying insects by diverting them away from the true exit. These structures appear to have a close analogue in the window-panes of the flower of the tropical aroid *Sauromatum guttatum* (Dakwale and Bhatnagar, 1985). Flies attracted by the smell, apparently rewarded neither by nectar nor pollen, enter the floral tube, are unable to escape, and hammer themselves into a state of exhaustion against the window-panes. If these windows are covered, the hammering ceases. Many insects die within the tube, but presumably a significant number escape after some 24 hours when the flower becomes flaccid.

Amongst the most spectacular of floral traps which are insect retainers is the Amazonian waterlily (*Victoria amazonica*). This flower is perhaps best considered as analogous to a snap-trap and is described on page 301.

No true carnivorous plant uses any floral part as a trap or part of a trap mechanism, although many of the features of the carnivorous syndrome, as we have seen, have analogues in flowers of non-carnivorous genera. Many flowers, too, kill insects indiscriminately (Eisner and Aneshansley, 1983; Vogel, 1981). The carnivorous plants have possibly exercised this 'restraint' in view of the apparent paradox of attracting and releasing insects for their respective pollination mechanisms (see Chs 15 and 16), but attracting and not releasing them for their nutrition (see Ch. 7). As we saw in Chapters 14, 15 and 16, there is the possibility that different

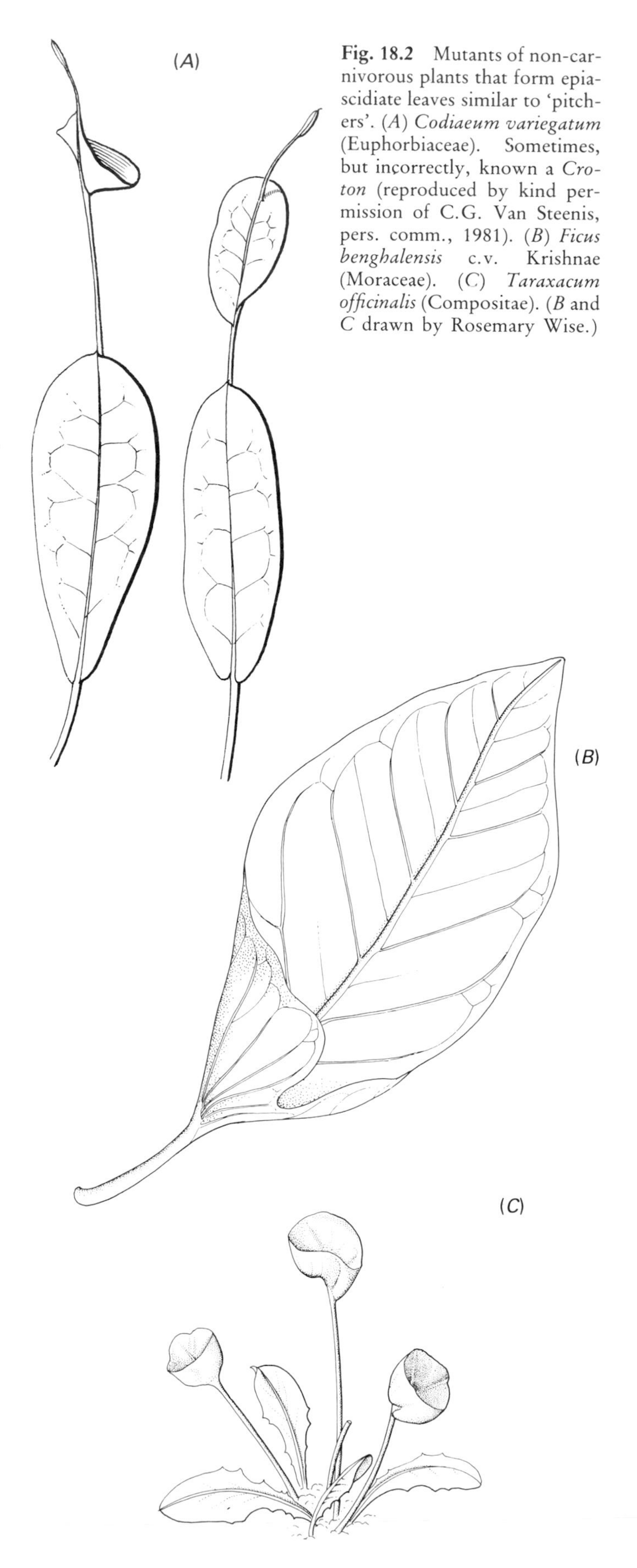

Fig. 18.2 Mutants of non-carnivorous plants that form epiascidiate leaves similar to 'pitchers'. (*A*) *Codiaeum variegatum* (Euphorbiaceae). Sometimes, but incorrectly, known a *Croton* (reproduced by kind permission of C.G. Van Steenis, pers. comm., 1981). (*B*) *Ficus benghalensis* c.v. Krishnae (Moraceae). (*C*) *Taraxacum officinalis* (Compositae). (*B* and *C* drawn by Rosemary Wise.)

populations of insects may possibly be employed for the different roles.

The Pitcher Forms of Leaves

The pitcher form of leaf, modified leaf or stem is also not uncommon. We find it in such primitive groups as the ferns (*Nephrolepis*, *Polypodium brunei* and *P. bifrons*; Bower, 1923), with many examples in the flowering plants. The tropical flowering plant genus *Dischidia* (Asclepiadaceae) (Huxley, 1986) has modified leaves of a pitcher form in which ants may temporarily live, insect frass and rain-water accumulate, and into which adventitious roots may penetrate (Juniper and Jeffree, 1983). There is no evidence that this 'pitcher' can absorb nutrients or that any positive attraction or trap mechanism operates. The ornamental Croton, *Codiaeum variegatum* (Euphorbiaceae), has an almost infinite flexibility of leaf form and pattern. According to van Steenis (pers. comm.) the tips of leaves can sometimes form almost perfect pitcher forms (Fig. 18.2*A*), but with no obvious purpose.

The tips of the leaves of the 'Krishnae' cultivar of *Ficus bengalensis* (Fig. 18.2*B*) can also form small pockets; revered in Hindu mythology, but of no other obvious value. So too can the mature leaves of a clone of dandelion (*Taraxacum officinale*) discovered by J. Adams in England (pers. comm., 1987; Fig. 18.2*C*). This latter mutant has now been discovered in considerable numbers suggesting that its (probably) apomictic 'seeds' are spreading the mutation widely. The normal imbricate leaf bases of *Dipsacus fullonum* (the European teazle), form another non-carnivorous pitcher and, as observed by Gerarde in 1597, catch considerable quantities of water (a single teazle may hold up to half a pint/250 ml). Both living and dead insects are often found in this liquid. The low surface tension of the fluid and hence its enhanced ability to drown insects is probably due to the presence of traces of soluble polysaccharides. The teazle may not derive any nutritional benefit from such insect deaths, although Christy (1923) considered it had all the features of a carnivore. Instead, the leaf-base pools may primarily serve to prevent climbing insects from reaching and damaging the flowers (cf. Kerner, page 293).

It is possible though that plants like *Dipsacus* absorb nutrients from this trapped water, presumably containing decomposing insects, in the same way that plants absorb nutrients from foliar feeding. *Sarracenia* and *Darlingtonia*, recognized as fully evolved carnivorous plants (see Fig. 1.1) contain within their pitchers dipteran larvae, protozoa and bacteria which break down insects. Similarly many 'tank' bromeliads obtain nutrients in just this manner. Could plants like these exhibit the most unsophisticated form of carnivory, a facultative primitive carnivory? We have already shown how a few species of Bromeliaceae appear to have become effectively fully carnivorous when growing in the right habitats.

Significant Features in the Ontogeny of a Pitcher

From the ontogenetic point of view (see Franck, 1975) the step forward in the evolution of pitchers was the adoption of a new sequence of events in the development of the leaf, based on a transverse zone of growth (Querzone) in the young leaf primordium (see page 56 and steps 1 and 2 in Fig. 17.4).

It is worth noting that all the carnivorous plant genera which form permanent digestive pools, i.e. *Brocchinia*, *Catopsis*, *Cephalotus*, *Darlingtonia*, *Genlisea*, *Heliamphora*, *Nepenthes*, *Sarracenia* and *Utricularia*, all form the epiascidiate type of a pitcher. (Franck, 1976). As we saw in Chapter 4, the term 'epiascidiate' was first used by de Candolle (1873) to describe a tubular organ whose inner surface represents the morphologically adaxial (upper) surface of a conventional leaf (Fig. 4.8). The morphological analogue of the epiascidiate leaf is the hypoascidiate appendage. Might the epiascidiate leaf form also be polyphyletic, or do Croizat's speculations (page 286) also apply? It is difficult to believe here in a common origin if only by a consideration of the Bromeliaceae. In one major sub-group, the Pitcairnioideae, most of the terrestrial examples do not apparently have effective absorbing trichomes and do not usually store water in tanks. The exception is *Brocchinia* which has both absorbing trichomes and a tank-forming habit (Smith and Downs, 1974 and Fig. 2.5).

In what is thought to be a more advanced sub-family, the Tillandsioideae, which are exclusively epiphytic and frequently form tanks, again just one exceptional genus, *Catopsis*, has recently been discovered to be carnivorous. It is difficult to sustain a common 'urceolate-tank' origin in these two anomalous genera of long-divergent sub-families.

No plant with hypoascidiate leaves is known to be an active carnivore. However, leaves of this type are just as common and are found in the genera *Dischidia* (Asclepiadaceae), *Celmisia* (Asteridaceae), *Cassiope* (Ericaceae) and *Tococa* and *Maieta* (Melastomaceae) (Troll, 1932).

The Combination of an Epiascidiate Leaf and a Glandular Surface

The evolution of the gland has already been considered (page 294 onwards). However, as already pointed out, if a gland-bearing leaf (e.g. with extrafloral nectaries) is transformed into a pitcher, it has the immediate advantage of having a closed water-pool and potentially a phytotelm (see Ch. 14) accessible to the leaf tissues *via* glands. It has already been shown that certain nectaries are not only capable of secreting nectar but are also capable of absorbing external matter such as amino acids (Ziegler and Lüttge, 1959). If this were true for primitive pitcher glands, the nectaries might have given the plant an immediate tool by which it could exploit occasional rainfall and debris. This view gains support from certain *Sarracenia* pitchers, which show a gradual change from nectaries to digestive glands along the pitcher wall from top to bottom, respectively. The development of special absorption systems is of an important adaptive value when the primitive pitcher collects rain-water and serves as a habitat for small animals. The absorptive epithelium in *Sarracenia* is one such system (Joel and Heide-Jørgensen, 1985) we now know; the development of glands is another. It is by now impossible to speculate which system developed first. However, the absence of a similar absorptive epithelium in *Heliamphora* (Sarraceniaceae), where glands are present (though quite few in number), might indicate that the glands were first to develop in this family.

The Non-Glandular Plant Surface and its Possible Evolution into Part of the Carnivorous Syndrome

The Evolution of the Trapping Zone in *Nepenthes*

When first clearly observed (Juniper and Burras, 1962) the trapping wax of *Nepenthes* was assumed to be a very special and perhaps unique case of a wax secretion adapted to insect retention. But these views are now changing (Juniper, 1986; Stork, 1986).

Within the pitcher of *Nepenthes*, and just below the nectar-secreting glands under the peristome lip, is the wax-secreting or conductive zone. The zone is highly variable in extent, sometimes extending far down the pitcher to the glandular zone; sometimes it is narrow or virtually absent (Kurata and Toyoshima, 1972). Its absence may govern the role of the pitcher as a phytotelm (see Ch. 14). Whatever its extent, its fine structure seems identical in all those species examined (Fig. 4.10). A full description of the trap's morphology and chemistry is given on page 113–116.

A Conventional View of Epicuticular Wax Development

Most epicuticular waxes in the angiosperms were, until recently, viewed as having mainly hydrophysiological roles (Pfeiffer *et al.*, 1957; Hall and Jones, 1961; Leyton and Juniper, 1963; Holloway, 1968; Holloway *et al.*, 1977; Martin and Juniper, 1970; Kolattukudy, 1976; Schönherr, 1976). Some waxes protect plants against high insolation (Barber and Jackson, 1957; Pearman, 1966); some may assist in dew condensation (Hull *et al.*, 1979), and a few protect against frost damage (Barber, 1955; Single and Marcellos, 1974; Thomas and Barber, 1974). A few protect against pathogens (Martin, 1964; Preece and Dickinson, 1971; Dickinson and Preece, 1976), although the evidence is very mixed (Schutt, 1971; Walla and Peterson, 1976; Hargreaves *et al.*, 1982). A few surfaces may give protection against insect predation (Anstey and Moore, 1954; Lupton, 1967; Stork, 1980a, b; Atkin and Hamilton, 1982; Woodhead, 1982; Juniper and Jeffree, 1983) but, again, the evidence is sometimes conflicting (Thompson, 1963; Way and Murdie, 1965; Southwood, 1986). Some may have a role as UV reflectors (Burkhardt, 1982).

The insect-related waxes, as in *Nepenthes* and *Iris pallida* (Knoll, 1914), were once considered as special evolutionary events of a late stage in angiosperm development. The more primitive plant groups, have all, or virtually all, of the chemistry and pathways of cutin and epicuticular waxes available to the angiosperms (Jeffree, 1986). If these ancient groups are studied it will be seen that extensive or thickened leaf-surface wax development is rare.

There are a few instances of hydrophysiologically related wax development on gymnosperm needles

(Leyton and Juniper, 1963) and waxes on surfaces in certain mosses appear to play a similar hydrophysiological role (Proctor, 1979). Epicuticular waxes are very rare in the ferns (see *Ceratopteris* and *Phlebodium* below). Yet all of these primitive groups penetrated, with only the chemistry of cutin to protect them (Jeffree, 1986), into virtually all the extreme habitats where the angiosperms are now also found.

Often the leaves and stems of these angiosperms have a generous epicuticular wax. Such features are held to account in part for the success of the flowering plants in exploiting extreme environments, and there is no doubt that wax mutants deficient in or totally lacking their epicuticular protection can be demonstrated to suffer adverse effects, in particular rapid water loss (Pfeiffer *et al.*, 1957; Hall and Jones, 1961; Hall, 1966; Grncarevic and Radler, 1967). This is the conventional view, but there is a danger in assuming that a selection pressure which now appears to have brought about a relevant environmental adaptation may have been that which initiated the evolutionary sequence. These wax forms may simply represent the fine tuning of the highly successful angiosperms in achieving the maximum penetration of habitats at minimal cost. The pressures are multiple, the response of the plant kingdom, shuffling its 'deck' of cards, almost infinite.

A New Appraisal of Epicuticular Wax Development

Let us assume for the sake of the present argument that the development of epicuticular wax upon a plant surface was an early evolutionary device to impede the movement or adhesion of phytophagous insects. The primitive groups of plants which still form part of our present flora do not present tempting targets to herbivorous insects. Many have other, different defence mechanisms. Some, such as the Equisetales, are rich in silica. Many are sclerophyllous or pachyphyllous in the sense used by Grubb (1986). The ferns and their allies are rich in such unpalatable compounds as phlobaphene, formed from tannins by oxidation and condensation (Sporne, 1975). Shikimic acid, an intermediate in phenolic synthesis, is common, as are the phytoecdysones in the Polypodiaceae which inhibit insect growth (Jizba *et al.*, 1967; Williams, 1970). In *Pteridium* there is even some as yet unidentified carcinogenic compound (Fletcher and Kirkwood, 1979). The significance of such a compound in defence terms is not clear.

Amongst the Pteridophyta (in the family Polypodiaceae) there are just two species with prominent wax develoment. One is the gold fern *Ceratopteris triangularis* (now more widely known as *Pityrogramma triangularis*) a native of the southwest United States. A yellowish, powdery wax, not so far identified, accumulates on the underside of the pinnae. This wax clearly has no role as a water-conservation device, but could be interpreted as a mechanism for restricting the adhesion of insects seeking to reach and attack the nutritionally rich sori.

Spore-eating was known to be widespread amongst the Carboniferous arthropods (Scott, *et al.* 1985) and the idea that defence mechanisms evolved in at least the Pteridophyta seems not unlikely. The second fern is *Phlebodium aureum*, a native of meso-America, and here the wax is concentrated on the petiole and rachis; again most readily interpreted as an insect-impedance device. Cultivars of this latter species are universally glaucous. It is common as a 'stove-plant' in cultivation and it is noted that the best glaucous condition is developed under cool, moist and shady conditions (Lowe, 1866), the opposite of what one would expect were the wax to have a hydrophysiological role. Is it possible then that waxes, *ab initio*, evolved as defences against insect predation and only later became hydrophysiological devices?

The Angiosperm's Floral 'Dilemma' and its Relation to Carnivory

The angiosperms, in concentrating reproductive features such as pollen, unfertilized eggs, tapetal cells, nectaries, petals and subsequently zygote, endosperm and fruit, group together a high proportion of protein, lipids, sugar and other soluble carbohydrates of the plant. This agglomeration tends to cluster much insect-attractive nutrient. The paradox of attracting the consistent, cross-pollinating hymenopteran and yet resisting the erratic phytophagous coleopteran immediately arises as we saw on page 273.

There can be no doubt that there were plenty of phytophagous insects of all types around in the latter half of the Cretaceous, ready to assault the emerging angiosperms at all points of their new life cycle (Gillet, 1962; Jarzembowski, 1984).

Anti-robbing Devices

The simplest device is to extend the floral mechanism on an elongated pedicel (see page 273). The second is to reinforce that pedicel in one of several ways: with rosette defences at the base (see page 273); by large-scale impedance devices on the main stem (Kerner, 1878; Sutherst and Wilson, 1986); by making the pedicel very thin, flexible and susceptible to wind movements (Curtis and Lesten, 1978; Southwood, 1986); or, most commonly, by coating the pedicel with crumbly, soft, non-adhesive epicuticular wax (Fig. 18.3). This last device is cheap and effective against a wide range of creeping, but not flying, phytophagous insects. That careful observer Rebecca Austin (1875D, Appendix 1) actually observed an ant to fall from a *Darlingtonia* flowering stem. The device can be combined with wind-whipping, as in many species of willow (*Salix*) which, although wind-pollinated may still be predated. The device might conceivably have a secondary advantage in providing an ultraviolet pointer. Many wax surfaces reflect strongly in the ultraviolet range (Clark and Lister, 1975; Burkhardt, 1982). Birds and some insects, particularly hymenoptera (Strong *et al.*, 1984) are known to have the capacity to detect UV. The erect pedicel might provide yet another guidance device of a linear kind for potential cross-pollinating insects.

Wax-coated pedicels and stems of the type suggested above are almost universal in the angiosperms. They can be seen at their most striking in the Iridaceae, as Knoll (1914) noted, in the Gramineae, Liliaceae, and Amaryllidaceae. In the latter family prominent examples are *Hippeastrum*, *Crinum*, *Haemanthus* and *Narcissus*. Far away, in phylogenetic terms, they can be seen in the genus *Dahlia* (Compositae) and they are not absent from tree genera such as *Salix* (Salicaceae). Interestingly, the genus *Sarracenia*, a well-attested carnivore, has a rich coating of wax on its pedicels although it uses a range of downward-pointing hairs in its pitcher (Chs 4 and 6) to retain insects. In the Droseraceae, *Dionaea* has a waxy-coated pedicel, but the closely related *Drosera* (see page 302 and Fig. 19.1) has a glandular flowering stem on which very small insects are occasionally seen to be trapped (an example of 'playing the cards' differently).

It is not difficult now to see how waxes on the one hand, or sticky glandular hairs (vide Kerner) on the other, might have evolved in pre-angiosperm eras to deflect or decoy insects. Now, with a similar chemistry and similar anatomy, they may be

Fig. 18.3 Epicuticular wax on the young flowering stems of *Salix alba argentea*. This soft, crumbly, wax may prevent phytophagous creeping insects from reaching the flowers by acting as a type of 'burglar-paint'. Liquid-nitrogen frozen SEM by C. Merriman, Oxford.

deployed as part of the carnivorous syndrome.

This 'card' concept of blocks of characters was developed, from earlier ideas, into the theory of 'reticulation' by Turrill (1950) to explain the relationship of features within the genus *Fritillaria*. What is proposed here is a tentative extension of the theory over the boundaries of families and orders. We have already seen from page 289 onwards how readily carnivorous plants may alter switch-mechanisms controlling whole blocks of genes relating to a single character to another, apparently irrelevant part of the plant. A similar transfer of, say, the whole wax-secreting trap system of *Nepenthes* from a conventional leaf surface to the inside of a pitcher seems to be no more improbable.

The Evolution of the Snap-Trap

Even amongst those humble members of the plant kingdom, the liverworts, there are some surprising morphological features which superficially bear a close resemblance to parts of a snap-trap carnivorous plant. In the families Lepidolaenaceae and Lejeuneaceae the 'leaf' underlobes may become modified into water-containing sacs or lobules. In some species they even have the capacity partially or completely to close like a *Utricularia* trap and, in some, the margins of these lobules bear soft mucilaginous teeth (Grolle, 1967). The whole structure resembles a miniature soft-spined *Dionaea* leaf, except that these leaf lobes are normally water-filled and may carry their own distinctive fauna of protozoa and nematodes: a microscopic phytotelm (Ch. 14). Decaying insect matter may add to the liverwort diet, but there is no evidence of designed carnivory.

Another curious snap-trap analogue, about which we find nothing more recent than Ludwig (1881) is the insect-catching grass *Molinia coerulea*. This common north-temperate grass can apparently catch small insects between its paleae, which act as the jaws of a spring trap analogous to *Dionaea*. During flowering, the paleae are forcibly separated by the swelling lodicules (a scale at the base of the ovule in grasses) and are held there for the duration on anthesis. The lodicules then shrink and allow the paleae to close. However, if during anthesis an insect, attracted by shining sappy turgid masses of the lodicules, attacks them by biting or puncturing, the collapse in turgor is sufficient to allow the paleae to close, which it does surprisingly rapidly (Ludwig, 1881). It is difficult to see what value to the *Molinia* accrues from trapping an indigestible insect, but it is not difficult to see such a mechanism leading on to carnivory, were the environmental situation favourable and following further mutation.

Victoria amazonica, the Amazonian Waterlily

The flowers of this species are amongst the most spectacular of floral traps which resemble the snap-traps of carnivorous plants. A group of flower buds clear of the water, begin to open simultaneously in the evening emitting a powerful pineapple scent. Leppard (1978) noticed that at 6.00 p.m. on the day of his observations the four prickly protective sepals folded back, closely followed by the petals. The flowers were completely open in two hours and then Dynastid beetles (cockchafers) entered the floral chamber and were trapped as the flowers slowly closed. The flowers were completely closed, trapping but not harming the beetles, by 10.00 a.m. of the following day. Around 4.00 p.m. of the second day the flowers opened again, releasing the pollen-coated beetles that flew off, presumably attracted again to first-day flowers. The original flowers closed on the third day and sank down below the surface of the river.

In *V. amazonica* stimulus-induced changes in the disposition of the floral parts bring about postural changes which temporarily trap the insects. But it is not yet clear what these stimuli are and whether the presence of the beetles or any of their activities is connected to these floral movements (Prance and Arias, 1975).

Caltha dionaeifolia

A species of *Caltha* from Tierra del Fuego (*C. dionaeifolia*) has leaf lobes which bear a striking resemblance to the paired lobes of *Dionaea* (Hooker, 1847; Joel, 1985). All the southern-hemisphere *Caltha* species bear this leaf form to a greater or lesser degree. The inner surface of each 'trap' is covered with numerous stomata which look to the naked eye like glands. Delpino (1874) suggested that this plant was analogous to *Dionaea*, and that its structure was adapted for the capture of insects (Hooker, 1875). However, neither digestive glands nor trigger hairs could be found on this leaf (Joel, 1985) nor is there any evidence of carnivory, even of an accidental nature. That there is no shortage of opportunity is demonstrated by the fact that this species of *Caltha* is growing in the same habitat as species of *Pinguicula* and *Drosera* which are obviously successful carnivores (D. Moore, pers. comm., 1982).

Snap trap analogues

Some features of the motor responses, to many the most fascinating feature of carnivorous plants, have analogues in other parts of the plant kingdom. The response to stimulus of the *Dionaea* or *Aldrovanda* leaf is not so very different from that of the leaf of *Mimosa pudica*. The movement of the trichomes of *Drosera* is not dissimilar to the movements of floral parts in the pollination mechanism of *Passiflora coerulea* or the pollination 'trigger' which operates in stamens of *Berberis*, the stigmas of *Mimulus*, the styles in *Sarothamnus* and the anthers in *Salvia*, all of which move relatively rapidly and in a predictable way on an insect's touch. There seems though to be no analogue elsewhere in the plant kingdom to the phenomenally rapid trap of *Utricularia*, apart perhaps from the column of the flower of *Stylidium* (Findlay and Findlay, 1975).

CHAPTER 19

The Evolution of Features of Carnivory as seen in the Current Flora

The Proposed Evolution of Specific Carnivorous Plants

In Chapters 17 and 18 we looked briefly at some of the separate components of the carnivorous syndrome and how they might have evolved. Is it possible, bearing in mind the vestigial fossil record, to integrate any of these isolated features into a synthesis bearing upon certain specific carnivorous plants?

The Evolutionary Path from *Drosophyllum* to *Dionaea*

The only possible sequence, following the lines of argument we have developed in Chapters 17 and 18 and with many reservations, may lie in the group of carnivores comprising *Drosophyllum*, *Drosera*, *Dionaea* and *Aldrovanda* (Fig. 19.1) (Williams, 1976). The pollen architecture (Kupriyanova, 1972) and the chromosome studies (e.g., Kondo, 1976; Kondo and Lavarack, 1984), as far as they go, suggest that these genera are closely related.

Let us speculate that some primitive form of carnivore which is analogous to a modern *Pinguicula*, but is not taxonomically related to it, arose by pressures similar to those envisaged by Kerner (1878) and Gillett (1962). *Drosophyllum* diverged early from this central stock, driven perhaps by gland pressures as set out on page 295, and is now pinned against the shore around the Iberian coastal fringe (see Ch. 3 and Fig. 3.6). Both its distribution and species number lend support to the idea that the central *Drosera* genus (see Fig. 19.1D) emerged before the major separation of the continents at the end of the Cretaceous (Speirs, 1981). This central *Drosera* group diversified over the whole globe in suitable habitats, particularly those both arid **and** disturbed (Figs 19.6 and 19.7), and is second only in species number to *Utricularia*. *Drosera* developed the nastic and tropic gland tentacle (Williams, 1976), which brought about economies in the quantities of both mucilage and enzymes required by bringing the gland head to the prey and not the prey to an already secreted pool. We can then speculate that evolution proceeded, via a species with types of leaf similar to those occasionally found in *Drosera erythrorhiza*; (Dixon and Pate, 1978, Dixon *et al.* 1980). In this species, lobe-pairs are occasionally found which totally lack the stalked glands and thereby the ability to absorb, but retain the sessile glands. (Dixon *et al.* (1980) do not state whether or not the slow leaf marginal rolling, which is a feature of stimulated *Drosera* leaves, is retained in these mutants. Nor do they indicate whether or not these sessile glands can be absorptive when stimulated.)

In plant evolutionary terms it would have been possible for the rapidly moving lobes of *Dionaea* to have evolved via the fusion of the tentacles of the margin of the leaf. However, a study of the leaf venation of *Dionaea* suggests that each lobe is much more likely to be an expanded and modified version of a *Drosera* leaf and that almost all the tentacles have been lost (Joel, 1982 and, unpub.; Figs 19.2 and 19.3).

We must then account for the retention of three (occasionally four or more) tentacles on each lobe,

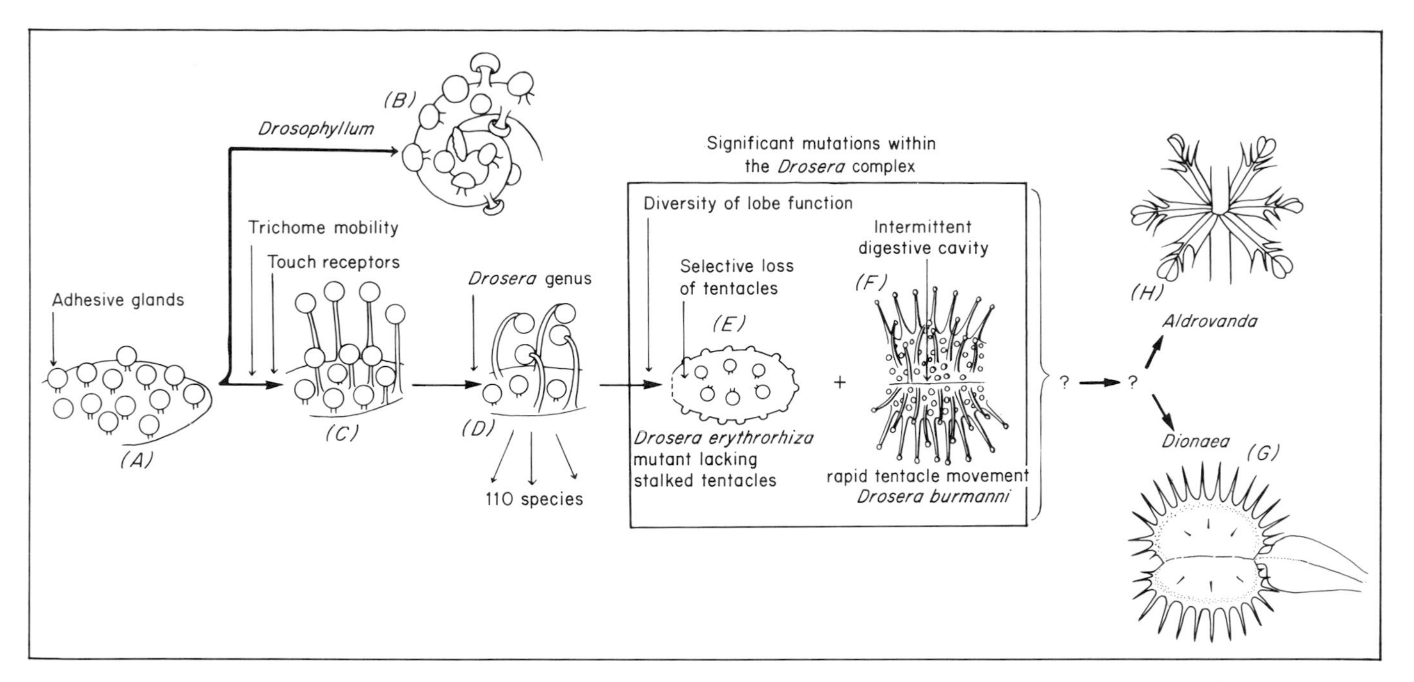

Fig. 19.1 A speculative scheme to illustrate the possible evolutionary path of a simple, single-glandular 'fly-paper' type carnivorous plant to a complex *Dionaea*/*Aldrovanda* type. (*A*) Simple, insect-defence-type glandular surface (see Fig. 18.1) evolving to basic carnivory using microorganism decay of prey. (*B*) One evolutionary path leads to *Drosophyllum*, with dimorphic glands, enzyme secretion, but no trichome movement. (*C*) Evolution of trichome mobility and touch-receptors leads to the central genus *Drosera* (*D*). Wide species diversity and diffuse habitat penetration. (*E*, *F*) Mutant forms in the *Drosera* complex lead to *D. erythrorhiza* types with reduced numbers of tentacles. Other diverse forms, e.g. *D. burmanni* (*F*) develop virtually non-glandular and very rapidly moving marginal tentacles leading to an intermittent digestive cavity. A combination of further glandular reduction and modification (see Figs 19.2–19.4) plus enhancement of the intermittent digestive cavity may have lead to a pre-*Dionaea* type of trap. A tolerance of flooding (see Ch. 3, page 30) leads to the divergence (*H*) of the ubiquitous aquatic *Aldrovanda* (see Figs 6.10, 6.11 and 6.23; and the localized *Dionaea* (*G*).

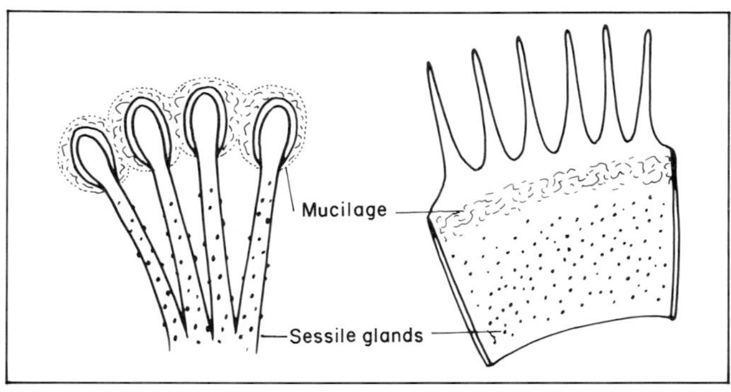

Fig. 19.2A The location of sessile and mucilage glands on *Drosera* (left) and *Dionaea* (right) leaves showing how fusion of the gland stalks of *Drosera* could bring about the pattern of lobe-surface sessile glands as seen on *Dionaea*; cf. Fig. 4.3.

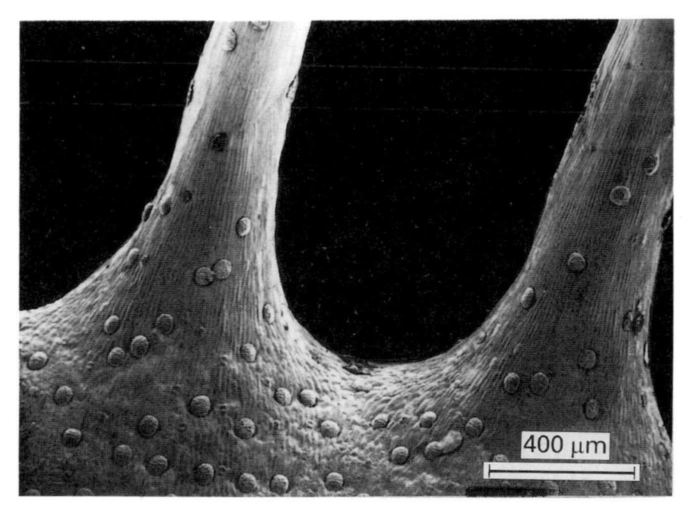

Fig. 19.2B The alluring glands on the margins of the trap lobes of *Dionaea*, continuing up the bases of the marginal teeth. (SEM by D. Kerr; from unpublished anatomical study by D.M.J.).

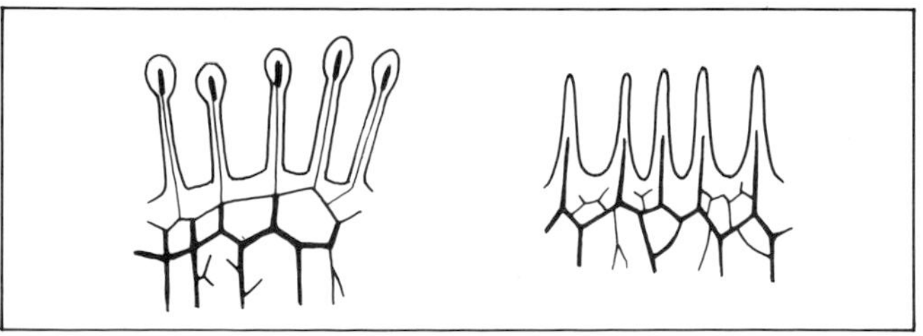

Fig. 19.3 The pattern of leaf venation in *Drosera* (left) and *Dionaea* showing the close similarity of the marginal venation.

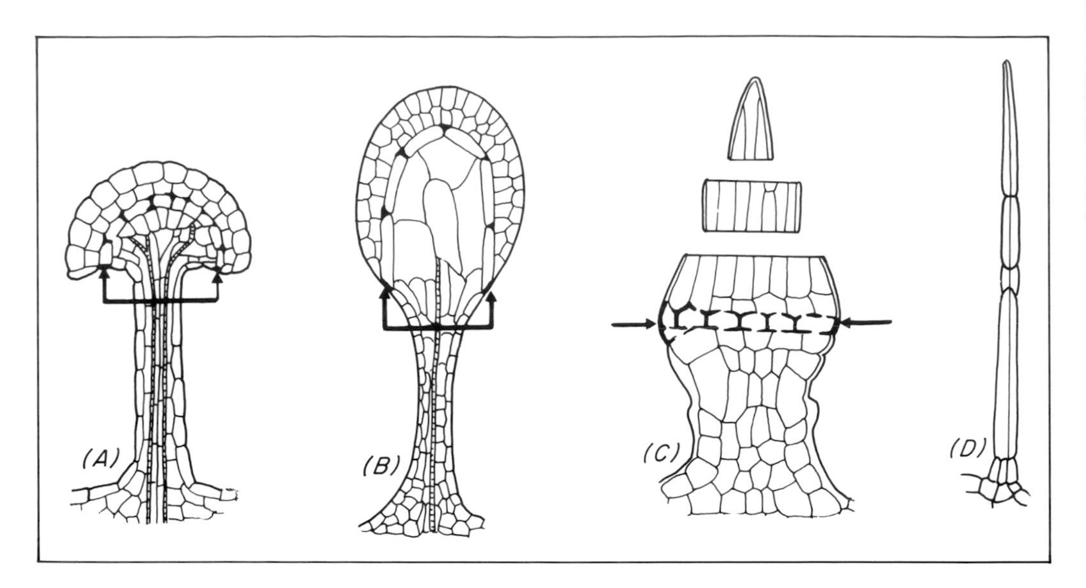

Fig. 19.4 A possible evolutionary sequence of the *Dionaea* trigger hair (*C*) via a *Drosophyllum* (*A*) and *Drosera* (*B*) stalked gland. The endodermoid layer, indicated by the broad arrows, is present in *A*, *B* and *C*, but missing in *Aldrovanda* (*D*). (Redrawn in part from Williams, 1976.)

which have become transformed into sensory hairs. The digestive glands of *Dionaea* thus appear, in evolutionary terms, to be derived from sessile glands equivalent to those now found on *Drosophyllum* and *Drosera*, a supposition supported by the close structural similarity of these organs.

As Williams (1976) has pointed out, reviewing earlier anatomical work at the same time, there is a striking resemblance between the stalked glands of *Drosophyllum*, the tentacles of *Drosera*, the trigger hairs of *Dionaea* and the trigger of *Aldrovanda* (Fig. 19.3). The first of these three genera still possess a recognizable endodermoid layer and the *Aldrovanda* bristles appear to be derived from endodermoid cells (Figs 6.5, 6.11, 6.17 and 19.4). Williams speculates that the endodermoid characteristics in *Dionaea* and *Aldrovanda* are non-functional relics, indicative of a secretory ancestry. In *Drosera*, the cells homologous to the sensory cells of the trigger hair of *Dionaea* would be the epidermal cells of the upper section of the tentacle stalk. Williams and Spanswick (1976) have shown, in confirmation, that the stalk epidermal cells in *Drosera* are excitable.

If *D. erythrorhiza* gives us some clue as to the possible path of evolution by selective loss of tentacles, we must assume that *some* of the marginal tentacles were retained, although losing all their secretory or receptor functions, and now form the marginal teeth around the edge of the *Dionaea* lobe (Fig. 19.4). We have a very suitable intermediate for this stage, the rare and very small *D. burmanni* from northeastern Australia (Fig. 19.1*E*). Not only do *Dionaea*-like tentacles develop around the margin of this species, it is also the fastest moving of all droseras: the peripheral glands moving visibly to the human eye (Nolan, 1978b). It is also tolerant of flooding and often goes on growing under water, provided the water is moving (Ashley, 1975c; cf. *Aldrovanda*).

However, the most important difference between *Drosera* and *Dionaea*/*Aldrovanda*, as Williams (1976) emphasizes, is the inability in *Drosera* of the action potential to pass from the stalked structure that has been stimulated. In *Dionaea*/*Aldrovanda* the entire leaf-blade can respond to stimulus at a single location (see page 93). A low-resistance circuit must therefore have evolved, the plasma-membrane perhaps, connecting all the potential motor cells of each lobe in a common circuit with each trigger hair. Only in this way could the proton-extrusion system leading to a synchronized release of stress in the lobe walls, so elegantly demonstrated by Williams and Bennett (1982), be achieved (see Ch. 6). No intermediate offers a separate clue as to the possible origin of *Aldrovanda*, but *Dionaea* often survives part of the year in the wild (see Ch. 3) totally submerged (Roberts and Oosting, 1958) and has even, in this state, been observed to catch aquatic insects (Schnell, 1976; cf. *D. burmanni*).

The Recent Evolutionary Path of the Sarraceniales

No other group of the carnivorous plants, apart from the proposed *Drosophyllum*–*Dionaea* progression, lends itself to detailed speculation as to an evolutionary path. As we saw, however, on page 296, the spontaneous evolution of a pitcher via an epiascidiate mutation is both widespread in the plant kingdom and frequent. There are sufficient 'borderline' carnivores such as *Dipsacus* and many bromeliads (Fig. 2.5) in this category to give us confidence both that this was the origin of the Sarraceniales and to make it likely, given the speed of modification of characters (page 288), and the development of vegetative means of spread (pages 41 and 297), that their primitive forerunners will never be found. The post-Cretaceous migratory path of the Sarraceniales may, however, be instructive.

The Possible Migratory Path of a *Sarracenia*-like Ancestor

The genus *Sarracenia* may have split off from a common stock with *Heliamphora* and *Darlingtonia* (Fig. 19.5; and see page 38). Because of its retention of certain primitive morphological and floral features it has been suggested that *Heliamphora* might represent something close to the ancestral form of this group. Maguire (1978) tabulates and discusses these features, but rejects any suggestion that this area of the Guyana Highlands (Fig. 19.5) was the point from which the migration of the three genera took place. Mellichamp (1983) believes these early pitcher-plants evolved in what is now the southeastern United States. These areas were, during the early Paleocene and Eocene (about 60–40 million years ago) moister and warmer, and such conditions may have prevailed across the entire southern United States. This favourable climate may have enabled *Darlingtonia* or its ancestors to migrate across to the west coast, before the rise of the Sierras and the Rockies split the continent. *Heliamphora* may have migrated south and is now pinned on the isolated sandstone 'tepui' of South America. *Darlingtonia* penetrated up the western sea-board of what is now the United States, and in its turn became

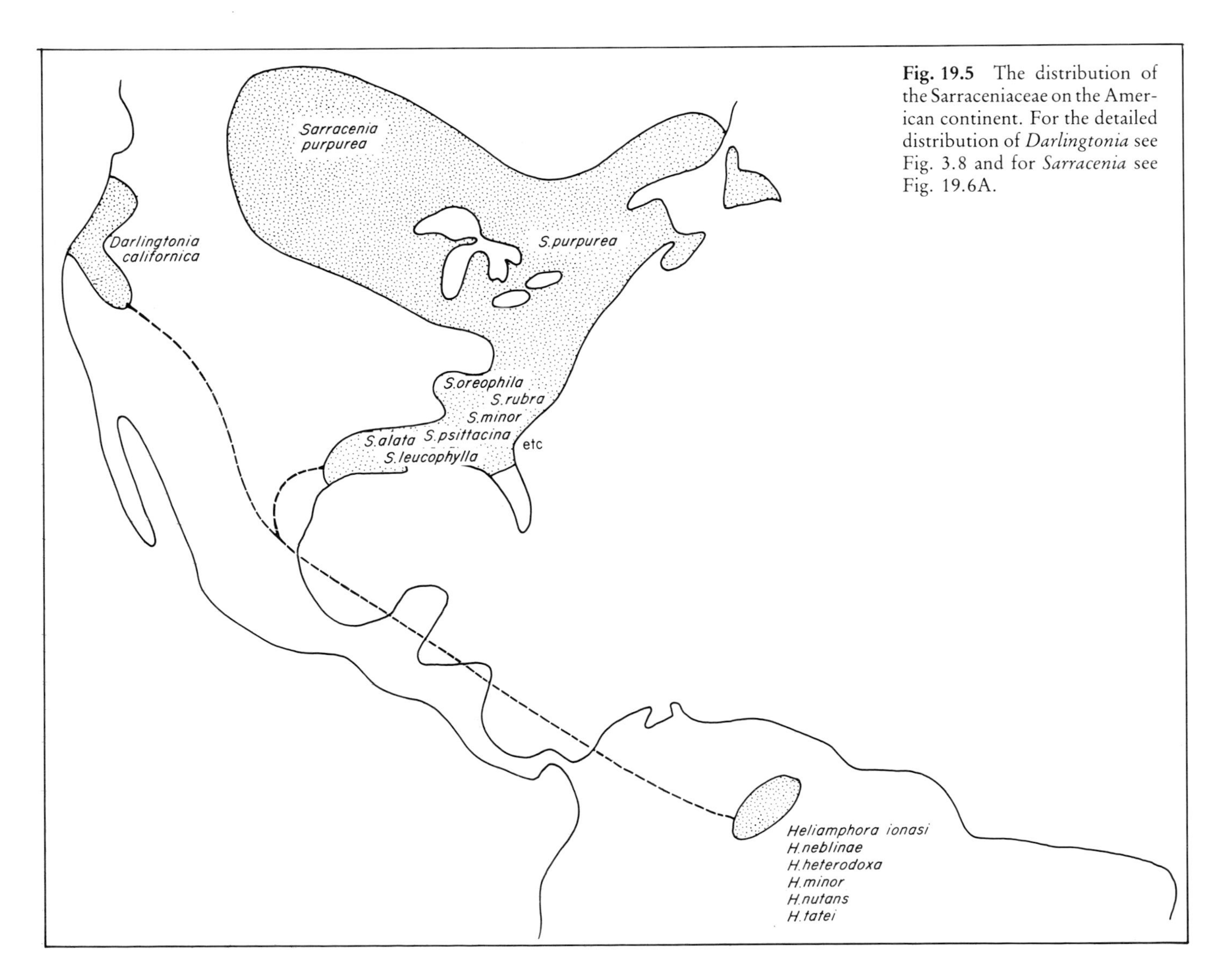

Fig. 19.5 The distribution of the Sarraceniaceae on the American continent. For the detailed distribution of *Darlingtonia* see Fig. 3.8 and for *Sarracenia* see Fig. 19.6A.

isolated and remained monotypic, principally upon the serpentine outcrops of the Tertiary formations of northern California and southern Oregon. The ecology of *Darlingtonia*'s distribution has been discussed in detail on page 38. What this facile interpretation of Sarraceniales wanderings does not explain is how *Heliamphora*, although apparently possessed of winged seeds capable of long-distance wind movement, seems to lack a dispersal ability (Maguire, 1978).

Sarracenia, or more likely its primitive forebear (Schnell and Krider, 1976) which is almost certainly extinct, colonized the eastern sea-board of the North American continent. Some of these migrations must have been very recent. *Sarracenia purpurea* can only have penetrated into the far north after the retreat of the continental glaciers some 10 000 years ago. Speciation, as well as migration into new habitats, is probably still going on (Figs 19.5 and 19.6). We can assume this to be the case from the taxonomic interest in the group, the debates over specific or sub-specific rank and the readiness with which many species form viable hybrids (see Ch. 3). Moreover, as Romeo *et al.* (1977) in their study of the amino acids and flavonoids of *Sarracenia* species point out, recent change seems likely. One cluster of *Sarracenia* species, the so-called '*rubra*-complex' virtually lacks the 3–0 glucosides and 3–0–galalactosides of quercetin and isorhamnetin. (Table 11.3). Such an absence, the authors suggest, in an otherwise general distribution of these compounds in the genus, implies the recent evolution of *S. rubra* and the subspecies *alabamensis* and *jonesii*.

Fig. 19.6 Two areas of very high numbers of species of carnivorous plants are located in the southeastern United States, on the Atlantic and Gulf coastal plains. The peninsula of Florida has only recently, in geological terms, emerged from the ocean, and this may explain its relative poverty.

(*A*) The continuous line enclosing the dotted area, which follows approximately the 'fall-line' of the mountains to the coastal plains, demarcates a region in which there are at least 1–3 species per county. Species number increases south and east and the diagonally hatched areas may have over 13 species per county. Most of the species of *Sarracenia* occur within the diagonally hatched areas – only *S. purpurea* extends significantly outside this zone. The approximate arcs or areas of each *Sarracenia* species are shown and, for illustration, the precise distributions of *S. leucophylla* and *S. rubra* are shown, in larger scale, below.

Of other N. American plants, *Drosera intermedia* has a distribution not unlike that of *S. purpurea*, *D. filiformis* and is almost entirely within the 'fall-line' boundary, and *D. capillaris*, and *D. brevifolia* completely within it. All of the southern butterworts – *P. pumila*, *P. lutea*, *P. caerulea*, *P. planifolia*, *P. primuliflora* and *P. ionantha* lie south and east of the fall-line, although the first three species extend down into southern Florida. Approximately thirteen of the eighteen broadly recognised N. American species of *Utricularia* are either restricted to, or have a significant portion of their ranges lying within the above zone. The detailed distribution of *Dionaea muscipula*, which is shown in Fig. 3.5 shows that it lies in the centre of the Atlantic coastal plain high-density area. (Data from Coker, 1928; Roberts and Oosting, 1958; Schnell, 1976; Gibson, 1983; Kral, 1983.)

(*B*) & (*C*) Detailed distribution of *Sarracenia leucophylla* and *S. rubra*. (Data from Schnell, 1976; cf. Fig. 19.6A.)

The possible enhancement of speciation in the genus *Sarracenia* by disturbance

After human or natural disturbance (see page 38) it has been noticed that natural hybrids of *Sarracenia* species are most prevalent. There is anecdotal evidence that the storm surges from the cyclones, which frequently batter the east coast of the United States, occasionally cause hybrid swarms. These swarms rarely survive long. Could it be that the more disturbed nature of North America's Atlantic coast has led to a greater speciation in *Sarracenia* than amongst its near relative *Darlingtonia* on the calmer Pacific rim?

It is interesting that, eastwards of the fall-line of the southeast corner of the United States, an area frequently devastated by storm surges, is one of the richest zones for a wide spectrum of carnivorous plant species. This huge coastal plain of ancient, arid soils is well-covered with large bogs, acid swamps and sandy heaths and, as Figure 19.6 and insets show, in the darker stippled areas there may be over thirteen species per county. The only other comparably rich area of the world, southwest Australia (Fig. 19.7), is also an area of ancient and arid soils within a cyclone belt.

The exploitation of the habitat by *Sarracenia* species

Except for there being an increase in general hardiness from south to north, the different species of *Sarracenia* do not seem to have exploited different habitats. In fact, several species of the genus may be found side by side in the same bog (see Ch. 3). However, it is noticeable that similar species, e.g. *S. flava* and *S. oreophila*, are rarely found in close association (McDaniel, 1971; Gibson, 1983a). This phenomenon is discussed in Chapter 7 in relation to prey-partitioning, a factor which we may find to be as potent an evolutionary pressure as the environment (Ch. 3) or commensal insects (Ch. 14).

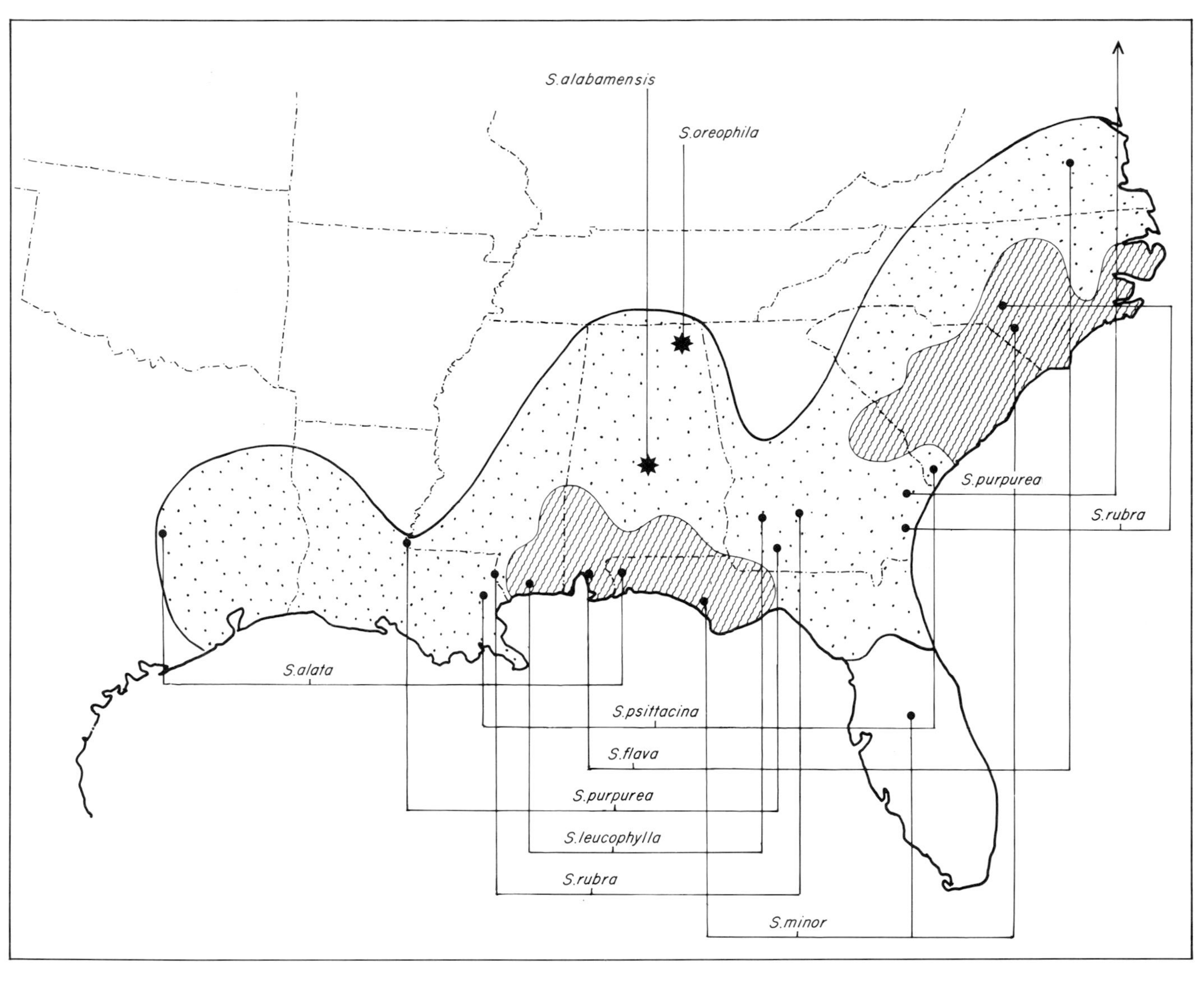
S.alabamensis
S.oreophila
S.purpurea
S.rubra
S.alata
S.psittacina
S.flava
S.purpurea
S.leucophylla
S.rubra
S.minor

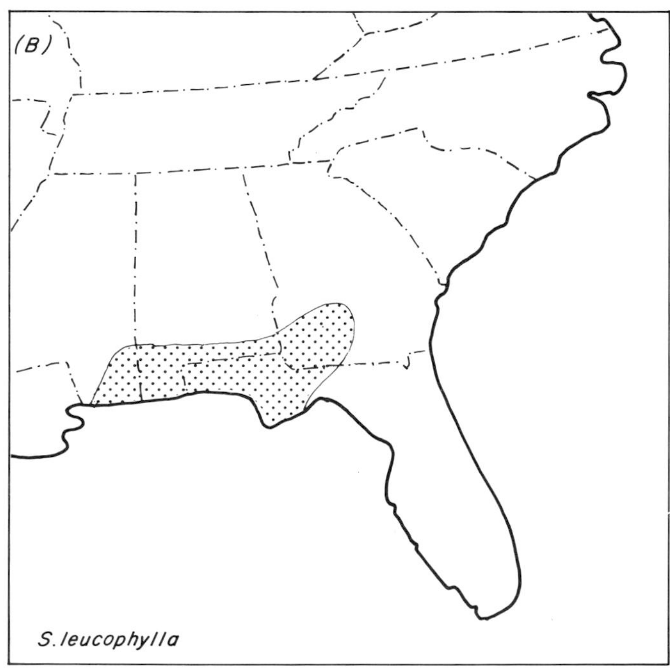
(B)
S. leucophylla

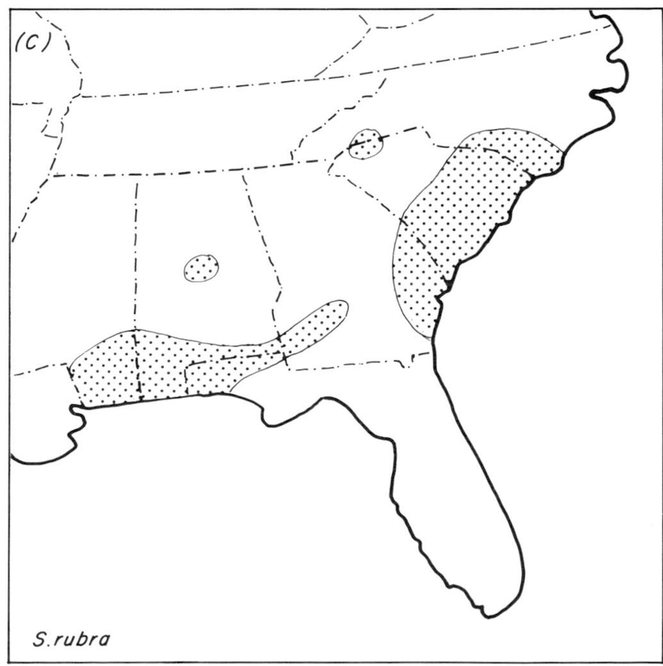
(C)
S. rubra

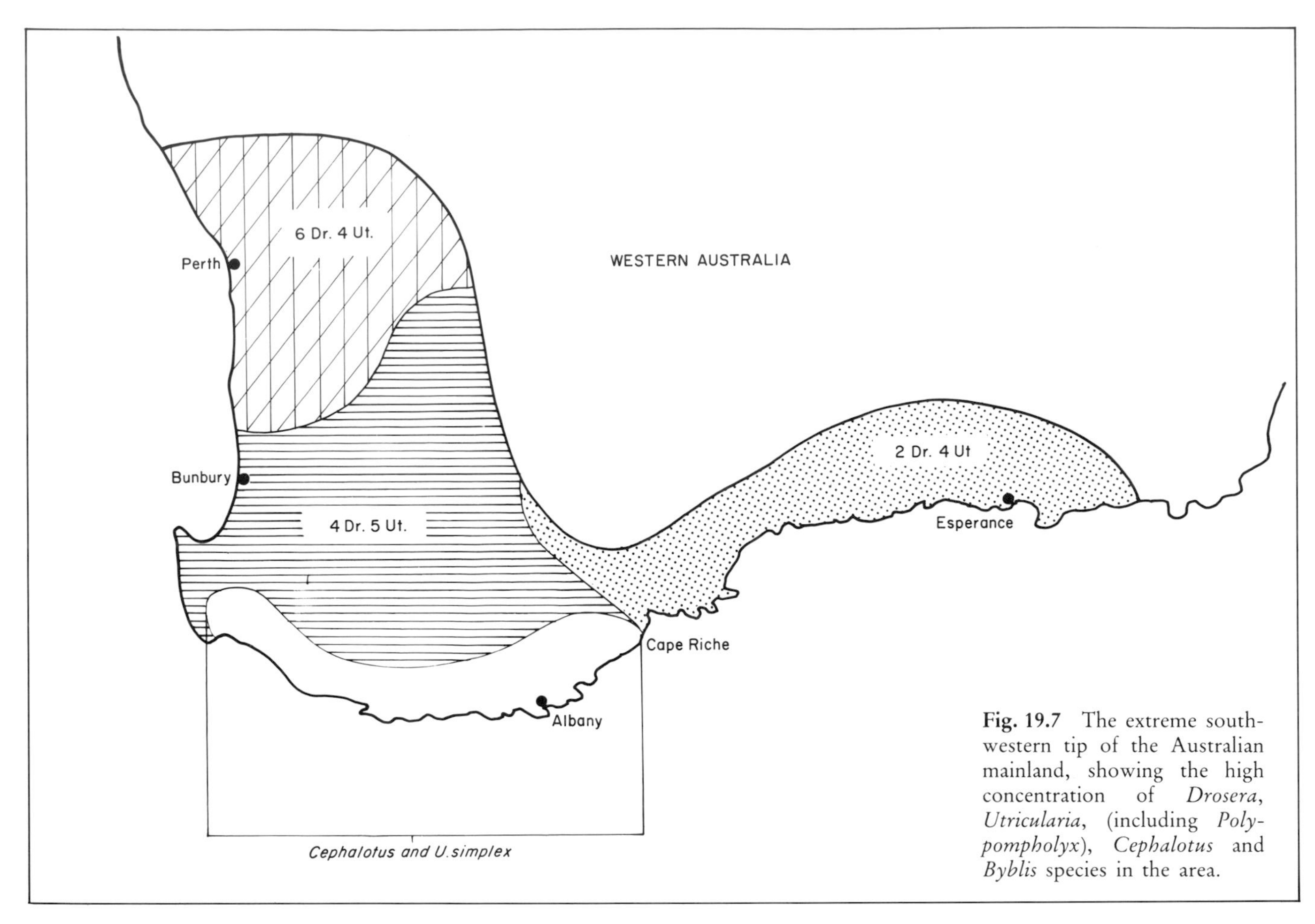

Fig. 19.7 The extreme south-western tip of the Australian mainland, showing the high concentration of *Drosera*, *Utricularia*, (including *Polypompholyx*), *Cephalotus* and *Byblis* species in the area.

Current Evolution Amongst Existing Genera

Evolution has not ceased and that of carnivorous plants, like most others, still continues. Thus a form of *Pinguicula vulgaris*, considered by some (Hadač, 1977; Studnicka, 1981) a different species (*P. bohemica*) but not so by others (Schlauer, 1986) probably arose as the result of the isolation of a small number of *P. vulgaris* in central Czechoslovakia. This strikingly tall representative of the genus, reaching up to 30 cm, has become adapted to the relatively warm lowland climate and a soil rich in mineral salts and breeds true, rarely hybridizing with *P. vulgaris*. Similarly, within the recent and very successful neotropical family of the Bromeliaceae, a few genera have evolved a carnivorous habit, probably in response to nutrient stress (see McWilliams, 1974; Smith and Downs, 1974; Givnish *et al.*, 1984).

There is evidence too that *Drosera anglica* of northern and central Europe may be a fairly recent amphidiploid hybrid between *D. rotundifolia* and *D. linearis* (Wood, 1955).

The Evolution out of Carnivory in Current Genera

In the evolving and often interlocked strategies of flowering plants we can see that there are several good candidates, given geological time and uninterrupted evolution, perhaps in the Bromeliaceae as we suggested above, for completely new carnivorous plants. At the same time there are members of well-authenticated carnivorous genera that would seem to be evolving away from the habit.

Nepenthes pervillei is restricted to two islands in the Seychelles. It roots into great cracks in the pre-Cambrian granite boulders and scrambles over the surrounding vegetation, mainly *Dianella ensifolia* and the fern *Gleichenia dichotoma*. Personal observation (B.E.J.) indicates that it catches few flies; flies of a suitable type are rare on windswept small islands. However, its nutritional needs may in part be met by conventional absorption through its roots, which penetrate deep into the granite cracks. These cracks are lined with a brown slime rich in the nitrogen-fixing blue-green alga *Llyngbia*. Since *N. pervillei* does not grow under a forest canopy, other nutritional needs may be met by the rain of guano from myriads of sea birds.

On Roraima, *Heliamphora* species also grow in what is, at least intermittently, a moist and warm atmosphere. They are not, apparently, very effective fly-catchers, either in the field or under greenhouse conditions. Could it be that the successive geological uplifts of their isolated tepuis have raised them beyond the range of rich fly populations? K. Burras (pers. comm., 1979) recorded few, if any, flies present in an *H. nutans* area. Again, living almost permanently in swampy, still muck, could they be using prokaryotic nitrogen fixation in the same way and short-circuiting the carnivorous syndrome?

The Evolutionary Future for Carnivorous Plants

Where should we look both for new carnivorous plants to add to the list and for the possibilities of future adaptation?

It seems likely that other tank-forming bromeliads, particularly the terrestrial forms like *Brocchinia reducta* (Fig. 2.5) will be found to possess all the syndrome. Perhaps, too, the desert fringe carnivorous plants like *Drosophyllum lusitanicum* on the one hand (Ch. 3) and *Ibicella lutea* (Fig. 2.3) and its relatives on the other, may not have received sufficient attention? As Cornelius Muller wrote (pers. comm., 1982): '. . . I have frequently noted numerous small dead insects, especially gnats and other small Diptera, stuck to the plants (*Proboscidea*). This phenomenon is not confined to the Martyniaceae. I have seen it on certain Nyctaginaceae and Compositae as well . . . If the capture, digestion, and absorption of nitrogenous prey were truly significant, the process could prove to be the basis of the common arenophilous habit of glandular non-leguminous herbs.'

Appendix 1

Letters of Rebecca Merritt Austin to Dr W.M. Canby of Wilmington, Delaware

From the transcripts by Frank Morton Jones of Wilmington, Delaware of the original letters made some time after 1920.

1875

A. February 9th. 1 page. *Darlingtonia* — perennial nature — vegetative reproduction — age of colonies.

B. March 2nd. 6 pages. *Darlingtonia* and *Drosera* — Intoxicating properties of *Darlingtonia* fluid — 'worms' on the captured flies — age of roots determined by numbers of whorls of fibrous roots — frequency of flowering — construction of the pitcher-tube, hood and 'windows' — persistence of the liquid in the pitcher even in the absence of rain — types of prey — moths, butterflies, beetles, flies and mosquitoes, but rarely bees — the direction of the twist of the pitcher — does the trap secrete a saccharine fluid to attract its prey? — the level of her scientific education and the problems connected with bringing up three children — a study of the 'snow-plant' (*Sarcodes sanguinea*) "... – send me a glass suitable for examining flowers."

C. May 3rd. 2 pages. *Darlingtonia* — method of aging individual plants — the prey of the pitcher plants — spiders' webs in the tubes — parasites on the old leaves.

D. May 15th. 9 pages. Determination of age of *Darlingtonia* — sends old plants to Canby — pollination of the 'snow-plant' — receives paper, magnifying glass and notes on *Sarracenia variolaris* — orientation of trap openings in relation to their position of emergence — all colonies observed grow on south-eastern slope — of the new leaves, the first two point nearly north and south — recently captured insects rapidly covered with white 'worms' — hundreds of such 'worms' in some tubes — absence of flying pollinators of *Darlingtonia* — two-winged insects and some spiders found inside the flowers — ant observed to fall off flowering stem — first pitcher always larger than subsequent leaves — many pitchers occupied by spiders and their webs — more flowers with spiders and webs within them — in fifty flowers there were thirty-five spiders — takes her sewing to *Darlingtonia* colony, but fails to observe any flying insect pollinating the flowers — observed soft bluish 'worm' about half an inch long, living in the pitchers.

E. June ? 2 pages. Sent vials of pitcher liquid to Canby — two types of 'worms' observed in the tubes, the larger located on the walls of the tubes and in the groove inside the hood, and the smaller in the liquid — larvae found in the tubes with closed hoods, the walls marked with spots as if pierced by some insect — the 'twist' appears designed to throw the leaves farther from the centre or axis of the plant — "the flower is fertilized by spiders, not cross-fertilized by insects." — "my notes on *Drosera rotundifolia* were published in our county paper."

F. July 7th. 3 pages. Worked with Mr. Lemmon and jointly studied *Darlingtonia* plants over three feet tall and the pitchers containing half a pint of liquid — larvae in the tubes only found in sunny locations — more larvae found in unopened tubes — increase of liquid in tubes as soon as insects are captured — nearly all of the flying insects that enter the pitcher must do so by night — feeding experiments on *Drosera* — leaves of fed plants fold over as readily on lettuce, radish, rice and potato as they do on beef, bread, butter, egg, cake and sugar — in thirty-six to forty-eight hours nothing remains of the lettuce and radish fibre—leaf closes on salt, but withers, as do adjacent leaves on the same side of the plant — meat, rice, cheese and bread in *Darlingtonia* show no sign of digestion, only decomposition — salt in *Darlingtonia* pitcher causes hood and fish-tail to wither — later upper, dotted part of tube withers also — adjacent tubes also dry and fall over — more larvae in unopened pitchers — the larvae use the captured insects as food.

G. July 17th–20th. 3 pages, incomplete. Salt passes to and damages neighbouring traps, but colouring matter does not appear to do so — more larvae in plants with closed hoods — notices attractive secretion on hood and fish-tail for the first time; flies attracted in great numbers — sweetness detectable on rim and inside hood — advice on growing live plants "give them plenty of *cold* water." — pays for a photograph of the plant to be presented to Canby and Gray — cannot believe the 'snow plant' (*Sarcodes*) to be a parasite — of three leaves of *Darlingtonia* fed with salt, two had fallen over and lay on the ground almost full of water, the third, away from the stream, was covered with salt crystals and two-thirds full of water — no change in leaves fed with various pigments.

H. August 1st. 2 pages. Noticed increase in 'honey' on the fish-tails — honey also on the outside of the hoods — honey-dew on some plants, where those plants grow in the sun, extends down the wing to the ground. Numerous black ants on the stems, fishtails and in the hoods — various spiders, wasps, hornets, ants, grass-hoppers, worms, butterflies, moths, flies, beetles and gnats caught in the traps — of a beetle, all but the hard shell consumed by the larvae living in the tubes — the fluid in the tubes changes the blue litmus paper slightly red.

I. August 8th. 1 page, incomplete — plants fed on bacon, raw and cooked mutton, and gingerbread. In all the tubes liquid level increased, but that fed with raw mutton least of all. After seven days gingerbread all gone and liquid level risen to within an inch of the orifice of the hood — honey greatly increased and hanging in drops on the end of the fish-tails. A finger inserted into the hood is coated with honey — two little 'cups' on either side of the insertion of the fishtails which catch the honey — small snails frequent in the tubes along with grasshoppers, spiders, ants, beetles.

J. August 23rd. 2 pages. Sends herbarium specimens of local flora other than *Darlingtonia* to Canby. Most specimens unnamed — dried hoods of *Darlingtonia* with abundant honey on them — stresses necessity of many seasons continuous work and much duplication to solve the various problems observed in *Darlingtonia* — sends thanks for promise of Darwin's new work.

K. September 30th. 2 pages. Honey on fishtail and in hood evenly spread and apparently not carried there by insects — on Sept. 24th noticed twelve honey-bees in the traps along with four green and yellow spotted 'worms' about an inch and a half long, snails, hairy flies, ants, various moths, red-coloured grasshoppers, 'katydids', yellow jackets, wasps, hornets, humble bees, spiders, beetles, white larvae in great numbers, and various smaller winged insects — water in tubes from one to six inches varying with quantity of insects captured — honey does not appear to possess any intoxicating property — liquid in the tube kills flies in three to five minutes.

L. November 19th. 1 page. Receives Darwin's 'Insectivorous Plants.' — found the lure on *Darlingtonia* more abundant on Oct. 26th than at any other time. Collected half a teaspoon to send you — "Where could I obtain Hooker's observations on *Nepenthes*?"

1876

A. January 17th. 1 page. Promises to send enough pure pitcher-liquid to experiment with, and half a dozen strong fresh plants — visits 'patch' once a week. White larvae abundant and active, notwithstanding the cold — liquid in tubes odourless and colourless and does not change the colour of litmus paper.

B. April 9th. 1 page. — commenced work this spring in real earnest, but with a sad heart as business troubles were not yet settled — hoped to sell specimens to pay for postage, buy books etc. — larvae still alive in pitchers on March 17th, but no change in litmus paper on March 25th, and no insects captured — April 10th insect capture begins — April 26th liquid in traps increases, more insects captured and action acid on litmus paper — flower stalks extending — no sign of lure on hoods or fishtails — May 9th liquid greatly increased and most hoods secreting the lure — flowers in full bloom.

C. May 20th. 1 page. Plans for two long excursions of forty and sixty miles — the lure now present on most of the perfect leaves, inner and rough portions of the hoods, fishtails, outside of hoods and inside, also two-thirds of the distance from the orifice down the wing to the ground.

Leaves Butterfly Valley — Moves to Fern Dell

D. June 11th. 2 pages. Concludes to camp in the wildwoods, with *Darlingtonia* about a mile from house, whilst husband goes elsewhere to seek a home — lure now abundant all over the last season's pitchers except those injured by the winter's snow and heavy freezing — more black ants in the tubes than any other insect — all kinds of insects in the tubes preserved in alcohol—pressed over one hundred *Darlingtonia* flowers last week; nearly all had webs and a great many with little spiders in them.

E. June 19th. 2 pages. Sees J.G. Lemmon's article in 'Scientific Press', very critical both of this paper and his earlier work in the 'Bulletin' — many factual errors; thus Lemmon writes that the plant produces four leaf stalks, whereas Austin records ten to eighteen new leaves each year, always in pairs, the first two the largest and always emerging between the two last of the year before — the first four large leaves stand far above all the others, marking the four cardinal points of the compass — more larvae and liquid in traps with closed hoods.

Moves from Fern Dell to American Valley

F. July 29th. 1 page. Chills every other day and too miserable for any exertion — sends a number of *Darlingtonia* at different stages of growth.

G. August 20th. 3 pages. Larvae of different sorts found in numerous leaves with closed hoods, along with up to twenty-five or thirty drops of liquid — largest *Darlingtonia* traps ever seen on Black Hawk creek — larvae are thought to be in tubes from four to six days *before* the hoods expand — was at *Darlingtonia* patch in severe thunderstorm, most of the water falls on the hoods and is drained off by the fishtails, little runs down the backs of the tubes and none enters the tubes — lure present on many surfaces of the plants not smeared but in the form of minute points — Prof. Edwards of the San Francisco Academy of Science says in an article, along with Dr. Gray in California Botany, that there are more Diptera caught than any other insects. This is not correct. Ants, spiders, beetles, butterflies and moths in the majority — found four long slim larvae, at first thought to be earth worms (*Lumbricus*), but found later on examination to be different.

H. December 18th. 1 page. Sends orchid specimens to Gray and Canby — J.D. Hooker also writes for a specimen of this orchid — recovering from the effects of chills and correspondence neglected.

1877

A. March 6th. 3 pages. Counted over one hundred white larvae in a single tube. The principal victims were moths, lepidoptera, grasshoppers and 'katydids', orthoptera, an occasional large hairy fly, a few mosquitoes, a dozen 'yellowjackets' in all and a few honey bees — picked pitcher on the table very attractive to the 'yellowjackets' — yellowjackets feed quietly on the nectar until satisfied, then attempt to escape through the skylights. When exhausted they fall back into the liquid and are soon powerless — the liquid seems to penetrate them as they touch it and destroys their vitality as if they were immersed in oil — if rescued, they recover — when feeding on the nectar they show no sign of stupidity or drunkenness and those that feed but do not enter the hoods fly away apparently unaffected — "when the leaves had been standing for three days I opened them to find so many yellowjackets that the liquid could not cover them."

Appendix 2

Some observations on *Nepenthes* in Papua New Guinea (1987) by Dr Matthew Jebb

Some Morphology Observations

Nepenthes plants exhibit a marked dimorphism, with Lower, or Rosette leaves, and Upper leaf forms (these forms are referred to in upper case throughout this appendix to distinguish them from their adjectival equivalent). Differences occur in the internode length, leaf blade shape, tendril coiling, and pitcher shape and disposition [Figure A2.1]. The Upper form of leaf is produced at the same time as the onset of flowering, and once reached these stems no longer bear the Lower leaf form. The morphological 'switch' appears to be reversible, with cuttings of upper stems reverting to producing Lower pitcher type leaves prior to the onset of flowering.

The regular pattern which this dimorphy takes has been appreciated by relatively few authors (Veitch, 1898; Danser, 1928). Veitch (1898) in his observations on cultivated plants believed that this change occurred very gradually. Likewise, Danser refers to Rosette pitchers, pitchers of the short shoots and Upper pitchers. My own observations on *N. ampullaria*, *N. maxima*, *N.. mirabilis* and

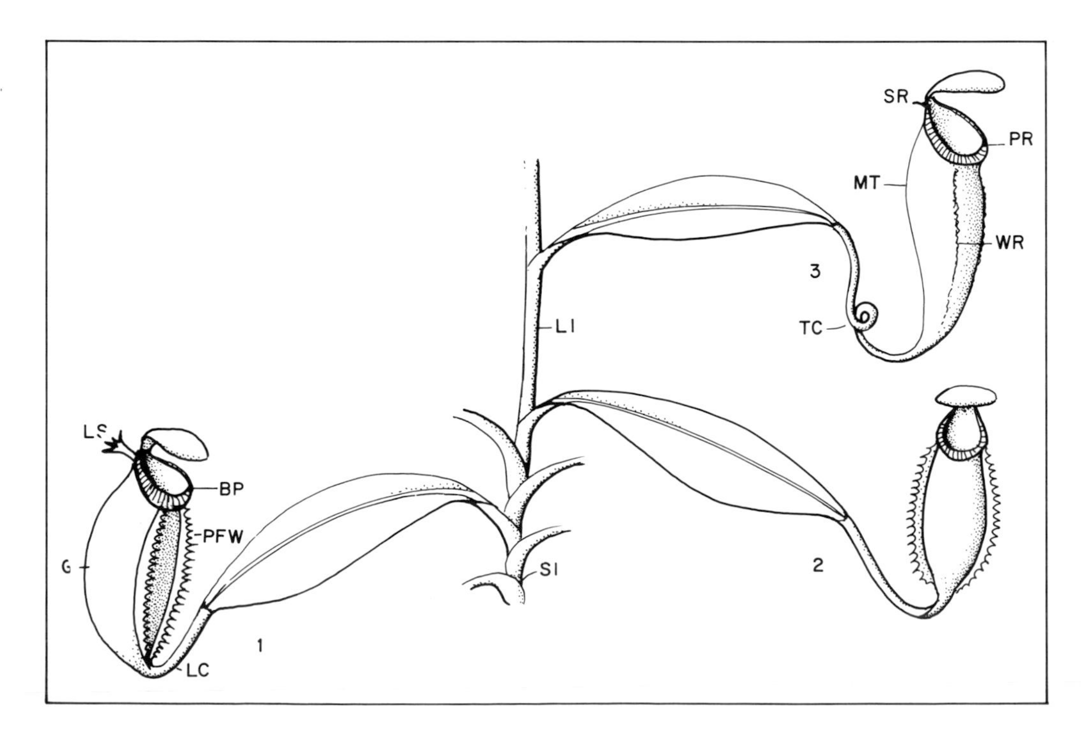

Fig. A2.1. Schematic drawing of terrestrial, intermediate and aerial forms of *Nepenthes* pitchers.
(1) Terrestrial: facing inwards, globose in form (G), large spur (LS), broad peristome (BP), petiole lacks coil (LC), wings are prominently fringed (PFW) and internodes are short (SI) (2) Intermediate form: pitcher arises sideways with respect to the main axis. (3) Aerial form: pitcher faces outwards, more tubular in form (MT), spur reduced (SR), peristome reduced (PR), petiole or tendril coils (TC), wings are reduced (WR) and internodes are long (LI). (From a sketch by, and with data from, Dr M.H.P. Jebb.)

N. neoguineensis growing in the wild suggests that the change is extremely rapid, and that a single intermediate type pitcher is only occasionally produced. *N. ampullaria* is remarkable in that it does not bear pitchers on its Upper leaves. The Lower leaves of this species are produced in compact rosettes, the leaf blade often being less than 2 cm in length, and the tendril 1 cm long, while the pitchers are highly urceolate. The Upper leaves on the other hand have long lanceolate leaf blades up to 25 cm in length with a coiling tendril to 15 cm in length. The tendril has a somewhat swollen tip, and this probably represents the vestigial pitcher.

The primary characteristic by which Upper and Lower pitcher forms are distinguished is in the origin of the pitcher in relation to the tendril. The Lower pitchers arise with their ventral side (mouth and wings) facing towards the tendril, while the Upper pitchers face away from the tendril [Figure A2.1]. In all plants, both wild and cultivated, that I have seen, this switch from one to the other is rapid, and only occasionally are single pitchers produced in which the tendril appears to arise from the 'side' of the pitcher. This change of axis is therefore a most useful standpoint by which to distinguish Upper and Lower leaf forms. Other features of the leaf may only change gradually, such as the shape and size of the pitcher, the degree of reduction in the ventral wings, and the size of the peristome, lid and spur.

The dimorphy of the pitchers apparently follows the same pattern in all species [Figure A2.1]. The stems are highly condensed in young plants, producing rosettes of leaves, often close to the ground. At, or shortly before the onset of flowering, a rapid morphological change takes place; the internodes become longer, the leaf blade may or may not change in size and shape, the tendril, between the leaf blade and pitcher, becomes both longer and coils readily. The Upper pitchers, which are produced at this point, differ from the Lower or Rosette pitchers in many respects; they are more tubular or infundibulate as opposed to the ovoid or globose lower pitchers; the mouth of the pitcher now faces away from the plant axis, and not towards it; the wings along the front of the pitcher become much reduced, in many species they are absent; the peristome is narrower; and the spur at the apex of the pitcher is now finer and more slender. At the point of transition a single, intermediate pitcher may be produced, but not invariably.

These many changes are associated with a change in growth form, from a short rosette plant to a climber; the internodes have lengthened, the tendrils are now coiled – for grasping other vegetation – the pitchers are less bulky, and face outwards. The difference between Rosette and Upper pitchers is most extreme in *N. ampullaria*, where the Upper pitchers have become reduced to mere swellings of the tendril tip. It may be that the extreme form of the urceolate pitchers in *N. ampullaria* prevents efficient aerial pitchers from being produced, since only a limited morphological amplitude may be available in this change from Lower to Upper pitcher. Alternatively, it may be that the habitat in which *N. ampullaria* lives has comparatively little prey above ground level, which makes the formation of such pitchers unnecessary. Ecological observations on this species would be of great interest.

A single *Nepenthes* plant will usually only bear some 4–8 functional pitchers. The leaves at the base of the plant dry out and become brittle, but do not abscise. Clearly the pitcher becomes more and more inefficient as chitin and other indigestible substances accumulate. *N. ampullaria* is again unusual in that the older stems continue to produce compact rosettes of Lower leaves. This difference may be to compensate for the fact that the Upper leaves do not produce pitchers.

Some Ecological Observations

Generally speaking, *Nepenthes* are characteristic of nutrient poor soils, often on heavily leached ridge tops. In New Guinea the most widely distributed and common species, *N. mirabilis*, is occasionally found in apparently nutrient rich situations, notably disturbed rain forest. *N. ampullaria* is also unusual in that it is often found in shady habitats, including *Araucaria* stands. However it does have long climbing stems which reach 15 metres or more in length. While other species may also climb in vegetation, they are apparently less tolerant of shade. The pitchers of this species are borne on very compact rosettes, and often become partially buried in leaf litter, their minute leaf blades, and urceolate form may be adaptations to this particular habitat. The blade is dispensed with since it is covered by leaf litter, while the pitcher has a relatively small mouth which may mean that it can operate as a pitfall trap even when buried by dead leaves. Furthermore the lid of this species is remarkably narrow and reflexed, and thus does not serve to cover the pitcher mouth in any way.

Examination of the contents of *Nepenthes* pitchers

Table A2.1. Trapping Spectra of *Nepenthes mirabilis*

	By Pitcher Type			
	18 lower traps	average	30 upper traps	average
Ants	**2573**	142.9	**1362**	45.4
Cockroaches	**32**	1.8	**24**	.8
Centipedes	**28**	1.6	**42**	1.4
Beetles	**8**	.4	**32**	1.2
Crickets	**6**		**8**	.3
Snails	**6**		**2**	
Spiders	**2**		**7**	
Others	**1**		**2**	
Flies	**23**	1.3	**94**	3.1
Ichneumon wasps	**12**	.7	**48**	1.8
Moths	**2**		**10**	.4
Wasps/Bees	—		**3**	.1

Bold = actual numbers
Medium = average: per pitcher [figures below .5 omitted]

suggests that there may be some correlation between pitcher type and captured prey. The contents of 52 pitchers of *N. mirabilis* were examined from a total of 20 plants, on Missima Island, Papua New Guinea. The results show that the lower pitchers trap greater numbers of ants, cockroaches, myriopods and spiders, while the aerial traps contain a greater number of flying prey such as dipteran flies, Ichneumon wasps and moths [Table A2.1]. Although the physical position of the pitcher is clearly a major factor influencing the trapped fauna, the pitcher morphology – large fringed wings, a broad peristome and a wide mouth for crawling creatures, as opposed to much reduced wings, a slender peristome and narrow mouth for flying insects – may increase the functional efficiency of the pitchers.

The decaying mass of trapped organisms may also contribute in luring necrophagous organisms such as beetles, flies and cockroaches. Older pitchers often contained a number of cockroaches which may have been attracted to the decaying contents. Some of the first captures of newly opened pitchers were found to be centipedes, while older pitchers rarely contained them. These active predators were either seeking refuge or prey in the pitcher, and not perhaps directly attracted by the pitcher.

References

Abderhalden, E. and Terruchi, Y. (1906). Vergleichende Untersuchungen über einige proteolytische pflanzlicher Herkunft (Comparative researches on proteolytic enzymes of plant origin from several sources). *Zeitschrift für Physiologische Chemie* **49**, 21–25.

Achterberg, C. van (1973). A study about the arthropoda caught by *Drosera* species. *Entomologischen Berichten* **33**, 137–140.

Adams, R.M. and Smith, G.W. (1977). An S.E.M. survey of the five carnivorous pitcher plant genera. *American Journal of Botany* **64**, 265–272.

Addicott, J.F. (1974). Predation and prey community structure: An experimental study of the effect of mosquito larvae on the protozoan communities of pitcher plants. *Ecology* **55**, 475–492.

Adowa, A.N. (1924). Zur Frage nach dem Fermenten von *Utricularia vulgaris* L. II. Der relative Gehalt der Blaschen und Zweige von *Utricularia vulgaris* an Proteoklastischen Fermenten (On the question of the enzymology of *U. vulgaris* II. The relative content of utricles and branches of *U. vulgaris* of proteolytic enzymes). *Biochemische Zeitung* **153**, 506–509.

Affolter, J.M. and Olivo, R.F. (1975). Action potentials in Venus's-flytraps: long-term observations following the capture of prey. *American Midland Naturalist* **93**, 443–445.

Ah-Lan, L. and Prakash, N. (1973). Life History of *Nepenthes gracilis*. *Malaysian Journal of Science* **2**, 45–53.

Airy-Shaw, H.K. (1951). On the Dioncophyllaceae, a remarkable new family of flowering plants. *Kew Bulletin* **6**, 327–347.

Åkerman, A. (1917). Untersuchungen über die Aggregation in den Tentakeln von *Drosera rotundifolia* (Observations on aggregation in the tentacles of *Drosera rotundifolia*). *Botanisker Notiser* 145–192.

Albrecht, R.M.. (1974). Microbial activity in *Sarracenia purpurea*. *Carnivorous Plant Newsletter* **3**, 32.

Aldenius, J., Carlsson, B. and Karlsson, B. (1983). Effects of insect trapping on growth and nutrient content of *Pinguicula vulgaris* L. in relation to the nutrient content of the substrate. *New Phytologist* **93**, 53–59.

Almborn, D. (1983). Flugtrumpet: *Sarracenia purpurea*, naturiliserad i Sverige (The pitcher-plant, *Sarracenia purpurea*, naturalized in Sweden). *Svensk Botanisk Tidskrift* **77**, 209–216.

Amagase, S. (1972). Digestive enzymes in insectivorous plants IV. Enzymic digestion of insects by *Nepenthes* secretion and *Drosera peltata* extract; Proteolytic and chitinolytic activities. *Journal of Biochemistry* **72**, 765–767.

Amagase, S., Nakayama, S. and Tsugita, A. (1969). Acid protease in *Nepenthes*: II. Study on the specificity of nepenthesin. *Journal of Biochemistry* **66**, 431–439.

Amagase, S., Mori, M. and Nakayama, S. (1972). Digestive enzymes in the insectivorous plants: III. Acid proteases in the genus *Nepenthes* and *Drosera peltata*. *Journal of Biochemistry* **72**, 73–81.

Angerilli, N.P.D. and Beirne, B.P. (1974). Influence of some fresh water plants on the development and survival of mosquito larvae in British Columbia, Canada. *Canadian Journal of Zoology* **52**, 813–815.

Angerilli, N.P.D. and Beirne, B.P. (1980). Influences of aquatic plants on colonization of artificial ponds by mosquitoes and their insect predators. *Canadian Entomologist* **112**, 793–796.

Anstey, T.H. and Moore, J.F. (1954). Inheritance of glossy foliage and cream petals in green sprouting broccoli. *Journal of Heredity* **45**, 39–41.

Arber, A. (1941). On the morphology of the pitcher-leaves in *Heliamphora*, *Sarracenia*, *Darlingtonia*, *Cephalotus* and *Nepenthes*. *Annals of Botany, New Series* **5**, 563–578.

Arisz, W.H. (1941). Transport door de tentakels van *Drosera* (Transport through the tentacles of *Drosera*). *Verslagen vergaderingen Nederlandse Akademie van Wetenschappen. Afdeeling Natuurkunde* **50**, 77–78.

Arisz, W.H. (1942a). Absorption and transport by the tentacles of *Drosera capensis*: I. Active transport of asparagine in the parenchyma cells of the tentacles. *Proceedings (Kon) Nederlandse Akademie van Wetenschappen* **45**, 2–8.

Arisz, W.H. (1942b). Absorption and transport by the tentacles of *Drosera capensis*: II. The activation of the transport of different substances by oxygen. *Proceedings (Kon) Nederlandse Akademie van Wetenschappen* **45**, 794–801.

Arisz, W.H. (1944a). Absorptie en transport door de tentakels van *Drosera capensis*: III. De absorptie van aminozuren en zouten door bindif aan het plasma (Absorption and transport by the tentacles of *D. capensis*: III. The absorption of amino acids . . .). *Verslagen vergaderingen Nederlandse Akademie van Wetenschappen Afdeeling Natuurkunde* **53**, 236–248.

Arisz, W.H. (1944b). Absorptie en transport door de tentakels van *Drosera capensis*: IV. Gelijktijdige absorpti van verschillende stoffen (Absorption and transport through the tentacles of *D. capensis*: IV. Parallel absorption of different substances). *Verslagen vergaderingen Nederlandse Akademie van Wetenschappen Afdeeling Natuurkunde* **53**, 249–260.

Arisz, W.H. (1953). Absorption and transport by the tentacles of *Drosera capensis*: V. Influence on the transport of substances inhibiting enzymic processes. *Acta Botanica Neerlandica* **2**, 74–106.

Arisz, W.H. and Oudman, J. (1937). On the influence of

aggregation on the transport of asparagine and caffeine in the tentacles of *Drosera capensis* L. *Proceedings (Kon) Nederlandse Akademie van Amsterdam* **40**, 431–439.

Asano, M. and Hase, J. (1943a) Hydroxyquinones VIII: The structure of droserone 2. *Journal of the Pharmaceutical Society of Japan* **63**, 83–90.

Asano, M. and Hase, J. (1943b). Hydroxyquinones IX: The structure of droserone 2. *Journal of the Pharmaceutical Society of Japan* **63**, 90–96.

Ashida, J. (1934). Studies on the leaf movement of *Aldrovanda vesiculosa* L. I. Process and mechanism of the movement. *Memoirs of the College of Science, Kyoto Imperial University* **49**, 141–246.

Ashida, J. (1935). Studies on the leaf movement of *Aldrovanda vesiculosa* L. II. Effects of mechanical, electrical, thermal, osmotic and chemical influences. *Memoirs of the College of Science, Kyoto Imperial University* **11**, 55–113.

Ashley, G. (1975). *Drosera burmanni* in Queensland, Australia. *Carnivorous Plant Newsletter* **4**, 47–48.

Ashley, T. and Gennaro, J.F. (1971). Fly in the sundew. *Natural History (New York)* **80**, 80–85, 102.

Atkin, D.S.J. and Hamilton, R.J. (1982). The effects of plant waxes on insects. *Journal of Natural Products (Lloydia)* **45**, 695–696.

Atkins, R.G. (1922). The hydrogen ion concentration of plant cells. *Proceedings of the Royal Dublin Society* **16**, 414–434.

Austin, R.M. (1875–1877). Selected Letters to W.M. Canby. *Society of Natural History of Delaware* (see Appendix 1).

Avadhani, P.N. and Higgs, R.E.A. (1977). C_4 photosynthesis among the 'Adinandra Belukar' species. *Plant Physiology:* Supplement to Vol. **59**. Session 25, Abstract 366.

Ayuga, C., Carretero, E. and Bermejo, P. (1985). Contribucion al estudie de flavonoides en *Drosera rotundifolia* (A contribution to the study of flavonoids in *D. rotundifolia*). *Annales de la Real Academie de Farmacie* **51**, 321–326.

Baker, H.G. and Baker, I. (1973). Amino acids in nectar and their evolutionary significance. *Nature (London)* **241**, 543–545.

Baker, J.G. (1877). 'Flora of Mauritius and the Seychelles', London.

Bakker, J.P. (1957). Quelques aspects du problème des sediments correlatifs en climat tropical humide (Some aspects of the problems of correlative sediments in a tropical humid climate). *Zeitschrift für Geomorphologie, New series* **1**, 3–43.

Balfour, T.A.G. (1875). Account of some experiments on *Dionaea muscipula* (Venus's Flytrap). *Transactions and Proceedings of the Botanical Society of Edinburgh* **12**, 334–369.

Balotin, N.M. and Di Palma, J.R. (1962). Spontaneous electrical activity of *Dionaea muscipula*. *Science* **138**, 1338–1339.

Barber, H.N. (1955). Adaptive gene substitution in Tasmanian Eucalypts 1. Genes controlling the development of glaucousness. *Evolution* **9**, 1–14.

Barber, H.N. and Jackson, W.D. (1957). Natural selection in action in *Eucalyptus*. *Nature* **179**, 1267–1269.

Barber, J.T. (1977). Mucilaginous seeds: Interactions with microorganisms. *Plant Physiology*: Supplement to vol. **59**, Abstract **6**.

Barber, J.T., Page, C.R. and Felsot, A.S. (1974). Interactions between mosquito larvae and mucilaginous plant seeds. I. Carbohydrate composition of mucilage in relation to entrapment of larvae. *Mosquito News* **34**, 394–398.

Barber, J.T., Page, C.R., Berger, A.I. and Hohenschutz, L.D. (1976). Interactions between mosquito larvae and mucilaginous plant seeds. III. Factors influencing attachment of larvae to seeds and their subsequent mortality. *Mosquito News* **36**, 301–307.

Barbour, M.G. and Major, J. (1977). 'Terrestrial Vegetation of California'. John Wiley, New York, London, Sydney and Toronto.

Barckhaus, R. and Weinert, H. (1974). Die fleischfressende Pflanze *Sarracenia purpurea*. Licht und elektron–mikroskopische Untersuchungen. (The carnivorous plant *S. purpurea*. Light and electron microscope investigations). *Mikrokosmos* **63**, 38–47.

Barr, A.R. and Chellapah, W.T. (1963). The mosquito fauna of pitcher plants in Singapore. *Singapore Medical Journal* **4**, 184–185.

Bartram, W. (1791). 'Travels in North and South Carolina, Georgia and Florida'. James and Johnson, Philadelphia.

Batalin, A. (1877). Mechanik der Bewegung der insektfressenden Pflanzen (Mechanics of the movement of carnivorous plants). *Flora* **60**, 145–154.

Batalin, A. (1880). Über die Function der Epidermis in den Schlauchen von *Sarracenia* und *Darlingtonia* (On the function of the epidermis in the pitchers of *Sarracenia* and *Darlingtonia*). *Acta Horti Petropolitani* **7**, 345–359.

Bate-Smith, E.C. (1984). Age and distribution of galloyl esters, iridoids and certain other repellents in plants. *Phytochemistry* **23**, 945–950.

Bauquis, P. and Miriamanoff, A. (1970). Observations d'ordres chimiques et ecologiques sur quelques espèces indigènes du genre *Pinguicula* (Chemical and ecological observations on some indigenous species of the genus *Pinguicula*). *Pharmaceutica Acta Helvetica* **45**, 122–131.

Beal, W.J. (1875). Carnivorous Plants. *Proceedings of the American Association for the Advancement of Science* **1875B**, 251–253.

Beaver, R.A. (1979a). A description of the male and larva of *Endonepenthia schuitemakeri* Schmitz from *Nepenthes* pitchers (Diptera, Phoridae). *Annales de la Société Entomologique de France (Nouvelle Serie)* **15**, 3–17.

Beaver, R.A. (1979b) Fauna and foodwebs of pitcher plants in West Malaysia. *The Malayan Nature Journal* **33**, 1–10.

Beaver, R.A. (1983). The communities living in *Nepenthes* pitcher plants: Fauna and food webs. *In* 'Phytotelmata: Terrestrial Plants as Hosts for Aquatic Insect Communities' (Eds. Frank, J.H. and Lounibos, L.P.) 125–159. Plexus, Medford, New Jersey, USA.

Beaver, R.A. (1985). Geographical variation in food web structure in *Nepenthes* pitcher plants. *Ecological Entomology* **10**, 241–248.

Beccari, O. (1886). *Nepenthes*. Plante Ospitatrici. (The *Nepenthes* pitcher as a habitat). Malesia Vol. 11. Tipografia del Reale Instituto Sordo muti, Genova.

Beckner, J. (1979). A method of growing the impossible bog orchids. *The Orchid Society Bulletin* **48**, 556–560.

Beebe, J.D. (1980). Morphogenetic responses of seedlings and adventitious buds of the carnivorous plant *Dionaea muscipula* in aseptic culture. *Botanical Gazette* **141**, 396–400.

Behre, K. (1928). Physiologische und cytologische Untersuchungen über *Drosera* (Physiological and cytological researches on *Drosera*). *Planta* **7**, 208–306.

Bell, C.R. (1949). A cytotaxonomic study of the Sarraceniaceae of N. America. *Elisha Mitchell Science Society* **65**, 137–164.

Bell, C.R. (1952). Natural hybrids in the genus *Sarracenia* L. *Elisha Mitchell Science Society* **68**, 55–80.

Bell, C.R. (1954). *Sarracenia leucophylla rafinesque*. *Elisha Mitchell Science Society* **70**, 57–60.

Bell, C.R. and Case, F.W. (1956). Natural hybrids in the genus *Sarracenia* II. Current notes on distribution. *Elisha Mitchell Science Society* **72**, 142–152.

Bell, E.A. (1980). The possible significance of secondary compounds in plants. *In*: 'Secondary Plant Products' (Eds. Bell, E.A. and Charlwood, B.V.), 11–21. Encyclopaedia of Plant Physiology, New Series **8**. Springer-Verlag, Berlin.

Bell, P.R. (1959). The ability of *Sphagnum* to absorb cations preferentially from dilute solutions resembling natural water. *Journal of Ecology* **47**, 351–355.

Bendz, G. and Lindberg, G. (1968). Naphthoquinones and anthocyanins from two *Drosera* species. *Acta Chemica Scandica* **22**, 2722–2723.

Bendz, G. and Lindberg, G. (1970). Pigments of some *Drosera* species. *Acta Chemica Scandica* **24**, 1082–1083.

Benolken, R.M. and Jacobsen, S.L. (1970). Response properties of sensory hairs excised from Venus's flytrap. *Journal of General Physiology* **56**, 64–82.

Bentham, G. (1840). On the *Heliamphora nutans*, a new pitcher plant from British Guiana. *Transactions of the Linnaean Society* **18**, 429–433.

Bentham, G. and Mueller, F. (1864). 'Flora Australiensis' Vol. II. Lovell, Reeve, London.

Bentley, B.L. (1977). Extrafloral nectaries and protection by pugnacious bodyguards. *Annual Review of Ecology and Systematics* **8**, 407–427.

Bentrup, F.W. (1979). Reception and transduction of electrical and mechanical stimuli. Physiology of Movement. *In*: Handbook of Plant Physiology, (Eds. Haupt, W. and Feinleib, M.E.). **7**, pp. 42–70, Springer-Verlag, Berlin.

Benzing, D.H. (1980). 'The Biology of the Bromeliads'. Mad River Press, California, USA.

Benzing, D.H. (1986). Foliar specializations for animal-assisted nutrition in *Bromeliaceae*. *In* 'Insects and the Plant Surface' (Eds. Juniper, B.E. and Southwood, T.R.E.). Edward Arnold, London.

Benzing, D.H. (1987). The origin and rarity of botanical carnivory. *Tree* **2**, 364–369.

Benzing, D.H., Henderson, K., Kessel, B. and Sulak, J. (1976). The absorptive capacities of bromeliad trichomes. *American Journal of Botany* **63**, 1009–1014.

Benzing, D.H., Givnish, T.J. and Bermudes, D. (1985). Absorptive trichomes in *Brocchinia reducta* (Bromeliaceae) and their evolutionary and systematic significance. *Systematic Botany* **10**, 81–91.

Benzing, D.H. and Renfrow, A. (1974). The mineral nutrition of Bromeliaceae. *Botanical Gazette* **135**, 281–288.

Bernon, G.L. (1968). Paper wasp nest in the pitcher-plant *Sarracenia purpurea* L. *Entomological News* **80**, 148.

Bezanger-Beauquesne, L. (1954). Sur le pigment jaune du *Drosera* (On the yellow pigment of *Drosera*). *Comptes rendus d'Académie des Sciences de Paris* **239**, 618–620.

Bezanger-Beauquesne, L. and Perrin, E. (1972). À propos du *Drosera* (Comments on *Drosera*) *Plant Médicinale et Phytothérapie* **6**, 183–193.

Bhattacharyya, J., Kokpol, U. and Miles, D.H. (1976). The isolation from *Sarracenia flava* and partial synthesis of Betulinaldehyde. *Phytochemistry* **15**, 432–433.

Bienenfeld, W. & Katzmeister, H. (1966). Flavanoide aus *Drosera rotundifolia* L. (Flavonoids from *D. rotundifolia* L.). *Archiv der Pharmazie (und Berichte der deutschen pharmazeutischen Gesellschaft)* **299**, 598–602.

Bird, D.F. and Kalff, J. (1986). Bacterial grazing by planktonic lake algae. *Science* **231**, 493–495.

Björklund, D. (1864). Untersuchungen aus dem Laboratorium der pharmaceutischen Gesellschaft in St. Petersburg. *Sarracenia purpurea* (a) Chemische Untersuchung des Wurzelstockes (Investigations from the laboratories of the pharmaceutical company in St. Petersburg. *S. purpurea* (a) Chemical investigations on rootstocks). *Archiv der pharmacie (Hannover)* **169**, 93–124.

Bogner, J. (1976). The natural habitat of *Heliamphora heterodoxa* Steyermark. *Carnivorous Plant Newsletter* **5**, 28.

Boller, T., Gehri, A., Mauch, F. and Vogeli, U. (1983). Chitinase in bean leaves: induction by ethylene, purification properties, and possible function. *Planta* **157**, 22–31.

Bonnet, M., Coumans, M., Hofinger, M., Ramaut, J.L. and Gaspar, Th. (1984). High performance gas chromatography of 1,4-naphthoquinones from Droseraceae. *Chromatographia* **18**, 621–622.

Bopp, M. and Weber, I. (1981). Studies on the hormonal regulation of the leaf blade movement of *Drosera capensis* L. *Physiologia Plantarum* **53**, 491–496.

Bosserman, R.W. (1985). Distribution of heavy metals in aquatic macrophytes from Okefenokee Swamp. *Symposia Biologica Hungarica* **29**, 31–40.

Botta, S.M. (1976). Sobre las trampas y las victimas o presas de algunas especies argentinas del genero *Utricularia* (The traps and their victims in some Argentine species of the genus *Utricularia*). *Darwiniana (Buenos Aires)* **20**, 127–154.

Boucher, D.H. (1985). The idea of mutualism, present and future. *In* 'The Biology of Mutualism, Ecology and Evolution' (Ed. Boucher, D.H.) Croom Helm, London.

Boucher, D.H., James, S. and Keeler, K.H. (1982). The ecology of mutualism. *Annual Review of Ecology and Systematics* **13**, 315–347.

Bouquet, A. (1970). Notes sur le *Drosera congolana* Taton (Notes on *D. congolana* Taton). *Plantes Médicinales et Phytothérapie* **4**, 221–222.

Bower, F.O. (1889a). On the morphology of *Nepenthes*: a study in the morphology of the leaf. *Annals of Botany* **3**, 239–252.

Bower, F.O. (1889b). On Dr Macfarlane's observations on pitchered insectivorous plants. *Annals of Botany* **4**, 165.

Bower, F.O. (1923). 'The Ferns' (Filicales) Vol. 1. Cambridge University Press, Cambridge.

Boyd, C.E. and Walley, W.W. (1972). Biogeography of Boron. I. Concentrations in surface waters, rainfall and aquatic plants. *American Midland Naturalist* **88**, 1–14.

Boyd, J.P. (Ed.) (1955). *Papers of Thomas Jefferson* **11**, 279–280.

Bradshaw, W.E. (1976). Geography of photo-periodic response in diapausing mosquito. Control of dormancy in *Wyeomyia smithii*. *Nature* **262**, 384–385.

Bradshaw, W.E. (1980a). Photoperiodism and the photic environment of the pitcher-plant mosquito *Wyeomyia smithii*. *Oecologia (Berlin)* **44**, 311–316.

Bradshaw, W.E. (1980b). Blood-feeding and capacity for increase in the pitcher-plant mosquito *Wyeomyia smithii*. *Environmental Entomology* **9**, 86–89.

Bradshaw, W.E. (1983). Interaction between the mosquito *Wyeomyia smithii*, the midge *Metriocnemus knabi*, and their carnivorous host *Sarracenia purpurea*. *In* 'Phytotelmata: Terrestrial Plants as Hosts for Aquatic Insect Communities' (Eds. Frank, J.H. and Lounibos, L.P.), 161–189. Plexus, Medford, New Jersey.

Bradshaw, W.E. and Creelman, R.A. (1984). Mutualism between the carnivorous purple pitcher plant and its inhabitants. *American Midland Naturalist* **112**, 294–304.

Bradshaw, W.E. and Lounibos, L.P. (1972). Photoperiodic control of development in the pitcher-plant mosquito *Wyeomyia smithii*. *Canadian Journal of Zoology* **50**, 713–719.

Bradshaw, W.E. and Lounibos, L.P. (1977). Evolution of dormancy and its photoperiodic control in pitcher-plant mosquitoes. *Evolution* **31**, 546–564.

Bradshaw, W.E. and Phillips, D.L. (1980). Photoperiodism and the photic environment of the pitcher-plant mosquito *Wyeomyia smithii. Oecologia (Berlin)* **44**, 311–316.

Brewer-Carias, C. (1973). *Utricularia humboldtii* grows in axils of bromeliad *Brocchinia. In*: 'Carnivorous plants of the Cerro de la Neblina'. *Natura (Buenos Aires)* **6**, 17–26.

Bringmann, G. (1985a). Acetogenine Isochinoline Alkaloide, 6–Aufbau und Cyclisierung zentral modifiziester β-Pentaketone: Synthese monocyclischer Isochinoalkaloid-Vorstufen (Acetogenin Isoquinoline Alkaloids, 6–Preparation and cyclization of centrally modified β-pentaketones: synthesis of monocyclic isoquinoline alkaloid precursors). *Liebigs Annales der Chemie*, 2105–2115.

Bringmann, G. (1985b). Acetogenine Isochinoline Alkaloide, 8–Biomimetische Synthesen beider Molekulhaften der *Ancistrocladus* und der *Triphyophyllum*. Alkaloide aus gemeinsamen Vorstufen (Acetogenin Isoquinoline Alkaloids 8–Biomimetic synthesis of both molecular moieties of *A.* and *T.* alkaloids from common precursors). *Liebigs Annales der Chemie*, 2126–2134.

Bringmann, G. (1986). The Naphthyl Isoquinoline Alkaloids. *In* 'The Alkaloids' (Ed. Brossi, A.), 141–184. Academic Press, Orlando, Florida.

Bringmann, G. and Jansen, J.R. (1984). A first and general route to naphthylisoquinoline alkaloids: the total synthesis of O-methyl-tetradehydro-triphyophylline. *Tetrahedron Letters* **25**, 2537–2540.

Bringmann, G. and Jansen, J.R. (1986). One pot preparation of 1,3-dimethyltetrahydroisoquinolines from their biosynthetic diketo precursors. *Heterocycles* **24**, 2407–2410.

Bristowe, W.S. (1939). 'The Comity of Spiders', Vol. 1. The Ray Society, London.

Brocher, F. (1911–12). Le problème de l'utriculaire (The problem of the bladderwort). *Annales de Biologie Lacustre (Bruxelles)* **5**, 33–46.

Brockenbro, T.W. (1981). News and views, *Drosera regia. Carnivorous Plant Newsletter* **10**, 60.

Broussaud, F. and Vintéjoux, C. (1982). Etudes ultrastructurales et cytoplasmiques des tissus superficiels places à l'entrée des urnes d'*Utricularia* (Lentibulariaceae) (Ultrastructural and cytochemical studies on the superficial tissues of the entrance of the *Utricularia* trap [Lentibulariaceael]). *Bulletin de la Société Botanique de France* **129**, 191–201.

Brower, J.H. and Brower, A.E. (1970). Notes on the biology and distribution of moths associated with the pitcher plant in Maine. *Proceedings of the Entomological Society of Ontario* **101**, 79–83.

Brown, N.E. (1880). New garden plants: *Nepenthes bicalcarata*, J.D. Hook. *Gardener's Chronicle*, 201–202.

Brown, N.E. (1901) Enumeration of the plants collected I. Spermatophyta. *Transactions of the Linnean Society of London* (Botany) 2nd series **6**, 1–108.

Brown, W.H. (1916). The mechanism of movement and duration of the effect of stimulation in leaves of *Drosera. American Journal of Botany* **3**, 68–90.

Browning, A.J. and Gunning, B.E.S. (1979a). Structure and function of transfer cells in the sporophyte haustorium of *Funaria hygrometrica* Hedw. I. The development and ultrastructure of the haustorium. *Journal of Experimental Botany* **30**, 1233–1246.

Browning, A.J. and Gunning, B.E.S. (1979b). Structure and function of transfer cells in the sporophyte haustorium of *Funaria hygrometrica* Hedw. II. Kinetics of uptake of labelled sugars and localisation of absorbed products by freeze-substitution and autoradiography. *Journal of Experimental Botany* **30**, 1247–1264.

Browning, A.J. and Gunning, B.E.S. (1979c). Structure and function of transfer cells in the sporophyte haustorium of *Funaria hygrometrica* Hedw. III. Translocation of assimilate into the attached sporophyte and along the seta of attached and excised sporophytes. *Journal of Experimental Botany* **30**, 1265–1273.

Bruce, A.N. (1905). On the glands of *Byblis gigantea* Lindl. *Notes from the Royal Botanical Society, Edinburgh* **16**, 9–14.

Bruneton, J., Bouquet, A., Fournet, A. and Cave, A. (1976). La triphyophylline, nouvel alcolide isole du *Triphyophyllum peltatum* (Triphyophylline, a new alkaloid isolated from *T. peltatum*). *Phytochemistry* **15**, 817–818.

Bryant, G.E. (1953). A new species of *Epitrix* (Coleoptera, Halticinae) from Western Australia. *Western Australian Naturalist* **4**, 8–9.

Buchen, B., Henzel, D. and Sievers, A. (1983). Polarity in mechanoreceptor cells of trigger hairs of *Dionaea muscipula* Ellis. *Planta* **158**, 458–468.

Buchen, B. and Schröder, W.H. (1986). Localization of calcium in the sensory cells of the *Dionaea* trigger hair by Laser Micro-Mass Analysis (LAMMA). *In* 'Molecular and Cellular Aspects of Calcium in Plant Development' (Ed. Trewavas, A.J.), 233–240. NATO ASI Series. Plenum Press, New York, London.

Buchmann, S.L. (1983). Buzz pollination in angiosperms. *In* 'Handbook of Experimental Pollination Biology' (Eds. Jones, C.E. and Little, R.J.) 73–113. Van Nostrand Reinhold, New York.

Buchmann, S.L. (1985). Bees use vibration to aid pollen collection from non-poricidal flowers. *Journal of the Kansas Entomological Society* **58**, 517–525.

Buffington, J. (1970). Ecological considerations of the cohabitation of pitcher-plants by *Wyeomyia smithii* and *Metriocnemus knabi. Mosquito News* **30**, 89–90.

Bünning, E. (1947). 'In der Wäldern Nord-Sumatra' (In the forests of Northern Sumatra), 113–114. Bonn.

Burbidge, F.W. (1880). *Nepenthes bicalcarata. Gardener's Chronicle. New series* **13**, 264–265.

Burbidge, F.W. (1882). Notes on the new *Nepenthes. Gardener's Chronicle. New series* **17**, 56.

Burbidge, F.W. (1896). *Nepenthes. Gardener's Chronicle. 3rd series* **20**, 105–106.

Burbidge, F.W. (1897). Note on *Nepenthes. Journal of the Royal Horticultural Society* **21**, 256–262.

Burdon Sanderson, J.S. (1873). Note on the electrical phenomena which accompany stimulation of the leaf of *Dionaea muscipula. Proceedings of the Royal Society* **21**, 495–496.

Burdon Sanderson, J.S. and Page, F.J.M. (1876). On the mechanical effects and on the electrical disturbance consequent on excitation of the leaf of *Dionaea muscipula. Proceedings of the Royal Society* **25**, 411–434.

Burdon Sanderson, J.S. (1911). The excitability of plants. *In* 'Sir John Burdon Sanderson, a memoir with selections from his Papers and Addresses'. Clarendon Press, Oxford. 172–198.

Burgess, L. and Rempel, J.G. (1971). Collection of the pitcher-plant mosquito *Wyeomyia smithii* (Diptera: Culicidae) from Saskatchewan. *Canadian Entomologist* **103**, 886–887.

Burkhardt, D. (1982). Birds, berries and UV: A note on some consequences of UV vision in birds. *Naturwissenschaften* **69**, 153–157.

Burkill, I.H. (1966). 'A Dictionary of the Forest Products of the Malay Peninsula', Vol. 2. Ministry of Agriculture and Cooperatives, Kuala Lumpur.

Burnell, J.N. and Anderson, J.N. (1973). Adenosine diphosphate sulfamylase activity in leaf tissue. *Biochemical Journal* **133**, 417–428.

Burnett, G.T. (1829). On the function and structure of plants, with reference to the adumbrations of a stomach in vegetals. *Quarterly Journal of Scientific Literature and Art; New Series* July–Dec, 279–292.

Bussy, J. (1974). Une station de *Drosera anglica* Huds. (= *Drosera longifolia* L.) sur tuf calcaire et milieu basique (A locality for *D. anglica* on calcareous tufa with a basic pH). *Bulletin de la Société Linnéenne de Lyon* **43**, 8.

Butler, D.E. (1985). Ecological adaptations of *Sarracenia purpurea* in Coastal Maine *Sphagnum* moss. *Carnivorous Plant Newsletter* **14**, 68–73.

Cajander, V.-R. and Ihantola, R. (1984). Mercury in some higher aquatic plants and plankton of the River Kokemaenjoki, southern Finland. *Annales Botanici Fennici* **21**, 151–156.

Cameron, C.J., Donald, G.L. and Paterson, C.G. (1977). Oxygen-fauna relationships in the pitcher-plant *Sarracenia purpurea* L. with reference to the chironomid *Metriocnemus knabi* Coq. *Canadian Journal of Zoology* **55**, 2018–2023.

Campbell, N. and Thomson, W.W. (1976). The ultrastructure of *Frankenia* salt glands. *Annals of Botany* **40**, 681–686.

Canby, W.M. (1874). *Darlingtonia californica*, an insectivorous plant. *American Association for the Advancement of Science. Section B. Natural History* **23**, 64–72.

Cannon, J.R., Lojanapiwatna, V., Raston, C.L., Sinchai, W. and White, A.H. (1980). The quinones of *Nepenthes rafflesiana*. The crystal structure of 2,5-dihydroxy-3,8-dimethoxy-7-methylnaphtho-1,4-quinone (nepenthone-E) and a synthesis of 2,5-dihydroxy-3-methoxy-7-methylnaphtho-1,4-quinone (nepenthone-C). *Australian Journal of Chemistry* **33**, 1073–1093.

Cantelo, W.W. and Jacobson, M. (1979). Phenylacetaldehyde attracts moths to bladder flower and to black light traps. *Environmental Entomology* **8**, 444–447.

Carlquist, S. (1976a). Wood anatomy of the Roridulaceae; ecological and phylogenetic implications. *American Journal of Botany* **63**, 1003–1008.

Carlquist, S. (1976b). Wood anatomy of the Byblidaceae. *Botanical Gazette* **137**, 35–38.

Carlquist, S. (1981). Wood anatomy of Cephalotaceae. *International Association of Wood Anatomists Bulletin* **2**, 175–178.

Carpenter, A. (1884). A carnivorous marine plant. *Nature* **30**, 289.

Carr, S.G.M. and Carr, D.J. (1976). The French contribution to the discovery of Australia and its flora. *Endeavour* **35**, 21–26.

Case, F.W. and Case, R.B. (1974). *Sarracenia alabamensis*: A newly recognised species from central Alabama. *Rhodora* **76**, 650–652.

Case, F.W. and Case, R.B. (1976). The *Sarracenia rubra* complex. *Rhodora* **78**, 270–325.

Casper, S.J. (1966). Monographie der Gattung *Pinguicula* (Monograph of the genus *Pinguicula*). *Bibliographica Botanica* **127–128**, 225.

Casser, M., Hodick, D., Buchen, B. and Sievers, A. (1985). Correlation of excitability and bipolar arrangement of endoplasmic reticulum during the development of sensory cells in trigger hairs of *Dionaea muscipula* Ellis. *European Journal of Cell Biology* **36**, supplement 7, 12.

Catesby, M. (1737–1754). 'The Natural History of Carolina, Florida and the Bahama Islands'. C. Marsh and T. Wilcox, London.

Chandler, G.E. and Anderson, J.W. (1976a). Studies on the nutrition and growth of *Drosera* species with reference to the carnivorous habit. *New Phytologist* **76**, 129–141.

Chandler, G.E. and Anderson, J.W. (1976b). Studies on the origin of some hydrolytic enzymes associated with the leaves and tentacles of *Drosera* species and their role in heterotrophic nutrition. *New Phytologist* **77**, 51–62.

Chandler, G.E. and Anderson, J.W. (1976c). Uptake and metabolism of insect metabolites by leaves and tentacles of *Drosera* species. *New Phytologist* **77**, 625–634.

Chapman, T.A. (1906). Observations on the life-history of *Trichoptilus paludum*. *Transactions of the Entomological Society of London*, 133–154.

Chapman, V.J. (1947). A new endemic species of *Nepenthes*. *Ceylon Journal of Science* **12**, 221–223.

Cheatham, N.H. (1976). Butterfly Valley botanical area. *Fremontia* **4**, 3–8.

China, W.E. (1953). Two new species of the genus *Cyrtopeltis* (Hemiptera) associated with sundews in Western Australia. *The Western Australian Naturalist* **4**, 1–9.

China, W.E. and Carvalho, J.C.M. (1951). A new ant-like mirid from Western Australia. *Annals and Magazine of Natural History 12th Series* **4**, 221–225.

Christen, K. von (1961). Beitrag zur Pharmakochemie und Pharmakologie des Fettkrautes (*Pinguicula vulgaris* L.) (Investigations into the chemical pharmacology of the butterwort). *Pharmazie* **16**, 92–102.

Christensen, N.L. (1976). The role of carnivory in *Sarracenia flava* L. with regard to specific nutrient deficiencies. *Journal of the Elisha Mitchell Scientific Society* **92**, 144–147.

Christy, M. (1923). The common teazel as a carnivorous plant. *Journal of Botany* **61**, 33–45.

Ciobanu, L.R. and Tăcină, F.L. (1973). L'ultrastructure des glandes digestives chez *Drosera capensis* (The ultrastructure of the digestive glands in *D. capensis*). *Review of Roumanian Biology: Series Botany* **18**, 249–253.

Clancy, F.G.A. and Coffey, M.D. (1977). Acid phosphatase and protease release by the insectivorous plant *Drosera rotundifolia*. *Canadian Journal of Botany* **56**, 480–488.

Clark, J.B. and Lister, G.R. (1975). Photosynthetic action spectra of trees. 2. The relationship of cuticle structure to the visible and ultra-violet spectral properties of needles from four conifer species. *Plant Physiology* **55**, 407–413.

Clautriau, G. (1899–1900) La digestion dans les urnes de *Nepenthes* (Digestion in the pitchers of *Nepenthes*). *Memoirs couronnes et autres memoires, publies des lettres et des Beaux Arts par Académie royale de Science Belgique.* **59**, 1–55, (Collection in 8, 3rd memoir).

Clusius (1601). 'Rariorum Plantarum Historia Book 3', 82.

Cohn, F. (1875). Ueber die Function der Blase von *Aldrovanda* und *Utricularia* (On the function of the traps of *Aldrovanda* and *Utricularia*). *Beiträge zur Biologie der Pflanzen* **1**, (Section III), 71–92.

Coker, W.C. (1928). The distribution of Venus's Flytrap (*Dionaea muscipula*). *Journal of the Elisha Mitchell Scientific Society* **43**, 222–228.

Collinson, P. (1765) *See* Rendle (1925) and Henrey (1975).

Colombo, P.M. and Rascio, N. (1977). Ruthenium red staining for electron microscopy of plant material. *Journal of Ultrastructural Research* **60**, 138–139.

Compton, R.H. (1909). The morphology and anatomy of *Utricularia brachiata* Oliver. *New Phytologist* **8**, 117–130.

Corbet, S.A. (1973). An illustrated introduction to the testate rhizopods in *Sphagnum* with special reference to the area around Malham Tarn, Yorkshire. *Field Studies* **3**, 801–838.

Corner, E.J.H. (1964). 'The Life of Plants', Natural History Series. Weidenfeld & Nicholson, London.

Cox, G.C. and Juniper, B.E. (1973). Autoradiographic evidence for paramural body function. *Nature, New Biology* **243**, 116–117.

Croat, T.B. (1986). Flowering behaviour of the neotropical

genus *Anthurium* (Araceae). *American Journal of Botany* **67**, 888–904.

Croizat, L. (1961). *See* Heads *et al.* (1984).

Cronquist, A. (1968). 'The Evolution and Classification of Flowering Plants'. Nelson, London.

Cronquist, A. (1981). 'An Integrated System of Classification of Flowering Plants.' Columbia University Press, New York.

Cummins, K.W. and Wuycheck, J.C. (1971). Caloric equivalents for investigations in ecological energetics. *International Vereinigen für theoretische und angewandte. Limnologie* **18**, 1–158.

Culpepper, N. (1813). 'English Physician and Complete Herbal' Vol. 1, 324.

Curran, P.F. and MacIntosh, J.R. (1962). A model system for biological water transport. *Nature* **193**, 347–348.

Curtis, J.D. and Lersten, N.R. (1978). Heterophylly in *Populus grandiflora*, Salicaceae, with emphasis on resin glands and extrafloral nectaries. *American Journal of Botany* **65**, 1003–1010.

Dafni, A. (1984). Mimicry and deception in pollination. *Annual Review of Ecology and Systematics* **15**, 259–278.

Dahlgren, R. (1975). A system of classification of the angiosperms to be used to demonstrate the distribution of characters. *Botanisker Notiser* **128**, 119–147.

Dahlgren, R., Jensen, S.R. and Nielsen, B.J. (1981). A revised classification of the Angiosperms with comments on correlations between chemical and other characters. *In* 'Phytochemistry and Angiosperm Phylogeny' (Eds. Young, D.A. and Siegler, D.S.), 149–199. Praeger, New York.

Dakin, W.J. (1919). The Western Australian pitcher plant *Cephalotus follicularis* and its physiology. *Journal of the Proceedings of the Royal Society of Western Australia* **4**, 37–53.

Dakwale, S. and Bhatnagar, S. (1985). Insect-trapping behaviour and diel periodicity in *Sauromatum guttatum* (Araceae). *Current Science* **54**, 699–702.

Damtoft, S., Jensen, S.R. and Nielsen, B.J. (1985). Iridoid glucosides from *Utricularia australis* and *Pinguicula vulgaris* (Lentibulariaceae). *Phytochemistry* **24**, 2281–2283.

Danser, B.H. (1928). The Nepenthaceae of the Netherlands Indies. *Bulletin du Jardin Botanique Buitenzorg* **9**, 249–438.

Danserau, P. and Segados-Vianna, F. (1952). Ecological study of the peat bogs of eastern N. America in the Laricetum phase of bog succession. *Canadian Journal of Botany* **30**, 490–520.

Darwin, C. (1875). 'Insectivorous Plants'. John Murray, London.

Darwin, C. (1877). On the nectar glands of the common brakefern. *Botanical Journal of the Linnaean Society* **15**, 407–409.

Darwin, C. (1887). 'The Life and Letters of Charles Darwin', 2nd edn, Vol. 3. John Murray, London.

Darwin, C. (1888). 'Insectivorous Plants', 2nd edn, revised by Francis Darwin. John Murray, London.

Darwin, F. (1876). The process of aggregation in the tentacles of *Drosera rotundifolia*. *Quarterly Journal of the Microscopical Society* **16**, 309–319.

Darwin, F. (1878). Experiments on the nutrition of *Drosera rotundifolia*. *Journal of the Linnaean Society of Botany* **17**, 17–32.

Daubenmire, R.F. (1974). 'Plants and environment: A Textbook of Plant Autecology', 3rd edn. John Wiley, New York.

Daumann, E. (1930). Das Blutennektarium von *Nepenthes* (The floral nectaries of *Nepenthes*). *Beihefte Botanisches Central blatt* **47**, 1–14.

Daumann, E. (1960). On the pollination Biology of *Parnassia* flowers. A new contribution to the experimental flower ecology. *Biologia Plantarum* **2**, 113–125.

Daumer, K. (1958). Blumenfarben, wie sie die Bienen sehen (Flower colours as seen by bees). *Zeitschrift für vergleichende Physiologie* **41**, 49–110.

De Buhr, L.E. (1973). Distribution and reproductive biology of *Darlingtonia californica*. M.A. Thesis, Claremont Graduate School, Claremont, California, May, 1973.

De Buhr, L.E. (1974). The distribution of *Darlingtonia californica*. *Carnivorous Plant Newsletter* **3**, 24–26.

De Buhr, L.E. (1975a). Observations on *Byblis gigantea* in Western Australia. *Carnivorous Plant Newsletter* **4**, 60–61.

De Buhr, L.E. (1975b). Phylogenetic relations of the Sarraceniaceae. *Taxon* **24**, 297–306.

De Buhr, L.E. (1976). Field notes on *Cephalotus follicularis* in Western Australia. *Carnivorous Plant Newsletter* **5**, 8–9.

De Buhr, L.E. (1977a). Wood anatomy of the Sarraceniaceae; Ecological and evolutionary implications. *Plant Systematics and Evolution* **128**, 159–169.

De Buhr, L.E. (1977b). Sectional reclassification of *Drosera* subgenus Ergaleium (Droseraceae). *Australian Journal of Botany* **25**, 209–218.

De Candolle, C.P. (1873). 'Prodromus; Systemis Naturalis Regni Vegetabilis'. S.G. Masson, Paris.

De Candolle, C.P. (1898–1899). Sur les feuilles peltées (On the formation of peltate leaves). *Bulletin des Travaux de la Société botanique de Genève* **9**, 1.

De Diaz, E.N.L. (1981). Desmidiaceae en *Utricularia foliosa* L. (Desmids in *Utricularia foliosa*). *Lilloa* **35**, 67–80.

Delaporte, F. (1982). 'Nature's Second Kingdom: Explorations of Vegetality in the 18th Century'. M.I.T. Press, Cambridge, Mass.

Delpino, F. (1874). Ulteriori osservazioni e considerazioni sulla dicogamia nel regno vegetale: 2. IV. Delle piante ziodifile (Further observations and considerations in the vegetable kingdom: 2. IV. of zidiophyllous plants). *Annali societa Italiana Naturale. Museo Civico Storia Naturale Milano* **16**, 151–349.

Dennis, W.M. (1980). *Sarracenia oreophila* (Kearny) Wherry in the Blue Ridge Province of North Eastern Georgia. *Castanea* **45**, 101–103.

Denoel, A. (1949a). Determination de l'activité des *Droseras* indigènes et de leurs teintures (Determination of the activity of the indigenous *Droseras* and their extracts). *Journal de Pharmacie Belgique* **4**, 3–19.

Denoel, A. (1949b). Les essais des *Drosera* indigènes et de leurs teintures (Essays on the indigenous *Droseras* and their extracts). *Journal Pharmacologie Belgique (new series 1)* **4**, 175–183.

Dernby, K.G. (1917a). Notiz betreffend die proteolitischen Enzyme der *Drosera rotundifolia* (Observations concerning the proteolytic enzymes in *D. rotundifolia*). *Biochemische Zeitschrift* **78**, 197–199.

Dernby, K.G. (1917b). Die proteolitischen Enzyme der *Pinguicula vulgaris* (The proteolytic enzymes of *P. vulgaris*). *Biochemische Zeitschrift* **80**, 152–158.

Desire, C. (1946). Action du rouge neutre sur les glandes de *Pinguicula vulgaris* (The action of neutral red on the glands of *P. vulgaris*). *Compte Rendu de la Société de Biologie* **140**, 265–267.

Dexheimer, J. (1972). Quelques aspects ultrastructuraux de la secretion de mucilage par les glandes digestives de *Drosera rotundifolia* (Some ultrastructural aspects of the secretion of mucilage by the digestive glands of *D. rotundifolia*). *Comptes Rendues de l'Académie des Sciences, Paris* **275D**, 1983–1986.

Dexheimer, J. (1976). Étude de la secretion de mucilage par les cellules des glandes digestives de *Drosera* (*D. rotundifolia* L. et *D. capensis* L.) Application de quelques techniques cytochimiques (Study of the secretion of mucilage in digesting gland cells of *Drosera*). *Cytobiologie* **13**, 307–321.

Dexheimer, J. (1978a). Study of mucilage secretion by the cells of the digestive glands of *Drosera capensis* L. using staining of the plasmalemma and mucilage by phosphotungstic acid. *Cytologia* **43**, 45–52.

Dexheimer, J. (1978b). Étude de la secretion de mucilage par les cellules des glandes digestives de *Drosera capensis* L. Localisation ultrastructurale des phosphatases neutres et de l'ATPase (Study of mucilage secretion by the cells of the glands of *D. capensis* L. Ultrastructural localisation of neutral phosphatases and of ATPase). *Zeitschrift für Pflanzenphysiologie* **86**, 189–201.

Dexheimer, J. (1978c). Localisation ultrastructurale des phosphatases acides (phosphomonoesterases acides) dans les cellules des glandes digestives du *Drosera capensis* L. pendant la synthèse de mucilage (Ultrastructural localisation of acid phosphatases (acid phosphomonoesterases) in the digestive gland cells of *D. capensis* during the synthesis of mucilage). *Revue de Cytobiologie et de Biologie Végétales* **1**, 49–57.

Dexheimer, J. (1979). Ultrastructural localisation of enzymatic activities in the cells of the digestive glands of *Drosera capensis* during the mucigenic phase: Detection of glucose-6-phosphatase activity. *Cytologia (Tokyo)* **44**, 153–160.

Di Gregorio, G.J. and Di Palma, J.R. (1966). Anthocyanin in *Dionaea muscipula* Ellis (Venus's Flytrap). *Nature (London)* **212**, 1264–1265.

Diannelidis, T. and Umrath, K. (1953). Aktionsstrome der Blasen von *Utricularia vulgaris* (Action potentials in the traps of *U. vulgaris*). *Protoplasma* **42**, 58–62.

Diaz-Gonzales, T.E., Guerra, J. and Nieto, J.M. (1982). Contribucion al Conocimieto de la clase Adiantetea BR–BL. 1942 en la peninsula Iberica (Contribution to the knowledge of the class Adiantetea BR–BL. 1942 in the Iberian peninsula). *Anales del Jardin Botanico de Madrid* **38**, 497–506.

Dickinson, C.H. and Preece, T.F. (1976). 'Microbiology of Aerial Plant Surfaces'. Academic Press, New York and London.

Didry, N., Pinkas, M. and Dubreuil, L. (1986). Activité antibactérienne de naphthoquinones d'origine végétale (Antibacterial activity of some naphthoquinones found in plants). *Annales Pharmaceutiques Françaises* **44**, 73–78.

Diels, L. (1906). Droseraceae. *In* 'Sarraceniales, Das Pflanzenreich' IV, 112 (Ed. Engler, A.), 1–137. Verlag von Wilhelm Engelmann, Leipzig.

Disney, R.H.L. (1981a). A new species of *Megaselia* from *Nepenthes* in Hong Kong, with re-evaluation of the genus *Endonepenthis* (Diptera: Phoridae). *Oriental Insects* **15**, 201–206.

Disney, R.H.L. (1981b). A new species of *Megaselia* (Diptera, Phoridae) that breeds in pitchers of *Nepenthes* in Sri Lanka. *Ceylon Journal of Science* **14**, 89–101.

Dixon, K.W. and Pate, S.J. (1978). Phenology, morphology and reproductive biology of the tuberous sundew, *Drosera erythrorhiza* Lindl. *Australian Journal of Botany* **26**, 441–454.

Dixon, K.W. and Pate, J.S. (1980). Biology of a Western Australian tuberous *Drosera*. *Carnivorous Plant Newsletter* **9**, 9–23.

Dixon, K.W., Pate, J.S. and Bailey, W.J. (1980). Nitrogen nutrition of the tuberous sundew *Drosera erythrorhiza* Lindl. with special reference to the catch of arthropod fauna by its glandular leaves. *Australian Journal of Botany* **28**, 283–297.

Dodge, H.R. (1947). A new species of *Wyeomyia* from the pitcher-plant (Diptera: Culicidae). *Proceedings of the Entomological Society of Washington* **49**, 117–122.

Dodonaeus (vulg. Dodoens), R. (1618). 'Cruydt-Boeck' (Ed., with preface, van Ravelingen, J.), 744. Leyden.

Dodson, C.H., Dressler, R.L., Hills, H.G., Adams, R.H. and Williams, N.H. (1969). Biologically active compounds in orchid fragrances. *Science* **164**, 1243–1249.

Dore Swamy, R. and Mohan Ram, H.Y. (1969). Studies on growth and flowering in axenic cultures of insectivorous plants I. Seed germination and establishment in cultures of *Utricularia inflexa*. *Phytomorphology*, **19**, 363–371.

Dore Swamy, R. and Mohan Ram, H.Y. (1971). Studies on growth and flowering in axenic cultures of insectivorous plants II. Induction of flowering and development of flower in *Utricularia inflexa*. *Zeitschrift für Pflanzenphysiologie* **65**, 315–325.

Dover, C. (1928). Notes on the fauna of pitcher-plants from Singapore Island. *Journal of the Malayan Branch of the Royal Asiatic Society* **6**, 1–27.

Dressler, R.L. (1968). Observations on orchids and euglossine bees in Panama and Costa Rica. *Revista Biologica Tropica* **15**, 143–183.

Duchartre, P. (1867). 'Elements de Botanique', 308, 358. J.B. Baillière, Paris.

Duffey, S.S. (1986). Plant glandular trichomes: their partial role in defence against insects. *In* 'Insects and the Plant Surface' (Eds. Juniper, B.E. and Southwood, T.R.E.), 151–172. Edward Arnold, London.

Dujardin-Neametz, G.E. and Egasse, E. (1889). 'Les Plantes Médicinales', 250–251.

Dupont, L.M. (1986). Temperature and rainfall variation in the Holocene, based on comparative palaeoecology and isotope geology of a hummock and a hollow (Bourtangerveen, the Netherlands). *Review of Palaeobotany and Palynology* **48**, 71–159.

Durand, R. and Zenk, M.H. (1971). Biosynthesis of plumbagin (5-hydroxy-2-methyl-1,4-naphthoquinone) via the acetate pathway in higher plants. *Tetrahedron Letters* **32**, 3009–3072.

Durand, R. and Zenk, M.H. (1974a). Homogentisate ring-cleavage pathway in the biosynthesis of acetate-derived naphthoquinones of the Droseraceae. *Phytochemistry* **13**, 1483–1492.

Durand, R. and Zenk, M.H. (1974b). Enzymes of the homogentisate ring-cleavage pathway in cell suspension cultures of higher plants. *FEBS Letters* **39**, 218–220.

Durand, R. and Zenk, M.H. (1976). The biosynthesis of the naphthoquinone 7-D methyljuglone. *Biochemie und Physiologie der Pflanzen* **169**, 213–217.

Dwyer, T.P. (1983). Seed structure of carnivorous plants. *Carnivorous Plant Newsletter* **12**, 8–23.

Eastop, V.F. (1986). Hooked epidermal hairs. *Botanical Society of the British Isles News* **44** (December), 14–15.

Edwards, H. (1876). *Darlingtonia californica* Torrey. *Proceedings of the Californian Academy of Science* **6**, 161–166.

Edwards, S. (1804). *Dionaea muscipula*. *Curtis's Botanical Magazine* **20**, 785.

Eichler, A.W. (1881). Über die Schlauchblätter von *Cephalotus follicularis* (On the trapping leaves of *C. follicularis*). *Jahrbuch des Berliner Botanish Garten* **1**, 193–197 (through Engler and Prantl).

Eisner, T. and Aneshausley, D.J. (1983). Adhesive strength of the insect-trapping glue of a plant (*Befaria racemosa*). *Annals of the Entomological Society of America* **76**, 295–298.

Eisner, T. and Shepherd, J. (1965). Caterpillar feeding on a sundew plant. *Science* **150**, 1608–1609.

Ekambaram, T. (1916). Irritability of the bladders in *Utricularia*. *Agricultural Journal of India* **11**, 72–79.

Eleuterius, L.N. and Jones, S.B. (1969). A floristic and ecological study of pitcher plant bogs in south Mississippi. *Rhodora* **71**, 29–34.

Ellis, J. (1770). 'Directions for Bringing over Seeds and Plants from the East Indies, and other Distant Countries in a State of Vegetation. To which is added, the figure and botanical description of a new plant, *Dionaea muscipula* or Venus's Flytrap'. L. Davis, London.

Erber, D. (1979). Untersuchungen zur Biozonos und Nekrozonos in Kannenpflanzen auf Sumatra (Investigations into the biocenosis and necrocenosis of pitcher plants in Sumatra). *Archiv für Hydrobiologie* **87**, 37–48.

Ernst, A. (1961). Revision der Gattung *Pinguicula* (A revision of the genus *Pinguicula*). *Botanische Jahrbucher für Systematik, Pflanzengeschichte und Pflanzengeographie* 145–194.

Erwin, D.H. and Valentine, J.W. (1984). 'Hopeful Monsters', transposons and Metazoan radiation. *Proceedings of the National Academy of Science, USA* **81**, 5482–5483.

Evans, K.W. and Brust, R.A. (1972). Induction and termination of diapause in *Wyeomyia smithii* (Diptera: Culicidae), and larval survival studies at low and subzero temperatures. *The Canadian Entomologist* **104**, 1937–1950.

Faber, F.C. von (1912). Das erbliche Zusammenleben von Bakterien und tropischen Pflanzen (The inherited states of symbiosis between bacteria and tropical plants). *Jahrbücher für wissenschaftliche Botanik* **51**, 285–375.

Fabian-Galan, G. and Sălăgeanu, N. (1968). Considerations on the nutrition of certain carnivorous plants (*Drosera capensis* and *Aldrovanda vesiculosa*). *Revue Roumaine de Biologie-Botanique* **13**, 275–280.

Fage, L. (1928). Notes on the fauna of pitcher plants IV Aranae. *Journal of the Malayan Branch of the Royal Asiatic Society* **6**, 13–19.

Fage, L. (1930). A sujet de deux araignées nouvelles trouvées dans les urnes de *Nepenthes* (Concerning two new spiders found in the pitchers of *Nepenthes*). *Treubia* **12**, 23–28.

Fahn, A. (1979). 'Secretory Tissues in Plants'. Academic Press, London and New York.

Fairbridge, R.W. Ed. (1975). 'The Encyclopedia of World Regional Geology, Part 1 Western Hemisphere (including Antarctica and Australia)'. Dowden, Hutchinson & Ross, Strondsberg, Pennsylvania.

Fashing, N.J. (1981). Arthropod associates of the cobra lily (*Darlingtonia californica*). 59th Annual Meeting of the Virginia Academy of Norfolk, Virginia USA May, 1981. *Virginia Journal of Science* **32**, 92.

Fashing, N.J. and O'Connor, B.M. (1984). *Sarraceniopsis* – A new genus for histiomatid mites inhabiting the pitchers of the Sarraceniaceae (Astigmata: Histiomatidae). *International Journal of Acarology* **10**, 217–227.

Fedotov, V.V. (1982). Dioncophyllites; a new genus in the Eocene flora of Raichika in Amur Oblast Russian–SFSR USSR (in Russian). *Botanicheski Zhurnal-Moskva (Leningrad)* **67** (7), 985–987.

Feinbrun, N. (1942). On the occurrence of *Drosera rotundifolia* L. in Lebanon. *Palestine Journal of Botany, Jerusalem, Series* **2**, 251–252.

Fenner, C.A. (1904). Beiträge zur Kenntnis der Anatomie, Entwicklungsgeschichte und Biologie der Laubblätter und Drüsen einiger Insektivoren (Contributions to the knowledge of the anatomy, the history of development and the biology of foliage leaves and glands of insectivorous plants). *Flora* **93**, 335–434.

Fermi, C. and Buscaglione (1899). Die proteolytischen Enzyme im Pflanzenreiche (Proteolitic enzymes in the Plant Kingdom). *Centralblatt für Bakteriologie, Parasitologie und Pflanzenkrankheit II*, **5**, 24–33, 63–66, 91–95, 125–134, 145–158.

Fernandes, A. (1941a). Morfologia e biologia das plantas carnivoras (Morphology and biology of the carnivorous plants). *Anuario da Sociedad Broteriana* **7**, 16–52.

Fernandes, A. (1941b). Morfologia e biologia das plantas carnivoras (Morphology and biology of the carnivorous plants). *Anuario da Sociedad Broteriana* **8**, 6–57.

Field, C. and Mooney, H.A. (1986). The photosynthesis–nitrogen relationship in wild plants. *In* 'On the Economy of Plant Form and Function' (Ed. Givnish, T.J.), 25–55. Cambridge University Press, Cambridge.

Fijalkowski, D. (1958). Badania nad rozmieszczeniem i ekologia aldrowandy pecherzykowatej (*Aldrovanda vesiculosa* L.) na Pojezierzu Leczynsko-Wlodawskim (The distribution and ecology of *Aldrovanda vesiculosa* L. in Leczyna-Wlodawa Lake). *Acta Societatis Botanicorum Poloniae*; 579–604.

Findlay, G.P. and Findlay, N. (1975). Anatomy and movement of the column of *Stylidium*. *Australian Journal of Plant Physiology* **2**, 597–621.

Findlay, G.P. and Findlay, N. (1981). Respiration dependent movement of the column of *Stylidium*. *Australian Journal of Plant Physiology* **8**, 1–12.

Fineran, B.A. (1980). Ontogeny of external glands in the bladderwort *Utricularia monanthos*. *Protoplasma* **105**, 9–25.

Fineran, B.A. (1985). Glandular trichomes in *Utricularia*: a review of their structure and function. *Israel Journal of Botany* **34**, 295–330.

Fineran, B.A. and Gilbertson, J.M. (1980). Application of lanthanum and uranyl salts as tracers to demonstrate apoplastic pathways for transport in glands of the carnivorous plant *Utricularia monanthos*. *European Journal of Cell Biology* **23**, 66–72.

Fineran, B.A. and Lee, M.S.L. (1974). Transfer cells in traps of the carnivorous plant *Utricularia monanthos*. *Journal of Ultrastructural Research* **48**, 162–168.

Fineran, B.A. and Lee, M.S.L. (1975). Organization of quadrifid and bifid hairs in the trap of *Utricularia monanthos*. *Protoplasma* **84**, 43–70.

Fineran, B.A. and Lee, M.S.L. (1980). Organization of mature external glands on the trap and other organs of the bladderwort *Utricularia monanthos*. *Protoplasma* **103**, 17–34.

Fish, D. (1976a). Structure and composition of the aquatic invertebrate community inhabiting epiphytic bromeliads in south Florida and the discovery of an insectivorous bromeliad. PhD Thesis, University of Florida, Gainesville, Florida, USA.

Fish, D. (1976b) Insect–plant relationships of the insectivorous pitcher plant *Sarracenia minor*. *The Florida Entomologist* **59**, 199–203.

Fish, D. (1983). Phytotelmata: Flora and Fauna. In 'Phytotelmata: Terrestrial Plants as Hosts for Aquatic Insect Communities' (Eds. Frank, J.H. and Lounibos, L.P.), 1–27. Plexus, New Jersey.

Fish, D. and Beaver, R.A. (1978). A bibliography of the aquatic fauna inhabiting bromeliads (Bromeliaceae) and pitcher plants (Nepenthaceae and Sarraceniaceae). *Proceedings of the Florida Anti-Mosquito Association*, 19th Meeting, April 1978, 11–19.

Fish, D. and Hall, D.W. (1978). Succession and stratification of aquatic insects inhabiting the leaves of the insectivorous pitcher plant, *Sarracenia purpurea*. *The American Midland Naturalist* **99**, 172–183.

Fisher, R.A. (1958). 'The Genetical Theory of Natural Selec-

tion'. Dover, New York, USA.
Fletcher, T.B. (1908). Description of a new plume-moth from Ceylon, with some remarks upon its life-history. *Spolia Zeylanica* **5**, 20–32.
Fletcher, W.W. and Kirkwood, R.C. (1979). The bracken fern (*Pteridium aquilinum*), its biology and control. *In* 'The Experimental Biology of the Ferns' (Ed. Dyer, A.F.) Academic Press, London and New York.
Folkerts, G.W. (1977). Endangered and threatened carnivorous plants of North America. *In* 'Extinction is Forever: the Status of Threatened and Endangered Plants of the Americas' (Eds. Prance, G.T. and Elias, T.S.), 301–313. New York Botanic Garden, Bronx, New York.
Folkerts, G.W. (1982). The gulf coast pitcher plant bogs. *American Scientist* **70**, 260–267.
Forsyth, A.B. and Robertson, R.J. (1975). K reproductive strategy and larval behavior of the pitcher-plant sarcophagid fly, *Blaesoxipha fletcheri*. *Canadian Journal of Zoology* **53**, 174–179.
Foss, P.J. and O'Connell, C. (1984). Further observations on *Sarracenia purpurea* in County Kildare. *Irish Naturalist's Journal* **21**, 264–266.
Fowlie, J.A. (1985). Malaya Revisited Part 29. Rediscovering the habitat of *Paphiopedilum dayanum* on serpentine cliffs on Mount Kinabalu in Eastern Malaysia (formerly North Borneo). *Orchid Digest* **49**, 125–129.
Fowlie, J.A. and Lamb, A. (1981). Malaya Revisited Part 25. *Paphiopedilum volonteanum* (Sand) Stein rediscovered in very peculiar habitats of North Borneo. *Orchid Digest* **48**, 164–169.
França, C. (1925). Recherches sur le *Drosophyllum lusitanicum*, Link et remarques sur les plantes carnivores (Researches on *D. lusitanicum* and comments on carnivorous plants). *Archives Portugaises des Sciences Biologiques* **1**, 1–30.
Franck, D.H. (1975). Early histogenesis of the adult leaves of *Darlingtonia californica* (Sarraceniaceae) and its bearing on the nature of the epiascidiate foliar appendages. *American Journal of Botany* **62**, 116–132.
Franck, D.H. (1976). Comparative morphology and early leaf histogenesis of adult and juvenile leaves of *Darlingtonia californica* and their bearing on the concept of heterophylly. *Botanical Gazette* **137**, 20–34.
Franco, J.D.A. (1971). 'Nova Flora de Portugal', Vol. 1, 244. Sociedade Astoria, Lisboa.
Frank, J.H. and Lounibos, L.P. (Eds.) (1983). 'Phytotelmata: Terrestrial Plants as Hosts for Aquatic Insect Communities'. Plexus, New Jersey.
Frank, J.H. and O'Meara, G.F. (1984). The bromeliad *Catopsis berteroniana* traps terrestrial arthropods but harbors *Wyeomyia* larvae (Diptera: Culicidae). *The Florida Entomologist* **67**, 418–42.
Franstadt, A. (1877). Anatomie der vegetativen Organe von *Dionaea muscipula* Ellis (Anatomy of the vegetative parts of *D. muscipula* Ellis). *Cohn's Beitrage Biologie der Pflanzen* **2**, 27–64.
Frenzer, R. (1980). 'Sonnentau' extract in 'Herbalpina'. *Carnivorous Plant Newsletter* **9**, 33–34.
Friis, E.M. and Skarby, A. (1981). Structurally preserved angiosperm flowers from the Upper Cretaceous of southern Sweden. *Nature* **191**, 484–486.
Fromm-Trinta, E. (1984). Genliseas Americanas. *Sellowia* **36**, 55–62.
Fromm-Trinta, E. (1986). *Genlisea pallidae* a new species for the genus *Genlisea* (Lentibulariaceae). *Bradea* **4**, 176–179.
Frost, T.M. (1976). Investigations of the Aufwuchs of freshwater sponges. Part 1. A quantitative comparison between the surfaces of *Spongilla lacustris* and three aquatic macrophytes. *Hydrobiologia* **50**, 145–149.
Garcia-Bellido, A. (1975). Genetic control of wing disc development in *Drosophila*. 'Cell Patterning'. *Ciba Foundation Symposium* **29**, 161–182.
Gardiner, W. (1885). On the phenomena accompanying stimulation of the gland-cells in the tentacles of *Drosera dichotoma*. *Proceedings of the Royal Society Series B. Biological Science* **39**, 229–234.
Garsault, F.P. de (1767). Description, vertus et usages de sept cent dix-neuf plantes (A description, properties and uses of seven hundred and nineteen plants) Tome 3, Planche 458; *Pinguicula*.
Gascoigne, R.M., Ritchie, E. and White, D.E. (1948). A survey of anthocyanins in the Australian flora. *Journal of the Proceedings of the Royal Society of New South Wales* **82**, 44.
Genéves, L. and Vintéjoux, C. (1967). Sur la présence et l'organisation en un reseau tridimensionnel d'inclusions de nature proteique dans les noyaux cellulaires des hibernacles d'*Utricularia neglecta* L. (Lentibulariaceae) (On the presence and organisation of a three-dimensional network of inclusions of a protein nature in the new cells of the utricles of *U. neglecta*). *Comptes Rendus de l'Academie des Sciences, Paris* **264D**, 2750–2753.
Gerard, J. (1597). 'The Herball or General History of Plants'. John Norton, London.
Gerard, J. (1633). 'The Herbal or General History of Plants', Revised and enlarged by Thomas Johnson. Norton & Whittakers, London.
Ghosh, E. (1928). Notes on the fauna of pitcher plants from Singapore. Part VIII. Protozoa. *Journal of the Malay Branch of the Royal Asiatic Society* **6**, 24–27.
Gibson, M. and Warren, K.S. (1970). Capture of *Schistosoma mansoni*, Miracidia and Cercariae by carnivorous aquatic vascular plants of the genus *Utricularia*. *Bulletin of the World Health Organization* **42**, 833–835.
Gibson, R.W. (1974). Aphid-trapping glandular hairs on hybrids of *Solanum tuberosum* and *S. berthaultii*. *Potato Research* **17**, 152–154.
Gibson, T.C. (1983a). Competition, disturbance and the carnivorous plant community in south eastern U.S. PhD Thesis, Dept. of Biology, University of Utah, December 1983.
Gibson, T.C. (1983b). On the cultivation of the giant Malaysian pitcher plant (*Nepenthes rajah*). *Carnivorous Plant Newsletter* **12**, 82–84.
Gilbert, L.E. (1971). Butterfly-plant coevolution: Has *Passiflora* won the selectional race with Heliconiine butterflies? *Science* **172**, 585–586.
Gilchrist, A.J. (1974). Absorption by the glands of insectivorous plants. PhD Thesis, Botany School, University of Oxford.
Gilchrist, A.J. and Juniper, B.E. (1974). An excitable membrane in the stalked glands of *Drosera capensis*. *Planta* **119**, 143–147.
Gillet, J.B. (1962). Pest pressure, an underestimated factor in evolution. *Systematics Association Publication* **4**, Taxonomy and Geography 37–46.
Gimingham, C.H. and Cormack, E. (1964). Plant distribution and growth in relation to aspect on hill slopes in north Scotland. *Transactions of the Botanical Society of Edinburgh* **39**, 525–538.
Givnish, T.J., Burkhardt, E.L., Happel, R.E. and Weintraub, J.D. (1984). Carnivory in the bromeliad *Brocchinia reducta*, with a cost/benefit model for the general restriction of carnivorous plants to sunny, moist, nutrient-poor habitats.

The American Naturalist **124**, 479–497.
Givnish, T.J., McDiarmid, R.W. and Buck, W.R. (1986). Fire adaptation in *Neblinaria celiae* (Theaceae), a high-elevation rosette shrub endemic to a wet equatorial tepui. *Oecologia (Berlin)* 70, 481–485.
Godfrey, R.K. and Stripling, H.L. (1961). A synopsis of *Pinguicula* (Lentibulariaceae) in the southern United States. *American Midland Naturalist* **66**, 395–409.
Goebel, K. (1891). 'Pflanzenbiologische Schilderungen' (Plant Biological Descriptions). Part 2V. Insectivoren. Marburg, Germany.
Goebel, K. (1893). 'Pflanzenbiologische Schilderungen (Plant Biological Descriptions) Elwert, Marburg, Germany.
Goins, A.E. (1977). Observations on the life history and ecology of the southern pitcher-plant mosquito, *Wyeomyia haynei* Dodge. MSc Thesis, Auburn University, Alabama, USA.
Goldschmidt, R. (1933). Some aspects of evolution. *Science* **78**, 539–547.
Goldschmidt, R. (1960) 'The Material Basis of Evolution', 390. Pageant Books, New Jersey.
Gonzales, T.E.D., Guerra, J. and Nieto, J.M. (1982). Contribucion al conocimiento de la clase Adiantetea BR–BL. 1942 en la peninsula Iberica (A contribution to the understanding of the class Adiantetea in the Iberian peninsula). *Annales del Jardin Botanico de Madrid* **38 (II)**, 497–506.
Gordonoff, T. (1951). Wie wirken Präparate aus 'insektenfressenden' Pflanzen bei Krampf- und Keuchhusten? (How do preparations from insectivorous plants work in cases of convulsive coughing and whooping-cough?) *Schweizerische Medizinische Wochenschrift* **81**, 111–113.
Gorup, E. von and Will, H. (1876). Fortgesetzte Beobachtungen über peptonbildende Fermente im Pflanzenreiche (Continued observations on the peptone-forming ferment of the plant kingdom). *Botanische Zeitung* 473–478.
Goss, R.C., Whitlock, L.S. and Westrick, J.P. (1964). Isolation and ecological observations of *Panagrodontus* sp. (Nematoda: Cephalobidae) in pitcher plants (*Sarracenia sledgei*). *Proceedings of the Helminthological Society of Washington* **31**, 19–20.
Gottwald, H. and Parameswaran, N. (1968). Das sekundare Xylem und die systematische Stellung der Ancistrocladaceae und Dioncophyllaceae (The secondary xylem and the systematic position of the Ancistrocladaceae and the Dioncophyllaceae). *Botanischer Jahrbücher* **88**, 49–69.
Gough, S.B. and Woelkerling, W.J. (1976). Wisconsin, USA desmids Part 2. Aufwuchs and plankton communities of selected soft water lakes, hard water lakes and calcareous spring ponds. *Hydrobiologia* **49**, 3–25.
Gowda, D.C., Reuter, G. and Schauer, R. (1982). Structural features of an acidic polysaccharide from the mucin of *Drosera binata*. *Phytochemistry* **21**, 2297–2300.
Gowda, D.C., Reuter, G. and Schauer, R. (1983). Structural studies of an acidic polysaccharide from the mucin secreted by *Drosera capensis*. *Carbohydrate Research* **113**, 113–124.
Green, S. (1967). Notes on the distribution of *Nepenthes* species in Singapore. *Gardens Bulletin, Singapore* **22**, 53–65.
Green, S., Green, T.L. and Heslop-Harrison, Y. (1979). Seasonal heterophylly and leaf gland features in *Triphyophyllum* (Dioncophyllaceae) a new carnivorous plant genus. *Botanical Journal of the Linnaean Society* **78**, 99–116.
Gregorio, G.J. Di and Palma, J.R. Di. (1966). See Di Gregorio, G.J. and Di Palma, J.R.
Gregory, P., Avé, D.A., Bouthyette, P.Y. and Tingey, W.M. (1986). Insect-defensive chemistry of potato glandular trichomes. *In* 'Insects and the Plant Surface' (Eds. Juniper, B.E. and Southwood, T.R.E.), 173–183. Edward Arnold, London.
Greshoff, M. (1909). Phytochemical Investigations at Kew. *Bulletin of Miscellaneous Information of the Royal Botanic Gardens, Kew* **1909 No. 10**, 397.
Grieve, B.J. (1961). The genus *Cephalotus*. *Australian Plants* **I**, 26–28.
Grieve, M. 1931 (reprinted 1959). 'A Modern Herbal', 640, 782. Ed. J. Cape, London.
Grigson, G. (1958). 'The Englishman's Flora', 192–193, 312–313. Phoenix House, London.
Grime, J.P. (1979). 'Plant Strategies and Vegetation Processes'. John Wiley, Chichester, New York, Brisbane, Toronto.
Grisley, G. (1661). 'Viridarium Lusitanicum'.
Grjebine, A. (1979). The mosquitoes living in the Malagasy pitcher-plants: new species of the genus *Uranotaenia* (Diptera: Culicidae). *Annales de la Société Entomologique de France* **15**, 53–74.
Grolle, R. (1967). Monographie der Lepidolaenaceae (Monograph on Lepidolaenaceae). *Journal of the Hattori Botanical Laboratory* **30**, 1–53.
Grncarevic, M. and Radler, F. (1967). The effect of wax components on cuticular transpiration-model experiments. *Planta* **75**, 23–27.
Grubb, P.J. (1986). Sclerophylls, pachyphylls and pycnophylls: the nature and significance of hard leaf surfaces. *In* 'Insects and the Plant Surface' (Eds. Juniper, B.E. and Southwood, T.R.E.), 137–150. Edward Arnold, London.
Grudger, E.W. (1947). The only known fish-eating plant *Utricularia* the bladderwort. *Scientific Monthly* **64**, 369–384.
Gunawardana, J. (1986). *Drosera indica*, a sundew of Sri Lanka. *Loris, The Journal of the Wildlife and Nature Protection Society of Sri Lanka* **17**, 130–131.
Gunning, B.E.S. (1978). Age-related and origin-related control of the numbers of plasmodesmata in cell walls of developing *Azolla* roots. *Planta* **143**, 181–190.
Gunning, B.E. and Pate, J.S. (1974). Transfer cells. *In* 'Dynamic Aspects of Plant Ultrastructure' (Ed. Robards, A.), 441–480. McGraw-Hill, Maidenhead.
Haberlandt, G.F.J. (1906). 'Sinnesorgane im Pflanzenreich' (Sense Organs in the Plant Kingdom). Leipzig, Germany.
Haberlandt, G. (1914). 'Physiological Plant Anatomy' (Translated from the fourth German edition by Drummond, M.), 591. Macmillan, London.
Hada, Y. (1930). The feeding habits of *Utricularia*. *Transactions of the Sapporo Natural History Society* **11**, 175–183.
Hadac, E. (1977). Poznamky o endemickych rostlinach Ceske socialisicke republicky (Observations on the endemic flora of the Czechoslovak Socialist Republic). *Zpravy Ceskoslovenske botanicke spolecnosti, Praha* **12**, 1–15.
Hall, D.M. (1966). A study of the surface wax deposits on apple fruit. *Australian Journal of Biological Science* **19**, 1017–1025.
Hall, D.M. and Jones, R.L. (1961). Physiological significance of surface wax on leaves. *Nature* **191**, 95–96.
Hall, D.W. and Fish, D.D. (1974). Baculovirus from the mosquito *Wyeomyia smithii*. *Journal of Invertebrate Pathology* **23**, 383–388.
Hallam, N.D. (1970). Leaf wax fine structure and ontogeny in *Eucalyptus* demonstrated by means of a specialized fixation technique. *Journal of Microscopy* **92**, 137–144.
Hanna, W. (1975). Brief observations on *Nepenthes mirabilis*. *Carnivorous Plant Newsletter* **4**, 31.
Harder, R. (1963). Über Blutenbildung durch tierische Zusatznahrung und andere Faktoren bei *Utricularia exoleta* R. Braun (On the flower formation as a result of animal

nutrition and other factors in *U. exoleta*). *Planta* **59**, 459–471.

Harder, R. (1967). Zur Frage der Ausscheidung antiseptischer Stoffe aus den Fangorganen von Carnivoren (On the question of excretion of antiseptic substances from the trapping organs of carnivores). *Review of Roumanian Biology: Series Botany* **12**, 159–162.

Harder, R. (1970a). *Utricularia* als Objekt für Heterotrophieuntersuchungen bei Blutenpflanzen (Wechselwirkung von Saccharose und Acetät) (*Utricularia* as a subject for the investigation of heterotrophic nutrition in flowering plants. The reciprocal action of sucrose and acetate). *Zeitschrift für Pflanzenphysiologie* **63**, 181–184.

Harder, R. (1970b). Einfluss von Daphniendekokt auf fünf Arten von *Utricularia* (Effect of *Daphnia* decoction on five species of *Utricularia*). *Beiträge zur Biologie der Pflanzen* **47**, 53–62.

Harder, R. and Zemlin, I. (1968). Blutenbildung von *Pinguicula lusitanica* in vitro durch Futterung mit Pollen (Flowering of *in vitro* cultures of *Pinguicula lusitanica* after feeding with pollen). *Planta* **78**, 72–78.

Hargreaves, J.A., Brown, G.A. and Holloway, P.J. (1982). The structural and chemical characteristics of the leaf surface of *Lupinus albus* in relation to the distribution of antifungal compounds. *In* 'The Plant Cuticle' (Eds. Cutler, D.F., Alvin, K.L. and Price, C.E.). Linnean Society Symposium Series No. 10. Academic Press, London and New York.

Harper, J.L. (1977). 'Population Biology of Plants'. Academic Press, London and New York.

Harshberger, J.W. (1925). Notes on the Portuguese insectivorous plant *Drosophyllum lusitanicum*. *Proceedings of the American Philosophical Society* **64**, 51–54.

Heads, M., Craw, R.C. and Gibbs, G.W. (Eds.) (1984). Croizat's Panbiogeography and Principia Botanica: Search for a novel biological synthesis. *Tuatara* **27**, 1–75.

Hegner, R.W. (1926a) The interrelationships of protozoa and the utricles of *Utricularia*. *Biological Bulletin* **50**, 239–270.

Hegner, R.W. (1926b). The protozoa of the pitcher-plant *Sarracenia purpurea*. *Biological Bulletin* **50**, 271–276.

Heide-Jørgensen, H.S. (1978). The xeromorphic leaves of *Hakea suaveolens R. Br.* 1. Structure of photosynthetic tissue with intercellular pectic strands and tylosoids. *Botanisk Tidsskrift* **72**, Bind.2.–3, 87–103.

Heide-Jørgensen, H.S. (1981). Parasitisme og Carnivoi (Parasitism and carnivory). *Kompendium, Institut for Planteanatomi og Cytologi, Kobenhavns Universitet* **61**.

Heide-Jørgensen, H.S. (1986). Kødaedende planter 1 (Carnivorous plants I: Introduction to the pitcher plants). *Naturens Verden* **3**, 81–102.

Heide-Jørgensen, H.S. (1987). Kødaedende planter 2 (Carnivorous plants 2: Active and passive sticky traps and snap traps). *Naturens Verden* **8**, 279–296.

Heinrich, G. (1984). LAMMA – Ionenspektren der Fangschleim carnivoren Pflanzen (Ionic spectra of the 'Fangschleim' of carnivorous plants). *Biochemistrie und Physiologie der Pflanzen* **179**, 129–143.

Helder, R.J. (1967). Transport of substances through the tentacles of *Drosera capensis*. *In* 'Translocation in Plants' (Ed. Ruthland, W.) Encyclopedia of Plant Physiology 13, 44–54. Springer-Verlag, Berlin.

Henrey, B. (1975). 'British Botanical and Horticultural Literature before 1800, Vol. II.' Oxford University Press, London.

Henry, Y. and Steer, M.W. (1985). Acid phosphatase localization in the digestive glands of *Dionaea muscipula* Ellis flytraps. *Journal of Histochemistry and Cytochemistry* **33**, 339–344.

Hepburn, J.S. (1918). Biochemical studies of the pitcher liquor of *Nepenthes*. *Proceedings of the American Philosophical Society* **57**, 112–129.

Hepburn, J.S., St. John, E.Q. and Jones, F.M. (1919). Biochemical studies of insectivorous plants. *Contributions of the Botanical Laboratory of the University of Pennsylvania* **4**, 419–463.

Hepburn, J.S., Jones, F.M. and St. John, E.Q. (1920). The absorption of nutrients and allied phenomena in the pitchers of the Sarraceniaceae. *Journal of the Franklin Institute* **189**, 147–184.

Hepburn, J.S., Jones, F.M. and St. John, E.Q. (1927). The biochemistry of the American pitcher plants: Biochemical studies of the North American Sarraceniaceae. *Transactions of the Wagner Free Institute of Science of Philadelphia* **11**, 1–95.

Heslop-Harrison, Y. (1970). Scanning electron microscopy of fresh leaves of *Pinguicula*. *Science* **167**, 172–174.

Heslop-Harrison, Y. (1975). Enzyme release in carnivorous plants. *In* 'Lysozymes in Biology and pathology', (Eds. Dingle, J.T. and Dean, R.T.), **4**, 525–578. North Holland Publishing Company, Amsterdam.

Heslop-Harrison, Y. (1976). Enzyme secretion and digest uptake in carnivorous plants. *In* 'Perspectives in Experimental Biology'. S.E.B. Symposial Volume **2**. Proceedings of the 50th Anniversary Meeting, Cambridge, 1974 (Ed. Sunderland, N.), 463–476. Pergamon Press, Oxford.

Heslop-Harrison, Y. (1978). Carnivorous Plants. *Scientific American* **February**, 104–115.

Heslop-Harrison, Y. and Heslop-Harrison, J. (1980) Chloride ion movement and enzyme secretion from the digestive glands of *Pinguicula*. *Annals of Botany* **45**, 729–731.

Heslop-Harrison, Y. and Heslop-Harrison, J. (1981). The digestive glands of *Pinguicula*: Structure and Cytochemistry. *Annals of Botany* **47**, 293–319.

Heslop-Harrison, Y. and Knox, R.B. (1971). A cytochemical study of the leaf-gland enzymes of insectivorous plants of the genus *Pinguicula*. *Planta (Berlin)* **96**, 183–211.

Hétet, F. (1879). Sur les principes qui donnent au *Sarracenia purpurea* ses propriétées therapeutiques (On the principles which give *Sarracenia purpurea* its therapeutic properties). *Comptes Rendues d'Académie des Sciences, Paris* **88**, 185.

Hevly, R.H. (1969). Nomenclatural history and typification of *Martynia* and *Proboscidea* (Martyniaceae). *Taxon* **18**, 527–534.

Hill, A.E. and Hill, B.S. (1976). Mineral ions. *In* 'Transport in Plants IIB. Tissues and Organs' (Eds. Lüttge, U. and Pitman, M.G.), 22–243. Springer-Verlag, Berlin.

Hill, B. and Findlay, G.P. (1981). The power of movement in plants: The role of osmotic machines. *Quarterly Review of Biophysics* **14**, 173–222.

Hill, J. (1756). 'The British Herbal', 107–108, 187–188. London.

Hilton, D.F.J. (1982). The biology of *Endothenia daeckeana* (Lepidoptera: Olethreutidae), an inhabitant of the ovaries of the northern pitcher plant *Sarracenia purpurea* (Sarraceniaceae). *The Canadian Entomologist* **114**, 269–274.

Hindley, K. (1980). The association of lady-slipper orchids and insectivorous plants Part 3 The association of *Cypripedium reginae* and *Cypripedium pubescens* with *Drosera rotundifolia* in bogs in Vermont. *Orchid Digest* **44**, 233–235.

Hirst, S. (1928). A new tyrogliphid mite (*Zwickia nepenthesiana* sp.n.) from the pitchers of *Nepenthes ampullaria*. *Journal of the Malayan Branch of the Royal Asiatic Society* **6**, 19–22.

Hodick, D. and Sievers, A. (1986). The influence of Ca^{++} on the action potential in mesophyll cells of *Dionaea muscipula* Ellis. *Protoplasma* **133**, 83–84.

Hodick, D. and Sievers, A. (1988). The action potential of *Dionaea muscipula* Ellis. Submitted for *Planta* (for preliminary report see Abstract from the 14th International Botanical Congress, Berlin, 1987).

Holloway, P.J. (1968). The effect of superficial wax on leaf wettability. *Annals of Applied Biology* **63**, 145–153.

Holloway, P.J., Brown, G.A., Baker, E.A. and Macey, M.J.K. (1977). Chemical composition and ultrastructure of the epicuticular wax in three lines of *Brassica napus*. *Chemistry and Physics of Lipids* **19**, 114–127.

Holter, H. and Linderstrom-Lang, K. (1933). Beiträge zur enzymatischen Histochemie III: Über die Proteinasen von *Drosera rotundifolia* (Investigations on enzyme histochemistry III: On the proteinases of *D. rotundifolia*). *Hoppe-Seyler's Zeitung für Physiologische Chemie* **214**, 223–240.

Holttum, R.E. (1954). 'Plant Life in Malaya'. Longman, Green, London.

Honsell, E. and Ghirardelli, L.A. (1972). Osservazioni sugli inclusi proteici nucleari nella cellule ghiandiolari degli ascidi di *Utricularia vulgaris* (Protein intra-nuclear inclusions in cell glands of ascidia in *Utricularia vulgaris* and their significance). *Giornale Botanicum Italia* **106**, 286.

Honsell, E., Ghirardelli-Gambardella, L. and Bole, E. (1975). Pinocitosi nelle cellule ghiandolari di *Utricularia vulgaris* (Pinocytosis in the glandular cells of *U. vulgaris*). *Informatore Botanico Italiano* **7**, 30–32.

Hooker, H.D. (1919). Notes on the life history of *Episilia monochromatea* Morr. (Lepidoptera. Noctuidae). *Entomological News* **30**, 61–63.

Hooker, J.D. (1847). 'Flora Antarctica', Vol. 2, 229–230. Reeve Brothers, London.

Hooker, J.D. (1859). On the origin and development of the pitcher of *Nepenthes*, with an account of some new Bornean plants of the genus. *Transactions of the Linnean Society* **22**, 415–424.

Hooker, J.D. (1869). *Drosophyllum lusitanicum*. *Curtis's Botanical Magazine* **25**, 3rd Series, Tab. 5796.

Hooker, J.D. (1875). Address to the Department of Botany and Zoology. Report of the 44th Meeting of the British Association for the Advancement of Science, Belfast 1874, 102–116.

Hooker, J.D. (1890). *Heliamphora nutans*, Native of British Guiana. *Curtis's Botanical Magazine* **46**, 3rd series (or Vol. 116 of the whole work), Tab. 7093.

Hooker, W.J. (1831). *Cephalotus follicularis*. *Curtis's Botanical Magazine* **58**, 3118–3119.

Hooker, W.J. (1858). *Nepenthes villosa* from Kina-Baloo, Borneo. *Curtis's Botanical Magazine* **14**, 3rd Series, 5080.

Hooper, S.N. and Chandler, R.F. (1984). Herbal remedies of the Maritime Indians: Phytosterols and triterpenes of 67 plants. *Journal of Ethnopharmacology* **10**, 181–194.

Horner, H.T. and Beltz, C.K. (1975). Ligule development and fine structure in two heterophyllous species of *Selaginella*. *Canadian Journal of Botany* **53**, 127–145.

Howard-Williams, C. and Junk, N.J. (1977). The chemical composition of Central Amazonian aquatic macrophytes with special reference to their role in the ecosystem. *Archives of Hydrobiology* **79**, 446–464.

Hughes, R.D. and Jackson, C.G. (1958). A review of the Anoetidae (Acari). *Virginia Journal of Science* **9**, 5–198.

Huie, L.M. (1897). Changes in the cell organs of *Drosera rotundifolia* produced by feeding with egg-albumin. *Quarterly Journal of Microscopical Science* **39**, 387–425.

Hull, H.M., Went, F.W. and Bleckman, C.A. (1979). Environmental modifications of epicuticular wax structure of *Prosopis* leaves. *Journal of the Arizona–Nevada Academy of Science* **236**, 39–42.

Hunter, P.E. and Hunter, C.A. (1964). A new *Anoetus* mite from pitcher-plants (Acarina: Anoetidae). *Proceedings of the Entomological Society of Washington* **66**, 39–46.

Hurst, E. (1942). 'Poisonous Plants of New South Wales', 136. Sydney.

Huxley, C.R. (1978). The ant-plants *Myrmecodia* and *Hydnophytum* (Rubiaceae), and the relationships between their morphology, ant occupants, physiology and ecology. *New Phytologist* **80**, 231–268.

Huxley, C.R. (1980). Symbiosis between ants and epiphytes. *Biological Reviews* **55**, 321–340.

Huxley, C.R. (1986). Evolution of benevolent ant-plant relationships. *In* 'Insects and the Plant Surface' (Eds. Juniper, B.E. and Southwood, T.R.E.). Edward Arnold, London, Victoria, Maryland.

Iijima, T. and Sibaoka, T. (1981). Action potential in the trap-lobes of *Aldrovanda vesiculosa*. *Plant and Cell Physiology* **22**, 1595–1601.

Iijima, T. and Sibaoka, T. (1982). Propagation of action potential over the trap lobes of *Aldrovanda vesiculosa*. *Plant and Cell Physiology* **23**, 679–688.

Iijima, T. and Sibaoka, T. (1983). Movement of K^+ during shutting and opening of the trap lobes in *Aldrovanda vesiculosa*. *Plant and Cell Physiology* **24**, 51–60.

Iijima, T. and Sibaoka, T. (1985). Membrane potentials in excitable cells of *Aldrovanda vesiculosa* trap-lobes. *Plant and Cell Physiology* **26**, 1–13.

Irwin, W.P. (1966). Geology of the Klamath Mountains Province. *In* 'Geology of Northern California'. *California Division of Mines and Geology Bulletin* **190**, 19–38.

Irwin, W.P. (1977). Ophiolitic terrains of California, Oregon and Nevada. *In* 'North American Ophiolites'. (Eds. Coleman, R.G. and Irwin, W.P.), 75–92. *Oregon Department of Geology & Mineral Industries, Bulletin* **95**.

Istock, C.A. (1978). Fitness variation in a natural population. *In* 'Evolution of Insect Migration and Diapause (Ed. Dingle, H.), 171–190. Springer-Verlag, New York.

Istock, C.A., Tanner, K. and Zimmer, H. (1983). Habitat selection by the pitcher-plant mosquito, *Wyeomyia smithii*: Behavioral and genetic aspects. *In* 'Phytotelmata: Terrestrial Plants as Hosts for Aquatic Insect Communities' (Eds. Frank, J.H. and Lounibos, L.P.), 191–204. Plexus, New Jersey.

Istock, C.A., Wasserman, S.S. and Zimmer, H. (1975). Ecology and evolution of the pitcher-plant mosquito. Part 1 Population dynamics and laboratory responses to food and population density. *Evolution* **29**, 296–312.

Istock, C.A., Zisfein, J. and Vavra, K.J. (1976a). Ecology and evolution of the pitcher-plant mosquito. Part 2 The substructure of fitness. *Evolution* **30**, 535–547.

Istock, C.A., Vavra, K.J. and Zimmer, H. (1976b). Ecology and evolution of the pitcher-plant mosquito. Part 3 Resource tracking by a natural population. *Evolution* **30**, 548–557.

Jacobson, S.L. (1974). Effect of ionic environment on the responses of the sensory hair of Venus's Flytrap. *Canadian Journal of Botany* **52**, 1293.

Jaffe, M.J. (1973). The role of ATP in mechanically stimulated rapid closure of the Venus's Flytrap. *Plant Physiology* **51**, 17–18.

Jahandiez, E. and Maire, R. (1931). 'Catalogue des Plantes du Maroc', 319. Alger Imprimerie Minerva En vente a Paris, Chez P. Lechevalier, Libraire 12, Rue de Tournou.

Jane, F.W. and Russell-Wells, B. (1935). Observations on the seeds and seedlings of *Utricularia vulgaris*, L. *Transactions of the Norfolk and Norwich Naturalists' Society* **14**, 31–54.

Janzen, D.H. (1985). The natural history of mutualism. *In* 'The Biology of Mutualism and Evolution' (Ed. Boucher, D.H.), 90–99. Croom Helm, London, Sydney.

Jarzembowski, E.A. (1984). Early Cretaceous insects from southern England. *Modern Geology* **9**, 71–93.

Jay, M. and Gonnet, J.F. (1973). Iso-Cutellareine isolated from *Pinguicula vulgaris*. *Phytochemistry* **12**, 953–954.

Jay, M. and Gonnet, J.F. (1974). Recherches chemotaxonomiques sur les plantes vasculaires XXX Les flavonoides des deux Lentibulariacées; *Pinguicula vulgaris* et *Utricularia vulgaris* (Chemotaxonomic studies on vascular plants Vol. XXX The flavonoids of two of the Lentibulariaceae; *P. vulgaris* and *U. vulgaris*). *Biochimica Systematica Ecologia* **2**, 47–51.

Jay, M. and Lebreton, P. (1972). Chemotaxonomic research on vascular plants 26. The flavonoids of the Sarraceniaceae, Nepenthaceae, Droseraceae and Cephalotaceae; a critical study of the order Sarraceniales. *Naturaliste Canadienne* **99**, 607–613.

Jebb, M. (1987). (See Appendix 2).

Jeffree, C.E. (1986). The cuticle, epicuticular waxes and trichomes of plants, with reference to their structure, functions and evolution. *In* 'Insects and the Plant Surface' (Eds. Juniper, B.E. and Southwood, T.R.E.), 23–64. Edward Arnold, London.

Jeffree, C.E., Baker, E.A. and Holloway, P.J. (1975). Ultrastructure and recrystallisation of plant epicuticular waxes. *New Phytologist* **74**, 539–549.

Jensen, S.R., Nielsen, J.B. and Dahlgren, R. (1975). Iridoid compounds, their occurrence and systematic importance in the angiosperms. *Botaniska Notiser* **128**, 181–197.

Jentsch, J. (1970). Probleme bei der Reinigung von *Nepenthes*-Kannensaft-Enzymen (Problems with the purification of *Nepenthes* pitcher-sap enzymes). *Berichte der Deutschen Botanischen Gesellschaft* **83**, 171–176.

Jentsch, J. (1972). Enzymes from carnivorous plants; *Nepenthes*. Isolation of the protease Nepenthacin. *FEBS Letters* **21**, 273–276.

Jepson, W.L. (1951). 'A Manual of Flowering Plants of California'. University of California Press, Berkeley, Los Angeles.

Jizba, J., Herout, V. and Sorm, F. (1967). Isolation of ecdysterone (crustecdysone) from *Polypodium vulgare* L. rhizomes. *Tetrahedron Letters* **18**, 1689–1691.

Joel, D.M. (1978). Light and electron microscopic studies on the secretory ducts of some Anacardiaceae species in relation to the process of secretion. PhD Thesis, The Hebrew University of Jerusalem, August.

Joel, D.M. (1982). How the bladderwort captures its prey. *Israel Land and Nature* **8**, 54–57.

Joel, D.M. (1985). Leaf anatomy of *Caltha dionaeifolia* Hooker (Ranunculaceae) – is this species carnivorous? *Botanical Journal of the Linnean Society* **90**, 243–252.

Joel, D.M. (1986). Glandular structures in carnivorous plants; their role in mutual exploitation of insects. *In* 'Insects and the Plant Surface' (Eds. Juniper, B.E. and Southwood, T.R.E.), 219–234. Edward Arnold, London.

Joel, D.M. (1988). 'Mimicry and Mutualism in Carnivorous Pitcher Plants. Pitchers' *Biological Journal of the Linnean Society,* **35**.

Joel, D.M. and Fahn, A. (1980). Ultrastructure of the resin ducts of *Mangifera indica* (Anacardiaceae) 2. Resin secretion in the primary stem ducts. *Annals of Botany* **46**, 779–783.

Joel, D.M. and Gepstein, S. (1985). Chloroplasts in the epidermis of *Sarracenia* (the American pitcher-plant) and their possible role in carnivory – an immunocytochemical approach. *Physiologia Plantarum* **63**, 71–75.

Joel, D.M. and Heide-Jørgensen, H.S. (1985). Ultrastructure and development of the pitcher epithelium of *Sarracenia*. *Israel Journal of Botany* **34**, 331–349.

Joel, D.M. and Juniper, B.E. (1982). Cuticular gaps in carnivorous plant glands. *In* 'The Plant Cuticle' (Eds. Cutler, D.F., Alvin, K.L. and Price, C.E.), 121–130. Linnaean Society Symposium No. 10, Academic Press, London and New York.

Joel, D.M. and Juniper, B.E. (1988). The external glands of some carnivorous plant traps (in press).

Joel, D.M., Juniper, B.E. and Dafni, A. (1985). Ultraviolet patterns in the traps of carnivorous plants. *New Phytologist* **101**. 585–593.

Joel, D.M., Rea, P.A. and Juniper, B.E. (1983). The cuticle of *Dionaea muscipula* Ellis. (Venus's Flytrap) in relation to stimulation, secretion and absorption. *Protoplasma* **114**, 44–51.

Johannsen, O.A. (1932). Tanypodinae from the Malayan subregion of the Dutch East Indies. *Archives of Hydrobiology suppl. bd.* **9**, 493–507.

Jones, C.E. and Buchmann, S.L. (1974). Ultraviolet floral patterns as functional orientation cues in Hymenopterous pollination systems. *Animal Behaviour* **22**, 481–485.

Jones, F.M. (1904). Pitcher-plant insects I. *Entomological News* **15**, 14–17.

Jones, F.M. (1907). Pitcher-plant insects II. *Entomological News* **18**, 413–420.

Jones, F.M. (1908). Pitcher-plant insects III. *Entomological News* **19**, 150–156.

Jones, F.M. (1916). Two insect associates of the California pitcher-plant, *Darlingtonia californica*. *Entomological News* **27**, 385–391.

Jones, F.M. (1918). *Dohniphora venusta* Coquillet (Dipt.) in *Sarracenia flava*. *Entomological News* **29**, 299–302.

Jones, F.M. (1920). Another pitcher-plant insect (Diptera; Sciarinae). *Entomological News* **31**, 91–94.

Jones, F.M. (1921). Pitcher-plants and their moths. *Natural History* **21**, 296–316.

Jones, F.M. (1923). The most wonderful plant in the world. *Natural History* **23**, 589–596.

Jones, K. (1964). Pollen structure and development in *Drosera*. *Journal of the Linnean Society of London, Series Botany* **59**, 81–87.

Josselyn, J. (1672). New-England's Rarities Discovered. *In* 'Birds, Beasts, Fishes, Serpents and Plants of that Country'. G. Widdowes, London.

Judd, W.W. (1959). Studies of the Byron Bog Southwestern Ontario X; Inquilines and victims of the pitcher-plant *Sarracenia purpurea* L. *Canadian Entomologist* **91**, 171–180.

Judd, W.W. (1969). Studies of the Byron Bog in Southwestern Ontario XXXIX; Insects trapped in the leaves of the sundew, *Drosera intermedia* Hayne, and *Drosera rotundifolia* L. *Canadian Field Naturalist* **83**, 233–237.

Jung, K-D. and Lüttge, U. (1980). Effects of fusicoccin and abscisic acid on sugar and ion transport from plant glands. *Annals of Botany* **45**, 339–349.

Jung, K-D., Ball, E. and Lüttge, U. (1980). Inhibition of sugar and salt elimination by glands with the amino-acid analog p-fluorophenyl alanine. *Annals of Botany* **45**, 351–356.

Juniper, B.E. (1986). The path to plant carnivory. *In* 'Insects and the Plant Surface' (Eds. Juniper, B.E. and Southwood,

T.R.E.), 195–218. Edward Arnold, London.

Juniper, B.E. and Barlow, P.W. (1969). The distribution of plasmodesmata in the root tip of maize. *Planta* **89**, 352–360.

Juniper, B.E. and Burras, J.K. (1962). How pitcher plants trap insects. *New Scientist* **13**, 75–77.

Juniper, B.E. and French, A. (1973). The distribution and redistribution of endoplasmic reticulum (ER) in geoperceptive cells. *Planta* **109**, 211–224.

Juniper, B.E. and Gilchrist, A.J. (1976). Absorption and transport of calcium in the stalked glands of *Drosera capensis* L. *In* 'Perspectives in Experimental Biology' (Ed. Sutherland, N.), 477–486. Pergamon Press, Oxford.

Juniper, B.E., Gilchrist, A.J. and Robins, R.J. (1977). Some features of secretory systems in plants. *Histochemical Journal* **9**, 659–680.

Juniper, B.E., Hawes, C.R. and Horne, J.C. (1982). The relationships between the dictyosomes and the forms of endoplasmic reticulum in plant cells with different export programs. *Botanical Gazette* **143**, 135–145.

Juniper, B.E. and Jeffree, C.E. (1983). 'Plant Surfaces'. Edward Arnold, London.

Juniper, B.E. and Southwood, T.R.E. (Eds.) (1986). 'Insects and the Plant Surface'. Edward Arnold, London.

Kandler, O. and Schmideder, H. (1952). Untersuchungen über die Geschwindigkeit der Fibrinverdauung bei *Nepenthes* (Investigations into the speed of fibrin digestion by *Nepenthes*). *Zeitschrift für Botanik* **40**, 317–326.

Kannowski, P.B. (1967). Colony populations of two species of *Dolichoderus* (Hymenoptera: Formicidae). *Annals of the Entomological Society of America* **60**, 1246–1252.

Karel, L. and Roach, E.S. (1951). 'A Dictionary of Antibiosis'. Columbia University Press, New York.

Karlsson, P.S. and Carlsson, B. (1984). Why does *Pinguicula vulgaris* trap insects? *New Phytologist* **97**, 25–30.

Kaul, R.B. (1982). Floral and fruit morphology of *Nepenthes lowii* and *N. villosa*, montane carnivores of Borneo. *American Journal of Botany* **69**, 793–803.

Kerner, A. (1878). 'Flowers and their Unbidden Guests'. C. Kegan Paul, London.

Kerner, A. and Oliver, F.W. (1895). 'The Natural History of Plants', **1**, 119–158. Blackie, London.

Kertland, M.P.H. (1968). *Sarracenia purpurea* as an introduced plant in Ireland. *Irish Naturalist's Journal* **16**, 50–51.

Kevan, P.G. and Baker, H.G. (1983). Insects as flower visitors and pollinators. *Annual Review of Entomology* **28**, 407–453.

Kiesel, A. (1924). Études sur la nutrition de l'*Utricularia vulgaris* (Studies on the nutrition of *U. vulgaris*). *Annales de l'Institut Pasteur* **38**, 879–891.

Kingsolver, J.G. (1979). Thermal and hydric aspects of environmental heterogeneity in the pitcher-plant mosquito. *Ecological Monographs* **49**, 357–376.

Kingsolver, J.G. (1981). The effect of environmental uncertainty on morphological design and fluid balance in *Sarracenia purpurea*. *Oecologia (Berlin)* **48**, 364–370.

Kircheimer, F. von (1941). Über ein Vorkommen der Gattung *Aldrovanda* Linné im Alttertiar Thüringens (On the occurrence of the taxon *Aldrovanda* in the Lower Tertiary of Thüringen). *Braunkohle* **40**, 308–311.

Kitching, R.L. and Pimm, S.L. (1986). The length of food-chains: phytotelmata in Australia and elsewhere. *Proceedings of the Ecological Society of Australia* **14**, 123–140.

Kitching, R. and Schofield, C. (1986). Every pitcher tells a story. *New Scientist* **109**, 48–50.

Knight, W., Knight, I. and Howell, J.T. (1970). A vegetation survey of the Butterfly Valley Botanical Area, California. *Wasman Journal of Biology* **28**, 1–43.

Knoll, F. (1914). Über die Ursache des Ausgleitens der Insectenbeine an wachsbedekten Pflanzentheilen (The reason why insects' feet slip on the waxy surface of plants). *Jahrbuch für wissenschaftliche Botanik* **54**, 448–497.

Kok, A.C.A. (1933). Uber den Transport körperfremder Stoffe durch parenchymatisches Gewebe (On the transport of alien substances through parenchymatous tissue). *Recueil des Travaux Botaniques Néerlandais* **30**, 23–139.

Kokpol, U. (1974). Constituents and antitumor activity of *Sarracenia flava*. Dissertation, Mississippi State University.

Kolattukudy, P.E. (1976). 'Chemistry and Biochemistry of Natural Waxes'. Elsevier, Amsterdam, Oxford and New York.

Kolesnikowa, T.D. (1961). K poznanju trieticznoj flory Zaobskowo Jara w Zapadnoj Sibiri (To an understanding of the Tertiary Flora of Zaobskiy Yar in Western Siberia). *Botanicheskiy Zhurnal* **46**, 125–130.

Kondo, K. (1972). A paper chromatographic comparison of *Utricularia cornuta* and *U. juncea*. *Phyton* **30**, 43–45.

Kondo, K. (1976). A cytotaxonomic study of some species of *Drosera*. *Rhodora* **78**, 532–541.

Kondo, K. (1984). Three new species of *Drosera* L. from Australia. *Boletim da Sociedade Broteriana* **57**, 51–60.

Kondo, K. and Lavarack, P.S. (1984). A cytotaxonomic study of some Australian species of *Drosera* L. (Droseraceae). *Botanical Journal of the Linnean Society* **88**, 317–333.

Kondo, K. and Yaguchi, Y. (1983). Stomatal responses to prey capture and trap narrowing in Venus's Flytrap (*Dionaea muscipula* Ellis). II. Effects of various chemical substances on stomatal responses and trap closures. *Phyton* **43**, 1–8.

Kondratjev, G.K. (1973). Pollen of a new species of *Aldrovanda* of the Eocene period in Priangarje (in Russian). *Botanicheski Zhurnal-Moskva* **58**, 878–879.

Konopeka, K.A. and Zeigenspeck, H. (1929). Der Kern des Droseratentakels und die Fermentbildung (The nucleus of *Drosera* tentacles and the development of fermentation). *Protoplasma* **7**, 62–71.

Koste, W. (1970). Das Radiertier-Porträt: *Colotheca trilobata* ein seltenes sessiles Radiertier (A portrait of the rotifer *Colotheca trilobata*: a rare sessile wheel-animalcule). *Mikrokosmos* **59**, 195–200.

Kovacs, M. (1978). The element accumulation in submerged aquatic plant species in Lake Balaton (Hungary). *Acta Botanica Academica Scientifica Hungary* **24**, 273–283.

Krahl, R. (1956). Ein wirksames Prinzip auf *Drosera rotundifolia* (An active principle from *D. rotundifolia*). *Arzneimittel-Forschung* **6**, 342–348.

Krahl, R. (1983). Sarraceniaceae. *United States Department of Agriculture Forest Service, Southern Region. Technical Publication R8-TP. Atlanta, Georgia. Papers* **197–199**, 549–560.

Kress, A. (1970). Zytotaxonomische Untersuchungen an einigen Insektenfängern (Droseraceae, Byblidaceae, Cephalotaceae, Roridulaceae and Sarraceniaceae) (Cytotaxonomic researches on some insect trappers). *Berichte der deutschen Botanischen Gesellschaft* **83**, 55–62.

Krishnamoorthy, V. and Thompson, R.H. (1969). New binaphthoquinone from *Drosera ramenatacea*. *Phytochemistry* **8**, 1591–1594.

Kristen, U. (1975). Feinstruktur Veränderungen in den submersen Laublattdrüsen von *Nomaphila stricta* Nees während der Seokretion (Alterations in fine structure of the submerged leaf glands of *N. stricta* during secretion) *Cytobiologie* **11**, 438–447.

Kristen, U. (1977). Granulocrine Ausscheidung von Narbensekret durch Vesikel des Endoplasmatischen Retikulums bei

Aptenia cordifolia (Granulocrine extrusion of stigmatic secretions via vesicles of the endoplasmic reticulum in *Aptenia cordifolia*). *Protoplasma* **92**, 243–251.

Kristen, U., Liebezeit, G. and Biedermann, M. (1982). The ligule of *Isoetes lacustris*: Ultrastructure, mucilage composition, and a possible pathway of secretion. *Annals of Botany* **49**, 569–584.

Krook, L. (see Fairbridge, 1975, p. 484).

Kruck, M. (1931). Physiologische und zytologische Studien über die Utriculariaceae (Physiological and cytological studies on the Utriculariaceae). *Botanisches Archiv* **33**, 257–309.

Kruckerberg, A.R. (1954). The ecology of serpentine soils III. Plant species in relation to serpentine soils. *Ecology* **35**, , 267–274.

Kucowa, I. (1955). *Droseraceae*. *Flora Polska* **7**, 17–21.

Kugler, H. (1956). Uber die optische Wirkung von Fliegenblumen auf Fliegen (On the visual effects of fly-flowers on flies). *Berichte der deutschen Botanische Gesellschaft* **69**, 387–398.

Kupriyanova, L.T. (1972). Morphological studies of *Drosera* pollen (in Russian). *Botanicheskii Zhurnal* **58**, 1155–1156.

Kurahashi, H. and Beaver, R.A. (1979). *Nepenthomyia malayana* gen.n.sp.n.: a new calliphorid fly bred from pitchers of *Nepenthes ampullaria* in West Malaysia. *Annales de la Société Entomologique de France* **15**, 25–30.

Kurata, S. and Toyoshima, M. (1972). Philippine species of *Nepenthes*. *Gardens' Bulletin, Singapore 26*, 155–158.

LaBilliardière, A.J.J. (1806). 'Novae Hollandiae Plantarum Specimen', Vol. 11, 7.

Laessle, A.M. (1961). A micro-limnological study of Jamaican bromeliads. *Ecology* **42**, 499–517.

Laird, M. (1988) 'The Natural History of Larval Mosquito Habitats'. Academic Press, London.

Lambert, G. (1902). A l'étude de la pharmacologie et de la physiologie du *Sarracenia purpurea* (On the study of the pharmacology and physiology of *Sarracenia purpurea*). *Annales de Hygeine et de Médécin Coloniale de Paris* **5**, 652–662.

Lambrecht, F.L. (1971a). Notes on the ecology of Seychelles mosquitoes. *Bulletin of Entomological Research* **60**, 513–532.

Lambrecht, F.L. (1971b). The pitcher plant. Carnivorous incubator and tool of natural selection. *The Garden Journal of the New York Botanic Garden* **21**, 100–105.

Lang, F.X. (1901). Untersuchungen über Morphologie, Anatomie und Samenentwicklung von *Polypompholyx* und *Byblis gigantea* (Research on the morphology, anatomy and seed development of *Polypompholyx* and *B. gigantea*). *Flora* **88**, 149–206.

Lang, J.T. (1978). Relationship of fecundity to the nutritional quality of larva and adult diets of *Wyeomyia smithii*. *Mosquito News* **38**, 396–403.

Lauchli, A., Kramer, D. and Stelzer, R. (1974). Ultrastructure and ion localization in xylem parenchyma cells in roots. *In* 'Membrane Transport in Plants' (Eds. Zimmerman, U. and Dainty, J.), 363–371. Springer-Verlag, New York.

Laundon, J.R. (1959). Droseraceae. *In* 'Flora of Tropical East Africa' (Eds. Hubbard, C.E. and Milne-Redhead, E.) Crown Agents, London.

Lavarack, P.S. (1979). Rainforest *Drosera* of north Queensland. *Carnivorous Plant Newsletter* **8**, 61–62.

Lavarack, P.S. (1981a). *Nepenthes mirabilis* in Australia. *Carnivorous Plant Newsletter* **10**, 69–75.

Lavarack, P.S. (1981b). The northern rainbow plant-*Byblis liniflora*. *Carnivorous Plant Newsletter* **10**, 102–103.

Lavault, M. and Bruneton, J. (1978). Sur trois nouveaux alcaloides de Dioncophyllaceae (On three new alkaloids from the Dioncophyllaceae). *Comptes Rendues de l'Académie des Sceinces, Paris* **287C**, 129–131.

Lavault, M., Kouhoun, M.T. & Bruneton, J. (1977). D-methyltriphyophylline et D-methyl-dehydro-1,2-triphyophylline, nouveaux alcaloides du *Triphyophyllum peltatum* (Dioncophyllaceae) (D-methyl-triphyophylline and D-methyl-dehydro-triphyophylline, new alkaloids from *T. peltatum*). *Comptes Rendues de l'Académie des Sciences, Paris* **285C**, 167–169.

Lawrence, P.A. and Morata, G. (1976). The compartment hypothesis. *In* 'Insect Development', (Ed. Lawrence, P.A.), 132–149. Symposium of the Royal Entomological Society of London No. 8. Blackwell Scientific, Oxford.

Lawton, J.R., Harris, P.J. and Juniper, B.E. (1979). Ultrastructural aspects of the development of fibres from the flowering stem of *Lolium temulentum* L. *New Phytologist* **82**, 529–536.

Lea, H.W. (1976). Muscle contracting substance from a plant's closing fly-trap. *Planta* **129**, 39–41.

Leavitt, R.G. (1905). Translocation of characters in plants. *Rhodora* **7**, 13–31.

LeClercq, J. and Angenot, L. (1984). A propos du *Drosera peltata* et de la standardisation de la teinture de *Drosera* (About *D. peltata* and the standardisation of tincture of *Drosera*). *Journal de Pharmacie Belgique* **39**, 269–274.

Leith-Ross, P. (1984). 'The Tradescants: Gardeners to the Rose and Lily Queen'. Peter Owen, London.

Leppard, M. (1978). The Amazonian waterlily. *The Garden Journal of the Royal Horticultural Society* **103**, 121–122.

Lever, R.J.A.W. (1950). Mosquitoes from pitcher-plants in the Cameron Highlands. *Malayan Nature Journal* **5**, 98–99.

Lever, R.J.A.W. (1956). Notes on some flies recorded from pitcher-plants. *Malayan Nature Journal* **10**, 109–110.

Levin, D.A. (1973). The role of trichomes in plant defence. *Quarterly Review of Biology* **48**, 3–15.

Levin, S.A. (1976). Population dynamic models in heterogeneous environments. *Annual Review of Ecology and Systematics* **7**, 287–310.

Levin, S.A. and Paine, R.T. (1974). Disturbance, patch formation, and community structure. *Proceedings of the National Academy of Sciences USA* **71**, 2744–2747.

Lewin, R. (1982). Can genes jump between eukaryotic species? *Science* **217**, 42–43.

Leyton, L. and Juniper, B.E. (1963). Cuticle structure and water relations of pine needles. *Nature* **198**, 770–771.

Lichtner, F.T. and Williams, S.E. (1977). Prey capture and factors controlling trap narrowing in *Dionaea* (Droseraceae). *American Journal of Botany* **64**, 881–886.

Lindley, J. (1846). 'The Vegetable Kingdom'. Bradbury and Evans, London.

Lindquist, J. (1975). Bacterial and ecological observations on the northern pitcher plant, *Sarracenia purpurea* L. MSc Thesis, University of Wisconsin–Madison, USA.

Lindsay, R. (1983). Blanket bogs: a cloak of peat. *The Living Countryside* **9**, 1996–2000.

Lindsay, R., Riggall, J. and Burd, F. (1985). The use of small-scale surface patterns in the classification of British peatlands. *Aquilo Series Botany* **21**, 69–79.

Link, H.F. (1824). 'Elementa Philosophiae Botanicae'. Berlin (cited by Troll, 1932).

Linnaeus, C. (1753). 'Species Plantarum, Exhibentes Plantas Rite Cognitas, ad Genera Relatas.'

Lloyd, F.E. (1932). The door of *Utricularia*, an irritable mechanism. *Canadian Journal of Research* **7**, 386–425.

Lloyd, F.E. (1934). Is *Roridula* a carnivorous plant? *Canadian Journal of Research* **10**, 780–786.

Lloyd, F.E. (1935). *Utricularia. Biological Review* **10**, 72–100.

Lloyd, F.E. (1942). 'The Carnivorous Plants'. Chronica Botanica, Vol. 9. Ronald Press, New York.

Lobareva, L.S., Rudenskaya, G.N. and Stepanov, V.M. (1973). Pepsinopodobnaya proteinaza nasekovnoyadnego rasteniya *Nepenthes* (Pepsin-like protease from the insectivorous plant *Nepenthes*). *Biokhimiya* **38**, 640–642 (*Biochemistry* **38**, 531–532, translation).

Loew, O. and Aso, R. (1907). Benzosäure in *Pinguicula vulgaris* (Benzoic acid in *P. vulgaris*). *Bulletin of the Agricultural College of Tokyo Imperial University* **7**, 411–412.

Lollar, A.Q., Coleman, D.C. and Boyd, C.E. (1971). Carnivorous pathway of phosphorus uptake by *Utricularia inflata*. *Archiv für Hydrobiologie* **69**, 400–404.

Lou, C.H. and Hsueh, Y.L. (1950). The time course of recovery from the trapping action of *Utricularia* bladder. *Chinese Journal of Experimental Botany* **3**, 63–66.

Lounibos, L.P. and Bradshaw, W.E. (1975). A second diapause in *Wyeomyia smithii*: a seasonal incidence and maintenance by photoperiod. *Canadian Journal of Zoology* **53**, 215–221.

Lounibos, L.P., Rey, J.R. and Frank, J.H. (1985). Ecology of Mosquitoes. Proceedings of a Workshop, Florida Medical and Entomological Laboratory.

Lounibos, L.P., Van Dover, C. and O'Meara, G.F. (1982). Fecundity, autogeny and larval environment of the pitcher-plant mosquito, *Wyeomyia smithii*. *Oecologia (Berlin)* **55**, 160–164.

Lowe, E.J. (1866). 'Ferns British and Exotic', Vol. 2. Groombridge, London.

Lowrie, A. (1978). *Cephalotus* hunting in the deep S.W. of Australia. *Carnivorous Plant Newsletter* **7**, 119–121.

Lucas, G. (1976). Observations of *Trichoptilus parvulus* and *Drosera*. *Carnivorous Plant Newsletter* **5**, 32–33.

Luckner, M. (1984). 'Secondary Metabolism in Microorganisms, Plants and Animals', 576. Springer-Verlag, Berlin.

Luckner, V.R. and Luckner, M. (1970). Naphthochinonderivaten aus *Drosera ramentacea* Burch. ex. Harv. et Sond (Naphthoquinone derivatives from *D. ramentacea*). *Pharmazie* **25**, 216–265.

Luckner, V.R., Winkler, K., Bessler, O., and Luckner, M. (1969). Vorschlag zum deutschen Anzeibuch, 7. Ausgabe Afrikanisches Sonnentaukraut (Proposal for a German report book 7. Distribution of African droseras). *Anzeimittelstandardisierung* **17**, 41–56.

Ludwig, F. (1881). *Molinia coerulea* als Fliegenfängerin (*Molinia coerulea* as a flycatcher). *Botanische Centralblatt* **8**, 87.

Luetzelberg, P. von (1910). Beiträge zur Kenntnis der *Utricularia* (Contributions to our understanding of *Utricularia*). *Flora* **100**, 145–212.

Lupton, F.G.H. (1967). The use of resistant varieties in crop protection. *World Review of Pest Control* **6**, 47–58.

Lüttge, U. (1963). Die Bedeutung des chemischen Reizes bei der Resorption von ^{14}C-Glutaminsaure, $^{35}SO_4$ und $^{45}Ca^{++}$ durch *Dionaea muscipula*. (The importance of chemical sequence in the resorbtion of ^{14}C-Glutamic acid, $^{35}SO_4$ and $^{45}Ca^{++}$ in *D. muscipula*). *Naturwissenschaften* **50**, 22.

Lüttge, U. (1964a). Untersuchungen zur Physiologie der Carnivoren-Drüsen (Investigations on the physiology of carnivorous-plant glands). *Berichte der deutschen Botanischen Gesellschaft* **77**, 182–187.

Lüttge, U. (1964b). Untersuchungen zur Physiologie der Carnivoren-Drüsen: I. Die den verdauungsvorgangen beteiligten Enzyme (Studies on the physiology of the glands of carnivorous plants. I. On the enzymes concerned with the digestive processes). *Planta (Berlin)* **63**, 103–107.

Lüttge, U. (1964c). Untersuchungen zur Physiologie der Carnivoren-Drüsen: III. Der Stoffwechsel der resorbierten Substanzen. (Studies on the physiology of carnivorous plant glands III. The metabolism of resorbed substances). *Flora (Jena)* **155**, 228–236.

Lüttge, U. (1965). Untersuchungen zur Physiologie der Carnivoren-Drüsen; II. Uber die Resorption verschiedener Substanzen (Studies on the physiology of carnivorous plant glands II. On the resorption of various substances). *Planta* **66**, 331–344.

Lüttge, U. (1966a). Untersuchungen zur Physiologie der Carnivoren-Drüsen: IV. Die Kinetik der Chloridsekretion durch das Drüsen-gewebe von *Nepenthes* (Studies on the physiology of carnivorous plant glands: IV. The kinetics of chloride secretion by the gland tissue of *Nepenthes*). *Planta* **68**, 44–56.

Lüttge, U. (1966b). Untersuchungen zur Physiologie der Carnivoren-Drüsen: V. Microautoradiographische Untersuchungen der Chloridsekretion durch das Drüsengewebe von *Nepenthes* (Studies on the physiology of carnivorous plant glands: V. Microautoradiographic studies of chloride secretion in glandular tissue of *Nepenthes*). *Planta* **68**, 269–285.

Lüttge, U. (1967). Drusenfunktionen bei fleischfressenden Pflanzen (Glandular function in carnivorous plants). *Umschau* **67**, 181–186.

Lüttge, U. (1971). Structure and function of plant glands. *Annual Review of Plant Physiology* **22**, 23–44.

Lüttge, U. (1983). Ecophysiology of carnivorous plants. *In* 'Physiological Plant Ecology III'. Encyclopaedia of Plant Physiology (New Series), Vol. 12C (Eds. Lange, O.L., Nobel, P.S., Osmond, C.B. and Zeigler, H.), 489–517. Springer Verlag, Berlin.

Lüttge, U. (1985). Les plantes carnivores (The carnivorous plants). *La Recherche* **16**, 1302–1313.

Lüttge, U. and Krapf, G. (1969). Die Ultrastruktur der *Nymphaea*-Hydropoten in Zusammenhang mit ihrer Funktion als Salz-transportierende (The ultrastructure of the hydropoten of *Nymphaea* in connection with their role in salt transport). *Cytobiologie* **1**, 121–131.

Lyte, H. (1578). 'A Nievve Herball or Historie of Plantes'. Gerard Drewes, London.

Mabberley, D.J. (1984). The optimistic in pursuit of the unrecognisable; a note on the origin of angiosperms. *Taxon* **33**, 77–79.

Mabberley, D.J. (1985). 'Jupiter Botanicus. Robert Brown of the British Museum'. Braunschweig, Verlag von J. Cramer. British Museum (Natural History), London.

MacBeth, A.K. Jr., Price, J.R. and Winzor, F.L. (1935a). The coloring matters of *Drosera whittakeri*. I. The absorption spectra and color reactions of hydroxynaphthaquinones. *Journal of the Chemical Society* **1**, 325–333.

MacBeth, A.K. and Winzor, F.L. (1935b). The coloring matters of *Drosera whittakeri* II. *Journal of the Chemical Society* **1**, 334–336.

Macbride, J. (1818). On the powers of *Sarracenia purpurea* to entrap insects. *Transactions of the Linnean Society of London* **12**, 48–52.

MacDougal, D.T. (1899). Symbiotic saprophytism. *Annals of Botany* **13**, 1–46.

Macfarlane, J.M. (1889). Observations on pitchered insectivorous plants Part I. *Annals of Botany* **3**, 253–265.

Macfarlane, J.M. (1892). Contributions to the history of *Dionaea muscipula* Ellis. *Contributions of the Botany Laboratory. Morris Laboratory, University Pa* **1**, 7–44.

Macfarlane, J.M. (1893). Observations on pitchered insectivorous plants Part II. *Annals of Botany* 7, 403–458.

Macfarlane, J.M. (1908a). Sarraceniaceae. *In* 'Sarraceniales, Das Pflanzenreich', IV. 110. (Ed. Engler, A.), 1–39. Verlag von Wilhelm Engelmann, Leipzig.

Macfarlane, J.M. (1908b). Nepenthaceae. *In* 'Sarraceniales, Das Pflanzenreich', IV. 111. (Ed. Engler, A.), 1–92. Verlag von Wilhelm Engelmann, Leipzig.

Macfarlane, J.M. and Steckbeck, D.W. (1933). *Sarracenia purpurea* var *stolonifera*, a noteworthy morphological and ecological type. *Bulletin of Miscellaneous Information, Kew*, **4**, 161–169.

Magnussen, B.C. (1982). An introduction to *Genlisea*. *Carnivorous Plant Newsletter* **11**, 13–15.

Maguire, B. (1978). The Botany of the Guyana Highlands Part 10: Sarraceniaceae. *Memoirs of the New York Botanical Garden* **29**, 36–62.

Maier, R. von (1973a). Wirkung von Trockenheit auf den Austrieb der Turionen von *Utricularia* L. (The effect of drought on the formation of turions in *Utricularia*). *Österreichische Botanische Zeitung* **122**, 15–20.

Maier, R. von (1973b). Das Austreiben der Turionen von *Utricularia vulgaris* L. nach verschiedenen langen Perioden der Austrocknung (The formation of turions in *U. vulgaris* after long periods of drought). *Flora (Jena)* **162**, 269–283.

Maier, R. von (1973c). Analysis of production and pigments in *Utricularia vulgaris*. *In* 'Oekosystemforschung. Ergebnisse von Symposien der Deutschen Botanischen Gesellschaft und der Gesellschaft für Angewandte Botanik'. (Ed. Ellenberg, H.), 87–101. Springer-Verlag, Berlin.

Mameli, E. (1916). Ricerche anatomiche, fisiologiche e biologiche sulla *Martynia lutea* Lindl. (Anatomical, physiological and biological research on *Martynia lutea*). *Atti dell'Universita di Pavia, Serie 2*, **16**, 137–188.

Mandossian, A.J. (1965). Plant associates of *Sarracenia purpurea* (pitcher plant) in acid and alkaline habitats. *Michigan Botanist* **4**, 107–114.

Mann, P. (1985). The Albany Area. *Carnivorous Plant Newsletter* **14**, 74–76.

Marburger, J.E. (1979). Glandular leaf structure of *Triphyophyllum peltatum* (Dioncophyllaceae): a 'fly-paper' insect trapper. *American Journal of Botany* **66**, 404–411.

Marchant, R. and Robards, A.W. (1968). Membrane systems associated with plasmalemma of plant cells. *Annals of Botany* **32**, 457–471.

Marco, J.L. (1985). Iridoid glucosides of *Pinguicula vulgaris*. *Journal of Natural Products* **48**, 338.

Markgraf, F. (1954). Les homologies des feuilles et les relations taxonomiques chez les Sarraceniales (Leaf homology and taxonomic relationships in the Sarraceniales) *8me Congrès International de Botanique, Paris, 1954. Rapports et communications, Section* **4**, 50–52.

Markgraf, F. (1955). Über Laubblatt – Homologien und verwandtschaftliche Zusammenhänge bei Sarraceniales (On the leaf blade of the Sarraceniales – homologies and relationships). *Planta* **46**, 414–446.

Marks, E.N. (1971). Mosquitoes that breed in pitcher plants. *News Bulletin of the Entomological Society of Queensland* **78**, 8–10.

Martin, J.T. (1964). The role of cuticle in the defence against plant disease. *Annual Review of Phytopathology* **2**, 81–100.

Martin, J.T. and Juniper, B.E. (1970). 'The Cuticles of Plants'. Edward Arnold, London.

Mather, T.N. (1981). Larvae of alderfly (*Megaloptera*: Sialidae) from a pitcher-plant. *Entomological News* **92**, 32.

Mather, T.N. and Catts, E.P. (1980). Seasonal dispersal of pitcher-plant *Sarracenia purpurea* mites (Acarina: Anoetidae). *Journal of the New York Entomological Society* **88**, 60–61.

Matile, L. (1979). *Xenoplatyura beaveri* n.sp. (Diptera: Mycetophiloidea) Keroplatidae nouveau de Malaisie inféode aux urnes de *Nepenthes (x. beaveri*, a new species of dipteran: Mycetophiliodea: Keroplatidae from the pitchers of *N. ampullaria* on the Island of Penang, Malaysia) *Annales de la Société Entomologique de France* **15**, 31–35.

Matthews, E.G. (1977). Signal-based frequency-dependent defense strategies and the evolution of mimicry. *American Naturalist* **111**, 213–222.

Matthews, R.E.F. (1960). A ribonuclease from *Nepenthes* spp. *Biochimica et Biophysica Acta* **38**, 552–553.

Mattingly, P.F. and Brown, E.S. (1955). The mosquitoes of the Seychelles. *Bulletin of Entomological Research* **46**, 69–110.

Mattson, W.J. (1980). Herbivory in relation to plant nitrogen content. *Annual Review of Ecology and Systematics* **11**, 119–161.

Mayr, E. (1983). 'The Growth of Biological Thought'. Harvard University Press.

Mazokhin-Porshnyakov, G.A. (1969). 'Insect Vision'. Plenum Press, New York.

Mazrimas, J.A. (1972). *Drosophyllum lusitanicum*. *Carnivorous Plant Newsletter* **1**, 5–6.

Mazrimas, J. (1975). Review of 'Carnivorous Plants of the Cerro de la Neblina' by Brewer-Carias, J. *Carnivorous Plant Newsletter* **4**, 36–37.

Mazrimas, J.A. (1979). Recent status of *Heliamphora*. *Carnivorous Plant Newsletter* **8**, 82–84.

McConnell, F.V., Quelch, J.J. and Brown, N.E. (1901). Report on two botanical collections made at Mount Roraima, British Guiana. *Transactions of the Linnean Society of London Series 2*, **6**, 1–108.

McConnell, R.B. (see Fairbridge, 1975, 318).

McDaniel, S. (1966). A taxonomic revision of *Sarracenia* (Sarraceniaceae). PhD Dissertation, Florida State University.

McDaniel, S. (1971). The genus *Sarracenia* (Sarraceniaceae). *Bulletin of the Tall Timbers Research Station* **9**, 1–36.

McEwan, S.F., Stewart, A. and Lee, V.E. (1988). Cytochemical localization of acid phosphatase activity in the stalked gland of *Drosera rotundifolia*. *Annals of Botany* (in press).

McGaha, Y.T. (1952). The limnological relations of insects to certain aquatic plants. *Transactions of the American Microscopical Society* **71**, 355–381.

McMath, V.E. (1966). Geology of the Taylorsville area, Northern Sierra Nevada. *In* 'Geology of Northern California' (Ed. Bailey, E.H.). *California Division of Mines and Geology Bulletin* **190**, 173–183.

McMillan, P. (1978). Some thoughts and observations on *Sarracenia*. *Carnivorous Plant Newsletter* 7, 105–107.

McWilliams, E.L. (1974). Evolutionary ecology. *In* 'Bromeliaceae (Pitcairnioideae)' (Eds. Smith, L.B. and Downs, R.J.), 40–45. *Flora Neotropica* Monograph **14**, Hafner, New York.

Meierhofer, H. (1902). Beiträge zur Kenntnis der Anatomie und Entwicklungsgeschichte der *Utricularia*-Blasen (Contributions to the knowledge of the anatomy and development of *Utricularia* traps). *Flora* **90**, 84–113.

Mellichamp, J.H. (1875). Notes on *Sarracenia variolaris*. *Proceedings of the American Association for the Advancement of Science, 23rd meeting* **1875B**, 113–133.

Mellichamp, T.L. (1979). Botanical history of carnivorous plants III. *Nepenthes*. *Carnivorous Plant Newsletter* **8**, 30–33.

Mellichamp, T.L. (1983). Cobras of the Pacific Northwest. *Natural History* **4**, 47–51.

Menninger, E.A. (1965). An African vine with three different leaves for three different jobs. *Garden Journal of the New York Botanical Garden:* **January–February**, 31–32.

Menzel, R. (1922). Beiträge zur Kenntnis der Mikroflora von Niederlandischen Ost-Indien; II. Uber den tierischen Inhalt der Kannen von *Nepenthes melamphora* Reinw. mit dem Berücksichtigung der Nematoden (Contributions to the knowledge of the microflora of the Netherlands East Indies; II. On the animal inhabitants of the pitchers of *N. melamphora* and identification of the nematodes). *Treubia* **3**, 116–122.

Menzel, R. and Micoletzky, H. (1928). Beiträge zur Kenntnis der Mikroflora von Niederlandische Ost-Indien; VII. *Anguillula nepenthicola* Menzel. aus Kannen von *Nepenthes gymnamphora* Nees bei Tjibodas (Contributions to the knowledge of the microflora of the Netherlands East Indies; VII. *A. nepenthicola* from the pitchers of *N. gymnamphora*). *Treubia* **10**, 285–295.

Menzel, R. (1979). Spectral sensitivity and colour vision in invertebrates. *In* 'Handbook of Sensory Physiology', Vol. VII/6A: Vision in invertebrates. A. Invertebrate photoreceptors. Springer-Verlag, Berlin, Heidelberg, New York.

Metcalfe, C.R. (1951). The anatomical structure of the Dioncophyllaceae in relation to the taxonomic affinities of the family. *Kew Bulletin* **6**, 351–368.

Meyer, A. and Dewèvre, A. (1894). Über *Drosophyllum lusitanicum* (On *D. lusitanicum*). *Botanisches Centralblatt* **60**, 33–41.

Meyers, D.G. (1982). Darwin's investigations of carnivorous aquatic plants of the genus *Utricularia*: Misconception, contribution and controversy. *Proceedings of the Academy of Natural Sciences of Philadelphia* **134**, 1–11.

Meyers, D.G. and Strickler, J.R. (1979). Capture enhancement in a carnivorous aquatic plant: Function of antennae and bristles in *Utricularia vulgaris*. *Science* **203**, 1022–1025.

Miles, D.H. and Kokpol, U. (1976). Tumor inhibitors 2. Constituents and antitumor activity of *Sarracenia flava*. *Journal of the Pharmaceutical Society* **65**, 284–285.

Miles, D.H., Kokpol, U. and Mody, N.V. (1975). Volatiles in *Sarracenia flava*. *Phytochemistry* **14**, 845–846.

Miles, D.H., Kokpol, U., Zalkow, L.H., Steindel, S.J. and Nabors, J.B. (1974). Tumor inhibitors 1. Preliminary investigation of antitumor activity of *Sarracenia flava*. *Journal of the Pharmaceutical Society* **63**, 613–61.

Miles, D.H., Kokpol, U., Battacharya, J., Attwood, J.L., Stone, K.E., Bryson, T.A. and Wilson, C. (1976). Structure of Sarracenin. An unusual enol diacetal monoterpene from the insectivorous plant *Sarracenia flava*. *Journal of the American Chemical Society* **98**, 1569–1573.

Minocha, S.C. (1985). In vitro propagation of *Dionaea muscipula*. *Horticultural Science* **20**, 216–217.

Minton, E.A. and Jeffrey, D.B. (1972). Modified floral parts of *Dionaea*. *Carnivorous Plant Newsletter* **1**, 45–47.

Miyoshi, E., Shizuri, Y. and Yamamura, S. (1984). Isolation and structures of Diomuscinone and Diomuscipulone from *Dionaea muscipula*. *Phytochemistry* **23**, 2385–2387.

Mody, N.V., Henson, R., Hedin, P.A., Kokpol, V. and Miles, D.H. (1976). Isolation of the insect paralysing agent coniine from *Sarracenia flava*. *Experimentia* **32**, 829–830.

Moeller, R.E. (1978). Carbon uptake by the submerged hydrophyte *Utricularia purpurea*. *Aquatic Botany* **5**, 209–216.

Moeur, J.E. and Istock, C.A. (1980). Ecology and evolution of the pitcher-plant mosquito: IV. Larval influence over adult reproductive performance and longevity. *Journal of Animal Ecology* **49**, 775–792.

Moeur, J.E. and Istock, C.A. (1982). Chromosome polymorphisms in the pitcher-plant mosquito, *Wyeomyia smithii*. *Chromosoma (Berlin)* **84**, 623–651.

Mogi, M. and Mokry, J. (1980). Distribution of *Wyeomyia smithii* (Diptera: Culicidae) eggs in pitcher plants in Newfoundland, Canada. *Tropical Medicine* **22**, 1–12.

Morgan, J.G. and Arnott, H.J. (1971). Endoplasmic reticulum in trichome glandular cells and its association with other cellular organelles. *Texas Reports on Biology and Medicine* **29**, 425 (abstract).

Morison, R. (1699). 'Plantarum Historiae Universalis Part 3', 533.

Morrissey, S.M. (1955). Chloride ions in the secretion of the pitcher plant. *Nature* **176**, 1220–1221.

Morrissey, S.M. (1960). The secretion of acid by *Nepenthes*. *Journal of Physiology (London)* **150**, 16–17.

Morrissey, S.M. (1963). Histamine and acid secretion in a tropical insectivorous plant (*Nepenthes*). *In* 2nd International Pharmacological Meeting, Prague. *Biochemical Pharmacology* **12**, Supplement 2D (abstract).

Morrissey, S.M. (1964). Carbonic anhydrase and acid secretion in a tropical carnivorous plant (*Nepenthes*). *International Union of Biochemistry* **32**, 660.

Mosto, P. (1979). Algas en trampas de *Utricularia oligosperma* St. Hill. (Algae in the traps of *U. oligosperma*). *Boletin de la Sociedad Argentina de Botanica* **18**, 89–99.

Moussli, Y. (1930). Étude chimique du *Drosera rotundifolia* et de ses éléments constituents (A chemical study of *D. rotundifolia* and its constituents). Doctoral Thesis (Pharmacology) Strasbourg University.

Muller, J. (1981). Fossil pollen records of extant angiosperms. *Botanical Review* **47**, 1–142.

Muma, M.H. and Denmark, H.A. (1967). Biological studies on *Macroseius biscutatus* (Acarina: Phytoseiidae). *Florida Entomologist* **50**, 249–255.

Munk, H. (1876). Die electrischen Reizmitteln und Bewegungserscheinungen am Blatt der *Dionaea muscipula* (The electrical stimulus and the appearance of movement in the leaves of *D. muscipula*). *Archiv für Anatomische, Physiologische und Wissenschaftliche Medizin* **1876**, 30–203.

Naeem, S. and Dushek, J. (1985). Plumbing the deathly depths of the California Pitcher plant. *Pacific Discovery*, **April–June**: 26–31. Cited in *Carnivorous Plant Newsletter* **15**, 25 (1986).

Nahrstedt, A. (1980). Absence of cyanogenesis from Droseraceae. *Phytochemistry* **19**, 2757–2758.

Nair, A.G.R. and Gunasegaran, R. (1982). Chemical investigations of certain in South Indian plants. *Indian Journal of Chemistry* **21B**, 979–980.

Nakayama, S. and Amagase, S. (1968). Acid protease in *Nepenthes*: Partial purification and properties of the enzyme. *Proceedings of the Japan Academy* **44**, 358–362.

Nash, R.C. (1973). Tuberous droseras in South Australia. *Carnivorous Plant Newsletter* **2**, 63–65.

Navale, G.K.B. and Misra, B.K. (1979). Some new pollen grains from Neyveli Lignite Tamil Nadu India. *Geophytology* **8**, 226–239.

Neamtu, G. and Bodea, C. (1972). Cercetari chemotaxonomice la plantele superioare IV. Pigmentiji carotinioidici din plante carnivore (Chemotaxonomic studies on higher plants IV. Pigments of carnivorous plants) *Studii si Cercetari de Biochimi* **15**, 181–185.

Nelson, E.C. and Seaward, M.R.D. (1981). Charles Darwin's correspondence with David Moore of Glasnevin on insectivorous plants and potatoes. *Biological Journal of the Linnean Society* **15**, 157–164.

Nemček, O., Sigler, K. and Kleinzeller, A. (1966). Ion transport in the pitcher of *Nepenthes henryana*. *Biochimica et Biophysica Acta* **126**, 73–80.

Nesbitt, H.H.J. (1979). A new anoetid (Acari) of the genus *Creutzeria* from the Seychelles. *Canadian Entomologist* **111**, 1201–1205.

Newman, M.C. and McIntosh, A.W. (1981). Lead in freshwater macrophytes, filamentous algae and 'aufwuchs'. *Bulletin of the New Jersey Academy of Science* **26**, 58.

Nicholls, K.W. (1985). Flavonoids and affinities of the Cephalotaceae. *Biochemical Systematics and Ecology* **13**, 261–263.

Nilsson, G. (1950). Nagot om baktereifloran; den svenska tatmjolken (The bacterial flora of the Swedish edible ropy milk). *Svenska Mejeritidmingen* **42**, 411–416.

Nitschke, T. (1860). Wachsthumverhältnisse des rundblattigen Sonnenthaues (Relative rates of growth of roundleaved sundews). *Botanische Zeitung* **18**, 57–69.

Nolan, G. (1978a). Vegetative reproduction in *Drosera montana*. *Carnivorous Plant Newsletter* **7**, 72.

Nolan, G. (1978b). On the foraging strategies of carnivorous plants: 2 Biological stimulus versus mechanical stimulus in the fast-moving periphery tentacles of the species *Drosera burmanni*. *Carnivorous Plant Newsletter* **7**, 79–81.

Nold, R.H. (1934). Die Funktion der Blase von *Utricularia vulgaris* (The function of the trap of *U. vulgaris*). *Beihefte zum Botanischen Zentralblatt* **52**, 415–448.

North, M. (1980). 'A Vision of Eden: The Life and Work of Marianne North'. In collaboration with the Royal Botanic Gardens, Kew. Webb & Bower, Exeter.

O'Connell, M. (1981). The phytosociology and ecology of Scragh Bog, Co. Westmeath. *New Phytologist* **87**, 139–187.

O'Neil, W. (1983). A preliminary report on the pollination of *Sarracenia purpurea* in a forest-swale ecotone. *Carnivorous Plant Newsletter* **12**, 60–62.

Obermayer, A.A. (1970). Roridulaceae. *In* 'Flora of Southern Africa' (Eds. Codd, L.E., De Winter, B., Killick, D.J.B. and Rycroft, H.B.), Vol. 13. The Government Printer, Pretoria.

Okahara, K. (1930). Physiological studies on *Drosera*: I. On the proteolytic enzyme of *Drosera rotundifolia*. *Scientific Reports of Tohoku Imperial University, fourth series Biology* **5**, 573–590.

Okahara, K. (1933). Physiological studies on *Drosera*: IV. On the function of microorganisms in the digestion of insect bodies by insectivorous plants. *Scientific Reports of Tohoku Imperial University, fourth series Biology* **8**, 151–168.

Oliver, F.W. (1944). A mass catch of cabbage whites by sundews. *Proceedings of the Royal Entomological Society of London* **19**, 5.

Oudemans, A.C. (1915). *Anoetus guentheri* nov.sp. *In* 'Die Lebende Bewohne der Insektfressenden Pflanze *Nepenthes distillatoria* auf Ceylon' (The living inhabitants of the insect-eating plant, *N. distillatoria*, in Ceylon). *Zeitschrift für wissenschaftliche Insektbiologie* **11**, 242–243.

Oudemans, A.C. (1924). Acaralogische Aanteekeningen LXXVI (Acaralogical drawings LXXVI). *Entomologische Berichte* **6**, 300–310.

Oudman, J. (1936). Über Aufnahme und Transport N-haltiger Verbindungen durch die Blätter von *Drosera capensis* L. (On the uptake and transport of N-containing compounds through the leaves of *D. capensis*). *Recueil des Travaux Botaniques Néerlandais* **33**, 351–433.

Outenreath, R.L. (1980). Ultrastructural and radioautographical studies of the developmental sequence, and of secretory modifications produced by stimulation of the digestive gland cells of *Drosera capensis* L. with reference to the golgi apparatus. PhD Thesis, University of Texas, Austin, Texas.

Outenreath, R. and Dauwalder, M. (1982). Ultrastructural and radioautographical studies of the digestive gland cells of *Drosera capensis* 1. Development and mucilage secretion. *Journal of Ultrastructural Research* **80**, 71–88.

Overbeck, C. (1982). 'Carnivorous Plants' (A Lerner Natural History Book). Lerner Publications, Minneapolis.

Oye, P. van (1921). Biologie der Kanne von *Nepenthes melamphora* Reinw. (Biology of the pitchers of *N. melamphora*). *Biologische Zentralblatt* **41**, 529–534.

Pagani, F. (1982). Fitocostituenti della *Martynia proboscidata* L. II. *Bolletin Chimica Farmacologica* **121**, 178–182.

Page, C.R. and Barber, J.T. (1975). Interactions between mosquito larvae and mucilaginous plant seeds. 2. Chemical attractions of larvae to seeds. *Mosquito News* **35**, 48–58.

Palczewska, I. (1966). Wystepowanie esteraz we wloskack gruzolowych i lisciach *Drosera rotundifolia* (On the presence of esterase in the glandular hairs and the leaves of *D. rotundifolia*). *Acta Societatis botanicorum Poloniae* **35**, 307–313.

Palmer, M.A., Goad, L.J., Goodwin, T.W., Copsey, D.B. and Boar, R.B. (1978). The conversion of 5α-lanost-24-ene-3β, 9α-diol and parleol into a poriferasterol by the alga *Ochromonas malhamensis*. *Phytochemistry* **17**, 1577–1580.

Panessa, B.J., Kuptsis, J., Piscopo-Rodgers, I. and Gennaro, J. (1976). Ion and uranium transport in the carnivorous pitcher plant: studies by SEM, TEM and X-ray microanalysis. *In* 'Proceedings, Workshop on Plant Science Applications of the SEM' (Ed. Johari, O.), 461–468. Illinois Institute of Technology, Chicago.

Panin, S. and Grishin, G. (1975). Sulfur content in some plants of the Semipalatinsk oblast of the Kazaskh SSR (Russian). *Rastitel Nye Resursy Akademi ja Nauk* **11**, 473–483.

Pant, D.D. and Bhatnagar, S. (1977). Morphological studies in *Nepenthes* (Nepenthaceae). *Phytomorphology* **27**, 13–34.

Paris, O.H. and Jenner, C.E. (1959). Photoperiodic control of diapause in the pitcher plant midge *Metriocnemus knabi*. *In* 'Photoperiodism and Related Phenomena in Plants and Animals' (Ed. Withrow, R.B.). *American Association for the Advancement of Science, Washington,* **55**, 601–624.

Paris, R. and Denis, J.C. (1957). Les Droseras; Leur caracterisation dans divers medicaments (The Droseraceae; Their characterisation in various medicines). *Annales Pharmacologies Français* **15**, 145–159.

Paris, R.R. and Delaveau, P. (1959). *Drosera*. Isolation of plumbagone from *Drosera auriculata* and ramentaceone from *D. ramentacea*. *Annales Pharmacologies Français* **17**, 585–592.

Paris, R. and Quevauvillier, A. (1947). Action de quelques drogues végétales sur les bronchospasmes histaminiques et acetycholiniques (The action of some plant drugs on the histamine and acetylcholine bronchospasm). *Thérapie* **2**, 69–72.

Parkes, D.M. (1980). Adaptive mechanisms of surfaces and glands in some carnivorous plants. MSc Thesis, Monash University, Clayton, Victoria, Australia.

Parkes, D.M. and Hallam, N.D. (1984). Adaption for carnivory in the West Australian pitcher plant *Cephalotus follicularis*. *Australian Journal of Botany* **32**, 595–604.

Parkinson, J. (1640). 'Theatrum Botanicum', 1235. London.

Pasteur, G. (1982). A classificatory review of mimicry systems. *Annual Review of Ecology and Systematics* **13**, 169–199.

Pate, J.S. and Dixon, K.W. (1978). Mineral nutrition of *Drosera erythrorhiza* Lindl. with special reference to its tuberous habit. *Australian Journal of Botany* **26**, 455–464.

Pate, J.S. and Dixon, K.W. (1982). 'Tuberous, Cormous and Bulbous Plants: Biology of an Adaptive Strategy in Western

Australia', 51–56. University of Western Australia Press.
Paterson, C.G. (1971). Overwintering ecology of the aquatic fauna associated with the pitcher-plant *Sarracenia purpurea*. *Canadian Journal of Zoology* **49**, 1455–1459.
Pearman, G.I. (1966). The reflection of visible radiation from leaves of some Western Australian species. *Australian Journal of Biological Sciences* **19**, 97–103.
Pennak, R.W. (1953) 'Fresh-water Invertebrates of the United States'. Ronald Press, New York.
Peyronel, B. (1932). Absence de mycorhizes chez les plantes insectivores et hemiparasites et signification probable de la mycorhize (The absence of mycorrhiza amongst carnivorous and hemiparasitic plants and its probable significance). *Bollettino Sezione Italiana Societa Internazionale di Microbiologia* **4**, 483–486.
Pfeiffer, R.K., Dewey, D.R. and Brunskill, R.T. (1957). Further investigations of the effect of pre-emergence treatment with trichloracetic acid and dichlorpropionic acids on the subsequent reactions of plants to herbicidal sprays. Chesterford Park Research Station Reports.
Pickett-Heaps, J.D. (1967). Further observations on the golgi apparatus and its functions in cells of the wheat seedling. *Journal of Ultrastructural Research* **18**, 287–303.
Pinner, R. (1967). Ungewöhnliche Raupe gedeiht auf dem fleischfressenden Sommerkraut (An unusual caterpillar that grows on *Drosera capillaris*). *Naturwissenschaftliche Rundschau* **20**, 479.
Pissarek, H.P. (1965). Beobachtungen über das Blühen von *Drosera rotundifolia* (Observations on the flowering of *D. rotundifolia*). *Schriften der Naturwissenschaftlichen Vereins für Schleswig-Holstein* **36**, 26–29.
Planchon, J.E. (1848). Sur la famille des Droseracées (On the family Droseraceae). *Annales des Sciences Naturelles de Botanique* **3 ser. 9**, 285–309.
Plummer, G.L. (1963). Soils of the pitcher-plant habitats in the Georgia Coastal plain. *Ecology* **44**, 727–734.
Plummer, G.L. and Jackson, T.H. (1963). Bacterial activities within the sarcophagus of the insectivorous plant *Sarracenia flava*. *American Midland Naturalist* **69**, 462–469.
Plummer, G.L. and Kethley, J.B. (1964). Foliar absorption of amino-acids, peptides and other nutrients by the pitcher-plant *Sarracenia flava*. *Botanical Gazette* **125**, 245–260.
Pocock, R.I. (1898). Spider and pitcher-plant. *Nature* **58**, 274–275.
Porcher, F.P. (1849). An examination into the medical and chemical properties of the *Sarracenia flava* and *variolaris* (side-saddle flower) fly-traps. *Charleston Medical Journal and Review* **4**, 1–13.
Prance, G.T. and Arias, J.RW. (1975). A study of the floral biology of *Victoria amazonica* (Poepp.) Sowerby. *Acta Amazonica* **5**, 109–139.
Preece, T.F. and Dickinson, C.H. (Eds.) (1971). 'Ecology of Leaf Surface Microorganisms'. Academic Press, London and New York.
Pringsheim, E.G. and Pringsheim, O. (1962). Axenic culture of *Utricularia*. *American Journal of Botany* **49**, 898–901.
Pringsheim, E.G. and Pringsheim, O. (1967). Kleiner Beitrag zur Physiologie von *Utricularia* (A small contribution towards the physiology of *Utricularia*). *Zeitschrift für Pflanzenphysiologie* **57**, 1–10.
Procter, J. (1974). The endemic flowering plants of the Seychelles: An annotated list. *Candollea* **29**, 345–387.
Proctor, M.C.F. (1979). Surface wax on the leaves of some mosses. *Journal of Bryology* **10**, 531–538.
Proctor, J. and Woodell, S.R.J. (1975). The Ecology of Serpentine Soils. *Advances in Ecological Research* **9**, 256–385.
Pullen, T. and Plummer, G.L. (1964). Floristic changes within pitcher plant habitats in Georgia. *Rhodora* **66**, 375–381.
Quintanilla, A. (1926). O problema das plantas carnivoras (On the problem of carnivorous plants). Dissertation, Coimbra. *Extract from Boletim da Sociedade Broteriana Second Series* **4**, 44–129.
Rachmilevitz, T. and Joel, D.M. (1976). Ultrastructure of the calyx glands of *Plumbago capensis* Thunb. in relation to the process of secretion. *Israel Journal of Botany* **25**, 127–139.
Ragetli, H.W.J., Weintraub, M. and Lo, E. (1972). Characteristics of *Drosera* tentacles: I. Anatomical and cytological detail. *Canadian Journal of Botany* **50**, 159–168.
Ramalingam, S. and Ramakrishnan, K. (1971). Redescription of *Aedes* (*Alanstonea*) *brevitibia* (Edwards) from Brunei, Borneo. *Proceedings of the Entomological Society of Washington* **73**, 231–238.
Ramanamanjary, W. and Botteau, P. (1968). Active protectrice du Matahanando, *Drosera ramentacea* Burch. vis-à-vis du bronchospasme (Protective activity of *D. ramentacea* with respect to a bronchospasm). *Comptes Rendues Hebdomadaires des Séances Académie des Sciences Naturelles Ser. D (Paris)* **266**, 1787–1789.
Rao, T.A., Shanware, P.G. and Tribedi, G.N. (1969). A note on the pitcher plant habit in Assam. *The Indian Forester* **95**, 611–613.
Rayle, D.L. and Cleland, R. (1977). Control of plant cell enlargement by hydrogen ions. *Current Topics in Developmental Biology* **11**, 187–214.
Rea, P.A. (1982a). Biochemical aspects of fluid secretion and digest absorption by the trap lobes of *Dionaea muscipula* Ellis (Venus's Flytrap). PhD Thesis, University of Oxford.
Rea, P.A. (1982b) Fluid composition and factors that elicit secretion by the trap lobes of *Dionaea muscipula* Ellis (Venus's Flytrap). *Zeitschrift für Pflanzenphysiologie* **108**, 255–272.
Rea, P.A. (1983a). The dynamics of H^+ efflux from the trap lobes of *Dionaea muscipula* Ellis (Venus's Flytrap). *Plant, Cell and Environment* **6**, 125–134.
Rea, P.A. (1983b). The influence of secretion elicitors and other effectors of H^+ efflux on carboxylate synthesis by the trap lobes of *Dionaea muscipula* Ellis (Venus's Flytrap). *Plant Science Letters* **32**, 159–167.
Rea, P.A. (1984). Evidence for the H^+ co-transport of D-alanine by the digestive glands of *Dionaea muscipula* Ellis (Venus's Flytrap). *Plant, Cell and Environment* **7**, 363–366.
Rea, P.A. and Whatley, F.R. (1983). The influence of secretion elicitors and external pH on the kinetics of D-alanine uptake by the trap lobes of *Dionaea muscipula* Ellis (Venus's Flytrap). *Planta* **158**, 312–319.
Rea, P.A., Joel, D.M. and Juniper, B.E. (1983). Secretion and redistribution of chloride in the digestive glands of *Dionaea muscipula* (Venus's Flytrap) upon secretion stimulation. *New Phytologist* **94**, 359–366.
Rees, M. and Will, H. (1875). Einige Bemerkungen über 'fleischressende' Pflanzen. *Botanische Zeitung* **33**, 713–718.
Reichle, D.E., Shanks, M.H. and Crossley, D.A. (1969). Calcium, potassium and sodium content of forest floor arthropods. *Annals of the Entomological Society of America* **62**, 57–62.
Reinert, G.W. and Godfrey, R.K. (1962). Reappraisal of *Utricularia inflata* and *U. radiata* (Lentibulariaceae). *American Journal of Botany* **49**, 213–220.
Reiskind, J. (1978). A crab-spider associate of *Nepenthes rafflesiana*. *Carnivorous Plant Newsletter* **7**, 77–78.
Rendle, A.B. (1925). An account of the introduction of American seeds into Great Britain by Peter Collinson. *The*

Journal of Botany **63**, 163–165.
Rennie, E.H. (1887). The colouring matter of *Drosera whittakeri*. *Journal of the Chemical Society* **51**, 371–377.
Richards, P.W. (1952). 'The Tropical Rain Forest'. Cambridge University Press, Cambridge.
Rickett, H.W. (1967). 'Wild Flowers of the United States', Vol. 2, Part 1: The Southeastern States. New York Botanical Garden. McGraw-Hill, New York.
Rickson, F.R. (1979). Absorption of animal tissue breakdown products into a plant stem – the feeding of a plant by ants. *American Journal of Botany* **66**, 87–90.
Ridley, H.N. (1906). Malay Drugs. *Agricultural Bulletin of the Straits and the Federated Malay States* **5**, 193–206.
Riley, C.V. (1873). Description and natural history of two insects which brave the dangers of *Sarracenia variolaris*. *Transactions of the Academy of Sciences of St. Louis* **3**, 235–240.
Riley, C.V. (1874). On the insects more particularly associated with *Sarracenia variolaris* (spotted trumpet-leaf). *Proceedings of the American Association for the Advancement of Science* **1874**, 18–25.
Roberts, M.L. (1972). *Wolffia* in the bladders of *Utricularia*: an 'herbivorous plant'? *Michigan Botanist* **11**, 67–69.
Roberts, P.R. and Oosting, H.J. (1958). Responses of Venus fly trap (*Dionaea muscipula*) to factors involved in its endemism. *Ecological Monographs* **28**, 193–218.
Robins, R.J. (1976). The nature of the stimuli causing digestive juice secretion in *Dionaea muscipula* Ellis (Venus's Flytrap). *Planta* **128**, 263–265.
Robins, R.J. (1978). Studies in secretion and absorption in *Dionaea muscipula* Ellis. PhD Thesis, University of Oxford.
Robins, R.J. and Juniper, B.E. (1980a). The secretory cycle of *Dionaea muscipula* Ellis. I. The fine structure and the effect of stimulation on the fine structure of the digestive glands. *New Phytologist* **86**, 279–296.
Robins, R.J. and Juniper, B.E. (1980b). The secretory cycle of *Dionaea muscipula* Ellis. II. Storage and synthesis of the secretory proteins. *New Phytologist* **86**, 297–311.
Robins, R.J. and Juniper, B.E. (1980c). The secretory cycle of *Dionaea muscipula* Ellis. III. The mechanism of release of digestive secretion. *New Phytologist* **86**, 313–327.
Robins, R.J. and Juniper, B.E. (1980d). The secretory cycle of *Dionaea muscipula* Ellis. IV. The enzymology of the secretion. *New Phytologist* **86**, 401–412.
Robins, R.J. and Juniper, B.E. (1980e). The secretory cycle of *Dionaea muscipula* Ellis. V. The absorption of nutrients. *New Phytologist* **86**, 413–422.
Robins, R.J. and Subramanyam, K. (1980). Scanning electron microscope study of the seed surface of some *Utricularia* (Lentibulariaceae) species from India. *Proceedings of the Indian National Science Academy* **46B**, 310–324.
Robinson, A.G. (1972). A new species of aphid (*Homoptera*: Aphididae) from a pitcher-plant. *Canadian Entomologist* **104**, 955–957.
Robinson, W.J. (1908). A study of the digestive power of *Sarracenia purpurea*. *Torreya* **8**, 181–194.
Robinson, W.J. (1909). Experiments on *Drosera rotundifolia* as to its protein digesting power. *Torreya* **9**, 109–114.
Roels, J.I., Pauwel, S.L., Evrard, C. and Pollak, H. (1968). Detection par activation neutronique de la teneur de manganese dans les plants Congolaises (Neutron activation detection of the level of manganese in Congolese plants). *Rapport de Recherches du Centre Nucléaire Trico, Kinshasa* **16**, 87–105.
Rolfe, W.D.I. and Ingham, J.K. (1967). Limb structure, affinity and diet of the Carboniferous centipede *Arthroplenia*. *Scottish Journal of Geology* **3**, 118–124.
Romeo, J.T., Bacon, J.D. and Mabry, T.J. (1977). Ecological considerations of amino acids and flavonoids in *Sarracenia* species. *Biochemical Systematics and Ecology* **5**, 117–120.
Rost, K. and Schauer, R. (1977). Physical and chemical properties of the mucin secreted by *Drosera capensis*. *Phytochemistry* **16**, 1365–1368.
Roth, I. (1953). Zur Entwicklungsgeschichte und Histogenese der Schlaublätter von *Nepenthes* (Developmental history and histogenesis of tubular leaves in *Nepenthes*). *Planta* **42**, 177–208.
Roth, I. (1954). Entwicklung und histogenetischer Vergleich der Nektar und Verdauungsdrusen von *Nepenthes* (Development and histogenetic comparison of the nectary and digestive glands of *Nepenthes*). *Planta* **43**, 361–378.
Roth, I. (1957). Relation between histogenesis of the leaf and its external shape. *Botanical Gazette* **118**, 237–245.
Rumphius, G.E. (1747). 'Herbarium Amboinense' – plurimas complectans, arbores, frutices, herbas, plantas, quae in Amboina. . . . In six parts 1741–1750, folio, Amsterdam. (A complete set is in the Bodleian Library, Oxford.)
Russell, M.C. (1953). Notes on insects associated with sundews (*Drosera*) at Lesmurdie. *The Western Australian Naturalist* **4**, 9–12.
Russell, M.C. (1958). Rediscovery of *Drosera zonaria* Planch. *The Western Australian Naturalist* **6**, 109–111.
Russell, M.C. (1959). Colouring matters from the western Australian sundews. II. The release of free pigments. *The Western Australian Naturalist* **7**, 30–34.
Ryabtsev, I.A. (1978). Observation of the relation of members of the food chain which are essential to wild water fowl and near-water animals (Russian). *In* 'Radioecology of Vertebrate Animals' (in Russian) (Ed. Il'Enko), 15–23. Academy Nauka, Moscow.
Rychnovska-Soudkova, M. (1953). Studie o mineralni vyzive rostliny *Drosera rotundifolia* L. I. Vliv kalcia jako dulezity a ekologiky cinitei (A study of the mineral nutrition of *D. rotundifolia*. 1. The physiological and ecological role of calcium). *Preslia* **25**, 51–66.
Rychnovska-Soudkova, M. (1954). Studie o mineralni vyzive rostliny *Drosera rotundifolia* L. 2. Koronova sorpce dusiku v anorganike forme (A study of the mineral nutrition of *D. rotundifolia*. 2. The absorption of inorganic nitrogen by the roots). *Preslia* **26**, 55–66.
Rymal, D.E. and Folkerts, G.W. (1982). Insects associated with pitcher-plants (*Sarracenia*: Sarraceniaceae), and their relationship to pitcher-plant conservation. *Journal of the Alabama Academy of Science* **53**, 131–151.
Sabalitschka, T. (1923). Ueber *Drosera rotundifolia* L. *Archiv der Pharmazie* **261**, 217–222.
Sachs, J. von (1887). 'Lectures on the Physiology of Plants' (translated by H. Marshall Ward), 655. Clarendon Press, Oxford.
Sampara-Rumantir, N. (1971). Rossoliside (Dutch). *Pharmaceutisch Weekblad voor Nederland* **106**, 653–664.
Sandved, K.B. and Prance, G.T. (1985). 'Leaves: the Formation, Characteristics, and Uses of Hundreds of Leaves Found in all parts of the World'. Crown, New York.
Sartory, A., Sartory, R. and Couderc, M. (1947). Contribution a l'étude des tannins du maté et du *Drosera*. *Comptes Rendues de l'Académie des Sciences, Paris* **224**, 1738–1740.
Sasago, A. and Sibaoka, T. (1985a). Water extrusion in the trap bladders of *Utricularia vulgaris*. I. A possible pathway of water across the bladder wall. *Botanical Magazine, Tokyo* **98**, 55–66.
Sasago, A. and Sibaoka, T. (1985b). Water extrusion in

the trap bladders of *Utricularia vulgaris*. II. A possible mechanism of water outflow. *Botanical Magazine, Tokyo* **98**, 113–124.

Sasaki, H., Yaguchi, H., Endo, T., Yosioka, I., Higashiyama, K. and Otomasu, H. (1978). The glycosides of *Martynia louisiana* a new phenyl propanoid glycoside martynoside. *Chemical and Pharmaceutical Bulletin (Tokyo)* **26**, 2111–2121.

Sattler, R. (1977). 'Fusion' and continuity in floral morphology. *Notes of the Royal Botanic Gardens Edinburgh* **36**, 397–405.

Scala, J., Iott, K., Schwab, D.W. and Semersky, F.E. (1969). Digestive secretion of *Dionaea muscipula* (Venus's Flytrap). *Plant Physiology* **44**, 367–371.

Scala, J., Schwab, D.W. and Simmons, E. (1968a). Plastids of the Venus's Flytrap (*Dionaea muscipula*). *Nature* **219**, 1183–1184.

Scala, J., Schwab, D.W. and Simmons, E. (1968b). The fine structure of the digestive gland of Venus's Flytrap. *American Journal of Botany* **55**, 649–657.

Scheele, G., Dobberstein, B. and Blobel, G. (1978). Transfer of proteins across membranes: biosynthesis in vitro of pretrypsinogen and trypsinogen by cell fractions of canine pancreas. *European Journal of Biochemistry* **82**, 593–600.

Schimper, A.F.W. (1882a). Notizen über insektenfressenden Pflanzen (Notes on insectivorous plants). *Botanische Zeitung* **40**, 225–234.

Schimper, A.F.W. (1882b). Notizen über insektenfressenden Pflanzen (Notes on insectivorous plants). *Botanische Zeitung* **40**, 241–248.

Schlauer, J. (1986). Nomenclatural synopsis of carnivorous phanerogamous plants. A world carnivorous plant list. *Carnivorous Plant Newsletter* **15**, 59–117.

Schmid, R. (1964). Die systematische Stellung der Dioncophyllaceen (The systematic position of the Dioncophyllaceae). *Botanische Jahrbücher* **83**, 1–56.

Schmid-Hollinger, R. (1979). *Nepenthes* Studien 5. Die Kannenformen der westlichen *Nepenthes* Arten (Studies on the genus *Nepenthes* 5. Pitcher formation in the westerly species). *Botanische Jahrbücher für Sytematik Pflanzengeschichte und Pflanzengeographie* **100**, 379–405.

Schmucker, Th. and Linnemann, G. (1959). Carnivorie In– 'Handbuch der Pflanzenphysiologie' (Encyclopedia of Plant Physiology) Vol. 11 Heterotrophy (Ed. Rühland, W.), 198–283. Springer-Verlag, Berlin.

Schnell, D.E. (1976). 'Carnivorous Plants of the United States and Canada' John F. Blair, Winston Salem, N. Carolina.

Schnell, D.E. (1978). Systematic flower studies of *Sarracenia*. *Castanea* **43**, 211–220.

Schnell, D. (1980). *Drosera linearis*. *Carnivorous Plant Newsletter* **9**, 16–17.

Schnell, D.E. (1980). Notes on the biology of *Sarracenia oreophila* (Kearney) Wherry. *Castanea* **45**, 166–170.

Schnell, D.E. (1982a). Effects of simultaneous draining and brush-cutting on *Sarracenia* L. population in a south-eastern North Carolina pocosin. *Castanea* **47**, 248–260.

Schnell, D.E. (1982b). Notes on *Drosera linearis* Goldie in northeastern Lower Michigan. *Castanea* **47**, 313–328.

Schnell, D.E. and Krider, D.W. (1976). Cluster analysis of the genus *Sarracenia* L. in the southeastern United States. *Castanea* **41**, 165–176.

Schnell, D.E., Leona, O. and Thayer, J.D. (1949). 'Antibiotic Activity of Plant Extracts' (unpublished).

Schnepf, E. (1960a). Zur Feinstruktur der Drüsen von *Drosophyllum lusitanicum* (On the fine structure of the glands of *Drosophyllum lusitanicum*). *Planta (Berlin)* **54**, 641–674.

Schnepf, E. (1960b). Kernstrukturen bei *Pinguicula* (Nuclear structures in *Pinguicula*). *Berichte der deutschen Botanische Gesellschaft* **73**, 243–245.

Schnepf, E. (1961a). Licht und elektronmikroskopische Beobachtungen an Insektivoren-Drüsen über der Secretion des Fangschleim (Light and electron microscopical observations on insectivorous glands and their secretion of 'fangschleim'). *Flora (Jena)* **151**, 73–87.

Schnepf, E. (1961b). Quantitative Zusammenhänge zwischen der Sekretion des Fangschleimes und den Golgi-Strukturen bei *Drosophyllum lusitanicum* (Quantitative relationships between the secretion of the viscid fluid of the tentacles and the Golgi structures of *D. lusitanicum*). *Zeitschrift für Naturwissenschaften* **16b**, 605–610.

Schnepf, E. (1963a). Zur Cytologie und Physiologie pflanzlicher Drüsen. 1. Teil. Über der Fangschleim der Insektivoren (On the cytology and physiology of plant glands. 1. On the 'fangschleim' of insectivorous plants). *Flora (Jena)* **153**, 1–22.

Schnepf, E. (1963b). Zur Cytologie und Physiologie pflanzlicher Drüsen. 2. Teil. Über die Wirkung von Sauerstoffentzung und von Atmungsinhibitoren auf die Sekretion des Fangschleimes von *Drosophyllum* und auf die Feinstruktur der Drüsenzellen (On the cytology and physiology of plant glands 2. On the effects of reduced oxygen supply and of respiratory inhibition on the secretion of the 'fangschleim' of *Drosophyllum* and on the fine structure of the gland cells). *Flora (Jena)* **153**, 23–48.

Schnepf, E. (1963c). Zur Cytologie und Physiologie pflanzlicher Drüsen. 3. Teil. Cytologische Veränderungen in den Drüsen von *Drosophyllum* während der Verdauung (On the cytology . . . 3. Cytological changes in the glands of *Drosophyllum* during digestion). *Planta* **59**, 351–379.

Schnepf, E. (1966). Die Morphologie der Sekretion in pflanzlichen Drüsen (The morphology of secretion by plant glands). *Berichte deutschen Botanische Gesellschaft* **78**, 478–483.

Schnepf, E. (1969). Sekretion und Exkretion bei Pflanzen (Secretion and excretion by plants). *Protoplasmatologia* **8**, 1–181.

Schnepf, E. (1972). Über die Wirkung von Hemmstoffen der Proteinsynthese auf die Sekretion des Kohlenhydrat-Fangschleims von *Drosophyllum lusitanicum* (The influence of inhibitors of protein synthesis on the secretion of the carbohydrate 'fangschleim' of *D. lusitanicum*). *Planta* **103**, 334–339.

Schnepf, E. (1974). Gland cells. *In* 'Dynamic Aspects of Plant Ultrastructure' (Ed. Robards, A.W.), 331–357. McGraw-Hill, Maidenhead.

Schönherr, J. (1976). Water permeability of isolated cuticular membranes: The effect of cuticular waxes on diffusion of water. *Planta* **131**, 159–164.

Schumacher, G.J. (1960). Further notes on the occurrence of desmids in *Utricularia* bladders. *Castanea* **25**, 62–65.

Schutt, P. (1971). Untersuchungen über den Einfluss von Cuticular-wachsen auf die Infektionsfähigkeit pathogener Pilzen. *Lophodermium pinastri* und *Botrytis cinerea* (Researches on the influence of cuticular waxes on the infection capacity of pathogenic fungi). *European Journal of Forest Pathology* **1**, 32–50.

Schwab, D.W., Simmons, E. and Scala, J. (1969). Fine structure changes during functioning of the digestive gland of Venus's Flytrap. *American Journal of Botany* **56**, 88–100.

Schwaegerle, K.E. (1983). Population growth of the pitcher-plant *Sarracenia purpurea* L. at Cranberry Bog, Licking County, Ohio. *Ohio Journal of Science* **83**, 19–22.

Schwartz, R. (1974). 'Carnivorous Plants' (Ed. Leavey, D.). Praeger, New York and Washington.

Scott, A.C. and Taylor, T.N. (1983). Plant/animal interactions during the upper Carboniferous. *The Botanical Review* **49**, 259–307.

Scott, A.C., Chaloner, W.G. and Paterson, S. (1985). Evidence of Pteridophyte-arthropod interactions in the fossil record. *Proceedings of the Royal Society of Edinburgh* **86B**, 133–140.

Sever, J.R., Lytle, T.F. and Thomas, F. (1972). Lipid geochemistry of a Mississippi coastal bog environment. *Contributions in Marine Science* **16**, 149–161.

Shetler, S.G. (1974a). Sarraceniales. Encyclopaedia Britannica 15th Edn, 252–256.

Shetler, S.G. (1974b). Nepenthales. Encyclopaedia Britannica 15th Edn. 958–962.

Shibata, C. and Komiya, S. (1972). Increase of nitrogen content in the leaves of *Drosera rotundifolia* fed with protein. *Japanese Bulletin of the Nippon Dental College, General Education* **1**, 55–75.

Shibata, C. and Komiya, S. (1973). Changes of nitrogen content in the leaves of *Drosera rotundifolia* during feeding with protein. *Japanese Bulletin of the Nippon Dental College, General Education* **2**, 89–100.

Shinonaga, S. and Beaver, R.A. (1979). *Pierretia urceola*: A new species of Sarcophagid fly found living in *Nepenthes* pitcher-plants in West Malaysia. *Annales de la Societée Entomologique de France* **15**, 37–40.

Shivas, R.J. (1983). *Nepenthes* of Gunung Ulu Kali. *Carnivorous Plant Newsletter* **12**, 65.

Sibaoka, T. (1966). Action potentials in plant organs. *Symposium for the Society for Experimental Biology* **20**, 49–74.

Sibaoka, T. (1980). Action potentials and rapid plant movements. *In* 'Plant Growth Substances 1979' (Ed. Skoog, F.). Springer-Verlag, Berlin, Heidelberg, New York.

Simms, J.E. (1884). *Utricularia vulgaris*: an enemy of fishes. *Nature* **30**, 295–296.

Simola, L.K. (1978). Dipeptides as nitrogen sources for *Drosera rotundifolia* in aseptic culture. *Physiologia Plantarum* **44**, 315–318.

Simola, L.K., Koskimes-Soininen, K. and Tomell, M. (1985). Glycopeptides of turions and leaves of *Utricularia vulgaris* at various stages of development. *Physiologia Plantarum* **65**, 23–26.

Single, W.V. and Marcellos, H. (1974). Studies on frost injuries to wheat IV. Freezing of ears after emergence from the leaf sheath. *Australian Journal of Agricultural Research* **25**, 679–686.

Slack, A.A.P. (1966). On the distribution in Britain of *Pinguicula lusitanica*. *Glasgow Naturalist* **18**, 438–444.

Slack, A.A.P. (1979). 'Carnivorous Plants', 240. Ebury Press, London.

Slack, A.A.P. (1986). 'Insect-eating Plants and How to Grow Them'. Alphabooks, Sherborne, Dorset.

Slater, F.M. (1981). The mineral contents of both peat and plants and their interrelationships at Borth Bog, Wales. Irish National Peat Communications. *Proceedings of the 7th International Peat Congress* **1**, 450–467.

Small, E. (1972). Photosynthetic rates in relation to nitrogen recycling as an adaptation to nutrient deficiency in peat bog plants. *Canadian Journal of Botany* **50**, 2227–2233.

Small, J.G.C. and Hendrikz, B. (1974). Evidence for nitrate reductase activity in the carnivorous plant *Drosera aliciae*. *South African Journal of Science* **70**, 156–157.

Small, J.G.C., Onraet, A., Grierson, D.S. and Reynolds, G. (1977). Studies on insect-free growth, development and nitrate-assimilating enzymes of *Drosera aliciae* Hamet. *New Phytologist* **79**, 127–134.

Smith, J.B. (1902). Life history of *Aedes smithii*. *Journal of the New York Entomological Society* **10**, 10–15.

Smith, L.B. and Downs, R.J. (1974). Pitcairnoideae (Bromeliaceae). 'Flora Neotropica', Monograph No. 14, Part I. Published for the Organization for Flora Neotropica, Hafner Press, New York.

Smith, L.B. and Downs, R.J. (1977). Tillandsioideae (Bromeliaceae). 'Flora Neotropica', Monograph No. 14, Part II. Published for the Organization for Flora Neotropica, Hafner Press, New York.

Smith, L.B. and Downs, R.J. (1979). Bromelioideae (Bromeliaceae). 'Flora Neotropica', Monograph No. 14, Part III. Published for the Organization for Flora Neotropica, Hafner Press, New York.

Smith, S.M. and Brust, R.A. (1971). Periodic control of the maintenance and termination of larval diapause in *Wyeomyia smithii* (Coq.) (Diptera: Culicidae) with notes on oogenesis in the adult female. *Canadian Journal of Zoology* **49**, 1065–1073.

Smith, Sir J.E. (1821). 'Correspondence of Linnaeus', Vol. 1, 69.

Smythies, B.E. (1963). The distribution and ecology of pitcher plants (*Nepenthes*) in Sarawak. Unesco Symposium in Kuching, July 1963, 'The Vegetation of the Humid Tropics', 170–178.

Sohma, K. (1975). Pollen morphology of the Japanese species of *Utricularia* L. and *Pinguicula* L. with notes on fossil pollen of *Utricularia* from Japan. *Journal of Japanese Botany* **50**, 164–209.

Sorenson, D.R. and Jackson, W.T. (1968). The utilization of *Paramecia* by the carnivorous plant *Utricularia gibba*. *Planta* **83**, 166–170.

Southwood, T.R.E. (1986). Plant surfaces and insects – an overview. *In* 'Insects and the Plant Surface' (Eds. Juniper, B.E. and Southwood, T.R.E.), 1–22. Edward Arnold, London.

Sowerby, J. (1790–1797). 'English Botany' (text J.E. Smith). London.

Speirs, D.C. (1981). The evolution of carnivorous plants. *Carnivorous Plant Newsletter* **10**, 62–65.

Sporne, K.R. (1975).'The Morphology of the Pteridophytes'. Hutchinson, London.

Stansell, V. (1980). *Darlingtonia californica* – Geographical distribution, habitat and threats. Dept. of the Interior U.S. Fish and Wildlife Service R-5 Special Report.

Steckelberg, R., Lüttge, U. and Weigl, J. (1967). Reinung der Proteinase aus *Nepenthes*-Kannensaft (Purification of the proteinase from the *Nepenthes* pitcher secretion). *Planta* **76**, 238–241.

Steenis, C.G.J. van (1957). Specific and intraspecific delimitation. *Flora Malesiana Series 1*, **5**, 167–234.

Steenis, C.G.J. van (1969). Plant speciation in Malesia with special reference to the theory of non-adaptive, saltatory evolution. *Biological Journal of the Linnean Society* **1**, 97–133.

Steenis, C.G.J. van (1977). Autonomous evolution in plants: Differences in plant and animal evolution. *The Gardens' Bulletin, Singapore* **29**, 103–126.

Steffan, W.A. and Evenhuis, N.L. (1982). *Toxorhynchites nepenthicola*: a new species from Papua New Guinea (Diptera: Culicidae). *Mosquito Systematics* **14**, 1–13.

Stellfeld, C. (1959). Contribuciones ao estudio da *Drosera* (Contributions to the study of *Drosera*). *Tribuna de Farmacia (Curitiba)* **27**, 57–61.

Stern, K. (1917). Beiträge zur Kenntnis der Nepenthaceen (Contributions to the knowledge of *Nepenthes*). *Flora (Jena)* **109**, 213–283.

Stern, K.G. and Stern, E. (1932). Über die Proteinasen insektivorer Pflanzen (On the proteinase from insectivorous plants). *Biochemische Zeitschrift* **252**, 81–96.

Stewart, W.N. (1983). 'Paleobotany and the Evolution of Plants'. Cambridge University Press, Cambridge.

Steyermark, J. and Maguire, B. (1984). Nuevos taxa de la Guyana Venezuelana (New taxa from Venezuelan Guyana). *Acta Botanica Venezuelica* **14**, 5–52.

Stork, N.E. (1980a). A scanning electron microscope study of tarsal adhesive setae in the Coleoptera. *Zoological Journal of the Linnean Society* **68**, 173–306.

Stork, N.E. (1980b). Role of wax blooms in preventing attachment to brassicas by the mustard beetle, *Phaedon cochleariae. Entomologica Experimentalis et Applicata* **28**, 100–107.

Stork, N.E. (1986). The form of plant waxes: a means of preventing insect attachment? *In* 'Insects and the Plant Surface' (Eds. Juniper, B.E. and Southwood, T.R.E.), 346–347. Edward Arnold, London.

Stowe, M.K. (1987). Chemical mimicry. *In* 'The Chemical Mediation of Co-evolution' (Ed. Spencer). Pegasus, New York.

Strong, D.R., Lawton, J.H. and Southwood, T.R.E. (1984). 'Insects on Plants: Community Patterns and Mechanisms'. Blackwell Scientific, Oxford.

Studnicka, M. (1981). The Czech butterwort: *Pinguicula bohemica. Carnivorous Plant Newsletter* **10**, 38–44.

Studnicka, M. (1982). The problem of carnivory in the common toothwort (*Lathraea squamaria*). *Carnivorous Plant Newsletter* **11**, 17–20.

Stuhlman, O. (1948a). A mechanical analysis of the closure movements of Venus's flytrap. *Physical Review* **74**, 119 (abstract).

Stuhlman, O. (1948b). A physical analysis of the opening and closing movements of the lobes of Venus's flytrap. *Bulletin of the Torrey Botanical Club* **75**, 22–44.

Sugden, A.M. and Robins, R.J. (1979). Aspects of the ecology of vascular epiphytes in Colombian cloud forest. 1. The distribution of epiphytic flora. *Biotropica* **11**, 173–188.

Sukul, N.C. (1971). *Paralongidorus droseri*: a new species of *Nematoda longidorae* associated with an insectivorous plant from West Bengal. *Bulletin of Entomology* **12**, 85–88.

Sutherst, R.W., Jones, R.J. and Schnitzerling, H.J. (1982). Tropical legumes of the genus *Stylosanthes* immobilize and kill cattle ticks. *Nature* **295**, 320–321.

Sutherst, R.W. and Wilson, L.J. (1986). Tropical legumes and their ability to immobilize and kill cattle ticks. *In* 'Insects and the Plant Surface' (Eds. Juniper, B.E. and Southwood, T.R.E.). Edward Arnold, London.

Swales, D.E. (1969). *Sarracenia purpurea* as a host and carnivore at Lac Carré, Terrebonne County, Quebec. *Naturaliste Canadienne* **96**, 759–763.

Swales, D.E. (1972). *Sarracenia purpurea* L. as host and carnivore at Lac Carré, Terrebonne County, Quebec, Part II. *Naturaliste Canadienne* **99**, 41–47.

Swales, D.E. (1975). An unusual habitat for *Drosera rotundifolia* L., its overwintering state and vegetative reproduction. *Canadian Field Naturalist 89*, 143–147.

Sydenham, P.H. and Findlay, G.P. (1973). The rapid movement of the bladder of *Utricularia* sp. *Australian Journal of Biological Sciences* **26**, 1115–1126.

Sydenham, P.H. and Findlay, G.P. (1975). Transport of solutes and water by resetting bladders of *Utricularia. Australian Journal of Plant Physiology* **2**, 335–351.

Syvanen, M. (1985). Cross-species gene transfer: Implications for a new theory of evolution. *Journal of Theoretical Biology* **112**, 333–343.

Szafer, W. (1953). Pleistocene stratigraphy of Poland from the floristic point of view. *Annals of the Geological Society of Poland* **22**, 1–99.

Tait, L. (1875). Insectivorous plants. Paper given to Birmingham Natural History Society, June 17, 1875. *Nature* **12**, 251–252.

Takahashi, K., Chang, W.J. and Ko, J.S. (1974). Specific inhibition of acid proteases from brain, kidney, skeletal muscle and insectivorous plants by diazoacetyl-DL-norleucine methyl ester and by pepstatin. *Journal of Biochemistry (Tokyo)* **76**, 879–899.

Takhtajan, A.L. (1969). 'Flowering Plants: Origin and Dispersal'. Oliver and Boyd, Edinburgh.

Takhtajan, A.L. (1980). Outline of the classification of flowering plants (Magnoliophyta). *The Botanical Review* **46**, 225–359.

Taylor, P. (1964). The genus *Utricularia* L. Lentibulariaceae in Africa (south of the Sahara) and Madagascar. *Kew Bulletin* **18**, 1–245.

Taylor, P. (1967). 'Botany of the Guiana Highlands' VII. Lentibulariaceae. *Memoirs of the New York Botanic Gardens* **17**, 201–228.

Taylor, P. (1976). Flora of Panama. Part 9. Lentibulariaceae. *Annals of the Missouri Botanic Garden* **63**, 565–580.

Taylor, P. (1988). The genus *Utricularia. Kew Bulletin, Additional Series* (in press).

Taylor, P. and Cheek, M. (1983). *Pinguicula agnata*, Lentibulariaceae. *Curtis's Botanical Magazine Tab. 874*, **184**, 159–160.

Thanikaimoni, G. (1966). Pollen morphology of the genus *Utricularia. Pollen et Spores* **8**, 265–284.

Thanikaimoni, G. and Vasanthy, G. (1972). Sarraceniaceae: Palynology and systematics. *Pollen et Spores* **14**, 143–155.

Thien, L.B., Heimermann, W.H. and Holman, R.T. (1975). Floral odors and quantitative taxonomy of *Magnolia* and *Liriodendron. Taxon* **24**, 557–568.

Thienemann, A. (1932). Die Tierwelt der *Nepenthes*-Kannen (The fauna of a *Nepenthes* pitcher). *Archiv für Hydrobiologie*, Supplement **11**, *Tropische Binnengewasser* **3**, 1–54.

Thienemann, A. (1934). Die Tierwelt der tropischen Pflanzengewasser (The fauna of a tropical phytotelm). *Archiv für Hydrobiologie,* Supplement **13**, 1–91.

Thomas, D.A. and Barber, H.N. (1974). Studies on leaf characteristics of a cline of *Eucalyptus urnigera* from Mount Wellington, Tasmania. 1. Water repellancy and the freezing of leaves. *Australian Journal of Botany* **22**, 501–512.

Thomas, D. and Gouranton, J. (1979). Ultrastructural and autoradiographic study of the intranuclear inclusions of *Pinguicula lusitanica* L. *Planta* **145**, 89–93.

Thomas, S.B. and McQuillin, J. (1953). Ropy milk organisms isolated from insectivorous plants. *Dairy Industries* **18**, 40–42.

Thompson, K.F. (1963). Resistance to the cabbage aphid, *Brevicoryne brassicae* in brassica plants. *Nature* **198**, 209.

Thompson, J.N. (1981). Reversed animal–plant interactions: The evolution of insectivorous and ant-fed plants. *Biological Journal of the Linnean Society* **16**, 147–155.

Thomson, R.H. (1949). Naturally occurring quinones: a synthesis of droserone. *Journal of the Chemical Society* **1949 (5)**, 1277–1278.

Thomson, R.H. (1971). 'Naturally Occurring Quinones', 2nd edn, 30, 31, 227–230, 234, 236, 237, 247, 254, 255, 668. Academic Press, London.

Thomson, R.H. and Wilkie, K.C.B. (1961). Some alleged naturally occurring quinones. *Chemistry and Industry* 1712–1713.

Thorn, R.G. and Barron, G.L. (1984). Carnivorous mushrooms. *Science* **224**, 75–78.

Thurston, E.L. and Seabury, F. (1975). A scanning electron microscope study of the utricle trichomes in *Utricularia*

biflora Lam. *Botanical Gazette* **136**, 87–93.
Tietze, L.F., Glusenkamp, K.H., Nakane, M. and Hutchinson, C.R. (1982). Stereocontrolled synthesis of (+)− Sarracenin via photochemical cycloaddition. *Angewandte Chemie* **21**, 70.
Tokés, Z.A., Woon, W.C. and Chambers, S.M. (1974). Digestive enzymes secreted by the carnivorous plant *Nepenthes macfarlanei* L. *Planta* **119**, 39–46.
Torrey, J. (1854). On the *Darlingtonia californica*, a new pitcher plant from Northern California. *Smithsonian Contributions to Knowledge*, **6**, 1–8 (Accepted for publication in 1850).
Tournefort, P. de (1689). Denombrement des plantes que iay trouve en Portugal en 1689 (A checklist of plants I found in Portugal in 1689). Manuscript in the Cabinet de Botanique, University of Coimbra, Portugal.
Tradescant, J. (Senior) (1618) 'Ashmole M.S.824', 175–184. Bodleian Library, Oxford.
Treat, M. (1875). Plants that eat animals. *Gardeners' Chronicle March 6th*, 303–304.
Treat, M. (1876). Is the valve of *Utricularia* sensitive? *Harper's New Monthly Magazine* **52**, 382–387.
Trécul, M.A. (1855). Feuilles du *Drosera rotundifolia* (The leaves of *D. rotundifolia*). *Annales des Sciences Naturelles* **4e series**, 303–311.
Treviranus, C.L. (1832–1833, October–April). Secretion of water by certain plants (Scientific Intelligence No. 19). *Edinburgh New Philosophical Journal* 195–197.
Troll, W. (1932). Morphologie der schildformigen Blätter (The morphology of shield-forming leaves). *Planta* **17**, 153–314.
Troll, W. (1939). 'Vergleichende Morphologie der Höheren Pflanzen (comparative morphology of the higher plants). Verlag von Gebruder Borntraeyer, Berlin.
Truswell, E.M. and Marchant, N.G. (1986). Early tertiary pollen of probable Droseracean affinity from central Australia. *Special Papers in Palaeontology* **35**, 163–178.
Tsang, P. (1980). A new *Drosera* from the top end of Australia. *Carnivorous Plant Newsletter* **9**, 46–48.
Turner (1568). 'A New Herball'. *Cited in*: 'The Vegetable Kingdom' Lindley, Jo (1846), 287–288. Bradbury and Evans, London.
Turrill, W.B. (1950). Character combinations and distributions in the genus *Fritillaria* and allied genera. *Evolution* **4**, 1–6.
Tutin, T.G. (1964). Family Droseraceae, genus *Drosophyllum*. *In* 'Flora Europaea' **1**, 350.
Uphof, J.C. (1936). Sarraceniaceae. *In* 'Die Natürlichen Pflanzenfamilien', **2**, Section 17b (Eds. Engler, A. and Prantl, K.). Wilhelm Engelman, Leipzig.
Van Achterberg, C. (1973). (See Achterberg, C. van).
Van Eseltine, G.P. (1929). A preliminary study of the unicorn plants (Martyniaceae). *New York State Agricultural Experimental Station, Technical Bulletin* **149**, 3–41.
Van der Pijl, L. (1953). On the flower biology of some plants from Java. With general remarks on fly-traps (species of *Annona*, *Ariocarpus*, *Typhonium*, *Gnetum*, *Arisaema* and *Abroma*). *Annales Bogorienses* **1**, 77–99.
Van der Pijl, L. (1982). 'Principles of Dispersal in Higher Plants'. Springer-Verlag, Berlin.
Vane-Wright, R.I. (1980). On the definition of mimicry. *Biological Journal of the Linnean Society* **13**, 1–6.
Vanhaelen, M. (1970). Identification par chromatographie sur couches minces de gel de silice de drogues végétales et se quelques-uns de leurs dérives galeniques. *Pharmaceutisch tijdschrift voor Belgie* **25**, 175–212.
Vassilyev, A.E. (1977). 'Functional Morphology of Plant Secretory Cells'. Nauka Publishing House, Leningrad (in Russian).
Vassilyev, A.E. and Muravnik, L.E. (1988a) The ultrastructure of the digestive glands in *Pinguicula vulgaris* L. (Lentibulariaceae) relative to their function. I. The changes during maturation. *Annals of Botany* (in press).
Vassilyev, A.E. and Muravnik, L.E. (1988b) The ultrastructure of the digestive glands in *Pinguicula vulgaris* L. (Lentibulariaceae) relative to their function. II. The changes on stimulation. *Annals of Botany* (in press).
Veitch, J.H. (1898). *Nepenthes. Journal of the Royal Horticultural Society* **21**, 1–30.
Veitch, J.H. (1906). 'Hortus Veitchii'. James Veitch, London.
Vesey-Fitzgerald, D. (1940). On the vegetation of the Seychelles. *Journal of Ecology* **28**, 465–483.
Villaneuva, V.R.,. Simola, L.K. and Mordon, M. (1985). Polyamines in turions and young plants of *Hydrocharis morsus-ranae* and *Utricularia intermedia*. *Phytochemistry* **24**, 171–172.
Vines, S.H. (1877). On the digestive ferment of *Nepenthes*. *Journal of the Linnean Society* **15**, 427–431.
Vines, S.H. (1897). The proteolytic enzymes of *Nepenthes* I. *Annals of Botany* **11**, 563–584.
Vines, S.H. (1898). The proteolytic enzymes of *Nepenthes* II. *Annals of Botany* **12**, 545–555.
Vines, S.H. (1901). The proteolytic enzymes of *Nepenthes* III. *Annals of Botany* **15**, 563–573.
Vinkenborg, J., Sampara-Rumantir, N. and Uffelie, O.F. (1969). De aawezigheid van hydroplumbagine-glucoside in *Drosera rotundifolia* L. (The presence of hydroplumbagin in *D. rotundifolia*). *Pharmaceutisch Weekblad voor Nederland* **104**, 45–49.
Vinkenborg, J., Sampara-Rumantir, N. and Uffelie, O.F. (1970). Rossoliside, een glucoside uit *Drosera rotundifolia* L. (Rossoliside, a new D-glucoside from *D. rotundifolia* L.). *Pharmaceutisch Weekblad voor Nederland* **105**, 414.
Vintéjoux, C. (1970). Compartement des inclusions intracellulaires cristallines dans les cellules meristématiques et différenciées au cours de la germination des hibernacles d'*Utricularia neglecta* L. (Lentibulariaceae) (Compartmentation of cristalline intracellular inclusions in meristematic cells and the differences in germination of the seeds of *U. neglecta*). *Comptes Rendues* **270D**, 2438–2440.
Vintéjoux, C. (1973a). Variations saisonnières des constituants ultrastructuraux dans les plastes foliares, chez l'*Utricularia neglecta* L. (Seasonal variations in the ultrastructural constituents in the leaf plastids of *U. neglecta*). *Comptes Rendues* **276D**, 1693–1696.
Vintéjoux, C. (1973b). Aspects ultrastructuraux de la sécrétion de mucilages chez une plante aquatique carnivore: l'*Utricularia neglecta* L. (Lentibulariacees) (Ultrastructural aspects of the secretion of mucilage from an aquatic carnivorous plant: *U. neglecta*) *Comptes Rendues* **277D**, 1745–1748.
Vintéjoux, C. (1973c). Études des aspects ultrastructuraux de certaines cellules glandulaires en rapport avec leur activité sécrétive chez l'*Utricularia neglecta* L. (Lentibulariaceae) (Studies of some ultrastructural aspects of certain gland cells and a report on their secretion activity in *U. neglecta*). *Comptes Rendues* **277D**, 2345–2348.
Vintéjoux, C. (1974). Ultrastructural and cytochemical observations on the digestive glands of *Utricularia neglecta* L. (Lentibulariaceae). Distribution of protease and acid phosphatase activities. *Portugaliae Acta Biologica Series A* **14**, 463–471.
Vintéjoux, C. (1976). Modifications ultrastructurales saisonnières dans les glandes externes d'*Utricularia neglecta* L. (Lentibulariacees) (Seasonal ultrastructural modifications in the external glands of *U. neglecta*). *Actes du 101 Congrès National des Sociétés Savantes, Lilles, Sciences*, **1**, 293–304.

Vintéjoux, C. (1979). Études ultrastructurales sur les phenomenes de secretion et d'absorption, chez une plante aquatique carnivore: *Utricularia neglecta* L. (Lentibulariaceae) (Ultrastructural studies on the phenomena of secretion and absorption in an aquatic carnivorous plant: *U. neglecta*). *Comptes rendus du 104 Congrès National des Sociétés Savantes, Sciences* **2**, 93–104.

Vintéjoux, C. and Prevost, M.C. (1976). Structure et réactivité des fermentations Golgiènnes, au cours de la sécrétion de mucilages polysaccharidiques dans les glandes externes de l'*Utricularia neglecta* L. (Lentibulariaceae) (Structure and activity of the Golgi during the secretion of polysaccharide mucilages by the external glands of *U. neglecta*). *Annales des Sciences Naturelles Botanique, Paris 12e ser. 17*, 375–394.

Voelcker, A. (1849). On the chemical composition of the fluid in the ascidia of *Nepenthes*. *Annals and Magazine of Natural History II* **4**, 128–136.

Vogel, S. (1960). Zur Feinstruktur der Drüsen von *Pinguicula* (On the fine structure of the glands of *Pinguicula*). *Beihefte zu den Zeitschriften des Schweizen den Forstvereins* **30**, 113–122.

Vogel, S. (1965). Kesselfallen-Blumen (Flowers that are insect traps). *Umschau* **65**, 12–16.

Vogel, S. (1981). Die Klebstoffhaare an den Antheren von *Cyclanthera pedata* (The adhesive hairs on the anthers of *C. pedata*). *Plant Systematics and Evolution* **137**, 291–316.

Vogel, S. (1983). Ecophysiology of zoophilic pollination. *In* 'Responses to the Chemical and Biological Environment' (Eds. Lange, O.L., Nobel, P.S., Osmund, C.B. and Ziegler, C.B.), Encyclopaedia of Plant Physiology: **12C**, 558–624, Springer-Verlag, Berlin.

Vos-Kelk, P. and Davids, C. (1977). A short note on *Lecane rotifera* found in *Utricularia* vegetation. *Hydrobiological Bulletin* **11**, 53–55.

Vries, H. de (1886). Über die Aggregation im Protoplasm von *Drosera rotundifolia* (On the aggregation in the protoplasm of *D. rotundifolia*). *Botanische Zeitung* **44**, 1–11.

Wagner, H., Kreher, B. and Jurcic, K. (1986). Immunological investigations of naphthoquinone – containing plant extracts, isolated quinones and other cytostatic compounds in cellular immunosystems. *Planta Medica*, 550.

Walker, T.J. and Hasler, A.D. (1949). Detection and discrimination of odours of aquatic plants by the bluntnose minnow *Hybrhynchus notatus*. *Physiological Zoology* **22**, 45–63.

Walla, J.A. and Peterson, G.W. (1976). *Dothistroma pini* and *Diplodia pinea* not affected by surface wax of pine needles. *Plant Disease Reporter* **60**, 1042–1046.

Wallace, R.L. (1977a). Substrate discrimination by larvae of the sessile rotifer *Ptygura beauchampii* Edmondson. *Freshwater Biology* **7**, 301–309.

Wallace, R.L. (1977b). Distribution of sessile rotifers in an acid bog pond. *Archiv für Hydrobiologie* **79**, 478–505.

Wallace, R.L. (1978). *Ptygura beauchampii*. *Ecology* **59**, 221–227.

Wallace, R.L. (1980). Ecology of sessile rotifers. *Hydrobiologia* **73**, 181–193.

Wan, A.G., Bexel, R.T., Ramsey, R.B. and Kholas, H.J.N. (1972). Sterols and triterpenes of the pitcher plant. *Phytochemistry* **11**, 456–461.

Watson, A.P., Matthiessen, J.N. and Springett, B.P. (1982). Arthropod associates and macronutrient status of the red-ink sundew (*Drosera erythrorhiza* Lindl.). *Australian Journal of Ecology* **7**, 13–22.

Way, M.J. and Murdie, G. (1965). An example of varietal variations in resistance of brussels sprouts. *Annals of Applied Biology* **56**, 326–328.

Weber, F. (1940). Vitamin-C-Gehalt gefütterten *Drosera*-blätter (The vitamin-C content of artificially fed *Drosera* leaves). *Berichte der deutschen botanische Gesellschaft* **58**, 370–373.

Wegmuller, S. (1972). Neue palynologische Ergebnisse aus den Westalpen (The latest results of pollen analysis of the western Alps). *Berichte der deutschen botanische Gesellschaft* **85**, 75–77.

Werle, E. (1955). Amine und betaine (Amines and betaines). 'Modern Methods of Plant Analysis **4**' (Eds. Paech, K. and Tracey, M.V.), 517–623.

Whalley, P. (1986). Bristly Oxtongue, *Picris echioides* L., a hazard for insects? *Botanical Society of the British Isles News*, **Sept. 1986**, 13–14.

Wherry, E.T. (1929). Acidity relations of the sarracenias. *Journal of the Washington Academy of Sciences* **19**, 379–390.

Whitaker, E.H. (1946). A study of the exudate of *Drosera longifolia* L. *American Journal of Botany* **33**, 838–839.

Whitaker, E.H. (1949). Physiological studies on two species of *Drosera* L. Doctoral Dissertation, Cornell University, Ithaca, New York.

White, J. (1911). The proteolytic enzymes of *Drosera*. *Proceedings of the Royal Society (London)* **83B**, 134–139.

Whitehead, B. (1973). Habitat of *Drosera peltata*. *Carnivorous Plant Newsletter* **2**, 61–62.

Whitesell, J.K., Matthews, R.S. and Wang, P.K.S. (1977). Studies in the total synthesis of mono- and sesquiterpenes. Antirride. *Synthetic Communications* **7**, 355–362.

Whitesell, J.K., Matthews, R.S., Minton, M.A. and Helbling, A.M. (1981). The total synthesis of sarracenin. *American Chemical Society* **103**, 3468–3472.

Wickler, W. (1968). 'Mimicry in Plants and Animals' (transl. Martin, R.D.). Weidenfeld & Nicholson, London.

Wieffering, J.H. (1966). Aucubinartige Glucoside (Pseudoindikene) und verwandte Heteroside als systematische Merkmale (Aucubin-type glucosides and related heterosides as taxonomic markers). *Phytochemistry* **5**, 1052–1064.

Wiens, D. (1978). Mimicry in plants. *Evolutionary Biology* **11**, 365–403.

Wijewanatha, R.T. (1952). Some preliminary observations on the genus *Nepenthes* in Ceylon. *Ceylon Journal of Science, Section A, Botany* **12**, 245–247.

Wilkinson, H.P. (1979). The plant surface. *In* 'Anatomy of the Dicotyledons' 2nd edn, **1** (Eds. Metcalfe, C.R. and Chalk, L.). 97–165 Oxford University Press, Oxford.

Williams, C.M. (1970). Hormonal interactions between plants and insects. *In* 'Chemical Ecology' (Eds. Sondheimer, E. and Simeone, J.B.), 103–132. Academic Press, New York and London.

Williams, M.E. and Mozingo, H.N. (1971). The fine structure of the trigger hair in Venus's flytrap. *American Journal of Botany* **58**, 532–539.

Williams, S.E. (1971). Rapid inflection of *Drosera* tentacles. Doctoral Dissertation, Washington University, St. Louis.

Williams, S.E. (1973a). The 'memory' of the Venus's Flytrap. *Carnivorous Plant Newsletter* **2**, 23–25.

Williams, S.E. (1973b). A salute to Sir John Burdon Sanderson and Mr. Charles Darwin on the centennial of the discovery of nerve-like activity in the Venus's Flytrap. *Carnivorous Plant Newsletter* **2**, 41.

Williams, S.E. (1976). Comparative sensory physiology of the Droseraceae – the evolution of a plant sensory system. *Proceedings of the American Philosophical Society* **120**, 187–204.

Williams, S.E. (1980). How Venus's Flytraps catch spiders and ants. *Carnivorous Plant Newsletter* **9**, 65–78.

Williams, S.E. and Bennett, A.B. (1982). Leaf closure in the

Venus flytrap: An acid growth response. *Science* **218**, 1120–1124.

Williams, S.E. and Bennett, A.B. (1983). Acid flux triggers the Venus's Flytrap. *New Scientist* **97**, 582.

Williams, S.E. and Pickard, B.G. (1969). Secretion, absorption and cuticle structure in *Drosera* tentacles. *Plant Physiology* **44**, Supplement 5, No. 22.

Williams, S.E. and Pickard, B.G. (1972a). Receptor potentials and action potentials in *Drosera* tentacles. *Planta* **103**, 193–221.

Williams, S.E. and Pickard, B.G. (1972b). Properties of action potentials in *Drosera* tentacles. *Planta* **103**, 222–240.

Williams, S.E. and Pickard, B.G. (1974). Connections and barriers between cells of *Drosera* tentacles in relation to their electrophysiology. *Planta* **116**, 1–16.

Williams, S.E. and Pickard, B.G. (1979). The role of action potentials in the control of capture movements of *Drosera* and *Dionaea*. *In* 'Tenth International Conference on Plant Growth Susbtances', 22–26. Madison, Wisconsin.

Williams, S.E. and Spanswick, R.M. (1972). Intracellular recordings of the action potentials which mediate the thigmonastic movements of *Drosera*. *Plant Physiology* **49**, 64.

Williams, S.E. and Spanswick, R.M. (1976). Propagation of the neuroid action potential of the carnivorous plant *Drosera*. *Journal of Comparative Physiology* **108**, 211–223.

Williamson, G.B. (1982). Plant mimicry – Evolutionary constraints. *Biological Journal of the Linnean Society* **18**, 44–58.

Willis, J.H. (1965). Historical notes on the Western Australian pitcher plant *Cephalotus follicularis* (including provenance of type material). *Western Australian Naturalist* **10**, 1–7.

Winston, R.D. and Gorham, P.R. (1979). Turions and dormancy states in *Utricularia vulgaris*. *Canadian Journal of Botany* **57**, 2740–2749.

Wirth, W. and Beaver, R.A. (1979). The *Dasyhelea* biting midges living in pitchers of *Nepenthes* in South East Asia (Diptera: Ceratopogonidae). *Annales de la Sociétee Entomologique de France* **15**, 41–52.

Withering, W. (1887). 'A Botanical Arrangement of British Plants'. Robinson & Robson, London; Balfour & Elliot, Edinburgh.

Woelkerling, W.J. (1976). Wisconsin, U.S.A. Desmids. Part 1 Aufwuchs and plankton communities of selected acid bogs, alkaline bogs and closed bogs. *Hydrobiologia* **48**, 209–232.

Wolfe, L.M. (1981). Feeding behavior of a plant: Differential prey capture in old and new leaves of the pitcher plant (*Sarracenia purpurea*). *American Midland Naturalist* **106**, 352–359.

Wood, C.E. (1955). Evidence for the hybrid origin of *Drosera anglica*. *Rhodora* **57**, 105–129.

Wood-Baker, C.S. (1972). A new food plant for *Acyrthosiphon pelargonii borealis* H.R.L. (Homoptera; Aphididae) in north Britain, with biometric data. *Entomologists Monthly Magazine* **108**, 95–97.

Woodhead, S. (1982). p-Hydroxybenzaldehyde in the surface-wax of sorghum: its importance in the seedling resistance to acridids. *Entomologica Experimentalis et Applicata* **81**, 296–302.

Wray, D.L. and Brimley, C.S. (1943). The insect inquilines and victims of pitcher plants in North Carolina. *Annals of the Entomological Society of America* **36**, 128–137.

Wurm, G., Grimm, H., Geres, U. and Schmidt, H. (1984). Plumbagin: Reactivität, Toxizität und antimikrobielle Aktivität des in *Drosera* und *Plumbago* Arten vorkommenden Naturstoffe (Reactivity, toxicity and antimicrobial activity of the natural compounds of *Drosera* and *Plumbago* species.) *Deutsche Apotheker Zeitung* **124**, 2128–2132.

Yaguchi, Y. and Kondo, K. (1981). Stomatal response to prey capture and trap narrowing in Venus's flytrap (*Dionaea muscipula*). *Phyton* **41**, 83–90.

Yeo, P.F. (1972). Floral allurements for pollinating insects. *In* 'Insect Plant Relationships' (Ed. Emden, H.F. van), 51–57. Blackwell Scientific, Oxford.

Zavortink, T.J. (1985). *Zingala*, a new subgenus of *Wyeomyia* with new species from pitcher-plants in Venezuela (Diptera, Culicidae, Sabethini). *Wasmann Journal of Biology* **43**, 46–59.

Zavortink, T.J. (1986). The occurrence of *Runchomyia frontosa* in carnivorous bromeliads in Venezuela, with notes on the biology of its immatures (Diptera, Culicidae, Sabethini). *Wasmann Journal of Biology* **44 (1–2)**, 127–129.

Zeeuw, J. de (1934). Versuche über die Verdauung von *Nepenthes*kannen (Experiments on the digestion of *Nepenthes* pitchers). *Biochemische Zeitschrift* **269**, 187–195.

Zeigler, H. and Lüttge, U. (1959). Über die Resorption von ^{14}C-Glutaminsäure durch sezierende Nektarien (On the uptake of C^{14}- labelled glutamic acid through dissected nectaries). *Naturwissenschaften* **46**, 176–177.

Zenk, M.H., Fürbringer, M. and Steglich, W. (1969). Occurrence and distribution of 7-methyl juglone and plumbagin in the Droseraceae. *Phytochemistry* **8**, 2199–2200.

Ziemer, R.R. (1973). Some field observations of *Darlingtonia* and *Pinguicula*. *Carnivorous Plant Newsletter* **2**, 25–27.

Zipperer, P. (1885). Beiträge zur Kenntis der Sarraceniaceen (Contribution to the knowledge of the Sarraceniaceae), 34. Inaugural Dissertation, Erlangen.

Zipperer, P. (1887). Beiträge zur Kenntis der Sarraceniaceen (Contributions to the knowledge of the Sarraceniaceae) *Botanisches Centralblatt* **29**, 358–359.

Index

Bold entries refer to figure captions and tables.

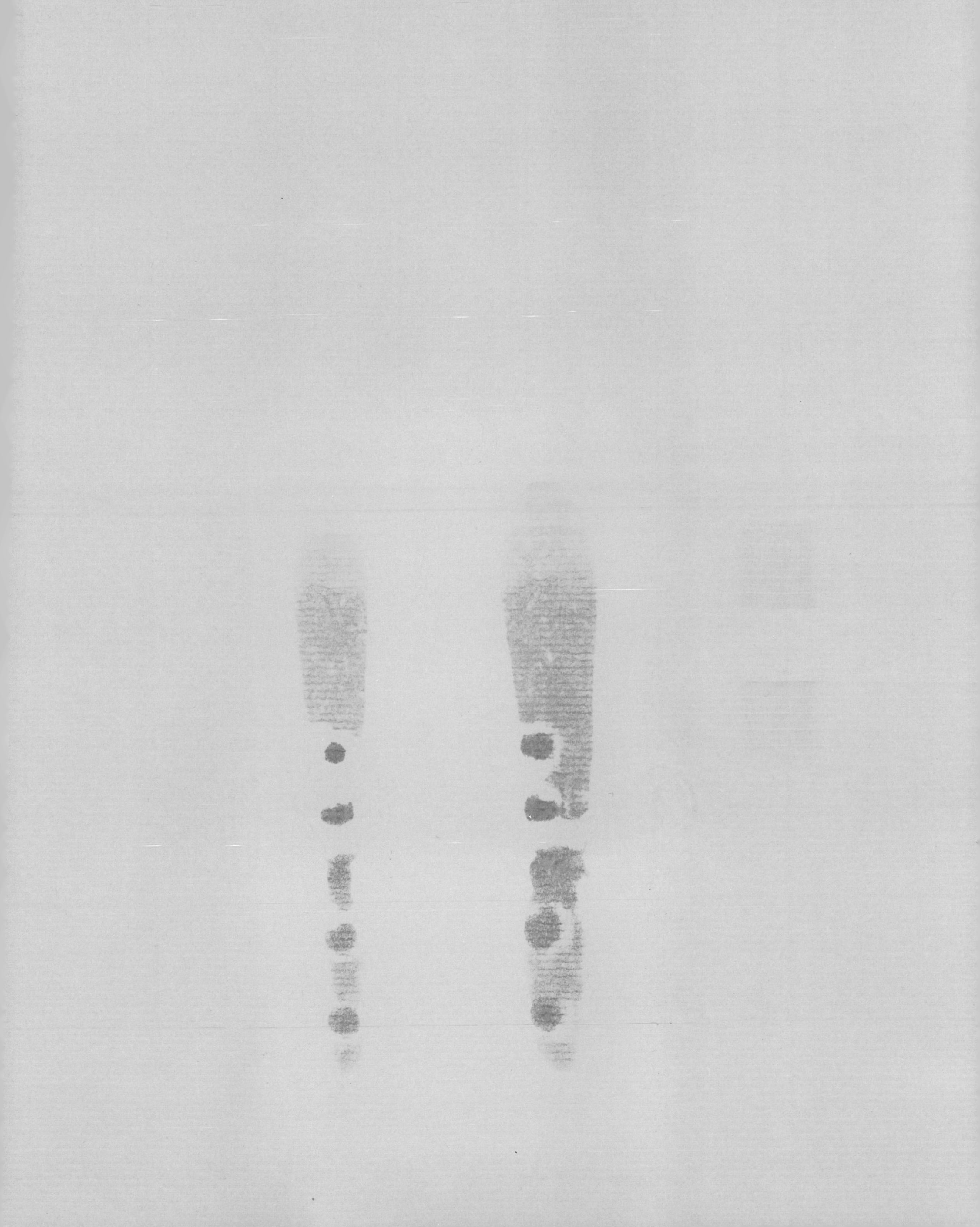

OVER
QK
917
.J854
1989

103081
Juniper, B. E.
The carniverous
plants.

DATE DUE

SUSAN COLGATE CLEVELAND
LIBRARY/LEARNING CENTER
COLBY-SAWYER COLLEGE
New London, New Hampshire 03257

GAYLORD

PRINTED IN U.S.A.